D0918724

The chemistry of
diazonium and diazo groups
Part 2

THE CHEMISTRY OF FUNCTIONAL GROUPS

A series of advanced treatises under the general editorship of Professor Saul Patai

The chemistry of alkenes (2 volumes)
The chemistry of the carbonyl group (2 volumes)
The chemistry of the ether linkage
The chemistry of the amino group
The chemistry of the nitro and nitroso groups (2 parts)
The chemistry of carboxylic acids and esters
The chemistry of the carbon–nitrogen double bond
The chemistry of amides
The chemistry of the cyano group
The chemistry of the hydroxyl group (2 parts)
The chemistry of the azido group
The chemistry of acyl halides
The chemistry of the carbon–halogen bond (2 parts)
The chemistry of the quinonoid compounds (2 parts)
The chemistry of the thiol group (2 parts)
The chemistry of the hydrazo, azo and azoxy groups (2 parts)
The chemistry of amidines and imidates
Supplement A: The chemistry of double-bonded functional groups (2 parts)
The chemistry of cyanates and their thio derivatives (2 parts)
The chemistry of diazonium and diazo groups (2 parts)

$-\overset{+}{N}\equiv N$ $\qquad$ $-CN_2$

The chemistry of diazonium and diazo groups

Part 2

Edited by
SAUL PATAI
The Hebrew University, Jerusalem

1978
JOHN WILEY & SONS
CHICHESTER — NEW YORK — BRISBANE — TORONTO
An Interscience ® Publication

Library of Congress Catalog Card No. 75-6913

ISBN 0 471 99492 8 (Pt. 1)
ISBN 0 471 99493 6 (Pt. 2)
ISBN 0 471 99415 4 (Set)

Printed in Great Britain by John Wright and Sons Ltd.,
at the Stonebridge Press, Bristol.

Contributing authors

W. Ando	Chemistry Department, The University of Tsukuba, Niiharigun, Ibaraki 300–31, Japan
D. A. Ben-Efraim	The Weizmann Institute of Science, Rehovot, Israel
C. F. Cooper	Department of Chemistry, University of Missouri-Rolla, Rolla, Mo. 65401, USA
A. J. Fry	Wesleyan University, Middletown, Connecticut, USA
A. F. Hegarty	Chemistry Department, University College, Cork, Ireland
E. S. Lewis	Department of Chemistry, Rice University, Houston, Texas, USA
G. Linstrumelle	Équipe de Recherche No. 12 du CNRS, Laboratoire de Chimie, École Normale Supérieure, Paris, France
J. F. McGarrity	Institute of Organic Chemistry, University of Lausanne, Lausanne, Switzerland
J. B. Moffat	Department of Chemistry, University of Waterloo, Waterloo, Ontario, Canada
H. M. Niemeyer	Institute of Organic Chemistry, University of Lausanne, Lausanne, Switzerland
M. Regitz	Department of Chemistry, University of Kaiserslautern, D-6750, Federal Republic of Germany
K. Schank	Fachbereich 14.1 Organische Chemie, Universität des Saarlandes, D-6600 Saarbrücken, Germany
R. Shaw	1162 Quince Avenue, Sunnyvale, California 94087, USA
P. J. Smith	Department of Chemistry and Chemical Engineering, University of Saskatchewan, Saskatoon, Saskatchewan, Canada
S. Sorriso	Istituto di Chimica Fisica, Università di Perugia, 06100 Perugia, Italy
V. Štěrba	Organic Chemistry Department, Institute of Chemical Technology, 532 10 Pardubice, Czechoslovakia
K. C. Westaway	Department of Chemistry, Laurentian University, Sudbury, Ontario, Canada
D. Whittaker	Department of Organic Chemistry, University of Liverpool, England
D. S. Wulfman	Department of Chemistry, University of Missouri-Rolla, Rolla, Mo. 65401, USA

Foreword

The present volume, 'The Chemistry of Diazonium and Diazo Groups' is, on the whole, organized and presented according to the general lines described in the 'Preface to the Series', printed on the following pages.

Some difficulty arose in the presentation owing to the fact that while the two groups treated, i.e. the diazo group and the diazonium group, are closely related and even occur in equilibrium with each other, their chemical behaviour and characteristics differ from each other considerably. Moreover, the material which had to be covered proved to be much more extensive than originally surmised. For these reasons, some of the subjects had to be divided into two or more chapters; for instance, the synthetic applications of diazonium and diazo groups are treated in two separate chapters and even so each of these turned out to be very large. Similarly, the syntheses of the different title compounds are discussed in three separate chapters.

The plan of the present volume also included a chapter on 'Biological and Pharmaceutical Effects' which, however, failed to materialize. It is hoped that this will appear in one of the supplementary volumes to the series.

Jerusalem, February 1977 SAUL PATAI

The Chemistry of Functional Groups
Preface to the series

The series 'The Chemistry of Functional Groups' is planned to cover in each volume all aspects of the chemistry of one of the important functional groups in organic chemistry. The emphasis is laid on the functional group treated and on the effects which it exerts on the chemical and physical properties, primarily in the immediate vicinity of the group in question, and secondarily on the behaviour of the whole molecule. For instance, the volume *The Chemistry of the Ether Linkage* deals with reactions in which the C—O—C group is involved, as well as with the effects of the C—O—C group on the reactions of alkyl or aryl groups connected to the ether oxygen. It is the purpose of the volume to give a complete coverage of all properties and reactions of ethers in as far as these depend on the presence of the ether group but the primary subject matter is not the whole molecule, but the C—O—C functional group.

A further restriction in the treatment of the various functional groups in these volumes is that material included in easily and generally available secondary or tertiary sources, such as Chemical Reviews, Quarterly Reviews, Organic Reactions, various 'Advances' and 'Progress' series as well as textbooks (i.e. in books which are usually found in the chemical libraries of universities and research institutes) should not, as a rule, be repeated in detail, unless it is necessary for the balanced treatment of the subject. Therefore each of the authors is asked *not* to give an encyclopaedic coverage of his subject, but to concentrate on the most important recent developments and mainly on material that has not been adequately covered by reviews or other secondary sources by the time of writing of the chapter, and to address himself to a reader who is assumed to be at a fairly advanced post-graduate level.

With these restrictions, it is realized that no plan can be devised for a volume that would give a *complete* coverage of the subject with *no* overlap between chapters, while at the same time preserving the readability of the text. The Editor set himself the goal of attaining *reasonable* coverage with *moderate* overlap, with a minimum of cross-references between the chapters of each volume. In this manner, sufficient freedom is given to each author to produce readable quasi-monographic chapters.

The general plan of each volume includes the following main sections:

(a) An introductory chapter dealing with the general and theoretical aspects of the group.

(b) One or more chapters dealing with the formation of the functional group in question, either from groups present in the molecule, or by introducing the new group directly or indirectly.

(c) Chapters describing the characterization and characteristics of the functional groups, i.e. a chapter dealing with qualitative and quantitative methods of determination including chemical and physical methods, ultraviolet, infrared, nuclear magnetic resonance and mass spectra: a chapter dealing with activating and directive effects exerted by the group and/or a chapter on the basicity, acidity or complex-forming ability of the group (if applicable).

(d) Chapters on the reactions, transformations and rearrangements which the functional group can undergo, either alone or in conjunction with other reagents.

(e) Special topics which do not fit any of the above sections, such as photochemistry, radiation chemistry, biochemical formations and reactions. Depending on the nature of each functional group treated, these special topics may include short monographs on related functional groups on which no separate volume is planned (e.g. a chapter on 'Thioketones' is included in the volume *The Chemistry of the Carbonyl Group*, and a chapter on 'Ketenes' is included in the volume *The Chemistry of Alkenes*). In other cases certain compounds, though containing only the functional group of the title, may have special features so as to be best treated in a separate chapter, as e.g. 'Polyethers' in *The Chemistry of the Ether Linkage*, or 'Tetraaminoethylenes' in *The Chemistry of the Amino Group*.

This plan entails that the breadth, depth and thought-provoking nature of each chapter will differ with the views and inclinations of the author and the presentation will necessarily be somewhat uneven. Moreover, a serious problem is caused by authors who deliver their manuscript late or not at all. In order to overcome this problem at least to some extent, it was decided to publish certain volumes in several parts, without giving consideration to the originally planned logical order of the chapters. If after the appcarance of the originally planned parts of a volume it is found that either owing to non-delivery of chapters, or to new developments in the subject, sufficient material has accumulated for publication of a supplementary volume, containing material on related functional groups, this will be done as soon as possible.

The overall plan of the volumes in the series 'The Chemistry of Functional Groups' includes the titles listed below:

The Chemistry of Alkenes (*two volumes*)
The Chemistry of the Carbonyl Group (*two volumes*)
The Chemistry of the Ether Linkage
The Chemistry of the Amino Group
The Chemistry of the Nitro and Nitroso Group (*two parts*)
The Chemistry of Carboxylic Acids and Esters
The Chemistry of the Carbon–Nitrogen Double Bond
The Chemistry of the Cyano Group
The Chemistry of Amides
The Chemistry of the Hydroxyl Group (*two parts*)
The Chemistry of the Azido Group
The Chemistry of Acyl Halides
The Chemistry of the Carbon–Halogen Bond (*two parts*)
The Chemistry of Quinonoid Compounds (*two parts*)
The Chemistry of the Thiol Group (*two parts*)
The Chemistry of Amidines and Imidates
The Chemistry of the Hydrazo, Azo and Azoxy Groups
The Chemistry of Cyanates and their Thio Derivatives
The Chemistry of Diazonium and Diazo Groups
Supplement A: The Chemistry of Double-Bonded Functional Groups (*two parts*)

Titles in press:

The Chemistry of the Carbon–Carbon Triple Bond
Supplement B: The Chemistry of Acid Derivatives

Future volumes planned include:

The Chemistry of Cumulenes and Heterocumulenes
The Chemistry of Organometallic Compounds
The Chemistry of Sulphur-containing Compounds
Supplement C: The Chemistry of Triple-bonded Functional Groups
Supplement D: The Chemistry of Halides and Pseudo-halides
Supplement E: The Chemistry of $—NH_2$, $—OH$, *and* $—SH$ *Groups and their Derivatives*

Advice or criticism regarding the plan and execution of this series will be welcomed by the Editor.

The publication of this series would never have started, let alone continued, without the support of many persons. First and foremost among these is Dr Arnold Weissberger, whose reassurance and trust encouraged me to tackle this task, and who continues to help and advise me. The efficient and patient cooperation of several staff-members of the Publisher also rendered me invaluable aid (but unfortunately their code of ethics does not allow me to thank them by name). Many of my friends and colleagues in Israel and overseas helped me in the solution of various major and minor matters, and my thanks are due to all of them, especially to Professor Z. Rappoport. Carrying out such a long-range project would be quite impossible without the non-professional but none the less essential participation and partnership of my wife.

The Hebrew University
Jerusalem, ISRAEL.

SAUL PATAI

Contents

CHAPTER **12**

Kinetics and mechanisms of reactions involving diazonium and diazo groups

A. F. HEGARTY
Chemistry Department, University College, Cork, Ireland

I. INTRODUCTION

Although diazoalkanes and alkane diazonium ions are in many cases interconvertible and thus share a common reaction pathway, the mechanistic aspects of diazoalkane and diazonium ion formation and reactions are treated separately in the present chapter. The main justification for this is that the bulk of the available kinetic data refers to reactions of arenediazonium ions, no doubt a reflection of their stability and the ease with which they can be handled (relative to their aliphatic analogues); diazoalkanes, on the other hand, show several reactions (e.g. cycloaddition, carbene formation) which are unique to this functional group.

The variety and complexity of behaviour shown by both groups is very great indeed and interpretation of their chemistry has been (and in some cases still is) the subject of renowned controversies. This has arisen since allowance was not always made for competing equilibria in solution and since several parallel reactions (e.g. ionic, free radical), whose relative importance can be changed by small variations in solvent or reaction conditions, can occur. Because of this and the probability that some of the reactions lie on a mechanistic borderline, it is more satisfactory to classify the reactions according to type (e.g. replacement of nitrogen in arenediazonium ions) rather than making the more usual classification based on mechanism.

Thus the main reaction types shown for arenediazonium ions are summarized in equations (1) to (4); these include replacement of nitrogen by a nucleophile (via phenyl cation, S_N2 or benzyne formation), reaction of a nucleophile at the terminal nitrogen (including reactions with activated aromatic systems), nucleophilic aromatic displacements activated by the strongly electron-withdrawing diazonium group and free radical reactions (where Y^- may be a metal ion or other electron donor).

$$Z-C_6H_4-N_2^+ + Y^- \longrightarrow Z-C_6H_4-Y + N_2 \quad (1)$$

$$Z-C_6H_4-N_2^+ + Y^- \longrightarrow Z-C_6H_4-N{=}N-Y \quad (2)$$

$$Z-C_6H_4-N_2^+ + Y^- \longrightarrow Y-C_6H_4-N_2^+ + Z^- \quad (3)$$

$$Z-C_6H_4-N_2^+ + Y^- \longrightarrow Z-C_6H_4^{\bullet} + N_2 + Y^{\bullet} \quad (4)$$

Diazoalkanes react with electrophiles (largely at carbon) and with nucleophiles (at nitrogen). Two further main reaction types (cycloaddition to unsaturated sites and carbene formation) are also observed (equation 5). Excellent discussions of

$$R_2\bar{C}-\overset{+}{N}{\equiv}N \longleftrightarrow R_2C{=}\overset{+}{N}{=}\bar{N}$$

$$\xrightarrow{E^+} R_2C(E)-N_2^+ \qquad \xrightarrow{Y^-} R_2\bar{C}-N{=}N-Y \qquad \xrightarrow{>C=C<} \text{pyrazoline } (R_2C-N{=}N \text{ ring}) \qquad \longrightarrow R_2C\colon + N_2 \quad (5)$$

earlier work are available in Zollinger's[1] and in Smith's[2] texts. Of necessity in such a wide field of interest some areas are dealt with either briefly (e.g. cycloadditions, alkanediazonium ions), or not at all (e.g. carbene and carbenoid formation). In the latter case recent comprehensive reviews are, however, available[3–5].

II. DIAZOTIZATION

The most direct route to aliphatic or aromatic diazonium salts is treatment of the corresponding primary amines (**1**) with nitrosating agents (Scheme 1). Although many reagents and conditions have been described for this conversion, only a fraction have been subjected to a detailed kinetic and mechanistic analysis. In many cases because of the solvents employed or the heterogeneity of the medium, a simple kinetic analysis is not possible. However, the variety of kinetic behaviour shown in the systems so far examined is enormous and has fascinated chemists for many decades. Several novel concepts now accepted in physical organic chemistry owe their origins to the challenge of these reactions.

$$\underset{(1)}{R{-}NH_2} \xrightleftharpoons{NOX} \underset{(2)}{R\overset{+}{N}H_2NO} \rightleftharpoons \underset{(3)}{R{-}NHNO} \rightleftharpoons \underset{(4)}{R{-}N{=}N{-}OH}$$

$$\underset{(4)}{R{-}N{=}N{-}OH} \xrightleftharpoons{HY} \underset{(6)}{RN_2^+\ Y^-}$$

$$\underset{(5)}{R^+} \longleftarrow \underset{(6)}{RN_2^+\ Y^-} \xrightarrow{-H^+} \underset{(7)}{R'N_2}$$

SCHEME 1

In spite of the complexity shown, several generalizations can be made that simplify the discussion which follows. Thus either the generation of the nitrosating agent or its attack on the amine can be rate determining. Normally the free amine is the reactive species even though its concentration may be very small in the acidic solutions used. Finally, the reaction rate can be independent of amine concentration if formation of the nitrosating agent is rate determining, and independent of amine structure if the nitrosating agent (for example NO^+) is so reactive that its reaction with the amine is diffusion controlled.

The most common nitrosating agents (NOX) have X = OH (nitrous acid), OR (alkyl or acyl nitrite esters), —ONO (nitrous anhydride which is normally formed *in situ* from nitrous acid), halogen (X = Cl, Br, I) together with nitrosonium salts (for example $NO^+ClO_4^-$, $NO^+BF_4^-$). Much of the mechanistic work has been directed toward the establishment of the relative reactivities of these reagents and determining which reagent provides the most important pathway under a given set of experimental conditions.

The diazonium ion (**6**), once formed, may be stable and thus is the 'product' of reaction. However, this may be labile (for example, heterocyclic or aliphatic diazonium salts undergo ready nitrogen loss to give **5** initially); similarly in reactions carried out at higher pH, diazohydroxides **4** (as their conjugate bases) or diazoalkanes **7** (if the group R contains an ionizable proton) may be formed. This diversity of products, however, does not normally complicate the study of diazonium ion formation since conditions can be usually be varied to make the subsequent reactions either markedly faster or slower than the nitrosation step. Another valuable tool is Ridd's[6] observation that secondary amines such as *N*-methylaniline (**8**) which form *N*-nitroso compounds (**9**) on reaction with nitrosating agents often give the same kinetic pattern as the corresponding primary amine. This has been used to great effect to show that the rate-determining steps are similar, ruling out slow proton transfers in the conversion of **3** to **6**.

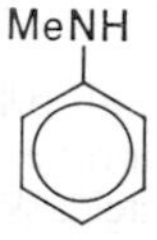

MeNH (**8**) MeN—N=O (**9**) R—(triazole N—N, N H)—NHNO (**10**)

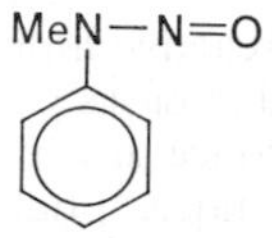

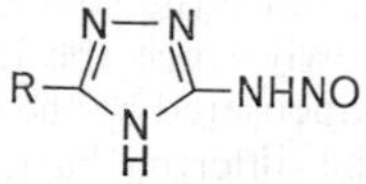

Although primary nitrosamines (**3**) are almost invariably invoked as reaction intermediates on the reaction pathway few examples of isolable materials are available. However, Muller[7] by carrying out the nitrosation of aniline in diethyl ether at −78 °C obtained a solution with the spectral characteristics of *N*-nitroso-aniline. The nitrosamine was stable at this temperature for several days. Attempts to isolate primary nitrosamines under normal diazotization conditions (at 0 °C in aqueous solution) have failed. Several heterocyclic primary nitrosamines (such as **10**) have however been described[8] and their stability has been attributed to, *inter alia*, the possibility of forming a strong intramolecular hydrogen bond to stabilize the N—H form. The nitrosamine **10** is only isolated when the reaction medium is weakly acidic and, to date, only when the heterocyclic ring is five- rather than six-membered.

A. Low Acidities ($[H^+] < 10^{-2}$ M)

At low acidities the diazotization of primary aromatic and aliphatic amines with nitrous acid in aqueous solution is independent of amine concentration and the kinetic law is of the form (6)[9]. Clearly in this case the slow step is formation of the reactive nitrosating species. This has been identified as nitrous anhydride formed as

$$\text{Rate} = k_{obs}[HNO_2]^2 \quad (6)$$

in equation (7)[10]. The best evidence for this comes from a study[11] in which the

$$2\,HONO \rightleftharpoons N_2O_3 + H_2O \quad (7)$$

equilibrium (7) was examined independently by following the rate of incorporation of ^{18}O-labelled H_2O into nitrous acid. It was shown that the rate of formation of N_2O_3 was the same as that of diazotization in weakly acidic media.

With amines carrying strongly electron-withdrawing groups, e.g. *p*-nitroaniline or 2,4-dinitroaniline, the nitrosation step is so slow that it becomes rate determining even under these mildly acidic conditions and equation (8) is followed[12].

$$\text{Rate} = k_{obs}[HNO_2]^2[ArNH_2] \quad (8)$$

Again the kinetic form implies that the anhydride is the reactive nitrosating species for these amines.

B. Intermediate Acidities (10^{-2}–10^{-1} M [H$^+$])

As the acidity of the medium is increased, the rate of diazotization of aniline passes through a region of intermediate order before becoming proportional to amine concentration [where equation (8) is followed]. For aniline the transition is complete at 0·1 M perchloric acid[13]. Again the anhydride N_2O_3 is implicated as the active nitrosating species; however, the observed rate of diazotization changes more slowly with acidity since the increasing concentration of the anhydride is partly compensated by the decrease in the reactive amine-free base.

At a given acidity the stronger amine bases (i.e. those with high pK_a's) in general react more slowly since the lower concentration of the free amine present is not altogether compensated by the greater nucleophilicity of the amine[12]. When allowance is made for the differing basicities of the amines, then there is a direct correlation between the electron-donating power of *meta* and *para* substituents of aromatic amines and rates of diazotization (e.g. *p*-chloroaniline is 3·4-fold less reactive than aniline)[9]. Secondary amines such as *N*-methylaniline are more reactive (*c*. 2-fold)[6] than primary amines of the same pK_a, the difference being similar to that observed in acylation of amines[14, 15]. However, aliphatic amines react surprisingly slowly—primary and secondary aliphatic amines are less reactive than aromatic amines whose pK_a's are up to 5 units lower[6].

As an alternative to nitrous anhydride as the active nitrosating species, Kenner suggested[16] that in this pH region the second molecule of HNO_2 was acting by a specific acid–general base mechanism by (a) protonating nitrous acid giving $H_2\overset{+}{O}NO$, and (b) removal of a proton (by NO_2^-) in the rate-controlling step [see equation (9)].

$$\text{HONO} \xrightleftharpoons{\text{H}^+} \text{H}_2\overset{+}{\text{O}}\text{NO} \xrightarrow{\text{ArNH}_2} \text{Ar}\overset{+}{\text{N}}\text{H}_2\text{NO} \xrightarrow{\text{NO}_2^-} \text{ArNHNO} \qquad (9)$$

However, this mechanism is unlikely in view of the fact that no other general bases have been found to catalyse diazotization under these conditions[17].

C. Moderate Acidities (0·1–6·5 M [H$^+$])

The order of reaction changes again as the acidity is increased so that the overall rate of reaction is given by equation (10). Clearly there is a change in the nitrosating

$$\text{Rate} = k_1[\text{ArNH}_2][\text{HNO}_2]h_0 + k_2[\text{Ar}\overset{+}{\text{N}}\text{H}_3][\text{HNO}_2]h_0 \qquad (10)$$

species since only one mole of HNO_2 is involved. The changeover from equation (8) to equation (10) is dependent on [HNO_2] but is complete for aniline in 1·0 M [$HClO_4$] when [HNO_2] is less than 10^{-3} M.

The first term in equation (10) is important only at relatively low acidities (h_0 is the Hammett acidity function) and the overall rate is independent of acidity (because of compensation between increasing h_0 and decreasing free amine $ArNH_2$). The separation of the two terms in equation (10) is greatly simplified by Ridd's observation[9] that k_1 does not vary greatly for amines which are more basic than *p*-nitroaniline and a value of 10^3 M^{-2} s^{-1} at 0 °C ($\mu = 3{\cdot}0$) has been used for k_1.

A plot of log k_{obs} for nitrosation of aniline against [$HClO_4$] up to 6·5 M or against $-H_0$ is linear, the latter having a slope close to unity. The ionic strength must be maintained constant (by the addition of $NaClO_4$) since the addition of perchlorate ion itself has a powerful catalytic effect at constant acidity, attributed to a salt effect[18].

This kinetic evidence for the second term in equation (10) was not expected, *a priori*, since it implies that the protonated amine is reactive towards an electrophilic species. However, Ridd[9] provides strong evidence that both anilinium ion and *N*-methylanilinium ion[6] do undergo reaction in this way.

The most compelling evidence comes from data measured at constant h_0 and $[HNO_2]$; the observed substituent effects are strikingly different from the well-established order of reactivity shown for the reaction of N_2O_3 with the free amine. Thus electron-donating substituents in the *meta* or *para* position generally increase the observed rate of nitrosation of anilines and *N*-methylanilines, which is inconsistent with the free amine reacting with a highly active nitrosating species.

The mechanism of equation (11) has been proposed[19] to account for these observations. The nitrosating species X^+ initially interacts with the π electrons of the aromatic ring. This allows removal of a proton by the weakly basic species (B) present at such acidities. Substituents in the *para* position will largely effect the initial equilibrium between the aromatic cloud and X^+, while *ortho* substituents would also be expected to influence the proton removal, as observed.

$$C_6H_5\overset{+}{N}H_3 \underset{}{\overset{X^+}{\rightleftharpoons}} [X^+ \cdots C_6H_5\overset{+}{N}H_3] \xrightarrow{B} [X^+ \cdots C_6H_5NH_2 \cdots \overset{+}{H} \cdots B] \longrightarrow C_6H_5\overset{+}{N}H_2X + \overset{+}{B}H \qquad (11)$$

The ratio of reactivity of $ArNH_2$ and $Ar\overset{+}{N}H_3$ towards X^+ ranges from 335 to 3×10^4. It is argued[19], however, that these relatively small numbers are reasonable on the basis of calculations involving simple electrostatic models. Moreover, the high reactivity of reagents is such that little discrimination is expected, especially in reactions of $ArNH_2$.

The active nitrosating species X^+ which in this acidity region reacts with either $ArNH_2$ or $Ar\overset{+}{N}H_3$ has the stoichiometric form (H^+) (HONO) and may be either nitrosonium ion (NO^+) or the hydrated form $(H_2ONO)^+$ [20]. However, the position of the equilibrium between these (equation 12) is not accurately known, although it is established that NO^+ is present in the more acidic solutions when $[HClO_4] > 6{\cdot}5$ M. It is likely that nitrosation by both species is extremely fast and occurs

$$H^+ + HONO \rightleftharpoons H_2\overset{+}{O}NO \rightleftharpoons H_2O + N\overset{+}{O} \qquad (12)$$

close to the diffusion limit even for quite deactivated amines such as *p*-nitroaniline[21]. Thus the actual nitrosating species undergoing reaction at these high acidities is simply the equilibrium-determined mixture of both reagents present under the given conditions.

The formation of nitrosonium ion (equation 12) is formally similar to that of nitronium ion (NO_2^+) from nitric acid at high acidities. However, NO_2^+ is a far more electrophilic species and undergoes electrophilic substitution at carbon *c.* 10^{14} times faster than NO^+ [22]; moreover, in these A-S_E2 reactions, deprotonation of the Wheland intermediate is usually rate determining[23].

The diazotization of some heterocyclic amines has been examined over the region 0·0025–5·0 M perchloric acid[24, 25]; a single mechanism is operative over the entire region for both 2- and 4-aminopyridine, being first order in amine, in acid and in nitrous acid. Unlike aromatic amines, (with pK_a's < 5) 4-aminopyridine (**11**) is

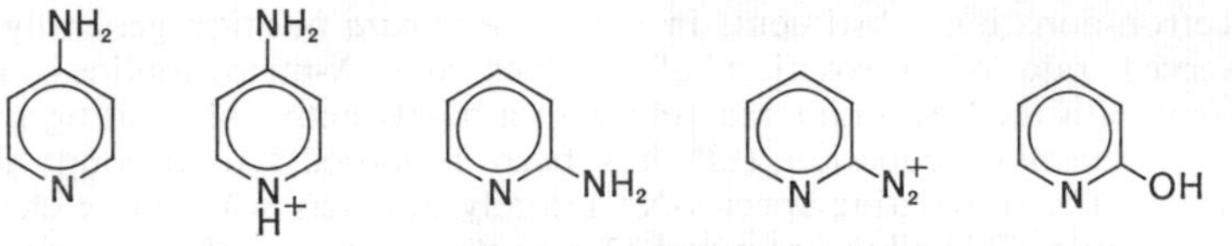

(**11**) (**12**) (**13**) (**14**) (**15**)

comparable in basicity to aliphatic amines (pK_a *c.* 10); however, protonation occurs on the ring nitrogen[25]. Protonation of the amino group of the pyridinium ion **12** is not expected to occur except at high acidities ($pK_a = -6{\cdot}55$). The mono-protonated species therefore behaves rather like an aromatic amine with a strongly electron-withdrawing substituent and reaction is thought to occur between this and nitrous acidium ion. The 2-amino isomer **13** follows a similar kinetic law[24], however the initial diazonium ion (**14**) formed reacts rapidly to give 2-hydroxypyridine **15**. The initial nitrosation step is also reversible in this case and satisfactory kinetics were obtained only by using a 5- or 10-fold excess of HNO_2.

D. Concentrated Acid

In the most acidic solutions the reactive species is NO^+ whose formation is favoured as the activity of water decreases (equation 12). At about 6 M-perchloric acid there is a sharp break in a k_{obs} vs. $[HClO_4]$ plot and the observed rate, instead of increasing with acidity (equation 10), decreases rapidly, following equation (13).

$$\text{Rate} = k[ArNH_3^+][HNO_2]h_0^{-2} \qquad (13)$$

This is consistent with a change in the rate-determining step and several pieces of evidence support the idea that trapping by a base of the species initially formed on reaction of the protonated amine is rate determining (equation 14). Thus a large

$$NO^+ + ArNH_3^+ \underset{}{\overset{\text{fast}}{\rightleftharpoons}} ArNH_2^+NO + H^+$$

$$ArNH_2^+NO + B \xrightarrow{\text{slow}} ArNHNO + BH^+ \qquad (14)$$

primary isotope effect is observed when hydrogen is replaced by deuterium, indicating proton transfer in the slow step[26]. Moreover, several aromatic amines were found to react at the same rate at high acidities; this is expected from equation (9) since the substituent effects in the initial equilibrium and the proton transfer should cancel.

One can rule out any change in the nitrosating species in this region since there is good evidence that NO^+ is the active reagent throughout[27]. Presumably the change in the rate-determining step to deprotonation of the intermediate is a consequence of two factors which reinforce each other: (a) *N*-nitrosation can be reversed at high acidities, and (b) the weakly basic species present at high acidities ensure that proton transfer from the *N*-nitrosoanilinium ion is thermodynamically unfavourable and slow.

E. Halide Ion Catalysis

At constant ionic strength (maintained by ClO_4^- or NO_3^-) it has been shown[28-32] that the rate of nitrosation of aniline is increased by added bromide, chloride or iodide ion (X^-) but not by fluoride ion (equation 15). Catalysis by halide is particularly important at low [HNO_2], which minimizes the concentration of N_2O_3, a competing nitrosation pathway.

$$\text{Rate} = k[\text{ArNH}_2][\text{H}^+][\text{HNO}_2][\text{X}^-] \tag{15}$$

In order to determine the reactive agent formed by added halide ion, Hughes and Ridd[30] used an elegant method, seeking conditions where the rate of formation of the reagent was the slow step. This was achieved at low acidities, low [HNO_2], and by using a reactive amine (*o*-chloroaniline). Under these conditions equation (16) is followed ($X^- = Br^-$ or I^-), i.e. the nitrosation step was rapid. This establishes that the reactive agent is the nitrosyl halide NOX (formed most likely by X^- attack on $(H_2ONO)^+$).

Schmid[30-32] calculated second-order rate constants for the reaction of aromatic and aliphatic amines with NOX, by correcting for the fraction of free amine and nitrosyl halide present under given conditions. It is clear from these results that all the aromatic amines studied react at much the same rate, *c.* 10^9 $M^{-1}\,s^{-1}$, which is close to the expected diffusion controlled limit. Moreover, both NOCl and NOBr

$$\text{Rate} = k'[\text{H}^+][\text{HNO}_2][\text{X}^-] \tag{16}$$

react at much the same rate, which is inconsistent with halide ion expulsion in the rate-determining step. Interestingly, the free aliphatic amines again (see Section II.B) react more slowly (*c.* 10^2-fold less than expected).

Because of the acidity dependence shown in equation (15) the rate of diazotization by HCl continues to increase up to *c.* 5·0 M. Under these conditions it has been shown that all the nitrous acid present is in the form of nitrosyl chloride[30].

F. Non-aqueous Solvents

Apart from the extensive work of Schmid and his coworkers[30-32] (who used methanol as solvent), little systematic study has been done on the mechanism of nitrosation in non-aqueous solvents. However, solvents such as benzene and ether, and reagents such as alkyl and acyl nitrites, have been used to achieve nitrosation under mild conditions. The use of some reagents, e.g. $NO^+BF_4^-$ requires dry organic solvents[33, 34]. However, kinetic analysis of such systems, which are often heterogeneous and involve highly associated substrates, is difficult.

In methanol, $H_2\overset{+}{O}\cdot NO$ is replaced by $Me\overset{+}{O}H\cdot NO$, which may be regarded as a solvated nitrosonium ion or protonated methyl nitrite. In the presence of HCl, equation (17) is followed, where f_{HCl} is the mean activity coefficient of HCl[35, 36].

$$\text{Rate} = kf_{\text{HCl}}^2[\text{PhNH}_3^+][\text{CH}_3\text{ONO}][\text{Cl}^-] \tag{17}$$

Schelly[37] has systematically varied the dielectric constant of the medium by the addition of CCl_4 as an inert cosolvent to methanol, and measured rate constants up to 60% CCl_4 when the dielectric constant (D) is 15. Correction was made for the degree of dissociation (α) of the ionic species present (e.g. $\alpha = 0{\cdot}6 \pm 0{\cdot}02$ for all electrolytes studied when $D = 19$). No diazotization occurs in the absence of added Cl^-, indicating that $CH_3\overset{+}{O}H\cdot NO$ and N_2O_3 do not nitrosate aniline, unlike their action in water. Although the active nitrosating agent is thought to be NOCl (as in pure methanol), shown by the rate dependence on $[NaNO_2]$, $[Cl^-]$ and $[H^+]$, its concentration is low due to the presence of a large excess of MeOH which pushes the equilibrium (18) to the right. Hence decreasing the dielectric constant of the medium (by the addition of CCl_4) actually increases the overall rate of diazotization.

$$CH_3OH + NOCl \rightleftharpoons CH_3ONO + H^+ + Cl^- \qquad (18)$$

The major effect operating is thought to be the decreasing activity of methanol which releases more NOCl (equation 18), although this is counteracted somewhat by the expected slowing of the nitrosation step with decreasing polarity of the medium due to (a) increased association and decrease in activity coefficients of the substrates, and (b) destabilization of the polar transition state. When $D < c.$ 15 it is suggested that tautomerization of the nitrosamine (PhNHNO → Ph—N=N—OH) is rate determining and this may be the slow step in aprotic media such as pure CCl_4. Data for this solvent were obtained by extrapolation, however, because of precipitation of the reagents[37].

III. REACTIONS OF ARENEDIAZONIUM IONS

A. Replacement of Nitrogen by Nucleophiles (Dediazoniation)

Three ionic pathways for the replacement of nitrogen from an arenediazonium ion (**16**) by a nucleophile Y^- (also termed dediazoniation[38]) are summarized in equations (19–21). Reaction (19) is analogous to the S_N1 mechanism and is

$$PhN_2^+ \ (\mathbf{16}) \xrightarrow[-N_2]{\text{slow}} Ph^+ \ (\mathbf{17}) \xrightarrow[\text{fast}]{Y^-} PhY \ (\mathbf{18}) \qquad (19)$$

$$\mathbf{16} \xrightarrow{Y^-} [\text{Ph}(\overset{\delta+}{N_2})(\overset{\delta-}{Y})] \ (\mathbf{19}) \xrightarrow{-N_2} \mathbf{18} \qquad (20)$$

$$\text{o-HC}_6H_4N_2^+ \ (+Y^-) \rightleftharpoons \text{o-}^-C_6H_4N_2^+ \ (\mathbf{20}) \longrightarrow C_6H_4 \ (\mathbf{21}) \xrightarrow{HY} \mathbf{18} \qquad (21)$$

characterized by a free phenyl cation (**17**) in the reaction pathway. Reaction (20) is a bimolecular nucleophilic aromatic substitution in which **19** can be either a transition state (synchronous loss of N_2 with attack by Y^-) or an intermediate; in the latter case either the formation or breakdown of **19** can be rate determining. The elimination–addition pathway (21) involves the formation of an aryne (**21**) followed by the addition of HY. Again any of the steps on this reaction sequence could conceivably be rate determining.

It must be remembered that these three pathways represent mechanistic extremes and that many reactions may be borderline and therefore have some of the characteristics of two pathways. Thus, for example, a strongly solvated phenyl cation **17** may be stabilized by Y^-; if such solvation is important in stabilizing the formation of **17** then the transition state will have S_N2 character (**19**). The benzyne pathway (equation 21) has not been observed in aqueous solution for normal diazonium salts in the absence of very strong bases; benzyne formation is common however under different conditions and is treated separately (see Section III.A.8). The distinction between mechanisms (19) and (20) remains an area of controversy and active research interest; it now seems certain however that path (19) is followed in water with most nucleophiles (H_2O, Cl^-, F^- and possibly Br^-) whereas path (20) may occur when stronger nucleophiles, e.g. NCS^-, are used particularly in solvents of lower ionizing power and when the arenediazonium ion has strongly electron-withdrawing substituents. An additional complication is that strong nucleophiles may react at the terminal nitrogen (see Section III.B) giving rise ultimately to free radical substitution of the aryl group; such alternative pathways must be rigorously excluded in any mechanistic study.

1. Phenyl cation pathway

Kinetic investigations by several groups have shown that the rate of decomposition of benzenediazonium ion in water is first order in **16** and shows little dependence on the concentration of added nucleophile[39–42]. Thus, for example, the rate of reaction in the presence of added HCl varies by less than 50% when the ratio of chlorobenzene to phenol formed as product changes from 0·05 to 3 [43, 44]. The simplest explanation for these observations (which have been confirmed when the ionic strength is maintained constant while the nucleophile concentration is varied[45]) is the formation of the phenyl cation **17** in the rate-determining step; **17** being relatively unstable is not very selective, reacting with the nucleophile present in excess[46]. Because of the low selectivity shown by **17**, displacements by this mechanism in water do not provide an efficient route to **18** except for phenol formation (**18**, Y = OH); alternative (catalysed) routes to **18** are however available (see section III.D).

Swain, Sheats and Harbinson[45] have presented several pieces of evidence supporting unimolecular reaction of **16**. Thus the rate of reaction is remarkably independent of solvent and nucleophile (see Table 1), e.g. there is a <2% change in rate when the solvent is changed from 80% to 105% H_2SO_4 in spite of the 10^3-fold change in a_{H_2O}. The entropy of activation (+10 e.u. at 25 °C) [43] is close to that for the solvolysis of *t*-butyl chloride[44] but distinctly different from the large negative entropies of activation shown by reactions in which a molecule of water is involved in the slow step. The reaction shows no solvent isotope effect ($k_{H_2O}/k_{D_2O} = 0{\cdot}98 \pm 0{\cdot}01$ [43, 45]; see Table 1) which rules out any mechanism involving the build-up of appreciable charge on oxygen in the transition state (as in equation (20), Y = OH_2).

Zollinger has demonstrated that benzenediazonium ion (**16**) labelled with ^{15}N in the β-nitrogen undergoes exchange with $^{14}N_2$ in trifluoroethanol under 300 atm nitrogen. Separate experiments also favour aryl cation formation in this solvent[48]. This result is particularly interesting both because of its analogy to the 'common ion effect', observed for the solvolysis of alkyl halides via the S_N1 mechanism[49], and the fact that this represents the first example of fixation of nitrogen at carbon. The extent of nitrogen incorporation is 2·46 ± 0·40% at 300 atm. Carbon monoxide gas, which is expected to be more nucleophilic than nitrogen, was also shown to react with the phenyl cation; in this case 2,2,2-trifluoroethylbenzoate was among the products[47].

TABLE 1. First-order rate constants for de-diazonization of 0·0003–0·0015 M $PhN_2^+BF_4^-$ at 25 °C [45]

Solvent	10^5k_1 (sec^{-1})
0·001% (0.001 M) H_2SO_4	4·59
0·10% (0·010 M) H_2SO_4	4·55
9·5% (1·0 M) H_2SO_4	4·12
23% (2·7 M) H_2SO_4	3·56
50% (6 M) H_2SO_4	2·68
80% (14 M) H_2SO_4	2·13
96% (18 M) H_2SO_4	2·11
105% (21 M) H_2SO_4	2·15
0·0001 M-D_2SO_4	4·76
0·010 M-D_2SO_4	4·71
1·0 M-D_2SO_4	4·11
100% CH_3CO_2H	4·26
100% CH_3CO_2H+0·13 M-LiCl	3·71
100% CH_3CO_2H+1·0 M-LiCl	4·51
CH_2Cl_2	2·20
3-Methylsulfolane	1·36
97% Sulfolane–3% $(C_2H_5)_2O$	1·24
Dioxane	1·15

2. Isotope effects

A large kinetic hydrogen isotope effect (k_H/k_D = 1·22) is observed for each hydrogen *ortho* to the diazonium group (equation 22)[50]. This indicates that a very electron-deficient species is being formed in the transition state (such as **22**). This

$$\xrightarrow{-N_2} \quad \text{(22)} \quad \longleftrightarrow \quad \text{(22a)} \qquad (22)$$

ortho effect is the largest secondary aromatic hydrogen isotope effect yet observed and is comparable to those observed for α-deuterium in reactions involving carbonium ion formation from tertiary aliphatic esters, and is taken as evidence for substantial hyperconjugative stabilization by the *o*-hydrogens of the phenyl cation (see **22a**).

The nitrogen isotope effect (replacement of the nitrogen attached to the ring in the diazonium group by ^{15}N), measured in 1% H_2SO_4 at 25 °C, is also consistent with this picture. The value obtained (1·038) is close to the calculated value (1·04–1·045) which would be observed for complete C—N bond cleavage in the transition state[51, 52]. Any smaller degree of C—N bond fission in the transition state would reduce the magnitude of this isotope effect so that any mechanism that does not allow for this (e.g. 20 with the first step being rate determining) can be confidently ruled out.

3. Selectivity and structure of phenyl cation

As mentioned above, the selectivity of the phenyl cation intermediate **17** is very low, comparable to that of the *t*-butyl cation (where $k_{Cl^-}/k_{H_2O} \sim 4$) [53]. The values reported for **17** (relative to 55 M-H_2O) are SO_4^{2-}, 1·4; Cl^-, 3; SCN^- (nitrogen), 3; SCN^- (sulphur), 6 [54]; Br^-, 6 [45].

The high reactivity of the phenyl cation probably arises from the location of the electron deficiency in an orbital (sp^2) with high s character[45]. The situation is formally similar to vinyl cation **23**.

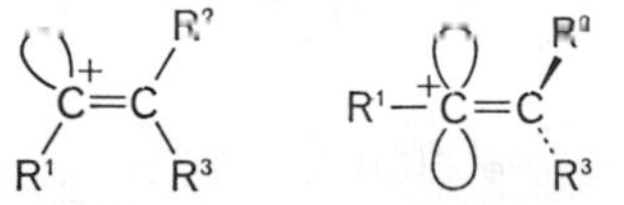

(23) **(24)**

Evidence for the formation of these unstable cations has recently been reported using several systems including electrophilic attack on acetylenes and solvolysis of vinyl halides[55, 56]. However, theoretical calculations (using *ab initio* methods)[55] indicated that the preferred structure is linear (**24**, favoured by 30 kcal mol^{-1} over the bent sp^2 structure **23**). Ring strain in the phenyl cation **17** would reduce any appreciable stabilization by structures analogous to **24** (similarly vinyl cation formation in cyclic systems is extremely slow)[55, 57].

A possible alternative structure for the (singlet) phenyl cation **17** is a triplet or biradical structure favoured by some authors[58–60] to explain *m*- and *p*-substituent effects on the rates of phenyl cation formation (see however Section III.A.4). On the other hand, molecular orbital calculations by Extended Hückel[61] and INDO[62] methods show that the singlet is favoured relative to the triplet by 146 kcal mol^{-1} [50]. But the preferred singlet structure predicted by this method is grossly distorted relative to that of benzene (e.g. the $C_{(1)}—C_{(2)}—C_{(6)}$ carbons are colinear)[50], which is surprising.

4. Substituent effects

The effect of *meta* and *para* substituents on the rate of dediazoniation of **16** is considerable (up to 10^5-fold rate difference) and although known for some time[43] has proved quite difficult to interpret. Thus all *para* substituents, including electron-withdrawing nitro and electron-donating alkyl, decrease the rate of nitrogen loss; in the *meta* position the strongly withdrawing NO_2, Br, Cl groups decrease the rate but *m*-MeO (which is normally electron withdrawing) and alkyl substituents increase the rate of reaction.

Any attempt to correlate these data using a simple Hammett equation (23) gives the scatter (Figure 1); similar results are also obtained with any other single

$$\log\frac{k}{k_{\mathrm{H}}} = \rho\sigma \qquad (23)$$

$$\log\frac{k}{k_{\mathrm{H}}} = f\mathscr{F} + r\mathscr{R} + i \qquad (24)$$

set of substituent constants. This arises since both the starting arenediazonium ion **25** and the aryl cation can be resonance stabilized by electron donation, but to different degrees. Excellent correlations are however obtained using Swain and Lupton's[63] four-parameter equation (24) (see Figure 2 for an example). The values of

(25) **(25a)**

$\mathscr{F}$ and $\mathscr{R}$ (the field and resonance contributions to substituent effects) were previously determined from other reaction series and the values of f and r which measure the sensitivity of the reaction to change in field and resonance effects calculated for the *meta* and *para* positions: $f_p = -2{\cdot}60$; $f_m = -2{\cdot}70$; $r_p = -3{\cdot}18$; $r_m = +5{\cdot}80$ [45].

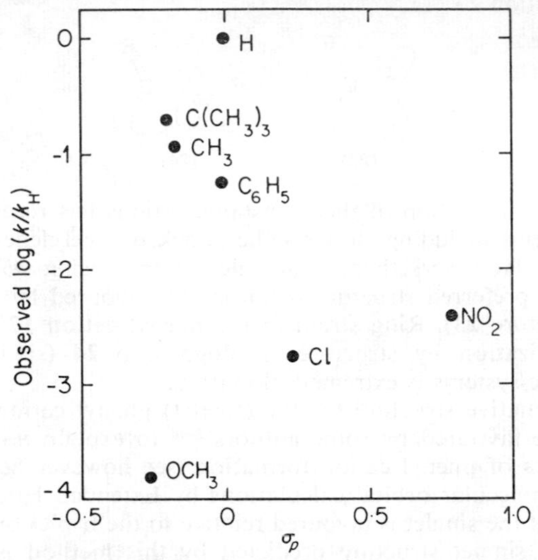

FIGURE 1. Hammett ($\rho\sigma$) plot of *para*-substituent effects on rate of dediazonization of **16**. Cl^- in 0·1 M-HCl at 25 °C [45].

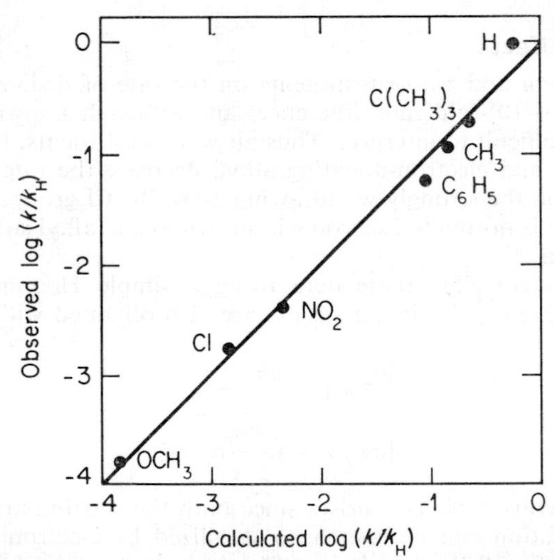

FIGURE 2. Dual substituent constant ($f\mathscr{F}+r\mathscr{R}+i$) plot of *para*-substituent effects on the rate of dediazonization of **16**. Cl^- in 0·1 M-HCl at 25 °C [45].

This implies that the inductive effect of *meta* and *para* substituents is similar but that the stabilization of the cation is most important for *meta* substituents while for *para* substituents the most important mechanism is stabilization of the starting benzenediazonium ion.

The excellent fit observed (Figure 2) using existing substituent constants is good evidence that a single mechanism (i.e. phenyl cation formation) is operative throughout the series. Moreover, this correlation does not involve a triplet structure for the cation[58–60], being consistent with singlet phenyl cation formation in each case.

5. Schiemann reaction

Fluoroarenes **27** are not prepared by the Sandmeyer reaction (Section III.D.1) since cuprous fluoride disproportionates to copper and cupric fluoride at room temperature. The Schiemann reaction which involves the thermal decomposition of arenediazonium fluoroborate however, provides an efficient route to these

$$\underset{\textbf{(26)}}{ArN_2^+BF_4^-} \xrightarrow{\Delta} \underset{\textbf{(27)}}{Ar{-}F} + BF_3$$

materials[64]. Mechanistic studies on this reaction[51] in methylene chloride at 25 °C have shown that the nucleophile involved is the BF_4^- rather than F^- and that reaction occurs via the rate-determining formation of a singlet aryl cation. Presumably the reactivity of the latter is sufficient to overcome the low nucleophilicity of BF_4^-. A secondary product noted in CH_2Cl_2 was ArCl formation and the relative amount of ArF formed was found to increase somewhat as **(26)** was increased; this was attributed to tighter ion-pair formation in the more concentrated solution.

6. Bimolecular mechanisms

The bimolecular nucleophilic displacement of nitrogen may be considered as a special case of nucleophilic aromatic substitution activated by the diazonium group (see Section III.C) at $C_{(1)}$. However, it is interesting to note that recent carbon-13 n.m.r. studies[65] have shown that there is a marked upfield shift of the carbon bearing the $-N_2^+$ group, indicating significant shielding, and that this effect is enhanced when a substituent Y on the ring is electron releasing. This emphasizes the importance of the contributing structures **28a** and **28b**. The increased electron density at $C_{(1)}$ would, of course, tend to disfavour direct nucleophilic attack at this position.

$$Y{-}C_6H_4{-}N_2^+ \longleftrightarrow Y{-}(C_6H_4)^{+}{}^{-}{-}N_2^+ \longleftrightarrow \overset{+}{Y}{=}C_6H_4{}^{-}{-}N_2^+$$

(28a) **(28b)**

Evidence has, however, been presented that the rate of dediazoniation of arenediazonium ions increases linearly with increasing bromide and thiocyanate ion[66]. The increase in rate however is quite small (and depends on a correction for the effect of added Na^+ on the activity of water) which makes an unambiguous distinction between the phenyl cation and bimolecular pathways inherently difficult. Thus the rate increases observed may have been due to specific medium effects (although the ionic strength was maintained constant in some studies)[67]. Clearly these data warrant reinvestigation in the light of Swain's results cited earlier. There seems also to be no good evidence[67] for the existence of a spirocyclic diazirine cation or excited diazonium ion on the reaction pathway for nucleophilic displacement of nitrogen[68].

Straightforward second-order kinetics have been observed for the heterolytic phenylation of a series of aromatic substrates in 2,2,2-trifluoroethanol. The second-order rate constants vary in the following direction: $C_6H_5CF_3 < C_6H_5OCH_3 < C_6H_5 \sim C_6H_5CH_3$ and a mechanism similar to equation (19) ($Y^- = C_6H_5X$) with considerable C—N bond cleavage in the transition state is proposed[69]. An alternative mechanism involving an encounter complex between the benzenediazonium ion and the aryl substrate is also consistent with the experimental data. However, direct displacement with the aryl substrate acting as a nucleophile can be ruled out since the strongly electron-withdrawing nitro group in the *para* position of the benzenediazonium ion does not increase the rate of arylation (as shown in other S_N—Ar displacements (Section III.C).

7. Nitrogen rearrangement accompanying hydrolysis

Lewis and coworkers[70–72] have demonstrated in an elegant series of experiments that the hydrolysis of diazonium salts is accompanied by a slower rearrangement in which the nitrogens reverse their positions (equation 25). This interesting rearrangement has attracted considerable attention (not always in agreement[73]) and is now supported by the work of several different groups[47, 74].

$$\mathrm{Ph{-}^{15}\overset{+}{\underset{\alpha}{N}}{\equiv}\underset{\beta}{N} \longrightarrow Ph{-}\overset{+}{N}{\equiv}^{15}N} \tag{25}$$

No rearrangement occurs when the solid diazonium salt is stored for long periods. The extent of rearrangement which occurs in solution is quite small, amounting to just 1·6% of the rate of hydrolysis[47, 74, 75]. The rate of rearrangement is first order in the arenediazonium ion, which rules out any mechanism involving exchange of nitrogen between two molecules of the arenediazonium ion. Moreover, the extent of rearrangement occurs in parallel to the hydrolysis of the arenediazonium ion, the percentage rearranged product increasing at the same rate as other hydrolysis products[74]. In fact the mechanism of both processes must be closely similar since the substituent effects on the rate of rearrangement show the same individualistic pattern as phenyl cation formation (see Section III.A.4). This is shown in Figure 3, where

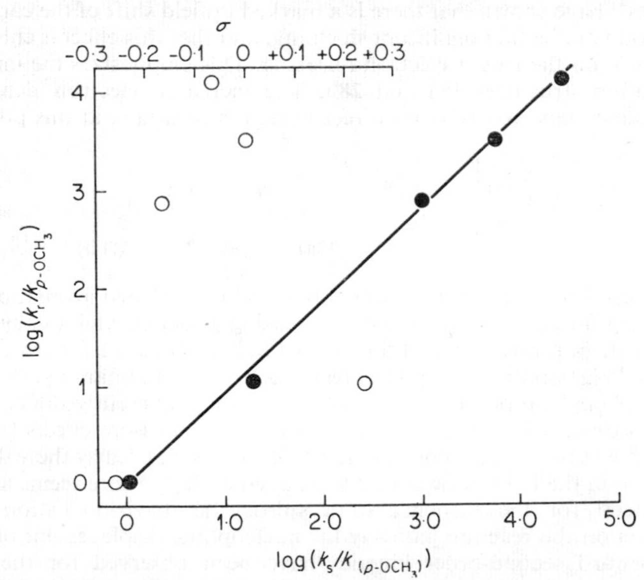

FIGURE 3. Relationship between rearrangement rates, k_r (relative to that of the *p*-methoxybenzenediazonium ion), and hydrolysis rates (k_s) also relative to that of the *p*-methoxy compound (●); lower scale (○), relation between k_r and Hammett σ.

a plot of the log of the relative rates of rearrangement of a series of aryl diazonium ions against the log of dediazonization rates is shown; the good correlation contrasts with that against the Hammett σ values.

Lewis originally proposed[70] that a stable spirodiazine cation **29** was the key intermediate in the rearrangement but later rejected this view[75] since the observed substituent effects (expected for the transition state **30**) are clearly different from the

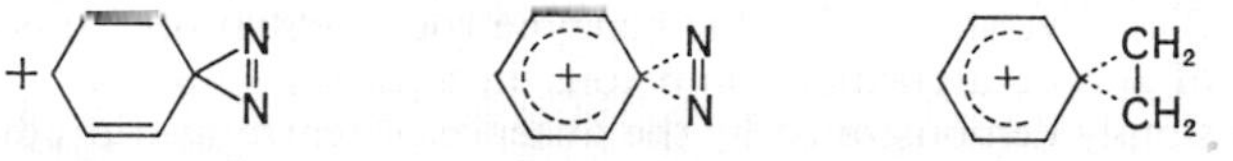

(29) (30) (31)

migratory aptitudes described by Cram[76] for phenonium ion formation. The rearrangement most likely involves a transition state close to the phenyl cation **32**; recapture of nitrogen can then occur before the species are separated by solvent **33**.

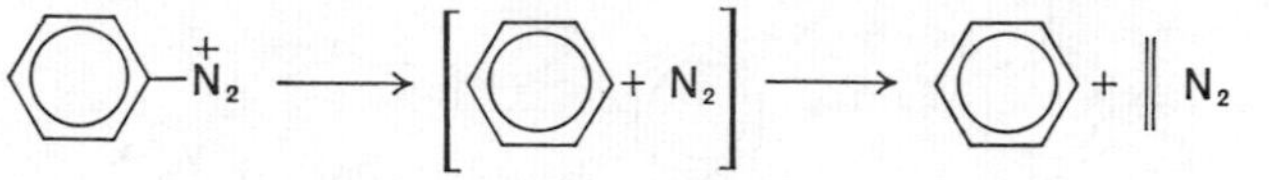

(32) (33)

The N_α–N_β rearrangement takes place at a higher rate in trifluoroethanol[47] (7·2% of the hydrolysis rate at 30 °C and 7·9% at 5 °C) and the possibility that it might occur via the mechanism of Scheme 2 was carefully investigated[48]. The cyclic intermediate **35** could give rise to the isomerized diazonium ion **36** but rearranged

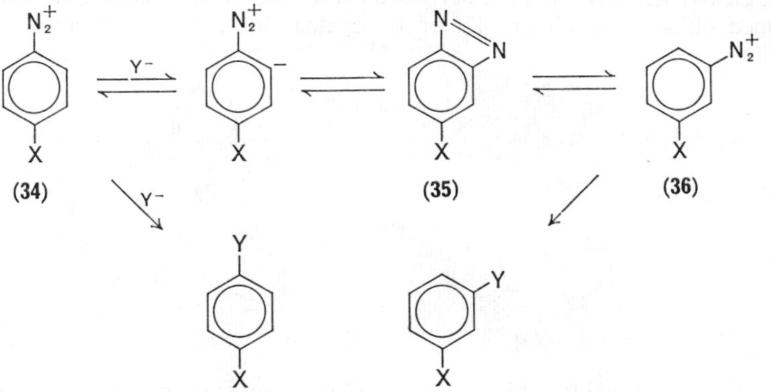

(34) (35) (36)

(37) (38)

SCHEME 2

products **38** would also result on reaction with the nucleophile Y^-. However, no such *m*-chloro products were detected (**38**, X = Cl) with *p*-chlorobenzenediazonium ion (**34**, X = Cl). It is clear therefore that the β-N becomes attached to the carbon vacated by the original α-N.

Addition of aryl radicals to the terminal nitrogen has been shown to yield the azobenzene radical cation **39**. By considering the initial reaction to be reversible,

$$Ar^1N_2^+ + Ar^\bullet \rightleftharpoons Ar^1-\dot{N}=\overset{+}{N}-Ar$$

(39)

^{15}N-scrambling (ArN_2^+ formation) can be explained[77]. However, kinetic studies argue against this mode of breakdown for **39** [78]; moreover such a radical mechanism is not consistent with the substituent effects observed for nitrogen reversal during solvolysis.

8. Benzyne formation

Several studies have shown that the benzyne route (equation 26) is not important in reactions of simple arenediazonium ions in aqueous solution at moderate pH. This is most simply demonstrated by the absence of rearranged products. Thus, for example, reaction of *o*-toluenediazonium chloride in water yields *o*-cresol but no *m*-cresol (<0·1% could be detected)[45]; another example is quoted in Section III.A.7.

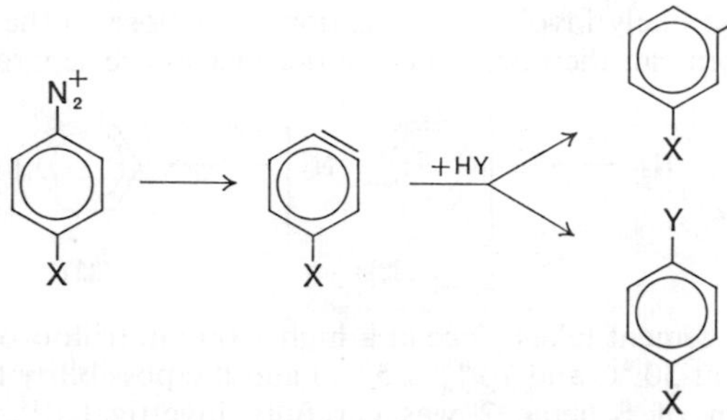

(26)

However, decomposition of benzenediazonium-2-carboxylate (**40**) in non-aqueous solution clearly leads to benzyne (**41**) formation since this intermediate can readily be trapped either as a dimer (**42**) or by cycloaddition to a 4-π donor (**43**) [79, 80].

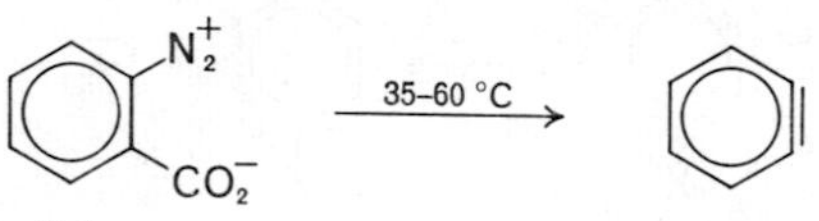

35–60 °C

(40) (41)

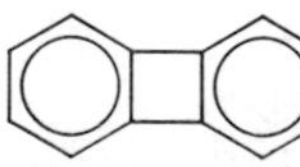

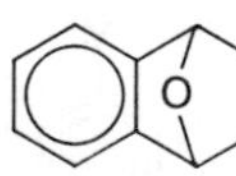

(42) (43)

Other *o*-diazonium acids which react similarly are benzenesulphinic (**44**) [81] and benzeneboronic (**45**) [82].

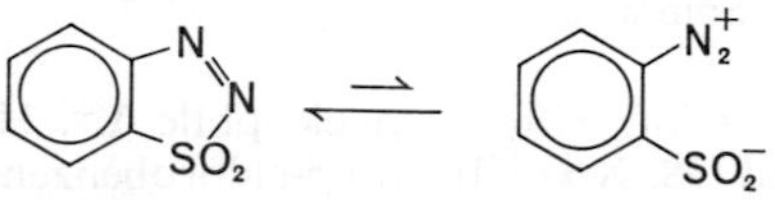

(44) 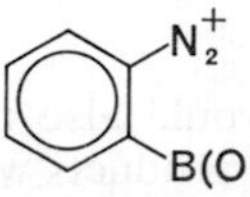 (45)

The formation of benzyne (**41**) from **40** could occur either synchronously (**46**) or asynchronously (**47** or **48**). In aqueous solution the formation of salicyclic acid (**49**) also occurs presumably by reaction with the zwitterion **47** [83]. Evidence is presented[84] that the formation of the aryne (trapped by furan) and salicylic acid (**49**) are formed via a common intermediate since the ratio of final products depends on the water

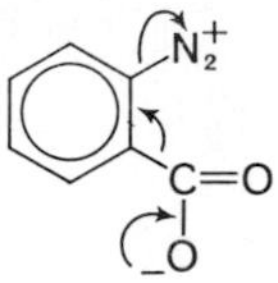

(46) 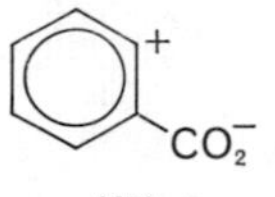 (47) 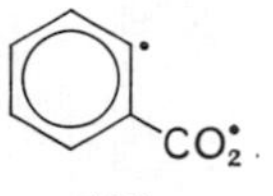 (48)

concentration but is independent of added furan. This is only consistent with a step-wise formation of benzyne involving preliminary loss of N_2 to give the zwitterion **47** (which may be in equilibrium with **50**).

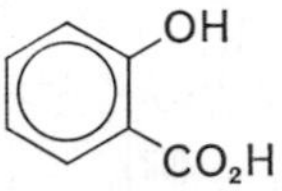

(49) 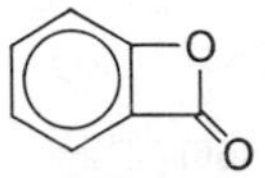 (50)

Cadogan[85-90], Ruchardt[91] and their coworkers have described another simple route to arynes (Scheme 3), involving acetate-induced elimination from benzenediazonium acetate (**53**). The simplest route to the acetate **53** is by heterolytic fission of the azoacetate **52**, which is formed on rearrangement of *N*-nitrosoacetanilide **51**.

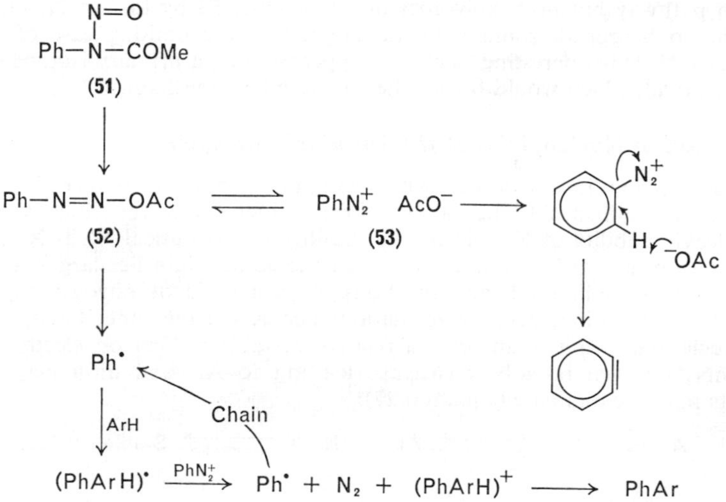

Ph—N(N=O)—COMe (51)

Ph—N=N—OAc (52) ⇌ PhN_2^+ AcO^- (53) ⟶ benzyne

(52) ⟶ $Ph^•$; $Ph^•$ + ArH ⟶ $(PhArH)^•$ $\xrightarrow{PhN_2^+}$ $Ph^•$ + N_2 + $(PhArH)^+$ ⟶ PhAr (Chain)

SCHEME 3

The azo compound **52** can also react by a competing free-radical mechanism (see Section III.E.1), but this route involves a chain reaction and can be suppressed by the addition of suitable traps such as 1,1-diphenylethylene[85]. Also of interest is the fact that the reactions of *N*-nitroso-*p*-chlorobenzanilides ($ArN(NO)COC_6H_4Cl$-*p*) in dry carbon tetrachloride give *p*-chlorobenzoic acid as a *primary* product, indicating that the abstraction of a proton as well as loss of nitrogen had occurred[86].

The formation of benzyne by acetate catalysis appears to be a concerted E2-type elimination (path (a), Scheme 4). Thus when **54** is labelled with deuterium in both positions *ortho* to the diazonium group, *c.* 50% is retained in the aryne **55** (which was trapped by anthracene as the adduct **57**)[92], contrary to an earlier report[93].

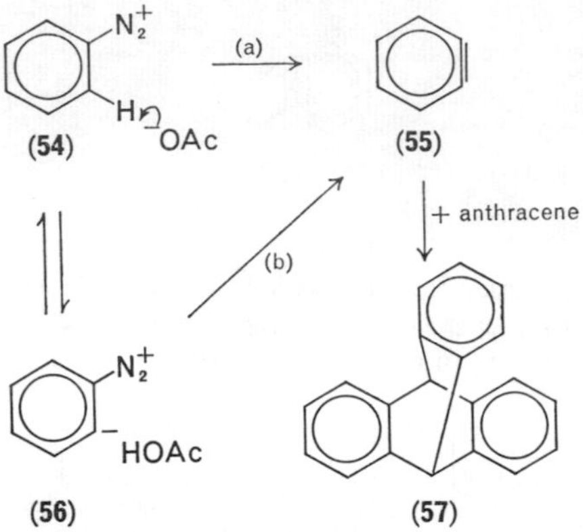

(54) (55) (56) (57)

SCHEME 4

Similarly there is no uptake of deuterium in the product **57** when the benzene-diazonium ion **54** is reacted in the presence of AcO^-–DOAc. This rules out an alternative pathway (b) via the betaine **56** which is in equilibrium with the starting diazonium ion **54**. However, the possibility exists that the betaine is in fact on the reaction pathway but invariably goes on to product **55** by loss of N_2 rather than returning to benzenediazonium ion (analogous to the limiting case of the ElcB mechanism)[94]. It is interesting that there appears to be a primary isotope effect for C—H removal, which would favour the concerted E2 mechanism[92].

B. *Reaction of Nucleophiles at the Terminal Nitrogen*

Arenediazonium ions may react with strong nucleophiles at the terminal nitrogen to give the azo adducts **58**. The stability of the adduct is however critically dependent on the leaving ability of X^- and on the stability of the radical $X^\bullet$. If X^- is a good leaving group (e.g. halide ion, acetate) then the equilibrium lies largely to the side of the arenediazonium ion and reactions characteristic of nitrogen replacement (aryl cation formation, benzyne formation) occur. On the other hand, if X^- is a good nucleophile which can form a relatively stable radical on electron transfer (e.g. PhN_2O^-) then homolytic cleavage leading to $Ar^\bullet$ and ultimately arylation products may predominate (equation 27).

$$\mathrm{Ar{-}\overset{+}{N}{\equiv}N + X^-} \rightleftharpoons \underset{(58)}{\mathrm{Ar{-}N{=}N{-}X}} \rightleftharpoons \text{Stable product}$$

$$\downarrow$$

$$\mathrm{ArN_2^{\bullet\bullet}X} \longrightarrow \mathrm{Ar^\bullet\ N_2\ X^\bullet} \qquad (27)$$

Stabilization of the adduct **58** can be achieved by using good nucleophiles which are relatively poor leaving groups (e.g. carbanions such as cyanide ion or phenoxide ion). An alternative method of stabilization of **58** is by conversion to a derivative which is more resistant to loss of X^-; two methods which have been observed are conversion to a conjugate base (e.g. diazotate formation) or isomerization of an initially formed *syn* isomer (where the Ar and X groups are *cis* to one another) to a more stable *anti* isomer.

The preferential formation of *syn* isomers, as the kinetically controlled product, has been noted in several instances with strong nucleophiles (such as ^{-}CN, SO_3^{2-}, MeO^- and HO^-). However, the *anti* isomer is generally more stable thermodynamically and thus the dilemma arises as to what features stabilize the transition state leading to the less stable isomer. Moreover with several nucleophiles [e.g. aromatic amines (reaction at nitrogen or carbon) and phenoxides (at carbon)] arenediazonium ions give only *anti* products.

This situation is, however, paralleled in the stereospecific reaction of nitrilium ions (**59**) with nucleophiles (e.g. AcO^-, MeO^-)[95, 96]. The exclusive formation of the *Z*-isomer (**60**) can be shown in all cases (R must however be chosen so that rapid

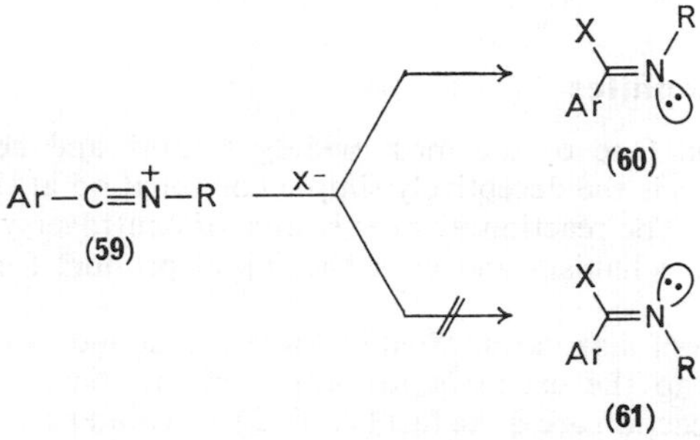

$$Ar{-}C{\equiv}\overset{+}{N}{-}R \ (\mathbf{59}) \xrightarrow{X^-} \mathbf{60} \ (Z) \quad \text{not} \quad \mathbf{61} \ (E) \qquad (28)$$

(59) **(60)** **(61)**

interconversion of the *Z* and *E* (**61**) isomers does not occur). However, when the two isomers are equilibrated at higher temperature the *E* isomer is usually present in larger amounts at equilibrium. The exclusive formation of **60** has been attributed to minimalization of interorbital repulsion between the incoming nucleophile and the forming lone pair on nitrogen and *ab initio* and CNDO/2 calculations support this picture. It is interesting that the isomer with the lone pair and group X *trans* to one another (i.e. **60**) should also lose X^- more rapidly to regenerate the ion **59** and this has been shown experimentally in one case (**60** and **61**, R = $OCH_2CH_2CH_3$; Ar = Ph; X = Cl)[97].

A similar reasoning has been applied to the preferential formation of the *syn* isomer **62** (which corresponds to the *E* isomer **60** since the electron pair on the α nitrogen and the nucleophile are *trans*)[98]. With the more reactive nucleophiles an

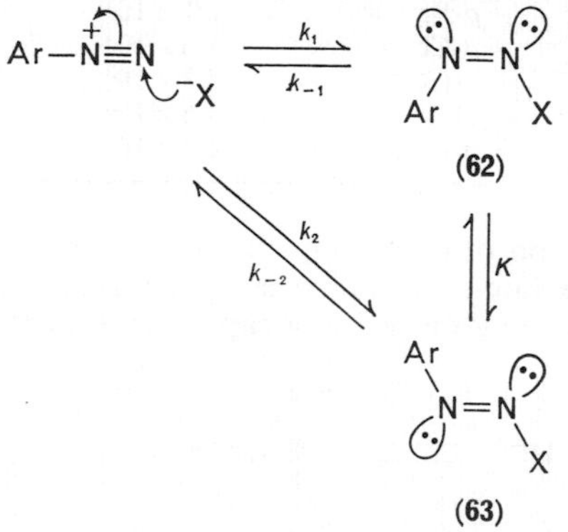

$$Ar{-}\overset{+}{N}{\equiv}N + {}^{-}X \underset{k_{-1}}{\overset{k_1}{\rightleftharpoons}} \mathbf{62}; \quad Ar{-}\overset{+}{N}{\equiv}N + {}^{-}X \underset{k_{-2}}{\overset{k_2}{\rightleftharpoons}} \mathbf{63}; \quad \mathbf{62} \overset{K}{\rightleftharpoons} \mathbf{63}$$

(62) **(63)**

early transition state (with relatively little N—X bond formation in the transition state) is implicated from substituent effects on the reactivity of the arenediazonium ion (see below); thus the various factors which help stabilize the *trans* relative to the *cis* azo linkage will be less important in the transition state. Zollinger[98] has also proposed that the observation of *anti* products (**63**) with some nucleophiles may be due to a changeover from an 'early' to 'late' transition state (where the greater stability of **63** would become important) and there is some evidence that the arenediazonium ion has lost a greater fraction of its charge in the transition state for these reactions. However in several instances, e.g. triazene formation and diazo coupling to phenolates, the initial coupling step is reversible. Hence the possibility arises that the *syn* isomer is in fact formed most rapidly in all cases (i.e. $k_1 > k_2$) but where the initial coupling step can be rapidly reversed (i.e. k_{-1} is large) then the product isolated is the *anti* isomer (**63**), since this reverts more slowly to the starting materials (k_{-2} small).

I. Oxygen nucleophiles

a. *Hydroxide ion.* One of the most widely studied and complex reactions of arenediazonium ions is the deceptively simple combination at the terminal nitrogen with hydroxide ion. The reaction was the source of controversy for many years and the details of the mechanism and even the initial product formed are not as yet universally agreed.

(i) Isomerization of diazotates. Most of the evidence indicates that the hydroxide ion reaction leads to the exclusive formation of *syn*-diazotates (**64**)[99]. The rate constants for this reaction are quite fast (Table 2) as would be expected for a reaction

TABLE 2. Rate constants for the reaction of arenediazonium ions with hydroxide ion in aqueous solution at 23 °C[100]

$$XC_6H_4N_2^+ + OH^- \overset{k}{\rightleftharpoons} syn\text{-}XC_6H_4N_2OH$$

X	k (l mol^{-1} sec^{-1})
p-NO_2	$5{\cdot}4 \times 10^5$
p-CN	$4{\cdot}2 \times 10^5$
m-CF_3	$1{\cdot}6 \times 10^5$
m-Cl	$6{\cdot}4 \times 10^4$
p-Br	$2{\cdot}1 \times 10^4$
p-Cl	$1{\cdot}6 \times 10^4$
p-CO_2^-	$5{\cdot}8 \times 10^3$
H	$4{\cdot}5 \times 10^3$
p-CH_3	$1{\cdot}2 \times 10^3$

involving the combination of oppositely charged ions[100, 101]. The kinetic behaviour clearly shows that the rate-determining step is formation of the diazohydroxide (**64**), the diazotate (**65**) being formed in a rapid subsequent equilibrium. These data

$$ArN_2^+ + HO^- \underset{k_{-1}}{\overset{k_1}{\rightleftharpoons}} \underset{(64)}{\overset{Ar\qquad OH}{N{=}N}} \overset{K_a}{\rightleftharpoons} \underset{(65)}{\overset{Ar\qquad O^-}{N{=}N}} \qquad (29)$$

give a good Hammett plot of $\log k_1$ against ordinary σ values[102] to give $\rho = +2{\cdot}61$. The overall equilibrium constant for diazotate formation ($K = [ArN_2O^-]/([ArN_2^+][HO^-]^2)$) is, as expected, more sensitive to the nature of the aryl substituent

(Table 3), giving a ρ value of 6·58 [100] (or 6·3 from earlier work)[99], since electron-withdrawing substituents which increase the reactivity of the diazonium ion will also increase the acidity constant K_a. The value of $\rho = 6{\cdot}58$ is therefore composite but reasonable on the basis of the ρ values for the individual steps (governed by k_1, k_{-1} and K_a) which have been separately determined (see below).

TABLE 3. Equilibrium constants for the reaction

$$XC_6H_4N_2^+ + 2\ OH^- \overset{K}{\rightleftharpoons} syn\text{-}XC_6H_4N_2O^-$$

in aqueous solution at 23 °C, $\mu = 0{\cdot}004$–$0{\cdot}03$ M

X	Buffer	K (M^{-2})
p-NO_2	Borate	$2{\cdot}0 \times 10^{10}$
p-CN	Borate	$3{\cdot}9\ (\pm 0{\cdot}3) \times 10^{9}$
m-CF_3	Borate, bicarbonate	$1{\cdot}0\ (\pm 0{\cdot}1) \times 10^{8}$
m-Cl	Bicarbonate	$1{\cdot}9\ (\pm 0{\cdot}3) \times 10^{7}$
p-Br	Bicarbonate, phosphate	$1{\cdot}4\ (\pm 0{\cdot}3) \times 10^{6}$
p-Cl	Bicarbonate, hydroxide	$6{\cdot}7\ (\pm 0{\cdot}8) \times 10^{5}$
p-CO_2^-	Bicarbonate, hydroxide	$1{\cdot}9\ (\pm 0{\cdot}2) \times 10^{5}$
H	Hydroxide	$2{\cdot}8\ (\pm 0{\cdot}7) \times 10^{4}$
p-CH_3	Hydroxide	$2{\cdot}2\ (\pm 0{\cdot}7) \times 10^{3}$

The order of reactivity shown by arenediazonium ions towards various nucleophiles ($N_3^- > HO^- \sim CH_3O^- > CN^-$) is the same as that shown in a more extended series of nucleophiles and in various solvents (methanol, water, dimethylsulphoxide, dimethylformamide) with carbonium ions as substrates[103]. Ritchie has used this to define constants N_+, which are characteristic only of the nucleophile and may be used to predict reactivity[104]. The invariant order of nucleophilicities is no doubt in part due to the absence of a leaving group in the defining reactions [as equation (29)].

Evidence for the presence of the *syn*-diazohydroxide **64** on the reaction pathway is difficult to obtain from the forward reaction ($ArN_2^+ + HO^-$) since at all pH's where the rate of formation of the diazohydroxide is appreciable, the —OH group is ionized (i.e. pH > pK_a). The concentration of free diazohydroxide present at equilibrium is therefore never greater than a few percent[105]. For example when p-nitrobenzenediazonium ion is titrated with HO^-, 2 moles base are consumed but only a single inflection point is noted with apparent pK_{app} *c.* 9·4. When 1 mole HO^- is used, 50% goes to the diazotate and 50% remains as the diazonium ion[106]. Such titrations are however complicated by isomerization of both the diazotate **65** and diazohydroxide **64**.

When an arenediazonium ion is dissolved in alkaline solution, an initial rapid reaction occurs (formation of the diazotate) followed by a slower reaction (with a half-life in the range of seconds to several hours at room temperature)[99, 107–109]. This was attributed by Lewis and Suhr[99] to the isomerization of the initially formed *syn*-diazotate **65** to the *anti*-diazotate.

Detailed studies by Štěrba and coworkers[107, 108] have determined that the rates of *syn–anti* isomerization of a series of diazotates follow equation (30) (k in s^{-1}). Possible mechanisms of isomerization which could explain the large degree of

$$\log k = 2{\cdot}1[\sigma + 2{\cdot}4(\sigma^- - \sigma)] - 2{\cdot}83 \tag{30}$$

resonance interaction found (2·4) include rotation about the N—N bond (**66**) or a lateral shift involving the N-aryl ring (**67**). The *anti*-diazotate isomer is normally the thermodynamically favoured isomer and hence the stable form in alkaline solution. A value of 600 has been estimated as the equilibrium constant for 4-nitrobenzenediazotate [110]. However, considerably smaller values have been reported for *ortho*-substituted benzenediazotates[107] (and in fact on this basis a proposal that the original structural assignments of *syn*- and *anti*-diazotates should be reversed has been made!)[107].

(66) **(67)**

Müller and coworkers[111] have reported the isolation of a relatively stable solid *syn*-diazotate at low temperature and determined its structure by X-ray crystallography; other *syn*-diazotates can be isolated if the arene group does not contain electron-donating groups. Several alkanediazotates have been reported to have the *syn* structure[112], including 1-phenylethanediazotate [113] and methanediazotate [114] (confirmed by X-ray)[115]. These *syn*-diazotates are usually prepared by basic cleavage of *N*-alkyl-*N*-nitrosourethanes, while the *anti*-diazotates can also be prepared by nitrosation of hydrazines[113, 116], a method originally used by Thiele and by Stolle[116] (equation 31).

$$\mathrm{RNHNH_2} \xrightarrow[\mathrm{NaOEt}]{\mathrm{EtONO}} \left[\begin{array}{c}\mathrm{ON\quad NO}\\ \mathrm{R-N-NH}\end{array}\right] \longrightarrow \mathrm{R{-}N{=}N{-}O^-} + \mathrm{N_2O} + \mathrm{EtOH} \qquad (31)$$

(ii) *Reaction of diazotates with acid.* The back reaction (conversion of arenediazotates to arenediazonium ions) has been investigated in detail by several groups

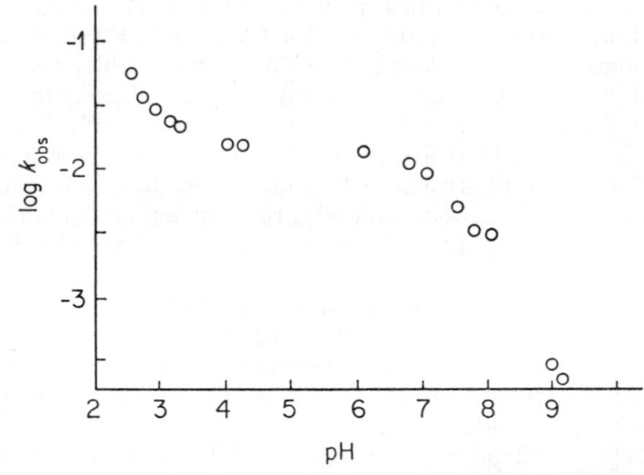

FIGURE 4. Dependence of rate of conversion of *anti*-benzenediazotate ion to diazonium ion on pH.

of workers. The substrate normally used in these studies is the more stable *anti*-diazotate and the observed pH–rate profile (see Figure 4 for a typical example) fits a two-component rate equation (32); the second term is only of importance at low

$$\text{Rate} = k[\text{Diazotate}\cdot\text{H}^+] + k'[\text{H}^+][\text{Diazotate}\cdot\text{H}^+] \qquad (32)$$

pH. Lewis and Hanson[117] have made a careful study of the basicity and equilibria involved for *anti*-diazotates and have shown that the kinetically determined pK_a's (from plots such as Figure 4) are consistent with estimates of diazotate pK_a's determined either spectrophotometrically or by titration (Table 4). The pK_a data fit the Hammett equation, $pK_a = 7{\cdot}3 - \rho\sigma$ with $\rho = 1{\cdot}45$, in agreement with other work[105, 118].

TABLE 4. pK_a's of *anti*-$ArN_2O^- \cdot H^+$ in water at 25·1 °C [117]

	pK_a		
Aryl group	Titration	Kinetic	Spectral
p-$CH_3C_6H_4$		7·40	
C_6H_5	7·25 ± 0·06	7·29	
p-ClC_6H_4	7·1 ± 0·1	6·95	
m-ClC_6H_4		6·76	
p$NO_2C_6H_4$	6·25 ± 0·05	6·13	6·36 ± 0·03
p-$N_2^+C_6H_4$		4·96	
2-Pyridyl		6·4	

The key questions which must be answered include the position of protonation of the diazotate, and whether the diazotate and/or the protonated species formed undergoes further isomerization before N—O bond cleavage occurs to give the arenediazonium ion.

The measured rate constant k (equation 32) could represent either (a) direct loss of HO^- from the protonated *anti*-diazotate (k_4, Scheme 5) or (b) reaction via the *syn*-diazohydroxide, with either its formation (k_1) or breakdown (k_3) as rate determining. Evidence that such a changeover in mechanism does occur comes from the data of Lewis and Hanson[117]. From Table 5 it is clearly seen that there is a minimum in a plot of $\log k$ against σ (or σ^-) [119], since both electron-donating and -withdrawing substituents can aid reaction.

TABLE 5. Rate constants for the first-order and acid-catalysed conversion of the conjugate acids of *anti*-diazotates to diazonium ions at 25·1 °C

Diazotate	σ	$k \times 10^2$ (sec^{-1})	k' ($l\ mol^{-1}\ sec^{-1}$)
p-$CH_3C_6H_4$	−0·069	5·0 ± 0·1	27 ± 4
C_6H_5	0·0	1·5 ± 0·1	13 ± 1
p-ClC_6H_4	+0·227	0·33 ± 0·01	7 +0·5
m-ClC_6H_4	+0·373	0·23 ± 0·01	1·3 ± 0·2
p-$NO_2C_6H_4$	+0·778	0·48 ± 0·01	0·053 ± 0·001
p-$N_2^+C_6H_4$	+1·91	35 ± 1	0·00195
2-C_5H_4N		5·8 ± 0·1	

It is proposed that with electron-donating substituents in the arenediazonium ion, k_4 is rate determining; the Hammett ρ value obtained (−2·6) is reasonable on the basis of the value (+2·61) already quoted for the back reaction between arenediazonium ions and hydroxide to give the *syn*-diazohydroxide (k_{-3}). When strongly electron-withdrawing substituents (e.g. *p*-NO_2, *p*-N_2^+) are present then k_4 is very much reduced; the rate of *anti* → *syn* isomerization is concomitantly increased so that reaction occurs via the more reactive *syn* isomer. From a two-point plot against σ^-, Lewis and Hanson[117] reported a ρ value of $+1{\cdot}1$ in this case. More extensive data have been presented by Štěrba and coworkers[107] using *para*-substituted *o*-nitrobenzenediazonium ions to give $\rho = 1{\cdot}0$ (with r, the degree of resonance interaction[120], $= 0{\cdot}65$). It is likely that k_1 rather than k_3 is rate determining under these conditions since such small positive ρ values are characteristic of isomerizations about the azo linkage.

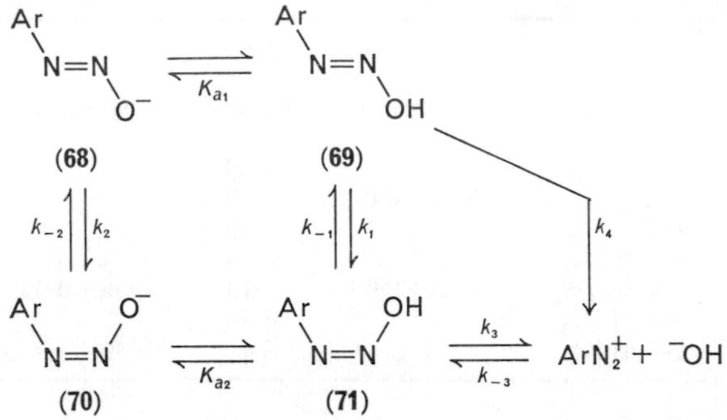

Ar–N=N–O⁻ (68) $\underset{K_{a_1}}{\rightleftharpoons}$ Ar–N=N–OH (69)

(68) $\underset{k_2}{\overset{k_{-2}}{\rightleftharpoons}}$ (70); (69) $\underset{k_1}{\overset{k_{-1}}{\rightleftharpoons}}$ (71); (69) $\xrightarrow{k_4}$ $ArN_2^+ + {}^-OH$

Ar, O⁻ N=N (70) $\underset{K_{a_2}}{\rightleftharpoons}$ Ar, OH N=N (71) $\underset{k_{-3}}{\overset{k_3}{\rightleftharpoons}}$ $ArN_2^+ + {}^-OH$

SCHEME 5

At high pH (*c.* 8) there is a further change in mechanism as shown in the inflection point in the pH–rate profile for the conversion of *anti*-2-nitro-4-chlorobenzenediazotate to the corresponding diazonium ions (Figure 5). This is attributed to the

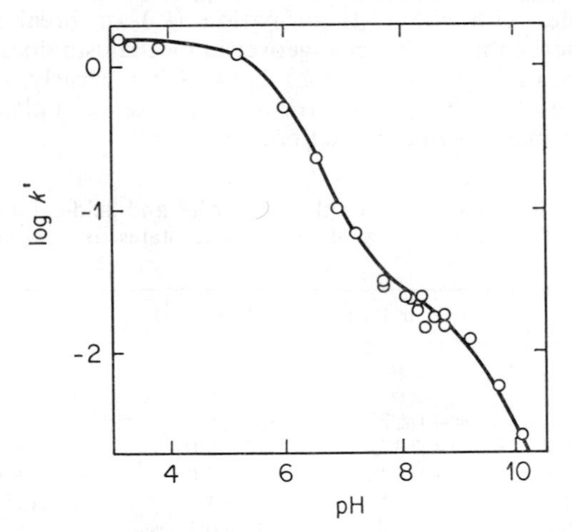

FIGURE 5. pH–rate profile for the inversion of *anti*-2-nitro-4-chlorobenzenediazotate into diazonium ions[107].

incursion of a rate-determining isomerization of the *anti*-diazotate to the *syn*-diazotate **70** (k_2). Since in this pH region pH > pK_{a_1} or pK_{a_2}, this interconversion is pH independent. However the observed rate of formation of arenediazonium ion decreases again at high pH since the back reaction of ArN_2^+ and HO^- becomes appreciable.

The pH profile for the acid-catalysed reaction of pyridine-2-diazotate (**72**) is broadly similar to that for other arenediazotates, but the 4-diazotate (**73**) shows

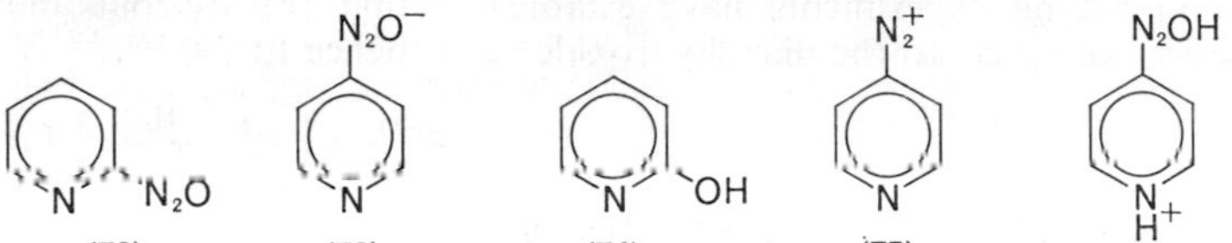

(**72**) (**73**) (**74**) (**75**) (**76**)

some interesting differences[121, 122]. The pH profile for **72** (Figure 6) is correlated by equation (31) with $k = 6 \times 10^{-2}$ sec^{-1} and pK_a of diazohydroxide = 6·2 (at 25 °C). The corresponding arenediazonium ion cannot be isolated and the product detected was 2-hydroxypyridine (**74**).

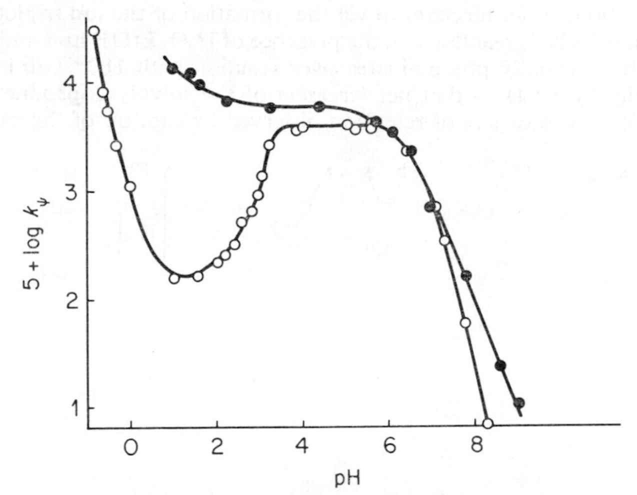

FIGURE 6. Variation of the rate constants for decomposition of pyridine-2-diazotate (**72**, solid circles) and pyridine-4-diazotate (**73**, open circles) as a function of pH.

The 4-diazotate **73** shows a distinctly different profile. In the pH region 7–8 the observed rate is proportional to $1/[HO^-]^2$, while there is a 'dip' in the pH–rate profile at *c.* pH 2. In this case an unstable intermediate was detected in the 'plateau' region (pH 4–6) and this was identified as the corresponding pyridine-4-diazonium ion (**75**) by trapping experiments. The back reaction of **75** with HO^-, which could be deduced from these trapping experiments, was insignificant in this pH region; however, this became kinetically important at high pH (7–8·5). The reduction in rate *c.* pH 2 was attributed to protonation of the pyridyl nucleus of the diazohydroxide, (**76**), which is expected to undergo a slower rate of HO^- loss than the corresponding free base. The results obtained by Bunton and coworkers[121] on this system do not require the inclusion of (or indeed rule out) steps involving the requirement of *syn–anti* isomerization before N—O bond cleavage; the initial material was assumed to be the more stable *anti*-diazotate since the same results were obtained when the diazotate was pretreated for 24 h in 0·1 M-$[HO^-]$ [121].

(iii) Alkanediazotates. The solvolysis of *syn*-alkanediazotates can take two routes (Scheme 6). When R is a secondary alkyl group then the formation of **79** is favoured.

At high pH both diazoalkane (**80**) and carbonium ion are formed[123, 124]. Stabilization of the α-carbon by conjugation with a vinyl or aryl group aids diazoalkane formation. Deuterium-labelling experiments have established that the diazoalkane **80** once formed does not revert to the diazohydroxide (and thence to **79**)[123].

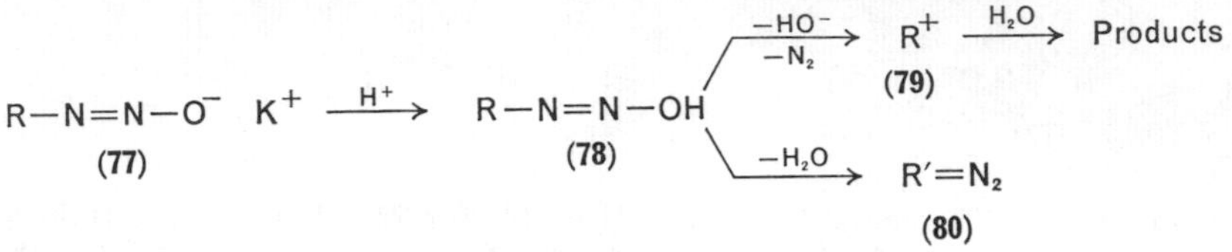

$$\text{R—N=N—O}^{-}\ \text{K}^{+} \xrightarrow{\text{H}^{+}} \text{R—N=N—OH} \quad \begin{cases} \xrightarrow[-\text{N}_2]{-\text{HO}^{-}} \text{R}^{+}\ \textbf{(79)} \xrightarrow{\text{H}_2\text{O}} \text{Products} \\ \xrightarrow{-\text{H}_2\text{O}} \text{R}'\text{=N}_2\ \textbf{(80)} \end{cases}$$

(77) (78)

SCHEME 6

Alkylation of the diazotate **81** with $Et_3O^+BF_4^-$ yields the corresponding ether with 70% net retention, and a mechanism via the formation of the ion triplet **82** has been proposed. In solvolysis reactions, in the presence of H_2O, EtOH and amines, reaction with the solvent usually predominates over reaction with HO^- within the solvent cage (typically by 3 : 1) so that net *inversion* of the solvolysis product is observed (equation 33)[112]. The degree of retention observed by capture of the carbonium ion

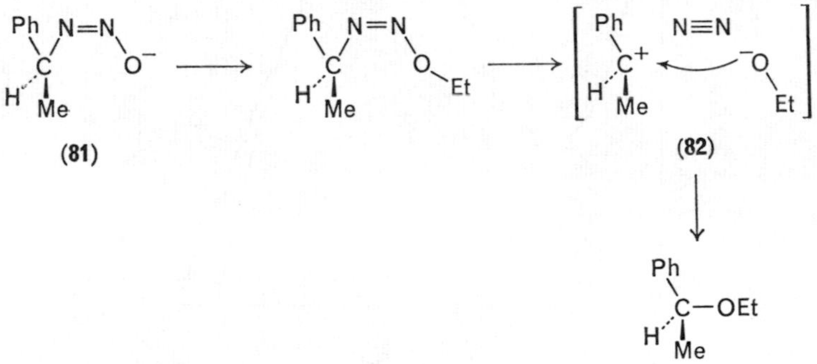

Ph(H)(Me)C—N=N—O⁻ (81) ⟶ Ph(H)(Me)C—N=N—O—Et ⟶ [Ph(H)(Me)C⁺ N≡N ⁻O—Et] (82) ⟶ Ph(H)(Me)C—OEt

formed by HO^- is increased as the stability of the carbonium ion is increased; it has been rationalized that the stability of the carbonium ion ensures that the two reactions (cleavage of N—O and N—C bonds in **83**) are more nearly concerted so that ^-OH is generated close to the carbonium ion[125].

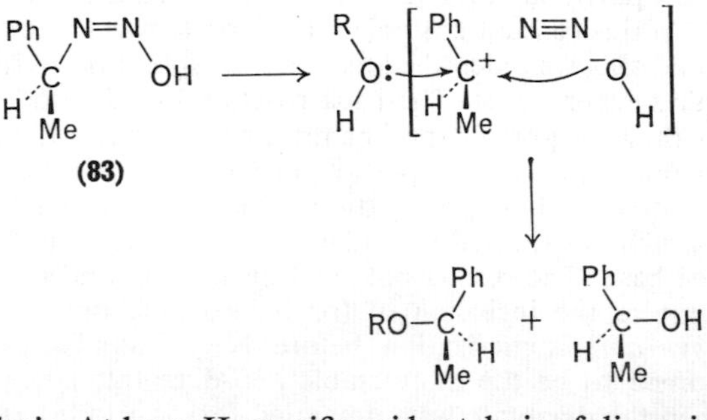

Ph(H)(Me)C—N=N—OH (83) ⟶ R(H)O: → [Ph(H)(Me)C⁺ N≡N ⁻O—H] ⟶ RO—C(Ph)(H)(Me) + Ph(H)(Me)C—OH (33)

(iv) Acid catalysis. At low pH, specific acid catalysis of diazonium ion formation from arenediazohydroxides is observed (see Figures 4 to 6). General acid catalysis may also be observed in this region in some instances, but this is weak (with a

Brönsted coefficient of *c.* 0·2)[105, 108]. The Hammett ρ value of $-2{\cdot}4$ observed[117] for specific acid catalysis in this region is consistent with protonation on oxygen of

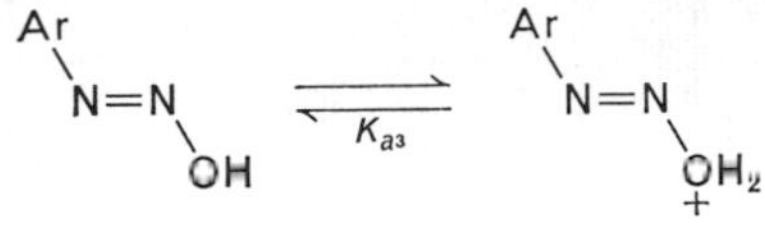

$$\mathrm{Ar{-}N{=}N{-}OH} \underset{K_{a3}}{\rightleftharpoons} \mathrm{Ar{-}N{=}N{-}\overset{+}{O}H_2}$$

(84)

the *anti*-diazohydroxide to give **84**, which undergoes loss of H_2O to give the arenediazonium ion. However protonation on the azo nitrogen to give **85** might be expected to facilitate *anti* → *syn* isomerization (by a rotation pathway).

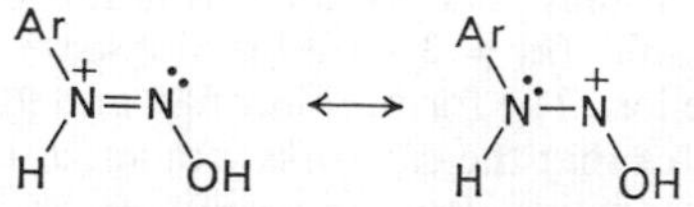

$$\mathrm{Ar(H)\overset{+}{N}{=}\ddot{N}{-}OH} \longleftrightarrow \mathrm{Ar(H)\ddot{N}{-}\overset{+}{N}{-}OH}$$

(85)

(v) Structure of 'diazohydroxide'. The position of protonation of the *anti*-diazotate (**68**) is unknown, although preferential oxygen protonation is generally assumed (to give **69**). In the absence of definitive evidence to the contrary, arguments have been presented[107, 117] that the initial position of protonation might equally be at nitrogen (to give the nitrosamine as the predominant tautomer). Because of

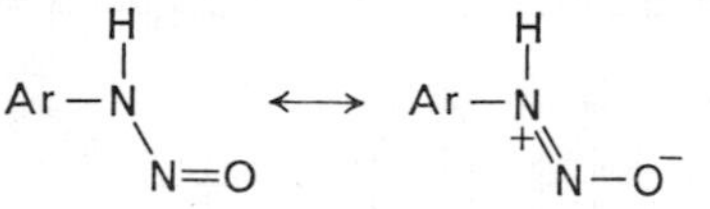

$$\mathrm{Ar{-}N(H){-}N{=}O} \longleftrightarrow \mathrm{Ar{-}\overset{+}{N}(H){=}N{-}O^-}$$

(86) (86a)

restricted rotation about the N—N bond in the nitrosamine (**86a**), which is also observed in *N*-nitroso secondary amines[126, 127], slow *syn–anti* isomerization in **86** (analogous to k_1 in Scheme 5) is also to be expected; loss of HO^- could then occur via the diazohydroxide tautomer.

b. *Alkoxide.* In spite of the well-established reaction of hydroxide ion with diazonium ion to give diazohydroxides, coupling with other oxygen nucleophiles rarely takes a simple course, possibly because an analogous conversion to the relatively stable diazotate is not possible. The initial product formed with other oxygen nucleophiles often undergoes homolytic fission to yield aryl radicals and ultimately arylation and/or reduction products.

Bunnett has shown[128, 129] that treatment of *p*-nitrobenzenediazonium ion (**87**) with sodium methoxide in methanol resulted in the formation of the reduction product, nitrobenzene (**90**), often in high yield. Kinetic studies have shown that an initial fast reaction occurs and that the *trans*-diazomethoxide (**89**) and nitrobenzene (**90**) are formed in approximately equal amounts. The *trans* isomer **89** is relatively stable and was isolated and characterized; in acid, **89** is slowly reconverted to the diazonium compound **87**. Interestingly the relative amounts of **89** and **90** formed were independent of [MeO^-]. This indicates that [MeO^-] is not involved in the rate-determining step for the formation of these two products and suggests that there is a common intermediate which is rapidly formed on the reaction pathway. This is most likely the *cis* material (**88**). The initial formation of **88** from *p*-nitrobenzenediazonium ion is very rapid ($k_1 = 3 \times 10^8$ l mol^{-1} sec^{-1} at 23 °C) [130], comparable to the rate with hydroxide ion. The formation of **89** and **90** initially in approximately equal amounts also requires that the *cis–trans* isomerization rate (k_2) is coincidently similar to that for the reductive decomposition of **88** (k_3). A free-radical chain mechanism is suggested for the latter rather than any aryldiazene-mediated reaction since deuterium is not incorporated when MeOD is used in place of MeOH [128].

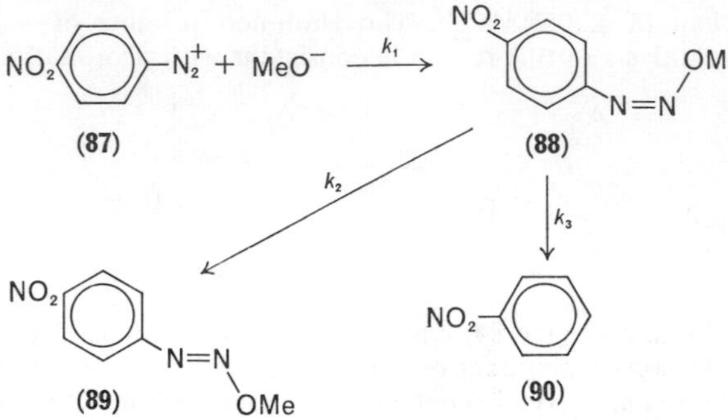

c. *Phenoxide*. The normal products of coupling between phenols and benzenediazonium ions are the corresponding *p*-azo materials (**93**). However, several groups of workers have described e.s.r. and CIDNP evidence for the presence of radical intermediates (which, of course, may not be the major pathway for the formation of **93**). A possible route to the electron transfer products **94** and **95** is via the covalent

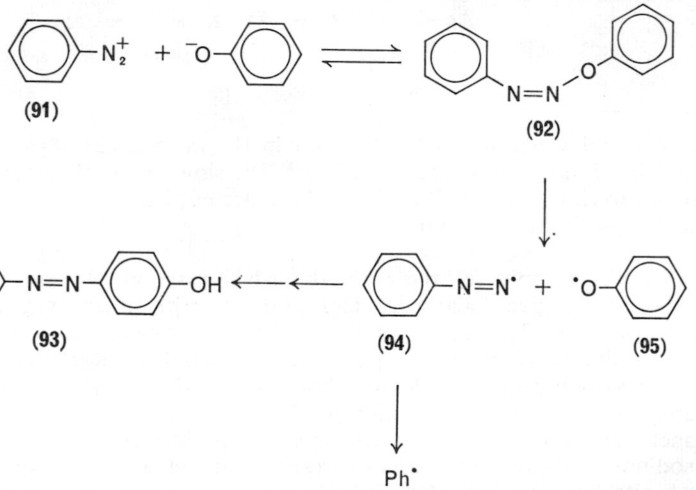

aryl diazoether **92** [98], formed by initial nucleophilic attack at nitrogen by the ionized phenol. With 2,4,6-tri-*t*-butylphenol the intermediate radical (analogous to **95**) can be detected by e.s.r.[131], while CIDNP evidence[132, 133] and the trapping[131] of the decomposition product Ph• are consistent with the presence of aryl diimine radicals (**94**). Although coupling of the radical species **94** and **95** to give **93** is possible, it is unlikely that this is a significant competing route to direct electrophilic reaction of the arenediazonium ion with phenolate[134].

d. *Acetate*. The equilibrium constant for the formation of a covalent azo compound **96** from acetate and benzenediazonium ion lies very much to the side of the starting ions[135]. An estimate of K as *c.* 10^{-5} (see equation 34) has been made[136]. In spite of the low concentration of **96**, however, this has been implicated as the reactive intermediate in radical formation (see Section III.E.1). Another important mode of reaction of benzenediazonium ions with acetate is benzyne formation, which is dealt with elsewhere (Section III.A.8).

$$\mathrm{Ar{-}N_2^+ + {}^-O{-}\overset{\overset{\displaystyle O}{\|}}{C}{-}CH_3 \;\underset{}{\overset{K}{\rightleftarrows}}\; Ar{-}N{=}N{-}O{-}\overset{\overset{\displaystyle O}{\|}}{C}{-}CH_3}\ \ \textbf{(96)}$$

$$\downarrow$$

$$\mathrm{Ar^{\bullet}\ N_2\ \ {}^{\bullet}O{-}\overset{\overset{\displaystyle O}{\|}}{C}{-}CH_3} \qquad (34)$$

e. *Diazotate*. A related reaction is the coupling of arenediazonium ions with diazotates in basic solution to form diazoanhydrides (**97**) [136–138]. Since the diazotate

$$\mathrm{Ar{-}N_2^+ + ArN_2O^- \rightleftarrows Ar{-}N{=}N{-}O{-}N{=}N{-}Ar}\ \ \textbf{(97)}$$

$$\downarrow$$

$$\mathrm{Ar\,N_2^{\bullet} + {}^{\bullet}ON_2Ar}\ \ \textbf{(97a)}\ \ \textbf{(97b)}$$

is also a good leaving group, the equilibrium lies to the left and the main utility is the homolytic decomposition to give **97a** and **97b** which can then give rise to chain initiation in free radical reactions[139].

2. Sulphur nucleophiles

Arenediazonium ions couple with thiophenols; the reaction occurs at nitrogen and involves the thiophenoxide anion (equation 35)[140]. The adducts are more stable than the corresponding oxygen analogues, because of the greater nucleophilicity of sulphur. A competing reaction is nitrogen loss to give disulphides (**98**) [141]. A more common mechanism for the reaction of the less nucleophilic neutral sulphur is

$$\mathrm{ArN_2^+ + Ar'S^- \longrightarrow Ar{-}N{=}N{-}SAr' + Ar{-}S{-}Ar'}\ \ \textbf{(98)} \qquad (35)$$

nitrogen displacement (e.g. thioureas yield *S*-arylthiouronium salts)[142]. The formation[143, 144] and decomposition[145] of diazosulphones (**99**) has been investigated. A ρ value of $+2{\cdot}40$ was reported[143] for substituents in the arenediazonium ion for sulphone (**99**) formation (in methanol), comparable to the value for reaction with HO^- or CN^- [100]. The thermal decomposition of **99** in protic, non-polar solvents occurs by a free radical pathway (equation 36) involving initial scission of the N—S bond. In the more polar acetonitrile, however, heterolysis occurs initially. Arenediazonium ions react with arenesulphonic acids (Ar^1SO_3H) to give charge-transfer complexes[98] (this is used industrially to stabilize diazonium ions) rather than diazosulphonates (ArN_2OSO_2Ar') as was previously thought[98].

$$ArN_2^+ + Ph{-}SO_2^- \longrightarrow Ar{-}N{=}N{-}SO_2{-}Ph \; \textbf{(99)}$$

$$\textbf{(99)} \longrightarrow [ArN{=}N^\bullet \; {}^\bullet SO_2Ph] \longrightarrow Ar^\bullet + Ph^\bullet + N_2 + SO_2 \qquad (36)$$

The reaction of arenediazonium ions with sulphite to give diazosulphonates (**100**) has been investigated by several groups of workers[146–148]. Only the sulphite (and not the bisulphite[146]) reacts in the pH range 5–9 to give the *syn* adduct (**100**) in an initial fast reaction (ρ = 5·5 for the equilibrium for the formation of **100**)[147].

$$ArN_2^+ + SO_3^{2-} \rightleftharpoons \underset{\textbf{(100)}}{(syn)\ Ar{-}N{=}N{-}SO_3^-} \xrightarrow{k_1} \underset{\textbf{(101)}}{(anti)\ Ar{-}N{=}N{-}SO_3^-}$$

$$\textbf{(100)} \longrightarrow \underset{\textbf{(102)}}{Ar{-}N(SO_3^-){-}NHSO_3^-}$$

Rearrangement to the more stable *anti*-sulphonate **101** then occurs ($k_1 \sim 2 \times 10^{-3}$ sec^{-1})[147] and this (unlike the isomerization of diazocyanates or diazotates) is relatively independent of the nature of Ar. In the presence of excess sulphite, a competing reaction is the formation of the hydrazine disulphonic acid (**102**) leading to reduction. The formation of the *anti* isomer **101** (but not **100**) is irreversible and the arenediazonium is not regenerated from **101** on acidification.

3. Carbon nucleophiles

a. *Cyanide ion*. The coupling of cyanide ions with arenediazonium ions initially gives orange *syn*-diazocyanides (**103**) which rearrange to the more stable red *anti* isomers (**104**)[149, 150]. Kinetic studies[100] on the initial reaction follow equation (37)

$$Ar{-}N_2^+ + {}^-CN \underset{k_{-1}}{\overset{k_1}{\rightleftharpoons}} \underset{\textbf{(103)}}{(syn)\ Ar{-}N{=}N{-}CN} \rightleftharpoons \underset{\textbf{(104)}}{(anti)\ Ar{-}N{=}N{-}CN}$$

$$\log k_1 = \rho\sigma + 2.32 \qquad (37)$$

with $\rho = +2{\cdot}31$. The equilibrium between **103** and the starting diazonium and cyanide ions gives a Hammett ρ of 3·53[100] (a higher value, 4·7, was given in a previous study)[150]. Thus considerable bond formation has already occurred in the transition state for the formation of **103**. The mechanism of *syn* → *anti* isomerization of the diazocyanides has been studied and a lateral shift, rather than rotation mechanism, has been proposed[151].

b. *Ketones and related compounds.* The coupling of benzenediazonium ions with aliphatic carbon activated by a neighbouring electron-withdrawing (usually acyl or nitro) group is known as the Japp–Klingemann synthesis (equation 38)[152–154]. The final product is usually the more stable hydrazone (**107**, **108**) formed either on tautomerization ($R^2 = H$) or removal of one of the acyl groups (R^2 = alkyl); however, the intermediate azo materials (**106**) are usually isolable.

The reaction of acetylacetone (**109**) with substituted benzene diazonium ions obeys the Hammett equation (39) ($\mu = 0{\cdot}1$, 20 °C)[155]. A plot of $\log k_{obs}$ against pH has a slope of $-1{\cdot}0$ (in the pH region *c.* 0), showing that the conjugate base

$R^1C(=O)-C^-(R^2)-C(=O)OR^3$ (**105**) + $\overset{+}{N_2}Ar$ ⟶ $R^1C(=O)-C(R^2)(N{=}N{-}Ar)-C(=O)OR^3$ (**106**)

106 —($R^2 = H$)→ $R^1C(=O)-C(=N{-}NH{-}Ar)-C(=O)OR^3$ (**107**)

106 —(R^2 = Alkyl)→ $R^1C(=O)-C(=N{-}NH{-}Ar)-R^2$ (**108**) (38)

$$\log k_{obs} = 3{\cdot}45\sigma + 5{\cdot}11 \qquad (39)$$

110 is the reactive species. With the more reactive 2,6-dichloro-4-nitrobenzene-diazonium ion, the rate of coupling becomes independent of acidity above $H_0 \sim -2$; from these data and the enol (**111**) content in aqueous solution, the relative reactivities of **110** and **111** are estimated as $10^9 : 1$.

$CH_3-C(=O)-CH_2-C(=O)-CH_3$ (**109**)

$CH_3-C(O^-)=CH-C(=O)-CH_3$ (**110**)

$CH_3-C(OH)=CH-C(=O)-CH_3$ (**111**)

The low acidity (pK_a *c.* 19) and enol content ($pK_T \sim 6$)[156] of acetone are partly compensated by the higher reactivity of the enolate anion, and rate constants for coupling with substituted benzenediazonium ions can be measured in phosphate and borate buffers (pH 6–9). The pH dependence shows that the enolate anion is the reactive species and that the second-order rate constants approach the diffusion-controlled limit; the Hammett ρ value is low ($+1{\cdot}89$, obtained for the substituted benzenediazonium ions with substituents less electron-withdrawing than *m*-Cl) as expected for such a reactive species[157]. Acetone actually reacts with 2 moles of diazonium ion under these conditions to give the formazan **112** (the hydrazone **113** is the final product under acidic conditions)[157]. The reaction with the first mole of diazonium ion is rate determining (the observed rate of coupling with the hydrazone is *c.* 10^7 times faster than with acetone itself)[157]. The reactivity of arylaldehyde

$$CH_3-\overset{O}{\overset{\|}{C}}-CH_3 \rightleftharpoons CH_3-\overset{O^-}{\overset{|}{C}}=CH_2 \xrightarrow[k_1]{ArN_2^+} CH_3-\overset{O}{\overset{\|}{C}}-CH_2-N=N-Ar$$

$$\rightleftharpoons CH_3-\overset{O}{\overset{\|}{C}}-CH=N-NH-Ar \quad \mathbf{(113)}$$

$$CH_3-\overset{O}{\overset{\|}{C}}-CH=N-NH-Ar \xrightarrow[fast]{ArN_2^+} CH_3-\overset{O}{\overset{\|}{C}}-\underset{N=N-Ar}{\underset{|}{C}}=N-NH-Ar \quad \mathbf{(112)}$$

hydrazones towards benzenediazonium ion has been investigated independently (equation 40)[159]. The initial product formed on reaction with the neutral hydrazone **114** is the unstable bis(arylazo)methane **(117)** which tautomerizes to the more stable formazan **116** in acidic or basic solution. Substituents in Ar′ have a larger effect than those in Ar, consistent with a transition state **(115)** in which most of the charge

$$PhN_2^+ + \underset{\mathbf{(114)}}{Ar-CH=N-NH-Ar'} \rightleftharpoons \underset{\mathbf{(115)}}{Ar-\overset{N=\overset{+}{N}H-Ar'}{\overset{|}{\underset{N=N-Ph}{\underset{|}{C}}}}-H} \xrightleftharpoons{-H^+} \underset{\mathbf{(117)}}{Ar-\overset{N=N-Ar'}{\overset{|}{\underset{N=N-Ph}{\underset{|}{C}}}}-H} \rightleftharpoons \mathbf{(116)} \qquad (40)$$

(116): $Ar-C(=N-N(Ar')\cdots H\cdots N(Ph)-N=)$ (cyclic hydrogen-bonded formazan)

is delocalized along the hydrazone chain. The bis(arylazo)methane **(117)** on treatment with strong acid can regenerate a hydrazone and arenediazonium ion (which may be different from those used in the initial coupling reaction).

Some of the results for the coupling of benzenediazonium ions with various carbon acids are summarized in Table 6. Of particular interest is the fact that the slow rate of reprotonation of nitroethane anion (attributed to the need for extensive rehybridization in the transition state) is not paralleled in the reaction with benzenediazonium ion.

TABLE 6. Rate and equilibrium constants for carbon acids[155]

Substrate	pK_a	$k_{H_3O^+}$ ($l\ mol^{-1}\ min^{-1}$)	$k_{PhN_2^+}$ ($l\ mol^{-1}\ min^{-1}$)	$\rho_{ArN_2^+}$
Nitroethane	8·6	9×10^2	$2{\cdot}3\times10^5$	2·88
Acetoacetanilide	10·7	8×10^9	5×10^6	3·06
Acetylacetone	8·9	1×10^0	$1{\cdot}2\times10^5$	3·45

c. *Other carbanions.* Reaction with organometallic reagents as potential sources of carbanions can yield azo compounds in good yield (equation 41)[159]. Reaction with acetylene initially gives **118** which, on reaction with water, gives an arylhydrazone of an α-keto-aldehyde (equation 42)[160].

$$ArN_2^+ + R_2Zn \xrightarrow{DMF} Ar{-}N{=}N{-}R \qquad (41)$$

$$R{-}C{\equiv}C^- + ArN_2^+ \longrightarrow R{-}C{\equiv}C{-}N{=}N{-}Ar \ \textbf{(118)} \xrightarrow{H_2O} R{-}\overset{O}{\overset{\|}{C}}{-}CH{=}N{-}NH{-}Ar \qquad (42)$$

d. *Aromatic substrates.* The electrophilic benzenediazonium ion reacts with activated aromatic substrates **119** to yield substitution products **121**. This reaction

$$X{-}C_6H_5\ \textbf{(119)} + ArN_2^+ \underset{k_{-1}}{\overset{k_1}{\rightleftharpoons}} \left[X{-}C_6H_5^+(H)(N{=}N{-}Ar)\right]\ \textbf{(120)} \xrightarrow[k_2]{-H^+} X{-}C_6H_4{-}N{=}N{-}Ar\ \textbf{(121)}$$

has been widely studied, because of its importance in the dye industry[161], and several of the parameters which influence reactivity are now clear.

In general the aromatic substrate **119** must be activated (X = NR_2 or OR) and when X = OH, the reactive form of the substrate is usually the conjugate base (X = O^-). The control of the pH is therefore important, especially since the concentration of the active diazonium ion decreases rapidly in basic solution (see Section III.B.1.a). The kinetic equation (43) is followed for coupling of benzene-diazonium ion to phenol, where $[PhOH]_T$ and $[PhN_2^+]_T$ refer to the total concentrations of phenol species (PhO^- + PhOH) and diazo species (PhN_2^+, PhN_2OH,

$$\text{Rate} = \frac{d[121]}{dt} = k[PhN_2^+][PhO^-] = k_{obs}[PhN_2^+]_T[PhOH]_T \qquad (43)$$

PhN_2O^-) in solution. The observed rate of coupling is at a maximum at *c.* pH 10 and then decreases in more acidic and basic solutions. In acid, k_{obs} is inversely proportional to $[H^+]$, due to the depletion of the reactive PhO^-; in base k_{obs} is inversely proportional to $[HO^-]$ due to the conversion of PhN_2^+ to the unreactive diazotate[162].

Either the initial electrophilic step (k_1) or the subsequent proton transfer (k_2) can be rate determining dependent on the substrate and the conditions used. When the second step (k_2) is rate determining, this is readily recognized by the appearance of general base catalysis and a primary isotope effect (large k_H/k_D value)[163, 164]. Thus *p*-methoxybenzenediazonium ion reacts with 1-naphthol-4-sulphonic acid and its 2-deutero analogue at the same rate. However, when a more electrophilic reagent is used (*p*-chlorobenzenediazonium ion), $k_H/k_D = 6{\cdot}5$ [165, 166]. Steric factors can also influence the magnitude of the isotope effect observed, and in a careful study by Zollinger[167] steric effects on the formation and base-catalysed breakdown of the

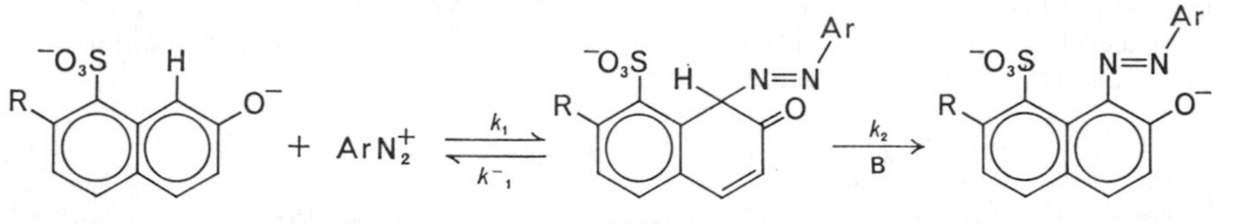

$$\textbf{(122)} + ArN_2^+ \underset{k_{-1}}{\overset{k_1}{\rightleftharpoons}} \textbf{(123)} \xrightarrow[B]{k_2} \text{product}$$

Wheland intermediate **123** have been examined. The substrates used were 2-naphthol-6,8-disulphonic acids and the kinetic equation followed was (44). The major influence of the butressing substituent R on k_1 was electronic, the size of the substituent being unimportant. The ratio k_2/k_{-1} however was sensitive to the size of R and it was

$$\frac{d[\mathbf{122}]}{dt} = [Ar'H]\,[ArN_2^+\cdot]\,\frac{k_1 k_2[B]/k_{-1}}{1+k_2[B]/k_{-1}} \qquad (44)$$

concluded that the major effect of increasing the size of substituents about the coupling site was to hinder the approach of the base (pyridine), i.e. to reduce k_2. The small effect on k_1 was attributed to the formation of a relatively unhindered intermediate **123** in which the electrophilic species was in a pseudo-axial position, while the proton to be expelled is in the same plane as the oxygen and the *peri*-sulphonic acid.

The magnitude of the isotope effect varies systematically with the pK_a of the catalysing base as shown in studies of reactive arenediazonium ions with 2-naphthol-5,8-disulphonic acid[168]. There is an apparent maximum in the plots of $\log(k_H/k_D)$ against pK_a of the catalysing base when the difference in pK_a's of the base and intermediate, akin to **123**, approach zero. However, the pK_a of the intermediate is low (*c.* 1) so that there are few bases on the descending limb ($pK_a < 1$) because the effect of these weak bases is overwhelmed by water itself acting as a general base.

(i) Site of coupling. The normally preferred site of coupling is *para* to the activating group and large amounts of *ortho* substitution are only rarely observed. The products are also *trans* about the azo linkage, probably arising from the reversible first step in the coupling reaction which results in a highly selective electrophile; base removal of the proton may also be facilitated in *para* attack.

However in naphthalene derivatives, substantial amounts of *ortho* substitution are observed when an OH group is present in the 1-position, which can be attributed to intramolecular general-base catalysis of proton removal by this group. Thus coupling of 1-naphthol (**124**) with benzenediazonium ion gives 2- and 4-substitution **128**, **127**. The formation of **127** is subject to general-base catalysis, but the formation of **128** is not, possibly due to intramolecular catalysis by the proximate ketonic group (**125**). The relative amounts of the two products **127** and **128** can therefore be changed by varying the concentration/nature of the general base present[169].

Deprotonation of the Wheland intermediate **125** can become rate determining when electron-donating substituents are present in the naphthol substrate[169].

The position of attack by benzenediazonium ion on **129** is also interesting[170]. At high pH (where the phenolate anion governs the position of attack), **130** is the major product. At lower pH, **131** is formed, attack occurring *para* to the dimethylamino group. Coupling with 2-naphthol (**132**) occurs at the 1-position to the

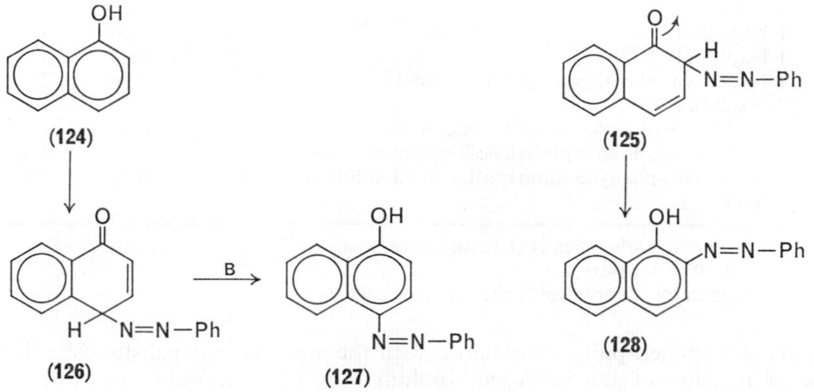

(124) (125) (126) (127) (128)

exclusion of 3- or 6-substitution. This is attributable to the highly selective nature of the diazonium ion since the aromatic structure of the unsubstituted ring in **132** remains relatively undisturbed only when attack occurs at the 1-position[171].

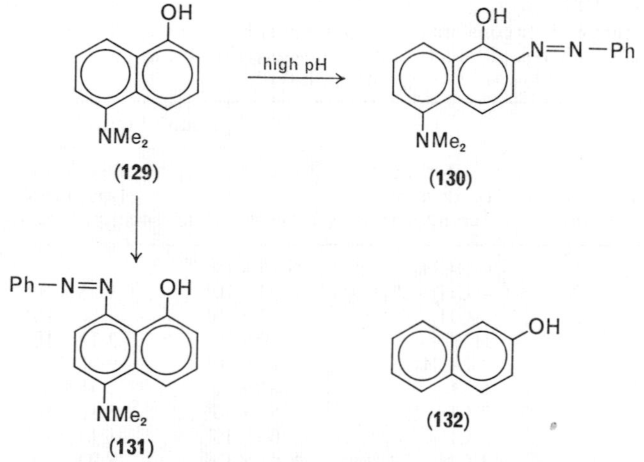

(129) (130) (131) (132)

(ii) Reactivity of arenediazonium ions. The reactivity of arenediazonium ions as electrophiles is increased by the presence of electron-withdrawing substituents which increase the electropositive nature of the diazonium ion. Good correlations of $\log k_{obs}$ vs. the σ value of the substituent in the arenediazonium group have been reported (Table 7) for a wide variety of substrates, with ρ values generally in the range 3·2–4·8. In general, the lower the reactivity of the substrate the higher the sensitivity to substituent effects (Table 7) in line with Brown's Reactivity–Selectivity principle[172].

TABLE 7. Hammett ρ values for the reaction of substituted benzenediazonium ions[173-177]

Substrate[a]	ρ	$\log k_0$[b]	Site
Phenol	4·20	4·43	4
4-Methylphenol	4·27	3·02	2
4-Methoxyphenol	4·09	4·50	2
1-Naphthol	4·15	7·50	4
1-Naphthol[c]	4·80	−1·90	4
1-Hydroxynaphthalene-4-sulphonic acid	3·94	4·56	2
2-Naphthol	3·20	6·26	1
2-Hydroxynaphthalene-6-sulphonic acid	3·18	4·76	1
1-Hydroxy-6-aminonaphthalene-3-sulphonic acid	4·04	2·25	5
1-Hydroxy-6-phenylaminonaphthalene-3-sulphonic acid	4·15	0·97	5

[a] The reactive species is the conjugate base.
[b] At 20 °C (μ = 0·3).
[c] The neutral naphthol is the reactive species.

Curved Hammett plots are obtained with the most reactive substrates[173] since the rate of reaction of the benzenediazonium ions with strongly electron-attracting groups may then approach the diffusion controlled limit (see Table 8 and Figure 7)[137]. However, a contributing factor in this case might be the inclusion of *ortho*-substituted derivatives or non-additivity of substituent effects in those ions with strongly electron-withdrawing groups.

TABLE 8. Rate constants of coupling of substituted benzenediazonium salts with 1-naphtholate and its sulpho derivative at 20 °C and ionic strength 0·05

Compound No.	Substituent on diazonium component	k_2 (l mol^{-1} min^{-1}) 1-Naphtholate	k_2 (l mol^{-1} min^{-1}) Dianion of 1-naphthol-4-sulphonic acid
1	4-OCH_3	$6{\cdot}15\times10^5$	—
2	4-CH_3	$4{\cdot}74\times10^6$	$1{\cdot}58\times10^3$
3	3-CH_3	$1{\cdot}22\times10^7$	$1{\cdot}56\times10^4$
4	H	$3{\cdot}39\times10^7$	$3{\cdot}14\times10^4$
5	3-OCH_3	$8{\cdot}15\times10^7$	$9{\cdot}02\times10^4$
6	4-Cl	$2{\cdot}63\times10^8$	$2{\cdot}45\times10^5$
7	2-Cl	$1{\cdot}55\times10^9$	$5{\cdot}16\times10^5$
8	3-Cl	$2{\cdot}06\times10^9$	$1{\cdot}14\times10^6$
9	3-CN	$8{\cdot}61\times10^9$	$8{\cdot}94\times10^6$
10	4-CN	$1{\cdot}64\times10^{10}$	$7{\cdot}18\times10^6$
11	3-NO_2	$2{\cdot}20\times10^{10}$	$2{\cdot}08\times10^7$
12	4-NO_2	$2{\cdot}76\times10^{10}$	$1{\cdot}65\times10^7$
13	2,5-Cl	$2{\cdot}48\times10^{10}$	$1{\cdot}78\times10^7$
14	2-Cl-4-NO_2	$1{\cdot}48\times10^{11}$	$3{\cdot}72\times10^8$
15	3-Cl-4-NO_2	—	$1{\cdot}78\times10^8$
16	4-Cl-3-NO_2	—	$5{\cdot}26\times10^7$
17	2,4,6-Cl_3	$7{\cdot}08\times10^{10}$	$2{\cdot}17\times10^7$
18	2,6-Cl_2-4-NO_2	$6{\cdot}31\times10^{11}$	$4{\cdot}45\times10^9$

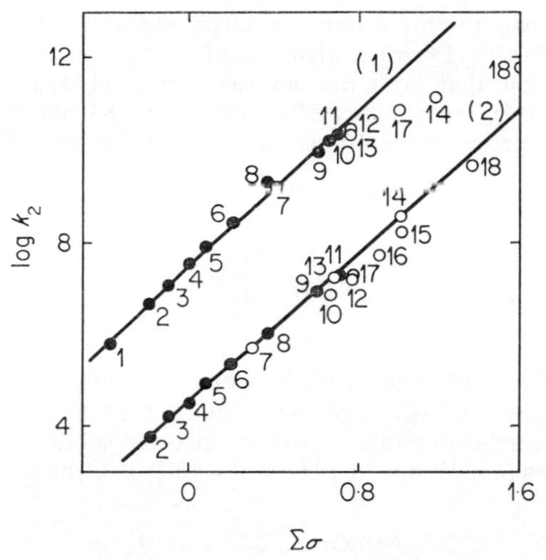

FIGURE 7. Hammett plots for the coupling of substituted benzenediazonium ions with 1-naphtholate (curve 1) and the dianion of 1-naphthol-4-sulphonic acid (curve 2); for substituents, see Table 8[137].

Some curvature was also noted[178, 179] in the reaction of substituted benzenediazonium ions with **133** (ρ = 3·90 for a plot of log k_{obs} vs. σ^+ when data for strongly electron-withdrawing substituents ($X = p$-NO_2, *m*-Cl) are not included). For the strongly electron-attracting substituents it is proposed that H_2O is *not* acting as a general base for proton removal in the transition state (see **134**). This is supported

(133) **(134)**

by the observation of a positive entropy of activation ($\Delta S = +17$ to 18 e.u. for X = *p*-NO_2, *m*-Cl), whereas ΔS is close to zero for the other substituted benzenediazonium ions. However, the incursion of diffusion-controlled kinetics could also explain these results.

(iii) Reactivity of substrate. The presence of substituents in the aryl group reacting with the diazonium ion has a dual effect: electron withdrawal increases the concentration (at pH's below the pK_a) but decreases the reactivity of the phenolate. With substituted anilines, unless studied in acidic solution (<pH 4), electron withdrawal decreases the nucleophilicity of the substrate as observed for other aromatic substituted reactions.

The reactivity order for benzene (or naphthalene) derivatives is $-O^- > NR_2 >$ NHR > OR ~ OH ≫ Me, as expected on the basis of their Hammett σ constants; the ionized phenolate is *c.* 10^{10}-fold more reactive than phenol itself, which is the reactive species only in acidic solution. Similarly 1-naphtholates are 7–9 orders of magnitude more reactive than the neutral naphthols[180].

The presence of two anionic groups increases reactivity of the substrate, but the effect is not additive[181]. Pyrroles also couple with benzenediazonium ion, and kinetic studies[182] show that both the neutral pyrrole (**135**) and the anion **136** (at high pH) are reactive[183]. The conjugate base of 3-methyl-1-phenyl-5-pyrazolone (**137**) is also the reactive species at the 4-position in neutral solution[184].

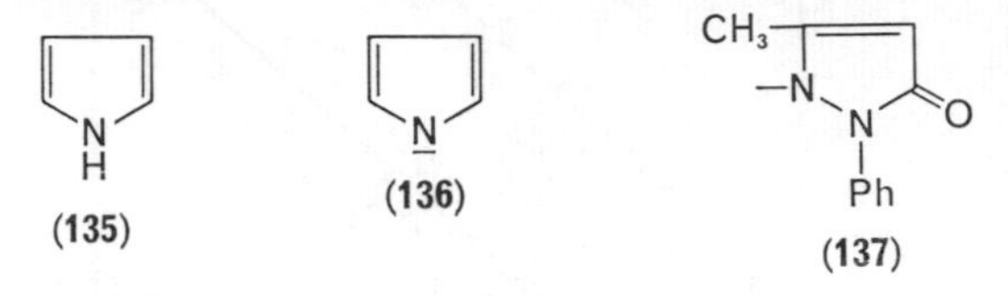

(**135**) (**136**) (**137**)

The reactivity of phenolate ion with benzenediazonium ion is about an order of magnitude greater than that with hydroxide ion. Bunton and coworkers[121] have used this to estimate second-order rate constants in basic solution using a competition method. Under these conditions at pH > 8 the rate of formation of **139** is given by

$$\underset{(\mathbf{138})}{ArN_2O^-} \underset{K_2}{\rightleftharpoons} ArN_2OH \underset{k_{-3}(HO^-)}{\overset{k_3}{\rightleftharpoons}} ArN_2^+ \xrightarrow[k_c]{Ar'O^-} \underset{(\mathbf{139})}{\text{Azo dye}}$$

$$ArN_2^+ \xrightarrow{k_4} ArOH \qquad Ar = \text{4-pyridyl}$$

equation (45). The major species present in this region is the diazotate **138** and since k_{-3} and k_4 are known, k_c can be determined by estimating the amount of azo dye

$$k_{obs} = \frac{k_c k_3[Ar'O^-]}{(k_{-3}[HO^-]+k_c[Ar'O^-])(K_2+[H^+])} \tag{45}$$

(**139**) formed (spectrophotometrically) as a function of [HO⁻] and [ArO⁻]. The results (Table 9) show that the more reactive anions react with pyridine-4-diazonium ion close to the diffusion-controlled limit.

TABLE 9. Second-order rate constants for azo coupling of pyridine-4-diazonium ion at 25 °C [121]

Coupling agent	k_c (1 mol^{-1} sec^{-1})
p-Hydroxybenzoate	$1{\cdot}3\times10^5$
o-Chlorophenol	$1{\cdot}0\times10^5$
Phenol	$1{\cdot}8\times10^6$
2-Naphthol	$1{\cdot}1\times10^7$
o-Methoxyphenol	$1{\cdot}2\times10^7$
o-Methoxyphenol	$1{\cdot}2\times10^8$
2,6-Dimethylphenol	$2{\cdot}4\times10^8$
1-Naphthol	$1{\cdot}9\times10^9$
2-Methyl-1-naphthol	$7{\cdot}2\times10^9$

(iv) Displacement of groups other than H⁺. Zollinger[184] has shown that benzenediazonium ion will displace SO_3 from 2-naphthol-1-sulphonic acid (**140**) to give **141**.

An initially formed π complex between the substrates precipitates from solution (in water). Displacement of Br— is also observed (equation 47), and this reaction is catalysed by thiosulphate ion, indicating that the loss of bromine is rate determining. A change in the rate-determining step to initial attack occurs in the displacements of Cl— and I— since no such catalysis is observed[185].

$$\textbf{(140)} \xrightarrow{PhN_2^+} \textbf{(141)} \qquad (46)$$

(140): 1-SO₃H-2-naphthol; (141): 1-(N=N—Ph)-2-naphthol

$$\text{1-Br-2-OH-6-}HO_3S\text{-naphthalene} \xrightarrow{PhN_2^+} \text{1-(N=N—Ph)-2-OH-6-}HO_3S\text{-naphthalene} \qquad (47)$$

Other groups which can readily be expelled as cationic species may also be displaced. Thus *para* substitution of **142** by diazotized sulphanilic acid occurs in preference to *ortho* substitution.

$$R{-}CH(OH){-}C_6H_4{-}NMe_2\ \textbf{(142)} + HO_3S{-}C_6H_4{-}N_2^+ \longrightarrow HO_3S{-}C_6H_4{-}N{=}N{-}C_6H_4{-}NMe_2 + RCHO \qquad (48)$$

4. Nitrogen nucleophiles

Ammonia and simple aliphatic primary amines couple with arenediazonium ions to give triazenes (**144**); in the presence of excess diazonium salt, pentazenes (**145**) may be formed. With secondary amines the reaction stops at the triazene stage. Since (a) the initial coupling product may readily tautomerize (to **148**), and

$$\underset{\textbf{(143)}}{ArN_2^+X^-} + RNH_2 \rightleftharpoons \underset{\textbf{(144)}}{Ar{-}N{=}N{-}NHR} + \underset{\textbf{(145)}}{Ar{-}N{=}N{-}N(R){-}N{=}N{-}Ar}$$

$$\textbf{(144)} \rightleftharpoons \textbf{(148)}; \qquad \textbf{(144)} \xrightarrow{R=H} \underset{\textbf{(149)}}{ArN_3}$$

$$\underset{\textbf{(146)}}{ArNH_2} + \underset{\textbf{(147)}}{RN_2^+} \rightleftharpoons \underset{\textbf{(148)}}{Ar{-}NH{-}N{=}N{-}R}$$

(b) the coupling is reversible, diazonium ion and amine exchange (to form **146** and **147**) may occur (especially in acid), together with isomerization of triazenes to form ring-substituted products. The intermediate triazene (**144**, R = H) can readily be oxidized to an aryl azide (**149**) by using arenediazonium tribromides in the initial coupling reaction (**143**, X = Br_3^-) [186]; by labelling the diazonium group and ammonia with ^{15}N, Clusius and coworkers[187, 188] have shown that the terminal nitrogen in phenyl azide (**149**, Ar = Ph) is derived from ammonia.

The isomerization of triazenes to *p*-amino azo compounds in acid (known as the diazoamino rearrangement, equation 49) has long been recognized as an intermolecular rearrangement.

$$Ar-N=N-NH-C_6H_5 \xrightarrow{H^+} Ar-N=N-C_6H_4-NH_2 \qquad (49)$$

The reaction occurs in a stepwise fashion, resulting in the formation of a free diazonium group which can be diverted and trapped by a more nucleophilic arene[189]. The rate of rearrangement is increased by added aromatic amine, suggesting a competing pathway in which the electrophilic species is the protonated triazene[190].

Hydrazine can couple with 1 or 2 moles of arenediazonium ion. With arylhydrazines (**150**), unsubstituted on the terminal nitrogen, elimination to give aryl azide and aryl amine occurs rapidly and the intermediate tetrazene **151** is not isolated. Because of rapid prototropy in the intermediate **151** four products, two

$$ArN_2^+ + Ar'NHNH_2\ \textbf{(150)} \longrightarrow Ar-N=N-NH-NH-Ar\ \textbf{(151)}$$

$$\textbf{(150)} \downarrow\ Ar-N=N-N(Ar')-NH_2\ \textbf{(152)} \qquad \textbf{(151)} \downarrow\ ArN_3 + Ar'N_3 + ArNH_2 + Ar'NH_2 \qquad (50)$$

amines and two azides, are formed (equation 50)[191, 192]; ^{15}N labelling has established that when Ar = Ar′ then equal amounts of all four products are formed. 1,3-Diaryltetrazenes (**152**) may also be formed (in sodium acetate buffer) and have been characterized by the formation of arylidene derivatives[193]. With hydrazine the major product isolated is an aryl azide, usually formed in >90% yield, together with a small amount of hydrazoic acid[194].

With hydroxylamine, reaction of arenediazonium ion at nitrogen takes a similar course. The intermediate hydroxytriazene is not normally isolated[195]; in base elimination yields amine **155** while at pH<7, decomposition to the azide **154** is observed[196]. The formation of N_2O in basic solution could also be the result of reaction of oxygen with the conjugate base of hydroxylamine.

$$Ar-N_2^+ + NH_2OH \longrightarrow [Ar-N=N-NHOH]\ \textbf{(153)} \underset{}{\overset{HO^-}{\rightleftharpoons}} ArNH-N=N-O^-$$

$$\textbf{(153)} \xrightarrow{H^+} ArN_3 + H_2O\ \textbf{(154)} \qquad ArNH-N=N-O^- \longrightarrow ArNH_2 + N_2O\ \textbf{(155)}$$

The reaction of arenediazonium salts with azide ion provides a general method for the introduction of the azido functional group (equation 51). Kinetic studies have shown that the rate of reaction with azide ion is far faster (*c.* 2×10^9-fold) [197] than the rate of phenyl cation formation and various labelling experiments (using ^{15}N) have confirmed that a direct displacement of N_2 by N_3^- does not occur.

$$ArN_2^+ \xrightarrow{N_3^-} Ar{-}N_3 + N_2 \tag{51}$$

The principal reaction involved is nucleophilic attack by azide ion on the terminal nitrogen to give, initially, a pentazene (equation 52). Ritchie and Wright[198] have investigated the kinetics of this process and obtained a ρ value of 3·2 for substituents in the arenediazonium ion. However several anomalies were apparent (e.g. for

$$ArN_2^+ + N_3^- \rightleftharpoons Ar{-}N{=}N{-}N{=}\overset{+}{N}{=}\overset{-}{N} \tag{52}$$

p-nitrobenzenediazonium ion, N_3^- was found to react faster than HO^- whereas the reverse is true for benzenediazonium ion) and this led to the conclusion[100] that the rates observed were for the subsequent decomposition of the diazoazide, and that N_3^- reacts faster than HO^- in the initial step in all cases. This is supported by the magnitude of the ρ value observed which is closer to that for the equilibrium formation of *syn*-diazocyanides (+3·53) than that for the rate of reactions of arenediazonium ions with cyanide ion (ρ = 2·31) [100].

Careful labelling experiments by Clusius, Huisgen, Ugi and their colleagues[199-204] have demonstrated that the initial reaction yields two products, the pentazene **157** and the cyclic pentazole **158** to the extent of 70% and 30% respectively (when Ar = Ph). Crystalline pentazoles have subsequently been isolated (**158**, Ar = *p*-$EtOC_6H_4$; electron-donating substituents stabilize the pentazole) and shown to decompose to aryl azide and nitrogen[202, 203]. The possibility that the pentazene **157** is also the precursor of the pentazole was eliminated by Ugi[204]. Loss of nitrogen from the pentazene **157** occurs more rapidly than from the pentazole **158**, thus when

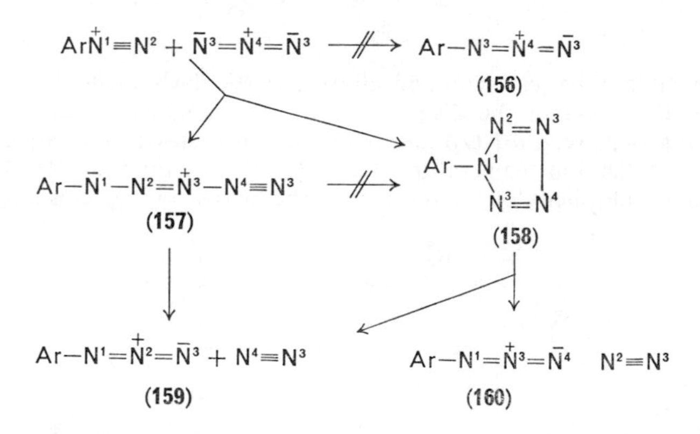

Ar = Ph there is an initial loss of about 70% N_2 at −40 °C the remaining nitrogen (from the pentazole) is evolved when the solution is heated to *c.* 0 °C [201].

Detailed studies on the rates of decomposition of phenylpentazole (**158**, Ar = Ph) in a variety of solvents show a relatively small effect (k_{obs} varies less than 10-fold between *n*-hexane and 1 : 1 methanol–water), typical of 1,3-dipolar cycloadditions[205]; formation of the pentazole from the diazonium ion and azide ion is thought to involve a similar concerted cycloaddition. The balance between pentazole and pentazene can be varied in the initial reaction; electron-withdrawing substituents increase the proportion of acyclic **159** product.

C. Nucleophilic Aromatic Substitution Activated by the Diazonium Group

The diazonium group is more strongly electron-withdrawing than the nitro or trimethylammonium group and its effect is approximately equivalent to two nitro groups. Lewis and Johnson[206] have determined Hammett σ_p and σ_m constants of +1·9 and +1·7 for the diazonium group from the ionization of benzoic and phenylacetic acids. These values have been successfully applied to other systems, e.g. the ionization of heterocyclic diazonium salts[207]; the diazonium group is therefore the most strongly electron-attracting group known. When direct resonance interaction between the diazonium group and the substituent is possible than the effective electron-withdrawing power of the diazonium group is greatly enhanced. Thus the pK_a of *p*-hydroxybenzenediazonium ion **(161)** is as low as 3·40 [208], due to the stabilization of the conjugate base by structures such as **162**. From these data and data for *p*-aminobenzenediazonium ions a σ_p^+ value of 3 has been calculated for the diazonium group[206, 209].

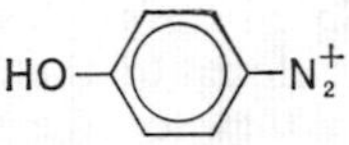

HO–C₆H₄–N_2^+ **(161)**   O=C₆H₄=$\overset{+}{N}=\overline{N}$ **(162)**

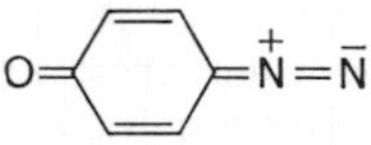

The electron-withdrawing character of the diazonium group activates the aromatic ring towards nucleophilic attack and displacement of suitable leaving groups can occur from the *ortho* and *para* positions. As leaving groups $I^- > Br^- > Cl^- > F^-$ [which suggests that the displacement of the leaving group from the Meisenheimer complex may be rate determining (equation 53)][200]. Other groups which may be

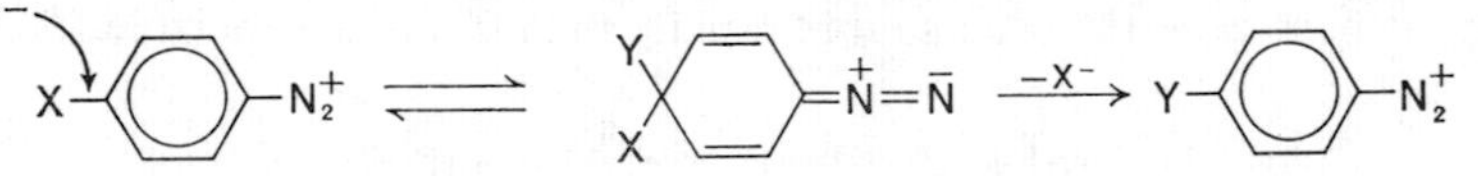

$$Y^- + X\text{–}C_6H_4\text{–}N_2^+ \rightleftharpoons (Y)(X)C_6H_4{=}\overset{+}{N}{=}\overline{N} \xrightarrow{-X^-} Y\text{–}C_6H_4\text{–}N_2^+ \quad (53)$$

displaced include nitro (as NO_2^-) and alkoxy. Particularly facile displacements can occur when the effect of the diazonium group is augmented by other electron-withdrawing groups (e.g. *o*-NO_2) and reaction can then occur in aqueous solution at normal diazotization temperatures (*c.* 0 °C) [210, 211] (equation 54). This however may lead to a multiplicity of products since the 'activating' groups may themselves

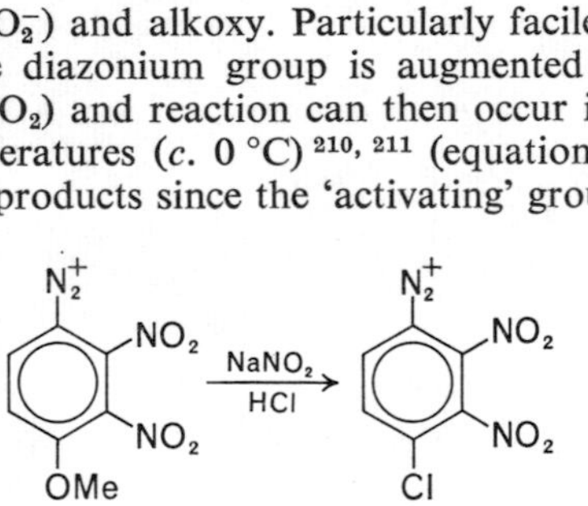

$$N_2^+\text{,}\ NO_2\text{,}\ NO_2\text{,}\ OMe\text{-benzene} \xrightarrow[HCl]{NaNO_2} N_2^+\text{,}\ NO_2\text{,}\ NO_2\text{,}\ Cl\text{-benzene} \quad (54)$$

$$2,6\text{-}Cl_2\text{-}4\text{-}NO_2C_6H_2N_2^+ \xrightarrow{HO^-} N_2\text{(Cl, O, NO}_2\text{)quinone diazide} + N_2\text{(Cl, Cl, O)quinone diazide} \quad (55)$$

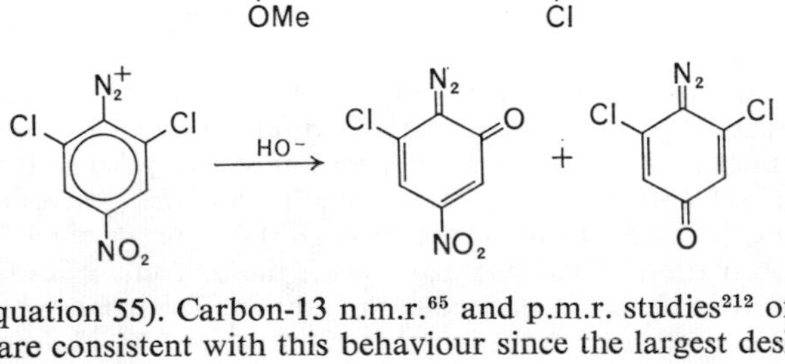

be displaced (equation 55). Carbon-13 n.m.r.[65] and p.m.r. studies[212] of the benzenediazonium ion are consistent with this behaviour since the largest deshielding effect is in the *para* position.

D. Metal-catalysed Reactions

The Sandmeyer (56, 57), Meerwein (58, 59) and related reductions to give biaryls (60) and azoarenes (61) and arenes (62) possibly share a common mechanism when catalysed by a metal or metal ion (most commonly Cu^{I}, Cu^{II} or metallic Cu

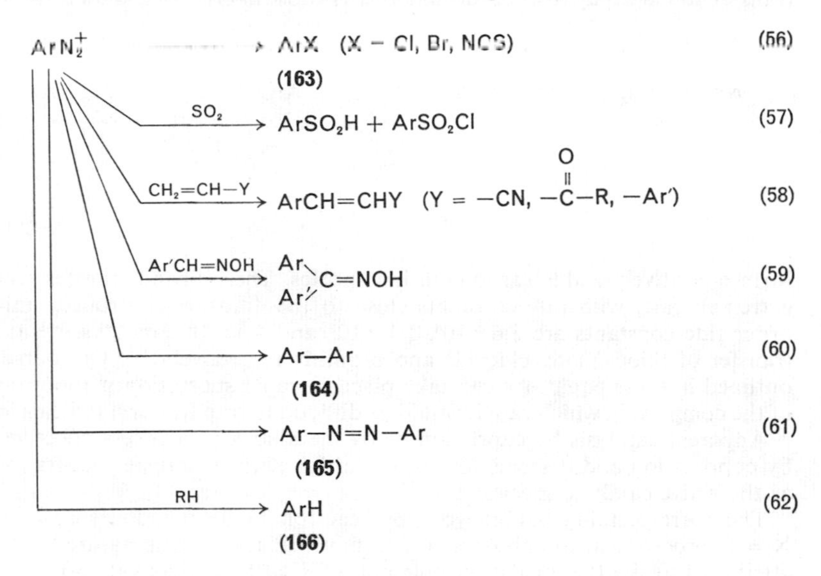

itself)[213, 214]. The existence of radical intermediates in these reactions[217] (at least under most conditions) is now generally accepted, but since their formation is often rate determining, it is difficult to obtain information on the role of various organo-copper complexes which may determine the product composition. The use of radical traps and other methods for the selective diversion of intermediates is widely employed.

1. Sandmeyer reaction

The rate of formation of **163** (X = Cl) catalysed by cuprous chloride is first order in diazonium ion and in catalyst but inversely proportional to $[Cl^-]$ [214]. The latter observation is explained by the conversion of $Cu^{I}Cl_2^-$ (the active form of the catalyst) to $Cu^{I}Cl_4^{3-}$. The rate-determining step is probably the initial coordination step; electron-withdrawing groups in Ar aid reaction. Zollinger[98] has pointed out the similarity of this sequence to electron transfer by an inner sphere mechanism, in

$$ArN_2^+ + ClCu^{I}Cl^- \xrightarrow{k_1} [Ar{-}N{\equiv}NClCu^{I}Cl]$$

$$\downarrow$$

$$Ar^{\bullet} + N_2 + CuCl_2 \longleftarrow [Ar{-}N{=}N^{\bullet}ClCu^{II}Cl]$$

which one of the ligands acts as a bridge between reductant and oxidant, proposed by Taube[218]. The electron-transfer sequence is also similar to the proposed mechanism of radical formation outlined in Section III.E.1; the same sequence has also been proposed for other catalytic reagents, e.g. $FeCl_2$ [219]. The relative reactivities shown in the Sandmeyer reaction (—CN>—I>—Br>—Cl) follow the same sequence followed in electron transfer by the inner sphere mechanism.

Extensive work by Kochi's group[220, 221] shows that aryl radical reaction with $Cu^{II}Cl$ can occur; an alternative oxidative substitution process involving electron transfer to Cu^{II} may also occur and this is most likely when the initial radical can

$$Ar^{\bullet} + Cu^{II}Cl_2 \longrightarrow [Ar\cdots Cl\cdots CuCl] \longrightarrow ArCl + Cu^{I}Cl$$

$$Ar^{\bullet} + Cu^{II}Cl_2 \rightleftharpoons ArCuCl_2 \rightleftharpoons ArCuCl^{+}Cl^{-} \rightleftharpoons Ar^{+}Cu^{I}Cl_2^{-} \longrightarrow ArCl + Cu^{I}Cl$$

form a relatively stable carbonium ion species. These ligand transfer reactions are extremely fast, with rate constants close to the diffusion-controlled limit (second-order rate constants are $3{\cdot}6 \times 10^8$, $1{\cdot}1 \times 10^9$ and $4{\cdot}3 \times 10^9$ l mol^{-1} sec^{-1} at 25 °C for transfer of thiocyanate, chloride and bromide respectively)[222]. Of course the steps outlined in these equations can take place in rapid succession without dissociation of the complex, in which case it would be difficult to trap free aryl radicals in solution.

Apparent catalysis by cupric ion or by metallic copper is less effective than that by cuprous ion and it seems likely that initial electron transfer occurs to form Cu^{I} as the active catalytic species[213].

The corresponding Sandmeyer-type reaction to form iodoarenes (equation 56, X = I) proceeds in the absence of a catalyst at room temperature. This has been attributed to the low oxidation potential of I^-, which allows it to reduce the aryl-diazonium ion, leading to aryl radical species.

2. Biaryl and azoarene formation

The formation of **164** and **165** catalysed by Cu^{I} (tetrakis(acetonitrile) copper(I) perchlorate in acetone, which gives a homogeneous solution) involves radical precursors and arylcopper intermediates. Thus using methyl iodide as a radical trap, *p*-nitroiodobenzene was formed from *p*-nitrobenzenediazonium ion and the yields of **164** and **165** were reduced to the same extent[223]. This rules out the non-radical pathway (Scheme 7) suggested by Cowdrey and Davies[213]. Direct recombination of aryl radicals to yield **164** [220] is also unlikely since the reaction order in $Ar^{\bullet}$ would then be second order whereas the estimated order is between one and two.

$$(ArN_2)_2CuCl_2^{+} + CuCl_2^{-} \longrightarrow Ar_2 + 2\,N_2 + 2\,CuCl_2$$

$$(ArN_2)_2CuCl_2^{+} + CuCl_2^{-} \longrightarrow Ar{-}N{=}N{-}Ar + N_2 + 2\,CuCl_2$$

SCHEME 7

A comprehensive set of pathways suggested by Cohen[223] is outlined in Scheme 8. The rate of formation of biaryl and azoarene is second order in Cu^{I} catalyst; low catalyst concentrations are therefore used in the Sandmeyer reaction to minimize these unwanted side-products. High concentrations of Cu^{I} also favour azoarene as opposed to biaryl formation.

Initial generation of $Ar^{\bullet}$ is suggested to occur as outlined for the Sandmeyer reaction. Analogies for most of the steps (Scheme 8) are available from other studies, e.g. Ar_2Cu^I reacts with ArN_2^+ to give azo compounds. However the organo-Cu^{II} and Cu^{III} species suggested probably have short lifetimes—analogous complexes have not as yet been isolated (although organo-Cu^I complexes are well known)[224]. Other observations are consistent with this scheme: Cu^{II} favours $Ar{-}N_2{-}Ar$ formation by reducing Ar_2Cu^{III}; electron-withdrawing groups in $Ar{-}N_2^+$ favour Ar_2 formation, consistent with the ArN_2^+ group acting as an oxidizing agent.

$$Ar^{\bullet} + Cu^{I} \rightleftharpoons ArCu^{II}$$

$$Ar^{\bullet} + ArCu^{II} \longrightarrow Ar_2Cu^{III}$$

$$Ar_2Cu^{III} \longrightarrow Ar_2 + Cu^{I}$$

$$\downarrow (Cu^{I} \rightarrow Cu^{II})$$

$$Ar_2Cu^{II} \xrightarrow{ArN_2^+} Ar_2Cu^{III} + Ar^{\bullet} + N_2$$

$$\downarrow (Cu^{I} \rightarrow Cu^{II})$$

$$Ar_2Cu^{I} \xrightarrow{ArN_2^+} ArCu^{I} + ArN_2Ar$$

$$ArCu^{I} \xrightarrow{ArN_2^+} ArN_2Ar + Cu^{I}$$

SCHEME 8

Reduction to ArH can occur when the Ar radical is trapped by a suitable hydrogen donor, e.g. an alcohol or ether. This pathway is also well recognized in reductions in the absence of metal ions (see Section III.G).

3. Meerwein reaction

Both addition and substitution products result in the catalysed arylation of unsaturated compounds. The reaction is first order in $Cu^ICl_2^-$ and ArN_2^+, but generally independent of the unsaturated substrate. This suggests that formation of $Ar^{\bullet}$ is rate determining; the further reactions (Scheme 9)[225] may however occur before this dissociates completely into solution since attempts to trap radical species or induce polymerization have not always been successful. The $Cu^{II}Cl_2$ acts as an oxidizing agent involving either hydrogen abstraction or chloride addition. The use of an excess of $Cu^{II}Cl_2$ is therefore advantageous in that the trapping of the radical is induced before polymerization can occur. Because of the similarity of the reaction conditions, the usual reduction products (**164–166**) are formed concomitantly in Meerwein reactions[214, 226–228].

$$Ar^{\bullet} + CH_2{=}CH{-}Y \longrightarrow ArCH_2{-}\dot{C}H{-}Y$$

$$\downarrow Cu^{II}Cl_2$$

$$CuCl + H^+ + Cl^- + ArCH{=}CH{-}Y \qquad Ar{-}CH_2CHCl{-}Y + CuCl$$

SCHEME 9

E. *Arylation*

1. Via aryl radicals

Using a variety of substrates of the general type **167** arylation of aromatic substrates can be achieved (this is known as the Gomberg–Bachmann synthesis)[229]. There is good evidence that these reactions involve a free radical chain, with the aryl radical ($Ar^{\bullet}$) as the reactive species.

Thus nitrobenzene is more readily arylated (*c.* threefold) than benzene, which rules out a highly electrophilic species such as Ar^+ [230]. Nitrobenzene is arylated in the *ortho* and *para* positions (with $<10\%$ *meta*) and the ratio of isomers formed is similar to that observed when phenyl radicals from other sources (e.g. the decomposition of benzoyl peroxide) are used[231, 232]. The electrophilicity of the radical is increased by the introduction of electron-withdrawing substituents (e.g. p-NO_2) but when $Ar^{\bullet} = p\text{-}MeC_6H_4^{\bullet}$ or $p\text{-}MeOC_6H_4^{\bullet}$ little electrophilic character is shown [233].

The reactions however show some unexpected features such as the absence of dimeric and disproportionation products (equation 63) which are normally formed in reactions involving phenyl radicals[234, 235]. Moreover, when $X = CH_3CO_2{-}$ acetic acid is formed rather than the expected[236] radical products from the decomposition of an acetoxy radical ($CH_3CO_2^{\bullet}$). Most of the controversy has centred on possible explanations for these observations. Clearly any mechanism must provide for efficient hydrogen abstraction from **168**.

$$\underset{\textbf{(167)}}{Ar{-}N{=}N{-}X} \longrightarrow Ar^{\bullet} \xrightarrow{Ar'H} Ar{-}Ar'$$

$$(X = MeCO_2{-},\ R_2N{-},\ ArN_2O{-})$$

$$2\ \underset{\textbf{(168)}}{\text{(Ar,H)}C_6H_5^{\bullet}} \longrightarrow Ar{-}C_6H_5 + \text{(Ar,H)}C_6H_5\text{(H,H)} \qquad (63)$$

The simplest route to **167** ($X = OCOCH_3$) is by preliminary rearrangement of *N*-nitrosoacetanilides (usually at room temperature in the aromatic compound as solvent). The acetate (**169**) can also be prepared by direct reaction between the diazonium ion and acetate but this method is less satisfactory and side-products may predominate. The initial rearrangement (equation 64) (which has been shown to be intramolecular)[237, 238] is rate determining for phenylation[239], so that most of the studies have concentrated on product analysis. Rüchardt and Freudenberg[240]

$$\text{Ph—N(N=O)—COMe} \longrightarrow \underset{\textbf{(169)}}{\text{Ph—N=N—OCOMe}} \rightleftharpoons \text{PhN}_2^+\bar{\text{O}}\text{COMe} \quad (64)$$

proposed that initiation of the radical reaction occurred through the formation and subsequent homolysis of a diazoanhydride **170** (see equations 65, 66). The phenyldiazotate radical **171** (originally given a π-type structure[241] which has subsequently

$$\text{Ph—N=N—OCOMe} + \text{MeCO}_2^- \longrightarrow \text{Ph—N=N—O}^- + (\text{MeCO})_2\text{O} \quad (65)$$

$$\text{Ph—N=N—O}^- + \text{Ph—N}_2^+ \rightleftharpoons \underset{\textbf{(170)}}{\text{Ph—N=N—O—N=N—Ph}} \longrightarrow \underset{\textbf{(171)}}{\text{Ph—N=N—O}^\bullet} + \text{N}_2 + \text{Ph}^\bullet \quad (66)$$

been modified to σ-type)[94] then played the key role of hydrogen abstractor in a subsequent step (equation 67) to give the diazohydroxide **172**. The acetic acid is then formed by reaction of acetate with **172** (equation 68) and the diazotate continues the chain process (equations 67, 68).

$$\text{Ph}^\bullet + \text{C}_6\text{H}_6 \longrightarrow \text{Ph(H)C}_6\text{H}_5^\bullet \xrightarrow{\textbf{171}} \text{Ph—Ph} + \underset{\textbf{(172)}}{\text{Ph—N=N—OH}} \quad (67)$$

$$\text{Ph—N=N—OH} + \text{CH}_3\text{CO}_2^- \rightleftharpoons \text{Ph—N=NO}^- + \text{CH}_3\text{CO}_2\text{H} \quad (68)$$

The presence of the iminoxy radical **(171)** has been detected using e.s.r. methods by Cadogan[88] using a wide variety of solvents. Hey, Perkins and their coworkers[242, 243] have also detected the presence of nitroxide radical **173** (probably formed as outlined in equation 69); however, this species is not present in all the solvents studied and it is questionable whether such a stable species would provide the rapid hydrogen abstractor required.

$$\text{Ph}^\bullet + \text{Ph—N(NO)—COMe} \longrightarrow \underset{\textbf{(173)}}{\text{Ph—N(—N(Ph)—O}^\bullet\text{)—COMe}} \quad (69)$$

The possibility arises that in spite of the detection of **171** (and in some solvents **173**), these may not be the principal species involved in chain propagation. In fact an alternative and simpler chain process was also proposed by Cadogan and coworkers[94] (Scheme 10) involving electron transfer between the radical **174** and benzenediazonium ion, followed by proton abstraction by acetate ion.

$$Ph^{\bullet} + PhN_2^+ \xrightarrow{C_6H_6} [\text{Ph(H)}C_6H_5^{+}] + Ph{-}N{=}N^{\bullet}$$

(174)

$$[\text{Ph(H)}C_6H_5^{+}] \xrightarrow{MeCO_2^-} Ph{-}Ph + MeCO_2H \qquad Ph{-}N{=}N^{\bullet} \rightarrow Ph^{\bullet} + N_2$$

SCHEME 10

Aryne and phenyl cation formation are of course competing reactions (see Section III.A.8) and the free-radical pathway can be suppressed by the addition of a suitable radical trap[85].

Only minor modifications of equations (66)–(68) are required to explain the formation of aryl radicals from diazonium ions in the presence of base. Again it is proposed that the diazoanhydride **170** is formed and that it is the active radical-producing species[245].

The absence of dimerization is explicable in terms of the high reactivity and low concentration of $Ar^{\bullet}$ in solution. The short lifetime of these radical species is emphasized by the interesting observation[246] that decomposition of optically active 2-methyl-6-nitrobiphenyl-2′-diazonium ion (in the presence of HO^-) gives an arylated product (**175**) in which there is >80% retention of configuration (equation 70).

$$\text{2-}CH_3\text{-6-}NO_2\text{-}C_6H_3\text{-}C_6H_4\text{-2'-}N_2^+ \xrightarrow[ArH]{HO^-} \text{2-}CH_3\text{-6-}NO_2\text{-}C_6H_3\text{-}C_6H_4\text{-2'-}Ar \qquad (70)$$

(175)

2. Via aryl cations

Small changes in solvent, reactants or reaction conditions or even in the presence or absence of oxygen[247] may be sufficient to change the mechanism of an arylation reaction from free radical to ionic and *vice versa*. Several groups of workers have described conditions where the aryl cation pathway is favoured[60, 248–257].

Kobayashi's[253–257] detailed work on the decomposition of benzenediazonium tetrafluoroborates shows that in aprotic solvents (MeCN, DMF and nitromethane), phenyl cations are produced which attack the aromatic substrate in the slow step; the partial rate factors for model free-radical and cationic phenylations of substituted benzenes were used to determine the mechanism. Ionic phenylation was also proposed using benzenediazonium 2,2,2-trifluoroacetate; but in methanol and pyridine, the reaction is free radical[256, 60].

Similar results have been reported by Zollinger's group[251]. Thus **176** in 2,2,2-trifluoroethanol (TFE) gives only the ether **178** and fluorobenzene **177**, by an ionic mechanism; the reaction is strictly first order in **176**. However, addition of even small amounts of pyridine decreases the yields of **177** and **178** and increases the amounts of benzene, biaryls and diazo-tars (typical homolytic products). The addition of pyridine actually increases the rate of disappearance of **176**, but good first-order kinetics are no longer obtained.

$$Ph{-}N_2^+\ BF_4^- \xrightarrow{TFE} Ph{-}F + Ph{-}OCH_2CF_3$$

(176) (177) (178) (179)

The radical-producing species which is formed on the addition of methanol, pyridine, *N,N*-diphenylhydroxylamine or nitrite ions, probably results from initial nucleophilic attack at the terminal nitrogen of the arenediazonium ion followed by homolytic cleavage (Section III.E.1); in the case of pyridine the radical species thus produced (**179**, termed a gegenradical[98]) is not particularly stable and the driving force in this case is thought to be the high nucleophilicity of pyridine itself.

The balance between the free radical and ionic pathways can also be shifted by substituents in the arene group. Hence under conditions where benzenediazonium ion reacts by a polar mechanism in DMSO (a poorly nucleophilic solvent), *p*-nitrobenzenediazonium ion gives typical homolytic products[249, 250]. This is explicable in terms of the relative stabilities of $Ar^{\bullet}$ and Ar^+ when Ar = Ph and $p\text{-}NO_2C_6H_4$ [98].

F. Intramolecular Reactions

Ring closure reactions involving displacement of the diazonium group from aromatic diazonium salts are known collectively as the Pschorr cyclization[258]. These cyclizations (equation 71 gives as an example the conversion of diazotized *o*-aminostilbene to phenanthrene) may be carried out under a variety of conditions in either acidic or basic solution in the presence or absence of a copper catalyst. Almost certainly both ionic (involving aryl cation intermediates) and radical pathways may be operative and since each of the pathways gives rise to side-products, Lewin and Cohen[259] have emphasized the importance of ensuring that the reaction conditions chosen favour just one mechanism.

HC=CH ; N_2^+ —Cu catalyst→ (71)

In general, in the absence of catalysts, cyclization in acidic solution occurs via aryl cation formation, particularly if the anion involved is not a good reducing reagent (e.g. BF_4^-; Cl^- on the other hand may give rise to radical reaction)[258, 260, 261]. Cyclization in basic solution is probably free radical in character[262] (via diazoanhydride formation, see Section III.E.1) whereas those catalysed by copper metal or cuprous ion are radical in character whether they are carried out in acidic or basic solution[263, 264]. These reactions are thus mechanistically analogous to intermolecular arylations (see Section III.D.2).

Thermal decomposition of 2-diazoniumbenzophenone **180** illustrates these points[259]. In the absence of copper catalysts the reaction is probably ionic (equation 72) and the products were fluoroenone (**181**, 65%) and 2-hydroxybenzophenone (**182**, 35%) [257]. Addition of a small amount of copper had little effect on the product composition or the reaction rate but the addition of cuprous oxide (or a large quantity of Cu) increased the rate of decomposition dramatically and the major product is the fluorenone; the formation of 2-hydroxybenzophenone (**182**) is virtually eliminated under these conditions and the only other product is benzophenone (**185**) formed in <10% yield[259]. The intermediacy of the radical **183** (which cyclizes to give the more stable radical **184**) is proposed. The radical intermediate

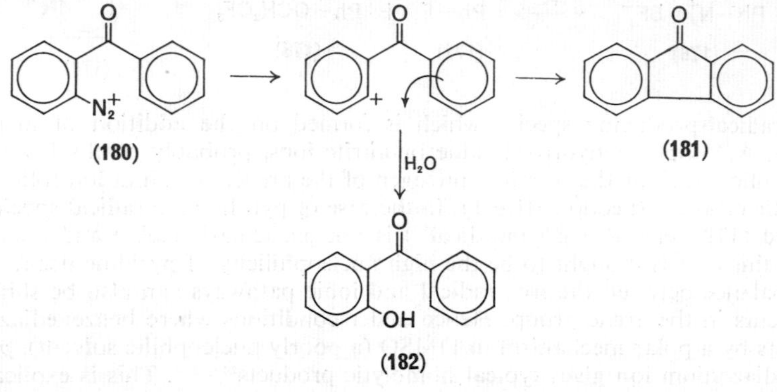

(72)

183 can be diverted to give entirely the reduction product benzophenone **185** in the presence of a suitable hydrogen source (e.g. RH = dioxan). Addition of a large excess (>300-fold) of $Cu(NO_3)_2$ results in the formation of *c.* 90% 2-hydroxybenzophenone (**182**). A possible mechanism involving electron transfer from the

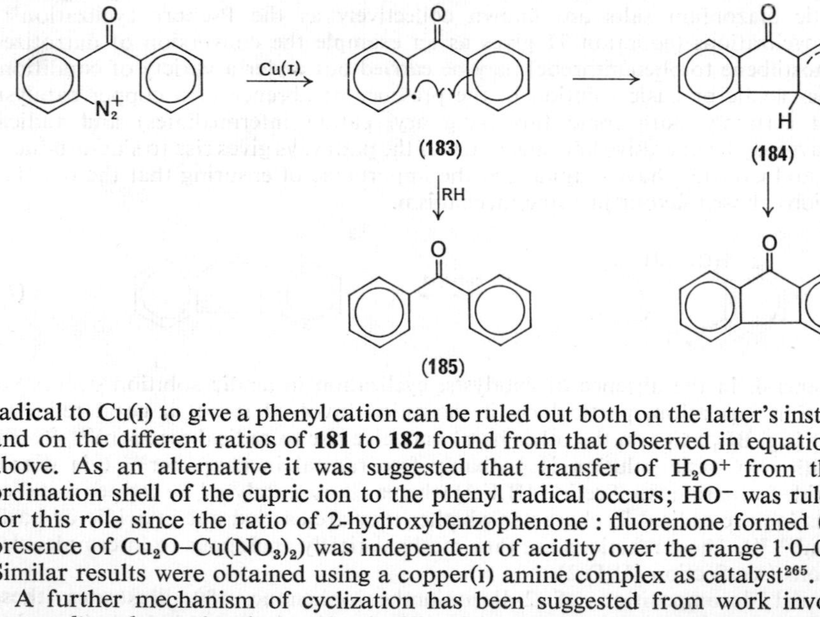

radical to Cu(I) to give a phenyl cation can be ruled out both on the latter's instability and on the different ratios of **181** to **182** found from that observed in equation (72) above. As an alternative it was suggested that transfer of H_2O^+ from the coordination shell of the cupric ion to the phenyl radical occurs; HO^- was ruled out for this role since the ratio of 2-hydroxybenzophenone : fluorenone formed (in the presence of Cu_2O–$Cu(NO_3)_2$) was independent of acidity over the range 1·0–0·01 M. Similar results were obtained using a copper(I) amine complex as catalyst[265].

A further mechanism of cyclization has been suggested from work involving, *inter alia*, electrochemical reduction[266]. This involves intramolecular reduction proceeding through an intramolecular charge-transfer complex in which electron transfer proceeds with simultaneous release of nitrogen within the solvent cage (equation 73).

A useful criterion was introduced by Huisgen and Zahler[267] to distinguish between radical and ionic pathways based on the very low selectivity of the phenyl cation (relative to the radical). The decomposition of diazotized 2-(*o*-aminobenzoyl)-naphthalene (**186**) yields two products, **187** and **188**, resulting from attack at the two sites adjacent to the acyl group. The ratio of **187** : **188** however varies with the conditions used for the decomposition from 2·4 in acid solution (where a phenyl

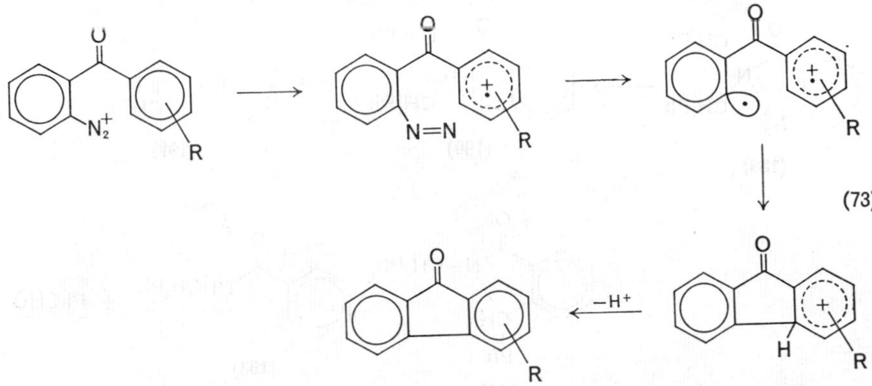

(73)

cation intermediate is implicated) to 9·5 in basic solution (free radical). The Cu_2O-catalysed cyclization also gives a ratio of *c*. 9, consistent with the mechanism suggested above[259]. The selectivity of the two intermediates decreases, so that the ratio of **187** : **188** decreases somewhat as temperature is increased.

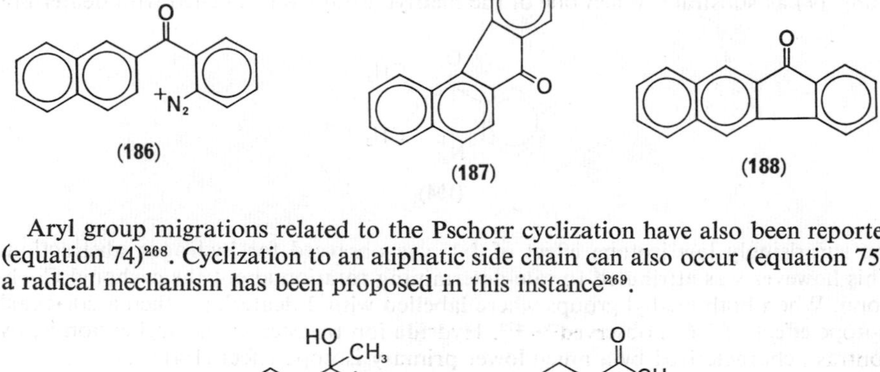

(186) (187) (188)

Aryl group migrations related to the Pschorr cyclization have also been reported (equation 74)[268]. Cyclization to an aliphatic side chain can also occur (equation 75); a radical mechanism has been proposed in this instance[269].

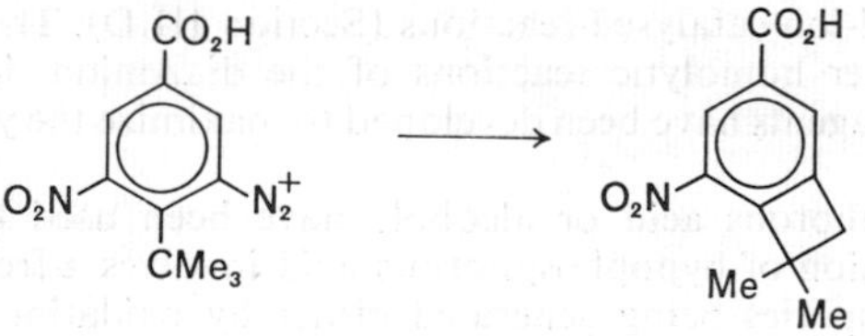

(74)

(75)

Intramolecular hydride ion transfer to a phenyl cation **190** (produced by thermal decomposition of the diazonium compound **189**) can also occur[270]. The driving force for this reaction is probably the formation of the stabilized azacarbonium ion **191**. On treatment with water the final products obtained are **193** and benzaldehyde.

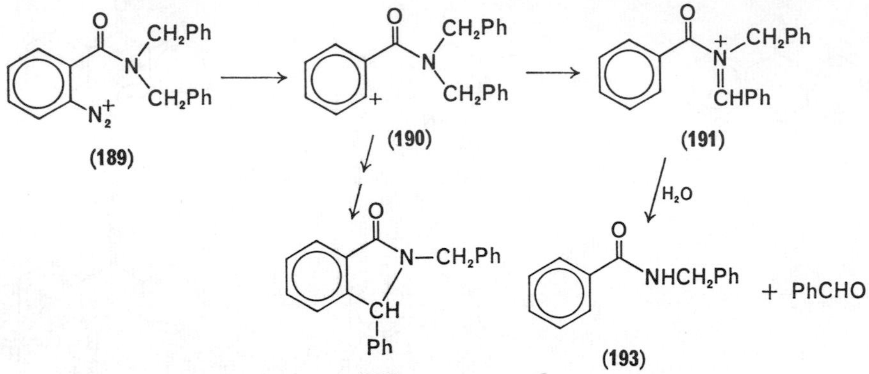

(189) (190) (191) (192) (193) + PhCHO

In addition 1-phenyl-2-benzylphthalimide (**192**) is formed and a mechanism involving cyclization of the phenyl cation **190** (rather than, say, cyclization of **191**) is favoured since **191**, prepared independently, was shown to be stable in an inert solvent with respect to ring formation. In the presence of Cu_2O, the reaction becomes free radical and 1,5-hydrogen atom transfer occurs rather than hydride ion transfer[271], using **194** as substrate. When one of the methyl groups was labelled with deuterium

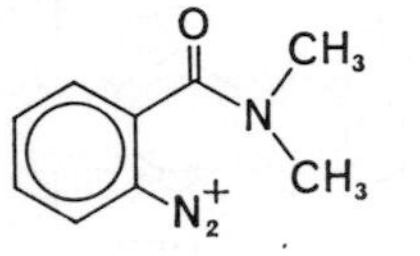

(194)

the surprisingly low isotope effect of 1·1 was observed for hydrogen abstraction. This however was attributed to rate-determining rotation about the carbonyl C—N bond. When both methyl groups where labelled with 2 deuteriums then a corrected isotope effect of 7·6 is observed[271, 272]. Hydride ion transfer to the aryl cation is, by contrast, characterized by a much lower primary isotope effect (1·4)[271].

G. Reduction

Reduction of diazonium ions to arenes, biaryls and azoarenes has already been referred to in metal-ion catalysed reactions (Section III.D). These are also common by-products in other homolytic reactions of the diazonium ion (Section III.E.1), although specific reagents have been developed to maximize the yields of the reduction products.

Thus hypophosphorous acid or alcohols have been used to form arenes. The mechanism of reaction of hypophosphorous acid involves a free radical chain, with the initial radical species being generated either by oxidation (traces of oxidizing agents have a catalytic effect)[273, 274] or by nucleophilic attack by $H_2PO_2^-$ on the diazonium ion followed by homolytic cleavage (equation 76).

$$ArN_2^+ + H_2PO_2^- \rightleftharpoons Ar{-}N_2OPOH_2 \rightleftharpoons Ar^{\bullet}N_2H_2PO_2^{\bullet}$$

$$Ar^{\bullet} + H_3PO_2 \longrightarrow ArH + H_2PO_2^{\bullet}$$

$$ArN_2^+ + H_2PO_2^{\bullet} \longrightarrow Ar^{\bullet} + N_2 + H_2PO_2^+$$

$$H_2PO_2^+ + H_2O \rightleftharpoons H_3PO_3 + H_3O^+ \qquad (76)$$

Both alcohols (particularly methanol) and ethers (usually dioxan) have also been used to achieve an efficient reduction, but the conditions, e.g. exclusion of O_2, use of buffered solutions (since aryl cation formation may predominate in acid), must be carefully controlled otherwise yields may be erratic[275]. In the presence of dioxan *and* D_3PO_2 hydrogen abstraction (to give ArH) occurs preferentially from the dioxan[276].

The kinetics of reduction of arenediazonium ions with ethers (equation 77) have been studied (by following the rate of nitrogen evolution) by Rüchardt and co-workers[277]. With *p*-chlorobenzenediazonium ion, buffered at pH 4·5, there is a considerable induction period. The induction period can be reduced by (a) an increase in pH, (b) the presence of oxygen and (c) addition of Cu^I, Fe^{II}, I^-. The

$$ArN_2^+X^- + RCH_2OR \longrightarrow N_2 + ArH + R{-}CH{=}\overset{+}{O}RX^- \qquad (77)$$

initiation step(s) (equations 78, 79) suggested involve electron transfer from diazotate (favoured by high pH) or from the metal ions or I^-. Oxygen is suggested to initiate

$$ArN_2^+ \longrightarrow \longrightarrow Ar^{\bullet} \qquad (78)$$

$$RCH_2OR \longrightarrow \longrightarrow R\dot{C}HOR \qquad (79)$$

via an alternative process involving the ether (equation 79). A free-radical chain reaction (Scheme 11) is also consistent with the observation that radical traps (e.g. $BrCCl_3$) inhibit arene formation.

$$Ar^{\bullet} + RCH_2OR \longrightarrow ArH + R\dot{C}HOR$$

$$R\dot{C}HOR + ArN_2^+ \longrightarrow ArN_2^{\bullet} + RCH{=}\overset{+}{O}R$$

$$ArN_2^{\bullet} \longrightarrow Ar^{\bullet} + N_2$$

SCHEME 11

Aryldiimine (ArN_2H) formation can be ruled out as an intermediate since this decomposes via a carbanionic pathway and the product would then contain deuterium (ArD) in the presence of D_2O, which was not observed. The rate of reduction varied with the nature of the ether (1,3-dioxan > THF > dioxan > glyme) and the arenediazonium ion (*p*-$NO_2C_6H_4$ > *p*-ClC_6H_4 > 1-naphthyl > *p*-anisyl), consistent with equation (78) and Scheme 11.

A similar mechanism has been proposed[139] for the reduction of *p*-phenylene-bisdiazonium ion (**195**) with alcohols in aqueous acid solution which follows equation (80). The inverse dependence on [H^+] is accounted for by the initiation mechanism (via the diazoanhydride or electron transfer from the diazotate). When propanol-2*d* was used, a primary isotope effect (k_H/k_D *c*. 6) was observed in product formation, but not on the overall rate; this is typical for such reactions[278] and

$$^{+}N_2-C_6H_4-N_2^{+}$$

(195)

$$-\frac{d[195]}{dt} = k\frac{[195]\,[ROH]}{[H^+]} \qquad (80)$$

indicates that hydrogen (deuterium) abstraction occurs at a rate which is less than diffusion controlled. With allyl alcohol the kinetic form followed and the isotope effects are different and product studies show considerable polymerization (possibly initiated by a Meerwein-type addition to the unsaturated centre)[139].

An interesting initiation process using irradiation has been described[279–282] (equation 81). The method is quantitative and removes some of the uncertainties

$$H_2O \rightsquigarrow e^-_{aq}, HO^\bullet, H^\bullet, H_2, H_2O_2 \qquad (81)$$

$$HO^\bullet \text{ (or } H^\bullet) + RCH_2OR' \longrightarrow R\dot{C}HOR + H_2O \text{ (or } H_2) \qquad (82)$$

$$e^-_{aq} + ArN_2^+ \longrightarrow Ar^\bullet + N_2 \qquad (83)$$

$$Ar^\bullet + RCH_2OR' \longrightarrow ArH + R\dot{C}HOR' \qquad (84)$$

$$R\dot{C}HOR + ArN_2^+ \longrightarrow Ar^\bullet + N_2 + RCH{=}\overset{+}{O}R \qquad (85)$$

$$Ar^\bullet + ArN_2^+ \longrightarrow Ar\overset{\bullet +}{N}NAr \qquad (86)$$

$$Ar\overset{\bullet +}{N}N{-}Ar + R\dot{C}HOR \longrightarrow ArN{=}N{-}Ar + RCH\overset{+}{O}R \qquad (87)$$

inherent in the thermal methods. Subsequent reactions (equations 82–87) yield the azoarene (the key step being aryl radical reaction with the arenediazonium ion[77], equation 86) and arene by hydrogen abstraction from the ether (or alcohol when R′ = H); aldehydes are produced quantitatively when alcohols are used as reducing agents. Packer and coworkers[280–282] show that with appropriate alcohols (e.g. ethanol) both α($CH_3\dot{C}HOH$) and β($\dot{C}H_2CH_2OH$) radicals are produced, but that only the α-radicals react with the diazonium ion according to equation (85). Beckwith and Norman[283] have shown that the e.s.r. signals due to α-radicals from ethanol and methanol quickly disappear on addition of ArN_2^+. The results also suggest that the possibility of termination by aryl radical dimerization (biaryl formation) proposed by Lewis[139] is remote.

In basic solution (anhydrous methanol containing methoxide ion) Bunnett and Takayama[129] have proposed that reduction occurs via two competing pathways: (a) ionic, in which the hydrogen on the arene (formed as product) comes from the methanol —OH, and (b) a radical chain mechanism in which the chain carrying species is $^\bullet CH_2O^-$ (rather than $^\bullet CH_2OH$ as proposed in acid). The radical $^\bullet CH_2O^-$ is expected to give a particularly efficient route to $Ar^\bullet$; after reaction with ArN_2^+, the leaving group for homolytic fission is formaldehyde.

Phenyldiazene (**196**) formation has been directly observed as a reduction product of arenediazonium ions with sodium borohydride[284, 285]. The diazene **196** is rapidly converted to arene in the presence of base via a mechanism involving loss of N_2 from the diazene anion[286]. The isolation of **196** is only possible under anaerobic conditions; however electron-withdrawing substituents (e.g. NO_2) stabilize **196** towards aerial oxidation[287].

$$X{-}C_6H_4{-}\overset{+}{N}{\equiv}N + BH_4^- \quad (X = H, NO_2)$$

$$\downarrow$$

$$\left[X{-}C_6H_4{-}N{=}N{-}H\right] \longrightarrow X{-}C_6H_4{-}H + N_2$$

(196)

Aryldiazene intermediates have also been proposed in the reduction of arenediazonium salts using stannous chloride and related reagents; the products are the corresponding hydrazines (equation 88). Reduction by bisulphite has been

$$ArN_2^+ \xrightarrow{SnCl_4} [Ar{-}N{=}N{-}H] \longrightarrow ArNHNH_2 \qquad (88)$$

shown[288b, 289] to occur via the *syn* diazosulphonic acid **197** (see Section III.B.2). The arylhydrazine sulphonic acid (**198**) is an intermediate and treatment with acid yields the arylhydrazine **199**.

$$ArN_2^+ \xrightarrow{SO_3^{2-}} \underset{\mathbf{(197)}}{\overset{Ar\quad\quad SO_3H}{N{=}N}} \xrightarrow{HSO_3H} \underset{\mathbf{(198)}}{Ar{-}NH{-}NH{-}SO_3H} \xrightarrow{H^+} \underset{\mathbf{(199)}}{ArNHNH_2}$$

IV. SYNTHESIS OF DIAZOALKANES

A. Nitrosation of Primary Amines

Direct reaction of primary aliphatic amines with nitrosating species in acidic solution only yields diazoalkanes when the group R (in **201**) is strongly electron withdrawing (e.g. R = CO_2Et, CF_3, SO_3H, COR)[288a]. Proton loss from the diazonium ion intermediate **201** to give the diazoalkane is competitive with N_2 loss. Generally

$$\underset{\mathbf{(200)}}{RCH_2NH_2} \longrightarrow \underset{\mathbf{(201)}}{RCH_2N_2^+} \longrightarrow RCH{=}N_2$$

the synthesis is only successful when an excess of acid is avoided (typically equimolar ratios of amine, $NaNO_2$ and HCl are used) and the diazoalkane is extracted into an organic layer (e.g. CH_2Cl_2) as soon as it is formed. α-Diazo esters which react relatively slowly in acid can be successfully prepared by this method[289]. Acidic conditions can be avoided by the choice of nitrosating species (see Section II.F), most typically by the use of NOCl [290]. When an arenediazonium salt has an acidic hydrogen then the corresponding diazoalkane may be formed on addition of base.

Equation (89) gives an example[291] and further examples involving heterocyclic diazonium salts are given in Reference 39.

$$\text{HO}-\text{C}_6\text{H}_4-\text{N}_2^+ \xrightarrow{:\text{B}} \text{O}=\text{C}_6\text{H}_4=\text{N}_2 \quad (89)$$

B. *From* N-*Nitrosamides*

The most widely used method for the synthesis of diazoalkanes is the treatment of *N*-nitrosamides (**202**) with base (usually KOH or KOEt in EtOH)[292, 293]; the diazoalkane is either directly distilled off or, more usually, extracted into an organic layer prior to distillation. The *N*-nitroso-*N*-alkyl compounds used include ureas, *p*-toluene sulphamides[294, 295], terephthalimides[296], nitroguanidines[297], oxamides[297] and sulpholanes[298]. The mechanisms which have been suggested are summarized in equations (90) to (93). Base attack at the carbonyl is favoured with urethanes (YH = OEt)[299] but with ureas direct attack at the carbonyl is so slow that the base reacts either to remove an ionizible proton (equation 92, YH = NH_2)[300, 301] or at the nitrosyl group (equation 93)[302–304]. In the latter case Y=C=O (e.g. Y = NH) is observed among the products. The ElcB mechanism (equation 92) is favoured by Hecht's observation[300, 301] that both NaH and triethylamine effect diazoalkane formation in dry solvents when YH = NH_2.

Preliminary rearrangement:

$$\underset{(\mathbf{202})}{\text{R}-\text{CH}_2-\text{N}(\text{N}=\text{O})-\text{C}(=\text{O})-\text{YH}} \longrightarrow \text{RCH}_2-\text{N}=\text{N}-\text{O}-\text{C}(=\text{O})-\text{YH} \xrightarrow{\text{HO}^-} \text{RCH}=\text{N}_2 + {}^-\text{O}_2\text{CYH} \quad (90)$$

Base attack at carbonyl:

$$\text{R}-\text{CH}_2-\text{N}(\text{N}=\text{O})-\text{C}(=\text{O})-\text{YH} \xrightarrow{\text{RO}^-} \text{RCH}_2-\text{N}(\text{N}=\text{O})-\text{C}(\text{O}^-)(\text{OR}')-\text{YH} \longrightarrow \text{RCH}_2-\text{N}=\text{N}-\text{O}^- \longrightarrow\longrightarrow \text{RCH}_2=\text{N}_2 \quad (91)$$

ElcB:

$$\text{RCH}_2-\text{N}(\text{N}=\text{O})-\text{C}(=\text{O})-\text{YH} \underset{}{\overset{\text{R}'\text{O}^-}{\rightleftharpoons}} \text{R}-\text{CH}_2-\text{N}(\text{N}=\text{O})-\text{C}(=\text{O})-\text{Y}^- \longrightarrow \text{R}-\text{CH}_2-\text{N}=\text{N}-\text{O}^- + \text{Y}=\text{C}=\text{O}; \quad \text{R}-\text{CH}_2-\text{N}=\text{N}-\text{O}^- \longrightarrow\longrightarrow \text{RCH}=\text{N}_2 \quad (92)$$

Attack at nitrosyl group:

$$RCH_2{-}N(N{=}O){-}C({=}O){-}YH \xrightarrow{R'O^-} R{-}CH_2{-}N({-}N(O^-){-}OR'){-}C({=}O){-}YH \longrightarrow RCH_2{-}N(HONOR'){-}C({=}O){-}Y^-$$

$$\xrightarrow{HO^-} R{-}CH_2N_2OR' \longrightarrow RCH{=}N_2$$

$$\xrightarrow{HO^-} R{-}CH_2{-}N{=}N{-}O^- \longrightarrow RCH{=}N_2 \qquad (93)$$

A special mechanism involving an internal acylation of the diazotate (equation 94) is indicated by Reimlinger's work which showed that when the oxygen of the *N*-nitroso group was labelled with ^{18}O, this was incorporated in the ethyl carbonate[305].

$$Me{-}C({=}O){-}N(N{=}O){-}CH_2CO_2Et \longrightarrow MeCO_2R + (EtO_2CCH_2){-}N{=}N{-}O^- \longrightarrow CH_2N_2 + O{=}C(O^-){-}OEt \qquad (94)$$

C. From Alkanediazotates

Solid potassium *syn*-diazotates (**203**) when dissolved in hydroxylic solvents give competitive reactions leading to diazoalkanes (**205**) and the normal products associated with alkylcarbonium ion (**204**) formation. The diazoalkane **205** predominates when R = H or when R is a group which can stabilize the forming negative charge on carbon (e.g. R = Ar, $CH_2{=}CH{-}$). Thus the methanolysis of

$$\underset{(203)}{R{-}CH_2{-}N{=}N{-}O^-} \rightleftharpoons RCH_2{-}N{=}N{-}OH \longrightarrow \underset{(204)}{R\overset{+}{C}H_2} + N_2 + HO^-$$

$$RCH_2{-}N{=}N{-}OH \xrightarrow{HO^-} \underset{(205)}{RCH{=}\overset{+}{N}{=}\overset{-}{N}} \longleftrightarrow \underset{(205a)}{R\overset{-}{C}H{-}\overset{+}{N}{\equiv}N}$$

substituted cinnamyl diazotates showed that electron-withdrawing substituents aid diazoalkane formation (relative to N_2 loss) while electron-donating substituents have the opposite effect[306]. When the group R can stabilize the carbonium ion **204** then this route predominates. Hence with secondary alkyl diazotates, diazoalkane formation is minimized[307]. In concentrated base (3·0 M HO^-), diazoalkane formation is first order in base while deuterium-labelling experiments have shown that **205**, once formed, cannot revert and give the carbonium ion **204** [112].

D. Other Methods

Oxidation of isolable hydrazones **206** with agents such as mercuric oxide, silver oxide and manganese dioxide in non-aqueous solvents has been described (equation 95). The action of trace amounts of base as catalyst suggests a mechanism involving electron transfer from the hydrazone anion[308–312]. The tendency of simple aldehydic

$$\underset{(\mathbf{206})}{R_2C{=}NNH_2} \rightleftharpoons R_2C{=}N{-}\bar{N}H \longrightarrow R_2C{=}N_2 \qquad (95)$$

hydrazones to disproportionate to symmetrical 2,3-diazabuta-1,3-dienes limits this method to ketone hydrazones (e.g. **206**, R = Ph). The diazabutadienes **207** can also be converted to diazoalkanes via the *N*-oxides **208**, followed by pyrolysis (equation 96)[313].

$$\underset{(\mathbf{207})}{R_2C{=}N{-}N{=}CR_2} \longrightarrow \underset{(\mathbf{208})}{R_2C{=}N{-}\overset{+}{N}(O^-){=}CR_2} \longleftrightarrow R_2C{=}N{-}N(O^-){-}\overset{+}{C}R_2 \xrightarrow{\Delta} R_2C{=}N_2 + O{=}CR_2 \qquad (96)$$

1,1-Elimination from sulphonyl hydrazones in the presence of base (known as the Bamford–Stevens reaction)[314, 315] also gives diazoalkanes (equation 97). An

$$Ph{-}CH{=}N{-}NHSO_2Ar \rightleftharpoons Ph{-}CH{=}N{-}\bar{N}{-}SO_2Ar \longrightarrow PhCH{=}N_2 + ArSO_2^- \qquad (97)$$

$$ArC(R){=}N{-}NH_2 + Ar'SCl \longrightarrow Ar{-}C(R){=}N{-}NH{-}SAr' \longrightarrow Ar{-}C(R){=}N_2 \qquad (98)$$

analogous reaction described by Anselme[316] uses *o*-nitrobenzene sulphenylhydrazones (formed in the presence of triethylamine) which undergo elimination in the presence of hydroxide (equation 98, Ar′ = $2\text{-}NO_2C_6H_4$).

An interesting reaction which probably involves a hydrazonyl chloride intermediate is the formation of diazomethane from chloroform and hydrazine (equation 99); the 1,3-dipolar ion **209** formed on elimination can undergo base-catalysed isomerization to the more stable diazoalkane[305].

$$HCCl_3 + NH_2NH_2 \xrightarrow{-2\,HCl} H(Cl)C{=}N{-}NH_2 \xrightarrow{-HCl} \underset{(\mathbf{209})}{H{-}\overset{+}{C}{=}N{-}\bar{N}{-}H} \rightleftharpoons H_2C{=}N_2 \qquad (99)$$

The Foster synthesis involves the reaction of an oxime with chloramine (X = Cl) or hydroxylamine *O*-sulphonic acid (X = OSO_3H, equation 100)[317, 318]. *N,N*-Dichloramines react similarly with hydroxylamine in the presence of methoxide ion (equation 101), but the yields are poor[319].

$$R_2C{=}NOH + NH_2{-}X \xrightarrow{HO^-} R_2C{=}N_2 + H_2O + X^- \quad (100)$$

$$R_2CH{-}NCl_2 + NH_2OH \longrightarrow [R_2CH{-}N(Cl){-}NHOH] \longrightarrow R_2CH{-}N{=}N{-}O^- \longrightarrow R_2CH{=}N_2 \quad (101)$$

The structure of simple diazoalkanes can also be modified without losing the diazo function, most simply by an electrophilic substitution at carbon; the reaction of acid chlorides with diazomethane to give α-diazoketones (see Section V.A.2.c) is an example of this.

V. REACTIONS OF DIAZOALKANES

A. With Electrophilic Species

1. Protic acids

a. *Pre-equilibrium protonation.* Diazoalkanes are generally stable in base but undergo rapid reaction in acid solution. The products formed (see Scheme 12) may be complex depending on the tendency of carbonium ion **212** to undergo further rearrangement and on the nature and number of nucleophilic species (X^-) present. Kinetic studies have shown[320] that proton transfer to the nucleophilic

$$R^1{-}\overset{R^2}{C}{=}\overset{+}{N}{=}\overset{-}{N}\ (\mathbf{210}) \leftrightarrow R^1{-}\overset{-}{C}(R^2){-}\overset{+}{N}{\equiv}N$$

$$\mathbf{210} \underset{-H^+}{\overset{+H^+}{\rightleftharpoons}} R^1{-}CH(R^2){-}N_2^+\ (\mathbf{211}) \longrightarrow R^1{-}\overset{+}{C}H(R^2)\ (\mathbf{212}) \longrightarrow \text{Rearrangement}$$

$$\mathbf{212} \xrightarrow{-H^+} \text{Alkenes}$$

$$\mathbf{211} \xrightarrow{X^-} R^1{-}CH(R^2){-}X\ (\mathbf{213}); \quad \mathbf{212} \xrightarrow{X^-} \mathbf{213}$$

$$\mathbf{212} \xrightarrow{R^1R^2C{=}N_2} R^1R^2C{=}CR^1R^2$$

SCHEME 12

carbon occurs followed by nitrogen loss and either of the steps can be rate determining. Diazoacetic esters (**210**, $R^1 = EtO_2C-$, R^2 = H), α-diazoketones (**210**, R^1 = R″CO, R^2 = H), α-diazosulphones[321] (**210**, $R^1 = R''SO_2$, R^2 = H) and diazomethane show the characteristics of a pre-equilibrium protonation with (a) specific (rather than general) acid catalysis, (b) faster reaction in D_2O than in H_2O (since D_3O^+ is a stronger acid than H_3O^+)[322] and (c) deuterium exchange (R^2 = H replaced by D) in the unreacted diazoalkane[323]. Whether or not the nucleophile

(X^-) reacts with the carbonium ion (**212**) or with the diazonium ion (**211**) has been a matter of controversy reminiscent of the corresponding dediazonization of arenediazonium ions (see Section III.A.1); several other routes to alkanediazonium ions (e.g. nitrosation of amines, rearrangements of *N*-nitrosamines, acidification of diazotates or alkyltriazenes) are available and the data for these reactions have been used as supporting evidence for nucleophilic involvement. Weak nucleophiles (e.g. $X^- = H_2O$) probably react with the carbonium ion **212**; however catalysis by Cl^-, Br^- and I^- has been observed for diazoacetic ester decomposition in acid and interpreted in terms of nucleophilic assistance in the transition state. However, the observed Swain–Scott[324] *s* value of *c.* 0·3 indicates a low sensitivity to the nature of the nucleophile (as expected for a reaction involving such a good leaving group as N_2)[325, 326].

b. *Proton transfer in slow step.* Rate-determining proton transfer to the diazoalkane **210** is observed with diaryldiazomethanes (**210**, $R^1 = R^2 = Ph$), aryldiazomethanes (**210**, $R^1 = Ar$, $R^2 = H$) and secondary diazoketones (**210**, $R^1 = MeCO$, $R^2 = Me$). The apparent anomaly that the rate-determining step shifts to protonation when strongly electron-withdrawing groups are *not* present in the diazoalkane is explicable in terms of the increased rate of N_2 loss to give the more stable species **212** (relative to H^+ loss to reform the diazoalkane). General acid catalysis is observed for diphenyldiazomethane with Bronsted α close to 0·5 [327]. Primary isotope effects (k_H/k_D = 3·5 and 3·6 for acetic acid and benzoic acids acting as general acid catalysts)[278, 328] clearly point to proton transfer in the rate-limiting step. Moreover, the reaction is faster in H_2O than in D_2O [329]. The (rapid) subsequent reaction of **211** ($R^1 = MeCO$, $R^2 = Me$) with nucleophiles Cl^-, Br^-, I^-, NCS^- follows the Swain–Scott relationship indicating nucleophilic participation in the product-determining step. Elimination (to give alkenes) shows a primary isotope effect (k_H/k_D = 2·4), which favours an E2-type mechanism for this step[330].

Because of the ease with which the reactions of diphenyldiazomethane with acid can be followed this reaction has been used to investigate in detail the mechanism of proton transfer and in the study of polar and steric effects in the catalysing acid[331, 332].

When $X^- = RCO_2^-$ (Scheme 12), then a relatively large proportion of the ester **213** is formed and this route is widely used for ester alkylation under mild conditions. Moreover addition of excess RCO_2^- (rather than RCO_2H) does not change the proportion of ester formed. Roberts[333, 334] explained this observation in terms of a cyclic transition state in which proton transfer from RCO_2H is concerted with the nucleophilic step by the forming RCO_2^-; the formation of a tight ion pair suggested by More O'Ferrall[320] is a limiting case of this mechanism (Scheme 13) and is more

$$Ph_2CN_2 + HX \longrightarrow Ph_2CHN_2^+ \longrightarrow Ph_2\overset{+}{C}HX^-N_2$$

$$\underset{(214)}{Ph_2CHOEt} \xleftarrow{EtOH} Ph_2CH^+ \parallel X^- \longrightarrow \underset{(215)}{Ph_2CH{-}X}$$

SCHEME 13

convincing on the basis of the evidence presented (e.g. the isotope effect, using $PhCO_2H(D)$ in ethanol was the same for the formation of **214** and **215**, whereas a different isotope effect might be expected for the product formed by a cyclic route), moreover it has been shown using ^{18}O-labelled *p*-nitrobenzoic acid that the same degree of ion return, scrambling, etc., is observed as in the solvolysis of benzhydryl *p*-nitrobenzoate (**215**, $X^- = p\text{-}NO_2C_6H_4{-}CO_2^-$)[335, 336].

c. *Weak acids*. With weak acids (ArOH, ROH) alkylation by diazomethanes is slow and the formation of ethers from aliphatic alcohols is normally catalysed by the addition of fluoroboric[337] or toluenesulphonic acids[338] or of aluminium alkoxides (equation 102)[339]. Ethers can also be alkylated (e.g. $Me_2O \rightarrow Me_3O^+$) as can

$$R\overset{+}{O}(H){-}Al(OR)_2 + CH_2N_2 \longrightarrow R{-}\overset{+}{O}(CH_3){-}Al(OR)_2 \rightleftharpoons ROCH_3 + Al(OR)_3 \quad (102)$$

ammonium ions (e.g. $R_3\overset{+}{N}H \rightarrow R_3\overset{+}{N}Me$)[339] using diazoalkane in the presence of an acid; the use of an acid with a counter ion of low nucleophilicity (e.g. ClO_4^-, BF_4^-) is essential in these reactions[341].

Alkylation of enolizable ketones or amides can occur on oxygen (nitrogen) or on carbon and at one time the ratios of O to C alkylation products formed were used to estimate equilibrium concentrations of the keto-to-enol tautomers. However, it is now clear that alkylation normally occurs at the same site from which the proton is removed and that the major products formed are kinetically controlled yielding the enol ethers (*O*-alkylation) in spite of the low equilibrium enol content. As expected for a reaction involving an ambident anion, however, the ratio of C to O alkylation can be sensitive to the solvent, ion pairing, etc.[342]

d. *Non-hydroxylic solvents*. In general, the reaction of diazoalkanes with (strong) acids follows the same course in non-hydroxylic solvents. For example, the reaction of Ph_2CN_2 with *p*-toluenesulphinic acid shows rate-determining proton transfer ($k_H/k_D = 3{\cdot}0$) in dichloromethane, benzene, acetonitrile, dioxan and DMSO [343]. The fastest rates are found in the dipolar aprotic DMSO and this is attributed to dedimerization of the acid which is not solvated by its conjugate base in this solvent. Both sulphinate (**217**) and sulphone (**216**) are observed among the products, but the sulphone **216** is the sole product in more dissociating DMSO.

$$ArSO_2H + Ph_2CN_2 \longrightarrow Ph_2CHN_2^+\ ^-O_2SAr$$

$$\downarrow$$

$$Ph_2\overset{+}{C}H \parallel {}^-O_2SAr \longleftarrow [Ph_2CH^+\ {}^-O_2SAr]$$

$$Ph_2\overset{+}{C}H \parallel {}^-O_2SAr \rightarrow Ph_2CHSO_2Ar\ (\mathbf{216}) \qquad [Ph_2CH^+\ {}^-O_2SAr] \rightarrow Ph_2CHOSOAr\ (\mathbf{217})$$

e. *Rearrangements involving alkanediazonium ions.* Alkanediazonium ions (**219**) formed either on protonation of diazoalkanes or nitrosation of primary amines (or other similar routes such as rearrangement of *N*-nitrosamides) undergo similar subsequent reactions[344, 345] including reaction with the solvent (to give **220**) or counter ion (**225**) and alkene **224** formation, often preceded by migration of a group α to the diazonium group. However, the characteristics of the carbonium ions

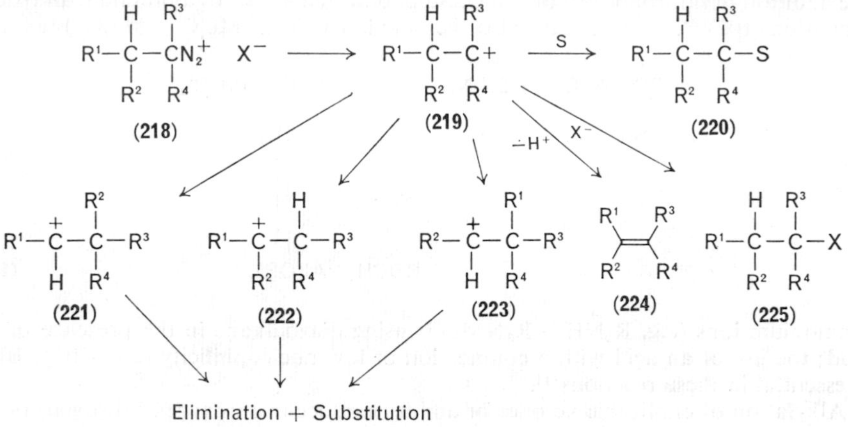

(**218**) (**219**) (**220**)

(**221**) (**222**) (**223**) (**224**) (**225**)

Elimination + Substitution

SCHEME 14

formed by this route are distinctly different from those formed by other (solvolytic) routes (this has been termed a 'memory effect')[346]. Thus the migratory aptitudes of the neighbouring groups (R^1, R^2 in **219**) may be close to unity, compared with values up to 10^3 in solvolysis[347]; reaction with the counter ion, X^-, usually predominates over reaction with other nucleophiles present in solution. These reactions have been the subject of several recent reviews and will be dealt with only briefly here[348–350].

Two alternative hypotheses have been put forward to explain these differences. Streitwieser[351, 352] has proposed that the rearrangements etc. occur, not from the carbonium ion (**219**), but from the diazonium ion (**218**). This would predict that migration of, say, R^2 is *concerted* with nitrogen loss. An alternative explanation offered by Huisgen and Rüchardt[353] and expanded on by Collins[349] and others is that the carbonium ion **219** actually lies on the reaction pathway; however, the ions formed are of high energy ('hot' carbonium ions[354] which are either vibrationally excited or unsolvated) and lack discrimination either for the group which migrates or between potential nucleophiles. Because of the particularly good leaving ability of nitrogen from **218** (the energy of activation is 3–5 kcal mol^{-1}), the carbonium ion **219** is formed with little solvent assistance (or solvent stabilization). The adjacent counter-ion X^- of the original diazonium ion retains its position on fast nitrogen loss, and reaction with X^- within the ion pair, rather than diffusion apart of the ions, is usually the major pathway.

A distinction between these mechanisms can be made by observing the stereochemistry of the migration process itself. Several groups of workers[349] have shown that when migration occurs, the configuration of the migration terminus is partially or predominantly retained. This would appear to rule out an S_N2-type mechanism for migration involving the diazonium ion **218**, which would lead to inversion.

The importance of ion pairing has been demonstrated[348] by using chiral diazonium ions (**218**) with ^{14}C-labelled acetate as the counter-ion. The product formed with retained configuration resulted predominantly from cation–anion recombination within the ion pair. Acetate ion from the solvent also attacks the intermediate ion pair. In this case the configuration is largely inverted indicating that the counter ion shields the side from which the nitrogen left. The lifetime of the ion pair is short since specifically O^{18}-labelled acetate counter ion reacts largely through the oxygen originally attached to nitrogen (in *N*-nitrosoamide decomposition). Cation–anion recombination within the ion pair can therefore compete even with reorientation of the counter ion.

2. Reaction with the carbonyl group

a. *Aldehydes and ketones.* Four products **226–229** are commonly formed in the reaction of diazoalkanes with aldehydes or ketones and the relative amounts are critically dependent on the reaction conditions. The reaction is catalysed by acid

$$RR^1C{=}O + CH_2N_2 \longrightarrow RC(=O)CH_2R^1\ \textbf{(226)};\quad RCH_2C(=O)R^1\ \textbf{(227)};\quad RR^1C(-O-CH_2-)\ \textbf{(228)};\quad R^1RC(OH)-CH{=}N_2\ \textbf{(229)}$$

but occurs in the absence of acid in hydroxylic solvents; catalysis by alcohols has also been noted[355]. The relative amounts of ketonic products are increased both by the presence of proton donors and Lewis acids, e.g. $AlCl_3$, BF_3; in fact in the presence of Lewis acids the sole products can be rearranged ketones[356].

Kinetic studies are consistent with the existence of two parallel reactions, initiated by cycloaddition of the diazoalkane to the ketone in the two possible orientations **230**, **232** [357]. Rapid ring opening with N_2 loss occurs to give **231** and **233** respectively (the direct formation of **231** and **233** with preliminary cycloaddition is also consistent with the data). The intermediate **233** or **235** can then rearrange (migration of R or R^1) to give **226** and **222** [359–362]. In the presence of a species (ROH, Lewis acid) capable of H-bonding or coordination to oxygen in **233**, the competing cyclization to give **228** is minimized. The alternative route via **231**, **234** gives solely the epoxide **228**. The rate of reaction is fastest with aliphatic aldehydes and in general aldehydes react more rapidly than ketones; substituent effects in the diazoalkane show that the rate is increased by an increase in the electrophilic character of carbon. The migratory aptitudes of the groups R and R^1 have been studied in detail and show that the group capable of stabilizing positive charge is most likely to migrate.

(230) (231) (228) (232) (233) (234) (235)

226 + 227

When R or R^1 is a chiral centre then migration occurs with retention of optical activity[363]. Steric factors have also been identified in determining the group which tends to migrate; thus the reaction of **236** with diazomethane gives largely **237**, i.e. the most highly substituted carbon migrates (electronic control). However, with

(236) $\xrightarrow{CH_2N_2}$ (237)

CH_3CHN_2, the less hindered carbon preferentially migrates[361]. This method of ring expansion, involving the conversion of a cyclic ketone to the next higher homologue, has been widely used[364]. A side reaction with ketones which are readily enolizable is *O*-alkylation.

Competing reactions occur with diphenylketen involving cycloaddition to the alkene and ketone groups; the products isolated are **238** and **239** [365]. α,β-Unsaturated ketones react somewhat similarly; in the absence of catalysts the major products

$Ph_2C{=}C{=}O + Ph_2C{=}N_2 \longrightarrow$ (238)

$Ph_2C{=}C(O^-){-}Ph_2C{-}N_2^+ \longleftrightarrow Ph_2\bar{C}{-}C{=}O\,(Ph_2C{-}N_2^+) \longrightarrow$ (239)

are the result of cycloaddition to the alkene. However, the addition of BF_3 diverts the reaction to give only homologous ketones[366].

The formation of the α-diazoalcohol (**229**) requires proton loss from the intermediate **233**, which competes with nitrogen loss. This is most likely when the diazoalkane has a strongly electron-withdrawing substituent (e.g. diazoacetic ester)[367].

b. *Esters*. Esters can be cleaved by diazoalkanes giving ester interchange (equation 103); apart from isolated exceptions, the normal products expected with ketones are not observed. It was originally thought that the ester was first cleaved by traces of H^+ or HO^- and that the acid formed then reacted with the diazoalkane. It is, however, clear that the rate of reaction excludes such a mechanism and it is proposed that the catalytic action of the diazoalkane is due to the increased nucleophilicity of oxygen because of coordination of the alcohol with the diazoalkane (**240**)[368].

$$R-C(=O)-OR' + CH_3OH \xrightarrow{CH_2N_2} R-C(=O)-OCH_3 + R'OH \qquad (103)$$

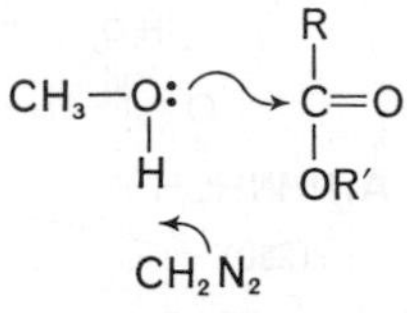

(**240**)

Epoxide formation is reported in the reaction of the lactone **241** with excess diazoethane: the NO_2 group increases the electrophilic character of the carbonyl carbon and 'solvates' the dipolar intermediate (see **233**)[369]. Intramolecular reaction of diazoalkanes with ester functions has also been reported in the conversion of **242** to **243** [370].

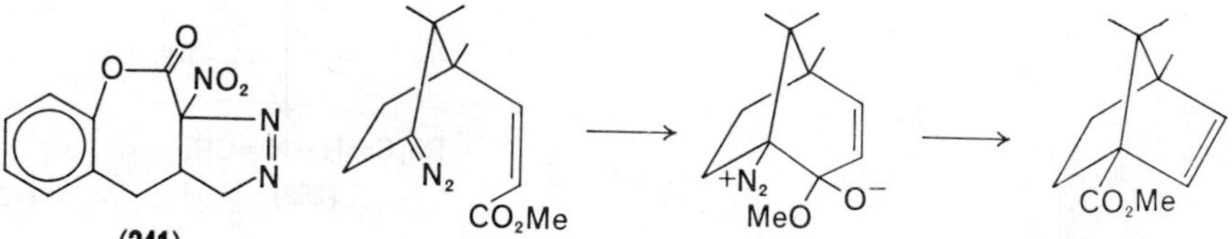

(**241**) (**242**) (**243**)

c. *Acid chlorides*. α-Diazoketones (**245**) are formed on reaction of acid chlorides or anhydrides with diazomethane[371]. This is the first step of the Arndt–Eistert

$$R-C(=O)-Cl + CH_2N_2 \longrightarrow R-C(O^-)(Cl)-CH_2N_2^+ \xrightarrow{R_3N} RC(=O)CHN_2 + R_3\overset{+}{N}HCl^-$$

(**244**) (**245**)

synthesis, a sequence which converts a carboxylic acid (via the acid chloride **244**) into its next higher homologue; the second step, which can be carried out *in situ*, is a Wolff rearrangement of the α-diazoketone (**245**) in the presence of Ag^+. Side products may occur in the initial reaction in the absence of a base; reaction of HCl with CH_2N_2 gives CH_3Cl while **245** gives $RCOCH_2Cl$ [372].

3. Arenediazonium ions

Electrophilic reaction of arenediazonium ions occurs at carbon to give unstable intermediates (**246**) which undergo further rearrangements and solvolysis. In the presence of Cl^-, hydrazonyl chlorides (**248**) are formed[373] (on tautomerization of **247**) which may be further solvolysed to the hydrazide (**250**)[358] or the hydrazonyl ether (**251**)[374]. The competing cyanamide (**249**) formation occurs via a rearrangement of the ArNH group to carbon as shown by ^{15}N-labelling[374].

$$[Ar{-}N{=}N{-}CH_2N_2^+] \xrightarrow{Cl^-} Ar{-}N{=}N{-}CH_2Cl$$

(**246**) → ArNH—C≡N (**249**)

(**247**) ⇌ Ar—NH—N=CHCl (**248**)

(**248**) $\xrightarrow{H_2O}$ ArNHNHCH(=O) (**250**)

(**248**) $\xrightarrow{MeOH}$ ArNH—N=CHOMe (**251**)

2,3-Diazabuta-1,3-diene ('azine', **252**) formation which is observed as a minor product in the reaction of diaryl diazomethanes in hydroxylic solvents and is promoted by the presence of Lewis acids (e.g. BF_3) probably occurs via a similar mechanism (equation 104)[375].

$$Ph_2C{=}N_2 \xrightarrow{BF_3} Ph_2C(BF_3^-){-}N_2^+ \xrightarrow{Ph_2CN_2} Ph_2C(BF_3^-){-}N{=}N{-}C(N_2^+)Ph_2 \xrightarrow{-N_2,\ -BF_3} Ph_2C{=}N{-}N{=}CPh_2\ (\mathbf{252}) \qquad (104)$$

4. Reaction with halogens

Halogens (Cl_2, Br_2, I_2) react with diazoalkanes or with α-diazoketones[376] to give *gem*-dihalides (equation 105) and this provides a simple route to these materials[377].

$$R_2CN_2 \xrightarrow{X_2} R_2C(X){-}N_2^+\ (\mathbf{253}) \xrightarrow{X^-} R_2CX_2$$

$$R_2CN_2 \xrightarrow{-N_2} R_2C{:}\ (\mathbf{254}) \xrightarrow{X_2} R_2CX_2 \qquad (105)$$

Although the mechanism of this reaction has not been studied in any detail, its rapidity in hydroxylic solvents[378] suggests an ionic mechanism with the halogen acting as a Lewis acid and the intermediacy of an α-halodiazonium ion. A possible alternative mechanism which may become operative in solvents of low ionizing power is via the carbene (**254**).

5. Carbonium ions

The stable triphenylmethyl carbonium ion reacts with diazomethane in ether at 0 °C to give the alkene **255** on rearrangement and nitrogen loss[375]. 1,2,3-Triphenylpropene is also formed by reaction of the intermediate carbonium ion with a further

$$Ph_3C^+ + CH_2N_2 \longrightarrow Ph_3C{-}CH_2N_2^+ \longrightarrow Ph_2C{=}CHPh + N_2$$

(255)

mole of CH_2N_2. A similar mechanism has been proposed for the conversion of xanthylium perchlorate to dibenzo[*b,f*]oxepine (the intermediate diazonium ion can be trapped before rearrangement by reaction with I^-)[379]. Azacarbonium ions (**256**) give aziridinium salts (**257**) on reaction with diazomethane; the aziridinium ion **258** is opened on reaction with alcohols[380].

$$Me_2C{=}\overset{+}{N}Me_2 \longrightarrow Me_2C(CH_2N_2^+){-}\ddot{N}Me_2 \longrightarrow \text{aziridinium } Me_2C(CH_2)\overset{+}{N}Me_2$$

(256) (257) (258)

6. Lewis acids

The catalysis by Lewis acids of the reactions of diazoalkanes with alcohols and ketones has already been referred to (Sections V.A.1.c. and V.A.2.a); the formation of azines **252** (in the absence of other electrophiles) is also observed.

The polymerization of diazoalkanes is initiated by boron halides (**259**, X = F, Cl, Br), boron alkyls (**259**, X = alkyl) and esters (**259**, X = OR) and a mechanism has been suggested involving initial reaction at carbon to give **260**, followed by N_2

$$CH_2N_2 + BX_3 \ (259) \longrightarrow CH_2{-}N_2^+ / {}^-BX_3 \ (260) \xrightarrow{X^-} CH_2X / BX_2 \ (261)$$

$$(260) \longrightarrow X_3\bar{B}CH_2^+ \ (262) \longrightarrow X_2BCH_2CH_2X \ (263)$$

$$(261) \xrightarrow{CH_2N_2} X_2\underline{B}(CH_2{-}X){-}CH_2N_2^+ \ (265) \longrightarrow X_2\underline{B}(CH_2{-}X){-}CH_2^+ \ (264) \longrightarrow X_2BCH_2CH_2X \ (263)$$

$$(263) \longrightarrow B[(CH_2)_aX][(CH_2)_bX][(CH_2)_cX] \ (266)$$

SCHEME 15

loss; in successive steps rearrangement of the growing chain to carbon is followed by further reaction with CH_2N_2. With diazomethane polymerization is very rapid and Davies has suggested[381] that the major pathway is via direct reaction of the carbonium ion **262**. The chain length is dependent on the diazoalkene : BX_3 ratio and the growing chain may be intercepted by the presence of a nucleophile.

Aluminium and silicon halides also promote polymerization and similar mechanisms have been proposed. Halomethyl derivatives are also obtained from anhydrous halides and alkyls including $SnCl_4$, $HgCl_2$, AlR_3 and $AsCl_3$ and the suggested mechanism is similar to the first step outlined in Scheme 15, giving **261** [382–384]. When the α-carbon carries a substituent then migration occurs, as observed with protic acids. However with BF_3 as catalyst, under anhydrous conditions, the normal migratory aptitudes (e.g. Ph > Me) are not observed and indiscriminate rearrangement occurs. This suggests a mechanism which is different from that with protic acids with the driving force being charge separation in the zwitterion **267** [304].

$$R{-}\overset{R^1}{\underset{R^1}{C}}{-}\overset{+}{C}H{-}\bar{B}X_3 \longrightarrow R{-}\overset{+}{\underset{R^1}{C}}{-}\overset{R^1}{C}H{-}\bar{B}X_3 + R^1{-}\overset{+}{\underset{R^1}{C}}{-}\overset{R}{C}H{-}BX^-$$

(267)

$$\downarrow \qquad\qquad \downarrow$$

$$(R)(R^1)C{=}C(R^1)(H) \qquad (R^1)(R^1)C{=}C(R)(H)$$

B. Cycloadditions

The mechanism of the 3+2 cycloadditions shown by diazoalkanes to a wide variety of unsaturated centres (e.g. alkenes, alkynes, azomethines and azo, nitroso and cyano groups) has been widely studied, particularly by Huisgen's group. These reactions are characterized by the retention of the alkene stereochemistry in the cycloadduct, high entropies[385] and a general insensitivity of the rate of reaction to the nature of the solvent used.

The experimental evidence in favour of a concerted ($\pi^4s+\pi^2s$) cycloaddition, which is thermally allowed by the Woodward–Hofmann rules, has been summarized[386]. Although an alternative reaction pathway via dipolar or diradical intermediates has been put forward[387] (and reiterated[388]) by Firestone, the weight of evidence supports the concerted mechanism; in particular the observed stereospecificity (observed not only with diazoalkanes but with all other 1,3-dipoles) is difficult to account for via the two-step mechanism.

The transition state for cycloaddition to diazomethane has been described as occurring in two planes (**268**) involving some bending of the C—N—N bond, but preserving the allyl anion orbital which makes contact with the dipolarophile, d—e. Calculations using the CNDO/2 method have shown that the initial bending of CH_2N_2 in this way to permit cycloaddition occurs with a low energy barrier[389].

A more recent theoretical study[390] using LCAO–SCF–MO methods of the reaction of diazomethane with ethylene has compared the energies of the transition state **268** and one in which all the heavy atoms are coplanar. It predicts that the 'all-coplanar' transition state is in fact the more probable pathway. The transition state is reached early on the reaction coordinate with a small degree of charge transfer from the dipole towards the alkene. Moreover (as Huisgen has often pointed out[386]), the degree of bond formation to nitrogen and carbon need not be exactly the same in the transition state. The study predicts that CC bond formation is slightly more advanced than C—N bond formation (see also below).

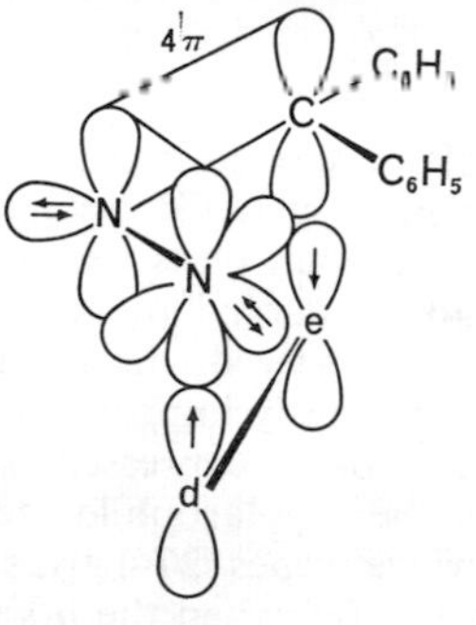

(268)

I. Orientation

Cycloaddition of a diazoalkene to an alkene such as **269** can give rise to two products, **270** and **271**. Many of the additions described originally as unidirectional

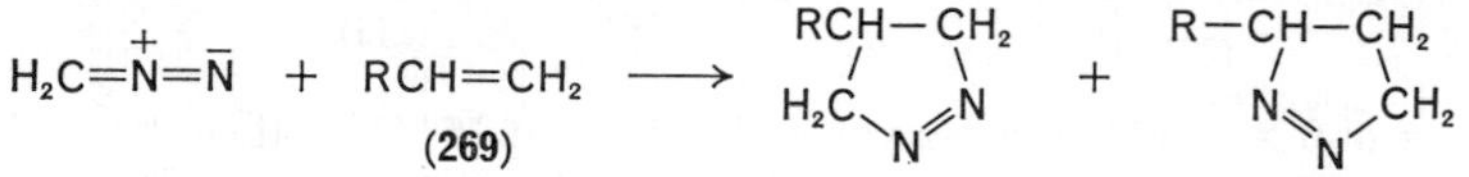

$$H_2C{=}\overset{+}{N}{=}\overset{-}{N} + RCH{=}CH_2 \ (\mathbf{269}) \longrightarrow (\mathbf{270}) + (\mathbf{271})$$

have later been shown to give mixtures. Several effects have been invoked to explain preferred orientation including electrostatic and steric effects and also the nucleophilic or electrophilic properties of the atoms undergoing reaction. The observed orientations for mono-, di- and tri-substituted alkenes and for mono- and di-substituted alkynes have been summarized by Bastide and coworkers[389]. The most important factor is the nucleophilic character of carbon in the diazoalkane (although the terminal nitrogen may carry more formal negative charge, this does not rule out carbon as the more nucleophilic centre[386]) which tends to react with the more electrophilic centre in the dipolarophile.

When the diazoalkane is itself substituted, then the formation of further isomeric products may occur. Thus in the addition of methyl diazoacetate to *trans*-ethyl cinnamate, two orientations are possible for the 'normal' pathway involving reaction of the diazoalkene carbon at the β-position of the unsaturated ester. *Syn* addition (**272**) is slightly favoured over *anti* (**273**) addition (k_{syn}/k_{anti}) = 1·5) when R = Me

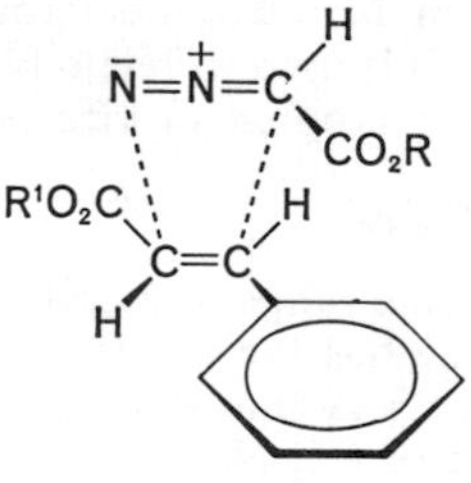

(272) 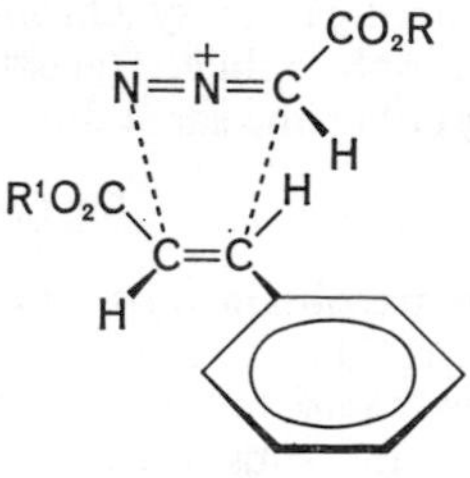 (273)

and the reverse is true when R = *t*-Bu (k_{syn}/k_{anti} = 0·47). Clearly both steric and electronic effects are operative, the preference for *syn* addition being attributed to π-overlap between the ester and aryl groups, which is counteracted by van der Waals' repulsions[391].

2. Substituent effects

In common with other dipolar cycloadditions, a variety of behaviour is shown when the nature of the substituent is changed systematically in either the dipolarophile or the dipole itself. Thus Hammett plots with positive or negative slopes have been reported or, more commonly, curved plots which are concave upwards are obtained. When curved plots are observed, the use of various alternative σ scales (σ^+, σ^-, σ^n, σ^0 etc.) does not significantly improve linearity[392].

Moreover, contrary to a view often expressed, some of the ρ values for the variation of the substituent in the dipolarophile are quite large. Thus the cycloaddition of diazomethane to arylacetylenes (**274**) gives a ρ value of +2·0 for electron-withdrawing substituents (ρ is *c.* +0·5 when the σ value of the substituents is <0) in DMF–ether at 25 °C[393]. Addition of diazomethane to the triple bond remote

$X-C_6H_4-C\equiv CH$ **(274)** $X-C_6H_4-C\equiv C-C\equiv C-C(=O)-CH_3$ **(275)**

from the group X in **275** gives a ρ value of +1·13 (at 25 °C in ether)[394] which implies the build-up of considerable negative charge on the dipolarophile in the transition state (when allowance is made for the normal attenuation factor due to the interpolation of the extra acetylenic group).

The observed curvature in Hammett type plots is dealt with (mainly by Sustmann and coworkers)[395–397] using a simple HOMO–LUMO model. A positive ρ value corresponds to the situation where $\Delta E_1 < \Delta E_2$ (where ΔE_1 is the difference in the energies between the HO of the dipole and the lowest unoccupied (LU) of the dipolarophile and ΔE_2 represents the interaction of the LU of the dipole and the HO of the dipolarophile for arylacetylenes and diazomethane). A minimum in a free-energy plot then results if there is an inversion in the relative values of ΔE_1 and ΔE_2 within a series. Equation (106), given by Sustmann[397], describes the variation of $\log k$ for cycloadditions involving the dipolar benzonitrile oxide.

$$\log k = A\beta^2(1/\Delta E_1 + 1/\Delta E_2) \quad (106)$$

Stephan[392] has attributed the non-linearity generally observed in simple Hammett correlations to a change in transition state structure involving asynchronous (two limiting structures) to synchronous bond formation between the termini of the dipole and dipolarophile. The extended equation (107) is used to relate two reaction series using either the same dipole and a variable dipolarophile or vice versa. Good

$$\log (k/k_0)_R = c\sigma + \log (k/k_0)_H \quad (107)$$

correlations are obtained using this equation for a wide range of reactions and it is concluded that, of the dipoles studied, diazomethane has the greatest nucleophilic character (as compared with, say, ArCNO or ArN_3)[392]. A similar conclusion arises from solvent and substituent effects on the addition of CH_2N_2 to styrenes where a concerted mechanism with considerable charge build-up on the carbon adjacent to the aryl ring is proposed (**276**)[398]. The reactivity of the diazoalkane which varies in the sequence $RCHN_2 > CH_2N_2 > Ar_2CN_2 > RO_2CCH_2N_2$ is consistent with this.

(**276**)

VI. REFERENCES

1. H. Zollinger, *Diazo and Azo Chemistry*, Interscience, New York, 1961.
2. P. A. S. Smith, *The Chemistry of Open-chain Nitrogen Compounds*, Benjamin, New York, 1961, Chaps 10, 11.
3. W. J. Baron, M. R. DeCamp, M. E. Hendrick, M. Jones, R. H. Lewin and M. B. Sohn, in *Carbenes*, Vol. 1 (Eds M. Jones and R. A. Moss), Interscience, New York, 1973, p. 1.
4. W. Kirmse, *Carbene Chemistry*, 2nd ed., Academic Press, New York, 1971.
5. D. Bethell, in *Organic Reactive Intermediates* (Ed. S. P. McManus), Academic Press, New York and London, 1973, Chap. 2.
6. E. Kalatzis and J. H. Ridd, *J. Chem. Soc.* (*B*), 529 (1966).
7. E. Muller and H. Haiss, *Chem. Ber.*, **96**, 570 (1963).
8. R. N. Butler, *Chem. Revs.*, **75**, 245 (1975).
9. J. H. Ridd, *Quart. Revs.*, **15**, 418 (1961).
10. L. P. Hammett, *Physical Organic Chemistry*, McGraw-Hill, New York, 1940, p. 294.
11. C. A. Bunton, D. R. Llewellyn and G. Stedman, *J. Chem. Soc.*, 568 (1959).
12. L. F. Larkworthy, *J. Chem. Soc.*, 3116 (1959).
13. E. D. Hughes, C. K. Ingold and J. H. Ridd, *J. Chem. Soc.*, 65, (1958).
14. T. C. Bruice and S. J. Benkovic, *Bioorganic Mechanisms*, Benjamin, New York, 1967, Chap. 1.
15. W. P. Jencks, *Catalysis in Chemistry and Enzymology*, McGraw-Hill, New York, 1969.
16. J. Kenner, *Chem. Ind.* (*London*), **19**, 443 (1941).
17. B. C. Challis and A. R. Butler, in *The Chemistry of the Amino Group* (Ed. S. Patai), Wiley–Interscience, New York, 1968, p. 305.
18. B. C. Challis and J. H. Ridd, *J. Chem. Soc.*, 5197 (1962).
19. E. C. R. de Fabrizio, E. Kalatzis and J. H. Ridd, *J. Chem. Soc.* (*B*), 533 (1966).
20. E. D. Hughes, C. K. Ingold and J. H. Ridd, *J. Chem. Soc.*, 83 (1958).
21. L. F. Larkworthy, *J. Chem. Soc.*, 3304 (1959).
22. B. C. Challis and R. J. Higgins, *J. Chem. Soc. Perkin II* 1498 (1975).
23. B. C. Challis, R. J. Higgins and A. J. Lawson, *J. Chem. Soc. Perkin II*, 1831 (1972).
24. E. Kalatzis and C. Mastrokalos, *J. Chem. Soc. Perkin II*, 498 (1974).
25. E. Kalatzis, *J. Chem. Soc.* (*B*), 273, 277 (1967).
26. B. C. Challis and J. H. Ridd, *Proc. Chem. Soc.*, 245 (1960).
27. N. S. Bayliss, R. Dingle, D. W. Watts and R. J. Wilkie, *Australian J. Chem.*, **16**, 933 (1963).
28. H. Schmid and G. Muhr, *Chem. Ber.*, **70**, 421 (1937).
29. E. D. Hughes and J. H. Ridd, *J. Chem. Soc.*, 82 (1958).
30. H. Schmid, *Monatsh. Chem.*, **85**, 424 (1954).
31. H. Schmid and C. Essler, *Monatsh. Chem.*, **88**, 1110 (1957).
32. H. Schmid and E. Hallaba, *Monatsh. Chem.*, **87**, 560 (1956).
33. G. Olah, N. A. Overchuk and J. C. Lapierre, *J. Amer. Chem. Soc.*, **87**, 5785 (1965).
34. P. B. Desai, *J. Chem. Soc. Perkin I*, 1865 (1973).
35. H. Schmid, *Chem. Ztg, Chem. App.*, **86**, 809 (1962).
36. H. Schmid and G. Muhr, *Monatsh. Chem.*, **93**, 102 (1962).
37. Z. A. Schelly, *J. Phys. Chem.*, **74**, 4062 (1972).
38. J. F. Bunnett, *J. Chem. Soc.*, 4717 (1954).
39. D. F. DeTar and A. R. Ballentine, *J. Amer. Chem. Soc.*, **78**, 3916 (1956).

40. D. F. DeTar and S. K. Wong, *J. Amer. Chem. Soc.*, **78**, 3921 (1956).
41. E. A. Moelwyn-Hughes and P. Johnson, *Trans. Faraday Soc.*, **36**, 948 (1940).
42. E. S. Lewis, *J. Amer. Chem. Soc.*, **83**, 4601 (1961).
43. M. L. Crossley, R. H. Kienle and C. H. Benbrook, *J. Amer. Chem. Soc.*, **62**, 1400 (1940).
44. E. S. Lewis, *J. Amer. Chem. Soc.*, **80**, 1371 (1958).
45. C. G. Swain, J. E. Sheats and K. G. Harbison, *J. Amer. Chem. Soc.*, **97**, 783 (1975).
46. S. Winstein and A. H. Fainberg, *J. Amer. Chem. Soc.*, **79**, 5940 (1957).
47. R. G. Bergstrom, C. H. Wahl and H. Zollinger, *Tetrahedron Letters*, 2975 (1974).
48. P. Burri, G. H. Wahl and H. Zollinger, *Helv. Chim. Acta*, **57**, 2099 (1974).
49. L. C. Bateman, M. G. Church, E. D. Hughes, C. K. Ingold and N. A. Taher, *J. Chem. Soc.*, 979 (1940).
50. C. G. Swain, J. E. Sheats, D. G. Gorenstein and K. G. Harbison, *J. Amer. Chem. Soc.*, **97**, 791 (1975).
51. C. G. Swain and R. J. Rogers, *J. Amer. Chem. Soc.*, **97**, 799 (1975).
52. P. Burri and H. Zollinger, *Helv. Chim. Acta*, **56**, 2204 (1973).
53. R. Huisgen, *Angew. Chem. Int. Ed.*, **9**, 751, 759 (1970).
54. E. S. Lewis and J. E. Cooper, *J. Amer. Chem. Soc.*, **84**, 3847 (1962).
55. G. Modena and U. Tonellato, *Adv. Phys. Org. Chem.*, **9**, 185 (1971).
56. R. H. Summerville, C. A. Senkler, P. V. R. Schleyer, T. E. D. Deuber and P. J. Stang, *J. Amer. Chem. Soc.*, **96**, 1100 (1974).
57. W. D. Pfeifer, C. A. Bahn, P. v. R. Schleyer, S. Bocher, C. E. Harding, K. Hummel, M. Hanack and P. J. Stang, *J. Amer. Chem. Soc.*, **93**, 1513 (1971).
58. R. J. Cox, P. Bushnell and E. M. Evleth, *Tetrahedron Letters*, 209 (1970).
59. R. W. Taft, *J. Amer. Chem. Soc.*, **83**, 3350 (1961).
60. R. A. Abramovitch and F. F. Gadallah, *J. Chem. Soc. (B)*, 497 (1968).
61. R. Gleiter, R. Hoffmann and W. D. Stohrer, *Chem. Ber.*, **105**, 8 (1972).
62. E. M. Evleth and P. M. Horowitz, *J. Amer. Chem. Soc.*, **93**, 5636 (1971).
63. C. G. Swain and E. C. Lupton, *J. Amer. Chem. Soc.*, **90**, 4328 (1968).
64. G. Balz and G. Schlemann, *Chem. Ber.*, **60B** 1186 (1927).
65. G. A. Olah and J. L. Grant, *J. Amer. Chem. Soc.*, **97**, 1546 (1975).
66. E. S. Lewis and W. H. Hinds, *J. Amer. Chem. Soc.*, **74**, 304 (1952).
67. E. S. Lewis, L. D. Hartung and B. M. McKay, *J. Amer. Chem. Soc.*, **91**, 419 (1969).
68. E. S. Lewis and J. E. Cooper, *J. Amer. Chem. Soc.*, **84**, 3847 (1962).
69. P. Burri and H. Zollinger, *Helv. Chim. Acta*, **56**, 2204 (1973).
70. E. S. Lewis and J. M. Insole, *J. Amer. Chem. Soc.*, **86**, 32, 34 (1964).
71. E. S. Lewis and R. E. Holliday, *J. Amer. Chem. Soc.*, **88**, 5043 (1966).
72. E. S. Lewis and P. G. Kotcher, *Tetrahedron*, **25**, 4873 (1969).
73. A. K. Bose and I. Kujayevsky, *J. Amer. Chem. Soc.*, **88**, 2325 (1966).
74. C. G. Swain, J. E. Sheats and K. G. Harbison, *J. Amer. Chem. Soc.*, **97**, 796 (1975).
75. E. S. Lewis and R. E. Holliday, *J. Amer. Chem. Soc.*, **91**, 426 (1969).
76. D. J. Cram, *J. Amer. Chem. Soc.*, **86**, 3767 (1964).
77. J. Bargon and K. G. Seifert, *Tetrahedron Letters*, 2265 (1974).
78. C. J. Heighway, J. E. Packer and R. E. Richardson, *Tetrahedron Letters*, 4441 (1974).
79. W. T. Ford, *J. Org. Chem.*, **36**, 3979 (1971).
80. M. Stiles, R. G. Miller and U. Burckhardt, *J. Amer. Chem. Soc.*, **85**, 1792 (1963).
81. S. Yaroslavsky, *Chem. Ind. (London)*, 765 (1965).
82. G. Wittig and R. W. Hoffmann, *Chem. Ber.*, **95**, 2718 (1962).
83. A. Hantzsch and W. B. Davidson, *Chem. Ber.*, **29**, 1522 (1896).
84. R. Gompper, G. Seybold and B. Schmolke, *Angew. Chem. Int. Ed.*, **7**, 398 (1968).
85. J. I. G. Cadogan, C. D. Murray and J. T. Sharp, *J. Chem. Soc., Chem. Commun.*, 901 (1974).
86. D. L. Brydon, J. I. G. Cadogan, D. M. Smith and J. B. Thompson, *Chem. Commun.*, 727 (1967).
87. D. L. Brydon, J. I. G. Cadogan, J. Cook, M. J. P. Harger and J. T. Sharp, *J. Chem. Soc. (B)*, 1996 (1971).
88. J. I. G. Cadogan, *Accounts Chem. Res.*, **4**, 186 (1971).
89. B. D. Baigrie, J. I. G. Cadogan, J. R. Mitchell, A. K. Robertson and J. T. Sharp, *J. Chem. Soc. Perkin I*, 2563 (1972).

90. J. I. G. Cadogan, A. G. Rowley, J. T. Sharp, B. Sledzinski and N. H. Wilson, *J. Chem. Soc. Perkin I*, 1072 (1975).
91. C. Ruchardt and C. C. Tan, *Angew. Chem. Int. Ed.*, **9**, 522 (1970).
92. J. I. G. Cadogan, R. M. Paton and C. Thomson, *Chem. Commun.*, 133 (1974).
93. P. C. Buxton and H. Heaney, *Chem. Commun.*, 545 (1973).
94. J. I. G. Cadogan, R. M. Paton and C. Thomson, *Chem. Commun.*, 614 (1969).
95. A. F. Hegarty and M. T. McCormack, *J. Chem. Soc. Chem. Commun.*, 168 (1975).
96. M. T. McCormack and A. F. Hegarty, *J. Chem. Soc. Perkin II*, 1976, in print.
97. D. McCarthy and A. F. Hegarty, *J. Chem. Soc. Perkin II*, 1701 (1976).
98. H. Zollinger, *Accounts Chem. Res.*, **6**, 335 (1973).
99. E. S. Lewis and H. Suhr, *Chem. Ber.*, **91**, 2350 (1958).
100. C. D. Ritchie and D. J. Wright, *J. Amer. Chem. Soc.*, **93**, 6574 (1971).
101. V. Beránek, V. Štěrba and K. Valter, *Coll. Czech. Chem. Commun.*, **38**, 257 (1973).
102. C. D. Ritchie and W. F. Sager, *Prog. Phys. Org. Chem.*, **2**, 323 (1964).
103. C. D. Ritchie and P. O. I. Virtanen, *J. Amer. Chem. Soc.*, **94**, 1589, 4966 (1972).
104. C. D. Ritchie, *Accounts Chem. Res.*, **5**, 348 (1972).
105. J. Jahelka, O. Macháčková and V. Štěrba, *Coll. Czech. Chem. Commun.*, **38**, 706 (1973).
106. Reference 1, p. 47.
107. J. Jahelka, O. Macháčková, V. Štěrba and K. Valter, *Coll. Czech. Chem. Commun.*, **38**, 3290 (1973).
108. O. Macháčková and V. Štěrba, *Coll. Czech. Chem. Commun.*, **37**, 3313 (1972).
109. V. A. Ketlinsku and I. L. Bagel, *Zh. Org. Khim.*, **9**, 1915 (1973).
110. J. S. Littler, *Trans. Faraday Soc.*, **59**, 2296 (1963).
111. E. Müller, W. Rundel, H. Haiss and H. Hagenmaier, *Z. Naturforsch.*, **15b**, 751 (1960).
112. R. A. Moss, *Accounts Chem. Res.*, **7**, 421 (1974).
113. E. H. White, T. J. Ryan and K. W. Field, *J. Amer. Chem. Soc.*, **94**, 1360 (1972).
114. H. Suhr, *Chem. Ber.*, **96**, 1720 (1963).
115. E. Müller, W. Hoppe, H. Hagenmaier, H. Haiss, R. Huber, W. Rundel and H. Suhr, *Chem. Ber.*, **96**, 1712 (1963).
116. J. Thiele, *Chem. Ber.*, **41**, 2808 (1908); R. Stolle, *Chem. Ber.*, **41**, 2811 (1908).
117. E. S. Lewis and M. P. Hanson, *J. Amer. Chem. Soc.*, **89**, 6268 (1967).
118. I. F. Gracher, *J. Gen. Chem. U.S.S.R.*, **17**, 1834 (1947).
119. J. Shorter, *Correlation Analysis in Organic Chemistry*, Clarendon Press, Oxford, 1973, Chap. 2.
120. M. Yoshito, K. Manamoto and T. Kubota, *Bull. Chem. Soc. Japan*, **35**, 1723 (1962).
121. C. A. Bunton, M. J. Minch and B. B. Wolfe, *J. Amer. Chem. Soc.*, **96**, 3267 (1974).
122. C. A. Bunton and B. B. Wolfe, *J. Amer. Chem. Soc.*, **96**, 7747 (1974).
123. R. A. Moss, *J. Org. Chem.*, **31**, 1082 (1966).
124. W. Kirmse and G. Wachterhausser, *Ann. Chem.*, **707**, 44 (1967).
125. M. C. Whiting, *Chem. Brit.*, **2**, 482 (1966).
126. R. K. Harris and R. A. Spragg, *Chem. Commun.*, **362** (1967).
127. D. J. Blears, *J. Chem. Soc.*, 6256 (1964).
128. W. J. Boyle, T. J. Broxton and J. F. Bunnett, *Chem. Commun.*, 1469 (1971).
129. J. F. Bunnett and H. Takayama, *J. Org. Chem.*, **33**, 1924 (1968).
130. C. D. Ritchie and P. O. I. Virtanen, *J. Amer. Chem. Soc.*, **95**, 1882 (1973).
131. A. Rieker, P. Niederer and H. B. Stegmann, *Tetrahedron Letters*, 3873 (1971).
132. J. Hollaender and W. P. Newmann, *Angew. Chem.*, **82**, 813 (1970).
133. N. N. Bubnov, K. A. Bilevitch, L. A. Poljakova and O. Y. Okhlobstin, *J. Chem. Soc. Chem. Commun.*, 1058 (1972).
134. E. Lippmaa, T. Pehk, T. Saluvere and M. Magi, *Org. Mag. Res.*, **5**, 441 (1973).
135. H. Suschitzky, *Angew. Chem. Int. Ed.*, **6**, 596 (1967).
136. E. Muller and H. Haiss, *Chem. Ber.*, **95**, 1255 (1962).
137. H. Kropacova, J. Panchartek, V. Štěrba and K. Valter, *Coll. Czech. Chem. Commun.*, **35**, 3287 (1970).
138. E. Bamberger, *Chem. Ber.*, **29**, 446 (1896); **33**, 3188 (1900).
139. E. S. Lewis and D. J. Chalmers, *J. Amer. Chem. Soc.*, **93**, 3267 (1971).
140. C. C. Price and S. Tsanawski, *J. Org. Chem.*, **28**, 1867 (1963).

141. H. van Zwet, J. Reiding and E. C. Kooyman, *Rec. Trav. Chim.*, **89**, 21 (1970).
142. A. Ginsberg and J. Goerdeler, *Chem. Ber.*, **94**, 2043 (1961).
143. C. D. Ritchie, J. D. Saltiel and E. S. Lewis, *J. Amer. Chem. Soc.*, **83**, 4601 (1961).
144. H. Meerwein, G. Dittmar, G. Kaufmann and R. Raue, *Chem. Ber.*, **90**, 853 (1957).
145. J. L. Kice and R. S. Gabrielsen, *J. Org. Chem.*, **35**, 1004, 1010 (1970).
146. R. Dijkstra and J. de Jonge, *Rec. Trav. Chim.*, **77**, 538 (1958).
147. E. S. Lewis and H. Suhr, *Chem. Ber.*, **92**, 3031 (1959).
148. A. Hantzsch and M. Schmiedel, *Chem. Ber.*, **30**, 71 (1897).
149. R. J. W. LeFevre and I. R. Wilson, *J. Chem. Soc.*, 1106 (1949).
150. E. S. Lewis and H. Suhr, *Chem. Ber.*, **92**, 3043 (1959).
151. P. Haberfield, P. M. Block and M. S. Lux, *J. Amer. Chem. Soc.*, **97**, 5804 (1975).
152. E. Enders and R. Putter, in *Methoden der Organischen Chemie*, Vol. X/3, Thieme, Stuttgart, 1965, pp. 467, 627.
153. H. C. Yao and P. Resnick, *J. Amer. Chem. Soc.*, **84**, 3514 (1962).
154. D. Y. Curtin and M. L. Poutsma, *J. Amer. Chem. Soc.*, **84**, 4887 (1962).
155. V. Macháček, J. Panchartek and V. Štěrba, *Coll. Czech. Chem. Commun.*, **35**, 844 (1970).
156. R. P. Bell and P. W. Smith, *J. Chem. Soc.* (*B*), 241 (1966).
157. V. Macháček, O. Macháčková and V. Štěrba, *Coll. Czech. Chem. Commun.*, **35**, 2954 (1970); **36**, 3187 (1971).
158. A. F. Hegarty and F. L. Scott, *J. Org. Chem.*, **32**, 1957 (1967).
159. D. Y. Curtin and J. L. Tveten, *J. Org. Chem.*, **26**, 1764 (1961).
160. A. D. Ainley and R. Robinson, *J. Chem. Soc.*, 369 (1937).
161. H. Zollinger, *Chem. Revs.*, **51**, 347 (1962).
162. Reference 106, p. 227.
163. R. Ernst, O. A. Stamm and H. Zollinger, *Helv. Chim. Acta*, **41**, 2274 (1958).
164. H. Zollinger, *Adv. Phys. Org. Chem.*, **2**, 163 (1964).
165. R. Ernst, O. A. Stamm and H. Zollinger, *Helv. Chim. Acta*, **41**, 2274 (1958).
166. H. Zollinger, *Helv. Chim. Acta*, **38**, 1597 (1955).
167. F. Snyckers and H. Zollinger, *Helv. Chim. Acta*, **53**, 1294 (1970).
168. S. B. Hanna, C. Jermini, H. Loewenschuss and H. Zollinger, *J. Amer. Chem. Soc.*, **96**, 7222 (1974).
169. S. Kishimoto, O. Manabe, H. Haciro and N. Hirao, *Nippon Kagaku Kaishi*, 2132 (1972); *Chem. Abstr.*, **78**, 70984 (1973).
170. R. Putter, *Angew. Chem.*, **63**, 188 (1951).
171. L. Pauling, *The Nature of the Chemical Bond*, 2nd ed., Cornell Univ. Press, New York, 1940, p. 142.
172. H. C. Brown, *Adv. Phys. Org. Chem.*, **1** (1963).
173. V. Štěrba and K. Valter, *Coll. Czech. Chem. Commun.*, **37**, 1327 (1972).
174. V. Štěrba and K. Valter, *Coll. Czech. Chem. Commun.*, **37**, 270 (1972).
175. H. Kropáčová, J. Panchartek, V. Štěrba and K. Valter, *Coll. Czech. Chem. Commun.*, **35**, 3287 (1970).
176. J. Kaválek, J. Panchartek and V. Štěrba, *Coll. Czech. Chem. Commun.*, **36**, 3470 (1971).
177. I. Dobas, J. Panchartek, V. Štěrba and M. Večeřa, *Coll. Czech. Chem. Commun.*, **35**, 1288 (1970).
178. B. Demain, *Tetrahedron Letters*, 3043 (1973).
179. B. Demain, *Bull. Soc. Chim. France*, 769 (1973).
180. H. Kropáčová, J. Panchartek, V. Štěrba and K. Valter, *Coll. Czech. Chem. Commun.*, **35**, 3287 (1970).
181. O. Macháčková, V. Štěrba and K. Valter, *Coll. Czech. Chem. Commun.*, **37**, 1851 (1972).
182. K. Mitsumura, Y. Hashida, S. Sekiguchi and K. Matsui, *Bull. Chem. Soc. Japan*, **46**, 1770 (1973).
183. R. M. Elofson, R. L. Edsberg and P. A. Mecherly, *J. Electrochem. Soc.*, **97**, 166 (1950).
184. P. B. Fischer and H. Zollinger, *Helv. Chim. Acta*, **55**, 2146 (1972).
185. P. B. Fischer and H. Zollinger, *Helv. Chim. Acta*, **55**, 2139 (1972).
186. J. H. Boyer and F. C. Canter, *Chem. Revs*, **54**, 1 (1954).

187. K. Clusius and H. Hurzeler, *Helv. Chim. Acta*, **37**, 383 (1954).
188. K. Clusius and M. Vecchi, *Ann. Chem.*, **607**, 16 (1957).
189. R. J. Friswell and A. G. Green, *J. Chem. Soc.*, **47**, 917 (1885); **49**, 746 (1886).
190. C. K. Ingold, *Structure and Mechanism in Organic Chemistry*, G. Bell and Sons, London, 1953, p. 610.
191. K. Clusius and H. Craubner, *Helv. Chim. Acta*, **38**, 1060 (1955).
192. P. Griess, *Chem. Ber.*, **9**, 1659 (1876).
193. J. P. Horwitz and V. A. Grakauskas, *J. Amer. Chem. Soc.*, **80**, 926 (1958).
194. T. Curtius, *Chem. Ber.*, **26**, 1263 (1893).
195. L. Gatterman and R. G. Ebert, *Chem. Ber.*, **49**, 2117 (1916).
196. J. Mai, *Chem. Ber.*, **25**, 372 (1892).
197. E. S. Lewis and M. D. Johnson, *J. Amer. Chem. Soc.*, **82**, 5408 (1960).
198. C. D. Ritchie and D. J. Wright, *J. Amer. Chem. Soc.*, **93**, 2429 (1971).
199. K. Clusius and H. Hürzeler, *Helv. Chim. Acta*, **37**, 798 (1954).
200. K. Clusius and M. Vecchi, *Helv. Chim. Acta*, **39**, 1469 (1956).
201. R. Huisgen and I. Ugi, *Angew. Chem.*, **68**, 705 (1956).
202. I. Ugi, R. Huisgen, K. Clusius and M. Vecchi, *Angew. Chem.*, **68**, 753 (1956).
203. I. Ugi, H. Perlinger and L. Behringer, *Chem. Ber.*, **91**, 2324 (1958); **92**, 1864 (1959).
204. I. Ugi, *Tetrahedron*, **19**, 1801 (1963).
205. R. Huisgen, *Angew. Chem.*, **75**, 742 (1963).
206. E. S. Lewis and M. D. Johnson, *J. Amer. Chem. Soc.*, **81**, 2070 (1959).
207. J. Vilarrasa, E. Melendez and J. Elguero, *Tetrahedron Letters*, 1609 (1974).
208. E. S. Lewis and H. Suhr, *J. Amer. Chem. Soc.*, **82**, 862 (1960).
209. J. D. Roberts, R. A. Clement and J. J. Srysdale, *J. Amer. Chem. Soc.*, **73**, 2181 (1951).
210. J. F. Bunnett and R. E. Zahler, *Chem. Revs*, **49**, 273 (1951).
211. R. A. Bolto, M. Liveris and J. Miller, *J. Chem. Soc.*, 750 (1956).
212. B. A. Porai-Koshits, *Russ. Chem. Rev.*, **39**, 283 (1970).
213. R. G. R. Bacon and H. A. O. Hill, *Quart. Revs*, **19**, 95 (1965).
214. W. A. Cowdrey and D. S. Davies, *Quart. Revs*, **6**, 358 (1952).
215. H. Meerwein, G. Dittmar, R. Gollner, K. Hafner, F. Mensch and O. Steinfort, *Chem. Ber.*, **90**, 851 (1957).
216. L. Gattermann, *Chem. Ber.*, **32**, 1136 (1899).
217. R. M. Elofson and F. F. Gadallah, *J. Org. Chem.*, **34**, 854 (1969).
218. H. Taube, *J. Chem. Ed.*, **45**, 452 (1968).
219. J. P. Snyder, R. J. Boyd and M. A. Whitehead, *Tetrahedron Letters*, 4347 (1972).
220. J. K. Kochi, *J. Amer. Chem. Soc.*, **79**, 2942 (1957).
221. C. L. Jenkins and J. K. Kochi, *J. Amer. Chem. Soc.*, **94**, 856 (1972).
222. J. K. Kochi and R. V. Subramanian, *J. Amer. Chem. Soc.*, **87**, 1508 (1965).
223. T. Cohen, R. J. Lewarchiek and J. Z. Tarino, *J. Amer. Chem. Soc.*, **96**, 7753 (1974).
224. D. Sutton, *Chem. Soc. Revs.*, **4**, 443 (1975).
225. J. K. Kochi, *Tetrahedron*, **18**, 483 (1962).
226. S. C. Dickerman, K. Weiss and A. K. Ingberman, *J. Amer. Chem. Soc.*, **80**, 1904 (1958).
227. C. S. Rondestvedt, *Org. Reactions*, **11**, 189 (1960).
228. W. F. Beech, *J. Chem. Soc.*, 1297 (1954).
229. W. E. Bachmann and R. A. Hoffmann, *Org. Reactions*, **2**, 224 (1941).
230. D. I. Davies, D. H. Heg and G. H. Williams, *J. Chem. Soc.*, 3112 (1961).
231. R. Huisgen and H. Nakaten, *Ann. Chem.*, **586**, 84 (1954).
232. D. H. Heg, C. J. M. Stirling and G. H. Williams, *J. Chem. Soc.*, 1475 (1956).
233. R. Ito, T. Migita, N. Morchaiva and O. Simamura, *Bull. Chem. Soc. Japan*, **36**, 992 (1963).
234. E. L. Eliel, M. Eberhardt and O. Simamura, *Tetrahedron Letters*, 749 (1962).
235. D. F. DeTar and R. A. J. Long, *J. Amer. Chem. Soc.*, **80**, 4742 (1958).
236. F. G. Edwards and F. R. Mayo, *J. Amer. Chem. Soc.*, **72**, 1265 (1950).
237. R. Huisgen and H. Nakaten, *Ann. Chem.*, **573**, 181 (1951).
238. D. H. Heg, J. Stuart-Webb and G. H. Williams, *J. Chem. Soc.*, 4657 (1952).
239. R. Huisgen and G. Horeld, *Ann. Chem.*, **562**, 181 (1951).
240. C. Rüchardt and B. Freudenberg, *Tetrahedron Letters*, 3623 (1964).

241. G. Binsch, E. Merz and C. Rüchardt, *Chem. Ber.*, **100**, 247 (1967).
242. G. R. Chalfont and M. J. Perkins, *J. Amer. Chem. Soc.*, **89**, 3054 (1967).
243. G. R. Chalfont, M. J. Perkins, D. H. Hey and K. S. Y. Liang, *Chem. Commun.*, 367 (1967).
244. S. Terabe and R. Konaka, *J. Amer. Chem. Soc.*, **91**, 5655 (1969).
245. C. Rüchardt and E. Merz, *Tetrahedron Letters*, 2431 (1964).
246. R. J. W. Le Fèvre and I. R. Wilson, *J. Chem. Soc.*, 1106 (1949).
247. T. J. Broxton, J. F. Bunnett and C. H. Pack, *Chem. Commun.*, 1363 (1970).
248. R. M. Cooper and M. J. Perkins, *Tetrahedron Letters*, 2477 (1969).
249. B. Gloor, B. L. Kaul and H. Zollinger, *Helv. Chim. Acta*, **55**, 1596 (1972).
250. B. L. Kaul and H. Zollinger, *Helv. Chim. Acta*, **51**, 2132 (1968).
251. P. Burri, H. Loewenschuss, H. Zollinger and G. K. Zwolinski, *Helv. Chim. Acta*, **57**, 395 (1974).
252. N. Kamigata, M. Kobayashi and H. Minato, *Bull. Chem. Soc. Japan*, **45**, 2047 (1972).
253. N. Kobori, M. Kobayashi and H. Minato, *Bull. Chem. Soc. Japan*, **43**, 223 (1970).
254. M. Kobayashi, H. Minato, E. Yamada and N. Kobori, *Bull. Chem. Soc. Japan*, **43**, 219 (1970).
255. M. Kobayashi, H. Minato and N. Kobori, *Bull. Chem. Soc. Japan*, **43**, 219 (1970).
256. K. Ishida, N. Kobori, M. Kobayashi and H. Minati, *Bull. Chem. Soc. Japan*, **43**, 285 (1970).
257. K. Kamigara, R. Hisada, H. Minato and M. Kobayashi, *Bull. Chem. Soc. Japan*, **46**, 1016 (1973).
258. D. F. DeTar, *Org. Reactions*, **9**, 409 (1957).
259. A. H. Lewin and T. Cohen, *J. Org. Chem.*, **32**, 3844 (1967).
260. A. H. Lewin, A. H. Dinwoodie and T. Cohen, *Tetrahedron*, **22**, 1527 (1966).
261. G. A. Olah and W. S. Tolgyesi, *J. Org. Chem.*, **26**, 2053 (1961).
262. G. H. Williams, *Homolytic Aromatic Substitution*, Pergamon Press, New York, 1960, p. 28.
263. R. A. Abramovitch, *Advan. Free Radical Chem.*, **2**, 87 (1967).
264. D. C. Nonhebel and W. A. Waters, *Proc. Roy. Soc.*, **A242**, 16 (1957).
265. A. H. Lewin and R. J. Michl, *J. Org. Chem.*, **38**, 1126 (1973).
266. F. F. Gadallah, A. A. Cantu and R. M. Elofson, *J. Org. Chem.*, **38**, 2386 (1973).
267. R. Huisgen and W. D. Zahler, *Chem. Ber.*, **96**, 736 (1963).
268. M. Stiles and A. J. Sisti, *J. Org. Chem.*, **24**, 268 (1959).
269. M. H. Knight, T. Putkey and H. S. Mosher, *J. Org. Chem.*, **36**, 1483 (1971).
270. T. Cohen and J. Lipowitz, *J. Amer. Chem. Soc.*, **86**, 2514, 2515 (1964), and previous papers in this series.
271. T. Cohen, C. H. McMullen and K. Smith, *J. Amer. Chem. Soc.*, **90**, 6866 (1968).
272. T. Cohen, K. W. Smith and M. D. Swerdolff, *J. Amer. Chem. Soc.*, **93**, 4303 (1971).
273. N. Kornblum, G. D. Cooper and J. E. Taylor, *J. Amer. Chem. Soc.*, **72**, 3063 (1950).
274. N. Kornblum, A. E. Kelley and G. D. Cooper, *J. Amer. Chem. Soc.*, **74**, 3074 (1952).
275. D. F. DeTar and M. N. Turetzky, *J. Amer. Chem. Soc.*, **78**, 3928 (1956).
276. D. V. Banthorpe and E. D. Hughes, *J. Chem. Soc.*, 3314 (1962).
277. R. Werner and C. Rüchardt, *Tetrahedron Letters*, 2407 (1969).
278a. J. D. Roberts and W. T. Moreland, *J. Amer. Chem. Soc.*, **75**, 2167 (1953).
278b. L. Melander, *Arkiv Kemi*, **3**, 525 (1951).
279. D. Schulte-Frohlinde and H. Blume, *Z. Phys. Chem.*, **59**, 282 (1968).
280. J. E. Packer, D. B. House and E. J. Rasburn, *J. Chem. Soc. (B)*, 1574 (1971).
281. J. E. Packer, R. K. Richardson, P. J. Soole and D. R. Webster, *J. Chem. Soc. Perkin II*, 1472 (1974).
282. J. E. Packer and R. K. Richardson, *J. Chem. Soc. Perkin II*, 751 (1975).
283. A. L. J. Beckwith and R. O. C. Norman, *J. Chem. Soc. (B)*, 403 (1969).
284. C. E. McKenna and T. G. Traylor, *J. Amer. Chem. Soc.*, **93**, 2313 (1971).
285. T. Severin, R. Schmitz, J. Loske and J. Hufnagel, *Chem. Ber.*, **102**, 4152 (1969).
286. P. Huang and E. M. Kosower, *J. Amer. Chem. Soc.*, **90**, 2354 (1968).
287. E. M. Kosower, P. C. Huang and T. Tsuji, *J. Amer. Chem. Soc.*, **91**, 2325 (1969).
288a. H. Gilman and R. G. Jones, *J. Org. Chem.*, **38**, 84 (1973).
288b. R. Huisgen and R. Lux, *Chem. Ber.*, **93**, 540 (1960).

289. N. E. Searle, *Org. Syntheses Coll.*, Vol. 4, 425 (1963).
290. E. Müller and W. Rundel, *Chem. Ber.*, **90,** 2673 (1957).
291. J. K. Stille and L. Plummer, *J. Amer. Chem. Soc.*, **85**, 1318 (1963).
292. W. D. McPhee and E. Klingsberg, *Org. Syntheses Coll.*, Vol. 3, 119 (1955).
293. R. Huisgen and J. Reinertshofer, *Ann. Chem.*, **575,** 174 (1952).
294. C. G. Overberger and J. P. Anselme, *J. Org. Chem.*, **28**, 592 (1963).
295. T. J. de Boer and H. J. Backer, *Org. Synthesis*, **36**, 16 (1956).
296. J. A. Moore and D. E. Reed, *Org. Synthesis*, **41**, 16 (1961).
297. H. Reimlinger, *Chem. Ber.*, **94**, 2549 (1961).
298. V. Horak and M. Prochazka, *Chem. and Ind.*, 472 (1961).
299. D. E. Applequist and D. E. McGreer, *J. Amer. Chem. Soc.*, **82**, 1965 (1960).
300. S. M. Hecht and J. W. Kozarich, *J. Org. Chem.*, **38**, 1821 (1973).
301. S. M. Hecht and J. W. Kozarich, *Tetrahedron Letters*, 5147 (1972).
302. W. M. Jones, D. L. Muck and T. K. Tandy, *J. Amer. Chem. Soc.*, **88**, 68 (1966).
303. W. M. Jones and D. L. Muck, *J. Amer. Chem. Soc.*, **88**, 3798 (1966).
304. G. W. Cowell and A. Ledwith, *Quart. Revs*, **24,** 119 (1970).
305. H. K. Reimlinger, L. Skattebol and F. Billiou, *Chem. Ber.*, **94**, 2429 (1961).
306. H. Hart and J. L. Brewbaker, *J. Amer. Chem. Soc.*, **91,** 716 (1969).
307. R. A. Moss, *J. Org. Chem.*, **31**, 1082 (1966).
308. C. D. Nenitzescu and E. Solomonica, *Org. Syntheses Coll.*, Vol. 2, 496 (1943).
309. R. Baltzly, N. B. Mehta, P. B. Russell, R. E. Brooks, E. M. Grivsky and A. M. Steinberg, *J. Org. Chem.*, **26**, 3669 (1961).
310. K. Nakagawa, H. Ondue and K. Minami, *Chem. Commun.*, 736 (1966).
311. A. C. Day, P. Raymond, R. M. Southam and M. C. Whiting, *J. Chem. Soc.* (*C*), 476 (1967).
312. J. R. Dyer, R. B. Randall and H. M. Deutsch, *J. Org. Chem.*, **29**, 3423 (1964).
313. L. Horner, W. Kirmse and H. Ferkeness, *Chem. Ber.*, **94**, 279 (1961).
314. D. G. Farnum, *J. Org. Chem.*, **28**, 870 (1963).
315. W. R. Bamford and T. S. Stevens, *J. Chem. Soc.*, 4735 (1952).
316. D. E. Dana and J. P. Anselme, *Tetrahedron Letters*, 1565 (1975).
317. J. Meinwald, P. G. Gassman and E. G. Miller, *J. Amer. Chem. Soc.*, **81**, 4751 (1959).
318. W. Rundel, *Angew. Chem.*, **74**, 469 (1962).
319. E. Bamberger and E. Renault, *Chem. Ber.*, **28**, 1682 (1895).
320. R. A. More O'Ferrall, *Adv. Phys. Org. Chem.*, **5**, 331 (1967).
321. B. Zwanenburg and J. B. F. N. Engberts, *Rec. Trav. Chim.*, **84**, 165 (1965).
322. P. Gross, H. Steiner and F. Krauss, *Trans. Faraday Soc.*, **32**, 877 (1936); **34**, 351 (1938).
323. J. D. Roberts, C. M. Regan and I. Allen, *J. Amer. Chem. Soc.*, **74**, 6779 (1952).
324. C. G. Swain and C. B. Scott, *J. Amer. Chem. Soc.*, **75**, 141 (1953).
325. W. J. Albery, J. E. C. Hutchins, R. M. Hyde and R. H. Johnson, *J. Chem. Soc.* (*B*), 219 (1968).
326. H. Dahn and H. Gold, *Helv. Chim. Acta*, **46**, 983 (1963).
327. G. Diderich and H. Dahn, *Helv. Chim. Acta*, **55**, 1 (1972).
328. R. A. More O'Ferrall, W. K. Kwok and S. I. Miller, *J. Amer. Chem. Soc.*, **86**, 553 (1964).
329. H. Dahn and G. Diderich, *Helv. Chim. Acta*, **54**, 1950 (1971).
330. G. Fierz, J. F. McGarrity and H. Dahn, *Helv. Chim. Acta*, **58**, 1058 (1975).
331. N. B. Chapman, J. R. Lee and J. Shorter, *J. Chem. Soc.* (*B*), 769 (1969), and other papers in this series.
332. W. J. Albery, *J. Chem. Soc. Perkin II*, 2180 (1972), and other papers in this series.
333. J. D. Roberts, W. Watanabe and R. E. McMahon, *J. Amer. Chem. Soc.*, **73**, 2521 (1951).
334. J. D. Roberts and C. M. Regan, *J. Amer. Chem. Soc.*, **74**, 3695 (1952).
335. H. L. Goering and J. F. Levy, *J. Amer. Chem. Soc.*, **84**, 3853 (1962).
336. A. F. Diaz and S. Winstein, *J. Amer. Chem. Soc.*, **88**, 1318 (1966).
337. M. Neeman and W. S. Johnson, *Org. Synthesis*, **41**, 9 (1961).
338. J. D. Roberts and W. Watanabe, *J. Amer. Chem. Soc.*, **72**, 4869 (1950).
339. H. Meerwein and G. Hinz, *Ann. Chem.*, **484**, 1 (1930).
340. R. Daniels and C. G. Kormenoy, *J. Org. Chem.*, **27**, 1860 (1962).

341. F. Klages and H. Meuresch, *Chem. Ber.*, **85**, 863 (1952).
342. N. Kornblum and G. P. Coffey, *J. Org. Chem.*, **31**, 3447 (1966).
343. E. T. Blues, D. Bryce-Smith, J. G. Irwin and I. W. Lawston, *J. Chem. Soc. Chem. Commun.*, 466 (1974).
344. C. J. Collins and B. M. Benjamin, *J. Amer. Chem. Soc.*, **85**, 2519 (1963).
345. L. Friedman and J. H. Baylen, *J. Amer. Chem. Soc.*, **91**, 1790 (1969).
346. J. A. Berson, *Angew. Chem.*, **80**, 765 (1968).
347. D. Y. Curtin and M. C. Crew, *J. Amer. Chem. Soc.*, **76**, 3719 (1954).
348. C. J. Collins, *Chem. Soc. Revs.*, **4**, 251 (1975).
349. C. J. Collins, *Accounts Chem. Res.*, **4**, 315 (1971).
350. E. H. White and D. J. Woodcock, in *The Chemistry of the Amino Group* (Ed. S. Patai), Wiley–Interscience, London, 1968, Chap. 8.
351. A. Streitwieser Jr, *J. Org. Chem.*, **22**, 861 (1957).
352. A. Streitwieser Jr and W. D. Schaeffer, *J. Amer. Chem. Soc.*, **79**, 2888 (1957).
353. R. Huisgen and C. Rüchardt, *Ann. Chem.*, **601**, 1 (1956).
354. D. Semenow, C. H. Ship and W. G. Young, *J. Amer. Chem. Soc.*, **80**, 5472 (1958).
355. J. N. Bradley, G. W. Cowell and A. Ledwith, *J. Chem. Soc.*, 4334 (1964).
356. W. S. Johnson, M. Neeman, S. P. Birkeland and N. A. Fedoruk, *J. Amer. Chem. Soc.*, **84**, 989 (1962).
357. J. N. Bradley, G. W. Cowell and A. Ledwith, *J. Chem. Soc.*, 4334 (1964).
358. R. Huisgen, *Angew. Chem.*, **67**, 439 (1955).
359. R. S. Bly, F. B. Culp and R. K. Bly, *J. Org. Chem.*, **35**, 2235 (1970).
360. T. D. Inch, G. J. Lewis, R. P. Pell, and N. Williams, *Chem. Commun.*, 1549 (1970).
361. N. J. Turro and R. B. Gagosian, *J. Amer. Chem. Soc.*, **92**, 2036 (1970).
362. J. Heiss, M. Bauer and E. Müller, *Chem. Ber.*, **103**, 463 (1970).
363. C. D. Gutsche and C. T. Chang, *J. Amer. Chem. Soc.*, **84**, 2263 (1962).
364. C. D. Gutsche and D. Redmore, *Carbocyclic Ring Expansion Reactions*, Academic Press, New York, 1968, p. 81.
365. P. Lipp, J. Buchkremer and H. Seeles, *Ann. Chem.*, **499**, 1 (1932).
366. W. S. Johnson, M. Neeman, S. P. Birkeland and N. A. Fedvruk, *J. Amer. Chem. Soc.*, **84**, 989 (1962).
367. H. Biltz and E. Kramer, *Ann. Chem.*, **463**, 154 (1924).
368. H. Bredereck, R. Sieber and L. Kamphenkel, *Chem. Ber.*, **89**, 1169 (1956).
369. F. M. Dean and B. K. Park, *J. Chem. Soc. Chem. Commun.*, 162 (1974).
370. E. H. Billett, I. Fleming and S. W. Hanson, *J. Chem. Soc. Perkin II*, 1658, 1669 (1973).
371. A. L. Wilds and A. L. Meader, *J. Org. Chem.*, **13**, 763 (1948).
372. W. Bradley and G. Schwarzenbach, *J. Chem. Soc.*, 2904 (1928).
373. R. Huisgen and H. J. Koch, *Ann. Chem.*, **591**, 200 (1955).
374. K. Clusius, H. Hürzeler, R. Huisgen and H. J. Koch, *Naturwissenschaften*, **41**, 213 (1954).
375. H. W. Whitlock, *J. Amer. Chem. Soc.*, **84**, 2807 (1962).
376. N. A. Preobrashenski and M. J. Kabatschnik, *Chem. Ber.*, **66**, 1542 (1933).
377. T. Curtius and A. Darapsky, *Chem. Ber.*, **39**, 1373 (1906).
378. A. F. Hegarty, J. A. Kearney, P. A. Cashell and F. L. Scott, *J. Chem. Soc. Perkin II*, 242 (1976).
379. H. W. Whitlock, *Tetrahedron Letters*, 593 (1961).
380. N. J. Leonard, J. V. Paukstelis and L. E. Brady, *J. Org. Chem.*, **29**, 3383 (1964).
381. A. G. Davies, D. G. Hare, O. R. Khan and J. Sikora, *J. Chem. Soc.*, 4461 (1963).
382. H. Hoberg, *Ann. Chem.*, **656**, 18 (1962).
383. G. Wittig and F. Wingler, *Ann. Chem.*, **656**, 18 (1962).
384. D. Seyferth and E. G. Rochav, *J. Amer. Chem. Soc.*, **77**, 907, 1302 (1955).
385. R. Huisgen, H. Stangl, H. J. Sturm and H. Wagenhofer, *Angew. Chem.*, **73**, 170 (1961).
386. R. Huisgen, *J. Org. Chem.*, **33**, 2291 (1968).
387. R. A. Firestone, *J. Org. Chem.*, **33**, 2285 (1968).
388. R. A. Firestone, *J. Org. Chem.*, **37**, 2181 (1972).
389. J. Bastide, N. El Ghandour and O. Henri-Rousseau, *Tetrahedron Letters*, 2979, 4225 (1972).
390. G. Leroy and M. Sana, *Tetrahedron*, **31**, 2091 (1975).

391. P. Eberhard and R. Huisgen, *Tetrahedron Letters*, 4337, 4343 (1971).
392. E. Stephan, *Tetrahedron*, **31**, 1623 (1975).
393. E. Stephan, L. Vo-Quang and Y. Vo-Quang, *Bulg. Soc. Chim.*, 2795 (1973).
394. P. Battioni, L. Vo-Quang and Y. Vo-Quang, *Tetrahedron Letters*, 4803 (1972).
395. R. Sustmann, *Tetrahedron Letters*, 2717 (1971).
396. R. Sustmann, *Tetrahedron Letters*, 963 (1974).
397. R. Sustmann and H. Trill, *Angew. Chem. Int. Ed.*, **11**, 838 (1972).
398. P. K. Kadaba and T. F. Colturi, *J. Heterocyclic Chem.*, **6**, 829 (1969).

CHAPTER 13

Rearrangements involving the diazo and diazonium groups

D. WHITTAKER
Department of Organic Chemistry, University of Liverpool, England

21

I. INTRODUCTION

Diazo compounds are highly reactive species which can behave, under suitable conditions, as electrophiles, nucleophiles, 1,3-dipoles or carbene sources. They can behave as bases, and those with a hydrogen on the diazo carbon atom can also behave as acids. Protonation of a diazo compound yields the even more reactive diazonium ion which readily decomposes to a carbonium ion. Consequently, diazo and diazonium groups are rich sources of rearrangement reactions.

A comprehensive survey of all the reactions involving diazo and diazonium groups which also involve rearrangement of the carbon skeleton of the substrate would exceed this book in size, so that this chapter is limited to a survey of the main types of reaction which involve rearrangements. Recent reviews provide examples of rearrangements involving addition reactions of diazo compounds[1–4], rearrangements involving carbenes[5–10], diazotization[11, 12], and rearrangements involving carbonium ions[13–15].

Since this chapter does not attempt to be comprehensive, examples of rearrangements have been chosen for their illustrative value without any consideration of priority of publication. It is intended, however, that all the main types of reaction which have been published up to the date of writing (July 1975) will be included.

II. ISOMERIZATION OF DIAZO AND DIAZONIUM COMPOUNDS

Diazo compounds are not usually thought of as readily undergoing isomerization reactions but it has recently been pointed out[16] that diazomethane is unique among small molecules in its number of known or suspected structural isomers.

$H_2C=\overset{+}{N}=\overset{-}{N}$ — Diazomethane

$H_2C\langle N=N \rangle$ (cyclic) — Diazirine

$H_2N-C\equiv N$ — Cyanamide

$^{-}C\equiv\overset{+}{N}-NH_2$ — Isocyanamide

$H-\overset{-}{C}=\overset{+}{N}=N-H$ — Nitrilimine

$HN=C=NH$ — Carbodiimide

The first three are well-characterized stable molecules[17, 18]; derivatives of the others are known. A diazomethane isomer, 'isodiazomethane', is also described in the literature, being obtained by treating the diazomethyl anion with potassium dihydrogen phosphate[19]. It was first thought to have the nitrilimine structure but recent spectroscopic evidence suggests the isocyanamide structure[20].

$$H_2C=\overset{+}{N}=N^{-} \rightleftharpoons H-\overset{-}{C}=\overset{+}{N}=N^{-} \rightleftharpoons H-\overset{-}{C}=\overset{+}{N}=N-H$$

$$\rightleftharpoons {}^{-}C\equiv\overset{+}{N}-\overset{-}{N}-H \quad N \rightleftharpoons {}^{-}C\equiv\overset{+}{N}-NH_2$$

Isocyanamide is isomerized to diazomethane in the presence of base; interconversion is suggested[16] to result from tautomerism of the anion. In support of this, the silver salt of diazomethane decomposed to yield only diazomethane[21].

$$Ag_2C{=}\overset{+}{N}{=}N^- \longrightarrow H_2C{-}\overset{+}{N}{-}N^-$$

Diazirine is isomerized to diazomethane on photolysis[22]. Although photolysis of both compounds ultimately yields a carbene, experiments in the gas phase in the presence of added nitrogen show that at least 20% of the primary decomposition of diazirine is isomerization to diazomethane[23]. However, the heats of formation of diazomethane and diazirine have been estimated[24] as 206 and 331 kJ/mol respectively, so that any diazomethane produced by diazirine isomerization cannot initially be in its electronic ground state. It has been suggested that the only possible pathway to diazomethane is intersystem crossing to yield excited triplet diazomethane, which has a long enough lifetime to lose its excess energy by collisions[25].

Substituted diazirines can undergo internal energy redistribution more efficiently than diazirine itself, and 3-phenyldiazirine is believed to isomerize to the diazo compound on photolysis[26]. Further substitution makes the rearrangement proceed so readily that 3,3-diphenyldiazirine has not been prepared[27]; attempted preparations have yielded only diphenyldiazomethane.

The reverse reaction, isomerization of diazo compounds into diazirines, is much less common, but the photochemical isomerization of α-diazoamides into the corresponding diazirinylamides in about 20% yield has been reported[28].

$$N_2CHCONC_5H_{10} \longrightarrow \text{diazirine } (N{=}N)C(H)CONC_5H_{10}$$

The reaction was not successful in producing disubstituted diazirines.

Nitrogen interchange within the diazonium group has only recently been detected. Experiments by Swain[29] on benzenediazonium fluoroborate labelled on the α-nitrogen atom with ^{15}N have shown a slow exchange between the α- and β-nitrogen atoms at a rate approximately 1·6% of the rate of decomposition of the diazonium ion. On the basis of substituent effects, the authors suggest that reaction occurs via an intermediate such as **1** rather than by fission and recombination.

$$C_6H_5{-}N{=}^{15}N \longrightarrow \mathbf{(1)} \longrightarrow C_6H_5{-}^{15}N{=}N$$

(1)

The existence of a fission and recombination route to nitrogen scrambling has been demonstrated by Zollinger[30], also working with benzenediazonium tetrafluoroborate. Decomposition of this substrate in the presence of labelled gaseous nitrogen under high pressure showed that the diazonium nitrogen exchanged with the gaseous nitrogen. Nearly 5% incorporation of external nitrogen was observed, providing the first evidence for the reaction of gaseous nitrogen with a purely organic reagent in solution.

Related to diazo and diazonium compounds are the diazotates, of the general type **2**, which undergo isomerization to **3**. When X is *t*-butyl sulphide[31], the reaction

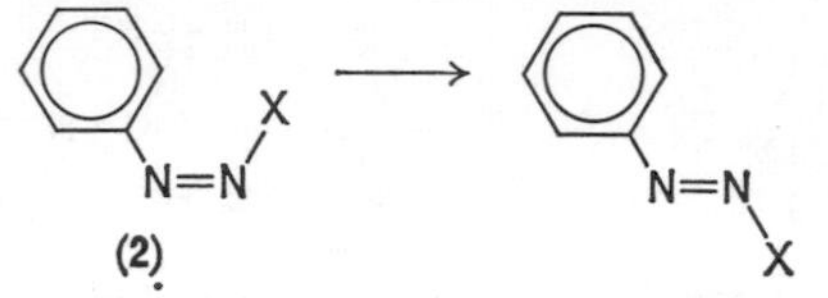

is believed to proceed by preliminary ionization to a diazonium thiolate ion pair. For X = OH, the mechanism is an acid-catalysed splitting and recombination[32]. Photochemical isomerization is also found when X = *t*-butyl sulphide or cyanide[33].

III. REARRANGEMENTS INVOLVING ADDITION REACTIONS OF DIAZO COMPOUNDS

A. Rearrangements Involving 1,3-Dipolar Addition

Diazoalkanes readily behave as 1,3-dipoles, and as such will undergo thermal cycloaddition reactions to give Δ^1-pyrazolines (**4**)[2].

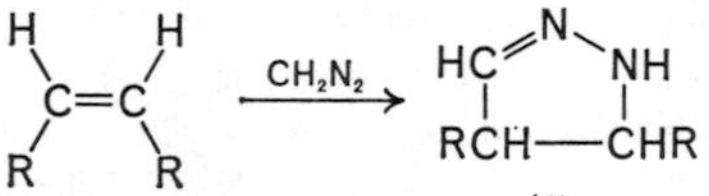

The mechanism is believed to be a 3+2 cycloaddition[34, 35] though a dissenting view favours a two-step mechanism involving biradical intermediates[36]. In addition to the thermal reaction, a report has appeared[37] of a photolytic addition reaction between diazofluorene and norbornadiene.

In many of these reactions, the Δ^1-pyrazolines are stable, or undergo rearrangement (possibly on work-up, e.g. alumina catalysis) to Δ^2-pyrazolines[38]; in others the pyrazolines cannot be detected and their presence is demonstrated only by isolation of their decomposition products. In a few cases, where formation of a pyrazoline would interfere with conjugation of the substrate, it is believed that the cycloaddition is never completed[39] as in the reaction of **5**.

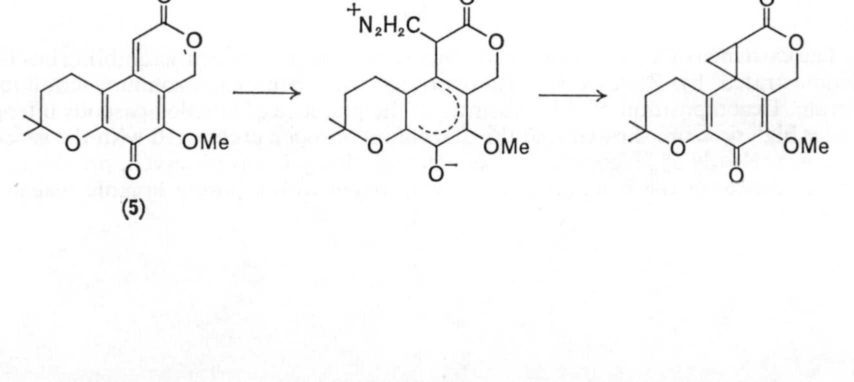

Pyrazoline derivatives of sufficient stability to permit their isolation can usually be decomposed either thermally or photolytically to yield cyclopropanes and alkylated olefins[40].

C C
R—C N
H N
→ H CH R + HC=C CH₂ R

If, however, one of the carbon atoms of the original double bond possesses a strongly electron-releasing substituent, formation of an alkylated olefin is favoured[41].

Ph—CH=C(CN)(CN) → Ph—CH—C(CN)(CN) H₂C N=N → Ph—C—C(CN)—CN H CH₂—N₂⁺
↓
Ph—C(CH₃)=C(CN)(CN)

Exceptionally, an alkyl shift can take place rather than a hydride shift; in cyclic systems, this can result in a ring expansion reaction[42].

Ph— =O N—N CH₃ → N=N Ph— =O N—N CH₃ —heat→ Ph— =O N—N CH₃

→ Ph Ċ=O N—N CH₃ → Ph— N—N O CH₃

Addition of diazoalkanes to acetylenes gives pyrazolenines which readily undergo pyrolysis or photolysis to yield the diazoalkene, and hence, via the carbene, a cyclopropane[43].

R C—N C N R¹ C —$h\nu$→ R C=N₂ R¹—C C— → R C: R¹—C C— → R R¹

Opening of the pyrazolenine ring is reversible, so that in some circumstances rearrangement to the more stable pyrazole is favoured over carbene formation[44].

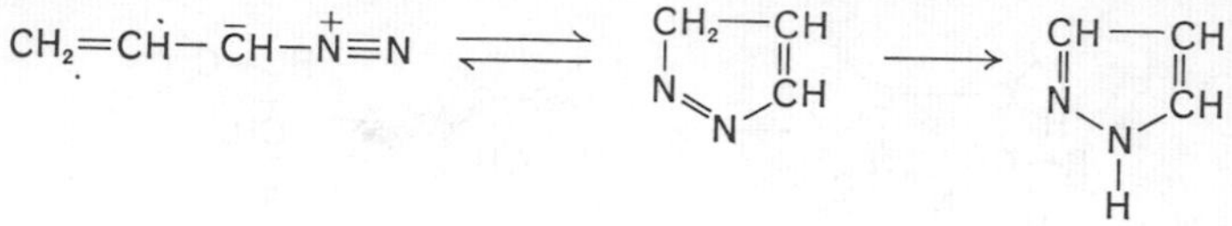

Substituents on the pyrazolenine ring can affect the course of the diazoalkene decomposition; an α-keto substituent, for example, gives rise to products of a competing Wolff rearrangement[45] (see p. 612).

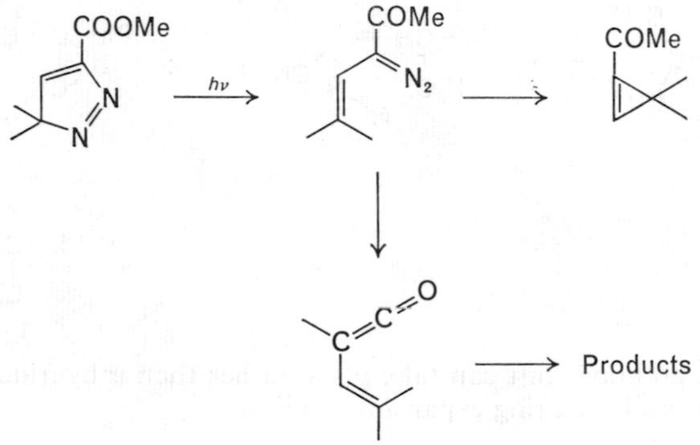

B. Rearrangements Involving Nucleophilic Addition

Diazomethane reacts readily with carbonyl compounds to yield epoxides and homologous carbonyl compounds. The mechanism of the reaction is believed to involve nucleophilic addition of the diazomethane to the carbonyl carbon atom to form a zwitterionic intermediate, followed by a 1,2-nucleophilic displacement of nitrogen by the electrons of a carbon–carbon bond.

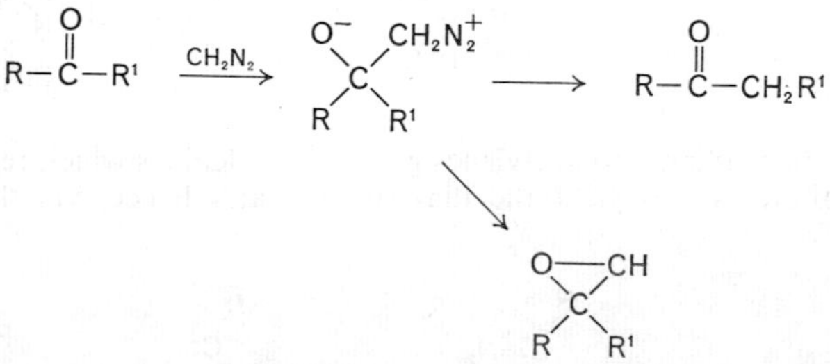

The overall reaction is basically similar to the Tiffenau–Demjanov rearrangement and is used as a method of ring expansion in cyclic ketones[46] (see p. 629).

The first step of the ring expansion reaction of ketones is addition of the diazoalkane, which takes place from the least hindered side of the molecule[47, 48]. This is followed by bond migration; in cases where either of the C—C=O bonds can migrate, then it is the more substituted carbon atom which is involved preferentially[49], i.e.

In cases where the diazo compound used is larger than diazomethane, such as diazoethane, then the direction of the bond migration is controlled by the conformation of the zwitterion[49].

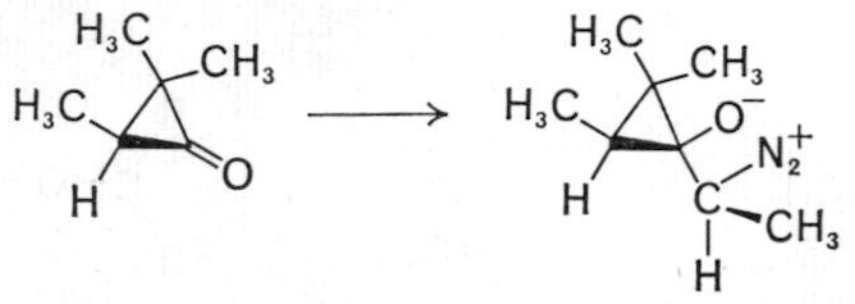

(6)

The most favourable conformation of the zwitterion is that shown in **6**. The next step of the reaction involves expulsion of the nitrogen by a bond shift displacing it

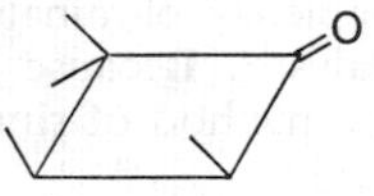

(7)

from behind, so that the main product is **7**. Other possible conformations of the diazoethane group contribute only traces of isomeric products.

The reaction can also be carried out intramolecularly, when the result is cyclization[50]. The same considerations of factors favouring bond migration apply in this case as in the previous one.

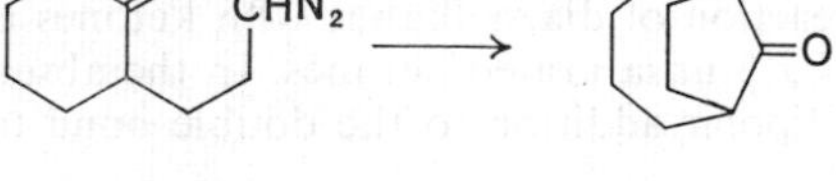

A theory that epoxides and ketones may not arise from the same intermediate was first put forward[51] in 1955. Since then, the accumulation of evidence from kinetic[52] and rearrangement product[50] studies has lead to the suggestion that ring expansion or insertion reactions result from nucleophilic attack of the diazo compound on the carbonyl carbon atom, while epoxide formation results from electrophilic attack of the diazo compound on the carbonyl oxygen atom, or 1,3-dipolar addition to the carbonyl group, i.e.

$$\begin{array}{c} RR^1C{=}O \xrightarrow{CH_2N_2} \end{array}$$

$$\text{(1,3-dipolar adduct: oxadiazoline ring } RR^1C{-}O{-}CH_2{-}N{=}N) \longrightarrow RR^1\overset{+}{C}{-}O\overset{-}{C}H_2$$

$$RR^1\overset{-}{C}{-}O{-}CH_2N_2^+ \longrightarrow RR^1\overset{-}{C}{-}O{-}CH_2^+$$

$$\longrightarrow \text{epoxide } RR^1C{-}O{-}CH_2$$

The existence of the second type of reaction is consistent with the observation that epoxide formation is almost completely eliminated if the reaction is carried out using boron trifluoride as a catalyst[53]. Because of this specificity, the catalysed reaction is now generally used as a method of ring homologation[54].

$$RR^1C{=}O + BF_3 \longrightarrow RR^1C{=}\overset{+}{O}{-}\overset{-}{B}F_3$$

$$\xrightarrow{CH_2N_2} RR^1C(CH_2N_2^+)(O\overset{-}{B}F_3) \longrightarrow R(R^1{-}CH_2)C{=}O\cdots BF_3\,.$$

The acid-catalysed reaction of diazoalkanes with ketones is particularly valuable in the homologation of α,β-unsaturated ketones. In the absence of the catalyst, the reagent undergoes 1,3-dipolar addition to the double bond to yield a pyrazoline[55].

$$Ph{-}CH{=}CH{-}C({=}O)CH_3 \longrightarrow Ph{-}CH{-}CH{-}C({=}O)CH_3 \text{ (with } H_2C{-}N{=}N \text{ ring)}$$

In the presence of fluoroboric acid, however, only the insertion product is observed[56].

$$Ph-CH=CH-C(=O)-CH_3 \longrightarrow Ph-CH=CH-CH_2-C(=O)-CH_3$$

Since decomposition of the 1,3-dipolar adduct can also lead to the insertion product, in many cases it is not possible to be sure which reaction is taken place.

Diazomethane can also be added to systems such as immonium salts, which behave in a manner similar to ketones, and are believed to react via nucleophilic attack of the diazo compound. The reaction yields a heterocyclic 3-membered ring, which opens readily to yield a ring-expanded product[57-59].

IV. REARRANGEMENTS INVOLVING CARBENES AND CARBENOIDS

Carbenes can be obtained by pyrolysis or photolysis of diazo compounds[7].

$$Ph_2CN_2 \longrightarrow Ph_2C{:} + N_2$$

Some diazo compounds are sufficiently stable to be isolated and characterized, but less stable diazo compounds are often prepared from toluene-*p*-sulphonyl hydrazones and decomposed *in situ.*

$$R'R''C=N-NH-SO_2C_6H_4CH_3 \longrightarrow R'RC=N-\bar{N}-SO_2C_6H_4CH_3$$

$$\longrightarrow R'R''C=N_2 \longrightarrow R'R''C{:} + N_2$$

In aprotic solvents, decomposition of the diazo compound yields only the carbene, but in proton-rich solvents protonation of the diazo compound yields the diazonium ion, and thence the carbonium ion. The borderline between these reactions is often ill defined.

In aprotic solvents, carbenes often decompose to yield olefins, the overall reaction being represented as follows:

$$RR'R''C{-}CH{=}N_2 \longrightarrow RR'R''C{-}\ddot{C}H \longrightarrow RR''C{=}CR'H$$

(8) (9)

It has been pointed out[6] that the evidence for involvement of free carbene (8) in these reactions is often flimsy, and that a displacement of the leaving group by migrating electrons of a C—C bond could give rise to the olefin (9) without intervention of the carbene. The same argument may be applied to reactions in which the toluene-*p*-sulphonyl hydrazone is the source of the diazo compound, except that in this case there is frequently no evidence of formation of either the diazo compound or the carbene.

In cases in which thermal decomposition of a diazo compound does yield a carbene, the carbene is probably formed as a singlet, that is, with a vacant orbital and an electron pair[7, 8]. It can then either react directly as the singlet, or form the triplet state in which it has two unpaired electrons and behaves as a diradical.

Rearrangements occur during carbene reactions either as a result of one carbene being transformed into another carbene, or during reaction of the carbene. Rearrangements of the first type are relatively uncommon; the overall reaction may be represented as:

$$R{-}C({=}X){-}\ddot{C}{-}R' \longrightarrow R{-}\ddot{C}{-}C({=}X){-}R'$$

The reaction involves the unusual feature that more than one bond must be broken and created during the rearrangement. Consequently, the energy barrier to rearrangement is relatively high and unless both carbenes are fairly stable, decomposition of the carbene will offer a lower energy pathway. For similar reasons, rearrangement by bond shift during carbene formation is unlikely, though one possible case of subsequent interaction of an electron-rich centre with the electron-deficient carbene has been reported[60].

Almost all rearrangements involving carbenes take place when the carbene reacts, and these rearrangements may be conveniently considered in terms of each of the reaction types discussed below.

A. Insertion into a C—H Bond

Consideration of spin conservation leads to the conclusion that the insertion of a singlet carbene into a C—H bond is a concerted process involving a three-centre transition state.

$$RR'C: + H{-}C\!\lt \longrightarrow \left[RR'C\cdots H \cdots C\!\lt \right] \longrightarrow RR'HC{-}C\!\lt$$

By similar reasoning, the insertion of a triplet carbene (diradical species) into a C—H bond is a two-step process involving radical formation and combination.

$$RR'\dot{\underset{\bullet}{C}} + H-\overset{/}{\underset{\backslash}{C}}- \longrightarrow RR'\dot{C}H + \dot{\overset{/}{\underset{\backslash}{C}}}- \longrightarrow RR'CH\overset{/}{\underset{\backslash}{C}}-$$

Insertion of a highly energetic carbene, such as methylene, takes place almost equally readily into primary, secondary and tertiary C—H bonds, but phenyl carbene is more selective, and the more stable chloromethylene even more selective, showing a selectivity ratio for secondary to primary bonds of 20 to 1, compared with 1·2 to 1 for methylene.

Insertion reactions of this type can only be regarded as giving rise to rearranged products when they proceed intramolecularly. Several examples of this type of reaction are known. In a simple case, isopropyl carbene, generated from the diazo compound either photolytically[61] or thermally[62], gives rise to approximately 35% of methylcyclopropane. The reaction proceeds even better in cyclic systems, where diazocamphane gives tricyclene in 100% yield[63].

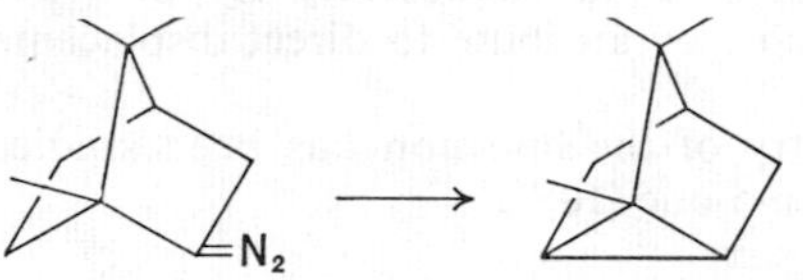

N_2

Long range insertion reactions are relatively uncommon and require special structural features of the substrate. Long range insertion can take place in carbenes which do not have C—H bonds α or β to the carbene centre[64], and in cases where the molecule is constrained by a cyclic structure[65].

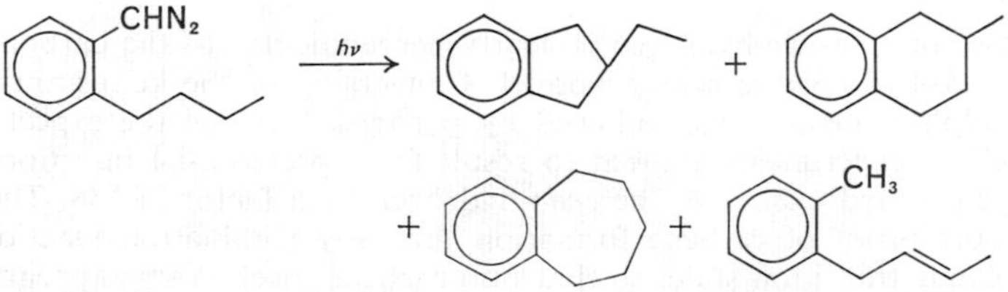

CHN_2 $\xrightarrow{h\nu}$ + + CH_3 +

A study of a number of insertion reactions in cyclic systems has shown that in all cases, insertion takes place into an axial C—H bond of the ring, to give a *cis*-bicyclo derivative. This is consistent with the view that carbene insertion reactions occur with retention of configuration.

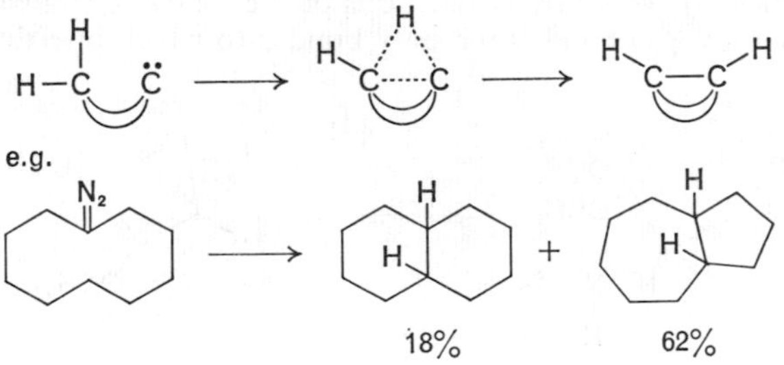

e.g.

N_2 ⟶ H H 18% + H H 62%

B. *Olefin-forming Insertions*

These reactions are of the general type:

$$R^3R^2R^1C{-}\ddot{C}{-}R^4 \longrightarrow R^3R^2C{=}CR^1R^4$$

The reaction consists of an intramolecular insertion of the carbene into one of the bonds to the α-carbon atom. These reactions are frequently but incorrectly described as '1,2-shift' reactions, by comparison with similar carbonium ion rearrangements, though this implies conversion of one carbene into another, which has not been observed in reactions of this type[66].

It is in the field of olefin-forming reactions of carbenes that evidence for the involvement of a carbene in the reaction is frequently thin. Reaction via a diazonium ion and a carbonium ion would often give similar products to a carbene reaction, and distinction between the two routes is not always possible. Evidence that alternative routes to olefin formation do exist has been obtained by Powell and Whiting[67], who have observed a stereoselectivity of reaction inconsistent with carbene formation, which they attribute to direct displacement of nitrogen to give the olefin.

The preferred geometry of the insertion has been studied by using the locked norbornyl system, brexan-5-one **(10)**[68].

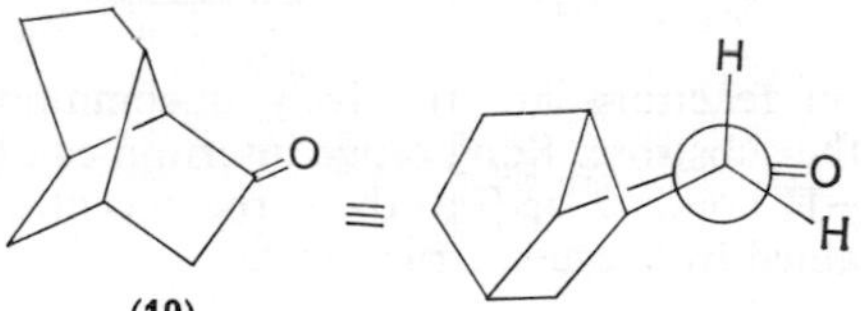

(10)

In this molecule, the *exo*-hydrogen is nearly perpendicular to the carbonyl group, while the *endo*-hydrogen is nearly parallel. Conversion of the keto group into the toluene-*p*-sulphonylhydrazone, followed by pyrolysis, yielded the expected olefin, which labelling experiments showed to result from preferential insertion into the bond to the *exo*-hydrogen over the *endo*-hydrogen by a factor of 138. The authors do not report proof of carbene formation, but any elimination reaction should favour a planar transition state, so that non-carbene mechanisms appear unlikely. Interestingly, photolysis of the toluene-*p*-sulphonyl hydrazone reduced the *exo*-to-*endo* ratio to only 4·8.

The relative ease of carbene insertion into neighbouring bonds has been measured, and it is found[69] that the order of ease of insertion is C—H > C—phenyl > C—methyl. Insertion into a C—O bond, though not uncommon in reactions of photochemically generated carbenes[70], is very rare in reactions of thermally generated carbenes, and occurs only when a C—O bond is the only bond into which insertion can possibly occur[71].

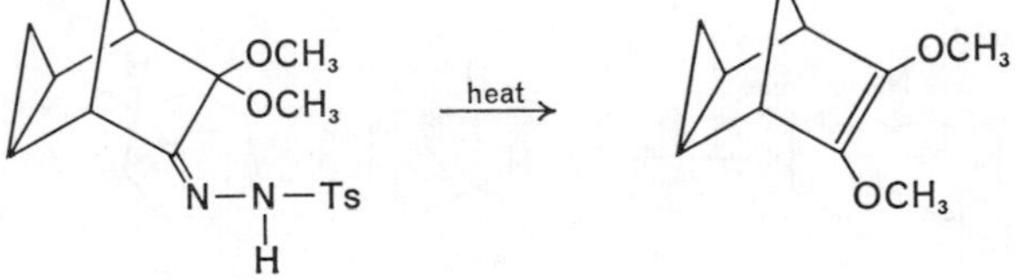

$$\text{(OCH}_3)_2\text{, N–NH–Ts} \xrightarrow{\text{heat}} \text{OCH}_3\text{, OCH}_3$$

Reactions of this type clearly produce skeletal rearrangements comparable to the 1,2-alkyl shifts observed during carbonium ion rearrangements; in acyclic[72] and bicyclic[73] systems, this can result in changes to the basic carbon skeleton of the molecule.

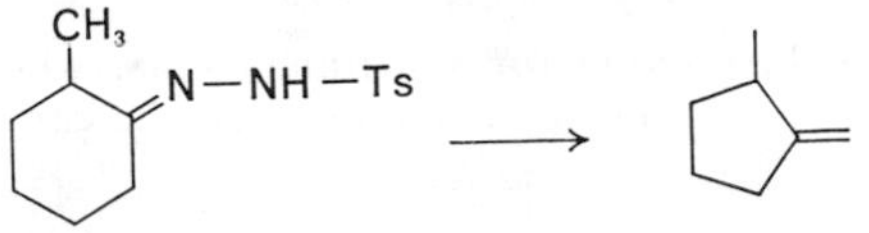

CH_3 N—NH—Ts

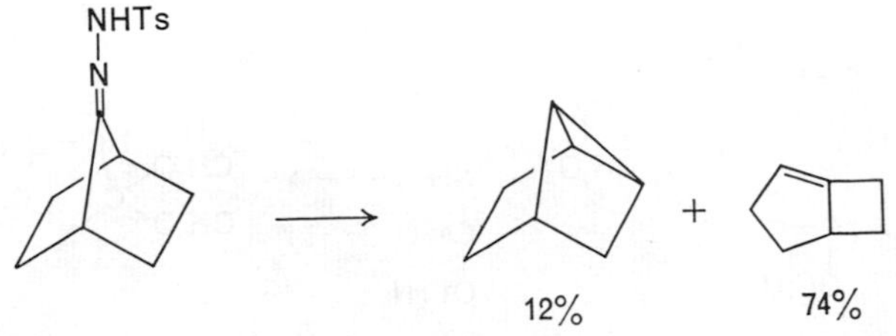

NHTs
N

12% 74%

Olefin-forming insertion reactions have recently been used to generate unstable olefins, such as those contravening the Bredt rule. These olefins are so highly reactive that they cannot be isolated, but decompose either by a retro Diels–Alder reaction or by dimerization. An example of the first type of reaction is that of the norbornyl derivative shown below[74].

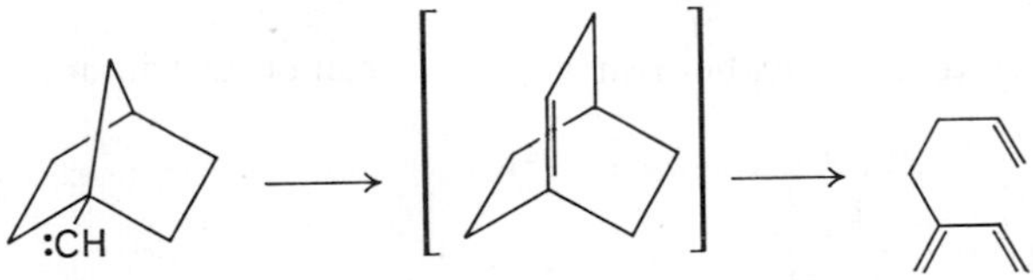

:CH

Olefins such as that generated from adamantyl carbene do not undergo a retro Diels–Alder reaction readily, so they decompose by dimerization[75].

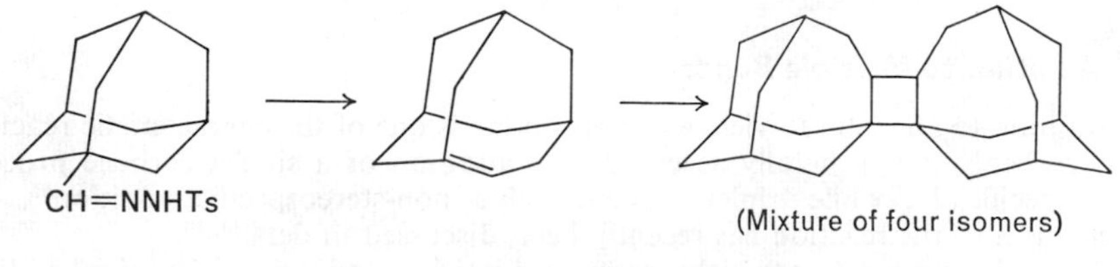

CH=NNHTs

(Mixture of four isomers)

C. Reaction with Nucleophiles

The reaction of a carbene with a hydroxyl group shows an overall similarity to insertion reactions of the type discussed in the previous sections, but the mechanism is different. From a detailed study of the reaction of diphenyl carbene with alcohols and with water, it was concluded[76] that the reaction involves attack of the carbene on the hydroxyl oxygen to give an ylide; this then undergoes prototropic shift to yield the alcohol or ether.

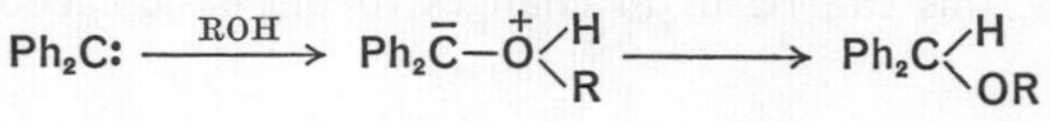

$$Ph_2C: \xrightarrow{ROH} Ph_2\bar{C}-\overset{+}{O}\langle^{H}_{R} \longrightarrow Ph_2C\langle^{H}_{OR}$$

Reactions of this type give little scope for rearrangements, which rarely occur; cyclization offers one possible rearrangement route when the carbene is generated in a suitable molecule[77].

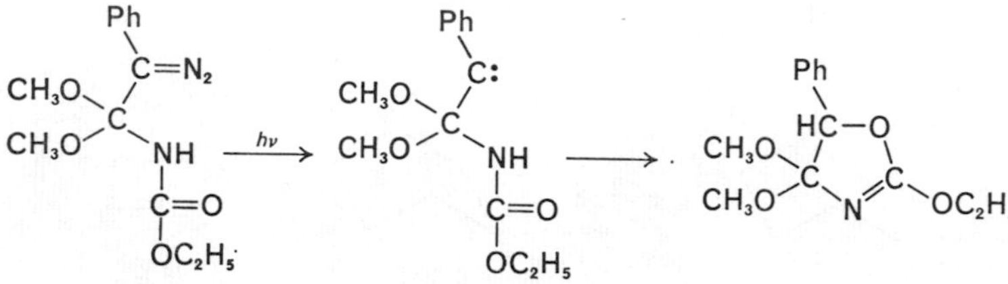

Attack of a carbene on an aziridine gives an unstable ylide, which fragments to give an olefin[78].

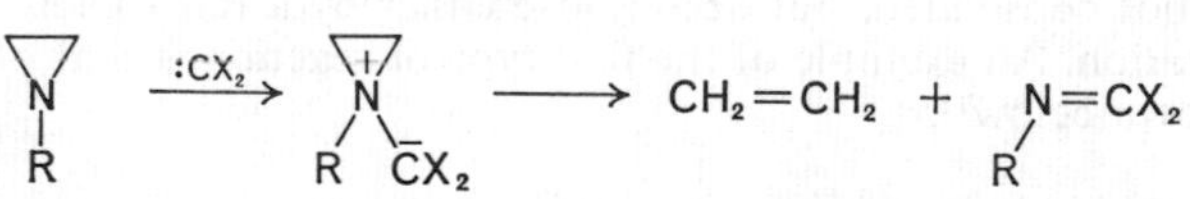

$$\xrightarrow{:CX_2} \longrightarrow CH_2{=}CH_2 + RN{=}CX_2$$

Episulphides cleave similarly, but rearrangement can occur within the ylide in a few special cases[79].

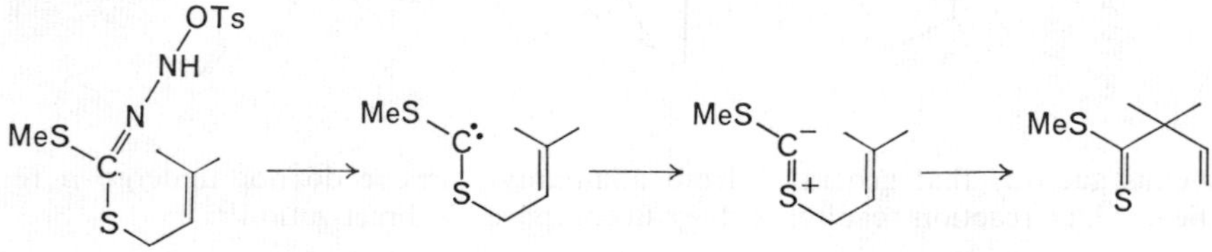

D. *Addition to Multiple Bonds*

Addition to an olefin to yield a cyclopropane is one of the characteristic reactions of a carbene. It is generally accepted that addition of a singlet carbene proceeds stereospecifically, while triplet carbenes give non-stereospecific addition[80]. The mechanism of the reaction has recently been discussed in detail[9, 10].

The addition reaction can yield rearranged products only when the initial addition product is unstable, or by an intramolecular reaction. An example of the first type of reaction is[81]:

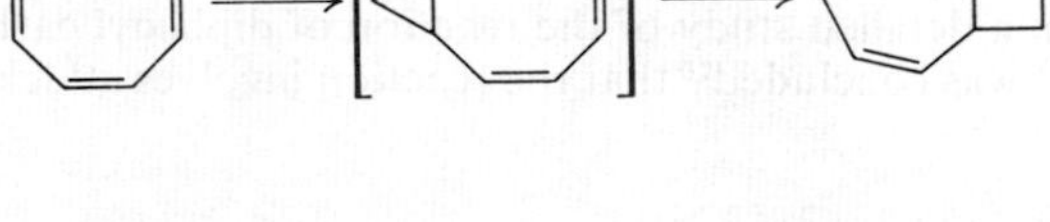

$$\xrightarrow{CH_2N_2}$$

A simple example[82] of an intramolecular carbene addition is the photolysis of allyldiazomethane at −78 °C:

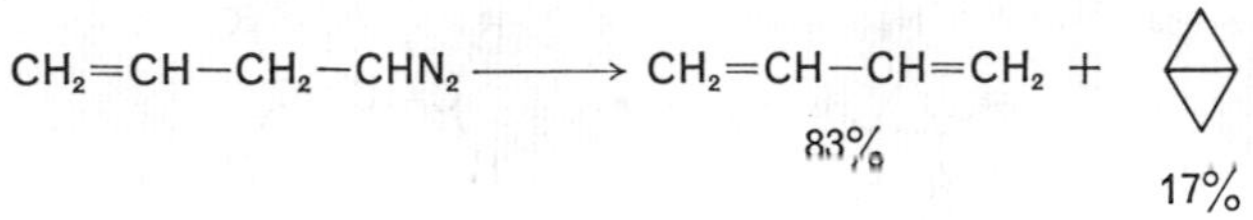

$$CH_2{=}CH{-}CH_2{-}CHN_2 \longrightarrow CH_2{=}CH{-}CH{=}CH_2 + \text{(bicyclobutane)}$$

83% 17%

The intramolecular addition reaction does not, however, take place in systems in which the separation of the diazo and olefinic functions exceeds one carbon atom. Thus, **11** cyclizes[83] while **12** does not[84].

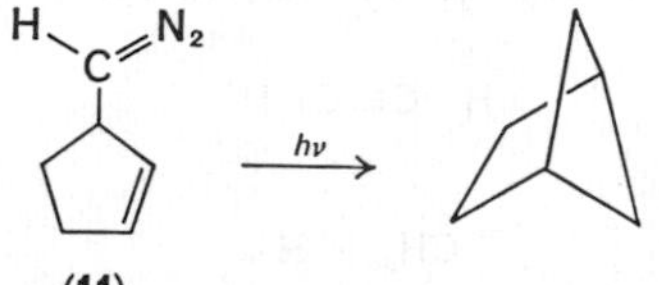

(11) $\xrightarrow{h\nu}$

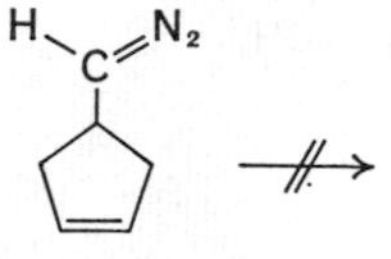

(12) $\not\rightarrow$

It has been suggested that this may be because the intermediate in intramolecular additions could be a 1,3-dipolar adduct rather than a carbene, in which case the reaction would be[85]:

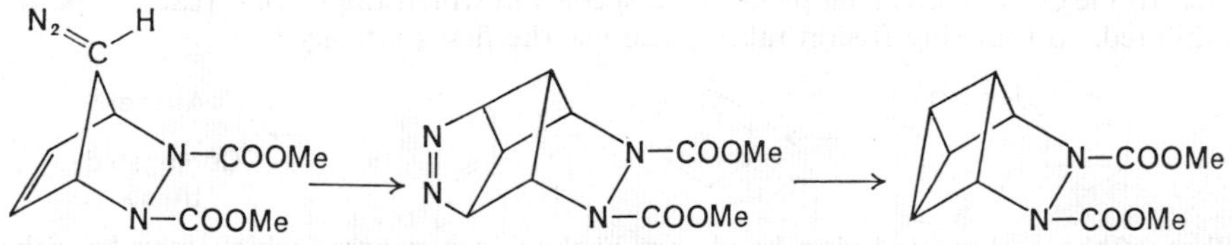

N−COOMe, N−COOMe ⟶ N−COOMe, N−COOMe ⟶ N−COOMe, N−COOMe

Such a reaction would be subject to more rigid stereochemical requirements than the carbene addition, and hence more easily inhibited by separation of the reacting groups.

E. *Fragmentation Reactions*

Cyclopropylidene (13) is readily generated by thermal decomposition of diazocyclopropane (14), and decomposes to yield allene (15). In a thorough study of the

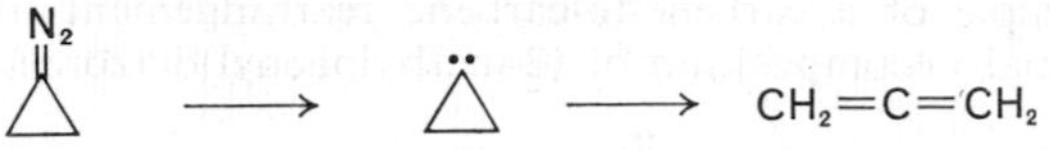

(14) ⟶ (13) ⟶ $CH_2{=}C{=}CH_2$ (15)

reaction, Jones and his coworkers[86] were able to prove the presence of both (13) and (14) by trapping experiments, so it is clear that (15) does arise from (14), in quantitative yield. Reaction of an optically active disubstituted diazo compound showed that the allene was formed with the retention of a high degree of rotation[87].

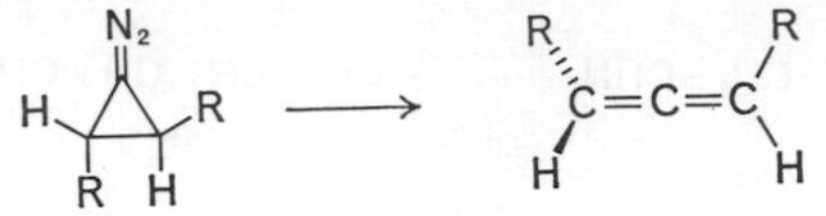

When a carbene is generated next to a cyclopropyl ring, as in cyclopropylcarbene, a completely different reaction is observed. Gas phase pyrolysis of *trans*-2,3-dimethyl cyclopropyl diazoethane gives a mixture of cleavage to olefin plus acetylene, and cleavage to the diene. The latter is formed via cyclobutane, generated by carbene insertion[88].

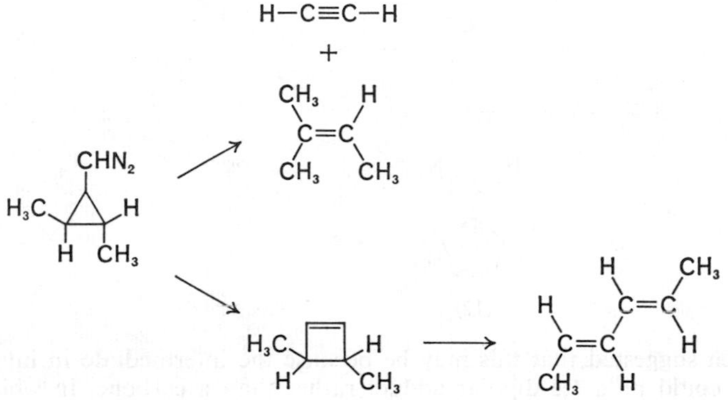

When the cyclopropyl ring is fused to a larger ring, generation of a carbene centre next to the cyclopropyl ring produced a species in which the second reaction path is inhibited, so that ring fission takes place via the first pathway[89].

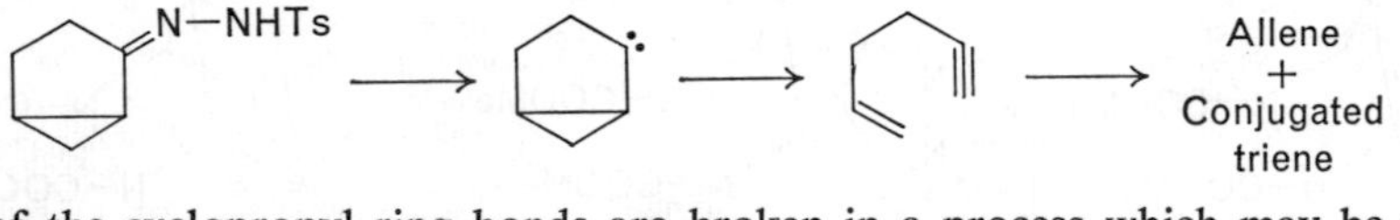

Allene + Conjugated triene

Two of the cyclopropyl ring bonds are broken in a process which may be either concerted or two step. Recent experiments favour the former, at least for reactions from a singlet ground state[90].

F. Carbene to Carbene Rearrangements

True carbene to carbene rearrangements are uncommon, since carbenes are highly reactive species which usually decompose more readily than they rearrange. Consequently, rearrangement offers a lower energy pathway than does reaction only when both the carbenes involved are unusually stable.

The first example of a carbene-to-carbene rearrangement to be observed was during the thermal decomposition of (2-methylphenyl)diazomethane.

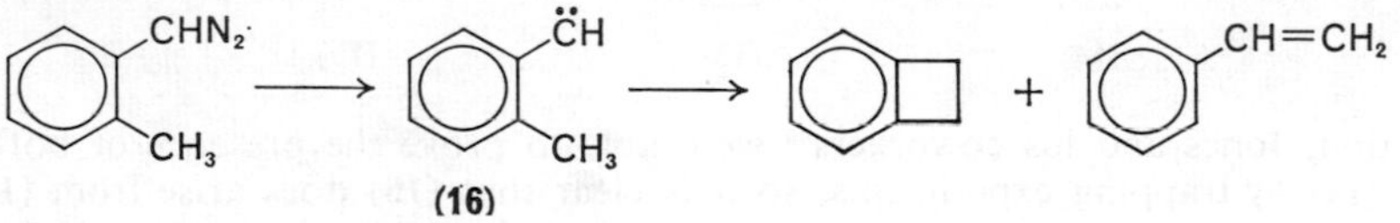

(16)

Observation of 9% styrene is inconsistent with reaction via the carbene (**16**), though the formation of benzocyclobutene is consistent with its presence. Similar reactions have been reported for the 3- and 4-methyl benzylidenes while reaction of the unsubstituted carbene, phenylcarbene, gave a quantitative yield of heptafulvalene[91].

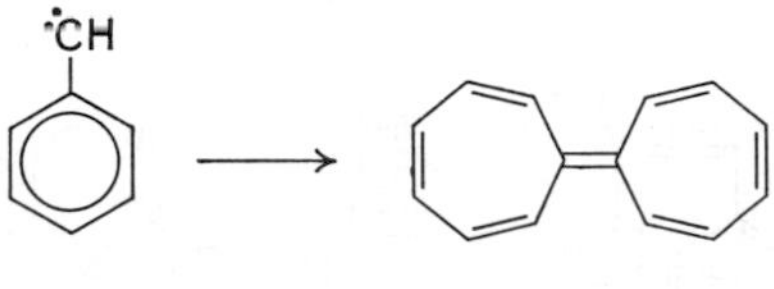

Trapping experiments subsequently confirmed that this reaction does proceed via cycloheptatrienylidene[92].

The interconversion of phenyl carbene and cycloheptatrienylidene (**17**) can be visualized as occurring by any of at least three mechanisms[93], but recent evidence[94] favours interconversion via the cyclopropene:

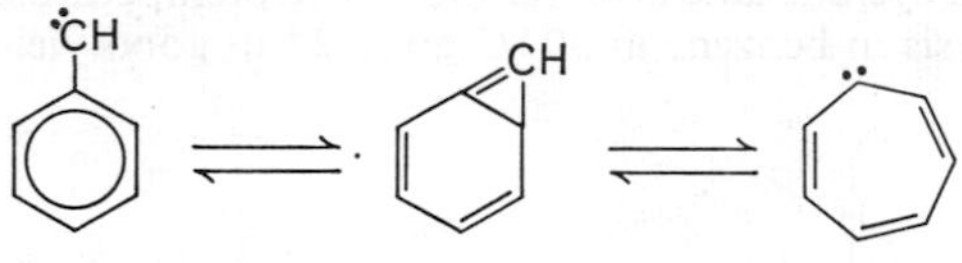

(**17**)

Since both steps of this reaction are reversible, it provides a mechanism for 'movement' of substituents in the ring.

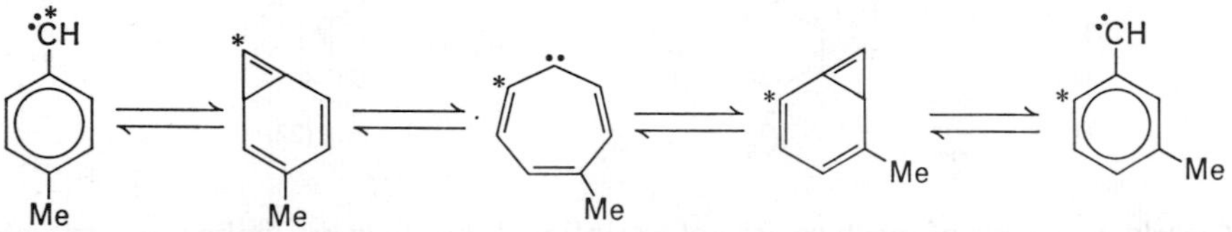

By a series of similar equilibria, 4-methylbenzylidenes can give rise to styrene; experiments on labelled materials gave styrene in which the position of the label was consistent with the proposed mechanism[95].

The cyclopropenyl intermediate involved in these reactions is analogous to the oxirene intermediate **18** (see later) in the Wolff rearrangement.

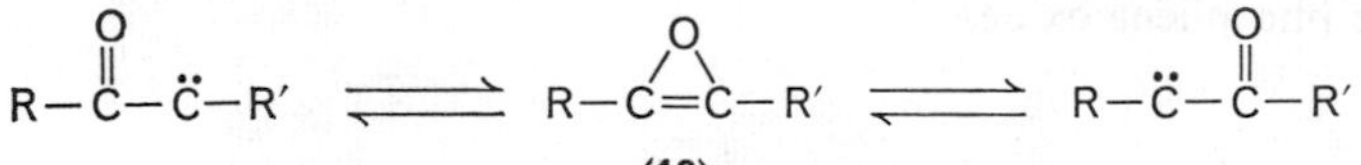

(**18**)

However, this heterocyclic intermediate probably draws extra stability from the presence of the hetero atom, since the 'carbene' species involved do not display such normal carbene reactions as addition to olefins. The difference may well be that the heterocyclic species is an intermediate, the carbocyclic a transition state.

Detailed discussions of the carbene-to-carbene rearrangements involving phenylcarbene have been given by several workers[91, 96, 97]. The mechanism seems to be a general one in systems of this type, as similar reactions have been observed in the gas-phase pyrolysis of diphenyldiazomethane[98], and between the acenaphthylcarbene **(19)** and phenylenylidine **(20)** species[99].

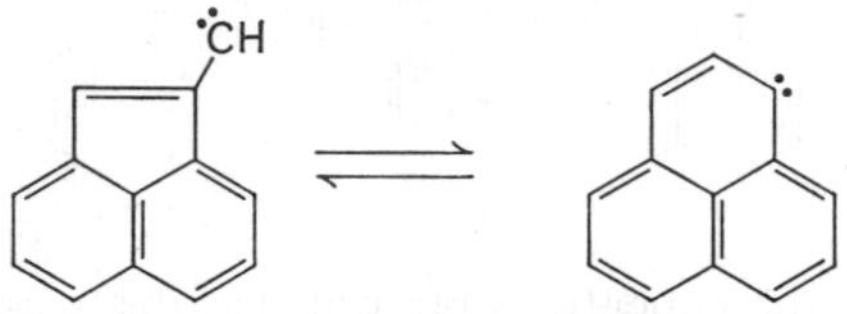

(19) **(20)**

All the above reactions take place under pyrolytic conditions (250–600 °C) in the gas phase, but a similar reaction has been observed which takes place in solvents such as benzene and cyclohexane at lower (30–80 °C) temperatures[100]. Pyrolysis of **21** at 80 °C or photolysis in benzene at 30 °C gives **22** in good yield.

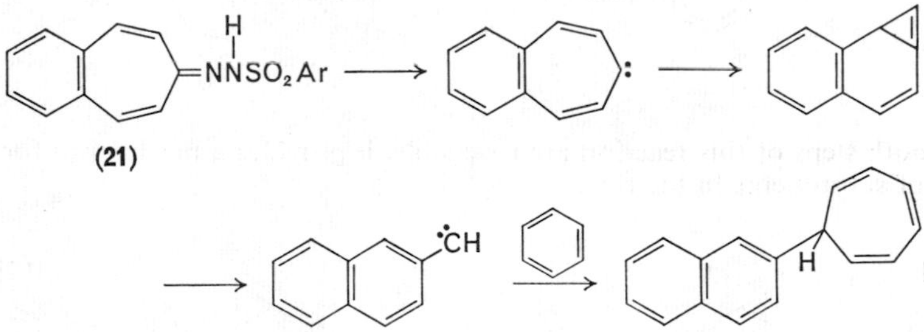

(21) **(22)**

Clearly, reactions of this type are not exceptional, but they are limited to very stable carbenes.

G. Rearrangements of Carbenes to Form Other Reactive Intermediates

When a carbene is generated in a system containing a suitable nitrogen atom, it may rearrange to yield a nitrene; a phosphorylcarbene can rearrange to yield a methylene phosphene oxide.

1. Carbene–nitrene rearrangements

This rearrangement shows mechanistic similarities to the carbene-to-carbene rearrangements described earlier. Like them, it is an uncommon reaction, and takes place only with very stable carbenes, since only under these conditions can it compete with decomposition of the carbene. A good example is the rearrangement of pyridyl carbene to phenyl nitrene; the pyridyl diazo compound is unstable, so is formed *in situ* by the pyrolysis shown[100a]:

N C=N N NH N → N C–H N_2 → N $\ddot{C}H$ ⇌ N ⇌ N:

It is suggested that the mechanism is closely analogous to that proposed for carbene–carbene rearrangements, i.e.

N $\ddot{C}H$ ⇌ N ⇌ N: ⇌ N ⇌ :N:

The equilibrium can also be approached from the phenyl nitrene side, and is strongly in favour of the nitrene[100b].

2. Phosphoryl carbene–methylene phosphene oxide rearrangements

Methylene phosphene oxides are short-lived phosphorus analogues of ketenes; this rearrangement is thus similar to the Wolff rearrangement. Unlike the Wolff rearrangement, the first step of the reaction is the formation of a phosphoryl carbene, and this may then react with a nucleophile directly by an insertion mechanism, or rearrange to a methylene phosphene oxide, then add the nucleophile[100c].

$Ph_2P(O)C(N_2)Ph \longrightarrow Ph_2P(O)\ddot{C}Ph \xrightarrow{H_2O} Ph_2P(O)CH(OH)Ph$

$\downarrow$

$O=P(Ph)=CPh_2 \xrightarrow{H_2O} PhP(O)(OH)CHPh_2$

Alternatively, the methylene phosphene oxide can be detected by trapping with a carbonyl compound to give a 1,2-oxaphosphetane[100d].

$O=P(Ph)=CPh_2 \xrightarrow{PhCHO}$ Ph(O=)P–C(Ph)(Ph)–C(Ph)(H)–O (ring)

Some 1,2-oxaphosphetanes, however, will undergo further rearrangement[100e].

H. The Wolff Rearrangement

Thermal or photolytic decomposition of a diazo compound which has a carbonyl group α to the diazo group does not proceed via a simple carbene reaction. The reaction does not involve a carbene which can be trapped by any conventional technique, and in the presence of water gives rise to a rearranged carboxylic acid. The reaction is known as the Wolff rearrangement[101].

$$R-C(=O)-C(=N_2)-R' \xrightarrow{H_2O} RR'CH-C(=O)-OH$$

The rearrangement did not attract much interest when it was first described in 1902; it was not until an efficient synthesis of α-diazoketones from acid chlorides and diazomethane was developed[102] that the synthetic uses of the reaction were expanded. In its original form, the reaction was carried out thermally; a photochemical analogue was demonstrated in 1944 [103] and a similar rearrangement catalysed by silver ions is now known[8, 104]. In contrast to this latter reaction, the decomposition of α-diazoketones in the presence of copper salts leads, in most cases, to addition and insertion reactions of carbonylcarbenes[104, 105]. Only a few cases of copper-catalysed Wolff rearrangements are known.

Despite extensive studies, the mechanism of the Wolff rearrangement is still in some doubt. The basic mechanism proposed by Wolff was via the route:

$$R-C(=O)-C(=N_2)-R' \longrightarrow R-C(=O)-\ddot{C}-R' \longrightarrow RR'C=C=O \xrightarrow{H_2O} RR'CH-C(=O)-OH$$

The intermediacy of ketenes seems well established, since ketenes or their decomposition products have been isolated from diazoketone thermolysis in aprotic solvents[106] and carbon monoxide has been isolated from photolysis of the ketene produced during α-diazoketone photolysis. It has been pointed out that rearrangement from an excited singlet state of the diazoketone is necessary to yield ketenes since the first excited state of ketenes lies at high energies, rendering triplet diazoketone to triplet ketene an endoenergetic process[108]. This is supported experimentally by showing that diazocyclohexanone underwent Wolff rearrangement on unsensitized irradiation via an intermediate which was not trapped by olefins, but that irradiation in the presence of benzophenone yielded an intermediate which did not undergo Wolff rearrangement, and could be trapped by olefins[10].

Three possible mechanisms for conversion of diazoketones into ketenes have been proposed;

(i) A concerted shift:

$$R^1-C(=O)-C(=N_2)-R^2 \longrightarrow O=C=C(R^2)(R^1)$$

(ii) Reaction via a carbonyl carbene:

$$R^1-C(=O)-C(=N_2)-R^2 \longrightarrow R^1-C(=O)-\dot{\dot{C}}-R^2 \longrightarrow O=C=C(R^2)(R^1)$$

(iii) Reaction involving an oxirene (**18**), probably formed from, or in equilibrium with, a carbonylcarbene:

$$R^1C(=O)C(=N_2)R^2 \longrightarrow R^1C(=O)\ddot{C}R^2 \rightleftharpoons \underset{(\mathbf{18})}{R^1C{=}CR^2\ (\text{oxirene})} \rightleftharpoons R^1\ddot{C}C(=O)R^2$$

Some evidence to support the first mechanism comes from the observation that most α-diazoketones have a *cis* structure; the only known exception to this rule does not undergo the Wolff rearrangement[109]. However, kinetic studies do not support the idea of a concerted mechanism. Electron-releasing substituents on a migrating aryl group strongly retard reaction[110] which is the reverse of the result expected for a concerted migration. Further evidence against a concerted migration has been obtained by showing a lack of kinetic isotope effects on the reaction[111].

Attempts to obtain proof of carbene intermediates in the reaction by trapping experiments have been generally unsuccessful, though the copper-catalysed reactions of the same diazo compounds, which do not give Wolff rearrangement products, readily yield trappable intermediates. Some support for the intermediacy of carbonyl carbenes has been obtained by generating these species via an alternative route; photolysis of **23** has produced an intermediate which behaves very like that produced by the photolysis of diazoacetophenone[112].

$$\underset{(\mathbf{23})}{Ph-CO-\overset{-}{C}H-\overset{+}{S}(CH_3)_2} \longrightarrow Ph-CO-CH{:} \xrightarrow{EtOH} PhCOCH_3$$

$$PhCOCHN_2 \longrightarrow Ph-CO-CH{:} \xrightarrow{EtOH} PhCH_2CO_2C_2H_5$$

Evidence for the intermediacy of oxirenes in the Wolff rearrangement has been obtained by labelling experiments.

$$R-\overset{*}{C}(=O)-C(=N_2)-R' \longrightarrow R\overset{*}{C}{=}CR'\ (\text{oxirene}) \longrightarrow RR'\overset{*}{C}{=}C{=}O + RR'C{=}\overset{*}{C}{=}O$$

The experiments of Strausz and his coworkers[113] using 3-diazo-2-propanone, 3-diazo-2-butanone, azibenzil ($PhCOCN_2Ph$) and α-diazoacetophenone showed that a ^{13}C label on the carbonyl carbon atom is scrambled over both carbon atoms of the corresponding ketene. This experiment affords clear proof of the existence of a symmetrical intermediate, most probably an oxirene. The authors found oxirene formation to be a characteristic of photolytic reactions, which is inconsistent with a thorough study of the decomposition of 3-diazoheptan-4-one by Sammes[114] during which the existence of oxirene participation was demonstrated during both thermal and photolytic decomposition but not during silver- or copper-catalysed decomposition.

An alternative approach to the oxirene question has been provided by the treatment of alkynes with *m*-chloroperoxybenzoic acid, a reaction which should give oxirenes directly[115]:

$$R-C\equiv C-R' \xrightarrow{[O]} R-\overset{\overset{O}{\frown}}{C=C}-R'$$

Oxidation of acetylenes was compared to the results of thermal decomposition of the appropriate α-diazoketones, and the products found to be qualitatively but not quantitatively similar.

Quantum mechanical calculations indicate that oxirenes are energetically accessible intermediates, and suggest that the carbonyl carbene to oxirene rearrangement is a process of little or no activation energy[116, 117].

On the basis of the above results, an overall scheme for the Wolff rearrangement may be written:

$$R-\overset{O}{\overset{\|}{C}}-\overset{N_2}{\overset{\|}{C}}-R \longrightarrow R-\overset{O}{\overset{\|}{C}}-\ddot{C}-R \rightleftharpoons R-\overset{\overset{O}{\frown}}{C=C}-R$$

$$R-\overset{O}{\overset{\|}{C}}-\ddot{C}-R \longrightarrow R_2C=C=O \longrightarrow \text{Products}$$

This scheme, however, has a number of exceptions in which a Wolff rearrangement takes place without any evidence of oxirene participation. Photolysis of 5-diazohomoadamantan-4-one proceeds without oxirene participation[118]; this may be due to its instability as part of a strained polycyclic system.

$$\text{5-diazohomoadamantan-4-one } (N_2, =O) \longrightarrow \text{adamantane-}CO_2H$$

Similarly, azibenzil[119] undergoes a thermal Wolff rearrangement without oxirene participation. The reasons for these exceptions are not known.

Related to the Wolff rearrangement is the photolysis of diazoacyl esters[120], which in methanol proceeds via two main competing pathways. One of these is the normal insertion of a carbene into the —OH bond of methanol, the other a path analogous to the Wolff rearrangement.

$$PhOC(=O)—CH(=N_2) \longrightarrow PhOC(=O)CH\!:$$

$$PhOC(=O)CH\!: \xrightarrow{CH_3OH} PhO—C(=O)CH_2OCH_3$$

$$PhOC(=O)CH\!: \longrightarrow PhOCH{=}C{=}O \xrightarrow{CH_3OH} PhOCH_2CO_2CH_3$$

However, a sulphur analogue of the above class of compounds decomposes only by a Wolff rearrangement pathway[121].

$$CH_3SC(=O)CH_2N_2 \xrightarrow[CH_3OH]{h\nu} CH_3SC(=O)\ddot{C}H \longrightarrow CH_3SCH{=}C{=}O$$

The presence of a second carbonyl function α to the diazo group does not modify the mechanism, but gives a mixture of the two possible rearrangement products[122].

The Wolff rearrangement is a very useful synthetic reaction, as it provides a valuable method of making strained cyclic systems by ring contraction[123], e.g.

$$\text{(bicyclic diazo ketone)}{=}N_2 \xrightarrow[CH_3OH]{h\nu} \text{(ring-contracted bicyclic)}—COOCH_3$$

It fails, however, in cases where contraction would result in an excessively strained system[124].

$$\text{(acenaphthenequinone monodiazide: O, } N_2\text{)}$$

For a detailed discussion of the uses of this reaction, the treatment by Kirmse[8] and references therein should be consulted.

J. Carbenoid Rearrangements

Carbenoids have been defined[125] as 'Intermediates which exhibit reactions qualitatively similar to those of carbenes without necessarily being free divalent carbon species'. Frequently the carbenoid species has a metal associated with the organic fragment, and carbenoid activity is often found in the reactive species formed when diazoalkanes decompose under the influence of metal salts such as $ZnCl_2$, $HgCl_2$, Cu(I) and Cu(II) chlorides, Cu(II) sulphate and tungsten(VI) chloride[126]. Recently, decomposition of a diazoalkane catalysed by a hydrocarbon, tetraphenylethylene, has been reported, but has not yet been sufficiently investigated for a useful comparison with metal-catalysed reactions to be made[127].

When a diazoalkane is decomposed in the presence of a copper salt, the carbenoid is more likely to undergo intermolecular reaction than would a photolytically generated carbene from the same diazoalkane.

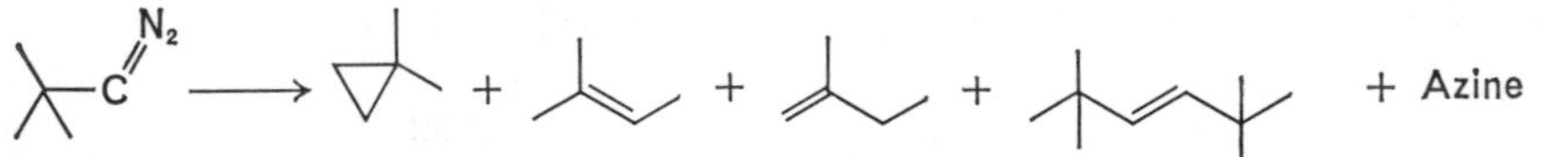

		+	+	+	+ Azine
$h\nu$	92	7	1	—	—
CuI (%)	7	12	6	62	13

In the presence of olefins, copper-catalysed decomposition of diazoalkanes gives a better yield of cyclopropanes than does the uncatalysed reaction. This is of value in such reactions as the formation of vinylcyclopropanes from vinyldiazomethane, where the photolytic or thermal reactions give poor yields due to pyrazoline formation or internal cyclization to pyrazolenines[128].

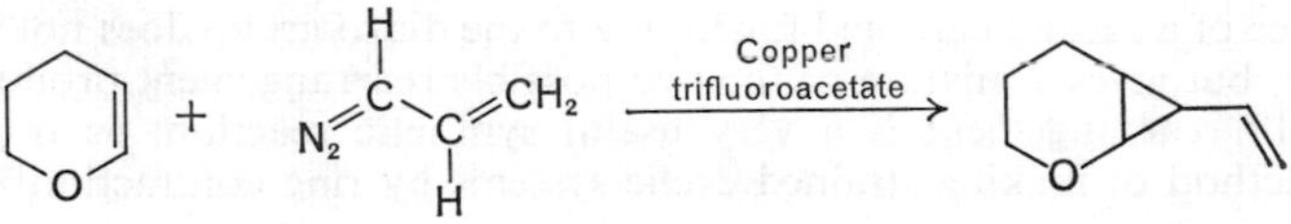

Copper trifluoroacetate

From the point of view of rearrangements, copper-catalysed reactions are important in that they suppress the Wolff rearrangement of α-diazoketones, and permit them to undergo addition and insertion reactions of normal carbenes[129]. In this, they differ from silver and platinum salts, which catalyse the Wolff rearrangement.

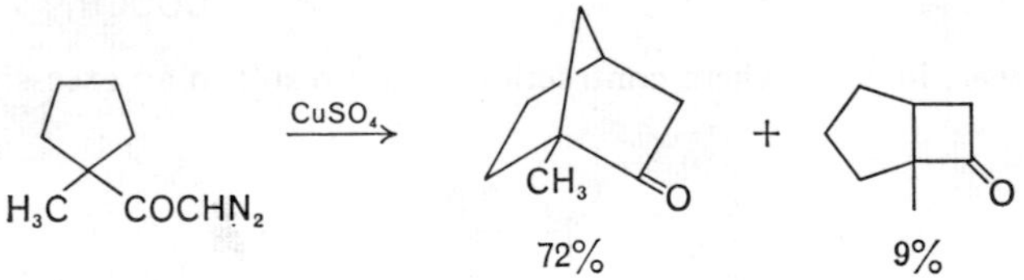

H_3C $COCHN_2$ $\xrightarrow{CuSO_4}$ 72% + 9%

Insertion into a C—C bond is also possible, as is illustrated by the novel rearrangement[130]:

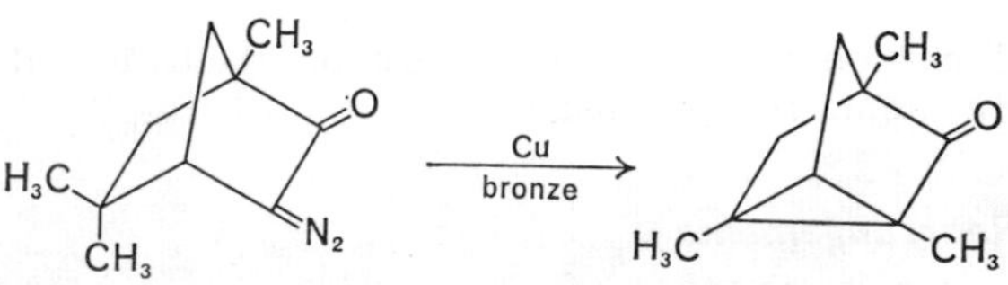

$\xrightarrow[\text{bronze}]{\text{Cu}}$

This reaction involves cyclopropane formation by insertion into a C—C bond, which is uncommon. The carbenoid from reaction of an α-diazoketone also undergoes addition to a double bond[131].

$$\text{cyclohexenyl}-\overset{O}{\overset{\|}{C}}-CHN_2 \longrightarrow \text{tricyclic ketone}$$

Copper carbenoids show a much greater tendency to dimerize than do 'free' carbenes, and this tendency can be used to obtain intramolecular cyclization products from the decomposition of *bis*-α-diazoketones. An example of this type of reaction is the synthesis of γ-tropolone[132].

$$Cl-CH(CH_2COCHN_2)_2 \longrightarrow \text{4-chlorocyclohept-2-ene-1,5-dione} \longrightarrow \gamma\text{-tropolone (OH)}$$

V. REARRANGEMENTS INVOLVING DIAZONIUM IONS

Reactions such as treatment of an aliphatic primary amine with nitrous acid give rise to a large number of rearranged products. These rearrangements have been listed in a number of comprehensive reviews[11–15, 46, 133], so that this section will consider the overall picture of the reaction, rather than attempt to give a comprehensive coverage of all known rearrangements.

Treatment of an aliphatic primary amine with nitrous acid is believed to yield the diazonium ion, which breaks down to the carbonium ion, and thence to products. However, the variety of products observed in these reactions exceeds by far that observed in formally similar carbonium ion reactions in which the ion is generated by ester or halide heterolysis, and has led to much investigation of why the ion involved is more reactive. This reactivity is a feature of all reactions of the diazonium ion, regardless of how it is obtained.

A. Formation of Diazonium Ions

The simplest possible route to an aliphatic diazonium ion is by treatment of a diazo compound with acid. The first step of this reaction is protonation of the diazo compound to give the diazonium ion, and this ion then decomposes to a carbonium ion, from which the products are obtained.

$$Ph_2CN_2 \xrightarrow{H^+} Ph_2CHN_2^+ \longrightarrow Ph_2\overset{+}{C}H + N_2$$

In practice, relatively few aliphatic diazo compounds have been prepared and purified, so protonation of the diazo compound is a little-used route to the diazonium ion. Several other routes are in more common use, and this variety of routes is important in mechanistic studies, since each route gives the diazonium ion associated to differing extents with different neutral and charged species. Comparison of the products of decomposition of formally similar species generated by different routes can thus provide useful information about the intermediates. Methods of generating the diazonium ion in common use are the following:

(a) Reaction of amines with nitrous acid. This reaction is often described as 'deamination' though Collins[133] has pointed out that this is not an appropriate description. Commonly, the amine is treated with nitrous acid, generated *in situ* from sodium nitrite and a mineral acid.

$$R{-}NH_2 + HNO_2 \longrightarrow RNHNO \longrightarrow RNNOH \longrightarrow RN_2^+ \longrightarrow R^+$$

The reagent is believed to be NO^+. Evidence in support of this mechanism has been obtained from the study of heterocyclic primary amines, where the intermediates are more stable than those obtained from primary aliphatic amines[12], and by using NOCl as a source of NO^+ under very mild conditions, when the diazotization of substituted anilines can be stopped at the nitrosamine stage[134]. A review of the processes involved has appeared[11].

The reaction can also be carried out in acetic acid, in which case the carbonium ion is generated as part of an ion pair[133].

$$2\,HNO_2 \longrightarrow H_2O + N_2O_3$$

$$N_2O_3 + CH_3COOH \longrightarrow CH_3COONO + NO_2^- + H^+$$

$$RNH_2 + CH_3C(=O)ONO \longrightarrow [R\overset{+}{N}H_2NO\ CH_3COO^-]$$

$$[R\overset{+}{N}H_2NO\ CH_3COO^-] \longrightarrow [R\overset{+}{N}H{=}NOH\ CH_3COO^-]$$

$$[R\overset{+}{N}H{=}NOH\ CH_3COO^-] \longrightarrow [RN_2^+\ H_2O\ CH_3CO\bar{O}]$$

$$[RN_2^+\ H_2O\ CH_3COO^-] \longrightarrow [R^+\ H_2O\ CH_3COO^-] + N_2$$

The water molecule and the acetate counterion strongly influence reactions of the carbonium ion.

(b) Thermal decomposition of *N*-alkyl-*N*-nitrosamides. In this reaction, the amine is converted first into an amide, then a nitrosamide, and then decomposed in a suitable solvent[135, 136].

$$RNH_2 \longrightarrow R{-}N(H){-}C(=O){-}R' \longrightarrow R{-}N(N{=}O){-}C(=O){-}R'$$

$$\longrightarrow [R{-}N{=}N{-}O{-}C(=O){-}R'] \longrightarrow [R^+\ \bar{O}OCR'] + N_2$$

The nitrosation process is best carried out with dinitrogen tetroxide. The reaction is reported to minimize rearrangements[13].

Variations on this process include the thermal decomposition of *N*-nitrosocarbamates[13], of *N*-nitrocarbamates[137] and *N*-nitroamides[138]. The nitro compounds decompose in the same way as the nitroso compounds, except that they give off NO_2.

(c) The triazine reaction. In this reaction, the aliphatic amine reacts with a diazonium salt to yield a triazine, which then undergoes acid-catalysed decomposition in a suitable solvent[139–141].

$$R{-}NH_2 \xrightarrow{ArN_2^+} RNH{-}N{=}N{-}Ar \xrightarrow{HX} R^+X^- + N_2 + ArNH_2$$

(d) The Bamford–Stevens reaction. This reaction consists of heating a toluene-*p*-sulphonylhydrazone with the sodium derivative of ethylene glycol in ethylene glycol[142]. Investigation of the reaction has shown it to consist of a unimolecular elimination from the anion of the sulphonylhydrazone, giving an aliphatic diazo compound[143].

$$R'R''C{=}N{-}NH{-}SO_2R \longrightarrow R'R''C{=}N{-}\bar{N}{-}SO_2R$$

$$\longrightarrow R'R''C{=}\overset{\delta+}{N}{\cdots}\overset{-}{N}{\cdots}\overset{\delta-}{SO_2R} \longrightarrow R'R''C{=}\overset{+}{N}{=}\bar{N} + {}^{-}SO_2R$$

In proton-rich solvents, the diazo compound forms the diazonium ion, and thence forms products of its decomposition; in the absence of a proton source, thermal decomposition of the diazo compound takes place[144].

The reaction can also be carried out photolytically[145], on either the salt or the free sulphonylhydrazone.

(e) Solvolysis of alkyl diazotates. Cleavage of *N*-alkyl-*N*-nitrosourethanes by the action of potassium *t*-butoxide yields the alkyl diazotate[146]:

$$R{-}N(NO){-}C(=O)OC_2H_5 + t\text{-}BuO^-K^+ \longrightarrow t\text{-}BuO{-}C(=O)OC_2H_5 + RN{=}NO^-K^+$$

In aqueous base, hydrolysis to the diazotic acid is almost instantaneous[147].

$$RN{=}NO^-K^+ + H_2O \rightleftharpoons RN{=}NOH + K^+OH^-$$

This then decomposes to yield typical carbonium ion products. In base, the leaving group is believed to be —N=NOH, yielding the carbonium ion directly, but in acid, reaction is probably via a diazonium ion.

B. Decomposition of Diazonium Ions

Theories of the mechanism of decomposition of diazonium ions are concerned with the variety of rearranged products which these reactions yield. The reaction was first postulated to take place via a carbonium ion, but the range of products was soon found to exceed that of the postulated ion generated by routes such as ester or halide heterolysis. To explain this, it was suggested[148] that the carbonium ion generated in deamination reactions was a high energy ion, sometimes described as a 'hot' ion, which had sufficient energy to cross barriers which blocked some reactions of the energetically normal species. The reason suggested for the greater energy of this species was the relatively low energy barrier (12–20 kJ/mol) to fission of the diazonium ion, relative to the energy of fission of the C—O bond of an ester (60–100 kJ/mol)[149].

Another theory was proposed by Huisgen[150], who suggested that in the decomposition of diazonium ion there is a compression of the energy profile relative to solvolysis reactions, leading to smaller differences in the energy of activation for several possible processes. The theory can, however, be regarded as replacing a 'high energy reactant' by a 'low energy reaction'.

The 'high energy ion' theory was challenged by Streitwieser[151], who pointed out that, on this hypothesis, diastereoisomeric cyclic amines should yield similar products, whereas different products were obtained on nitrous acid decomposition. He therefore proposed that the diazonium ion rather than the carbonium ion was the branching point for competing reactions. Reactions of the diazonium ion proceeded via a low energy route, since loss of nitrogen was assisted by the reagent.

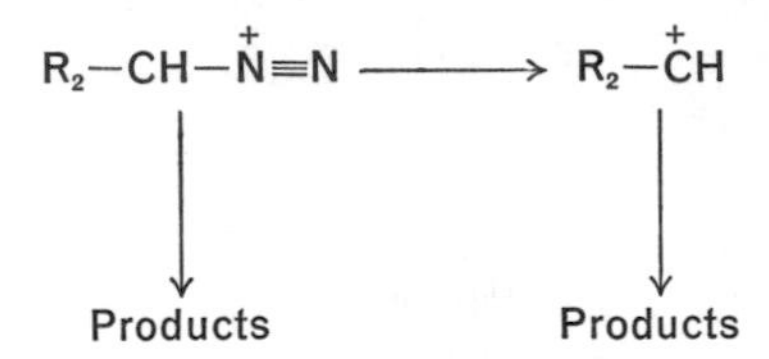

In support of this, he pointed out that the 1-butyl-1-*d*-acetate obtained from the reaction of **24** with nitrous acid showed 69% inversion of configuration[152].

$$CH_3CH_2CH_2C(NH_2)(H)(D)$$

(24)

An elegant test of this theory was devised by Whiting[141, 153]. He pointed out that the limiting cases of reaction via the diazonium ion and via the carbonium ion may be written as follows:

$$R{-}N{=}N{-}X \longrightarrow R{-}\overset{+}{N}{\equiv}N + X^-$$

$$R{-}N{=}N{-}X \longrightarrow R^+ + N_2 + X^-$$

In the first case, the diazonium ion is generated by removal of X^- well away from the centre of positive charge; if the diazonium ion has a finite lifetime, then complete separation of X^- will be probable, so that when the diazonium ion decomposes to the carbonium ion, the latter will not be associated with the counterion, X^-. In the other limiting case, where the diazonium ion represents a transition state rather than an intermediate, the carbonium ion and X^- will be generated simultaneously, and hence formation of the ion pair is likely, particularly in solvents of low ionizing power. Thus, if the diazonium ion is formed by a number of routes, each involving a different counterion, reactions in which the diazonium ion has a long life should yield the same product mixture in which counterion capture is absent, while reactions in which the diazonium ion has a short life should give product mixtures with varying amounts of products of capture of the different counterions.

Experimentally, Whiting found that reactions of 1-octylamine by different routes gave essentially similar products, suggesting that in this case the diazonium ion has a long lifetime, but reactions of 4-octylamine gave substantial yields of the products of ion pair collapse, suggesting that the diazonium ion had a lifetime which was short relative to the rate of molecular diffusion. This result is consistent with Streitwieser's theory, formation of a primary carbonium ion from a diazonium ion being slow while the more stable secondary carbonium ion is formed rapidly from its diazonium ion.

The interpretation of Whiting's results has recently been challenged by White[154]. He suggests that Whiting's picture is oversimplified since it does not allow for the exchange of counterion with the solvent, and that a full picture of the reaction of an amine, RNH_2, in solvent HY is

$$
\begin{array}{l}
R{-}NH_2 \longrightarrow R{-}\overset{+}{N}{\equiv}NX^- \longrightarrow R^+X^- \longrightarrow R^+Y^- \\
R{-}\overset{+}{N}{\equiv}NX^- \longrightarrow X{-}R \\
R{-}\overset{+}{N}{\equiv}NX^- \longrightarrow Y{-}R \\
R{-}\overset{+}{N}{\equiv}NX^- \longrightarrow R{-}\overset{+}{N}{\equiv}NY^- \\
R{-}\overset{+}{N}{\equiv}NY^- \longrightarrow R^+Y^- \\
R{-}\overset{+}{N}{\equiv}NY^- \longrightarrow Y{-}R
\end{array}
$$

White studied the decomposition of the methylnitrocarbamates of the primary, secondary and tertiary butyl alcohols, carrying out the reactions in ethanol in order to reduce counterion exchange. The results obtained were the reverse of those of Whiting; the product of capture of the carbonium ion by the counterion was 23% of the total in the primary case, 15% in the secondary and 2% in the tertiary.

A possible explanation of this discrepancy is that the initial assumption that a diazonium ion pair would separate relatively rapidly is incorrect. Formation of a diazonium ion pair, which can lose nitrogen to give a carbonium ion pair, provides a satisfactory explanation of White's results, and differing rates of exchange of the counterion with solvent would explain Whiting's results. To date, however, the difficulty of quantitative treatment of the complex mixtures of products involved in the reactions has prevented clear resolution of the problem.

C. External Stabilization of Carbonium Ions Formed from Diazonium Ions

1. Neutral species

Before the role of the counterion in diazonium ion decomposition is considered, it should be realized that the carbonium ion and counterion are not formed in close contact, but are separated by at least one neutral molecule. Decomposition of the diazonium ion produces a molecule of nitrogen, so that the carbonium ion and counterion are, in all cases, initially separated by this molecule. This species, described as an 'inert-gas-separated ion pair', is believed to be enclosed within a solvent cage, and to be in a state of considerable disorder[155, 156]. The mixture of species covered by the term 'inert-gas-separated ion pair' probably ranges from species in which the nitrogen atom separates the ion pair, so that the carbonium ion can react with solvent, to species in which the nitrogen molecule is off the line between the ions, which makes them more likely to react by collapse of the ion pair.

Even more complex is the situation during reaction of a primary amine with nitrous acid in acetic acid, where the species involved consists of an ion pair separated by a molecule of nitrogen and a molecule of water, presumably within a solvent cage.

$$
R{-}NH_2 \longrightarrow R{-}N{=}N{-}OH \xrightarrow{CH_3COOH} R{-}N{=}N{-}OH \cdots CH_3COOH \longrightarrow RN_2^+ \;\; CH_3COO^- \cdots H{-}OH \longrightarrow R^+ \;\; {}^-OCOCH_3 \;\; H_2O \;\; N_2
$$

Capture of R^+ by nitrogen is unlikely to be important[30], but reaction with either water or acetate is probable.

The importance of the water molecule in deamination in solvents such as anhydrous acetic acid has been neglected by many workers in the field. Although as long ago as 1963, White[157] pointed out that deamination in acetic acid of RNH_2 could yield ROH, RONO, $RONO_2$ and $ROCOCH_3$, many examples exist in which the experimenters have simply reduced the product mixture with lithium aluminium hydride and obtained only alcohols. Since each individual component of the mixture has its own mode of formation, this procedure destroys much of the evidence of the reaction, and can be misleading.

In the few cases in which alcohols have been isolated from the reaction of amines with nitrous acid in anhydrous acetic acid, they have been found to be formed with approximately 80% retention of configuration[157, 158]. This is consistent with the water molecule occupying a position close to that vacated by the leaving group, nitrogen. The association between the carbonium ion and the water molecule seems to be strong, since it can remain in place through a hydride shift and a Wagner–Meerwein shift[159].

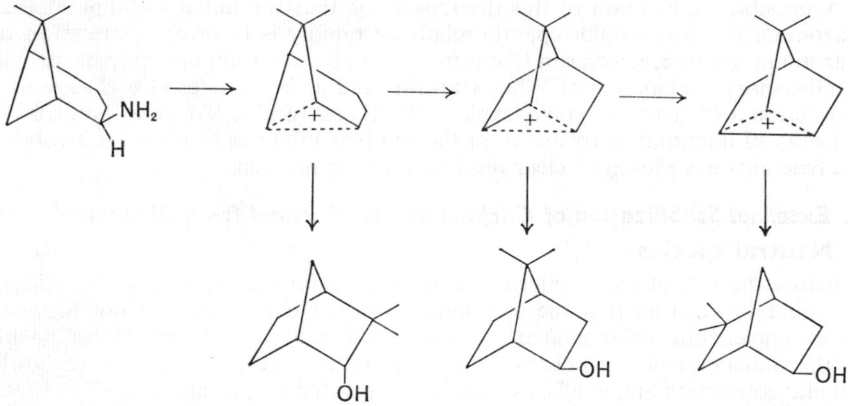

NH_2 H + + + OH OH OH

This type of reaction is detectable even in aqueous conditions. Deamination of cyclohexylamine in water, using nitrous acid labelled with ^{18}O, showed that approximately 10% of the oxygen in the product was derived from the nitrous acid[160].

2. Charged species

The first evidence of the importance of ion pairs in nitrous acid deamination comes from the work of Ott[161] in 1931. He showed that reaction of optically active α-phenylethylamine with nitrous acid in acetic acid gave an acetate which yielded an alcohol of partially retained configuration, while a similar reaction in water gave alcohol of partially inverted configuration. Retention of configuration must have resulted from ion pair collapse.

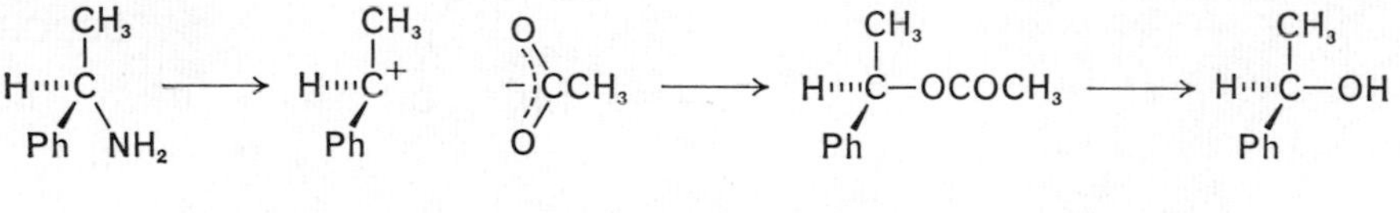

$$H\cdots C(CH_3)(Ph)NH_2 \longrightarrow H\cdots C^+(CH_3)(Ph)\ \ {}^-OC(O)CH_3 \longrightarrow H\cdots C(CH_3)(Ph){-}OCOCH_3 \longrightarrow H\cdots C(CH_3)(Ph){-}OH$$

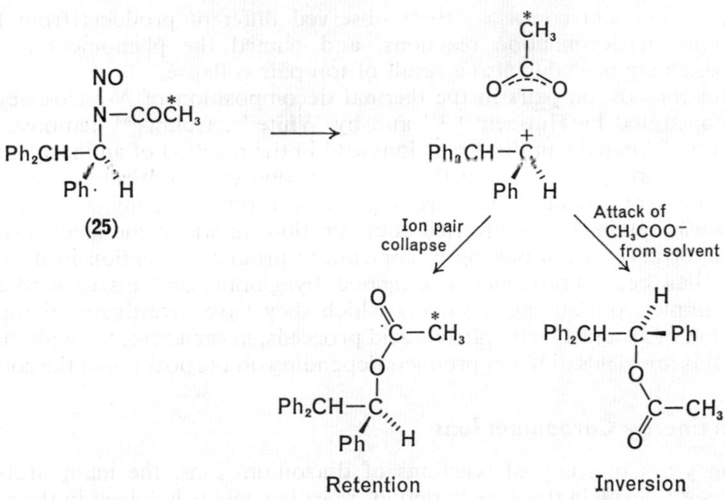

(25)

Ion pair collapse

Attack of CH_3COO^- from solvent

Retention

Inversion

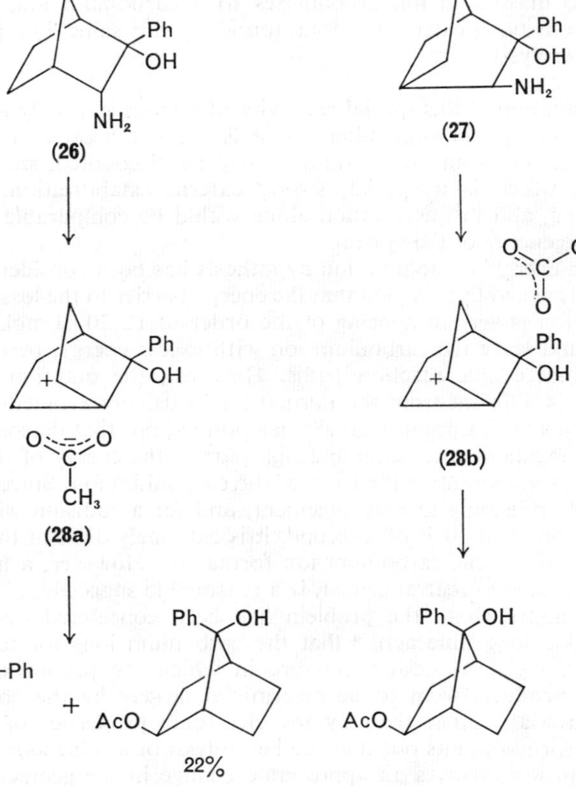

(26)

(27)

(28a)

(28b)

19%

22%

Later, Berson and coworkers[162, 163] observed different products from formally similar ions in deamination reactions, and named the phenomenon 'Memory effects'; these are probably also a result of ion pair collapse.

The presence of ion pairs in the thermal decomposition of *N*-nitrosoacylamines was demonstrated by Huisgen[158, 164] and by White[165]. Collins[166] demonstrated the importance of ion pairs in these reactions and in the reaction of amines with nitrous acid. Collins' group showed that the decomposition of an optically active, acetate-labelled substrate (**25**) gave a mixture of product of retained configuration, in which the label was retained, and inverted configuration, in which the label was lost.

The importance of ion pairing in controlling product formation in deamination reactions has been thoroughly investigated by Collins and his co-workers[133]. A single example from among the many which they have investigated[167] shows that the reaction of **26** and **27** with nitrous acid proceeds, in each case, through the ion **28**, but that this ion yields different products depending on the position of the counterion.

D. High Energy Carbonium Ions

During years of study of reactions of diazonium ions, the main problem has always been to explain the wide variety of rearrangements involved in the reactions. There remain only two basic theories:

(a) Rearrangements occur via the diazonium ion, not the carbonium ion.
(b) The diazonium ion decomposes to a carbonium ion, which has special reactivity, greater than that found in the same ion generated by ester heterolysis.

The explanation of this special reactivity of the ion is usually given by attributing to it a high energy, or suggesting that it is reactive because it is unsolvated. The view that the ion is unsolvated may clearly be discounted, since the effect of the counterion, which is a special, strong external stabilization, has clearly been demonstrated, and this association alone would be comparable to solvation in its effect on the energy of the system.

The 'high energy' carbonium ion hypothesis has been considered by a number of authors. It is generally accepted that the energy barrier to the loss of nitrogen from a diazonium ion is very low, being of the order of 12–20 kJ/mol, and that this low barrier should leave the carbonium ion with excess energy, permitting it to follow otherwise inaccessible reaction paths. However, the question of how the 'high energy' ion is different from the normal ion is usually avoided. In one of the few considerations of this point, Corey[168] has pointed out that diazonium ion decomposition is an exothermic reaction and that part of the energy of the reaction may be released as excess vibrational energy of the carbonium ion. Since the times required for a vibration leading to rearrangement, and for a collision with the solvent, are both of the order of 10^{-13} of a second, it is extremely difficult to predict the consequences of exothermic carbonium ion formation. However, a higher than normal incidence of internal rearrangement is a reasonable suggestion.

Another approach to the problem has been considered by Kirmse[169]. It was pointed out a long time ago[170] that the carbonium ions formed from diazonium ions normally give a product mixture in which the proportion of the different rearranged products seem to be determined largely by the conformation of the starting materials rather than by the electronic properties of the groups which rearrange. Kirmse points out that the heterolysis of a toluene-*p*-sulphonate to yield a carbonium ion involves an appreciable change in the geometry of the reactant.

Solvolysis, with its transition state 'late' on the reaction coordinate, favours rearrangements which achieve a minimum energy path. Decomposition of a diazonium ion, on the other hand, starts from a high energy species, and therefore passes its energy maximum early on the reaction coordinate and without significant distortion of nuclear positions. Consequently, deamination can produce cations which are bypassed in solvolysis, or which have a slightly different geometry from the formally similar ions produced in solvolysis reactions.

However, the distinction between displacement of the diazonium nitrogen by attack of an internal or external nucleophile (e.g. C—C bond electrons) and spontaneous loss of nitrogen to give an ion which reacts with a nucleophile with very little movement of nuclear positions is a very fine one.

The work of Whiting and of White demonstrates clearly the differences in mechanistic pathways between the reactions of primary, secondary and tertiary aliphatic primary amines. The rate of formation of a carbonium ion from a diazonium ion will depend on the energies of both species, and since the primary carbonium ion has yet to be demonstrated to take part in solvolysis reactions, it seems reasonable that it will be formed slowly, if at all, in diazonium ion decomposition.

On the basis of results obtained to date, it is impossible to exclude either reactions of diazonium ions or reactions of high energy carbonium ions as the source of the rearrangements peculiar to diazonium ion decomposition. It is not in fact necessary to exclude either, since rearrangement by nucleophilic attack or decomposition to a high energy carbonium ion may well be competing pathways, in which the outcome of the competition depends on the stabilities of the diazonium ion and the carbonium ion. The problem is certainly made more complicated by the fact that the reactive species involved are firmly attached to a counterion for most of their lives, and may have this attachment complicated by having to share a solvent cage with one or two neutral species. It is clear that the last word on the intermediates involved in diazonium ion decomposition has not yet been written.

E. Rearrangements Accompanying Diazonium Ion Decomposition

The wide variety of rearrangements which accompany diazonium ion decomposition can be divided into a number of main classes, which are considered below.

I. Semi-pinacol rearrangements

1,2-Aminoalcohols (semi-pinacols) undergo the pinacol rearrangement on treatment with nitrosating reagents. The reaction is similar to the acid-catalysed rearrangement of the 1,2-diol, and the ring opening of epoxides and hydrolysis of chlorohydrins.

Under similar conditions, the first three all gave similar pinacol-to-pinacolone ratios[171].

The product composition of the products of the semi-pinacol reaction does not necessarily reflect the ground state conformation of the starting material, as was demonstrated by the work of Collins and his coworkers[172–174], who showed that the migrating group can undergo a 1,2-shift with either retention or inversion of configuration at the migration terminus.

The carbonium ion generated from **29** must have sufficient lifetime to permit rotation of the central C—C bond of the molecule.

In the above reaction, only the *p*-tolyl group migrates, though since the molecule has sufficient lifetime to permit rotation around the central C—C bond, there is no stereochemical reason why the methyl group should not migrate. Even in a high energy system, relative migratory aptitudes are still of some importance though they would be expected to be reduced over the values measured in normal energy ions. Some attempts have been made to measure migratory aptitudes in the semi-pinacol reaction[175], and results suggest that they are probably lower than in the corresponding pinacol reactions, but the problems of measuring a true migratory aptitude, in which the result is in no way influenced by the ground state conformation, remain formidable[176] and the results obtained have probably only qualitative significance.

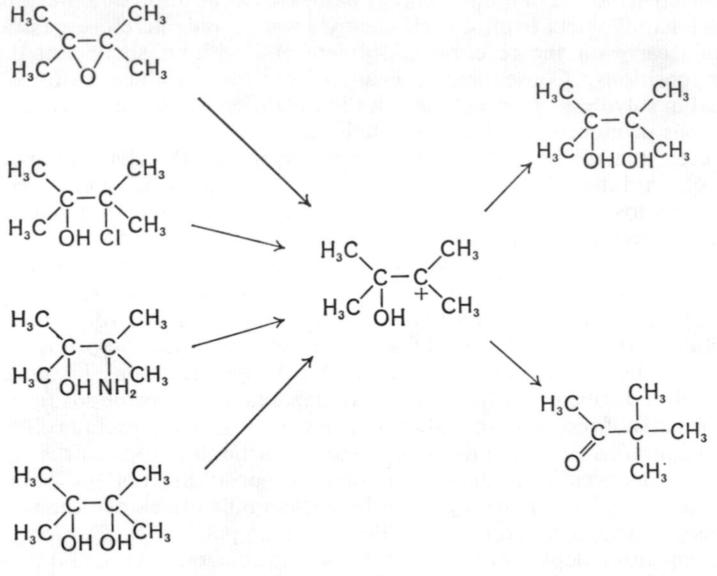

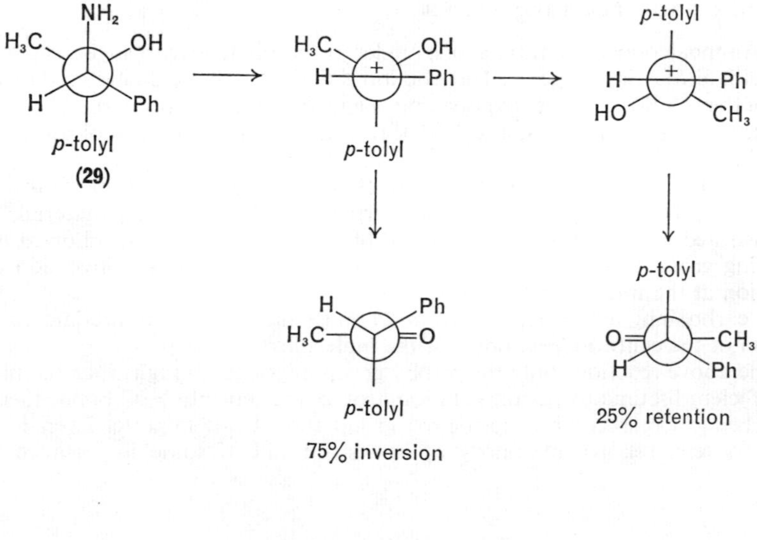

(29)

75% inversion

25% retention

When the reaction is carried out on a molecule of fixed conformation, it is found that, as expected, the product composition can be related to the ground state conformation[177].

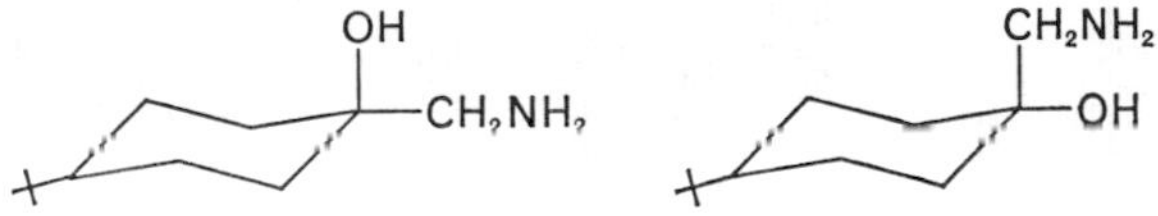

OH —CH₂NH₂ CH₂NH₂ —OH

Product composition

80%	*t*-Butylcycloheptanone	89%
1%	*t*-Butylcyclohexanone	5%
19%	Epoxide	6%

The small amount of C—C bond fission reported in the above reaction parallels a similar reaction observed in nitrosation of 1,3-amino alcohols[178]:

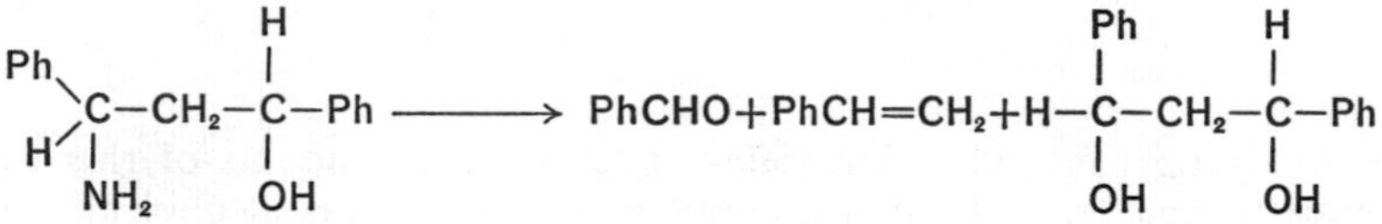

$$\text{Ph(H)C(NH}_2\text{)—CH}_2\text{—CH(OH)—Ph} \longrightarrow \text{PhCHO} + \text{PhCH=CH}_2 + \text{H—C(Ph)(OH)—CH}_2\text{—CH(OH)—Ph}$$

The mechanism of this reaction is presumably similar to that of fission of a 1,3-diol in acid proposed earlier[179].

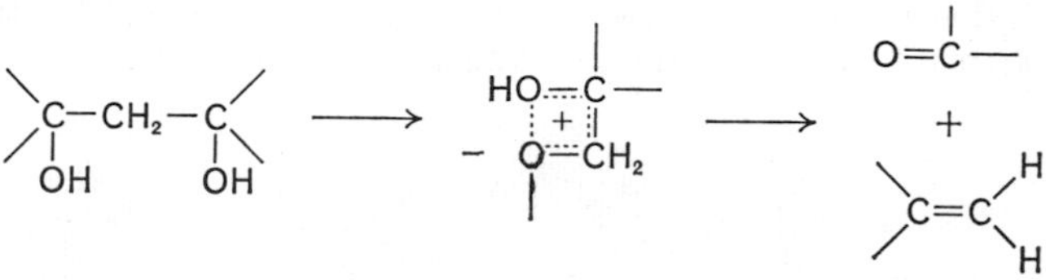

$$\text{>C(OH)—CH}_2\text{—C(OH)<} \longrightarrow \text{HO}\!\!=\!\!\text{C—/O}\!\!=\!\!\text{CH}_2\ (+) \longrightarrow \text{O=C<} + \text{>C=CH}_2$$

The higher energy of the reaction of the 1,2-amino alcohol presumably permits operation of a similar mechanism via an epoxide-like intermediate.

2. Ring expansion reactions

The driving force in a carbonium ion rearrangement is formation of a more stable ion from a less stable ion; this may be the result of release of ring strain or steric interactions but is more commonly electronic in origin, resulting from formation of a tertiary or secondary ion from a secondary or incipient primary ion[46], e.g.

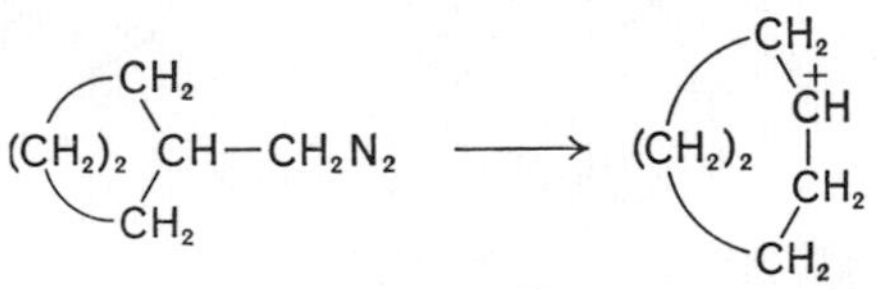

$$(\text{CH}_2)_2\text{(CH}_2)_2\text{CH—CH}_2\text{N}_2 \longrightarrow (\text{CH}_2)_2\text{—CH}_2\text{—}\overset{+}{\text{CH}}\text{—CH}_2\text{—CH}_2 \text{ (ring)}$$

In this case, the unstable primary ion is probably not even formed.

Ring expansion involves breaking a C—C bond, a process with a fairly high energy barrier. Consequently, a 1,2 hydride shift is often preferred. The hydride shift forms a more stable ion without breaking a C—C bond, so is energetically favoured even though it does not release any ring strain. For these reasons, a high energy carbonium ion reaction would be expected to favour ring expansion, which is in fact observed[180], e.g.

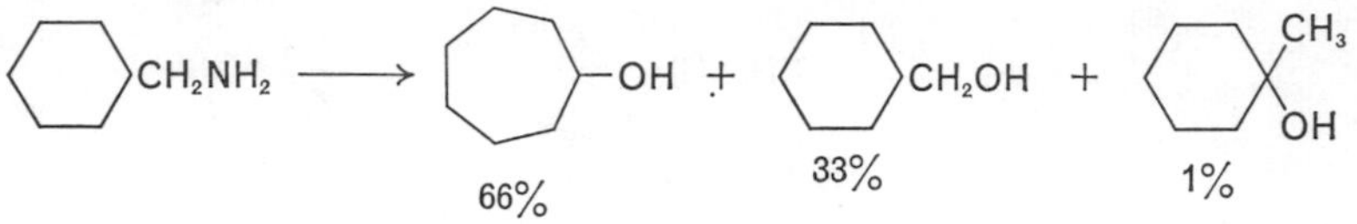

CH_2NH_2 ⟶ OH (66%) + CH_2OH (33%) + CH_3 / OH (1%)

An exception to the general rules of ring expansion reactions is provided by the cyclopropyl methyl system, which does not proceed cleanly to the cyclobutyl system[181].

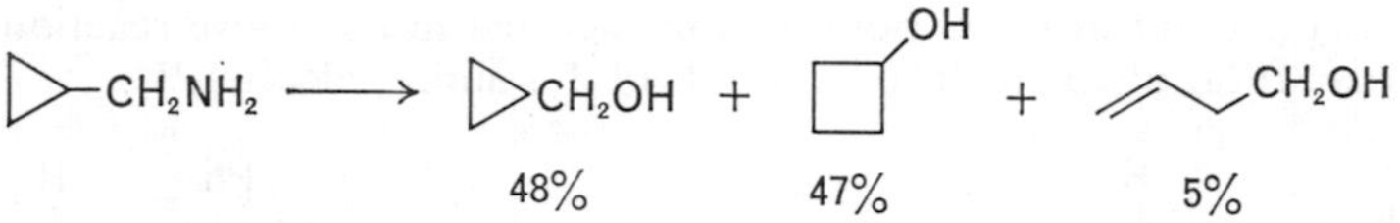

CH_2NH_2 ⟶ CH_2OH (48%) + OH (47%) + CH_2OH (5%)

The cyclopropyl methyl cation yields roughly equal amounts of ring expanded and unrearranged products. Any attempt to equilibrate the alcohol mixture leads to ring opening to give the thermodynamically stable allylic alcohol[182]. The reaction is, however, exceptional in that the carbonium ion centre of the cyclopropylmethyl cation draws some stabilization from the cyclopropane ring.

Ring expansion reactions of bicyclic systems are more favourable than those of monocyclic systems, since the release of ring strain is greater[162].

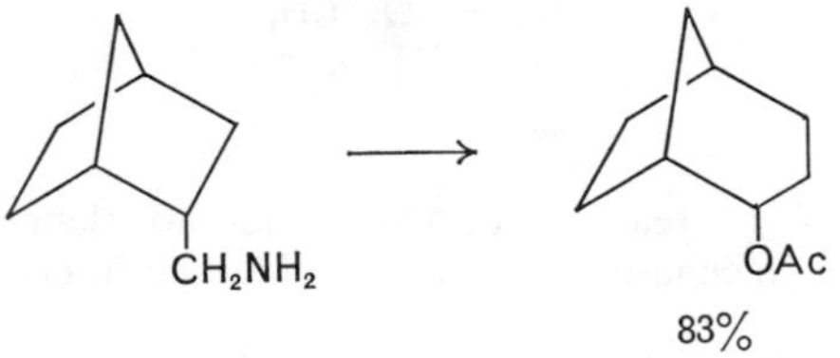

CH_2NH_2 ⟶ OAc (83%)

Surprisingly, however, the presence of neighbouring substituents inhibits the ring expansion[183, 184].

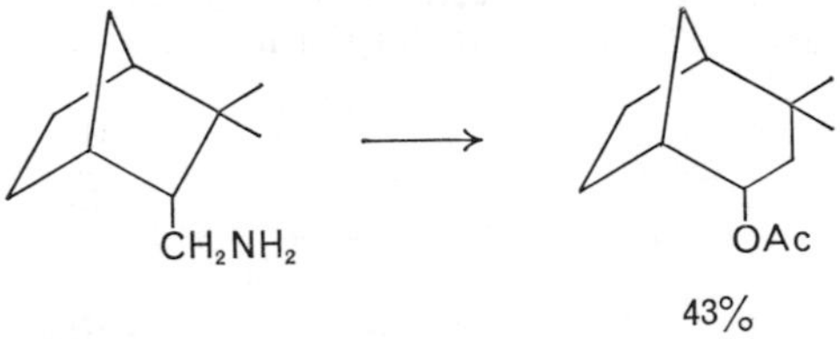

CH_2NH_2 ⟶ OAc (43%)

The reason for this inhibition is not known, though it has been observed in both solvolysis and amine nitrosation reactions.

The main alternate pathway to ring expansion in most systems is hydride shift, particularly in solvolysis reaction, where it can accompany departure of the leaving group[185]. In deamination it offers a lower energy pathway which should predominate, though to a lesser extent than in solvolysis. Replacement of hydrogen by a substituent should increase ring expansion[186], but opposed to this electronic effect is the conformational preference to put the more bulky substituent in a position *anti* to the amino group, hence favouring alkyl migration over ring expansion[187].

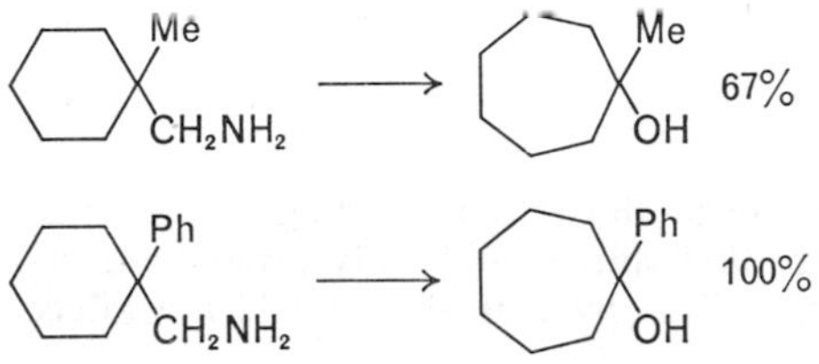

Me, CH_2NH_2 → Me, OH 67%

Ph, CH_2NH_2 → Ph, OH 100%

The ideal substituent to favour ring expansion should be small and yet readily able to stabilize a carbonium ion; hydroxyl is ideal. The semi-pinacol reaction is thus much used for ring expansion. 1,2-Amino alcohols are readily prepared from carbonyl compounds, the whole prodecure, which is known as the Demjanov–Tiffenau reaction, being outlined below[188].

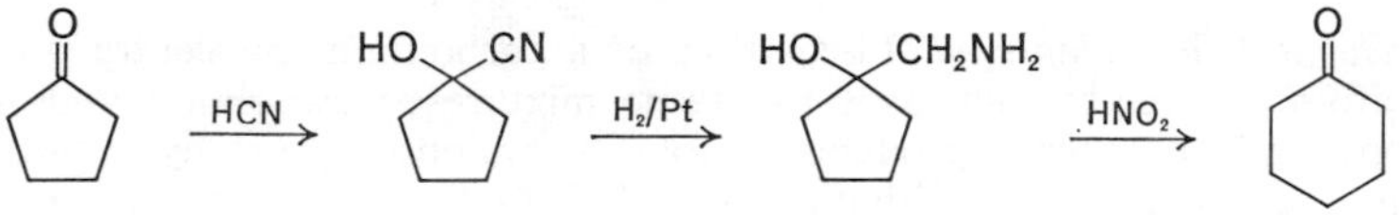

O —HCN→ HO, CN —H_2/Pt→ HO, CH_2NH_2 —HNO_2→ O

Asymmetrically substituted rings usually expand to give a mixture of products whose proportions depend on the conformation of the original ring[188].

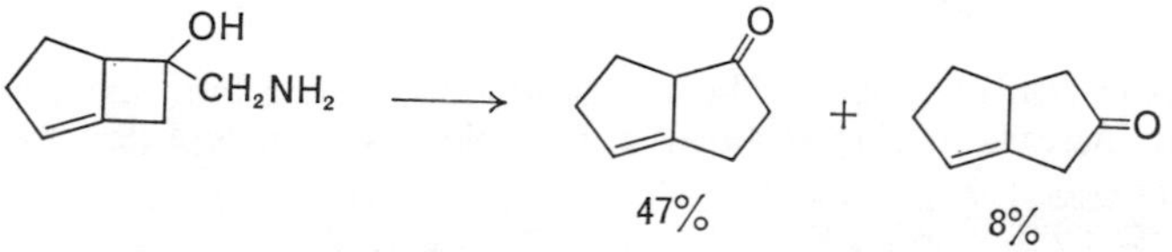

OH, CH_2NH_2 → O (47%) + =O (8%)

In view of the high energy which carbonium ions derived from diazonium ions are believed to have, it is not surprising that two of the three known examples of ring expansion by 1,3-alkyl shift in carbonium ions occur during amine nitrosation. Reaction of 2-cyclopropylethylamine **(30)** with nitrous acid gave 9% of cyclopentanol, 39% product of a hydride shift, and 52% of unrearranged product[189].

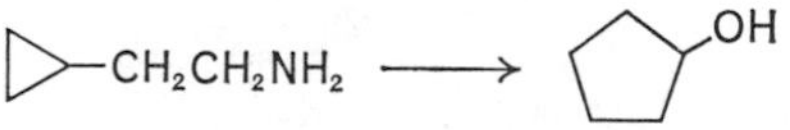

$-CH_2CH_2NH_2$ **(30)** → OH

A similar reaction, in which a 1,3-alkyl shift was observed despite competing 1,2-alkyl and 1,2-hydride shift pathways, was observed on reaction of *cis*-myrtanylamine **(31)** with nitrous acid[190].

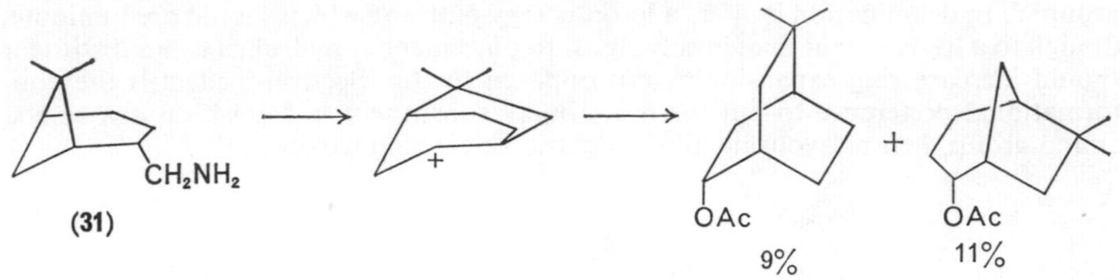

(31) CH_2NH_2 → + → OAc 9% + OAc 11%

In this case, there is substantial release of ring strain on expansion of the four-membered ring; a 1,2-alkyl shift would only expand the six-membered ring to a seven-membered ring, while a hydride shift would not affect the carbon skeleton.

3. Ring contraction reactions

Ring contraction reactions generally follow the same mechanistic pathways as ring expansion reactions and are driven by the same force, that is, formation of a more-stable from a less-stable carbonium ion. They are usually opposed by increase of steric and ring strain, but the electronic effects can overcome these. The whole field of ring contraction reactions has recently been reviewed[191].

a. *Cyclobutyl to cyclopropyl.* Formation of a carbonium ion centre on a ring carbon atom of cyclobutane gives the same mixture of cyclobutyl and methylcyclopropyl and ring-opened products as does the cyclopropylcarbinyl cation. Thus, treatment of cyclobutylamine with nitrous acid yields 48% of cyclopropylcarbinol[181].

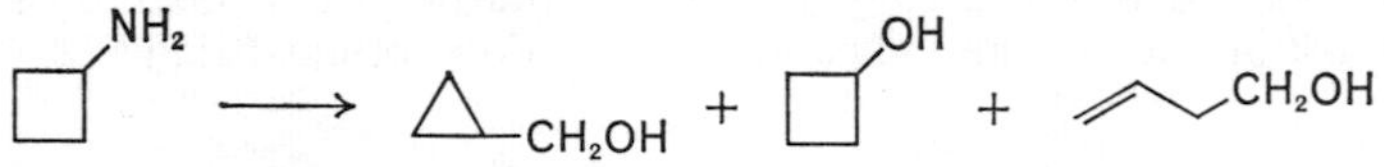

NH_2 → CH_2OH + OH + CH_2OH

b. *Cyclopentyl to cyclobutyl.* These reactions are uncommon as they involve a considerable increase in ring strain. An example has been reported by Collins[192]:

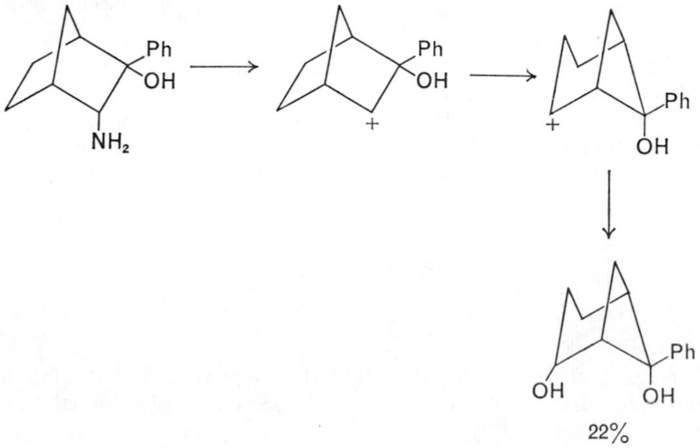

Ph OH NH_2 → Ph OH + → + Ph OH → OH Ph OH 22%

Similar results have been reported by Kirmse from studies of bornyl, fenchyl and related systems[193, 194]. These results may be summarized:

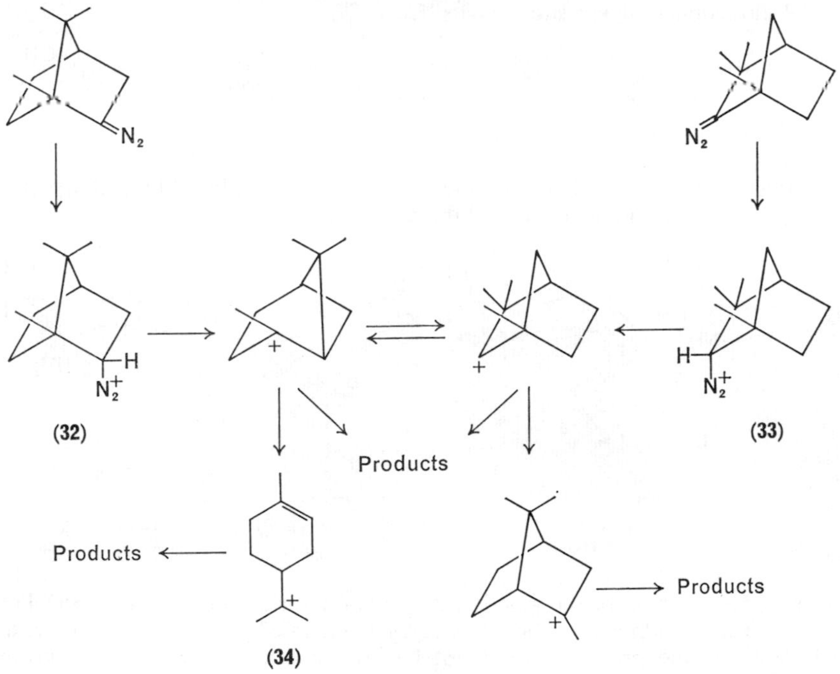

N_2 N_2

H N_2^+ + + H N_2^+

(32) (33)

Products

Products + Products

+

(34)

Kirmse obtained products of decomposition of **34** in 10% yield, showing that the bicyclo[2.2.1]heptane ring was contracted to a bicyclo[3.1.1]heptane ring. The diazo compounds used were generated by photolysis of camphor benzenesulphonylhydrazone and fenchone toluene-*p*-sulphonylhydrazone; their results should be compared with those of earlier workers, who generated formally similar intermediates by treatment of the appropriate amines with nitrous acid[195]. Reactions of **35** and **36** with nitrous acid gave the monocyclic product in yields of up to 39%; these reactions should clearly proceed via the species **32** and **33** postulated by Kirmse but the product mixtures obtained are different; also, reaction of diazocamphane with acetic acid, which should yield **32** directly[195], proceeded without ring opening. The reasons for these differences are not known; a possible explanation is that the reactions of Kirmse could involve a photo-excited diazo compound or diazonium ion, rather than a carbonium ion.

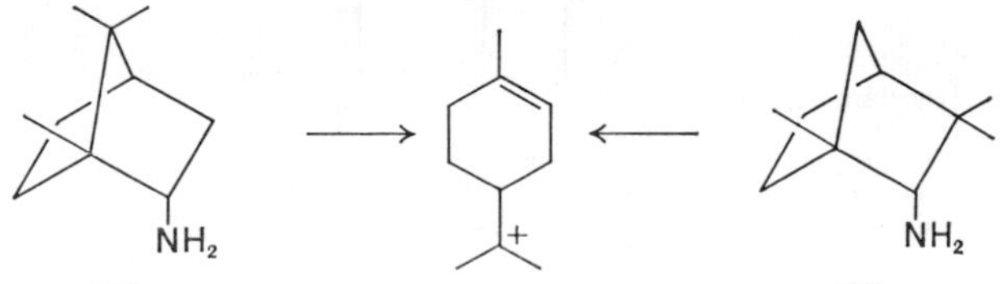

NH_2 + NH_2

(35) (36)

c. *Cyclohexyl to cyclopentyl.* Contraction of a cyclohexyl ring to a cyclopentane ring, though nominally less unfavourable energetically than the previous ring contraction, takes place only when the cyclopentyl carbonium ion is stabilized by powerful electron-supplying substituents[196].

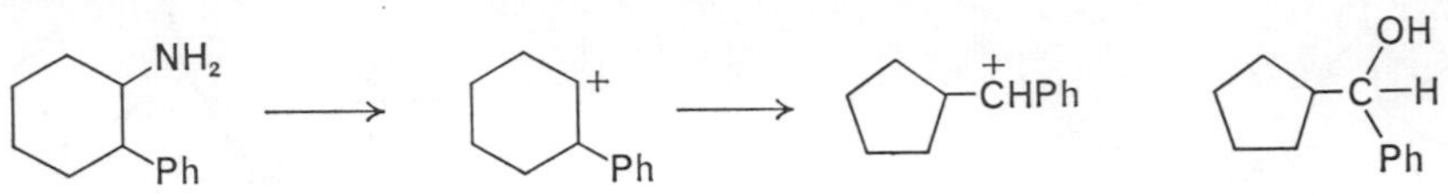

NH_2 Ph → + Ph → $\overset{+}{C}HPh$ OH C—H Ph

The stereochemistry of this reaction was subsequently explored by a study of the reactions of the 2-amino-4-*t*-butyl cyclohexanols[197].

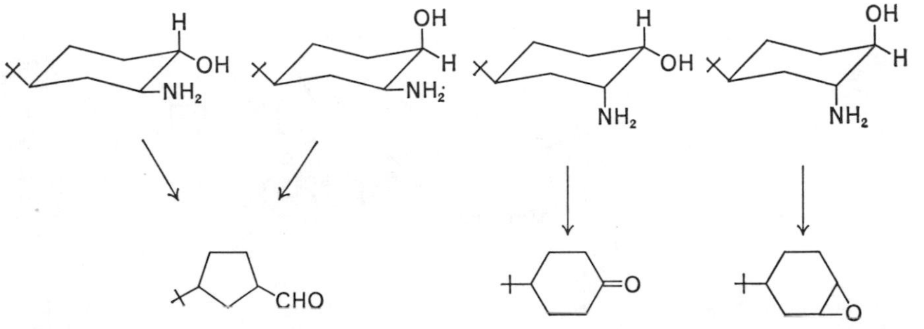

H OH NH_2 OH H NH_2 H OH NH_2 OH H NH_2

CHO =O O

Clearly, the outcome of the reaction depends on which group is anti-periplanar to the amino group. However, when the work was extended to cover conformationally mobile systems, the products could not be related quantitatively to the ground state conformations of the amines, but corresponded to the calculated ground state conformations of the diazo-hydroxide, suggesting that this intermediate may have sufficient lifetime to control the overall stereochemistry of the reaction.

4. Cyclization and ring opening

As might be expected, a high energy carbonium ion is more likely to undergo ring fission, and less likely to undergo ring closure, than is a normal ion in a similar system. Examples quoted earlier show significant amounts of ring opening, such as in the reaction of bornylamine with nitrous acid[195], whereas solvolysis of the corresponding chloride[198] does not show this reaction.

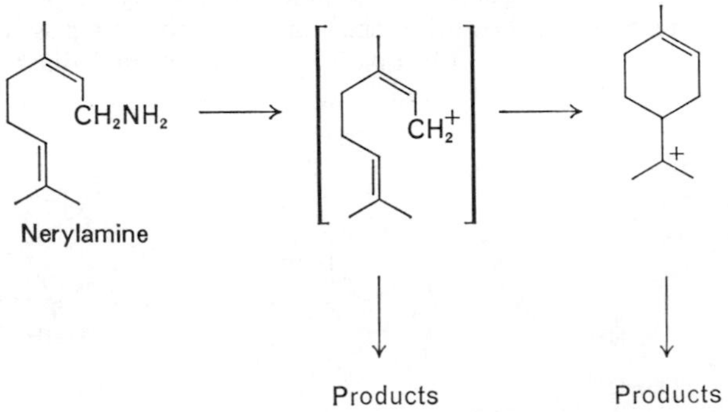

CH_2NH_2 → [CH_2^+] → +

Nerylamine

Products Products

An example of decreased cyclization in deamination reactions relative to solvolysis is provided by the cyclization of nerol derivatives[199]; where solvolysis of the chloride gives 77% cyclization, and reaction via the diazonium ion gives only 42% of cyclic products.

Cyclization is less important in deamination than in solvolysis because π-participation, which is important in formation of the transition state for solvolysis, is unnecessary when nitrogen is lost from the diazonium ion.

An exception to the general rules of cyclization and ring opening during reactions of diazonium ions is formation of cyclopropanes, by both intramolecular and intermolecular routes. The intramolecular route was at first suggested to be a carbene mechanism, since cyclopropanes are readily formed during carbene reactions, but an ingenious labelling experiment showed that the mechanism was cationic[200].

$$(H_3C)_2CH{-}CD_2NH_2 \xrightarrow{\text{Cationic route}} \text{cyclo-}[C(CH_3)(H){-}CH_2{-}CD_2]$$

$$(H_3C)_2CH{-}CD_2NH_2 \xrightarrow{\text{Carbene route}} (H_3C)_2CH{-}\ddot{C}D \longrightarrow \text{cyclo-}[C(CH_3)(H){-}CH_2{-}CHD]$$

Most of the product contained two deuterium atoms, showing that the reaction did involve the cationic route.

Experiments in which the solvent was varied showed that aprotic solvents favoured cyclopropane formation[201]. In an aprotic solvent, even the cyclopropylmethyl diazonium compound underwent cyclization to bicyclobutane[202].

$$\text{cyclo-}[H_2C{-}H_2C{-}C(H)]{-}CH_2NH_2 \longrightarrow C_4H_6 \text{ (bicyclobutane)}$$

Since these reactions are taking place in aprotic solvents, reaction with solvent no longer provides a pathway for decomposition of the ion, and elimination or cyclopropane formation are the most likely to take place. Elimination is the lower energy pathway, and is usually followed by ions generated solvolytically but diazonium ion decomposition is more likely to follow the high energy pathway.

An intermolecular acid-catalysed cyclopropane formation reaction has been reported by Closs[203].

$$ArCHN_2 + C{=}C \xrightarrow{HX} \text{cyclopropane (Ar, H on ring carbon)}$$

The reaction between aryl diazomethanes and *cis*-2-butene, using trifluoroacetic acid as catalyst, gave a 43% yield of cyclopropane. The authors have eliminated the obvious possibility of carbene formation, and suggest that the reaction involves the olefin and the diazonium ion.

$$ArCHN_2 + HX \xrightarrow{slow} ArC(=N_2)(H)\cdots H\cdots X$$

$$\xrightarrow{C=C} Ar{-}C(N_2)(H\cdots X){-}H \cdots (C{=}C) \longrightarrow \text{cyclopropane (Ar, H)}$$

It is notable that this reaction is not observed in the decomposition of the diazo compound derived from nerol. Possibly the greater stability of the arylmethyl diazonium ion gives it sufficient lifetime to permit attack of the π-bond electrons of the olefin.

5. Transannular interactions

A high energy ion would be well able to overcome the energy barriers to transannular reactions in cases where the conformation (and hence the entropy change) is favourable. Several examples of this type of reaction exist, a typical one being that below[204].

$$\text{OH, H, } CH_2NH_2 \longrightarrow \text{OH, H, } CH_3, + \longrightarrow \text{OH}^+, \text{H, } CH_3$$

6. Delocalized ions

The low energy barrier to the loss of nitrogen from a diazonium ion should preclude the need for any σ or π electron interaction with the leaving group. However, there is clear evidence that rearrangements of the type usually considered to proceed through delocalized ions under solvolytic reactions proceed in a similar manner when the carbonium ion is formed from a diazonium ion[205, 206].

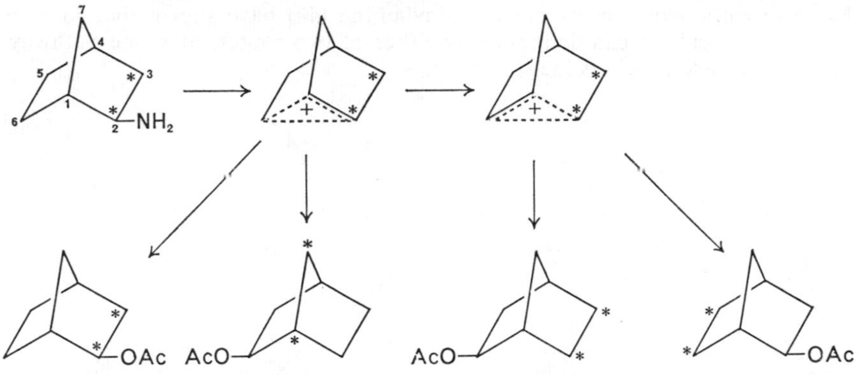

7 4 5 3 1 6 2 $-NH_2$

OAc AcO AcO* *OAc

Clearly, transfer of the label from C2 and C3 to C5 and C6 involves a rearrangement similar to that postulated to occur during solvolysis reactions. Similarly, deamination of the *syn-* and *anti*-7-norbornenylamines shows a stereospecificity inconsistent with undelocalized ions[207].

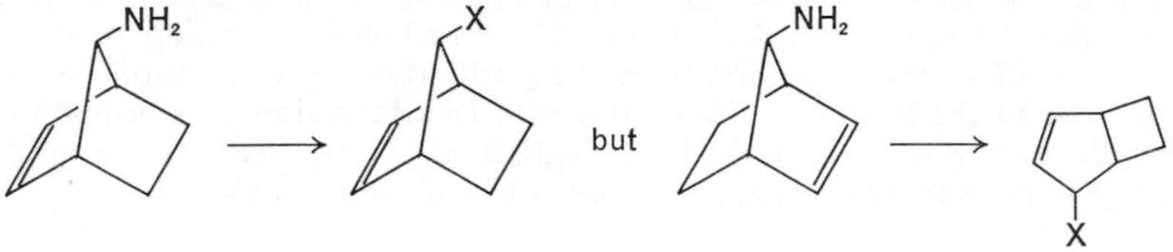

NH_2 X but NH_2 X

X = OH, 77%
X = OAc, 23%

X = OH, 24%
X = OAc, 72%

These and other results, suggesting the existence of delocalized ion in deamination reactions, were investigated by a detailed study of the reaction of *endo*-norbornylamine (**37**). Berson[208] found that reaction of the amine with nitrous acid gives *exo*-norbornyl acetate with 23% retention of optical activity, in contrast to acetolysis of the ester, which proceeds almost entirely through a symmetrical ion (**38**) to give a racemic product. Corey[168] found that the *exo* isomer also reacts with partial retention of optical activity, and concludes that the first step of the reaction is formation of a classical ion (**39**) which then delocalizes to the ion **38**.

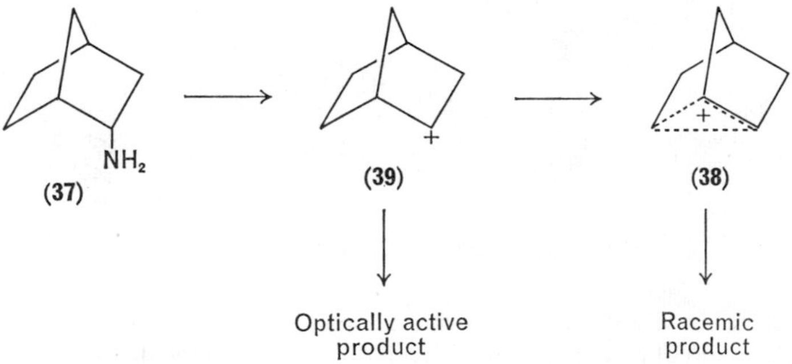

NH_2
(37)

(39)

Optically active product

(38)

Racemic product

More recently, experiments on α-nopinylamine (**40**) have shown that in a case where the classical ion can delocalize by either of two routes, only one pathway is followed when acetic acid is the solvent[159].

(40) → (43) → (41) → Bicyclic products

(43) → (42) → Monocyclic + Bicyclic products

Reaction of **40** with nitrous acid in acetic acid gives only products derived from **41**; when water is the solvent, a small amount of reaction through **42** is observed. It seems probable, then, that the first step of the reaction is formation of the undelocalized ion **43** as part of an ion pair, the position of the counterion being, as expected, close to the site of the leaving group. Delocalization of a σ-bond then displaces this counterion, giving only the delocalized ion **41**; in water, some separation of the ion pair prior to delocalization permits delocalization to yield **42**.

7. Allylic rearrangements

Formation of an allylic ion, like the formation of any other delocalized ion, is not favoured by formation of a carbonium ion from a diazonium ion. The reaction is again believed to consist of formation of an undelocalized ion, followed by π-electron interaction with the positive centre. Consequently, the reaction yields a mixture of products of the delocalized and undelocalized ions as in the case of geranylamine deamination[199].

	CH_2OH (geraniol)	CH_2OH (nerol)	OH (tertiary alcohol)	+ Cyclic products
Chloride hydrolysis	10%	—	81%	
Deamination in water	51%	15%	29%	
Deamination in HOAc+	59%	23%	14%	

(Products are alcohol + acetate mixtures)

Solvolysis of geranyl chloride yields the tertiary alcohol as the main product, as expected. Reaction of geranylamine with nitrous acid gives the unrearranged product in good yield, even though a small amount of the tertiary alcohol shows that delocalization can take place. The presence of water increases reaction at the tertiary centre so it seems reasonable to postulate that the first step of the reaction is formation of the undelocalized ion as part of an ion pair, separation of which is accompanied by delocalization.

Formation of an allylic ion by electron interaction following a C—C bond shift greatly favours the latter pathway[209].

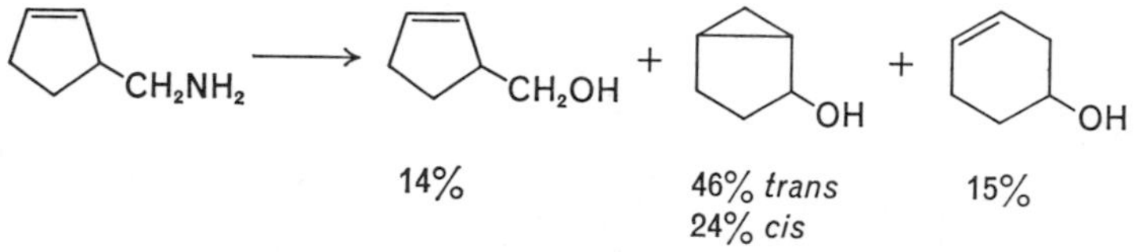

CH_2NH_2 ⟶ CH_2OH + OH + OH

14% 46% *trans* 24% *cis* 15%

In a suitable system formation of a homoallylic ion subsequent to loss of nitrogen has been observed[210].

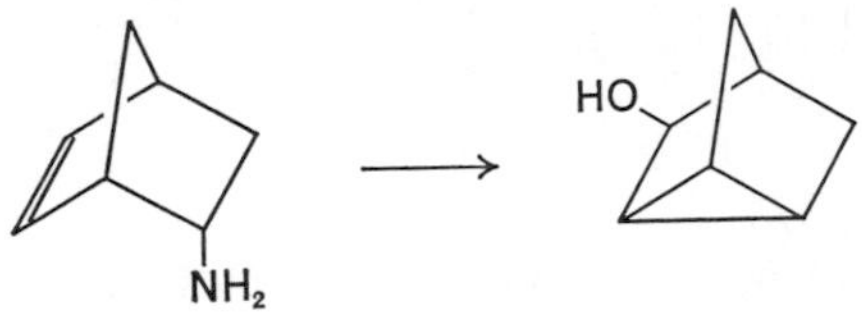

NH_2 ⟶ HO

However, in a more flexible system reaction proceeds without any interaction between the double bond and the cationic centre[211].

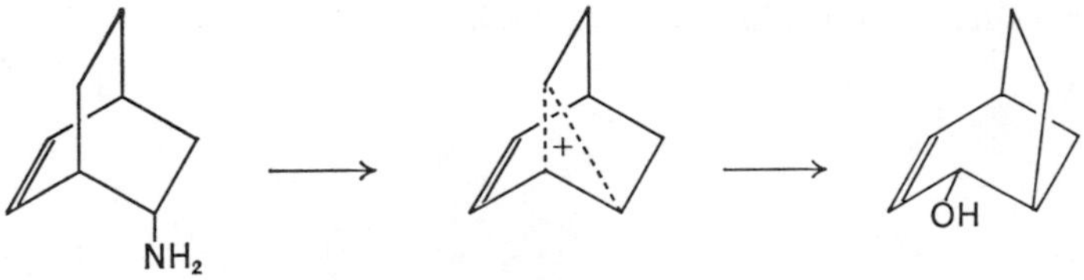

NH_2 ⟶ + ⟶ OH

F. Rearrangements of α-Ketodiazonium Ions

The presence of an α-keto group would be expected to increase the stability of a diazo group and reduce the stability of a carbonium ion, relative to the unsubstituted species. Consequently, the acid-catalysed decomposition of α-ketodiazo compounds is much slower than that of aliphatic diazo compounds and the reaction is amenable to kinetic study. A good review of the reaction has been published[15].

A kinetic study of a simple reaction, the acid-catalysed reaction of diazoacetic ester[212], showed that the reaction involves a pre-equilibrium proton transfer from acid to ester; in support[213] the solvent deuterium isotope effect is 2·9 at 25 °C. Displacement of nitrogen is then rate determining. Whether this is spontaneous or is assisted, by either an external nucleophile or neighbouring group participation, is still in dispute, though the bulk of the evidence favours participation[15]. Rearrangements during the acid-catalysed decomposition of α-diazoketones result, mainly

from participation of neighbouring group electrons, in loss of nitrogen as shown in the example below[214].

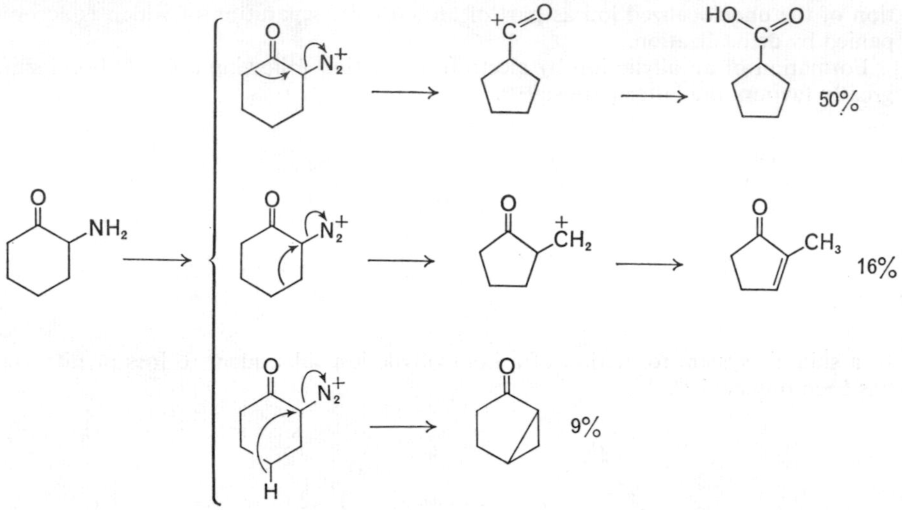

Extension of this work to bicyclic systems illustrates the importance of the direction of protonation on the stereochemistry of the overall reaction[215]. Protonation of 3-diazonorcamphor (44) is expected to be from the *exo* side to yield 45 which then decomposes to yield the delocalized ion 46. Products are then formed by attack of this ion at C3 or C4, or by loss of a proton. Retention of the stereochemistry at C3 in 47 demonstrates that 46 is probably delocalized and eliminates the possibility that 47 arises from nucleophilic attack on 45.

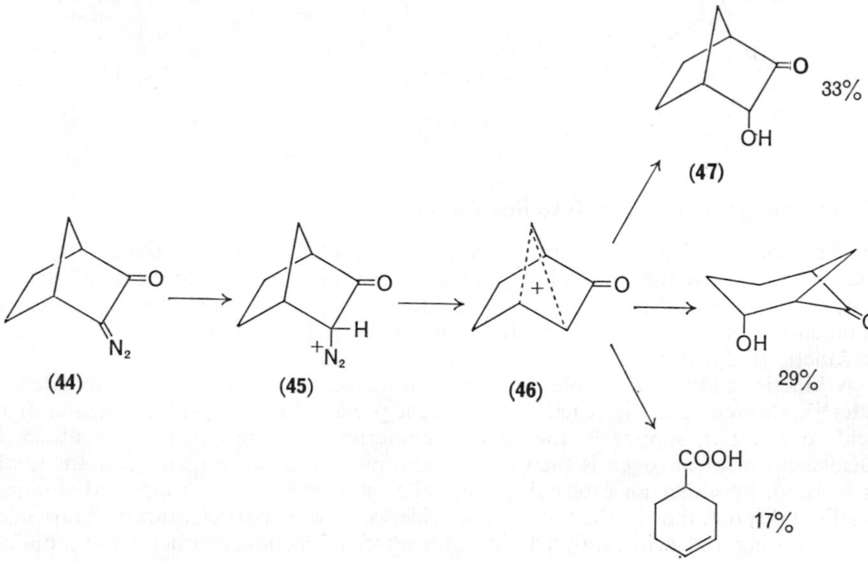

It should be noted that protonation is a reversible reaction, and that the stereochemistry of the reaction depends on which form is the more stable, rather than from which direction protonation most readily occurs. Thus, 3-diazocamphor (**48**) is most readily protonated from the *endo* side, but protonation from the *exo* side yields the more stable ion and the reaction product is a complex mixture of products of reaction of both **49** and **50**.

(**48**) (**49**) (**50**)

G. Rearrangements of Aromatic Diazonium Ions

Aromatic diazonium ions are much more stable than their aliphatic counterparts, but some evidence of the interconversion of *ortho*, *meta* and *para* isomers has been found during replacement of the diazonium group with halide ions[216]. Reaction of the *ortho* methyl, nitro, and trifluoromethyl derivatives of benzenediazonium salts with halide ions in pyridine/hydrogen fluoride showed that nucleophilic attack could occur at several sites.

	ortho-F	meta-F
X = CH_3	100%	0%
X = NO_2	0%	100%
X = CF_3	8%	91%

Meta and *para* isomers gave similar results; other halide ions reacted with these substrates to give mixtures of isomers. The results have been discussed in terms of the ^{13}C nuclear magnetic resonance spectra of the benzenediazonium ions, which show a spreading of the charge of the diazonium group around the ring which is consistent with the observed rearrangements[217].

These observations tend to support the theory that bimolecular attack of the nucleophile on the diazonium ion is the rate-determining step of the reaction[218].

VI. REFERENCES

1. C. W. Cowell and A. Ledwith, *Quart. Rev.*, **24**, 119 (1970).
2. R. Huisgen, R. Grashey and J. Sauer, in *The Chemistry of Alkenes* (Ed. S. Patai), Interscience, London, 1964, p. 739.
3. F. M. Dean, in *Some Recent Developments in the Chemistry of Natural Products* (Eds S. Rangaswami and N. V. Subba Rao), Prentice Hall of India, New Delhi, 1972, p. 1.
4. B. W. Peace and D. S. Wulfman, *Synthesis*, 137 (1973).
5. G. L. Closs, *Topics Stereochem.*, **3**, 193 (1968).
6. T. L. Gilchrist and C. W. Rees, *Carbenes, Nitrenes and Arynes*, Nelson, London, 1969.
7. D. Bethell, *Adv. Phys. Org. Chem.*, **7**, 153 (1969).

8. W. Kirmse, *Carbene Chemistry*, Academic Press, London and New York, 1971.
9. D. Bethell, in *Organic Reactive Intermediates* (Ed. S. P. McManus), Academic Press, New York, 1973, p. 61.
10. M. Jones Jr and R. A. Moss, in *Carbenes* (Eds M. Jones Jr and R. A. Moss), Vol. 1, Interscience, New York, 1973.
11. J. H. Ridd, *Quart. Rev.*, **15**, 418 (1961).
12. R. N. Butler, *Chem. Rev.*, **75**, 241 (1975).
13. R. H. White and D. J. Woodcock, in *The Chemistry of the Amino Group* (Ed. S. Patai), Interscience, London, 1968, p. 407.
14. D. V. Banthorpe, in *The Chemistry of the Amino Group* (Ed. S. Patai), Interscience, London, 1968, p. 585.
15. R. A. More O'Ferral, *Adv. Phys. Org. Chem.*, **5**, 331 (1967).
15a. L. Friedman, in *Carbonium Ions*, Vol. II (Eds G. A. Olah and P. von R. Schleyer), Wiley, New York, 1970.
16. B. T. Hart, *Aust. J. Chem.*, **26**, 461 (1973).
17. P. A. S. Smith, *Open Chain Nitrogen Compounds*, Vol. 1, Benjamin, New York, 1965.
18. S. R. Paulsen, *Angew. Chem.*, **72**, 781 (1960); E. Schmitz and R. Ohme, *Chem. Ber.*, **94**, 2166 (1961).
19. J. P. Anselme, *J. Chem. Educ.*, **43**, 596 (1966).
20. E. Müller, R. Beutler and B. Zeeh, *Ann. Chem.*, **719**, 72 (1968).
21. E. T. Blues, D. Bryce-Smith, J. G. Irwin and I. W. Lawston *J. Chem. Soc. Chem. Commun.*, 466 (1974).
22. M. J. Amrich and J. A. Bell, *J. Amer. Chem. Soc.*, **86**, 292 (1964).
23. C. B. Moore and G. C. Pimentel, *J. Chem. Phys.*, **41**, 3504 (1964).
24. G. S. Paulett and R. Ettinger, *J. Chem. Phys.*, **39**, 825, 3534 (1963); *J. Chem. Phys.*, **41**, 2557 (1964).
25. H. M. Frey, *Advan. Photochem.*, **4**, 225 (1966).
26. C. G. Overberger and J. P. Anselme, *J. Org. Chem.*, **29**, 1188 (1964).
27. C. G. Overberger and J. P. Anselme, *Tetrahedron Lett.*, 1405 (1963).
28. G. Lowe and J. Parker, *Chem. Commun.*, 1135 (1971).
29. C. G. Swain, J. E. Sheats and K. G. Harbison, *J. Amer. Chem. Soc.*, **97**, 783 (1975).
30. R. G. Bergstrom, G. H. Wahl Jr and H. Zollinger, *Tetrahedron Lett.*, 2975 (1974).
31. J. Brokken-Zijp and H. van den Bogaert, *Tetrahedron*, **29**, 4169 (1973).
32. O. Macháčková and V. Štěrba, *Coll. Czech. Chem. Commun.*, **37**, 3467 (1972).
33. J. Brokken-Zijp, *Tetrahedron Lett.*, 2673 (1974).
34. R. Huisgen, *Angew. Chem. Internat. Edn.*, **7**, 321 (1968).
35. R. Huisgen, *J. Org. Chem.*, **33**, 2291 (1968).
36. R. A. Firestone, *J. Org. Chem.*, **33**, 2285 (1968).
37. N. Filipescu and J. R. DeMember, *Tetrahedron*, **24**, 5181 (1968).
38. W. C. Howell, M. Ktenas and J. M. MacDonald, *Tetrahedron Lett.*, 1719 (1964).
39. D. H. R. Barton and J. B. Hendrickson, *J. Chem. Soc.*, 1208 (1956).
40. T. V. Van Auken and K. L. Rinehart, *J. Amer. Chem. Soc.*, **84**, 3736 (1962).
41. J. B. Bastus, *Tetrahedron Lett.*, 955 (1963).
42. M. Franck-Neumann and G. LeClerc, *Tetrahedron Lett.*, 1063 (1969).
43. G. L. Closs and W. A. Boll, *Angew. Chem. Internat. Edn.*, **2**, 399 (1963); *J. Amer. Chem. Soc.*, **85**, 3904 (1963).
44. A. C. Day and M. C. Whiting, *J. Chem. Soc.* (*C*), 1719 (1966); A. Ledwith and D. Parry, *J. Chem. Soc.* (*B*), 41 (1967).
45. A. C. Day, A. N. McDonald, B. F. Anderson, T. J. Bartczak and C. J. R. Hodder, *J. Chem. Soc. Chem. Commun.*, 247 (1973).
46. C. D. Gutsche and D. Redmore, *Carbocyclic Ring Expansion Reactions*, Suppl. I to *Advances in Alicyclic Chemistry* (Eds E. Hart and G. J. Karabatsos), Academic Press, New York, 1968, p. 81.
47. R. G. Carlson and N. S. Behn, *J. Org. Chem.*, **33**, 2069 (1968).
48. J. B. Jones and P. Price, *J. Chem. Soc. Chem. Commun.*, 1478 (1969).
49. N. J. Turro and R. B. Gagosian, *J. Chem. Soc. Chem. Commun.*, 949 (1969).
50. C. D. Gutsche and J. E. Bowers, *J. Org. Chem.*, **32**, 1203 (1967).
51. C. D. Gutsche and H. H. Peter, *J. Amer. Chem. Soc.*, **77**, 5971 (1955).

52. J. N. Bradley, G. W. Cowell and A. Ledwith, *J. Chem. Soc.*, 4334 (1964).
53. H. O. House, E. J. Grubbs and W. F. Gannon, *J. Amer. Chem. Soc.*, **82**, 4099 (1960).
54. E. Müller, H. Kessler and B. Zeeh, *Fortschr. Chem. Forsch.*, **7**, 128 (1966).
55. L. I. Smith and K. L. Howard, *J. Amer. Chem. Soc.*, **65**, 165 (1943).
56. W. S. Johnson, M. Neeman, S. P. Birkeland and N. A. Fedoruk, *J. Amer. Chem. Soc.*, **84**, 989 (1962).
57. N. J. Leonard and K. Jann, *J. Amer. Chem. Soc.*, **84**, 4806 (1962).
58. N. J. Leonard, K. Jann, J. V. Paukstelis and C. K. Steinhardt, *J. Org. Chem.*, **28**, 1499 (1963).
59. D. R. Crist and N. J. Leonard, *Angew. Chem. Internat. Edn.*, **8**, 962 (1969).
60. S.-I. Murahashi, K. Okumura, T. Kubota and I. Moritani, *Tetrahedron Lett.*, 4197 (1973).
61. W. Kirmse, H. D. von Scholz and H. Arold, *Ann. Chem.*, **711**, 22 (1968).
62. L. Friedman and H. Shechter, *J. Amer. Chem. Soc.*, **81**, 5512 (1959).
63. J. W. Powell and M. C. Whiting, *Tetrahedron*, **7**, 305 (1959); R. H. Shapiro, J. H. Duncan and J. C. Clopton, *J. Amer. Chem. Soc.*, **89**, 1442 (1967).
64. C. D. Gutsche, G. L. Bachmann and R. S. Coffey, *Tetrahedron*, **18**, 617 (1962).
65. L. Friedman and H. Shechter, *J. Amer. Chem. Soc.*, **83**, 3159 (1961).
66. W. E. Slack, C. G. Mosley, K. A. Gould and H. Shechter, *J. Amer. Chem. Soc.*, **96**, 7596 (1974).
67. J. W. Powell and M. C. Whiting, *Tetrahedron*, **12**, 168 (1961).
68. A. Nickon, F.-Chih Huang, R. Weglein, K. Matsuo and H. Yagi, *J. Amer. Chem. Soc.*, **96**, 5264 (1974).
69. H. Phillip and J. Keating, *Tetrahedron Lett.*, 523 (1961).
70. D. C. Richardson, M. E. Hendrick and M. Jones, *J. Amer. Chem. Soc.*, **93**, 3790 (1971).
71. P. G. Gassman and X. Creary, *Tetrahedron Lett.*, 4407 (1972).
72. J. W. Wilt and W. J. Wagner, *J. Org. Chem.*, **29**, 2788 (1964).
73. R. A. Moss and J. R. Whittle, *Chem. Commun.*, 341 (1969).
74. A. D. Wolf and M. Jones, *J. Amer. Chem. Soc.*, **95**, 8209 (1973).
75. M. Fărcasiu, D. Fărcasiu, R. T. Conlin, M. Jones and P. von R. Schleyer, *J. Amer. Chem. Soc.*, **95**, 8207 (1973).
76. D. Bethell, A. R. Newall, G. Stevens and D. Whittaker, *J. Chem. Soc. B*, 749 (1969); D. Bethell, A. R. Newall and D. Whittaker, *J. Chem. Soc. B*, 23 (1971).
77. M. L. Graziano, R. Scarpati and D. Tafuri, *Tetrahedron Lett.*, 2469 (1972).
78. Y. Hata and M. Watanabe, *Tetrahedron Lett.*, 3827, 4659 (1972); Y. Hata, M. Watanabe, S. Inoue and S. Oae, *J. Amer. Chem. Soc.*, **97**, 2553 (1975).
79. J. E. Baldwin and J. A. Walker, *J. Chem. Soc. Chem. Commun.*, 354 (1972).
80. P. S. Skell and R. C. Woodworth, *J. Amer. Chem. Soc.*, **78**, 4496 (1956).
81. W. R. Roth, *Ann. Chem.*, **671**, 10 (1964).
82. D. M. Lemal, F. Menger and G. W. Clark, *J. Amer. Chem. Soc.*, **85**, 2529 (1963).
83. D. M. Lemal and K. S. Shim, *Tetrahedron Lett.*, 3231 (1964).
84. A. Viola, S. Madhavan and R. J. Proberb, *J. Org. Chem.*, **39**, 3154 (1974).
85. B. M. Trost and R. M. Cory, *J. Amer. Chem. Soc.*, **93**, 5572 (1971); B. M. Trost, R. M. Cory, P. H. Scudder and H. B. Neubold, *J. Amer. Chem. Soc.*, **95**, 7813 (1973).
86. W. M. Jones and M. H. Grasley, *Tetrahedron Lett.*, 927 (1962); W. M. Jones and J. M. Walbrick, *J. Org. Chem.*, **34**, 2217 (1969).
87. W. M. Jones and J. W. Wilson Jr, *Tetrahedron Lett.*, 1587 (1965).
88. A. Guarino and A. P. Wolf, *Tetrahedron Lett.*, 655 (1969).
89. P. K. Freeman and D. G. Kuper, *J. Org. Chem.*, **30**, 1047 (1965).
90. S. S. Olin and R. M. Venable, *J. Chem. Soc. Chem. Commun.*, 273 (1974).
91. W. M. Jones, R. C. Joines, J. A. Myres, T. Mitsuhashi, K. E. Krajca, E. E. Waali, T. L. Davies and A. B. Turner, *J. Amer. Chem. Soc.*, **95**, 826 (1973).
92. E. E. Waali and W. M. Jones, *J. Amer. Chem. Soc.*, **95**, 8114 (1973).
93. T. Mitsuhashi and W. M. Jones, *J. Amer. Chem. Soc.*, **94**, 677 (1972).
94. T. T. Coburn and W. M. Jones, *J. Amer. Chem. Soc.*, **96**, 5218 (1974).
95. W. J. Barron, M. Jones and P. P. Gaspar, *J. Amer. Chem. Soc.*, **92**, 4739 (1970).
96. W. D. Crow and M. N. Paddon-Row, *Aust. J. Chem.*, **26**, 1705 (1973).
97. C. Wentrup, *Tetrahedron*, **30**, 1301 (1974).

98. J. A. Myers, R. C. Joines and W. M. Jones, *J. Amer. Chem. Soc.*, **92**, 4740 (1970).
99. T. T. Coburn and W. M. Jones, *Tetrahedron Lett.*, 3903 (1973).
100. K. E. Krajca, T. Mitsuhashi and W. M. Jones, *J. Amer. Chem. Soc.*, **94**, 3661 (1972).
100a. W. D. Crow and C. Wentrup, *Tetrahedron Letters*, 6149 (1968); C. Wentrup, *J. Chem. Soc. Chem. Commun.*, 1386 (1969).
100b. C. Wentrup, *Tetrahedron*, **30**, 1301 (1974).
100c. M. Regitz, A. Liedhegener, W. Anschütz and H. Eckes, *Chem. Ber.*, **104**, 2177 (1971).
100d. H. Eckes and M. Regitz, *Tetrahedron Letters*, 447 (1975).
100e. M. Regitz, *Angew. Chem. Int. Ed.*, **14**, 222 (1975).
101. L. Wolff, *Ann. Chem.*, **325**, 129 (1902); *Ann. Chem.*, **394**, 23 (1912).
102. W. Bradley and R. Robinson, *J. Chem. Soc.*, 1310 (1928).
103. O. Süs, *Ann. Chem.*, **556**, 65 (1944).
104. R. Casanova and T. Reichstein, *Helv. Chim. Acta*, **33**, 417 (1950).
105. P. Yates and R. J. Crawford, *J. Amer. Chem. Soc.*, **88**, 1562 (1966); P. Yates and J. Fugger, *Chem. Ind. (London)*, 1511 (1957).
106. G. Schroeter, *Ber. Deut. Chem. Ges.*, **42**, 2336 (1909).
107. I. G. Csizmadia, J. Font and O. P. Strausz, *J. Amer. Chem. Soc.*, **90**, 7360 (1968).
108. M. Jones Jr and W. Ando, *J. Amer. Chem. Soc.*, **90**, 2200 (1968).
109. F. Kaplan and G. K. Meloy, *J. Amer. Chem. Soc.*, **88**, 950 (1966).
110. Y. Yukawa, Y. Tsuno and T. Ibata, *Bull. Chem. Soc. Jap.*, **40**, 2613, 2618 (1967); W. Bartz and M. Regitz, *Chem. Ber.*, **103**, 1463 (1970).
111. Y. Yukawa and T. Ibata, *Bull. Chem. Soc. Jap.*, **42**, 805 (1969).
112. B. M. Trost, *J. Amer. Chem. Soc.*, **88**, 1587 (1966); *J. Amer. Chem. Soc.*, **89**, 138 (1967).
113. J. Fenwick, G. Frater, K. Ogi and O. P. Strausz, *J. Amer. Chem. Soc.*, **95**, 124 (1973).
114. S. A. Matlin and P. G. Sammes, *J. Chem. Soc. Perkin I*, 2623 (1972).
115. P. W. Concannon and J. Ciabattoni, *J. Amer. Chem. Soc.*, **95**, 3284 (1973).
116. A. C. Hopkinson, *J. Chem. Soc. Perkin II*, 794 (1973).
117. M. J. S. Dewar and C. A. Ramsden, *J. Chem. Soc. Chem. Commun.*, 688 (1973).
118. Z. Majerski and C. S. Redvanly, *J. Chem. Soc. Chem. Commun.*, 694 (1972).
119. V. Franzen, *Ann. Chem.*, **614**, 31 (1958).
120. H. Chaimovich, R. J. Vaughan and F. H. Westheimer, *J. Amer. Chem. Soc.*, **90**, 4088 (1968).
121. K.-P. Zeller, H. Meier and E. Müller, *Tetrahedron*, **28**, 5831 (1972).
122. S. S. Hixon and S. H. Hixon, *J. Org. Chem.*, **37**, 1279 (1972).
123. T. Gibson and W. F. Erman, *J. Org. Chem.*, **30**, 3028 (1966).
124. W. Ried and H. Lohwasser, *Ann. Chem.*, **683**, 118 (1965).
125. G. L. Closs and R. A. Moss, *J. Amer. Chem. Soc.*, **86**, 4042 (1964).
126. E. Müller, H. Kessler and B. Zeeh, *Fortschr. Chem. Forsch.*, **7**, 128 (1966).
127. Chi-Tang Ho, R. T. Conline and P. P. Gaspar, *J. Amer. Chem. Soc.*, **96**, 8109 (1974).
128. R. G. Salomon, M. F. Salomon and T. R. Heyne, *J. Org. Chem.*, **40**, 756 (1975).
129. E. Wenkert, B. L. Mylari and L. L. Davies, *J. Amer. Chem. Soc.*, **90**, 3870 (1968).
130. P. Yates and S. Danishefsky, *J. Amer. Chem. Soc.*, **84**, 879 (1962).
131. W. von E. Doering, E. T. Fossel and R. L. Kaye, *Tetrahedron*, **21**, 25 (1965).
132. J. Font, J. Valls and F. Serratosa, *Tetrahedron*, **30**, 455 (1974).
133. G. J. Collins, *Accounts Chemical Research*, **4**, 315 (1971).
134. E. Mueller and H. Haiss, *Chem. Ber.*, **96**, 570 (1963).
135. R. Huisgen and H. Reimlinger, *Ann. Chem.*, **599**, 161, 183 (1956).
136. E. H. White and C. A. Aufdermarsh, *J. Amer. Chem. Soc.*, **83**, 1174, 1179 (1961); E. H. White, *J. Amer. Chem. Soc.*, **77**, 6011 (1955).
137. E. H. White and L. A. Dolak, *J. Amer. Chem. Soc.*, **88**, 3790 (1966); E. H. White, M. C. Chen and L. A. Dolak, *J. Org. Chem.*, **31**, 3038 (1966).
138. E. H. White and D. W. Grisley Jr, *J. Amer. Chem. Soc.*, **83**, 1191 (1961).
139. E. H. White and H. Scherrer, *Tetrahedron Lett.*, 758 (1961).
140. E. H. White and M. Schroeder, *Abstr. Pap. Amer. Chem. Soc.*, 149th Meeting, Detroit, 37P.
141. H. Maskill, R. M. Southam and M. C. Whiting, *Chem. Commun.*, 496 (1965).
142. W. H. Bamford and T. S. Stevens, *J. Chem. Soc.*, 4735 (1952).
143. J. W. Powell and M. C. Whiting, *Tetrahedron*, **7**, 305 (1959).

144. L. Friedman and H. Shechter, *J. Amer. Chem. Soc.*, **81**, 5512 (1959).
145. W. G. Dauben and F. G. Willey, *J. Amer. Chem. Soc.*, **84**, 1497 (1962).
146. R. A. Moss, *J. Org. Chem.*, **31**, 1082 (1966).
147. R. A. Moss and S. M. Lane, *J. Amer. Chem. Soc.*, **89**, 5655 (1967).
148. D. J. Cram and J. E. McCarty, *J. Amer. Chem. Soc.*, **79**, 2866 (1957).
149. D. Semenow, C. H. Shih and W. G. Young, *J. Amer. Chem. Soc.*, **80**, 5472 (1958).
150. R. Huisgen and Ch. Ruchardt, *Ann. Chem.*, **601**, 1 (1956).
151. A. Streitwieser Jr, *J. Org. Chem.*, **22**, 861 (1957).
152. A. Streitwieser Jr and W. D. Schaeffer, *J. Amer. Chem. Soc.*, **79**, 2888 (1957).
153. M. C. Whiting, *Chem. Brit.*, **2**, 482 (1966).
154. E. H. White and K. W. Field, *J. Amer. Chem. Soc.*, **97**, 2148 (1975).
155. E. H. White, H. P. Tiwari and M. J. Todd, *J. Amer. Chem. Soc.*, **90**, 4734 (1968).
156. E. H. White, R. H. McGirk, C. A. Aufdermarsh Jr, H. P. Tiwari and M. J. Todd, *J. Amer. Chem. Soc.*, **95**, 8107 (1973).
157. E. H. White and J. E. Stuber, *J. Amer. Chem. Soc.*, **85**, 2168 (1963).
158. R. Huisgen and Ch. Ruchardt, *Ann. Chem.*, **601**, 21 (1956).
159. H. Indyk and D. Whittaker, *J. Chem. Soc. Perkin II*, 646 (1974).
160. D. L. Boutle and C. A. Bunton, *J. Chem. Soc.*, 761 (1961).
161. E. Ott, *Ann. Chem.*, **488**, 186 (1931).
162. J. A. Berson and P. Reynolds-Warnhoff, *J. Amer. Chem. Soc.*, **84**, 682 (1962); *J. Amer. Chem. Soc.*, **86**, 595 (1964).
163. J. A. Berson and D. Willner, *J. Amer. Chem. Soc.*, **84**, 675 (1962); *J. Amer. Chem. Soc.*, **86**, 609 (1964).
164. R. Huisgen and H. Reimlinger, *Ann. Chem.*, **599**, 161, 183 (1956).
165. E. H. White, *J. Amer. Chem. Soc.*, **77**, 6011, 6014 (1955); E. H. White and C. A. Aufdermarsh Jr, *J. Amer. Chem. Soc.*, **80**, 2597 (1958).
166. C. J. Collins, W. A. Bonner and C. T. Lester, *J. Amer. Chem. Soc.*, **81**, 466 (1959); C. J. Collins and J. B. Christie, *J. Amer. Chem. Soc.*, **82**, 1255 (1960); C. J. Collins, J. B. Christie and V. F. Raaen, *J. Amer. Chem. Soc.*, **83**, 4267 (1961).
167. C. J. Collins, V. F. Raaen and M. D. Eckart, *J. Amer. Chem. Soc.*, **92**, 1787 (1970); C. J. Collins, B. M. Benjamin, V. F. Raaen, I. T. Glover and M. D. Eckart, *Ann. Chem.*, 739, 7 (1970); C. J. Collins, I. T. Glover, M. D. Eckart, V. F. Raaen, B. M. Benjamin and B. S. Benjaminov, *J. Amer. Chem. Soc.*, **94**, 899 (1972).
168. E. J. Corey, J. Casanova Jr, P. A. Vatakencherry and R. Winter, *J. Amer. Chem. Soc.*, **85**, 169 (1963).
169. W. Kirmse and G. Voigt, *J. Amer. Chem. Soc.*, **96**, 7598 (1974).
170. J. G. Burr and L. S. Ciereszko, *J. Amer. Chem. Soc.*, **74**, 5426, 5431 (1952).
171. Y. Pocker, *Chem. and Ind.* (*London*), 332 (1959).
172. B. M. Benjamin, P. Wilder Jr and C. J. Collins, *J. Amer. Chem. Soc.*, **83**, 3654 (1961).
173. B. M. Benjamin and C. J. Collins, *J. Amer. Chem. Soc.*, **83**, 3662 (1961).
174. C. J. Collins, M. M. Staum and B. M. Benjamin, *J. Org. Chem.*, **27**, 3525 (1962).
175. D. Y. Curtin and M. C. Crew, *J. Amer. Chem. Soc.*, **77**, 354 (1955).
176. D. Bethell and V. Gold, *Carbonium Ions*, Academic Press, London, 1967, p. 22.
177. H. Farre and D. Gravel, *Canad. J. Chem.*, **41**, 1452 (1963).
178. J. English Jr and A. D. Bliss, *J. Amer. Chem. Soc.*, **78**, 4057 (1956).
179. H. E. Zimmerman and J. English Jr, *J. Amer. Chem. Soc.*, **76**, 2285, 2291, 2294 (1954).
180. P. A. S. Smith and D. R. Baer, *J. Amer. Chem. Soc.*, **74**, 6135 (1952).
181. J. D. Roberts and R. H. Mazur, *J. Amer. Chem. Soc.*, **73**, 2509 (1951).
182. R. A. Daby and R. E. Lutz, *J. Org. Chem.*, **22**, 1353 (1957).
183. P. I. Meikle and D. Whittaker, *J. Chem. Soc. Perkin II*, 322 (1974).
184. T. Gibson, *J. Org. Chem.*, **37**, 700 (1972).
185. P. I. Meikle, J. R. Salmon and D. Whittaker, *J. Chem. Soc. Perkin II*, **23**, (1973).
186. J. Diamond, W. F. Bruce and F. T. Tyson, *J. Org. Chem.*, **30**, 1840 (1965).
187. R. Kotani, *J. Org. Chem.*, **30**, 350 (1965).
188. P. A. S. Smith and D. R. Baer, *Org. Reactions*, **11**, 157 (1960).
189. G. E. Cartier and S. C. Bunce, *J. Amer. Chem. Soc.*, **85**, 932 (1963).
190. P. I. Meikle and D. Whittaker, *J. Chem. Soc. Perkin II*, 318 (1974).

191. D. Redmore and C. D. Gutsche, *Carbocyclic Ring Contraction Reactions* in *Advances in Alicyclic Chemistry*, Vol. 3 (Eds H. Hart and G. J. Karabatsos), Academic Press, London, 1973, p. 1.
192. C. J. Collins, V. F. Raaen, B. M. Benjamin and I. T. Glover, *J. Amer. Chem. Soc.*, **89**, 3940 (1967).
193. W. Kirmse and G. Arend, *Chem. Ber.*, **105**, 2738, 2746 (1972).
194. W. Kirmse and R. Siegfried, *Chem. Ber.*, **105**, 2754 (1972).
195. D. V. Banthorpe, D. G. Morris and C. A. Bunton, *J. Chem. Soc.*, (*B*), 687 (1971); W. Huckel and H. J. Kern, *Ann. Chem.*, **728**, 49 (1969).
196. D. V. Nightingale, J. D. Kerr, J. A. Gallacher and M. Maienthal, *J. Org. Chem.*, **17**, 1017 (1952).
197. M. Chérest, H. Felkin, J. Sicher, F. Šipoš and M. Tichý, *J. Chem. Soc.*, 2513 (1965).
198. P. Beltrame, C. A. Bunton, A. Dunlop and D. Whittaker, *J. Chem. Soc.*, 658 (1964).
199. C. A. Bunton, D. L. Hachey and J. P. Leresche, *J. Org. Chem.*, **37**, 4036 (1972).
200. A. T. Jurewicz and L. Friedman, *J. Amer. Chem. Soc.*, **89**, 149 (1967).
201. J. H. Bayless, F. D. Mendicino and L. Friedman, *J. Amer. Chem. Soc.*, **78**, 5790 (1965).
202. J. H. Bayless, L. Friedman, J. A. Smith, B. C. Cook and H. Shechter, *J. Amer. Chem. Soc.*, **87**, 661 (1965).
203. G. L. Closs and S. H. Goh, *J. Org. Chem.*, **39**, 1717 (1974).
204. T. A. Wnuk, J. A. Tonnis, M. J. Dolan, S. J. Padegimas and P. Kovacic, *J. Org. Chem.*, **40**, 444 (1975).
205. J. D. Roberts, C. C. Lee and W. H. Saunders, *J. Amer. Chem. Soc.*, **76**, 4501 (1954).
206. J. D. Roberts, C. C. Lee and W. H. Saunders, *J. Amer. Chem. Soc.*, **77**, 3034 (1955).
207. H. Tanida, T. Tsuji and T. Irie, *J. Org. Chem.*, **31**, 3941 (1966).
208. J. A. Berson and D. A. Ben-Efraim, *J. Amer. Chem. Soc.*, **81**, 4094 (1959).
209. M. Hanack and H. J. Schneider, *Tetrahedron*, **20**, 1863 (1964).
210. W. Parham, W. T. Hunter and R. Hanson, *J. Amer. Chem. Soc.*, **73**, 5068 (1951).
211. W. C. Wildman and D. R. Saunders, *J. Amer. Chem. Soc.*, **76**, 946 (1954).
212. J. D. Roberts, C. M. Regan and I. Allen, *J. Amer. Chem. Soc.*, **74**, 3679 (1952).
213. P. Gross, H. Steiner and F. Krauss, *Trans. Faraday Soc.*, **32**, 877 (1936); *Trans. Faraday Soc.*, **34**, 351 (1938).
214. O. E. Edwards and M. Lesage, *Canad. J. Chem.*, **41**, 1592 (1963).
215. P. Yates and R. J. Crawford, *J. Amer. Chem. Soc.*, **88**, 1561 (1966).
216. G. A. Olah and J. Welch, *J. Amer. Chem. Soc.*, **97**, 208 (1975).
217. G. A. Olah and J. L. Grant, *J. Amer. Chem. Soc.*, **97**, 1546 (1975).
218. H. Zollinger, *Accounts Chem. Res.*, **6**, 335 (1973).

CHAPTER **14**

Preparation of diazonium groups

K. SCHANK†

Fachbereich 14.1 Organische Chemie, Universität des Saarlandes, D-6600 Saarbrücken, Germany

I. INTRODUCTION

The most usual method for the preparation of the diazonium[1] group has been found to be the diazotization of primary amines with NO^+ donors:

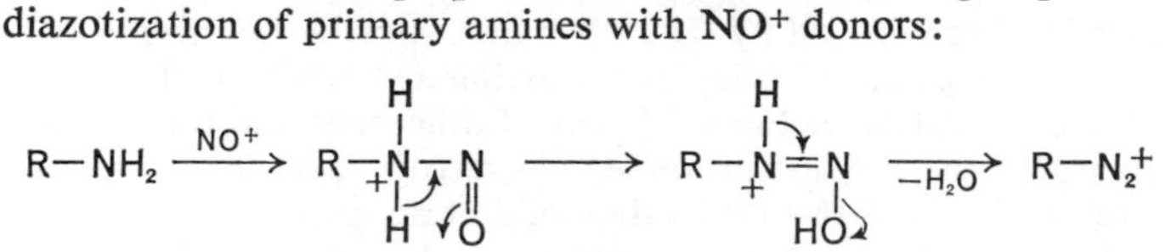

$$R{-}NH_2 \xrightarrow{NO^+} R{-}\overset{+}{N}H_2{-}N{=}O \longrightarrow R{-}\overset{+}{N}H{=}N{-}OH \xrightarrow{-H_2O} R{-}N_2^+$$

† The author gratefully acknowledges the permission of G. Thieme, publishers, and of Academic Press Inc., to use details of his contribution published in *Methodicum Chimicum.*

Since molecular nitrogen is an extremely good leaving group, diazonium ions lose it very easily and consequently methods of preparation must allow for this. The stability depends on the character of R to which nitrogen is attached, and may be achieved through the electronic interaction between the diazonium group and the bonding atom of R, that is in the first instance by the presence of free π-electron pairs. Consequently, fluorine-[2] and *N*-diazonium[3] ions have been described, but in this summary special attention will be given to carbon-bonded diazonium groups. Since aliphatic diazonium ions cannot delocalize the positive charge at nitrogen on the attached substituent R, they are unstable and immediately lose molecular nitrogen yielding reactive carbonium ions[4]. Thereby, the bonding carbon changes from the non-planar sp^3 state to the planar sp^2 hybridization state accompanied by shortening of the remaining three bonds. Prevention of this transhybridization by involving the bridge-head carbon of a bicyclic system with fixed non-planar arrangement could trap such aliphatic diazonium ions by azo coupling[5]. Olefinic diazonium salts, however, in which nitrogen is attached to a sp^2 carbon atom of an olefin, have been prepared easily from tosylazo olefins and Lewis acids[6]. In accordance with the rule of Staudinger–Schmidt concerning the strength of bonds at olefins, these compounds were found to be rather stable. A similar stabilization may be observed if the sp^2 carbon bears a negative charge as in the case of diazo compounds which may be considered as zwitterionic diazonium carbeniates. The most stable and most important diazonium compounds, however, are the aromatic (and heteroaromatic) diazonium salts in which the diazonium nitrogen is attached to an aromatic (or heteroaromatic) moiety. The two species last mentioned are included in the group of so-called *quinone diazides* which may be considered as cyclic vinylogues of α- or γ-diazo carbonyl compounds, and which may be written as diazonium phenolates:

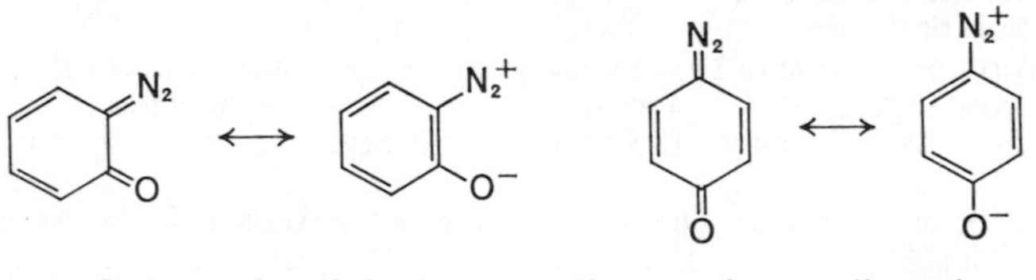

Another group of connecting links between diazo and arenediazonium compounds represents the new class of crystalline BF_2-chelate diazonium fluoroborates which have been synthesized by treatment of open-chain α-diazo-β-dicarbonyl compounds with BF_3 [7]:

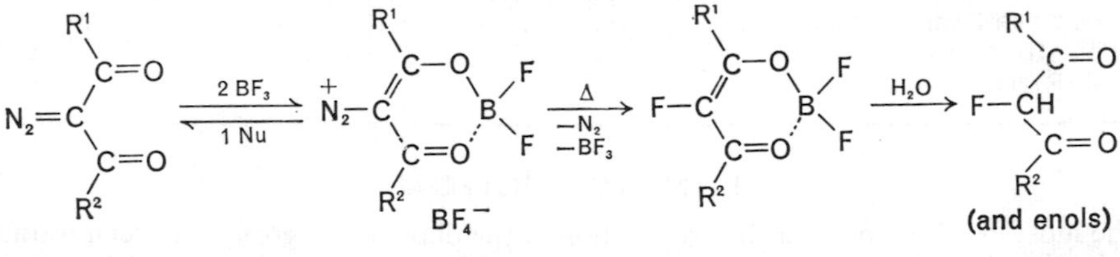

(and enols)

Their structure has been proved by ^1H-, ^{19}F-n.m.r. spectra, by inverse reactions with nucleophiles such as ether, yielding the starting materials, and also by a Balz–Schiemann-like degradation yielding 2-fluoro-β-diketones and their enols.

In consequence of the easy loss of molecular nitrogen, the simple arenediazonium salts are more or less *explosive* (!) in the solid state, especially the most common diazonium chlorides. The explosive character is enhanced by oxidizing substituents on the aromatic nucleus as well as by oxidizing anions; it is diminished by higher molecular weight and large complex anions. Diazonium chlorides are usually easily soluble in water, and complex stable diazonium salts can normally be precipitated from concentrated aqueous solutions of diazonium chlorides and sodium salts of complex acids. The stability of arenediazonium ions towards nucleophilic attack, i.e. the hydrolysis in aqueous solutions yielding nitrogen and phenols, is enhanced by electron-donating substituents[8].

II. FORMATION OF AROMATIC DIAZONIUM SALTS FROM THE CORRESPONDING AMINES BY DIAZOTIZATION

Diazotization of primary aromatic amines by NO^+-donors proceeds in three steps[9]:

$$(1)\quad Ar{-}NH_2 + NO^+ \xrightarrow{\text{slow}} Ar{-}\underset{H^+}{\overset{H}{N}}{-}NO$$

$$(2)\quad Ar{-}\underset{H^+}{\overset{H}{N}}{-}NO \xrightarrow{\text{fast}} Ar{-}\underset{H}{N}{-}NO + H^+$$

$$(3)\quad Ar{-}\underset{H^+}{N}{-}NO + H^+ \xrightarrow{\text{fast}} Ar{-}\overset{+}{N}{\equiv}N + H_2O$$

Recently, the intermediate step of a primary nitrosamine could be isolated in the course of the diazotization of an appropriate amine[10]. Since most diazotizations are performed in aqueous medium, protonated nitrous acid (from sodium nitrite and acids[11]) is usually used; free NO^+ should only appear in concentrated acids and in the course of conversions with complex NO^+-salts. Reactivity decreases under normal conditions in the following sequence: $NO^+ > H_2O^+{-}NO >$ $NCS{-}NO > Br{-}NO > Cl{-}NO > O_2N{-}NO$.

Kinetic measurements have dealt with the dependence of the reaction on pH, on anions, on solvents, etc.[9], as well as with the stability of standard solutions for common use[12].

A. Preparation Procedures

1. In aqueous medium

Normal diazotizations of aromatic amines with nitrous acid are carried out in dilute aqueous mineral acids and usually yield the diazonium salts in solution. The direct method starts from solutions or suspensions of the amine in dilute hydrochloric acid (approx. 2·5 mol acid per mol amine) which are treated with the molar amount of conc. aqueous sodium nitrite solution at 0–10 °C. The conversions proceed rapidly and almost quantitatively[13]; the end-point can be determined by numerous methods, and an excess of nitrous acid can easily be removed by adding urea or sulphamic acid[14]. If chloride ions are not desirable in view of subsequent reactions, other mineral acids such as sulphuric acid, phosphoric acid, etc., may be used but with the disadvantage of reduced solubility of the generated diazonium salts. The indirect method of diazotization is applied to amines which contain strongly acidic groups and therefore appear as sparingly soluble betains. In these cases, a solution of the amine and sodium nitrite in dilute alkali is slowly added under vigorous stirring to the acid.

2. In concentrated acids[15]

Weakly basic amines bearing strongly electron-attracting substituents on the aromatic (or heteroaromatic) nucleus are diazotized preferentially in concentrated acids, since hydrolysis[16] of the generated diazonium compounds occurs in aqueous solutions with rising dilution. Good results have been described with conc. sulphuric acid[17] alone as well as when used with glacial acetic acid or phosphoric acids; the use of conc. fluoroboric acid is winning increasing popularity whereas conc. or anhydrous hydrofluoric acid is principally used in connection with the preparation of organic fluorine compounds. Concentrated nitric acid has been used occasionally having the advantage of higher reaction rates than are observed with conc. sulphuric acid; it has, however, the disadvantage of undesired nitrations and oxidations as well as the enormously high danger of explosion of diazonium nitrates which are similar to the corresponding perchlorates[18]. Nitrosating agents in these cases are sodium nitrite, nitrosyl halides or complex salts.

3. In organic solvents[19]

Usually, in order to prepare solid diazonium salts, diazotizations of aromatic amines are carried out in organic solvents (glacial acetic acid[20], methanol, ethanol, formamide, DMF, acetone and others) by means of nitrous acid esters (preferentially pentyl nitrite). The explosive chlorides can be stabilized as double salts with heavy metal chlorides, which are commercially available in many cases, as well as the relatively stable diazonium salts of organic sulphonic acids and of tetrafluoroboric acid.

III. FORMATION OF AROMATIC DIAZONIUM SALTS STARTING FROM AROMATIC DIAMINES[21]

A common characteristic of aromatic diazonium ions is the high electrophilicity of their β-nitrogen which effects easy azo-coupling reactions with appropriate nucleophiles. Thus, 1,2-phenylenediamine and its homologues are diazotized in dilute acids yielding intermediate mono-diazonium ions which suffer an immediate ring closure to 1,2,3-benzotriazoles; diazotization in concentrated acids, however, prevents such a ring closure and enables bis-diazotization to occur:

NH_2 / NH_2 $\xrightarrow{NO^+}$ [N_2^+ / NH_2]

Dilute acids, $-H^+$ → N, N, N–H (1,2,3-benzotriazole)

NO^+ Conc. acids → N_2^+ / N_2^+

Similarly, aromatic 1,3- and 1,4-diamines must be diazotized in concentrated acids to obtain the corresponding bis-diazonium salts. The mono-diazonium ion from 1,3-phenylenediamine undergoes an *inter*molecular coupling reaction yielding the azo dye 'Bismarck Braun'. Aromatic bicyclic diamines like 1,8-naphthalenediamines and benzidine have been found to show similar properties depending on the diazotization conditions.

IV. DIAZOTIZATION OF HETEROAROMATIC AMINES

Normally, heteroaromatic amines behave in a similar fashion to aniline derivatives in the course of diazotization provided there exists no essential interaction between heteroatoms of the nucleus and the nitrogen of the amino group, i.e. 3-aminopyridine yields the corresponding diazonium ion[22] without difficulty:

$$\text{3-aminopyridine (NH}_2\text{)} \xrightarrow{HNO_2/HX} \text{3-pyridinediazonium } N_2^+ X^-$$

However, 2- and 4-aminopyridines may be regarded as amidines and their vinylogues, and are rapidly hydrolysed to hydroxy compounds (similarly to carbonamides which are converted to carboxylic acids by nitrous acid):

$$\text{2-aminopyridine (NH}_2\text{)} \xrightarrow[H_3O^+X^-]{HNO_2} \left[\text{2-pyridinediazonium } \overset{+}{N}_2 \; X^- \right] \longrightarrow \text{2-hydroxypyridine (OH)} \rightleftharpoons \text{2-pyridone (N–H, C=O)}$$

Either by diazotization in concentrated acids[23] or by conversion first into *N*-oxides with subsequent diazotization[24], these heteroaromatic amines can be easily converted to the corresponding diazonium species. Even 2-aminoimidazoles (as formal guanidines) could be diazotized by nitrosyl sulphate in conc. sulphuric acid[25]:

$$H_3C\text{-benzimidazole(1-}CH_3\text{)-2-C–}NH_2 \xrightarrow[\text{Conc. } H_2SO_4]{NOHSO_4} H_3C\text{-benzimidazolium(N–H, N–}CH_3\text{, +)-2-C–}N_2^+$$

However, the more electron-attracting substituents are attached to the aromatic or heteroaromatic nucleus of the diazonium ion, the more easily is the diazonium group hydrolysed by-passing the stage of nitrosamines[26]. If the primary amino group is directly attached to a nitrogen of the heteroaromatic nucleus, an *N*-diazonium ion is generated as an intermediate which decomposes with ultimate deamination and formation of dinitrogen monoxide[27]:

$$\text{1,2,4-triazol-4-yl–}NH_2 \xrightarrow[[H^+]]{HNO_2} \text{1,2,4-triazole (NH)} + N_2O + H_2O$$

V. FORMATION OF ARENEDIAZONIUM COMPOUNDS FROM OTHER ARENEDIAZONIUM COMPOUNDS

A. By rearrangement

The most surprising rearrangement observed with aromatic diazonium compounds is the exchange of the two nitrogen atoms of the diazonium group, as evidenced by isotope labelling[28]:

$$R\text{–}C_6H_4\text{–}\overset{+}{^{15}N}{\equiv}N \xrightarrow[\text{reaction}]{\text{First-order}} R\text{–}C_6H_4\text{–}\overset{+}{N}{\equiv}{^{15}N}$$

Another rearrangement involving a fluctuating 1,2,3-thiadiazol ring has been found after diazotization of 7-amino-6-substituted 1,2,3-benzothiadiazoles[29]:

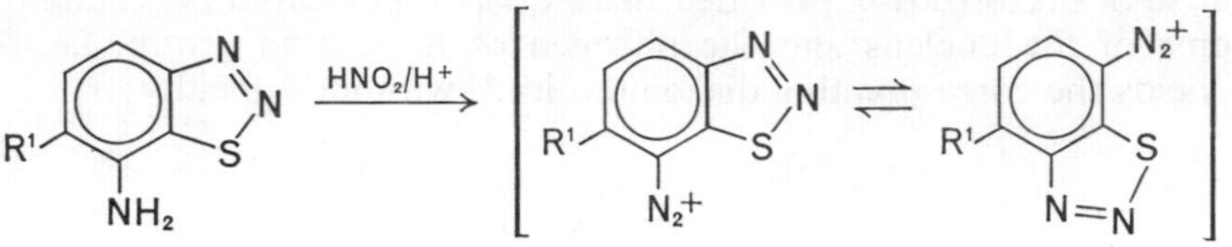

The change of the position of the diazonium function has been proved by the Sandmeyer reaction and by reduction.

B. By substitution

The diazonium group, as a strongly electron-attracting substituent, retards electrophilic substitution on aromatic nuclei much more even than the nitrogroup[30]. Therefore, quinone diazides[31] or their imines[32] can be protonated only by strong mineral acids yielding hydroxy or substituted amino aryl diazonium ions:

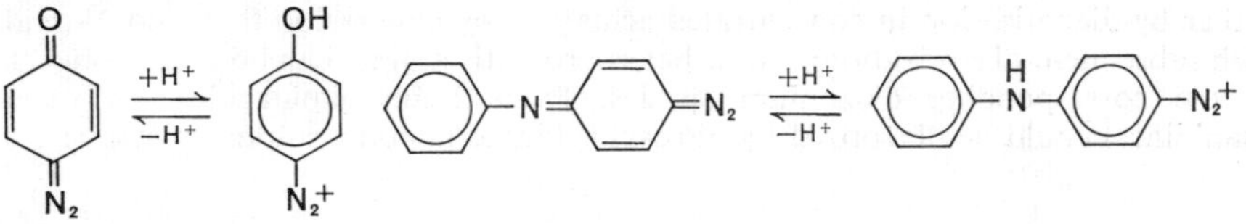

Similarly acylations have been carried out with quinone diazides[33] (as well as the reversal of this reaction[34]):

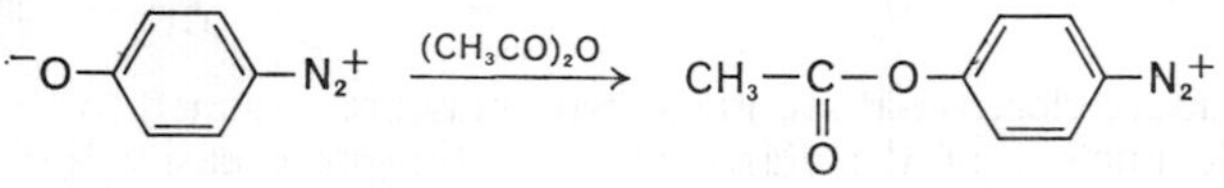

However, the quinone diazide oxygen as a strong electron donor compensates for the effect of the diazonium group and enables electrophilic substitutions again:

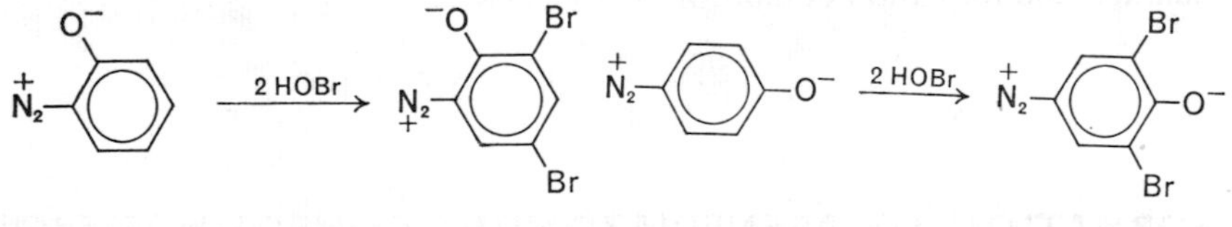

A tabular survey on S_E reactions of quinone diazides is given in the literature[35]. Similarly, Friedel–Crafts-like condensations of substituted amino phenyl diazonium compounds with formaldehyde have been described[36]. A quite interesting reaction is the immediate oxidation of diazotized α-aminonaphthalene by means of alkaline potassium hexacyanoferrate(III)[37]:

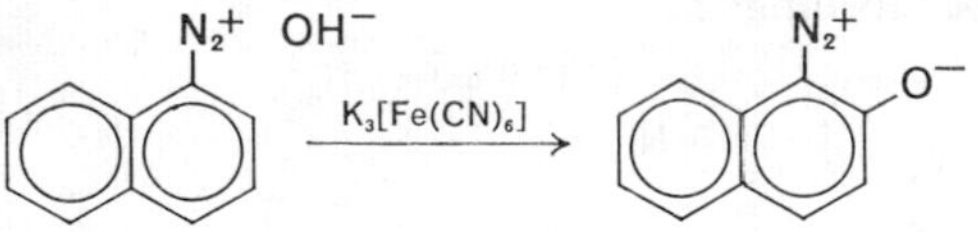

Since electron withdrawal from arene nuclei by the diazonium group is extremely effective, substituents in the *ortho* and *para* positions which are good leaving groups can be easily substituted by an S_Nar mechanism[38]:

$$\text{2,6-Cl}_2\text{-4-NO}_2\text{C}_6\text{H}_2\text{N}_2^+ \xrightarrow[-\text{HNO}_2]{\text{Conc. HCl}} \text{2,4,6-Cl}_3\text{C}_6\text{H}_2\text{N}_2^+ \xleftarrow[-\text{HNO}_2]{\text{Conc. HCl}} \text{2,4-Cl}_2\text{-6-NO}_2\text{C}_6\text{H}_2\text{N}_2^+$$

Starting from strongly activated 2- or 4-nitroarylamines this type of reaction enables the immediate formation of 2- or 4-chloroaryl diazonium salts on warming with hydrochloric acid[39]. In general, appropriate leaving groups in the *ortho* or *para* position to the diazonium group are easily substituted by more effective nucleophiles[40].

VI. FORMATION OF AROMATIC DIAZONIUM SALTS BY TRANSDIAZOTIZATION

Related to the hydrolysis of very reactive diazonium ions, prepared from weakly basic amines, is the aminolysis by aryl amines of high basicity, which effects a transfer of the diazo function, resulting in transdiazotization[41]. This method proved to be very useful for synthesizing benzidine monodiazonium salt either from its bis-diazonium salt or from the *p*-nitrophenyl diazonium salt[42]:

$$H_2N\text{-}C_6H_4\text{-}C_6H_4\text{-}NH_2 \xrightarrow[-H^+]{+\ O_2N\text{-}C_6H_4\text{-}N_2^+} H_2N\text{-}C_6H_4\text{-}C_6H_4\text{-}NH\text{-}N{=}N\text{-}C_6H_4\text{-}NO_2 \xrightarrow{HCl(aq)}$$

$$H_2N\text{-}C_6H_4\text{-}C_6H_4\text{-}N_2^+\ Cl^- + H_2N\text{-}C_6H_4\text{-}NO_2$$

VII. FORMATION OF AROMATIC DIAZONIUM SALTS BY AZO DECOUPLING

The N–N cleavage of the triazene in acidic medium described above may be regarded as a reversal of an azo coupling at nitrogen. Such decoupling reactions have also been observed at other centres. The reversible cleavage of diazo sulphonates[43] and diazo sulphones[44] by mineral acids, yielding diazonium ions, is well known; less known is the related alkali- or acid-catalysed rearrangement of aryl azoxy sulphones[45] yielding diazonium sulphonates:

$$C_6H_5\text{–}N(\rightarrow O)\text{=}N\text{–}SO_2\text{–}C_6H_5 \xrightarrow[\text{[OH}^-]]{[H^+]\text{ or}} C_6H_5\text{–}N_2^+ \;\; C_6H_5\text{–}SO_3^-$$

C–N cleavages of azo coupling products of CH-activated compounds under various conditions appear to be of some interest. Azo dyes from methine-activated compounds and diazonium salts showed reversible decoupling under the influence of mineral acids, boron trifluoride or bromine[46]. A similar fission could also be observed in the course of the reaction of chlorine with a phenylhydrazone[47]:

$$\text{indane-1,3-dione-2-}(=N\text{–}NH\text{–}C_6H_5) \xrightarrow[-HCl]{+2\,Cl_2} \text{2,2-dichloroindane-1,3-dione} + C_6H_5\text{–}N_2^+\,Cl^-$$

VIII. FORMATION OF AROMATIC DIAZONIUM COMPOUNDS FROM NITROSO ACYL AMINES

Acylated aromatic amines Ar—NH—X, where X may be COR, SO_3H or NO_2, are readily nitrosated by NO^+ donors; in the two latter cases immediate rearrangements occur yielding diazonium sulphates or nitrates. However, nitroso carbonamides can be isolated and undergo an intramolecular rearrangement by a so-called 'uncoiling mechanism' [48] yielding diazonium carboxylates. Under the influence of bases[49] diazotate anions are generated from nitroso acyl amines accompanied by elimination of the acyl group. Between diazotate anions and diazonium cations an equilibrium exists which depends on the pH value of the corresponding aqueous solution[50], and recently the kinetics and mechanism of such conversions have been studied[51]. Since sydnones are derivatives of *N*-nitroso amines, the intermediate formation of diazonium ions during hydrolytic ring cleavage in the presence of strong acids[52] is not surprising.

IX. FORMATION OF AROMATIC DIAZONIUM SALTS BY NITROSATION OF AROMATIC IMINES

Treatment of azomethines with NOX (X = Cl [53], NO_3 [54], BF_4 [55]) seems to be a useful method for preparing crystalline diazonium salts:

$$R^1\text{–}C_6H_4\text{–}CH\text{=}N\text{–}C_6H_4\text{–}R^2 \xrightarrow{NOBF_4} R^1\text{–}C_6H_4\text{–}CHO + R^2\text{–}C_6H_4\text{–}N_2^+\,BF_4^-$$

In a similar way, phenyl isocyanate[56] and aromatic sulphinylimines[57] are converted to diazonium salts by NOX with simultaneous elimination of carbon dioxide or sulphur dioxide.

X. FORMATION OF AROMATIC DIAZONIUM SALTS BY NITROSATION OF AROMATIC NITROSO COMPOUNDS AND RELATED SPECIES

N-Phenylhydroxylamine can be nitrosated readily to form *N*-nitroso-*N*-phenylhydroxylamine; an excess of nitrous acid, however, leads to benzene diazonium nitrate[58]:

$$C_6H_5\text{—NHOH} \xrightarrow[-H_2O]{+HNO_2} C_6H_5\text{—N(N{=}O)(O—H)} \xrightarrow[-H_2O]{+HNO_2} C_6H_5\text{—}N_2^+\ NO_3^-$$

Similarly, the following diazonium salt has been obtained from the dioxime of anthraquinone:

$$\text{anthraquinone dioxime (NOH, NOH)} \xrightarrow{\text{(a) } N_2O_4/-50\ ^\circ C \quad \text{(b) } HBF_4} \text{10-nitro-9-anthracenediazonium } N_2^+\ BF_4^-\ (NO_2)$$

Aromatic nitroso compounds, as the next higher oxidation stages, undergo redox diazotizations with excess of nitrous acid[59]:

$$Ar\text{—}N{=}O + 3\ HNO_2 \xrightarrow[-H_2O,\ -HNO_3]{} Ar\text{—}N_2^+\ NO_3^-$$

Since electron-rich aromatic compounds are easily nitrosated to form nitroso compounds, both reactions can be combined to synthesize diazonium nitrates in one step[60]:

$$R_2N\text{—}C_6H_5 + 4\ HNO_2 \xrightarrow[-2\ H_2O,\ -HNO_3]{} R_2N\text{—}C_6H_4\text{—}N_2^+\ NO_3^-$$

Less electron-rich aromatic compounds need more reactive nitrosating agents or metal catalysis preferentially by Hg^{2+} [61]. The use of nitrosyl chloride sometimes effects further substitution of the aromatic nucleus by chlorine[62]:

$$\text{1-nitrosonaphthalene (NO)} \xrightarrow[-HNO_3]{+2\ NOCl} \text{4-chloro-1-naphthalenediazonium } N_2^+\ Cl^-\ (Cl)$$

XI. FORMATION OF AROMATIC DIAZONIUM IONS BY ALDOL-LIKE CONDENSATIONS OF NITROSO COMPOUNDS AND OF NITROBENZENE

Reaction of nitrosobenzene (and derivatives) with hydroxylamine in presence of sodium carbonate gives diazonium ions in good yields[63]:

$$C_6H_5{-}N{=}O + H_2NOH \xrightarrow{Na_2CO_3} C_6H_5{-}N_2^+ \; OH^- + H_2O$$

Correspondingly, nitrobenzene as the next higher oxidized species reacts with ammonia as the next lower oxidized species in the presence of alkali to give the diazonium (diazotate) compound in poor yield[64]:

$$C_6H_5{-}NO_2 + NaNH_2 \xrightarrow[-H_2O]{} C_6H_5{-}N{=}N{-}O^- \; Na^+$$

XII. FORMATION OF AROMATIC DIAZONIUM SALTS FROM ARYLHYDRAZINES AND THEIR DERIVATIVES[65]

As mentioned before (Section VII) phenylhydrazones can be chlorinated yielding benzenediazonium chloride. Similarly, bromine may be used and, instead of phenylhydrazones, thionylphenylhydrazine and other derivatives may be applied. Phenylhydrazine itself can be oxidized by different reagents yielding the diazonium step; these methods, however, have not reached particular preparative importance. The reaction of phenylhydrazine with nitrous acid, depending on reaction conditions, seems to be quite interesting[66]:

$$C_6H_5{-}NH{-}NH_2 + HNO_2 \xrightarrow[-2\,H_2O]{(b)} C_6H_5{-}N_3$$

$$C_6H_5{-}NH{-}NH_2 + HNO_2 \xrightarrow[(a)]{-H_2O} C_6H_5{-}N(NO){-}NH_2 \xrightarrow[HX]{HNO_2} C_6H_5{-}N_2^+ \; X^-$$

$$C_6H_5{-}N(NO){-}NH_2 \xrightarrow{[H^+] \text{ or } [OH^-]} C_6H_5{-}N_3$$

In mineral acid solution with an excess of phenylhydrazine, nitrous acid effects conversion to phenyl azide. If nitrous acid is in excess, the formation of the diazonium salt predominates. *N*-Nitroso-*N*-phenylhydrazine, the intermediate product preferentially formed in weakly acid medium, can either be converted into azide by strong acids and strong bases or into the diazonium ion by treatment with nitrous acid. The latter reaction corresponds to the related conversion of *N*-nitrosophenylhydroxylamine with nitrous acid (Section X).

XIII. MISCELLANEOUS

The syntheses of tritium-labelled diazonium salts[67] and of polymeric diazonium salts[68] derived from 3-vinylaniline have been described. Immediate introduction of the diazonium group by tosyl azide[69] or by tosylhydrazine[70] requires special starting materials and cannot be generalized. These methods are of great importance in the series of aliphatic diazo compounds. Recently an interesting synthesis of diazonium salts has been achieved by using the fragmentation of *N*-chloro azocarbonamides[71]:

$$C_6H_5-N=N-C(=O)-NHR \xrightarrow{(CH_3)_3COCl} \left[C_6H_5-N=N-C(=O)-N(R)-Cl + (CH_3)_3COH\right] \longrightarrow C_6H_5-N_2^+\,Cl^- + RNH-CO_2C(CH_3)_3$$

XIV. REFERENCES

1. K. Schank, *Methodicum Chimicum*, Vol. 6 (Ed. F. Zymalkowski), Academic Press, New York, San Francisco, London, 1975, p. 159; Georg Thieme Publishers, Stuttgart, 1975.
2. A. V. Pankratov and N. I. Savenkova, *Zh. Neorg. Khim.*, **13**, 2610 (1968); H. W. Roesky, D. Bormann and O. Glemser, *Kurznachr. Akad. Wiss. Göttingen, Sammelh.*, **2**, 51 (1966); *Chem. Ber.*, **99**, 1589 (1966); J. K. Ruff, *Inorg. Chem.*, **5**, 1791 (1966); D. Moy and A. R. Young, *J. Amer. Chem. Soc.*, **87**, 1889 (1965).

3a. M. P. Doyle, J. G. Kalmbacher, W. Wierenga and J. E. DeBoer, *Tetrahedron Lett.*, 1455 (1974).

3b. R. Kreher, *Angew. Chem.*, **83**, 915 (1971).

3c. G. F. Terescenko, G. I. Koldobskij and L. I. Bagal, *Zh. Org. Khim.*, **6**, 1132 (1970).

3d. A. Schmidt, *Chem. Ber.*, **99**, 2976 (1966).

3e. K. Bott, *Angew. Chem.*, **77**, 683 (1965).

4. W. Kirmse, *Angew. Chem.*, **88**, 273 (1976).
5. D. Y. Curtin, B. H. Klandermann and D. F. Tavares, *J. Org. Chem.*, **27**, 2709 (1962).
6. K. Bott, *Chem. Ber.*, **108**, 402 (1975); *Synthesis*, 161 (1973).
7. L. Prim and K. Schank, to be published.
8. H. Salkowski, *Ber. dt. chem. Ges.*, **7**, 1008 (1874).
9. Reference 1, p. 163.
10. G. N. Dorofeenko, Y. P. Andreichikov and G. E. Trukhan, *Khim. Geterotsikl. Soedin.*, 1344 (1974).
11. D. H. Wilcox, Jr, *Amer. Dyest. Rept.*, **55**, 891 (1966).
12. A. Spevak and M. Matrka, *Coll. Czech. Chem. Commun.*, **37**, 2397 (1972); H. Zollinger, *Accounts Chem. Res.*, **6**, 335 (1973).
13. R. Pütter, *Methoden der Organischen Chemie* (Houben–Weyl–Müller), Vol. X, Part 3, 4th ed., Georg Thieme Verlag, Stuttgart, 1965, p. 16.
14. Reference 1, p. 164.
15. Reference 13, p. 22.
16. M. Matrka, Z. Sagner, V. Chmatal, V. Štěrba and M. Vesely, *Coll. Czech. Chem. Commun.*, **32**, 1462 (1967); Reference 13, p. 12, note 6.
17. H. E. Fierz-David and L. Blangley, *Grundlegende Operationen der Farbenchemie*, 8th ed., Springer Verlag, Wien, 1952, p. 244; Reference 13, p. 23.
18. Cf. Reference 1, p. 165.
19. Reference 13, p. 28.

20. M. Colonna, L. Greci and P. Bruni, *Gazz. Chim. Ital.*, **102**, 527 (1972).
21. Reference 1, p. 166.
22. P. Tomasik, E. Kucharzewska-Rusek and A. Thomas, *Roczniki Chem* **44**, 1131 (1970); cf. Reference 1, p. 169.
23. E. Koenigs, G. Kinne and W. Weiss, *Ber. dt. chem. Ges.*, **57**, 1172 (1924).
24. E. Ochiai, *J. Org. Chem.*, **18**, 534 (1953).
25. S. N. Kolodyazhnaya, A. M. Simonov and I. G. Uryukina, *Khim. Geterotsikl. Soedin.*, 1690 (1972).
26. Reference 13, p. 55.
27. Th. Curtius, A. Darapsky and E. Müller, *Ber. dt. chem. Ges.*, **40**, 836 (1907); cf. Reference 3a.
28. E. S. Lewis and R. E. Holiday, *J. Amer. Chem. Soc.*, **91**, 426, 430 (1969); **88**, 5043 (1966); G. W. Van Dine and R. Hoffmann, *J. Amer. Chem. Soc.*, **90**, 3227 (1968); A. K. Bose and I. Kugajewsky, *J. Amer. Chem. Soc.*, **88**, 2325 (1966).
29. E. Haddock, P. Kirby and A. W. Johnson, *J. Chem. Soc. C*, 2514 (1970).
30. Hammett substituent constant of the diazonium group: S. C. Gardner and E. C. Lupton, Jr, *J. Amer. Chem. Soc.*, **90**, 4328 (1968).
31. Reference 13, pp. 8, 51.
32. Reference 13, p. 44.
33. *DRP* 206 455 (1907), Farbenf. Bayer; *Fortschritte der Teerfabrikation und verwandter Industriezweige*, **9**, 300 (1911).
34. D. J. Triggle and S. Vickers, *Chem. Commun.*, 544 (1965); S. Vickers, D. J. Triggle and D. R. Garrison, *J. Chem. Soc. C*, 632 (1968).
35. Reference 13, p. 89.
36. V. V. Kozlov and S. K. Eremin, *Zh. Org. Khim.*, **7**, 1697 (1971); *Teor. Eksp. Khim.*, **6**, 795 (1970); cf. Reference 13, p. 89.
37. J. Sauer and R. Huisgen, *Angew. Chem.*, **72**, 303 (1960); E. Bamberger, *Ber. dt. chem. Ges.*, **55**, 3383 (1922).
38. B. Andersson and B. Lamm, *Acta Chem. Scand.*, **23**, 2983 (1969).
39. Tabular survey: cf. Reference 13, pp. 97, 98.
40. Reference 13, pp. 99–109; Reference 1, p. 172.
41. Reference 1, p. 173.
42. Reference 13, p. 46.
43. Reference 13, p. 576.
44. Reference 13, p. 584.
45. Reference 13, p. 586.
46. Reference 1, p. 174.
47. W. M. Moon, *J. Org. Chem.*, **37**, 383, 386 (1972).
48. R. Huisgen, *Liebigs Ann. Chem.*, **573**, 163 (1951); **574**, 157, 171 (1951); cf. Reference 13, p. 68; E. H. White and C. A. Elliger, *J. Amer. Chem. Soc.*, **89**, 165 (1967).
49. E. Bamberger, *Ber. dt. chem. Ges.*, **27**, 914 (1894); **30**, 366 (1897); cf. R. Huisgen, *Liebigs Ann. Chem.*, **573**, 163 (1951).
50. Reference 13, p. 552.
51. O. Machackova and V. Štěrba, *Coll. Czech. Chem. Commun.*, **37**, 3467 (1972); J. Jahelka, O. Machackova and V. Štěrba, *Coll. Czech. Chem. Commun.*, **38**, 706 (1973).
52. L. E. Cholodov and V. T. Jasunskij, *Zh. obshch. Chim.*, **37**, 670 (1967); G. S. Puranik and H. Suschitzky, *J. Chem. Soc.*, 1006 (1967).
53. J. Turcan, *Bull. Soc. Chim. France*, 627 (1935).
54. R. M. Scribner, *J. Org. Chem.*, **29**, 3429 (1964).
55. M. P. Doyle, W. Wierenga and M. A. Zaleta, *J. Org. Chem.*, **37**, 1597 (1972).
56. G. B. Bachmann and W. Michalowicz, *J. Org. Chem.*, **23**, 1800 (1958).
57. K. Bott, *Angew. Chem.*, **77**, 132 (1965).
58. Reference 13, p. 71.
59. Reference 13, p. 73.
60. C. Sellers and H. Suschitzky, *J. Chem. Soc.*, 6186 (1965).
61. Reference 1, p. 176.
62. *Jap. Patent* 10 343 (1966) (C1.23 D 01), 4.6.63, Appl. 27.12.63, Toyo Rayon Co., Ltd, Inv.: S. Torimitsu and M. Ohno; *Chem. Abstr.*, **65**, 15 548f (1966).

63. Reference 13, p. 86.
64. E. Bamberger and A. Wetter, *Ber. dt. chem. Ges.*, **37**, 629 (1904); F. W. Bergstrom and J. S. Buehler, *J. Amer. Chem. Soc.*, **64,** 19 (1942).
65. Reference 1, p. 177.
66. K. Clusius and K. Schwarzenbach, *Helv. Chim. Acta*, **42**, 739 (1959).
67. P. S. Traylor and S. J. Singer, *Biochemistry*, **6**, 881 (1967).
68. C. L. Arcus and R. H. Still, *J. Chem. Soc.*, 1340 (1964).
69. J. M. Tedder and B. Webster, *J. Chem. Soc.*, 4417 (1960).
70. W. R. Bamford and T. S. Stevens, *J. Chem. Soc.*, 4735 (1952).
71. R. Ohme and H. Preuschhof, *J. Prakt. Chem.*, **313**, 642 (1971).

CHAPTER 15

Synthesis of diazoalkanes

M. REGITZ

Department of Chemistry, University of Kaiserslautern, D-6750 Kaiserslautern, Federal Republic of Germany

I. INTRODUCTION

Diazoalkanes have been used as synthetic reagents in organic chemistry for nearly 100 years, and during that time they have lost none of their original attraction or significance. During the past decade it has been principally carbene chemistry, predominantly accessible via diazoalkanes, which has furthered the development and stimulated the production of a seemingly endless variety of new representatives of this class. We can expect the future to provide us with yet further surprises in this field.

Syntheses of diazoalkanes can be systematically classified in the following way:

(1) Condensation of two compounds which each possess a nitrogen-containing functional group, as in the diazotization of amines or the Forster reaction.
(2) Conversion of compounds containing a functional group with two N atoms into diazoalkanes by dehydrogenation (hydrazones) or cleavage (tosylhydrazones, *N*-acyl-*N*-nitrosoamines).
(3) Transfer of a diazo group from a donor (usually tosyl azide) to an acceptor (diazo group transfer).
(4) Substitution reactions of diazoalkanes which leave the N_2 group unchanged.

Disregarding the synthesis of acetylated diazoalkanes, as in the present section, the method listed under (2) is of predominant importance. The preparation of diazoalkanes has been the subject of several recent surveys[1,2,3]; hence we shall confine our attention mainly to recent work.

Even though the properties of aliphatic diazo compounds are considered elsewhere in this monograph, it is nevertheless appropriate to mention the ready decomposition or explosive nature of various members of this class. Such properties should not be overemphasized; while they can admittedly hamper preparative work, this drawback is of minor importance in relation to the enormous synthetic value of aliphatic diazo compounds. Caution must be exercised on purification of diazoalkanes by distillation, and the use of large amounts should be avoided. Working with diazomethane and its homologues, which are particularly susceptible to trace-catalysed decomposition, becomes almost unproblematical when used only in solution. The homologues in particular are powerful poisons—their inhalation and contact with skin should be strictly avoided.

The following discourse will consider the advantages and disadvantages of individual methods and their practical scope. It is restricted to diazoalkanes which do not bear CO, PO or SO_2, i.e. strong electron acceptors, at the diazo carbon atom. Compounds of that kind are considered in a separate Chapter.

II. SYNTHETIC METHODS

A. Diazotization of Amines

The diazotization of primary amines leads to diazoalkanes only if elimination of H^+ at the diazonium ion stage (**1** → **2**) can successfully compete with the usual decomposition of the aliphatic diazonium ion (**1** → **3**). This is the case if proton-activating groups are present on the diazonium carbon atom, if the entire process can be performed in alkaline medium, or if the diazo function to be formed is part of a system capable of conjugation. Since these conditions are seldom met, amine diazotization has only limited scope for the preparation of non-acylated diazoalkanes.

$$R_2CH{-}NH_2 \xrightarrow[-2\,H_2O]{H^+,\ HNO_2} \underset{(1)}{R_2CH{-}N_2^+} \begin{cases} \xrightarrow{-H^+} \underset{(2)}{R_2C{=}N_2} \\ \xrightarrow{-N_2} \underset{(3)}{R_2\overset{+}{C}{-}H} \longrightarrow \text{(further reactions)} \end{cases}$$

1. Diazoalkanes

Diazotization reactions of amines are generally carried out in acid media, in which diazoalkanes are subject to elimination of nitrogen; in special cases, however, this reaction can also be accomplished in a basic medium, as in the case of diazotization of methylamine with nitrosyl chloride (equation 1)[4, 5, 6]. The nitrosamine (**4**) can be

$$H_3C{-}NH_2 \xrightarrow{NOCl,\ ether,\ -80\,°C} \underset{(4)}{H_3C{-}NH{-}NO} \begin{cases} \xrightarrow[-C_2H_5OH]{C_2H_5OK} \underset{(5)}{H_3C{-}N{=}N{-}O^-\ K^+} \xrightarrow{KOH,\ H_2O} H_2C{=}N_2 \\ \xrightarrow{KOH,\ H_2O} \underset{(6)}{H_2C{=}N_2} \end{cases} \quad (1)$$

detected by u.v. methods[5], and potassium methyldiazotate (**5**) can even be isolated after addition of potassium ethoxide; it is designated 'stabilized diazomethane'[5]. Both **4** and **5** yield diazomethane (55–60 and 74–78%, respectively) on alkaline hydrolysis.

An analogous synthesis of 1-diazooctane from 1-octylamine has been reported[7].

2. Fluorodiazoalkanes

Diazotization of methylammonium salts with trifluoromethyl or other highly fluorinated alkyl groups attached to the amino carbon atom normally proceeds as expected because the fluorinated alkyl groups effect deprotonation of the diazonium intermediates (equation 2).

$$R^1{-}CF_2{-}\overset{R^2}{\overset{|}{C}H}{-}\overset{+}{N}H_3\ Cl^- \xrightarrow[-NaCl,\ -2H_2O]{NaNO_2} \underset{(6)}{R^1{-}CF_2{-}\overset{R^2}{\overset{|}{C}}{=}N_2} \quad (2)$$

a: $R^1 = F$, $R^2 = H$ **d**: $R^1 = CHF_2$, $R^2 = H$

b: $R^1 = F$, $R^2 = CH_3$ **e**: $R^1 = CF_2{-}CF_3$, $R^2 = H$

c: $R^1 = F$, $R^2 = CF_3$ **f**: $R^1 = F$, $R^2 = C_6H_5$

1-Diazo-2,2,2-trifluoroethane (**6a**) (70%)[8, 9], 2-diazo-1,1,1-trifluoropropane (**6b**) (no yield given)[10], 2-diazohexafluoropropane (**6c**) (49%)[11], 1-diazo-2,2,3,3-tetrafluoropropane (**6d**) (45%)[12], 1-diazo-2,2,3,3,4,4,4-heptafluorobutane (**6e**) (58%)[13] and 1-diazo-2,2,2-trifluoro-1-phenylethane (**6f**) (17%)[14] are instances of successful preparation of fluorinated diazoalkanes by diazotization of amines. Remarkably, this method fails on attempted synthesis of 1-diazo-2,2-difluoroethane[12] and diazopentafluorophenylmethane (equation 3)[15]. In the latter case, the elimination of N_2 from

$$F_5C_6CH_2{-}NH_2 \xrightarrow{\text{diazotization}} F_5C_6CH_2{-}N_2^+ \xrightarrow[-N_2]{} F_5C_6CH_2^+ \xrightarrow[-H^+]{H_2O} F_5C_6CH_2OH \quad (3)$$

the diazonium ion intermediate is seen as a consequence of the pronounced stability of the pentafluorobenzyl cation. Nor is 1,1,1-trichloro-2-diazoethane accessible in this way; nitrogen elimination and chlorine migration in the diazonium ion intermediate are completely dominant over the desired deprotonation (equation 4)[16]. The 1,2 Cl shift to the diazonium carbon atom is important for all the products (**7, 8** and **9**); direct attack by chloride ion appears unlikely because not even the solvent water exploits such a possibility (no 2,2,2-trichloroethanol is formed).

$$Cl_2C(Cl){-}CH_2{-}N_2^+ \xrightarrow[-N_2]{} \underset{(7)}{Cl_3C{-}CH_2{-}Cl} \xrightarrow{\text{Hydrolysis}} \underset{(8)}{HOOC{-}CH_2{-}Cl}$$

$$Cl_2C(Cl){-}CH(H){-}N_2^+ \xrightarrow[-N_2,\ -HCl]{} \underset{(9)}{Cl_2C{=}CHCl} \quad (4)$$

3. Diazocycloalkadienes

The synthesis of diazocyclohexadiene (**11**) from 4-aminophenylmalononitrile is reminiscent of the diazotization of aminophenols to give quinone diazides[17]; this reaction is unusual in that deprotonation of the diazonium ion intermediate (**10**)

$$4\text{-}H_2N{-}C_6H_4{-}CH(CN)_2 \xrightarrow{\text{Diazotization}} \underset{(10)}{4\text{-}N_2^+{-}C_6H_4{-}CH(CN)_2} \xrightarrow{-H^+} \underset{(11)}{4\text{-}N_2^+{-}C_6H_4{-}\ddot{C}^-(CN)_2 \text{ or } (NC)_2C{=}C_6H_4{=}N_2}$$

does not occur at the reaction site but in position 4 relative to that site. In spite of the pronounced acceptor properties of the cyano groups, the quinonoid resonance formula best represents the electron distribution, as shown by the longwave diazo stretching absorption (2128 cm^{-1})[17]. Finally, diazotetracyanocyclopentadiene (88%) which is formed on diazotization of the tetraethylammonium salt of aminotetracyanocyclopentadiene (equation 5) should also be included here[18]. The high frequency of the diazo stretching absorption (2252 cm^{-1}) and the magnitude of the dipole moment ($\mu = 11{\cdot}44$ D)[18] justify the betaine-type formulation.

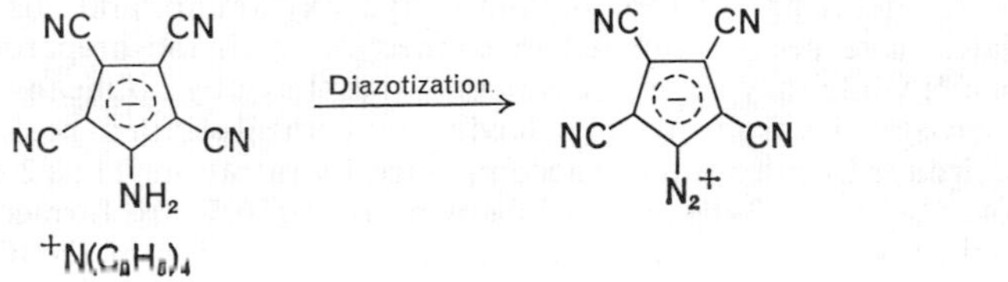

$$\text{(NC)}_4\text{C}_4^{-}\text{-NH}_2 \ {}^{+}\text{N(C}_6\text{H}_5)_4 \xrightarrow{\text{Diazotization}} \text{(NC)}_4\text{C}_4^{-}\text{-N}_2^{+} \quad (5)$$

4. Diazoheteroaromatics

The diazotization of primary heterocyclic amines also displays certain parallels with the corresponding reaction of aminophenols. Heterocyclic diazonium salts are formed first, there being usually no need to isolate them, and are deprotonated directly with weak bases to give diazoheterocycles. This reaction appears plausible if it is borne in mind that heterocyclic diazonium salts represent very strong acids: 1*H*-2-imidazolediazonium chloride has a pK_a value of 2·6, 1*H*-3-(1,2,4-triazole)-diazonium chloride one of 0·3 (water)[19]. The production of 3-diazo-3*H*-pyrazole can be cited as a model example in which the diazonium salt was also isolated (equation 6)[20]. Diazo derivatives of pyrroles[21], indazoles[22, 23], other pyrazoles[24],

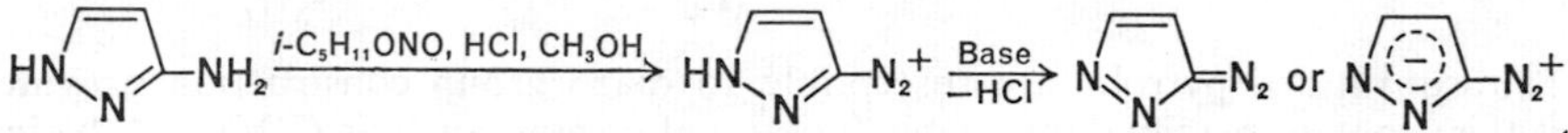

$$\text{HN-N=C-NH}_2 \xrightarrow{i\text{-C}_5\text{H}_{11}\text{ONO, HCl, CH}_3\text{OH}} \text{HN-N=C-N}_2^{+}\ \text{Cl}^{-} \xrightarrow[-\text{HCl}]{\text{Base}} \text{N=N-C=N}_2 \text{ or } \text{N}^{-}\text{-N-C-N}_2^{+} \quad (6)$$

imidazoles[25–28], 1,2,3-triazoles[27, 29], 1,2,4-triazoles[30] and of tetrazole[31], have also been prepared in the same way.

B. Construction of Diazo Groups by Nitrosation

Pyrroles and pyrazoles which are not completely substituted are transformed into the corresponding diazoheterocycles on treatment with sodium nitrite in acid media; in special cases the same reaction is used to prepare diazonium salts of aromatic compounds[32]. Formation of the diazo group by nitrosation proceeds in a clear-cut manner if the heterocyclic compound possesses only one reaction site, as does 2,3,5-triphenylpyrrole or 3,5-dimethylpyrazole; both give the expected diazoheterocycles **12** [33] and **13** [34].

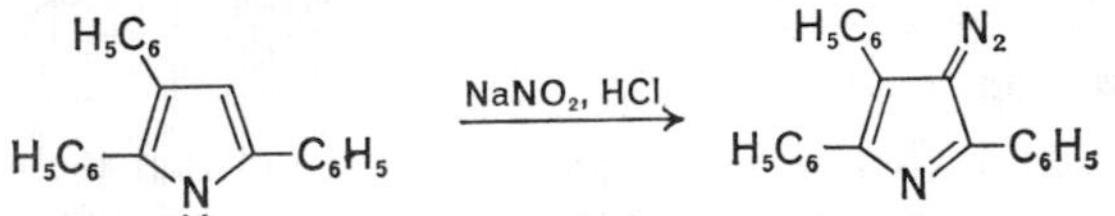

$$\text{2,3,5-}(\text{C}_6\text{H}_5)_3\text{-pyrrole} \xrightarrow{\text{NaNO}_2\text{, HCl}} \textbf{(12)}$$

$$\text{3,5-(CH}_3)_2\text{-pyrazole} \xrightarrow{\text{NaNO}_2\text{, HCl}} \textbf{(13)}$$

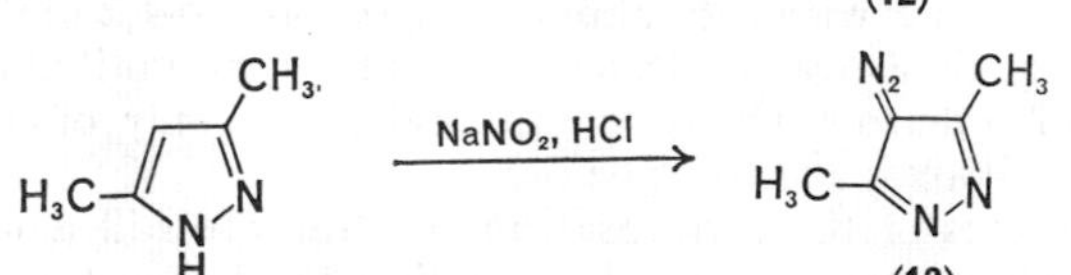

Presumably, nitroso compounds are intermediates since both 3-nitroso-2,5-diphenylpyrrole[33] and its isomer 2-nitroso-3,5-diphenylpyrrole[35], as well as 5-methyl-3-nitroso-2-phenylpyrrole[33] are transformed into the corresponding diazopyrroles by nitrous fumes in chloroform or acetic ester. More sophisticated mechanistic concepts involving nitroso, oximino, hydroxylamino and diazohydroxide intermediates have been postulated in connection with the preparation of an α-diazo lactone[36]; while they appear reasonable, they have by no means been confirmed.

When starting from pyrroles having two unsubstituted positions, formation of diazo groups by nitrosation is accompanied by nitration. Thus 2-diazo-3,5-diphenylpyrrole (**14**)[35] and 3-diazo-2,5-diphenylpyrrole (**16**)[33] are formed together with the diazonitropyrroles **15** and **17**. Overall, the introduction of the diazo group appears to proceed faster than nitration.

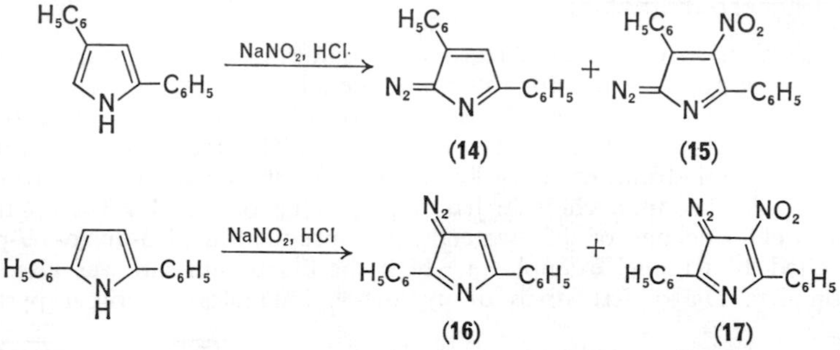

$NaNO_2$, HCl → (**14**) + (**15**)

$NaNO_2$, HCl → (**16**) + (**17**)

Tetrasubstituted pyrroles are susceptible to diazo group construction provided that they contain readily displaceable functional groups, such as COOH. A typical example is shown in equation (7)[35]. The same kind of reaction is observed with aromatic *o*- and *p*-hydroxy carboxylic acids which are transformed into diazonium salts with loss of CO_2 [37]. The diazopyrrole was not isolated but identified by azo coupling with β-naphthol[35].

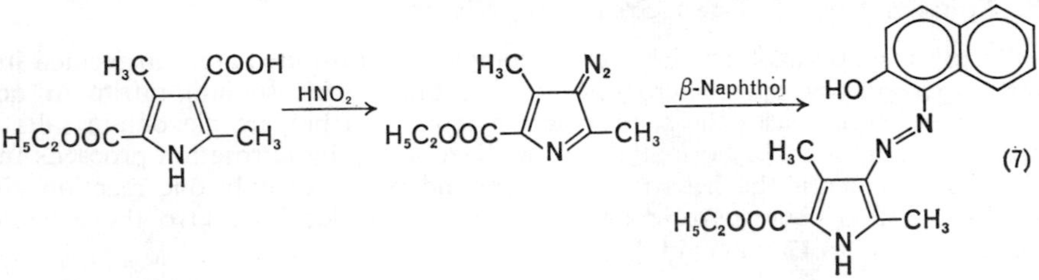

HNO_2 → β-Naphthol → (7)

C. Forster Reaction

This reaction was discovered by Forster[38] who found that the configurationally isomeric oximes of benzil react in aqueous alkaline solution with sodium hypochlorite and ammonia, i.e. with chloramine, to give 1-diazo-1,2-diphenyl-2-ethanone ('azibenzil'). This synthesis was neglected for many years, until adopted mainly for the preparation of α-diazo carbonyl compounds; it is only of limited utility for non-acylated diazoalkanes, as shown below.

The individual steps of the condensation reaction are still a matter of conjecture and the reaction course shown in equation (8) has not been substantiated by experiment[39].

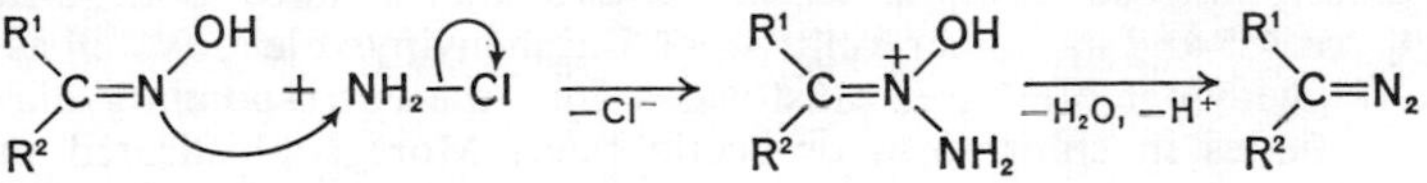

$$R^1R^2C{=}N{-}OH + NH_2{-}Cl \xrightarrow{-Cl^-} R^1R^2C{=}\overset{+}{N}(OH)NH_2 \xrightarrow{-H_2O,\ -H^+} R^1R^2C{=}N_2 \qquad (8)$$

1. Diazomethane

The action of a chloramine solution on the sodium salt of formaldehyde oxime furnishes diazomethane in satisfactory yield (70–75%) (equation 9)[40]. On use of hydroxylamine-*O*-sulphonic acid in place of chloramine for condensation with the oxime salt, the acid strength of the former reagent adversely affects the yield of the proton-sensitive compound diazomethane[40]. Application of the Forster reaction to the oximes of alkylated ketones has met with little success[40].

$$H_2C{=}N{-}O^-\ Na^+ + H_2NCl \xrightarrow{\text{ether, methanol, 0°C}} H_2C{=}N_2 + NaCl + H_2O \qquad (9)$$

2. Aryldiazomethanes

Although 9-diazofluorene (**18**), diazodiphenylmethane (**19**), 1-diazo-1-phenylethane (**20**) and diazophenylmethane (**21**) are accessible by the Forster reaction of the corresponding oximes with chloramine or hydroxylamine-*O*-sulphonic acid[39],

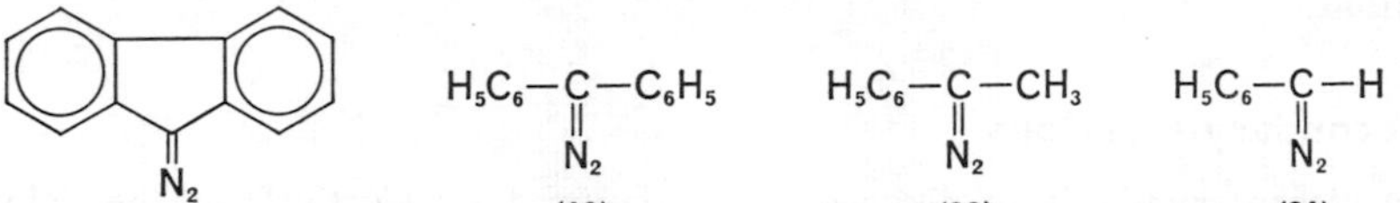

N_2 **(18)** $H_5C_6{-}C({=}N_2){-}C_6H_5$ **(19)** $H_5C_6{-}C({=}N_2){-}CH_3$ **(20)** $H_5C_6{-}C({=}N_2){-}H$ **(21)**

the yields are far exceeded by reactions to be considered later. 1-(2-Diazoethylidene)-indene (**22**) has so far only been prepared by the Forster reaction (63%)[41]. Condensation of 9-oximino-10-anthrone with chloramine is not uniform because the

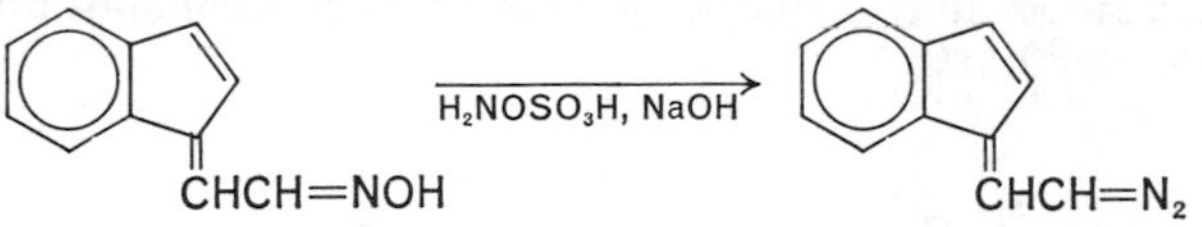

CHCH=NOH $\xrightarrow{H_2NOSO_3H,\ NaOH}$ CHCH=N_2 **(22)**

product, 9-diazo-10-anthrone, is accompanied by 9-imino-10-anthrone (equation 10)[42]. Branching of the reaction is assumed to occur at the 9-amino-9-nitroso-10-anthrone stage[42], the rather obvious alternative of 9-diazo-10-anthrone undergoing Cl/N insertion with chloramine to give 9-amino-9-chloro-10-anthrone apparently being neglected; elimination of hydrogen chloride from the last-named compound would afford 9-imino-10-anthrone.

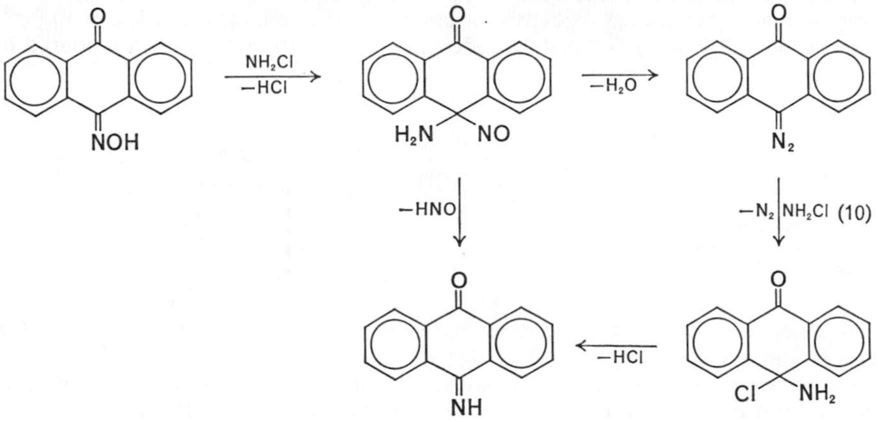

O, NOH $\xrightarrow[-HCl]{NH_2Cl}$ O, H_2N NO $\xrightarrow{-H_2O}$ O, N_2

$-HNO$ ↓ $\qquad$ $-N_2$ NH_2Cl ↓ (10)

O, NH $\xleftarrow{-HCl}$ O, Cl NH_2

D. Dehydrogenation of Hydrazones

Dehydrogenation of hydrazones to diazoalkanes (equation 11) is one of the oldest diazoalkane syntheses[43] known. While mercuric oxide was the dominant dehydrogenation reagent, silver oxide and manganese dioxide have recently found wide

$$R^1R^2C{=}N{-}NH_2 \xrightarrow{-2\,H} R^1R^2C{=}N_2 \qquad (11)$$

application and provided fresh incentive for revival of the method. Far less frequent use is made of mercuric trifluoro-acetate and -acetamide, and of iodine, phenyliodosoacetate, nickel peroxide, hydrogen peroxide, peroxyacetic acid, calcium hypochlorite or even atmospheric oxygen. Before discussing the individual dehydrogenation reagents and their scope, we shall first consider a number of secondary reactions.

I. Secondary reactions

An elegant synthesis of alkynes was discovered by Curtius[43] in the dehydrogenation of benzil bis(hydrazone) with mercuric oxide; diphenylacetylene was formed instead of the initially expected 1,2-bis(diazo)-1,2-diphenylethane (equation 12). There is no reason to doubt the intermediacy of the bis(diazo)alkane, even if no such compound has yet been isolated. Dehydrogenation with silver trifluoroacetate has been successfully employed for the production of substituted diphenylacetylenes (yields ⩾80%)[44].

$$H_5C_6{-}C({=}N{-}NH_2){-}C({=}N{-}NH_2){-}C_6H_5 \xrightarrow[-2\,Hg,\ -2\,H_2O]{2\,HgO} \left[H_5C_6{-}C({=}N_2){-}C({=}N_2){-}C_6H_5\right] \xrightarrow{-2\,N_2} C_6H_5{-}C{\equiv}C{-}C_6H_5 \qquad (12)$$

This reaction acquired particular significance in the synthesis of highly strained cycloalkynes (**23**). Whereas the existence of cyclopentyne (**23**, $n = 5$), cyclohexyne (**23**, $n = 6$) and cycloheptyne (**23**, $n = 7$) is only surmised from the products of trapping reactions with 2,5-diphenyl-3,4-benzofuran[45], cyclooctyne (**23**, $n = 8$)[45, 46], cyclononyne (**23**, $n = 9$)[47, 48] and cyclodecyne (**23**, $n = 10$)[49] have been prepared in isolable form.

$$(CH_2)_{n-2}\,[C({=}N{-}NH_2){-}C({=}N{-}NH_2)] \xrightarrow[-2\,Hg,\ -2\,H_2O]{2\,HgO} \left[(CH_2)_{n-2}\,[C({=}N_2){-}C({=}N_2)]\right] \xrightarrow{-2\,N_2} (CH_2)_{n-2}\,[C{\equiv}C]\ \ (\mathbf{23})$$

$n = 5, 6, 7, 8, 9, 10$

On dehydrogenation of hydrazones with an α-azomethine group, the resulting diazoalkanes undergo very fast 1,5-cyclization. Thus reaction of pyridine-2-aldehyde with silver oxide or potassium hexacyanoferrate(III) directly affords triazolopyridine (**25**), without detectable formation of diazo-2-pyridylmethane (**24**)[50,51].

Dehydrogenation → (24) — 1,5-Ring closure → (25)

Evidence that the 1,5-cyclization **(24)** → **(25)** is very fast can be gathered from the product distribution obtained on dehydrogenation of the bishydrazone **(26)**. Acetylene formation **(27)** → **(28)** is completely overshadowed by the cyclization reaction **(27)** → **(29)**[51].

(26) → (27) → (28) (not formed); (27) → (29)

2. Dehydrogenation with mercuric oxide

Dehydrogenation of hydrazones with mercuric oxide is usually carried out in solvents such as benzene, toluene, chloroform or light petroleum with admixture of compounds capable of binding water, like sodium sulphate, and catalysed by an ethanolic solution of potassium hydroxide. The dehydrogenations of benzophenone hydrazone and fluorenone hydrazone to give diazodiphenylmethane[52] and 9-diazofluorene[53], respectively, can be regarded as model reactions. The joint action of mercuric oxide and alkali hydroxide in the dehydrogenation reaction can be

$$(H_5C_6)_2C{=}N{-}NH_2 \xrightarrow{HgO} (H_5C_6)_2C{=}N{-}\overset{+}{N}H_2{-}HgO^- \xrightarrow{H^+\sim} (H_5C_6)_2C{=}N{-}NH{-}Hg{-}OH \tag{13}$$

$$\xrightarrow[-H_2O]{OH^-} (H_5C_6)_2C{=}N{-}\bar{N}{-}Hg{-}OH \xrightarrow[-Hg,\ -OH^-]{} (H_5C_6)_2C{=}N_2$$

envisaged as shown for diazodiphenylmethane in equation (13)[52]. Since no intermediates have been isolated in this reaction and oxygen dehydrogenations apparently proceed via hydrazone anions[54], the function of the hydroxyl ions can justifiably be assumed to lie in primary deprotonation.

In place of secondary diazoalkanes, azines are sometimes formed in dehydrogenation reactions with mercuric oxide[55–59]. This is more probably due to the instability of the diazoalkanes rather than a consequence of, say, a mercury-catalysed decomposition reaction. Experience gained so far indicates that the formation of mercury-bis(diazoalkanes) on dehydrogenation of hydrazones is to be expected[60] only if the group R is a powerful proton activator (equation 14).

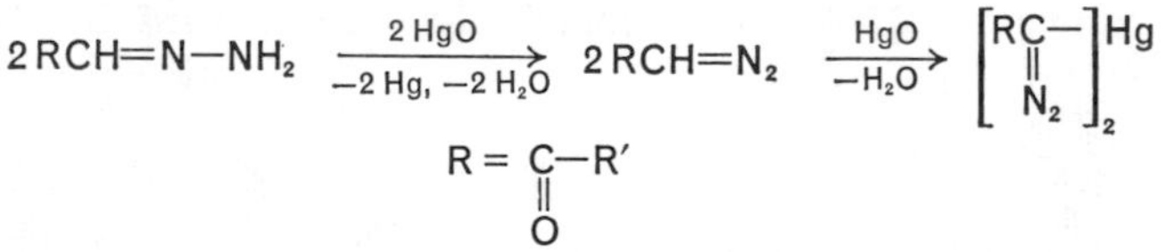

$$2\,RCH{=}N{-}NH_2 \xrightarrow[-2\,Hg,\ -2\,H_2O]{2\,HgO} 2\,RCH{=}N_2 \xrightarrow[-H_2O]{HgO} \left[\begin{matrix}RC{-}\\ \| \\ N_2\end{matrix}\right]_2 Hg \qquad R = \underset{\underset{O}{\|}}{C}{-}R' \tag{14}$$

The effect of solvent on success or failure of hydrazone dehydrogenations can be seen in the dehydrogenation of benzaldehyde hydrazone. While the aryldiazoalkanes are formed in benzene, light petroleum, dioxane or ether, reaction in 1,2-dimethoxyethane gives arylnitriles; this is ascribed to the high solvent polarity (equation 15)[61].

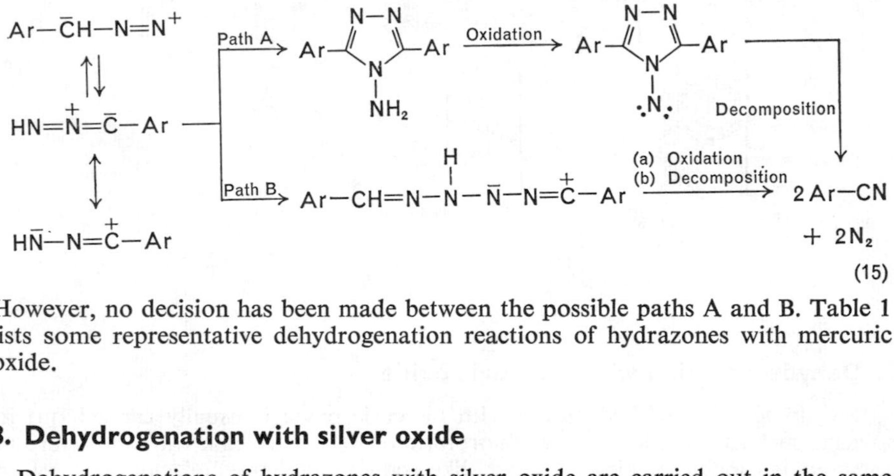

$$\begin{matrix} Ar{-}\bar{C}H{-}N{=}N^{+} \\ \uparrow\downarrow \\ HN{=}\overset{+}{N}{=}\bar{C}{-}Ar \\ \updownarrow \\ H\bar{N}{-}N{=}\overset{+}{C}{-}Ar \end{matrix}
\begin{matrix} \xrightarrow{\text{Path A}} \text{3,5-diaryl-4-amino-1,2,4-triazole (N-NH}_2) \xrightarrow{\text{Oxidation}} \text{3,5-diaryl-1,2,4-triazol-4-yl nitrene } (N{-}\ddot{N}{:}) \xrightarrow{\text{Decomposition}} \\ \xrightarrow{\text{Path B}} Ar{-}CH{=}N{-}\overset{H}{\overset{|}{N}}{-}\bar{N}{-}N{=}\overset{+}{C}{-}Ar \xrightarrow{\text{(a) Oxidation, (b) Decomposition}} 2\,Ar{-}CN + 2\,N_2 \end{matrix} \tag{15}$$

However, no decision has been made between the possible paths A and B. Table 1 lists some representative dehydrogenation reactions of hydrazones with mercuric oxide.

3. Dehydrogenation with silver oxide

Dehydrogenations of hydrazones with silver oxide are carried out in the same solvents as the above reactions with mercuric oxide; once again base catalysis is just as necessary as the addition of reagents for binding water. The rates of dehydrogenation appear relatively high, thereby favouring unstable diazoalkanes. Thus even at low temperature, dehydrogenation of cyclohexanone hydrazone with Ag_2O still yields diazocyclohexane at an adequate rate, whereas the HgO reaction merely gives azine over a wide temperature range (equation 16)[58]. Silver carbonate dehydrogenates hydrazones very fast, but precisely that reagent also seems to be

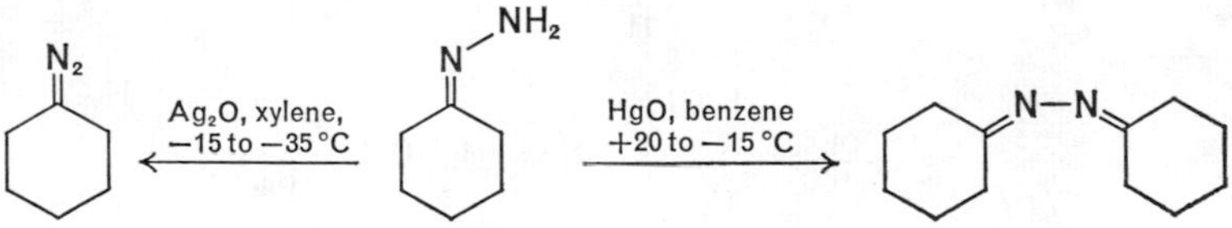

$$\text{diazocyclohexane } (C_6H_{10}{=}N_2) \xleftarrow[-15\text{ to }-35\,^{\circ}C]{Ag_2O,\ \text{xylene}} C_6H_{10}{=}N{-}NH_2 \xrightarrow[+20\text{ to }-15\,^{\circ}C]{HgO,\ \text{benzene}} C_6H_{10}{=}N{-}N{=}C_6H_{10} \tag{16}$$

TABLE 1. Aryl-, heteroaryl- and alkyl-diazomethanes by dehydrogenation with HgO

Diazoalkane	Solvent	Yield (%)	Reference
$H_5C_6CH{=}N_2$	Petroleum ether	45–56	57
	Pentane	38	62
	Ether[a]	≥54	63
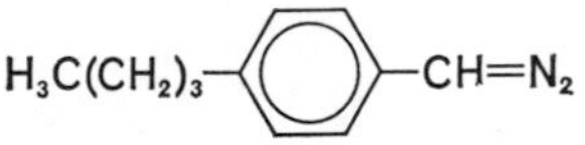 $H_3C(CH_2)_3-C_6H_4-CH{=}N_2$ (para)	Ether[a]	35–45	64
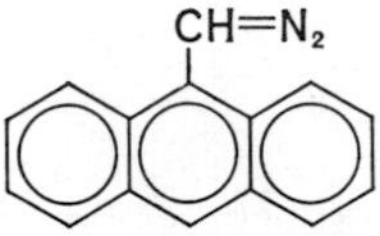 9-anthryl–$CH{=}N_2$	Ether[a]	65	65
$(H_5C_6)_2C{=}N_2$	Petroleum ether	85–98	66, 67
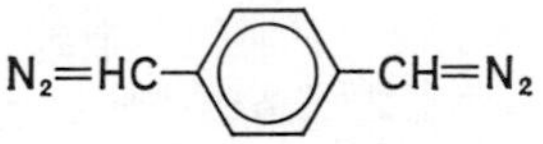 $N_2{=}HC-C_6H_4-CH{=}N_2$ (para)	Benzene[a]	—	68
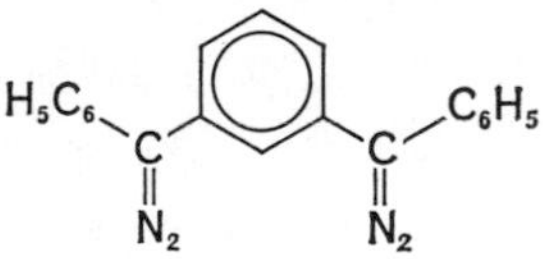 $H_5C_6-C(=N_2)-C_6H_4-C(=N_2)-C_6H_5$ (meta)	Ether[a]	80	69
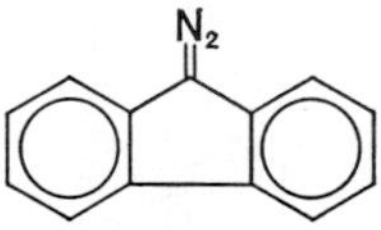 9-diazofluorene (N_2)	Ether[a]	fast 100	53, 55
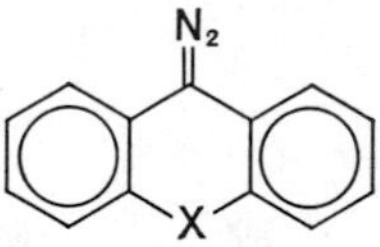 9-diazo tricycle, X = CO	Tetrahydrofuran	96	42
X = S	Ether	50	70
X = O	Ether[a]	—	71
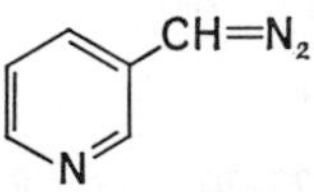 3-pyridyl–$CH{=}N_2$	Ether	68	72
$(H_3C)_2C{=}N_2$	Xylene	—	57, 73
	Ether[a]	70–90	59, 74
$H_5C_6-C(=N_2)-CH_3$	Petroleum ether	—	57
	Ether[a]	60	75
$H_5C_6-C(=N_2)-CF_3$	Ether[a]	84	14

[a] KOH-catalysis.

TABLE 2. Aryl- and alkyl-diazomethanes by dehydrogenation with Ag_2O

Diazoalkane	Solvent	Yield (%)	Reference
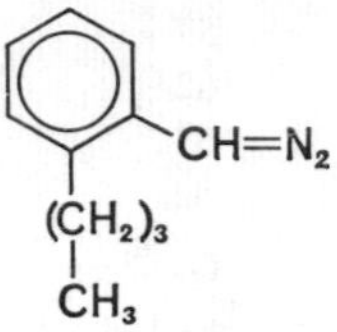 $CH{=}N_2$; $(CH_2)_3$; CH_3	Ether	99	77
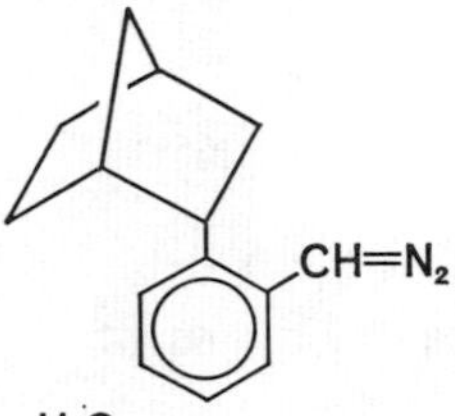 $CH{=}N_2$	Cyclohexane	77	78
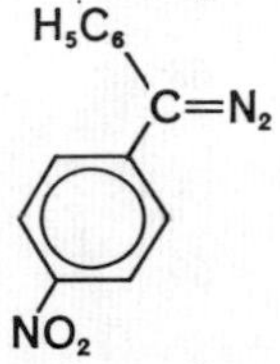 H_5C_6; $C{=}N_2$; NO_2	Petroleum ether	95	79
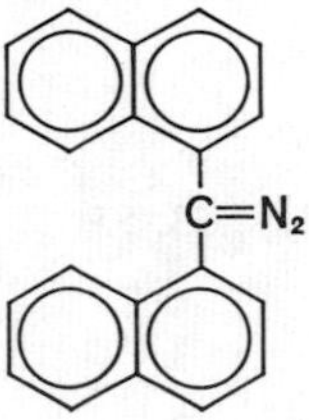 $C{=}N_2$	Benzene	—	80
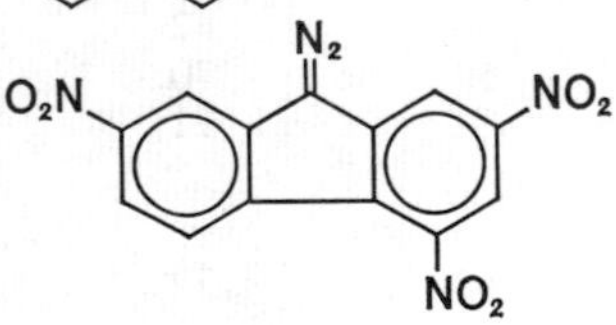 N_2; O_2N; NO_2; NO_2	Tetrahydrofuran	89	81
$(H_3C)_2C{=}N_2$	Ether	20–30	82
$H_3C{-}C({=}N_2){-}CF_3$	Ether[a]	—	83
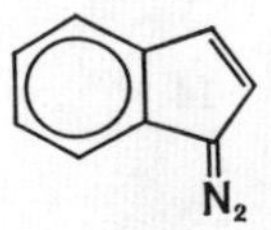 N_2	Ether[a]	93	84

[a] KOH-catalysis.

responsible for secondary reactions, hence brief contact with the dehydrogenating reagent gives diazodiphenylmethane in 88% yield, while prolonged reaction leads to

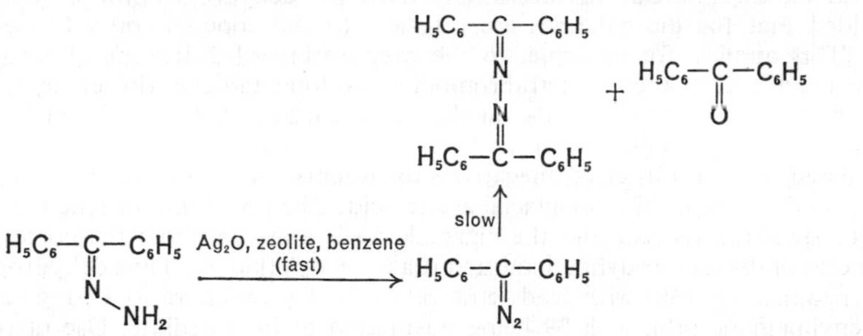

$$\mathrm{H_5C_6{-}C({=}N{-}NH_2){-}C_6H_5} \xrightarrow[\text{(fast)}]{\mathrm{Ag_2O}\text{, zeolite, benzene}} \mathrm{H_5C_6{-}C({=}N_2){-}C_6H_5} \xrightarrow{\text{slow}} (\mathrm{H_5C_6})_2\mathrm{C{=}N{-}N{=}C}(\mathrm{C_6H_5})_2 + \mathrm{H_5C_6{-}C({=}O){-}C_6H_5} \quad (17)$$

benzophenone azine (49%) and benzophenone (48%) (equation 17)[76]. Further examples are listed in Table 2.

4. Dehydrogenation with manganese dioxide

Dehydrogenation of hydrazones is carried out with activated manganese freshly prepared from potassium permanganate and manganese(II) sulphate tetrahydrate[85]. As found in the synthesis of (4-chlorphenyl)diazophenylmethane[86], use of manganese dioxide prepared in this way obviates the need for large excesses of dehydrogenation reagent. The most commonly used solvent is chloroform[87, 88]. In the preparation of diazophenyl(triphenylsilyl)methane (**30**) manganese dioxide

$$(\mathrm{H_5C_6})_3\mathrm{Si{-}C({=}N{-}NH_2){-}C_6H_5} \xrightarrow{\mathrm{MnO_2}} (\mathrm{H_5C_6})_3\mathrm{Si{-}C({=}N_2){-}C_6H_5}$$

(**30**) 95%

is superior to both mercuric oxide and silver oxide[89]. The preparative value of manganese dioxide in the dehydrogenation of hydrazones is seen, *inter alia*, in the production of 1,4-bis(diazobenzyl)benzene (**31**)[90], 1,3-bis(diazobenzyl)benzene (**32**)[68], diazo-2-furylmethane (**33a**)[91], diazo-2-thienylmethane (**33b**)[91], diazo-(phenyl-4-pyridyl)methane (**34**)[92], diazo(4-pyridyl)methane (**35**)[93], 4-diazomethyl-1,5-diphenyl-1,2,3-triazole (**36**)[94] and bis(1-adamantyl)diazomethane (**37**)[95].

$\mathrm{H_5C_6{-}C({=}N_2){-}C_6H_4{-}C({=}N_2){-}C_6H_5}$ (**31**) 92%

$\mathrm{H_5C_6{-}C({=}N_2){-}C_6H_4{-}C({=}N_2){-}C_6H_5}$ (1,3) (**32**) 100%

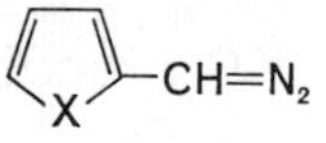

(**33**) 2-($\mathrm{CH{=}N_2}$)-heterocycle, X in ring
a: X = O, 50%
b: X = S, 69%

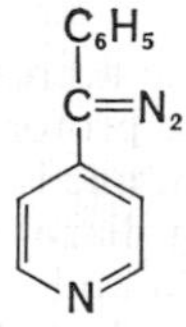

(**34**) 80% — 4-pyridyl–$\mathrm{C(C_6H_5){=}N_2}$

(**35**)— — 4-pyridyl–$\mathrm{CH{=}N_2}$

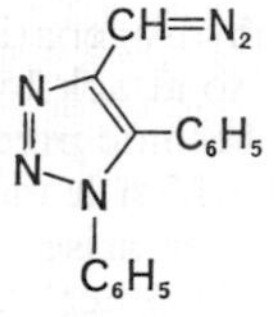

(**36**) 48% — 1,2,3-triazole with $\mathrm{CH{=}N_2}$, $\mathrm{C_6H_5}$, N–$\mathrm{C_6H_5}$

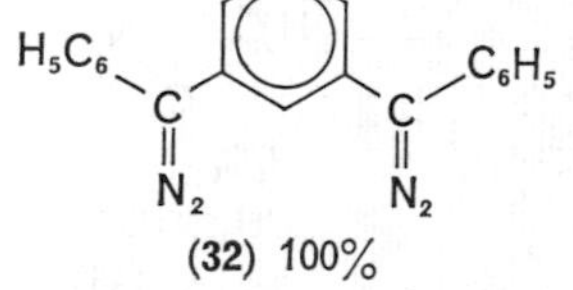

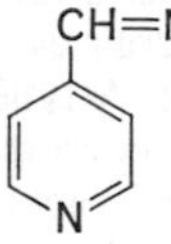

$(1\text{-Ad})_2\mathrm{C} = \mathrm{N_2}$
1-Ad = (adamantyl)
(**37**) 100%

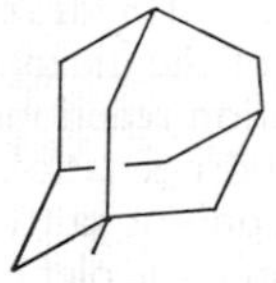

5. Dehydrogenation with lead tetraacetate

Lead tetraacetate can be successfully used for dehydrogenation of hydrazones provided that the diazoalkanes are resistant to the concomitantly formed acetic acid. This applies, for example, to the preparation of 2-diazohexafluoropropane (in benzonitrile)[96] and of tetrabromodiazocyclopentadiene (in ether)[97]. In the former case the electron density on the diazo carbon atom is reduced by the −I effect of the CF_3 groups and in the latter case by charge delocalization in the five-membered ring with its electronegative substituents. In both instances this prevents attack and decomposition by glacial acetic acid. The possibility of reaction between dehydrogenation reagent and the diazoalkane is apparent from the corresponding syntheses of diazodiphenylmethane and diazophenylethane[98]. Thus dehydrogenation of benzophenone (**38**) with lead tetraacetate in the molar ratio 1 : 1 gives diazodiphenylmethane (**40**), with **39** being postulated as intermediate. Use of twice the molar amount of dehydrogenation reagent, however, leads to complete decomposition of **40** by acetic acid (**40** → **44**) and by lead tetraacetate, product **41** being formed via the presumed intermediates **43** and **42**[98, 99]. These secondary reactions have already been observed to accompany azine formation at molar ratios of 1 : 1 on dehydrogenation of acetophenone hydrazone[98].

$$(H_5C_6)_2C{=}N{-}NH_2\ \textbf{(38)} \xrightarrow[-CH_3COOH]{Pb(OCOCH_3)_4,\ benzene} (H_5C_6)_2C{=}N{-}NH{-}Pb(OCOCH_3)_3\ \textbf{(39)}$$

$$\textbf{(39)} \xrightarrow[-CH_3COOH]{-Pb(OCOCH_3)_2} (H_5C_6)_2C{=}N_2\ \textbf{(40)}$$

$$\textbf{(40)} \xrightarrow[-N_2]{CH_3COOH} (H_5C_6)_2CHOCOCH_3\ \textbf{(44)}$$

$$\textbf{(40)} \xrightarrow{Pb(OCOCH_3)_4} (H_5C_6)_2C(N_2^+)Pb(OCOCH_3)_3\ CH_3COO^-\ \textbf{(43)} \xrightarrow{-N_2} (H_5C_6)_2C(OCOCH_3)Pb(OCOCH_3)_3\ \textbf{(42)} \xrightarrow{-Pb(OCOCH_3)_2} (H_5C_6)_2C(OCOCH_3)_2\ \textbf{(41)}$$

The utility of hydrazone dehydrogenation with lead tetraacetate decreases as the sensitivity of the diazoalkanes to attack by acid increases, because proton-catalysed decomposition reactions then become predominant. A typical example is to be seen in the dehydrogenation of **45** which leads via the corresponding diazo compound to the diazonium salt (**46**): thence arise the epimeric acetates **47a** and **47b**, on the one hand, and the olefin **48**, on the other[100]. 3β-Hydroxy-24-anosten-7-one hydrazone behaves in an essentially similar manner[101].

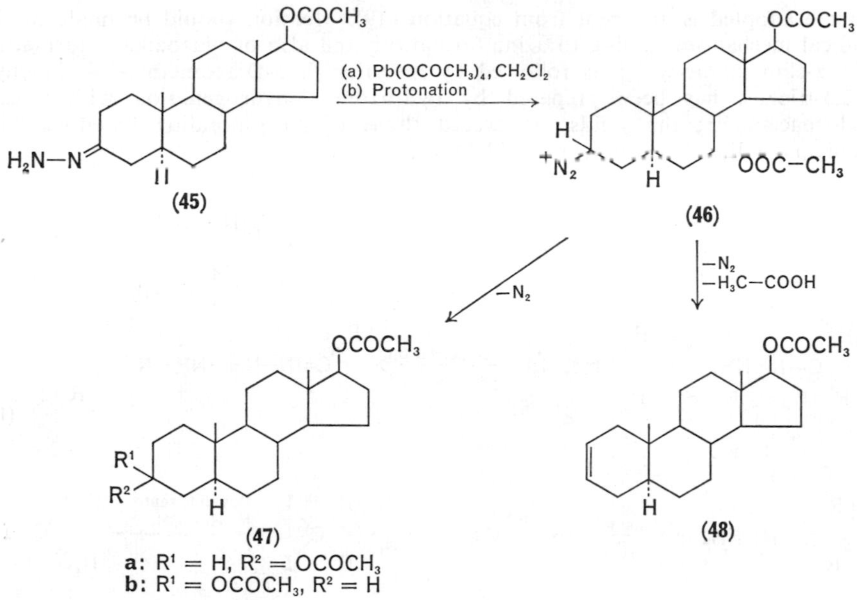

$OCOCH_3$ $OCOCH_3$

$H_2N{-}N$ (a) $Pb(OCOCH_3)_4, CH_2Cl_2$
(b) Protonation
H $^{+}N_2$ H $OOC{-}CH_3$

(45) (46)

$-N_2$ $-N_2$ $-H_3C{-}COOH$

$OCOCH_3$ $OCOCH_3$

R^1 R^2 H H

(47) (48)

a: $R^1 = H, R^2 = OCOCH_3$
b: $R^1 = OCOCH_3, R^2 = H$

6. Dehydrogenation with iodine

Diazodiphenylmethane[102] and tetrachlorodiazocyclopentadiene[103] have been prepared by dehydrogenation of hydrazones with iodine; however, both reactions require the presence of a base (triethylamine) to neutralize the hydrogen iodide. Among the possible secondary reactions of diazoalkane formation are iodine insertion and azine formation, which are observed in the absence of base[102, 104]. Under suitable conditions both these secondary reactions are encountered with 3β-hydroxy-5α-pregnan-20-one hydrazone. The geminal diiodo compound undoubtedly formed initially undergoes elimination of HI, apparently under the influence of the base (equation 18)[102]. The effect of adding base upon the reaction

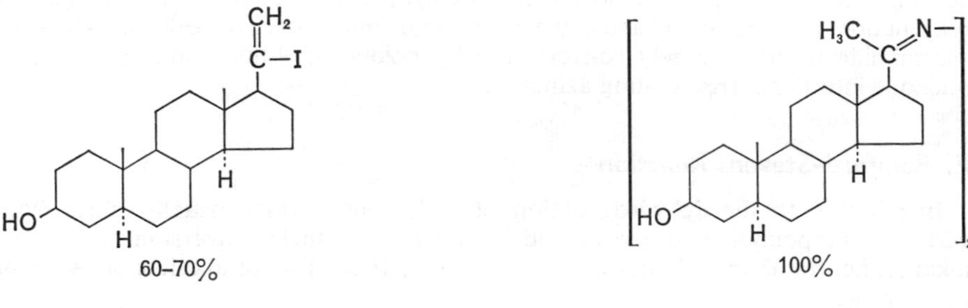

$H_3C{-}C{=}N{-}NH_2$

$I_2, N(C_2H_5)_3$, THF, 20 °C I_2, THF, 20 °C

HO H H (18)

$CH_2{=}C{-}I$ $[H_3C{-}C{=}N{-}]_2$

HO H H HO H H

60–70% 100%

course adopted is apparent from equation (19). Mention should be made of the radical mechanism leading to azine formation, and also of diazoalkane formation via α-elimination which is followed by insertion[101]. 5-Diazomethyl-1,4-diphenyl-1,2,3-triazole has been prepared by hydrazone dehydrogenation with phenyl iodosoacetate[105]; the yields far exceed those of corresponding reactions with manganese dioxide or mercuric oxide[106].

$$R_2C{=}N{-}NH_2 \xrightarrow{I_2} R_2C{=}N{-}\dot{N}H \xrightarrow{\text{Dimerization}} R_2C{=}N{-}NH{-}NH{-}N{=}CR_2 \xrightarrow{-H_2,\ -N_2} R_2C{=}N{-}N{=}CR_2 \qquad (19)$$

$$R_2C{=}N{-}NH_2 \xrightarrow{I_2\text{(base)}} R_2C{=}N{-}NHI \xrightarrow[-HI]{\alpha\text{-Elimination}} R_2C{=}N_2 \xrightarrow[-N_2]{I_2} R_2CI_2 \xrightarrow{\text{Possible route for } R = CH_3} H_2C{=}C(R){-}I$$

7. Miscellaneous

In the presence of ethanol/sodium ethoxide, 9-fluorenone hydrazone is dehydrogenated by oxygen to 9-diazofluorene[107]; the reaction can be generalized insofar as various hydrazones metallated with methyllithium react analogously with oxygen. Peroxy intermediates are assumed to precede diazoalkane formation (equation 20)[54].

$$R^1R^2C{=}N{-}NH_2 \xrightarrow{CH_3Li,\ THF} R^1R^2C{=}N{-}\bar{N}H \xrightarrow{O_2} R^1R^2C{=}N{-}N(H){-}O{-}O^- \qquad (20)$$

$$\xrightarrow{H\sim} R^1R^2C{=}N{-}\bar{N}{-}O{-}O{-}H \xrightarrow{-^-OOH} R^1R^2C{=}N_2 \qquad \begin{matrix} R^1 = R^2 = C_6H_5 \\ R^1 = C_6H_5,\ R^2 = CH_3 \end{matrix}$$

Nickel peroxide[108] has proved of value in the preparation of diazodiphenylmethane (100%), 1-diazo-1-phenylethane (56%), diazophenylmethane and 9-diazofluorene (92%)[109].

Finally, the dehydrogenation of tetrachlorocyclopentadienone hydrazone can be accomplished with sodium hypochlorite in ethyl acetate/methanol[110], although this reagent does not appear to have any particular advantage over the others. *N*-Bromosuccinimide in ether merely converts the hydrazone of 9-fluorenone and benzophenone into the corresponding azines[111].

E. Bamford–Stevens Reaction

In addition to the dehydrogenation of hydrazones, which usually starts from carbonyl compounds, another method is available for their conversion into diazoalkanes, i.e. the Bamford–Stevens reaction[112, 113]. It consists of alkaline cleavage of tosylhydrazones (equation 21), which are normally synthesized from tosylhydrazine and suitable CO compounds. This reaction has proved of synthetic

$$R^1R^2C{=}N{-}NH{-}SO_2{-}C_6H_4{-}CH_3 \underset{BH,\ -B^-}{\overset{^-B,\ -BH}{\rightleftharpoons}} R^1R^2C{=}N{-}\ddot{N}^-{-}SO_2{-}C_6H_4{-}CH_3 \longrightarrow R^1R^2C{=}N_2 + {}^-O_2S{-}C_6H_4{-}CH_3 \quad (21)$$

utility not only in diazoalkane chemistry but also by virtue of various secondary and alternative reactions, of which the carbene and carbonium ion reactions seen in equation (22) are probably the most important; the former occurs in aprotic

$$R_2C{=}N_2 \xrightarrow[-N_2]{\Delta\text{ (aprotic solvent)}} R_2C{:} \longrightarrow \text{Consecutive reactions}$$

$$R_2C{=}N_2 \xrightarrow{\text{(protic solvent)}} R_2CH{-}N_2^+ \xrightarrow{-N_2} R_2CH^+ \longrightarrow \text{Consecutive reactions} \quad (22)$$

media, and the latter in protic media at elevated temperature[114, 115, 116]. With regard to carbene chemistry, the significance of the Bamford–Stevens reaction lies in the intentional *in situ* generation of diazoalkanes with subsequent elimination of N_2 [115, 116, 117]; it plays a comparable role in carbonium ion chemistry[116–121].

Admixture of at least a stoichiometric amount of base is essential for the success of the Bamford–Stevens reaction; otherwise unreacted hydrazone acting as a proton donor promotes decomposition of the diazoalkane by the carbonium pathway (equation 23). *N*-Alkylated tosylhydrazones are then obtained as reaction products[122–126].

$$R^1R^2C{=}N{-}N{-}Tos \xrightarrow{-H^+} R^1R^2C{=}N{-}\bar{N}{-}Tos$$

$$R^1R^2C{=}N_2 \xrightarrow[-N_2]{H^+\text{(tosylhydrazone)}} R^1R^2CH^+$$

$$\longrightarrow R^1R^2C{=}N{-}N(Tos){-}CHR^1R^2 \quad (23)$$

I. Alternative and secondary reactions

Bases such as sodium hydride, sodium amide or organometallic reagents (twofold molar amount) can also deprotonate tosylhydrazones in the α-position to the azomethine group, if this is structurally permitted. Olefins are then formed without there being any need for the intermediacy of diazoalkanes or carbenes[114, 127, 128, 129].

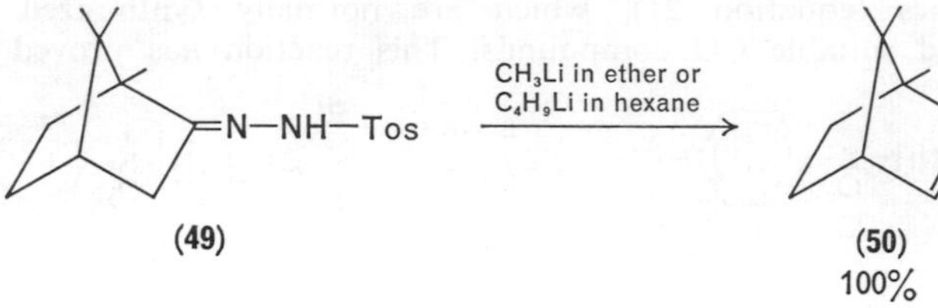

=N—NH—Tos

CH$_3$Li in ether or
C$_4$H$_9$Li in hexane

(49) → (50) 100%

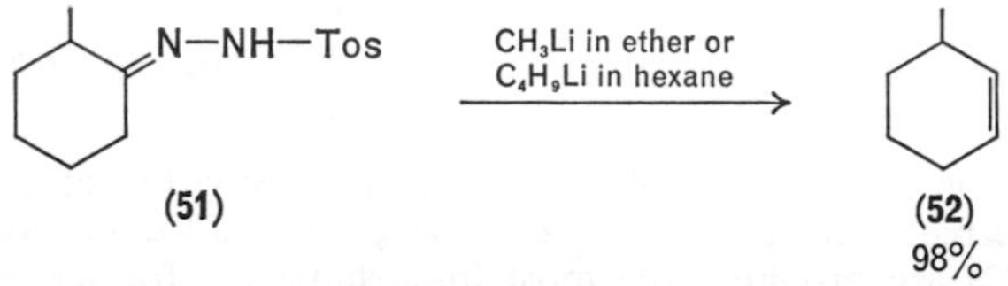

N—NH—Tos

CH$_3$Li in ether or
C$_4$H$_9$Li in hexane

(51) → (52) 98%

Supporting evidence comes from the corresponding reactions of camphor- and 2-methyl-cyclohexanone tosylhydrazone (**49** → **50** and **51** → **52**)[128]. Deuteriation experiments substantiate the mechanism shown in equation (24)[128]. Azine formation is only rarely observed in the Bamford–Stevens reaction[130–132]; thus cyclododecanone tosylhydrazone yields up to 60% of azine together with *cis*- and *trans*-cyclododecene. Conventional mechanisms of formation starting from the diazoalkane have to give way here to other concepts[130, 131].

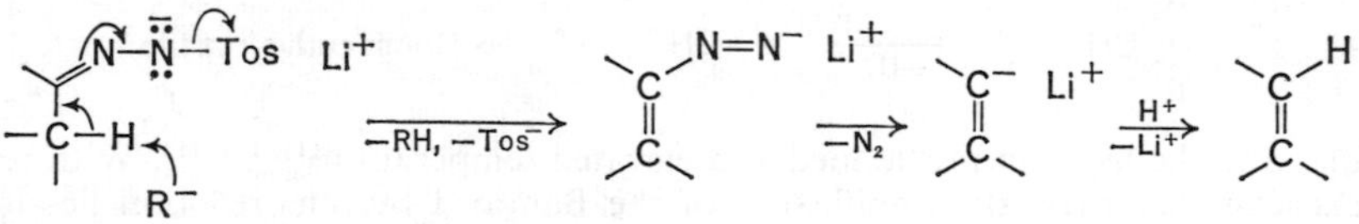

N—N̈—Tos Li$^+$ / —C—H / R$^-$ —(−RH, −Tos$^-$)→ C(N=N$^-$ Li$^+$)=C —(−N$_2$)→ C$^-$ Li$^+$=C —(H$^+$, −Li$^+$)→ C(H)=C (24)

Attempts to isolate 1,2-bis(diazo)alkanes on alkaline cleavage of 1,2-bis(tosylhydrazones) are just as unsuccessful as on dehydrogenation of hydrazones. Starting from benzil bis(tosylhydrazone) (**53a**), diphenylacetylene is formed via the hypothetical species **54**[113], whereas cleavage of 2,3-butanedione bis(tosylhydrazone) (**53b**) via the α-diazoimine (**55**) with subsequent 1,5-cyclization yields the triazole (**56**)[113, 133].

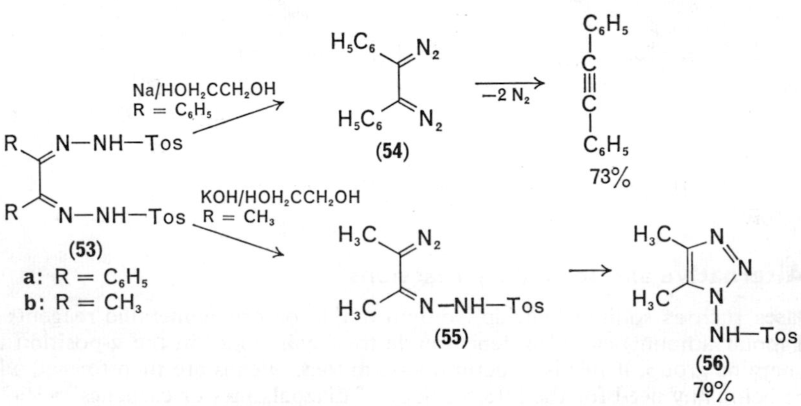

R—C(=N—NH—Tos)—C(=N—NH—Tos)—R (**53**)
a: R = C$_6$H$_5$
b: R = CH$_3$

Na/HOH$_2$CCH$_2$OH, R = C$_6$H$_5$ → H$_5$C$_6$C(N$_2$)—C(N$_2$)C$_6$H$_5$ (**54**) —(−2 N$_2$)→ C$_6$H$_5$—C≡C—C$_6$H$_5$ 73%

KOH/HOH$_2$CCH$_2$OH, R = CH$_3$ → H$_3$C—C(N$_2$)—C(=N—NH—Tos)—CH$_3$ (**55**) → triazole H$_3$C, H$_3$C, N, N, N—NH—Tos (**56**) 79%

Like the bis(tosylhydrazone) (**53a**), α-hydroxy- and α-acyloxytosylhydrazones of type **57** give diphenylacetylene on Bamford–Stevens reaction; however, in the case of the latter type a diazo and a diazonium intermediate have been postulated (**58** and **59**)[113, 134].

A highly interesting fragmentation reaction affording acetylenes, carbonyl compounds, sulphinates and nitrogen was discovered in the basic cleavage of α,β-epoxy ketone tosylhydrazones (equation 25)[135–137]; this reaction is also suitable

(25)

for preparation of cycloalkynes[135]. In conclusion, the formation of alcohols (**60** and **61**), for example, from 1-norbornanaldehyde tosylhydrazone in *N*-methyl-2-pyrrolidone[138–141] warrants mention; the occurrence of a skeletal rearrangement in the case of **60** is of no relevance to formation of the alcohol function. The role of

N-methyl-2-pyrrolidone in the reaction is shown in equation (26)[139, 141]. The basic cleavage of methylsulphinylhydrazones[142] and of *o*-nitrophenylsulphenylhydrazones[143] can be regarded as variants of the classical Bamford–Stevens reaction; no advance has come through either of these reactions.

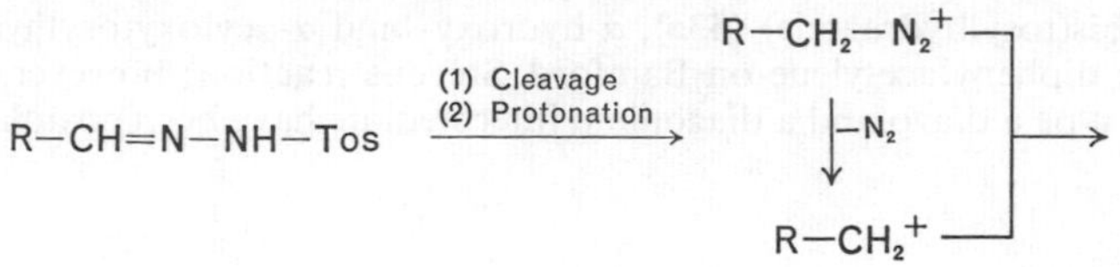

$$R{-}CH{=}N{-}NH{-}Tos \xrightarrow[\text{(2) Protonation}]{\text{(1) Cleavage}} R{-}CH_2{-}N_2^+ \xrightarrow{-N_2} R{-}CH_2^+ \longrightarrow \quad (26)$$

$$R{-}CH_2{-}O{-}SO{-}C_6H_4{-}CH_3 \xrightarrow{\Delta,\ \text{N-methylpyrrolidone}} R{-}CH_2OH$$

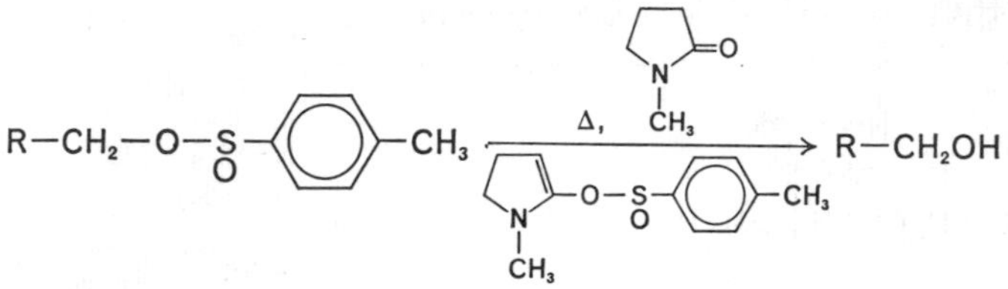

2. Aryldiazoalkanes and alkyldiazoalkanes

The tosylhydrazones of arylated and heteroarylated aldehydes and ketones undergo basic cleavage in the manner of a diazoalkane synthesis merely upon warming; the natural limitations of the methods are set by the thermal stability of the diazo compounds. Solvent/base systems such as ethanol/sodium ethoxide[113], triethylene glycol/sodium methoxide[144], dimethylformamide/diethylamine[145], and pyridine/sodium[146], etc., have proved suitable for arylated diazo compounds. Diazophenylmethane (**62a**)[113, 144, 146], diazo(4-nitrophenyl)methane (**62b**)[145], diazo-(4-methoxyphenyl)methane (**62c**)[146], diazodiphenylmethane (**62d**)[113], 1-diazo-1-phenylethane (**62e**)[113], 1-diazo-2,2,2-trifluoro-1-phenylethane (**62f**)[147], 5-diazo-dibenzo[*a*,*d*]cycloheptatriene (**63**)[148], 5-diazo-10,1-dihydrodibenzo[*a*,*d*]cyclohepta-triene (**64**)[149], 9-diazo-1,2-benzofluorene (**65**)[92], diazo(3-pyridyl)methane (**66**)[72] and 5-diazomethyl-1,4-diphenyl-1,2,3-triazole (**67**)[94] all testify to the broad scope of the reaction

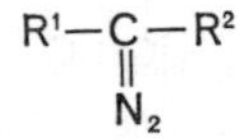

$R^1{-}C({=}N_2){-}R^2$ (**62**)

a: $R^1 = C_6H_5$; $R^2 = H$ (55–70%)
b: $R^1 = O_2N{-}C_6H_4{-}$; $R^2 = H$ (65–70%)
c: $R^1 = H_3CO{-}C_6H_4{-}$; $R^2 = H$ (59%)
d: $R^1 = R^2 = C_6H_5$ (58%)
e: $R^1 = C_6H_5$; $R^2 = CH_3$ (27–32%)
f: $R^1 = C_6H_5$; $R^2 = CF_3$ (≅30%)

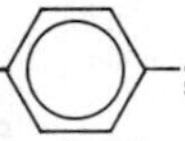

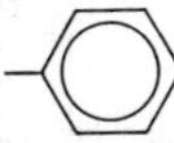

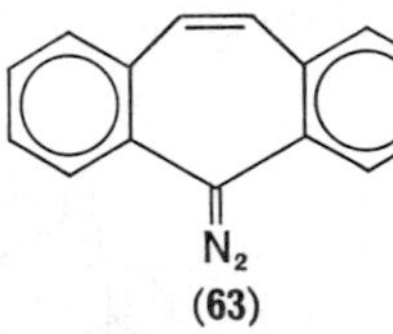

N_2 (**63**) 25%

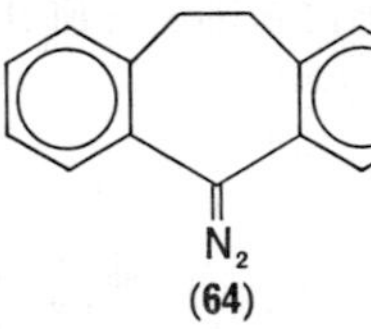

N_2 (**64**) 76%

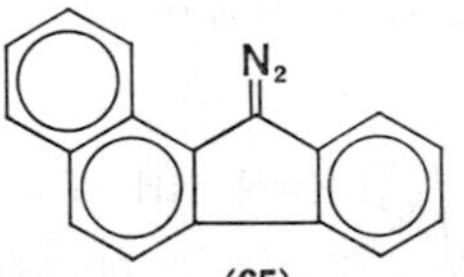

N_2 (**65**) 90%

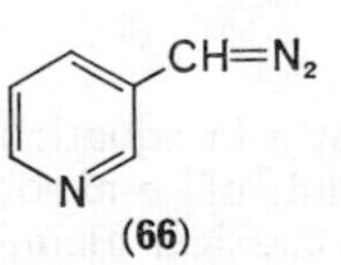

$CH{=}N_2$ (**66**) 60–80%

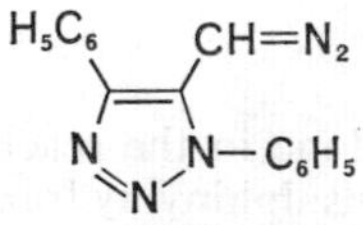

H_5C_6 $CH{=}N_2$ N N N$-C_6H_5$ (**67**) 55%

2-Allyloxyphenyldiazomethane, which is accessible by Bamford–Stevens reaction in the usual way, rapidly undergoes intramolecular 3+2 cycloaddition according to equation (27)[150]. 1,5-Cyclization of diazophenyl-2-pyridylmethane[50, 51, 92], a reaction already mentioned in connection with hydrazone dehydrogenation, is incomparably faster. Understandably, the isomeric 3- and 4-pyridyl derivatives do not undergo ring closure[72]. The same applies to 2-diazomethylpyridine *N*-oxide[151, 152].

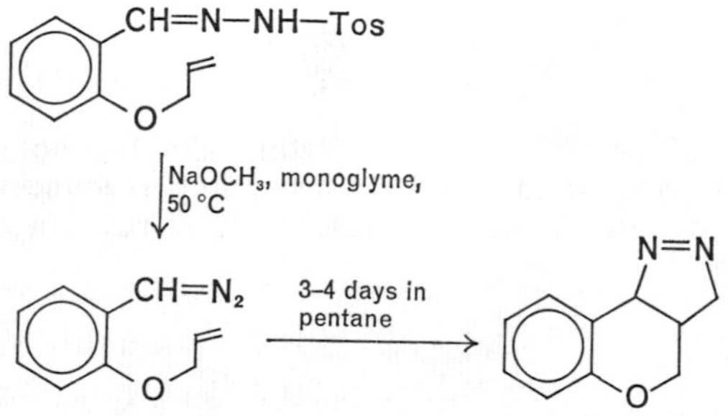

CH=N—NH—Tos

$NaOCH_3$, monoglyme, 50 °C

CH=N_2 → 3–4 days in pentane → N=N (27)

Alkyldiazoalkanes are extremely proton sensitive and cannot be prepared under the conditions described at the beginning of this section. Even those protons which arise from neutralization of tosylhydrazones are responsible for decomposition. In contrast they are very stable thermally, and can therefore be prepared by vacuum pyrolysis of alkali salts of tosylhydrazone (equation 28)[153]. In isolated cases (2-diazopropane, diazocyclopentane, diazocyclohexane) minimal yields ($\leqslant 5\%$) are unavoidable if the diazo compounds are unable to stand up to thermal stress[153]. Some aryldiazoalkanes are also accessible without solvent in this way[154].

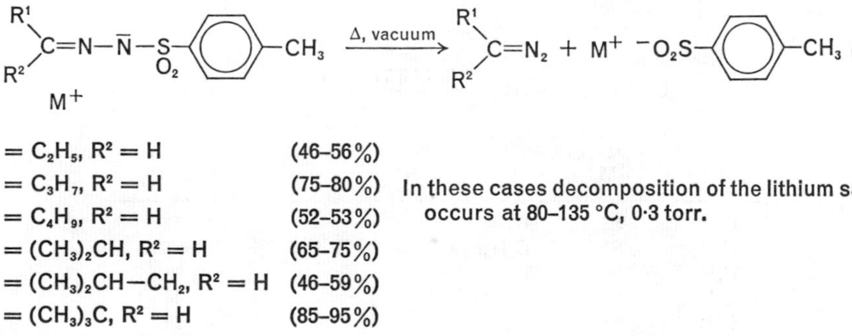

$$R^1R^2C{=}N{-}\bar{N}^-{-}SO_2{-}C_6H_4{-}CH_3\ M^+ \xrightarrow{\Delta,\ \text{vacuum}} R^1R^2C{=}N_2 + M^+\ {}^-O_2S{-}C_6H_4{-}CH_3 \quad (28)$$

$R^1 = C_2H_5$, $R^2 = H$ (46–56%)
$R^1 = C_3H_7$, $R^2 = H$ (75–80%)
$R^1 = C_4H_9$, $R^2 = H$ (52–53%)
$R^1 = (CH_3)_2CH$, $R^2 = H$ (65–75%)
$R^1 = (CH_3)_2CH{-}CH_2$, $R^2 = H$ (46–59%)
$R^1 = (CH_3)_3C$, $R^2 = H$ (85–95%)

In these cases decomposition of the lithium salts occurs at 80–135 °C, 0·3 torr.

More complex diazo compounds such as 2-diazo-1,1,3,3-tetramethylcyclobutane (**68**)[153], cyclopropyldiazophenylmethane (**69**)[153], 4-diazomethylcyclohexene (**70**)[155], 3-diazomethyl-3-methylcyclopentene (**71**)[156], 1-adamantyldiazomethane (**72**)[157] and diazohomoadamantane (**73**)[158] are also accessible in satisfactory yields by vacuum

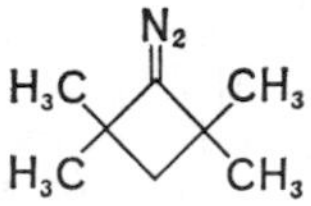

(68) 72%　　**(69)** 70%　　**(70)** 80%

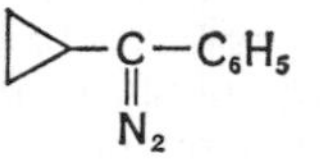

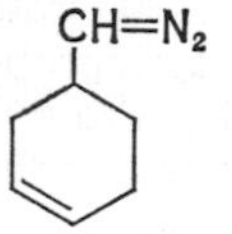

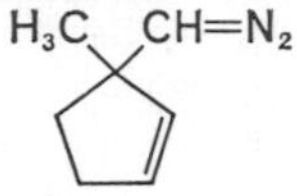

H_3C $CH{=}N_2$

(71)

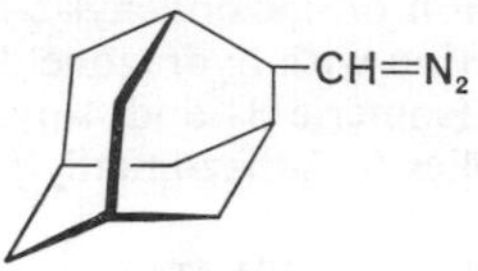

$-CH{=}N_2$

(72)
85%

N_2

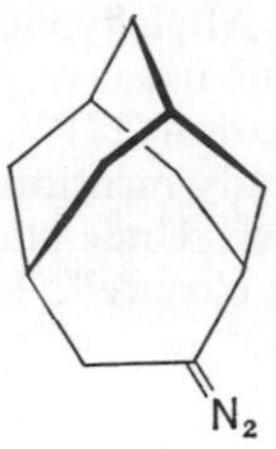

(73)

pyrolysis, usually of the sodium tosylhydrazone salts. It is not surprising that tosylhydrazones having alkenyl and alkyl groups form pyrazolines on alkaline cleavage without the diazo isomers being isolable[159, 160]. A relevant example is given in equation (29)[159].

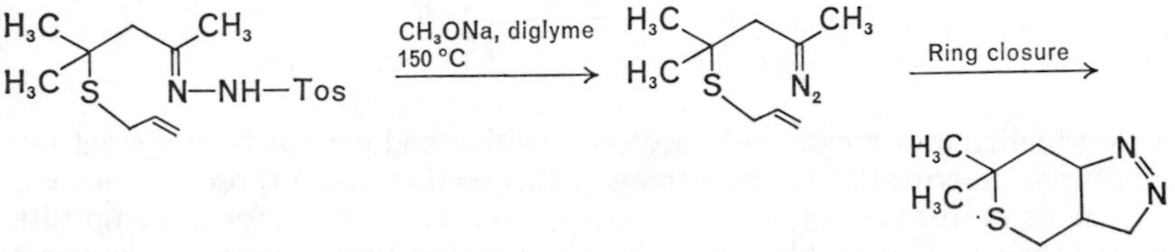

H_3C CH_3 / H_3C S $N{-}NH{-}Tos$ —(CH_3ONa, diglyme, 150 °C)→ H_3C CH_3 / H_3C S N_2 —(Ring closure)→ H_3C / H_3C S N N (29)

3. Silyldiazoalkanes and germyldiazoalkanes

Several silyldiazoalkanes and one representative of the germyldiazomethanes are preparable in good yield by Bamford–Stevens reaction at surprisingly low temperature; they are highly stable[161].

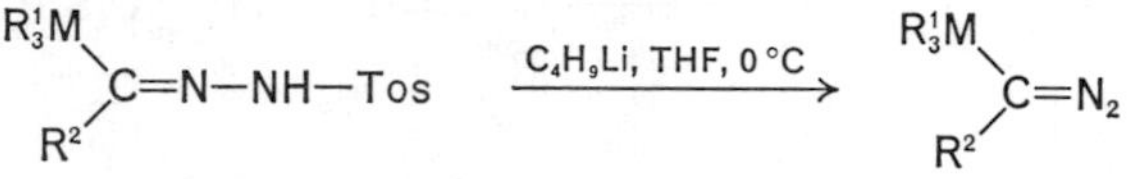

$$R^1_3M(R^2)C{=}N{-}NH{-}Tos \xrightarrow{C_4H_9Li,\ THF,\ 0\,°C} R^1_3M(R^2)C{=}N_2$$

R^1	M	R^2	Yield (%)
C_6H_5	Si	C_6H_5	73
C_6H_5	Si	$C_6H_4{-}X$ (X = F, Cl, Br)	56–71
C_6H_5	Si	CH_3	82
C_6H_5	Si	$CH_2{-}C_6H_5$	48
CH_3	Si	C_6H_5	73
C_6H_5	Ge	C_6H_5	65

4. α,β-Unsaturated diazoalkanes

Two secondary reactions can sometimes greatly complicate the preparation of α,β-unsaturated diazoalkanes by the Bamford–Stevens reaction: one is the formation of cyclopropenes via the sequence **74** → **75** → **78** [162–164] and the other is the 1,5-cyclization **74** → **76** which may also be associated with a [1,5]-sigmatropic substituent shift (**76** → **77**). Disregarding the photochemical decomposition of tosylhydrazones and their salts[165], the temperature necessary for occurrence of the Bamford–Stevens reaction should not be exceeded. This is clear from equation (30)[163].

In contrast to the alkaline cleavage of mesityl oxide tosylhydrazone, in which all the products derive from the corresponding carbene[163], Bamford–Stevens reaction of the similar compound 2,4-diphenyl-2-buten-4-one tosylhydrazone only gives products in which the nitrogen is completely retained according to the general formulas **74**, **76** and **77**[166].

a: $R^1 = CH_3$, $R^2 = H$; b: $R^1 = H$, $R^2 = CH_3$ (30)

Many cases are known in which 3*H*-pyrazoles (**76**) are formed on cleavage of corresponding tosylhydrazones; they can be transformed photochemically into cyclopropenes with loss of N_2[167,168]. Specific $n \rightarrow \pi^*$-excitation of the heterocycles can, however, also lead to isolation of the isomeric α,β-unsaturated diazoalkanes, as shown in equation (31)[169]. The α,β,-unsaturated diazoalkanes arising from cycloalkanone tosylhydrazones with an exocyclic carbon double bond cyclize according to **74** → **76** → **77**[170], just like those whose double bond is incorporated into, e.g., a tropone ring[171].

350–410 nm (31)

New approaches to the synthesis of heterocycles are based on the intermediate generation of doubly conjugated diazoalkanes; a C=C double bond may also be an integral part of an aromatic ring[170]. Thus the tosylhydrazone (**79**) has been shown to yield initially the α,β;γ,δ-doubly unsaturated diazo compound **80** which undergoes 1,7-cyclization to give **82**, the terminal diazo nitrogen attacking a benzene ring. Sigmatropic [1,5]-H shift concludes the reaction sequence with formation of **81** [170].

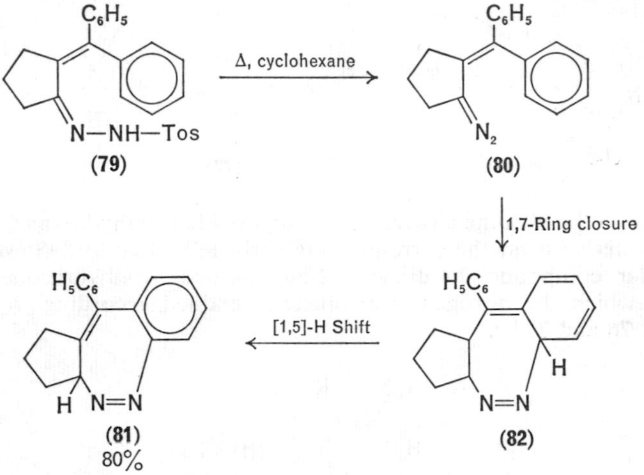

C_6H_5 N—NH—Tos (**79**) —Δ, cyclohexane→ C_6H_5 N_2 (**80**)

1,7-Ring closure

H_5C_6 H N=N (**81**) 80% ←[1,5]-H Shift— H_5C_6 H N=N (**82**)

This section will be closed by a mention of some vinyldiazoalkanes whose structures preclude subsequent cyclization reactions. For instance, 1-diazo-2,3,4-triphenylcyclopentadiene (**84**) is formed from the cyclopentenone (**83**) on heating with tosylhydrazine[172]; 1-diazo-4,4-dimethylcyclohexadiene (**85**)[173] and the tricyclic diazocyclohexenone (**86**)[174] are further examples.

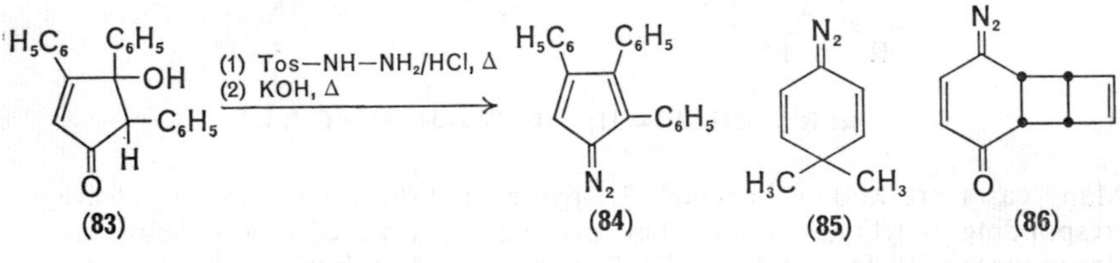

H_5C_6 C_6H_5 OH C_6H_5 H O (**83**) —(1) Tos—NH—NH_2/HCl, Δ (2) KOH, Δ→ H_5C_6 C_6H_5 C_6H_5 N_2 (**84**) N_2 H_3C CH_3 (**85**) N_2 O (**86**)

F. *Cleavage of β-(N-Alkyl-N-nitrosoamino) Ketones and Sulphones*

Although the method to be considered under this heading gives very good results in some cases, it is seldom used and has not been further developed in recent years.

β-(*N*-Alkyl-*N*-nitrosoamino) ketones are prepared in two facile synthetic steps: addition of suitable alkylamines to 2-methyl-2-penten-4-one ('mesityl oxide') is followed by nitrosation[175, 176]. The ensuing alkoxide cleavage is probably to be interpreted as a carbanion elimination initiated by deprotonation of the α-position; this is followed by decomposition to the alkyldiazotate and the starting material. Transformation of the former into diazoalkanes is treated at length in the next Section; the α,β-unsaturated ketone can re-enter the reaction (equation 32).

$$(CH_3)_2C{=}CH{-}CO{-}CH_3 \xrightarrow[\text{(2) Nitrosation}]{\text{(1) }H_2N\text{-Alkyl}} H_3C{-}C(CH_3)(N(NO)CH_2{-}\text{Alkyl}){-}CH_2{-}CO{-}CH_3 \xrightarrow[ROH]{OR^-} H_3C{-}C(CH_3)(N(N{=}O)CH_2{-}\text{Alkyl}){-}\bar{C}H{-}CO{-}CH_3 \quad (32)$$

$$\rightarrow\ {}^-O{-}N{=}N{-}CH_2{-}\text{Alkyl} + (CH_3)_2C{=}CH{-}CO{-}CH_3$$

$$^-O{-}N{=}N{-}CH_2{-}\text{Alkyl} \xrightarrow[-H_2O,\ -RO^-]{ROH} N_2{=}CH{-}\text{Alkyl}$$

Apart from sodium isopropoxide[175, 176], use of sodium cyclohexanolate in cyclohexanol/ether[176–178] furnishes yields of up to 85% of diazomethane. Interference arising from a possible 3+2 cycloaddition of diazomethane with 'mesityl oxide'[177] does not detract from the utility of the method.

Many examples demonstrate the value of this synthetic method: diazomethane (50%)[177, 179], 1-diazopropane (44–47%)[177], 1-diazobutane (41%)[179], (diazomethyl)-cyclopropane (—)[180], 1-diazohexane (23%)[179], 1-diazooctane (16%)[179] as well as the unsaturated diazo compounds 3-diazo-1-propene (41%)[179] and 1-diazo-2-butene (12%)[179] round off the spectrum.

A systematic classification will also assign a place in this Section to the KOH cleavage of 3-(*N*-methyl-*N*-nitrosoamino)sulpholane which leads to diazomethane (equation 33)[181, 182]. There have been no reports concerning generalization of this synthetic variant.

$$\text{3-(N(NO)CH}_3\text{)-sulpholane }(SO_2) \xrightarrow{\text{KOH in water, 60 °C}} \text{3-sulpholene }(SO_2) + CH_2{=}N_2 \quad (33)$$

G. *Acyl Cleavage of N-Alkyl-N-nitrosamides*

The synthesis of diazomethane from *N*-methylnitrosourethane by von Pechmann[183, 184] not only imparted the decisive impetus to the development of acyl cleavage of *N*-alkyl-*N*-nitrosamides but also to the study of diazoalkane chemistry. The predominant use of the method lies in the synthesis of diazomethane and its homologues, although acyldiazomethanes have recently also become accessible in this way (see the last chapter in this volume).

1. Variants

As already mentioned, the alkaline cleavage of *N*-methyl-*N*-nitrosourethanes (**87a**) takes historic precedence; although the suitability of *N*-alkyl-*N*-nitrosocarboxamides (**87b**) had long been known[183, 184] practical application came very late.

N-Alkyl-*N*-nitrosoureas (**87c**), *N*-alkyl-*N'*-nitro-*N*-nitrosoguanidines (**87d**) and *N*-alkyl-*N*-nitroso-*p*-toluenesulphonamides (**87e**) complete the picture. After a consideration of mechanistic aspects, the advantages and disadvantages of the individual variants will be discussed together with their scope.

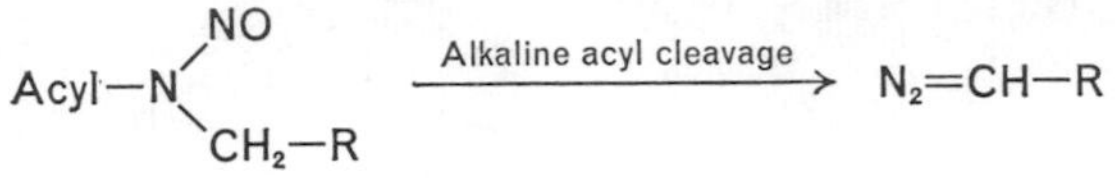

$$\text{Acyl—N(NO)—CH}_2\text{—R} \xrightarrow{\text{Alkaline acyl cleavage}} N_2\text{=CH—R}$$

(87)

a: Acyl = ROCO **b**: Acyl = R—CO **c**: Acyl = R_2N—CO
d: Acyl = O_2N—NH—C(=HN) **e**: Acyl = Ar—SO_2

2. Reaction mechanisms

Alkanediazotates[185] are nowadays unchallenged as intermediates of alkaline cleavage of *N*-alkyl-*N*-nitrosamines (**87**). Those which have been isolated and structurally verified—either by physicochemical methods or by their reactions—are: potassium methanediazotate (**88**)[186–189], potassium phenylmethanediazotate (**89**[186], potassium octanediazotate (**90**)[190] and potassium cyclopropylmethanediazotate (**91**)[191] (all from the corresponding urethanes **87a**). The extensive independence of diazotate formation of the reactants is apparent from the fact that **88** is also accessible from *N*-methyl-*N*-nitrosoacetamide (**87b**)[192] and *N*-methyl-*N*-nitroso-*p*-toluenesulphonamide (**87e**)[187, 188]. Formation of pyrrolidinium 2,2-diphenylcyclopropanediazotate (**92**) from the corresponding urea derivative (**87c**) can also be understood in this way[193].

H_3C—N=N—O^- K^+
(88) (as *syn* isomer)

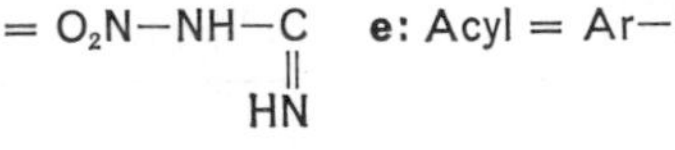

C_6H_5—CH_2—N=N—O^- K^+
(89)

C_8H_{17}—N=N—O^- K^+
(90)

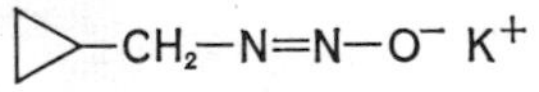

C_3H_5—CH_2—N=N—O^- K^+
(91)

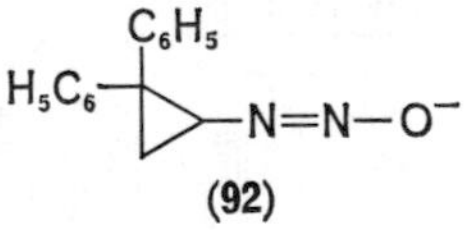

2,2-$(C_6H_5)_2C_3H_3$—N=N—O^- (pyrrolidinium, H_2N^+)
(92)

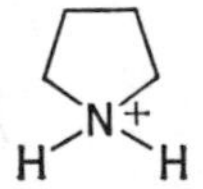

In the case of *N*-alkyl-*N*-nitrosourethanes and *N*-alkyl-*N*-nitrosocarboxamides nucleophilic attack of the base, i.e. alkoxide, at the carbonyl carbon atoms is assumed to account for diazotate formation according to equation (34); the carboxylic esters formed as a consequence to this reaction course are frequently isolable[190, 194–197]. The COOR group is of course incorporated directly into the reaction products[198] in the case of nitrosated lactams[189, 199].

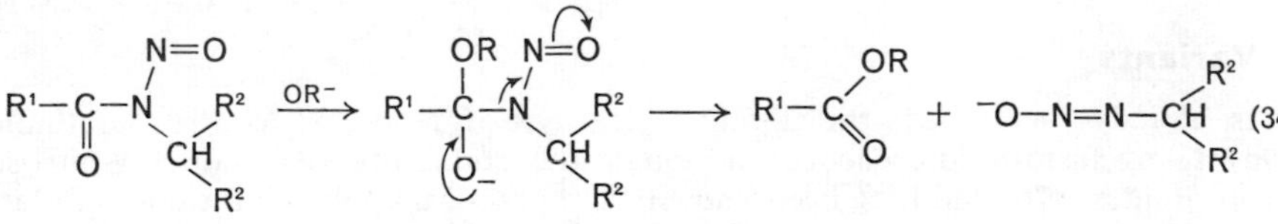

$$R^1\text{—C(=O)—N(N=O)—CH}R^2_2 \xrightarrow{OR^-} R^1\text{—C(OR)(O}^-\text{)—N(N=O)—CH}R^2_2 \longrightarrow R^1\text{—C(=O)OR} + {}^-\text{O—N=N—CH}R^2_2 \quad (34)$$

The deviating primary reaction in the cleavage of urea derivatives will be considered later. In general, the multistep conversion of the diazotate intermediate into diazoalkanes can be envisaged as involving initial protonation to give diazo hydroxides which transform via aliphatic diazonium ions into diazoalkanes. The yields will depend upon the extent to which the diazonium ions are converted into carbenium ions, i.e. lead to undesired secondary reactions (equation 35). The overall product distribution, in which secondary reactions of the diazonium ions themselves must also be considered[197, 200], is extremely sensitive to solvent and substituent effects[185, 186, 190, 197, 200–202].

$$R^2_2CH{-}N{=}N{-}O^- \underset{-H^+}{\overset{H^+}{\rightleftharpoons}} R^2_2CH{-}N{=}N{-}OH \underset{H_2O,\,-H^+}{\overset{H^+,\,-H_2O}{\rightleftharpoons}} R^2_2CH{-}N_2^+$$

$$R^2_2CH{-}N_2^+ \xrightarrow{-N_2} R^2_2\overset{+}{C}H \longrightarrow \text{Consecutive reactions}$$

$$R^2_2CH{-}N_2^+ \underset{+H^+}{\overset{-H^+}{\rightleftharpoons}} R^2_2C{=}N_2 \qquad (35)$$

While the intermediacy of alkanediazotates in the alkaline cleavage of *N*-alkyl-*N*-nitrosoureas is undisputed, there exists convincing evidence that reaction according to equation (36) is initiated by deprotonation of the non-nitrosated amide group. The further course of the reaction which is accompanied by 'transprotonation' leads to diazoalkanes and cyanate ions (Scheme 36)[197, 203–206]. It is

$$R{-}N(N{=}O){-}C({=}O){-}NH_2 \xrightarrow[-HB]{B^-} R{-}N(N{=}O){-}C({=}O){-}\bar{N}H \longrightarrow RN{=}N{-}O^- + HNCO$$

$$RN{=}N{-}O^- + HNCO \rightleftharpoons RN{=}N{-}OH + NCO^- \xrightarrow{-H_2O} R{-}CH{=}N_2 \qquad (36)$$

extremely doubtful whether alkoxide attack at the NO group of *N*-alkyl-*N*-nitrosoureas also occurs as a primary step in the cleavage reaction[196, 206].

3. *N*-Alkyl-*N*-nitrosourethanes

N-Alkyl-*N*-nitrosourethanes are generally accessible from reaction of chloroformic esters with alkylamines and subsequent nitrosation (methyl derivative[207]). Their skin-irritant properties indicate caution during preparative work.

a. *Diazoalkanes*. Decomposition of *N*-methyl-*N*-nitrosourethane can be accomplished in alcoholic solution with catalytic amounts of potassium carbonate; this is held responsible for the formation of small amounts of alkoxide which effect the cleavage[198]. This method is mainly employed for *in situ* generation of diazomethane (equation 37)[208], but has also given satisfactory results (41–47% yield), for example, in the ring expansion of cyclohexanone to give 2-phenylcycloheptanone with diazophenylmethane[209, 210].

$$H_5C_2OOC{-}N(NO)(CH_3) \xrightarrow{ROH/K_2CO_3} (RO)(H_5C_2O)C{=}O + CH_2{=}N_2 + H_2O \quad (37)$$

In other cases, solutions of diazomethane, e.g. in ether, are obtained by cleavage of urethanes with methanolic caustic alkali and subsequent distillation[183, 184, 211]; in order to prepare non-alcoholic solutions cleavage is often performed in high-boiling alcohols and the diazomethane conducted out of the reaction vessel in a stream of N_2 [212]. Diazoethane[213–215] and 1-diazopropane[215] are synthesized in analogous manner. Table 3 lists some representative examples and the relevant reaction conditions. The synthetic limits of the reaction, which are also affected to some extent by the instability of the respective diazoalkanes, can be seen in the following examples.

TABLE 3. Diazoalkanes via *N*-Alkyl-*N*-nitrosourethanes

Diazoalkanes	Reaction conditions	Yield (%)	Reference
Bis(2,4,6-trimethylphenyl)diazomethane ($(2,4,6\text{-}(H_3C)_3C_6H_2)_2C{=}N_2$)	Potassium *t*-butanolate, *t*-butanol, Δ	5–15	216
$(H_5C_6)_3CCH{=}N_2$	Sodium ethanolate, ethanol/ether, −15 °C	100	217
$(H_3C)_3CCH{=}N_2$	Sodium glycolate, glycol, distillation	41	218
$H_3C{-}C(CH_3)(C_6H_5){-}CH{=}N_2$	Sodium ethanolate	—	219
cyclopropyl$-CH{=}N_2$	Sodium triethylene glycolate, triethyleneglycol, −25 °C	⩾24	191
2-diazo-1,7,7-trimethylbicyclo[2.2.1]heptane (H_3C, CH_3, CH_3, $=N_2$)	Sodium 2-ethoxy-1-ethanolate, 2-ethoxy-1-ethanol, 20 °C	—	118
$H_3COOC{-}(CH_2)_n{-}CH{=}N_2$, $n = 1–5$	Sodium hydroxide, water, ether, 0 °C	~40	220

The fluorinated nitrosourethane (**93**) does not give (**96**) in an alkaline medium but is apparently transformed into the diazonium ion (**94**), which decomposes to difluorocarbene and diazomethane via **95** [12]. Formation of the carbene from **94** reveals a formal resemblance to α-elimination of haloforms. Production of diazotrimethylsilylmethane by the urethane route follows a normal course on cleavage with aqueous caustic soda/ether (≈20 °C), while at higher temperature in the absence of organic solvent diazomethane is formed[221].

$$F_2CH-CH_2-N(NO)(COOC_2H_5)\ \textbf{(93)} \xrightarrow{KOH,\ ether/n\text{-}propanol} F_2CH-CH_2-N_2^+\ \textbf{(94)} \xrightarrow{-H^+} F_2\bar{C}-CH_2-N_2^+\ \textbf{(95)}$$

$$\textbf{(94)} \not\xrightarrow{-H^+} F_2CH-CH=N_2\ \textbf{(96)} \qquad \textbf{(95)} \longrightarrow F_2C: + {}^-CH_2-N_2^+$$

Cleavage of the cyclic *N*-nitrosourethanes **(97)**[222] and **(98)**[223, 224] with lithium ethoxide follows a different course not leading to the expected diazoalkanes; nor does that of the bisurethane **(99)**[225, 226] with sodium methoxide.

(97) **(98)** **(99)**

b. *α,ω-Bis(diazo)alkanes.* α,ω-Bis(diazo)alkanes become more and more stable the farther apart the diazo groups are. In full accord with this statement, bis(diazo)-methane has so far escaped direct detection; however, formation of formaldehyde acetal according to equation (38) suggests their intermediate occurrence[227]. 1,2-Bis-(diazo)ethane also escapes detection by decomposing into nitrogen and acetylene[228]; it can, however, be trapped[229]. 1,3-Bis(diazo)propane is tolerably stable in solution[229]; it decomposes in cyclohexane at 25 °C to nitrogen and pyrazole according to equation (39)[230]. Higher homologues, i.e. 1,4-bis(diazo)butane up to 1,8-bis(diazo)-octane, have been detected directly or characterized by their reactions[228–231].

$$(ON)(H_5C_2OOC)N-CH_2-N(NO)(COOC_2H_5) \xrightarrow{Base} [N_2=C=N_2] \xrightarrow[-2\,N_2]{2\,C_2H_5OH} H_2C(OC_2H_5)_2 \quad (38)$$

$$N_2=CH-CH_2-CH=N_2 \longrightarrow :CH-CH_2-CH=N_2 \longrightarrow \text{3H-pyrazole} \xrightarrow{H\sim} \text{pyrazole (NH)} \quad (39)$$

c. *Unsaturated diazoalkanes.* Cleavage of *N*-allyl-*N*-nitrosourethane with potassium hydroxide or sodium methoxide expectedly affords 3-diazo-1-propene (vinyldiazomethane) which, however, slowly undergoes 1,5-cyclization to the pyrazole isomer (up to 80%) according to equation (40)[179, 232–234]. The rate constants

$$H_2C=CH-CH_2-N(NO)-COOC_2H_5 \xrightarrow{KOH\ or\ NaOCH_3} H_2C=CH-CH=N_2 \xrightarrow{Cyclization} \text{pyrazole (NH)} \quad (40)$$

of 1,5-ring closure have been determined for methyl- and phenyl-substituted vinyldiazomethanes such as *trans*-1-diazo-2-butene and *trans*-1-diazo-3-(3-nitrophenyl)-propene[235].

Some β,γ-unsaturated diazo compounds have also been prepared by way of nitrosourethanes. 4-Diazo-1-butene (allyldiazomethane)[236], 1-chloro-4-diazo-2,3-dimethyl-1-pentene[237], 3-diazomethyl-1,2-diphenylcyclopropene[238] and 3-diazomethyl-1-cyclopentene[239] demonstrate the scope of the method. In conclusion, mention should be made of the synthesis of 1,2,3-tri(*t*-butyl)-1-diazomethylcyclopropene (**101**) which is formed together with the cyclobutene (**102**) from the urethane (**100**) and provides an entry to stable cyclobutadienes[240].

$$\text{(100)}\ \xrightarrow[-30\,^{\circ}C]{CH_3ONa,\ CH_3OH}\ \text{(101)}\ +\ \text{(102)}$$

(100): tri-*t*-butylcyclopropenyl–CH_2–N(NO)–$COOC_2H_5$; (101): tri-*t*-butylcyclopropenyl–CH=N_2; (102): $OH(OCH_3)$-substituted tri-*t*-butylcyclobutene

4. *N*-Alkyl-*N*-nitrosoamides

N-Methyl-*N*-nitrosoacetamide[241] is cleaved both by methanolic potassium hydroxide[242, 243] and by methyl- or phenyl-lithium[192] to form diazomethane; in the latter case it is possible to isolate lithium ethoxides and lithium methanediazotate thus confirming base attack at the CO group (equation 41). The cleavage of

$$H_3C{-}C(=O){-}N(NO){-}CH_3 \xrightarrow{KOH,\ CH_3OH} H_2C{=}N_2 + H_3C{-}COO^- K^+ + H_2O \tag{41}$$

$$H_3C{-}C(=O){-}N(NO){-}CH_3 \xrightarrow{2\ RLi,\ ether} H_3C{-}CR_2{-}O^-\ Li^+ + H_3C{-}N{=}N{-}O^-\ Li^+ \xrightarrow{Hydrolysis} H_2C{=}N_2$$

$R = CH_3, C_6H_5$

N,N'-dimethyl-*N,N'*-dinitrosooxalamide (**103**) with methylamine is of interest in that the diamide (**104**) can be subjected to the same reaction after nitrosation[244]. This variant is also suitable for *in situ* generation of diazomethane, as demonstrated by the homologation of cyclohexanone to give cycloheptanone (72%)[245].

$$\underset{(103)}{ON(H_3C)N{-}C(=O){-}C(=O){-}N(CH_3)NO} \underset{As_2O_3,\ HNO_3}{\overset{2H_3C-NH_2,\ ether}{\rightleftharpoons}} \underset{(104)}{H_3CHN{-}C(=O){-}C(=O){-}NHCH_3} + 2\,H_2C{=}N_2 + 2\,H_2O$$

N,N'-Dimethyl-*N,N'*-dinitrososuccinamide[246] and *N,N'*-dimethyl-*N,N'*-dinitrosoterephthalamide[247] do not seem to have any advantages for generation of diazomethane.

Diazoethane[245], 1-diazopropane[243, 245], 1-diazobutane[243, 245], 3-chloro-1-diazopropane[248], diazophenylmethane[243] and diazocyclohexane[58] have been prepared partly via the oxalamide pathway but mainly by the acetamide route.

This variant has found only limited application to the synthesis of bis(diazo)-alkanes, e.g. of 1,3-bis(diazo)propane according to equation (42); the potassium bis(diazotate) can be isolated[202].

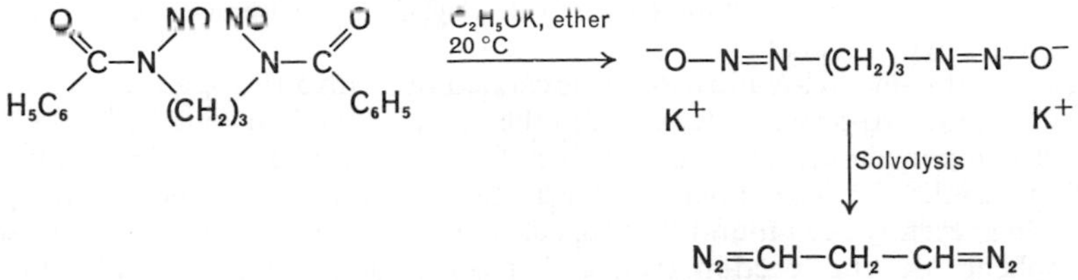

$$H_5C_6-C(=O)-N(NO)-(CH_2)_3-N(NO)-C(=O)-C_6H_5 \xrightarrow[20\,°C]{C_2H_5OK,\ ether} {}^-O-N{=}N-(CH_2)_3-N{=}N-O^-\ K^+\ K^+ \xrightarrow{Solvolysis} N_2{=}CH-CH_2-CH{=}N_2 \quad (42)$$

An elegant pathway leading to 1,6-bis(diazo)hexane consists of the nitrosation of 'nylon-66' to give **105** followed by alkaline cleavage[202].

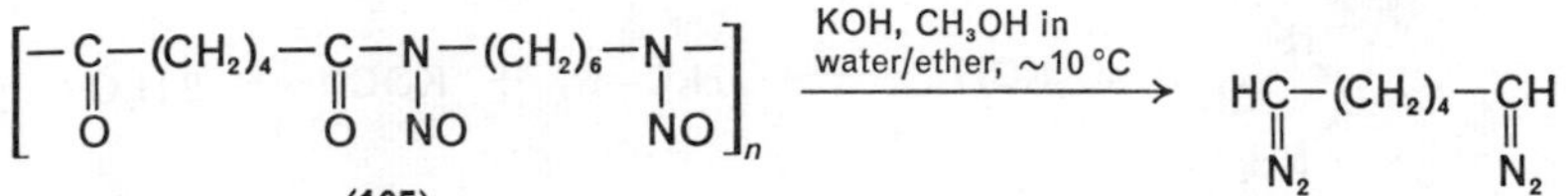

$$\left[-C(=O)-(CH_2)_4-C(=O)-N(NO)-(CH_2)_6-N(NO)-\right]_n \xrightarrow[water/ether,\ \sim 10\,°C]{KOH,\ CH_3OH\ in} HC(=N_2)-(CH_2)_4-CH(=N_2)$$

(105)

Cyclic nitrosated carboxamides are of course transformed into diazo compounds with ring cleavage and incorporation of the nucleophile employed. Such reactions are known in the case of *N*-nitroso-ε-caprolactam[199] and **106**; the latter compound furnishes methyl 2-diazomethylbenzoate (**107**) in high yield[249].

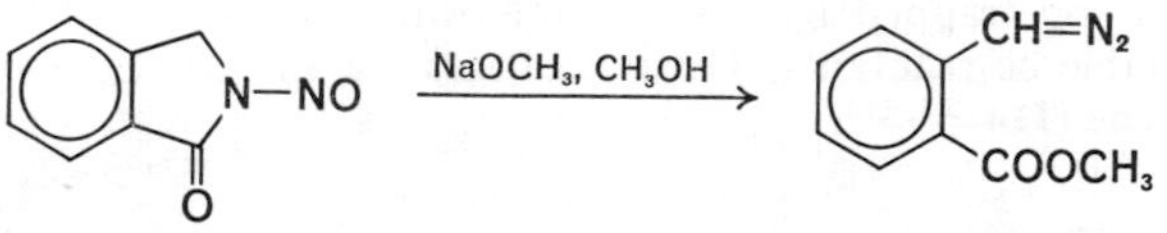

$$\textbf{(106)}\ \xrightarrow{NaOCH_3,\ CH_3OH}\ C_6H_4(CH{=}N_2)(COOCH_3)\ \textbf{(107)}$$

(106) **(107)**

Diazopropyne (**109**) is directly accessible from the nitrosated carboxamide (**108**)[250, 251], but can also be prepared via the trinitroso compound **110** [251]; it appears reasonable to assume the occurrence of tris(diazo)propane (**111**) as an unstable intermediate which is transformed into **109** by loss of 2 moles of N_2.

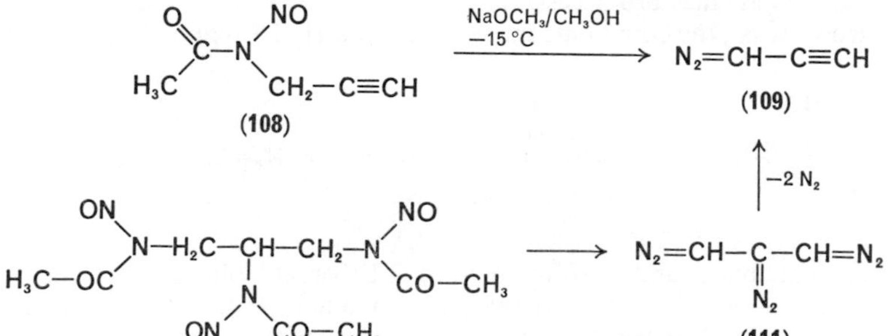

$$H_3C-C(=O)-N(NO)-CH_2-C\equiv CH\ \textbf{(108)} \xrightarrow[-15\,°C]{NaOCH_3/CH_3OH} N_2{=}CH-C\equiv CH\ \textbf{(109)}$$

$$H_3C-OC-N(NO)-H_2C-CH[N(NO)-CO-CH_3]-CH_2-N(NO)-CO-CH_3\ \textbf{(110)} \longrightarrow N_2{=}CH-C(=N_2)-CH{=}N_2\ \textbf{(111)} \xrightarrow{-2\,N_2} \textbf{(109)}$$

5. *N*-Alkyl-*N*-nitrosoureas

N-Methyl-*N*-nitrosourea is one of the most commonly used starting materials for the preparation of diazomethane; it is accessible by reaction of potassium cyanate or urea with methylamine and subsequent nitrosation[205, 252–256]. Its limited thermal stability is a drawback; spontaneous decomposition to nitrogen, water, and methyl isocyanate takes place above 20–30 °C [205, 253, 254, 257, 258]. A reaction of a pyrocatechol with diazomethane (from *N*-methyl-*N*-nitrosourea) has been reported in which methyl isocyanate also participates directly[259].

Cleavage of *N*-methyl-*N*-nitrosourea is carried out with 50–70% aqueous potassium hydroxide in a two-phase system, with ether as preferred solvent[205, 252, 260, 261]. The organic component becomes dispensible if diazomethane is distilled out of the reaction vessel[262, 263]. The stoichiometry of the cleavage is apparent from equation (43). It has recently been found that this cleavage can be accomplished with sodium 4-nitrophenoxide, the reaction then giving both sodium cyanate and 4-methoxynitrobenzene (from diazomethane and 4-nitrophenol)[264]. Time will tell whether this modification acquires synthetic importance as a methylation reaction.

$$O{=}C(NH_2){-}N(NO){-}CH_3 \ + \ KOH \longrightarrow H_2C{=}N_2 \ + \ KOCN \ + \ 2\,H_2O \qquad (43)$$

Simple homologues of diazomethane such as diazoethane[205], 1-diazopropane[265], 1-diazobutane[205, 266] or 1-diazododecane[267] are obtained in the same way as the rather labile cyclic representatives diazocyclobutane[268] and diazocyclohexane[58]. This applies in particular to 1-diazo-2,2-diphenylcyclopropane (**113**) obtained on lithium ethoxide cleavage of **112**; it can be recognized by an intermediate yellow to red colouration and trapped by 3+2 cycloaddition with diethyl fumarate. In the absence of suitable co-reactants, **113** rearranges via the corresponding carbene to 1,1-diphenylallene (**114**)[269, 270].

$$\underset{(\mathbf{112})}{(H_5C_6)_2C_3H_3{-}N(NO){-}C({=}O){-}NH_2} \xrightarrow{C_2H_5OLi} \underset{(\mathbf{113})}{(H_5C_6)_2C_3H_2{=}N_2} \xrightarrow[-N_2]{\text{via carbene}} \underset{(\mathbf{114})}{(H_5C_6)_2C{=}C{=}CH_2}$$

α,ω-Bis(diazo)alkanes are accessible by acyl cleavage of suitable polymethylene-bis(*N*-nitrosoureas), the comments made in Section II.G.3.b applying to the stability

$$H_2N{-}C({=}O){-}N(NO){-}CH_2{-}(CH_2)_n{-}CH_2{-}N(NO){-}C({=}O){-}NH_2 \longrightarrow N_2{=}CH{-}(CH_2)_n{-}CH{=}N_2 \qquad (44)$$

$$n = 1,2,3,4$$

of bis(diazo)methane and bis(diazo)ethane. Ethereal solutions of the α,ω-bis-(diazo)alkanes shown in equation (44) are obtained[231, 271]. A further insight into the scope of urea cleavage as a method of preparing diazoalkanes is given in Table 4.

TABLE 4. Diazoalkanes via *N*-alkyl-*N*-nitrosoureas

Diazoalkanes	Reaction conditions	Yield (%)	References
$H_5C_6{-}CH{=}N_2$	C_2H_5OLi, C_2H_5OH, −20 °C	79	272
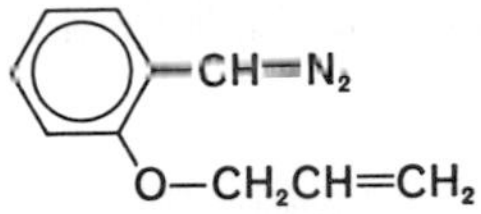 —CH=N_2 (ring with O—$CH_2CH{=}CH_2$)	KOH, monoglyme, pentane, 0 °C	21	150
$(H_3C)_2CHCH{=}N_2$	KOH, ether, −12 to −15 °C	—	273
$H_3COCH_2CH_2CH{=}N_2$	KOH, benzene, 5 to 10 °C	82	274
$Cl{-}CH_2CH_2CH{=}N_2$	KOH, benzene, 5 to 10 °C	47	274
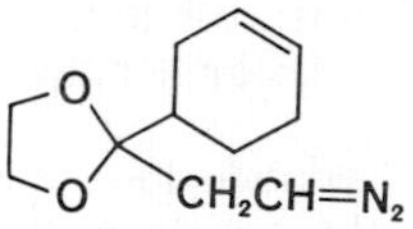 (O, O ring) $CH_2CH{=}N_2$	NaOH, hexane, 0 °C	—	275
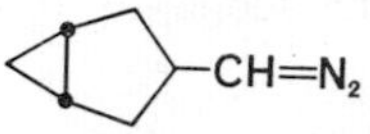 —CH=N_2 (*cis*- and *trans*-CH=N_2 group)	KOH, pentane, 0 °C	85–90	276
$(H_3C)_3Si{-}CH{=}N_2$	KOH, pentane, 20 °C	64	277, 278, 279
$H_2C{=}CH{-}CH{=}N_2$	NaOH, ether, 20 °C	19	280
	KOH, methanol, THF, 0 °C	65–70	281
$HC{\equiv}C{-}CH{=}N_2$	No detailed information	—	282

6. *N*-Alkyl-*N*′-nitro-*N*-nitrosoguanidines

Alkaline cleavage of *N*-alkyl-*N*′-nitro-*N*-nitrosoguanidines is strongly reminiscent of the corresponding reaction of nitrosoureas. They have the advantage of greater thermal stability than the latter, but are skin irritants as are nitrosourethanes. Cleavage of the reactants presumably occurs according to equation (45), the potassium salt of nitrocyanamide being isolable[283]. Although no mechanistic studies have been published, it is tempting to assume that deprotonation at the imido nitrogen is the initial reaction step, which is followed by cleavage according to equation (45).

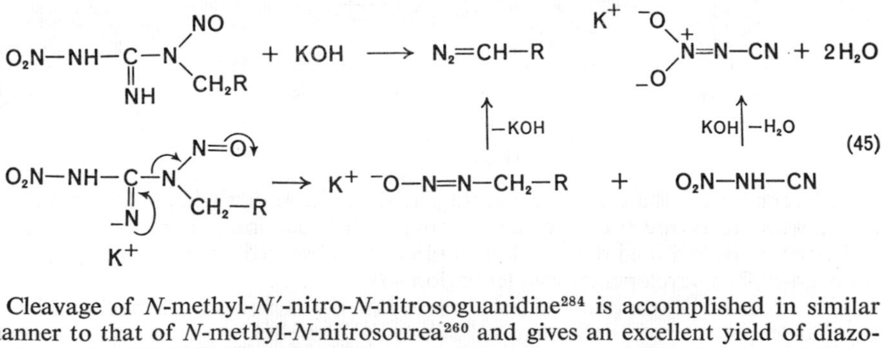

$$O_2N{-}NH{-}C({=}NH){-}N(NO)CH_2R + KOH \longrightarrow N_2{=}CH{-}R \quad K^+\ {}^-O(O^-)\overset{+}{N}{=}N{-}CN + 2H_2O$$

$$O_2N{-}NH{-}C({=}N^-\,K^+){-}N({=}O\text{ as }N{=}O)CH_2{-}R \longrightarrow K^+\ {}^-O{-}N{=}N{-}CH_2{-}R \;(\xrightarrow{-KOH} N_2{=}CH{-}R) + O_2N{-}NH{-}CN \;(\xrightarrow{KOH,\ -H_2O}) \qquad (45)$$

Cleavage of *N*-methyl-*N*′-nitro-*N*-nitrosoguanidine[284] is accomplished in similar manner to that of *N*-methyl-*N*-nitrosourea[260] and gives an excellent yield of diazomethane (73–93%)[283, 285]. Diazoethane, 1-diazopropane, 1-diazobutane, 1-diazopentane and diazophenylmethane (55–76%) reveal the versatility of the method[283]; other syntheses were thwarted by the inaccessibility of the nitroso compounds[283].

7. *N*-Alkyl-*N*-nitroso-*p*-toluenesulphonamides

A recent trend seems to be the increasing use of acyl cleavage of *N*-methyl-*N*-nitroso-*p*-toluenesulphonamide—accessible by the classical method of reacting tosyl chloride and methylamine with subsequent nitrosation[286]—for preparation of diazomethane (equation 46)[287, 288]. The thermal conditions of acyl cleavage (50–75 °C) do not seriously impair the yield of diazomethane since it is stable towards the effects of temperature in the absence of catalyst.

$$H_3C-C_6H_4-SO_2-N(NO)(CH_3) + ROH \xrightarrow{KOH} H_2C{=}N_2 + H_2O + H_3C-C_6H_4-SO_2-OR \quad (46)$$

The advantages of preparing diazomethane from *N*-methyl-*N*-nitroso-*p*-toluenesulphonamide include commercial availability and long storage life at room temperature and the fact that no irritant properties have been reported.

Further published applications (diazocyclohexane (94%)[58], diazophenylmethane (60%)[289], alkoxydiazoalkanes[290]) convey the impression that the scope of this method has not yet been exploited to the full.

H. Diazo Group Transfer

The term 'diazo group transfer' is applied to those reactions in which a 'complete' N_2 group is transferred from a donor (azides, diazoalkanes) to an acceptor (CH-acid compounds, double and triple bond systems) by exchange or addition[291–294]. The principal use of this reaction lies in the synthesis of diazomethanes bearing electron-acceptor groups (in this connection see the last chapter of this volume).

I. Cyclopentadienes and cyclohexadienes

Cyclopentadienes and cyclohexadienes possess sufficient proton activity to undergo diazo group transfer with tosyl azide, the most commonly employed transfer reagent. Thus cyclopentadienyllithium affords diazocyclopentadiene, probably via

$$C_5H_5^- Li^+ \xrightarrow{Tos-\bar{N}-N{=}\overset{+}{N}} C_5H_5-N{=}N-N^-(Tos)\ Li^+ \ (115) \xrightarrow{-Tos-NH^-Li^+} C_5H_4{=}N_2 \text{ or } C_5H_4^--N_2^+$$

the triazene intermediate **115** [295]; working in acetonitrile/diethylamine[296] or diethylamine without solvent[297] considerably simplifies the reaction conditions. Formation of isomeric disubstituted diazocyclopentadienes is observed on diazo group transfer onto 1,4-diphenylcyclopentadiene (equation 47)[296, 298].

This product distribution is a consequence of the mesomeric carbanion with its two nucleophilic centres. Similar behaviour is observed on diazo transfer onto methyl or trimethylsilyl cyclopentadiene-1-carboxylate[299]. Numerous other substituted diazocyclopentadienes, such as 5-diazo-1,2,3-triphenylcyclopentadiene (piperidine/acetonitrile, 97%)[296], 1-benzyl-5-diazo-2,3,4-triphenylcyclopentadiene (piperidine/acetonitrile, 60%)[300], 5-diazo-1,2,3,4-tetraphenylcyclopentadiene (piperidine/acetonitrile, 95%)[296] and 1-diazoindene (diethylamine/no solvent, 15–25%)[297, 301], are readily accessible by diazo group transfer.

$$\text{H}_5\text{C}_6\text{-C}_5\text{H}_4\text{-C}_6\text{H}_5 \xrightarrow[-\text{H}^+]{\text{CH}_3\text{CN, piperidine}} [\text{H}_5\text{C}_6\text{-C}_5\text{H}_3^-\text{-C}_6\text{H}_5] \xrightarrow[-\text{Tos-NH}^-]{\text{Tos N}_3} \text{diazo products: } 58\% + 30\% \quad (47)$$

9-Diazo-10-anthrone (**116a**)[302] and 9-diazothioxanthene *S,S*-dioxide (**116b**)[302, 303] may be cited as formal representatives of the diazocyclohexadienes; both readily undergo base-catalysed azine formation, which will clearly affect the choice of reaction conditions for diazo group transfer[302, 304, 305].

$$\xrightarrow[-\text{TosNH}_2]{\text{TosN}_3} \; (\mathbf{116}) \; \xrightarrow{\text{Decomposition}} \text{azine}$$

(**116**)
a: X = CO; C_2H_5OH/piperidine, 94%[302]
b: X = SO_2; C_2H_5OH/C_2H_5ONa, 98%[302]

2. Enamines

Transfer of diazo groups onto enamines constitutes a significant methodical extension in that the CH acidity of the diazo group acceptor becomes unimportant. The reaction is usually carried out with sulphonyl azides; it is associated with cleavage of the enamine C=C double bond and assumed to follow the course indicated in equation (48)[306]. Enamines display a great proclivity for azide addition[307]; this proceeds as shown in equation (48)[308], triazolines being isolable in some instances[308, 309]. Diazomethane has been prepared on various occasions with moderate success ($\leqslant$40%) from enamines of type **117**, reaction also yielding the amidines **118** (**117a**[307], **117b**[310], **117c**[311], **117d**[306], **117e**[306] and **117f**[306]). In the reactions of **117d**–**117f** the structural components of enamines, i.e. acetaldehyde and secondary amine, can react directly with tosyl azide in benzene[306]. Diazo group transfer onto 2-methylene-1,3,3-trimethyl-2,3-dihydroindole ('Fischer base') according to equation (49) leaves little room for improvement, giving yields of 81–87%[306]. It has not yet been established why the enamines **119** and **120** give such different diazomethane yields on diazo group transfer[306].

$$R^1R^2C{=}C(R^3)NR_2 + TosN_3 \longrightarrow \text{triazoline} \xrightarrow{\text{Ring opening}} R^1R^2C(N_2^+)\text{–}C(R^3)(\underline{N}\text{–}Tos)NR_2 \longrightarrow R^1R^2C{=}N_2 + R^3\text{–}C(=N\text{–}Tos)NR_2 \quad (48)$$

$$R^1R^2C{=}CH_2 \; (117) + TosN_3 \longrightarrow R^1R^2C{=}N\text{–}Tos \; (118) + H_2C{=}N_2$$

a: $R^1 = C_6H_5$, $R^2 = N(CH_3)(C_6H_5)$
b: $R^1 = CH(CH_3)_2$, R^2 = N-morpholino
c: $R^1 = CH_3$, R^2 = N-morpholino
d: $R^1 = H$, $R^2 = N(C_2H_5)_2$
e: $R^1 = H$, R^2 = N-piperidino
f: $R^1 = H$, R^2 = N-morpholino

$$\text{1,3,3-trimethyl-2-methyleneindoline} + N_3\text{–}SO_2\text{–}C_6H_4\text{–}X \longrightarrow \text{indoline-2-}{=}N\text{–}SO_2\text{–}C_6H_4\text{–}X + H_2C{=}N_2 \quad (49)$$

X = OCH_3, CH_3, NO_2

(119) $H_2C{=}N_2$: 73–75%; Amidine: 67–72%

(120) $H_2C{=}N_2$: 5–9%; Amidine: 9%

3. Enol ethers

Compared with enamines, diazo group transfers with enol ethers play only a minor role. Modest yields of diazomethane and imino ethers are obtained from alkoxyethylene and phosphoryl azides[312].

$$H_2C{=}CH{-}OR^1 + R^2_2P(O){-}N_3 \longrightarrow H_2C{=}N_2 + HC(OR^1){=}N{-}P(O)R^2_2$$

$$R^1 = CH(CH_3)_2,\ R^2 = OC_2H_5 \text{ or } C_6H_5$$

Phenylketene dimethyl acetal reacts with ethyl azidoformate after the manner of a diazo group transfer to give **121** [310, 314, 315]. In solution an equilibrium is observed with the isomeric triazoline[314, 315], via which **121** is undoubtedly formed.

$$H_5C_6(H)C{=}C(OCH_3)_2 + N_3{-}COOC_2H_5 \longrightarrow \text{triazoline } [(H_3CO)_2C,\ C_6H_5,\ N{-}COOC_2H_5] \rightleftarrows C_6H_5{-}C({=}N_2){-}C(OCH_3)_2{-}NH{-}COOC_2H_5$$

(121)

4. Acetylenes

By far the greatest interest is shown in diazo group transfers on to ynamines[316], which are dealt with in the last chapter of this volume; nevertheless, some reactions appear to deserve mention here. Thus cyanogen azide and acetylene afford a triazole/diazo imine equilibrium mixture[317], deduced from the pronounced changes in the ^{1}H-n.m.r. spectrum between -60 and $+80$ °C. Apart from ring/chain isomerism, *syn/anti* isomerization at the azomethine group and rotation about the C—C bond all have a role to play (equation 50).

$$HC{\equiv}CH + NC{-}N_3 \quad \text{1-cyano-1,2,3-triazole} \rightleftarrows NC{-}N{=}CH{-}CH{=}N_2 \rightleftarrows N(CN){=}CH{-}CH{=}N_2$$
$$\text{(each diazo imine} \rightleftarrows \text{its C—C rotamer: } NC{-}N{=}CH{-}CH{=}N_2\text{)} \qquad (50)$$

One of the relatively few diazo group transfers involving diazoalkanes is initiated by the [3+2] cycloaddition of 2-diazopropane to diphenylacetylene; the resulting 3*H*-pyrazole (**122**) can be isomerized photochemically to vinyldiazoalkane (**123**). However, the yields suffer somewhat from cyclopropene formation[318]. Other reactions of this type leading to acyldiazoalkanes are dealt with in the last chapter.

$$H_5C_6{-}C{\equiv}C{-}C_6H_5 + (H_3C)_2C{=}N_2 \longrightarrow \underset{(122)}{\text{3,3-dimethyl-4,5-diphenyl-3}H\text{-pyrazole}} \longrightarrow \underset{\substack{(123)\\ 90\%}}{H_3C{-}C(CH_3){=}C(C_6H_5){-}C(C_6H_5){=}N_2} + \underset{10\%}{\text{3,3-dimethyl-1,2-diphenylcyclopropene}}$$

I. Substitution Reactions

Although isolated substitution reactions at the diazo carbon atom, such as mercuration[319] or acylation[320], have been known for a very long time, the full preparative significance of this synthetic principle has only recently been recognized. The diazo groups are found to be surprisingly stable, even under extreme conditions. This approach has only limited synthetic utility, however, for diazoalkanes without electron-acceptor substituents.

1. Halogenation

Diazomethane can be chlorinated with *t*-butyl hypochlorite to give chlordiazomethane, which is moderately stable only at temperatures below −40 °C [321]. The same principle is applicable to bromodiazomethane[321] (equation 51).

$$H_2C{=}N_2 \xrightarrow{(H_3C)_3C-O-X,\ n\text{-}C_5H_{12}/CF_3Cl,\ -100\,°C} X{-}CH{=}N_2 \qquad (51)$$

$$X = Cl, Br$$

1-Bromo-1-diazopropane cannot be prepared in this way: however, the decomposition products obtained do indicate its intermediate occurrence[322]. With *N*-bromosuccinimide, on the other hand, complete substitution of diazocyclopentadiene occurs to give the stable compound tetrabromodiazocyclopentadiene[323].

2. Metallation

Exchange of hydrogen for lithium in diazomethane proceeds without complication with methyl- or phenyllithium in ether[324], or also with lithium *N*-methyl-*N*-trimethylsilylamide[325]. A completely independent synthesis of diazomethyllithium starts from nitrous oxide and methyllithium[326] (equation 52). Sodium can be

$$H_2C{=}N_2 \xrightarrow[H_3CNSi(CH_3)_3Li\ \text{in ether}]{CH_3(C_6H_5)Li\ \text{or}} LiCH{=}N_2 \xleftarrow{CH_3Li\ \text{in ether}} N_2O \qquad (52)$$

introduced into diazomethane with the aid of triphenylmethylsodium[327]. Both alkali metal derivatives are highly explosive in the absence of solvent.

The ability of alkali-metal-substituted diazomethanes to undergo further metallation reactions is demonstrated by the preparation of cadmiobis(diazomethane) **(124)** from diazomethyllithium and cadmium chloride[328].

$$2\,LiCH{=}N_2 + CdCl_2 \xrightarrow{\text{THF or ether at } -70\,°C} \underset{\mathbf{(124)}}{Cd(CH{=}N_2)_2} + 2\,LiCl$$

An elegant method for introducing organometallic groups into diazomethyl compounds consists of their reaction with suitably substituted metal amides. It is utilized in the synthesis of zinco- [329], cadmio- [329] and mercurodiazomethane[329] **(125a–c)**; all these diazo derivatives are highly explosive. The mercury derivative

$$M\left[N\left(Si(CH_3)_3\right)_2\right]_2 \xrightarrow[-2\,HN(Si(CH_3)_3)_2]{H_2C{=}N_2,\ \text{ether}} \underset{\mathbf{(125)}}{MC{=}N_2}$$

a: M = Zn (polymer);
b: M = Cd;
c: M = Hg

(**125c**) is presumably also formed together with other products from diazomethane and mercury acetate[330]. Mercurobis(diazomethane) (**124**, Hg instead of Cd) is synthesized by Li/Hg exchange[328]. In this context, the mercuration of diazocyclopentadiene with mercuric acetate also deserves mention even though substitution cannot take place at the diazo carbon atom (equation 53)[323]. Double metallation of

$$\text{C}_5\text{H}_4\text{=N}_2 \xrightarrow[\text{(2) NaI, C}_2\text{H}_5\text{OH, H}_2\text{O}]{\text{(1) Hg(OCOCH}_3)_2\text{, DMSO, 25 °C}} \text{IHg–C}_5\text{H}_2\text{(=N}_2\text{)–HgI} \tag{53}$$

diazomethane is possible with silver acetate according to equation (54); pyridine can be removed from the adduct[331]. Silylation of diazomethane is accomplished by first metallating with lithium and treating the product with chlorotrimethylsilane[332].

$$3\,H_2C{=}N_2 + 2\,AgOCOCH_3 \xrightarrow{\text{Ether, pyridine, }-50\,°C} Ag{-}C({=}N_2){-}Ag\cdot C_5H_5N + 2\,CH_3COOCH_3 + 2\,N_2 \tag{54}$$

Diazobis(trimethylgermyl)methane[329, 333], diazobis(trimethylstannyl)methane[333] and diazobis(trimethylplumbyl)methane[333, 334] have been prepared by direct metallation of diazomethane by the 'metal amide method' (equation 55).

$$2\,(H_3C)_3M{-}N(CH_3)_2 + H_2C{=}N_2 \longrightarrow (H_3C)_3M{-}C({=}N_2){-}M(CH_3)_3 + 2\,HN(CH_3)_2 \tag{55}$$

$$M = Ge, Sn, Pb$$

Another variant for the formation of bisgermylated and bisstannylated diazomethanes consists of reaction of trisubstituted germane or stannane with diazomethyllithium. The second mole of diazomethyllithium acts as a metallating agent for intermediate monogermyl- or monostannyl-diazomethane (equation 56)[333].

$$(H_5C_6)_3MCl \xrightarrow[-\text{LiCl}]{\text{LiCH=N}_2} (H_5C_6)_3M{-}CH({=}N_2) \xrightarrow[-\text{H}_2\text{C=N}_2]{\text{LiCH=N}_2} (H_5C_6)_3M{-}C({=}N_2){-}Li \xrightarrow[-\text{LiCl}]{(\text{H}_5\text{C}_6)_3\text{MCl}} \left[(H_5C_6)_3M\right]_2CN_2 \tag{56}$$

$$M = Ge, Sn$$

Since the second substitution step fails to occur in the analogous preparation of bis(silyl)diazomethanes, the alternative procedure of, e.g., first metallating diazotrimethylsilyl methane at −90–100 °C with butyllithium and then introducing the second silyl group with chlorotrimethylsilane[335] may be adopted.

The exceptional performance of the 'metal amide method' is also manifested in the preparation of arsenic-, antimony-, and bismuth-substituted diazomethanes which were entirely unknown until very recently[336]. In the case of diazobis-(dimethylarsino)methane, catalysis by chlorotrimethylstannane is necessary, but not for the Sb and Bi analogues[336].

3. Aldol addition

Aldol-type additions of diazomethyl compounds require a certain degree of CH acidity of the diazomethane hydrogen, which is most pronounced in CO- and PO-substituted diazomethanes. Nevertheless, unsubstituted diazomethane does add to carbonyl compounds having a highly electrophilic CO carbon atom. Thus the tolerably stable diazomethane adducts **126** [337] and **127** [338] were isolated on reaction with 3,3-dibromo-1,2-indandione and 1-phenyl-1,2,3,4-tetrahydro-2,3,4-quinolinetrione.

$H_2C{=}N_2$; OH; $CH{=}N_2$; Br Br; C_6H_5

(126) **(127)**

With various isatin derivatives the occurrence of diazomethane-aldol adducts can be established either by their detection as dioxoles after reaction with tetrabromo-*o*-quinone or by direct trapping in a 3+2 cycloaddition[337]. The latter technique can also be employed to confirm the formation of an aldol adduct from chloral and diazomethane by addition to dimethyl acetylendicarboxylate (equation 57)[339]. The overwhelming importance of electron-acceptor substituents of the

$$Cl_3C{-}CHO + H_2C{=}N_2 \longrightarrow Cl_3{-}CH(OH){-}CH{=}N_2 \xrightarrow{H_3COOC{-}C\equiv C{-}COOCH_3} \text{pyrazole: } Cl_3C{-}CH(OH),\ COOCH_3,\ COOCH_3,\ N{-}NH \quad 37\% \tag{57}$$

carbonyl component in aldol-type additions follows from the above examples; as shown in equation (58) it also applies to additions involving the C=N double

$$Cl_3C{-}CH{=}N{-}Tos \xrightarrow{H_2C=N_2} Cl_3C{-}CH(N(CH_3){-}Tos){-}CH{=}N_2 \text{ (with } N\text{-methylation)} + Cl_3C\text{-aziridine}$$

$$(F_3C)_2C{=}N{-}C(O){-}R \xrightarrow{H_2C=N_2} (F_3C)_2C(NH{-}CO{-}R)(CH{=}N_2) + \text{triazoline: } F_3C,\ CF_3,\ N{-}C(O){-}R,\ N{=}N \tag{58}$$

R = $C(CH_3)_3$, $CH(CH_3)_2$, $-C_6H_4-OCH_3$

bond[340]. Aldol additions of lithiodiazotrimethylsilylmethane, whose diazo carbon atom is decidedly nucleophilic to 'non-activated' aldehydes and ketones (benzaldehyde, acetone, acetophenone), have recently also been reported[341].

4. Cleavage reactions

Diacylated diazomethanes are generally susceptible to ready basic cleavage to give acyldiazomethanes[342]. The only attractive synthetic example yet discovered in the field of non-acylated diazomethanes is the alkaline cleavage of 'azibenzil' to form diazophenylmethane (equation 59)[343].

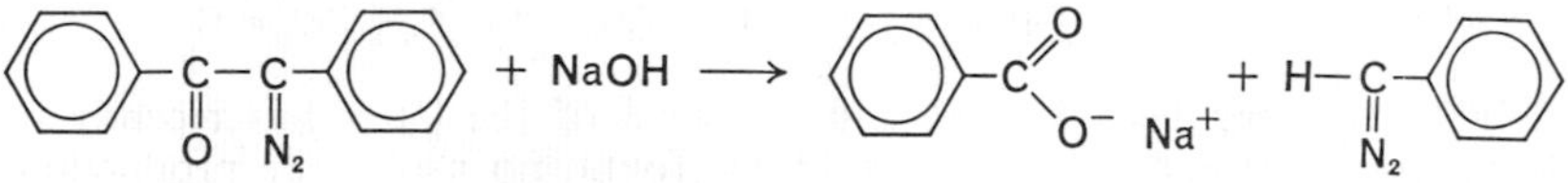

$$C_6H_5-C(=O)-C(=N_2)-C_6H_5 + NaOH \longrightarrow C_6H_5-C(=O)O^- \; Na^+ + H-C(=N_2)-C_6H_5 \qquad (59)$$

5. Miscellaneous

Compared with substitutions at the diazo atom, reactions at other positions of diazoalkanes are quite rare. Thus in diazocyclopentadiene the two positions adjacent to the diazo group are alkylated by 5-benzylidene-2,2-dimethyl-1,3-dioxane-4,6-dione (equation 60)[344].

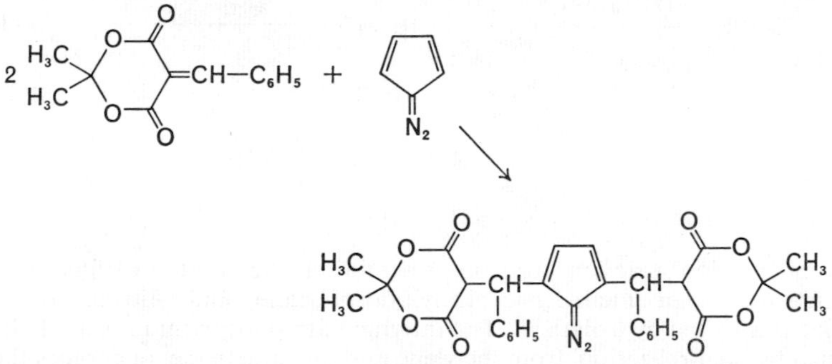

$$2\ (H_3C)_2C(OCO)_2C{=}CH-C_6H_5 + C_5H_4N_2 \longrightarrow \text{bis-adduct: } C_5H_2(=N_2)[CH(C_6H_5)-CH(COO)_2C(CH_3)_2]_2 \qquad (60)$$

In conclusion, mention is made of some nucleophilic substitution reactions occurring with displacement of chloride from 4,6-dichloro-2-diazomethyl-1,3,5-triazine; the nucleophiles may contain N, O or S (equation 61)[345, 346].

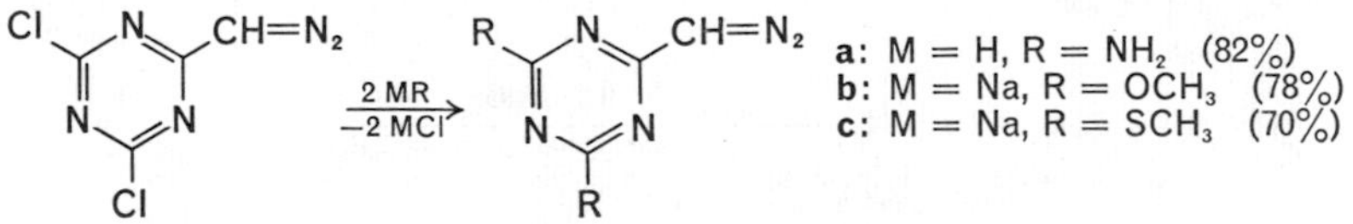

$$Cl_2C_3N_3-CH{=}N_2 \xrightarrow[-2\,MCl]{2\,MR} R_2C_3N_3-CH{=}N_2 \qquad (61)$$

a: M = H, R = NH_2 (82%)
b: M = Na, R = OCH_3 (78%)
c: M = Na, R = SCH_3 (70%)

J. Special Methods

In the course of time, various reactions have been reported in which the intermediacy of diazoalkanes can be detected or is very probable, but cannot be utilized synthetically; they will not be considered in detail[347–349]. In the following a few reactions are considered which could acquire preparative interest.

Alkaline hydrolysis of *N*-methoxypyridazinium salts affords α,β-unsaturated diazoalkanes[351], which have recently been attracting interest[350]; they undergo 1,5-cyclization on warming (equation 62)[351]. The two condensation reactions (63)[352]

$$\text{1-methoxypyridazinium salt } (R^1, R^2; OCH_3\ X^-) \xrightarrow{KOH/H_2O} N_2{=}C(R^1){-}CH{=}CH{-}C(O){-}R^2\ (70\text{–}90\%) \longrightarrow \text{3-acyl-5-}R^1\text{-pyrazole } (N{-}H) \quad (62)$$

$R^1 = R^2 = H$; $R^1 = C_6H_5$, $R^2 = H$; $R^1 = R^2 = CH_3$; $R^1 = C_6H_5$, $R^2 = CH_3$

and (64)[353] have long been known, and the yield of the latter has recently been greatly improved[354]; and yet they have so far found no use in the production of

$$H_3C{-}NCl_2 + H_2NOH \cdot HCl + 3\ H_3CONa \longrightarrow H_2C{=}N_2 + 3\ NaCl + 3\ H_3COH + H_2O \quad (\sim 18\%) \quad (63)$$

$$CHCl_3 + H_2N{-}NH_2 + 3\ KOH \longrightarrow H_2C{=}N_2 + 3\ KCl + 3\ H_2O \quad (64)$$

diazomethane. The conversion of dimesityl ketone imine into diazodimesitylmethane is dominated by a condensation step (equation 65)[355].

$$Mes_2C{=}NH \xrightarrow[H_3C-COONa]{N_2O_4,\ CCl_4} Mes_2C{=}N{-}NO \xrightarrow[-30\ °C]{LiAlH_4,\ ether} Mes_2C{=}N_2\ (53\%) \quad (65)$$

Mes = 2,4,6-trimethylphenyl (H_3C, CH_3, H_3C)

A condensation reaction formally resembling the Wittig olefination affords diazomethane from methylenetriphenylphosphorane and nitrous oxide after subsequent alkaline hydrolysis[356]. The intermediate phosphonium salt (**128**), which is formed by 'neutralization' from the ylene and diazomethane, undergoes the above hydrolysis.

$$(H_5C_6)_3P{=}CH_2 + N_2O \xrightarrow{Ether,\ 0\ °C} (H_5C_6)_3P{=}O + H_2C{=}N_2$$

$$(H_5C_6)_3P{=}CH_2 \xrightarrow{H_2C=N_2} (H_5C_6)_3\overset{+}{P}{-}CH_3\ \ {}^{-}CH{=}N_2\ (\mathbf{128}) \xrightarrow[-(H_5C_6)_2(H_3C)P=O,\ -C_6H_6]{KOH/H_2O} H_2C{=}N_2$$

III. REFERENCES

1. B. Eistert, M. Regitz, G. Heck and H. Schwall, in *Methoden der Organischen Chemie* (*Houben–Weyl–Müller*), Vol. X/4, 4th ed., p. 557, G. Thieme-Verlag, Stuttgart, 1968, p. 557.
2. M. Regitz, in *Methodicum Chimicum*, Vol. 6, 1st ed., G. Thieme-Verlag, Stuttgart/Academic Press, New York, 1974, p. 211; Engl. ed. Vol. 6, p. 206.

3. M. Regitz, *Diazoalkane*, G. Thieme-Verlag, Stuttgart, in the press.
3. M. Regitz, *Aliphatic Diazo Compounds*, G. Thieme-Verlag, Stuttgart, in the press.
4. E. Müller and W. Rundel, *Chem. Ber.*, **91**, 466 (1958).
5. E. Müller, H. Haiss and W. Rundel, *Chem. Ber.*, **93**, 1541 (1960).
6. Phrix-Werke, A.G. (Inv. E. Müller and W. Rundel), *DBP* 1033 671; *Chem. Abstr.*, **54**, 10861 (1960); Phrix-Werke A.G. (Inv. E. Müller, H. Haiss and W. Rundel), *DBP* 1104 518; *Chem. Abstr.*, **56**, 11445 (1962).
7. J. Bakke, *Acta Chem. Scand.*, **22**, 1833 (1968).
8. H. Gilman and R. G. Jones, *J. Amer. Chem. Soc.*, **65**, 1458 (1943).
9. B. L. Dyatkin and E. P. Mochalina, *Izv. Akad. Nauk S.S.S.R., Ser. Khim.*, 1225 (1964); *Chem. Abstr.*, **61**, 11881f (1964).
10. R. A. Shepard and P. L. Sciaraffa, *J. Org. Chem.*, **31**, 964 (1966).
11. E. P. Mochalina and B. L. Dyatkin, *Izv. Akad. Nauk S.S.S.R., Ser. Khim.*, 926 (1965); *Chem. Abstr.*, **63**, 5515c (1965).
12. J. H. Atherton, R. Fields and R. N. Haszeldine, *J. Chem. Soc. C*, 366 (1971).
13. L. C. Krogh, T. S. Reid and H. A. Brown, *J. Org. Chem.*, **19**, 1124 (1954).
14. R. A. Shepard and S. E. Wentworth, *J. Org. Chem.*, **32**, 3197 (1967).
15. J. M. Birchall and R. N. Haszeldine, *J. Chem. Soc.*, 3722 (1961).
16. A. Roedig and K. Grohe, *Tetrahedron*, **21**, 2375 (1965).
17. H. D. Hartzler, *J. Amer. Chem. Soc.*, **86**, 2174 (1964).
18. O. W. Webster, *J. Amer. Chem. Soc.*, **88**, 4055 (1966).
19. J. Villarrasa, E. Meléndez and J. Elguero, *Tetrahedron Lett.*, 1609 (1974).
20. H. Reimlinger, A. van Overstraeten and H. G. Viehe, *Chem. Ber.*, **94**, 1036 (1961).
21. J. M. Tedder and B. Webster, *J. Chem. Soc.*, 3270 (1960).
22. A. Hantzsch, *Ber. Dt. chem. Ges.*, **35**, 888 (1902).
23. U. Simon, O. Süs and L. Horner, *Liebigs Ann. Chem.*, **697**, 17 (1966).
24. D. G. Farnum and P. Yates, *J. Amer. Chem. Soc.*, **84**, 1399 (1962).
25. Y. F. Shealy, C. A. Krauth and J. A. Montgomery, *J. Org. Chem.*, **27**, 2150 (1962).
26. Y. F. Shealy, *J. Org. Chem.*, **26**, 2396 (1961).
27. J. W. Jones and R. K. Robins, *J. Amer. Chem. Soc.*, **82**, 3773 (1960).
28. W. A. Sheppard and O. W. Webster, *J. Amer. Chem. Soc.*, **95**, 2696 (1973).
29. D. Stadler, W. Anschütz, M. Regitz, G. Keller, D. van Asche and J. P. Fleury, *Liebigs Ann. Chem.*, 2159 (1975).
30. A. N. Frolov, M. S. Pevzner, J. N. Shokhor, A. G. Gal'kovskaya and L. I. Bagal, *Khim. Geterotsikl. Soedin.*, **705** (1970); *Chem. Abstr.*, **73**, 45420k (1970).
31. J. Thiele, *Liebigs Ann. Chem.*, **270**, 59 (1892); J. Thiele and J. T. Marais, *Liebigs Ann. Chem.*, **273**, 147 (1893); J. Thiele and H. Ingle, *Liebigs Ann. Chem.*, **287**, 235 (1895).
32. J. M. Tedder, *Tetrahedron*, **1**, 270 (1957); *J. Chem. Soc.*, 4003 (1957); *J. Amer. Chem. Soc.*, **79**, 6090 (1957); J. M. Tedder and G. Theaker, *J. Chem. Soc.*, 4008 (1957); *Chem. Ind. (London)*, 1485 (1957); *J. Chem. Soc.*, 2573 (1958); *Tetrahedron*, **5**, 288 (1959); *J. Chem. Soc.*, 257 (1959).
33. J. M. Tedder and B. Webster, *J. Chem. Soc.*, 3270 (1960).
34. H. P. Patel, J. M. Tedder and B. Webster, *Chem. Ind. (London)*, 1163 (1961).
35. J. M. Tedder and B. Webster, *J. Chem. Soc.*, 1638 (1962).
36. S. Torii, S. Endo, H. Oka, Y. Kariya and A. Takeda, *Bull. Chem. Soc. Jap.*, **41**, 2707 (1968).
37. J. M. Tedder and G. Theaker, *J. Chem. Soc.*, 257 (1959).
38. M. O. Forster, *J. Chem. Soc.*, 260 (1915).
39. J. Meinwald, P. G. Gassmann and E. G. Miller, *J. Amer. Chem. Soc.*, **81**, 4751 (1959).
40. W. Rundel, *Angew. Chem.*, **74**, 469 (1962).
41. T. Severin, H. Krämer and P. Adhikary, *Chem. Ber.*, **104**, 972 (1971).
42. J. C. Fleming and H. Schechter, *J. Org. Chem.*, **34**, 3962 (1969).
43. T. Curtius, *Ber. dt. chem. Ges.*, **22**, 2161 (1889).
44. M. S. Newman and D. E. Reid, *J. Org. Chem.*, **23**, 665 (1958).
45. G. Wittig and A. Krebs, *Chem. Ber.*, **94**, 3260 (1961).
46. A. T. Blomquist and L. H. Liu, *J. Amer. Chem. Soc.*, **75**, 2153 (1953).
47. A. T. Blomquist, L. H. Liu and J. C. Bohrer, *J. Amer. Chem. Soc.*, **74**, 3643 (1952).
48. V. Prelog, K. Schenker and W. Küng, *Helv. Chim. Acta*, **36**, 471 (1953).

49. A. T. Blomquist, R. E. Burge and A. C. Sucsy, *J. Amer. Chem. Soc.*, **74**, 3636 (1952).
50. J. D. Bower and G. R. Ramage, *J. Chem. Soc.*, 4506 (1957).
51. J. H. Boyer, R. Borgers and L. T. Wolford, *J. Amer. Chem. Soc.*, **79**, 678 (1957).
52. J. B. Miller, *J. Org. Chem.*, **24**, 560 (1959).
53. A. Schönberg, W. I. Awad and N. Latif, *J. Chem. Soc.*, 1368 (1951).
54. W. Fischer and J.-P. Anselme, *J. Amer. Chem. Soc.*, **89**, 5312 (1967).
55. H. Staudinger and O. Kupfer, *Ber. dt. chem. Ges.*, **44**, 2197 (1911).
56. H. Staudinger and J. Goldstein, *Ber. dt. chem. Ges.*, **49**, 1923 (1916).
57. H. Staudinger and A. Gaule, *Ber. dt. chem. Ges.*, **49**, 1897 (1916).
58. K. Heyns and A. Heins, *Liebigs Ann. Chem.*, **604**, 133 (1957).
59. A. C. Day, P. Raymond, R. M. Southam and M. C. Whiting, *J. Chem. Soc. C*, 467 (1966).
60. P. Yates and F. X. Garneau, *Tetrahedron Lett.*, 71 (1967).
61. D. B. Mobbs and H. Suschitzky, *Tetrahedron Lett.*, 361 (1971).
62. W. M. Jones and W. T. Tai, *J. Org. Chem.*, **27**, 1324 (1962).
63. J.-P. Anselme, *Organic Preparations and Procedures*, **1**, 73 (1969).
64. C. D. Gutsche, G. L. Bachmann and R. S. Coffey, *Tetrahedron*, **18**, 617 (1962).
65. T. Nakaya, T. Tomomoto and M. Imoto, *Bull. Chem. Soc. Jap.*, **40**, 691 (1967).
66. H. Staudiner, E. Enthes and F. Phenninger, *Ber. dt. chem. Ges.*, **49**, 1928 (1916).
67. L. I. Smith and K. L. Howard, *Org. Syn. Coll.*, Vol. III, 351 (1955).
68. R. W. Murray and A. M. Trozzolo, *J. Org. Chem.*, **29**, 1268 (1964).
69. S. I. Murahashi, Y. Yoshimura, Y. Yamamoto and I. Moritani, *Tetrahedron*, **28**, 1485 (1972).
70. A. Schönberg and M. M. Sidky, *J. Amer. Chem. Soc.*, **81**, 2259 (1959).
71. N. Latif and I. Fathy, *Can. J. Chem.*, **37**, 863 (1959).
72. B. Eistert, W. Kurze and G. W. Müller, *Liebigs Ann. Chem.*, **732**, 1 (1970).
73. P. C. Guha and D. K. Sankaran, *Ber. dt. chem. Ges.*, **70**, 1688 (1937).
74. S. D. Andrews, A. C. Day, P. Raymond and M. C. Whiting, *Org. Syn.*, **50**, 27 (1970).
75. G. L. Closs and J. J. Coyle, *J. Org. Chem.*, **31**, 2759 (1966).
76. M. Fetizon, M. Golfier, R. Milcent and I. Papadakis, *Tetrahedron*, **31**, 165 (1975).
77. C. D. Gutsche, E. F. Jason, R. S. Coffey and H. E. Johnson, *J. Amer. Chem. Soc.*, **80**, 5756 (1958).
78. C. D. Gutsche, G. L. Bachmann, W. Udell and S. Bäuerlein, *J. Amer. Chem. Soc.*, **93**, 5172 (1971).
79. R. Hüttel, J. Riedl, H. Martin and K. Franke, *Chem. Ber.*, **93**, 1425 (1960).
80. V. Franzen and H. J. Joschek, *Liebigs Ann. Chem.*, **633**, 7 (1960).
81. A. K. Colter and S. S. Wang, *J. Org. Chem.*, **27**, 1517 (1962).
82. D. E. Applequist and H. Babad, *J. Org. Chem.*, **27**, 288 (1962).
83. R. A. Shepard and P. L. Sciaraffa, *J. Org. Chem.*, **31**, 964 (1966).
84. R. A. Moss and J. D. Funk, *J. Chem. Soc. C*, 2026 (1967).
85. J. Attenburrow, A. F. B. Cameron, J. H. Chapman, R. M. Evans, B. A. Hems, A. B. A. Janson and T. Walker, *J. Chem. Soc.*, 1094 (1952).
86. W. Schroeder and L. Katz, *J. Org. Chem.*, **19**, 718 (1954).
87. S. Hauptmann and H. Wilde, *J. Prakt. Chem.*, **311**, 604 (1969).
88. H. Morrison, S. Danishefsky and P. Yates, *J. Org. Chem.*, **26**, 2617 (1961).
89. K. D. Kaufmann, B. Auräth, P. Träger and K. Rühlmann, *Tetrahedron Lett.*, 4973 (1968).
90. R. W. Murray and A. M. Trozzolo, *J. Org. Chem.*, **26**, 3109 (1961).
91. J. B. F. N. Engberts, G. van Bruggen, J. Strating and H. Wynberg, *Rec. Trav. Chim. Pays-Bas*, **84**, 1610 (1965).
92. H. Reimlinger, *Chem. Ber.*, **97**, 3493 (1964).
93. H. G. Biedermann and H. G. Schmid, *Z. Naturforsch. B*, **28**, 378 (1973).
94. P. A. S. Smith and J. G. Wirth, *J. Org. Chem.*, **33**, 1145 (1968).
95. J. H. Wieringa, H. Wynberg and J. Strating, *Tetrahedron*, **30**, 3053 (1974).
96. D. M. Gale, W. J. Middleton and C. G. Krespan, *J. Amer. Chem. Soc.*, **87**, 657 (1965); E. I. du Pont de Nemours (Inv. C. G. Krespan and W. J. Middleton, *U.S. Patent* 3242 166 (1966); *Chem. Abstr.*, **64**, 17422h (1966).
97. E. T. McBee and K. J. Sienkowski, *J. Org. Chem.*, **38**, 1340 (1973).

98. A. Stojiljković, N. Orbović, S. Sredojević and M. L. Mihailović, *Tetrahedron*, **26**, 1101 (1970).
99. R. Hensel, *Chem. Ber.*, **88**, 257 (1955).
100. M. Debono and R. M. Molloy, *J. Org. Chem.*, **34**, 1454 (1969).
101. D. H. R. Barton, P. L. Blatten and J. F. McGhie, *J. Chem. Soc. Chem. Commun.*, 450 (1969).
102. D. M. Barton, R. E. O'Brien and S. Sternhell, *J. Chem. Soc.*, 470 (1962).
103. Hooker Chemical Corporation (Inv. D. Knutson), *U.S. Patent* 3422 158 (1969); *Chem. Abstr.*, **70**, 67877e (1969).
104. H. Wieland and A. Roseeu, *Liebigs Ann. Chem.*, **381**, 229 (1911).
105. J. G. Shorfkin and H. Satzman, *Org. Syn.*, **43**, 62 (1963).
106. P. A. S. Smith, *J. Org. Chem.*, **39**, 1047 (1974).
107. H. Staudiner and A. Gaule, *Ber. dt. chem. Ges.*, **49**, 1951 (1916).
108. K. Nakagawa, R. Konaka and T. Nakata, *J. Org. Chem.*, **27**, 1597 (1962).
109. K. Nakagawa, H. Onoue and K. Minami, *J. Chem. Soc. Chem. Commun.*, 730 (1966).
110. H. Disselnkötter, *Angew. Chem.*, **76**, 431 (1964); *Angew. Chem., Int. Ed.*, **3**, 379 (1964).
111. M. Z. Barakat, M. F. A. El-Wahab and M. M. El-Sadr, *J. Amer. Chem. Soc.*, **77**, 1670 (1955).
112. W. Borsche and R. Frank, *Liebigs Ann. Chem.*, **450**, 75 (1926).
113. W. R. Bamford and T. S. Stevens, *J. Chem. Soc.*, 4735 (1952).
114. W. Kirmse, B. G. von Bülow and H. Schepp, *Liebigs Ann. Chem.*, **691**, 41 (1966).
115. W. Kirmse, *Carbene, Carbenoide und Carbenanaloge*, 1st ed., Verlag Chemie, Weinheim, 1969, p. 137.
116. W. Kirmse, *Carben Chemistry*, 2nd ed., Academic Press, New York, 1971, p. 29.
117. J. Casanova and B. Waegell, *Bull. Soc. Chim. Fr.*, 922 (1975).
118. J. W. Powell and M. C. Whiting, *Tetrahedron*, **7**, 305 (1959).
119. J. W. Powell and M. C. Whiting, *Tetrahedron*, **12**, 168 (1961).
120. L. Friedman and H. Shechter, *J. Amer. Chem. Soc.*, **81**, 5512 (1959).
121. C. H. De Puy and D. M. Froemsdorf, *J. Amer. Chem. Soc.*, **82**, 634 (1960).
122. D. M. Lemal and A. J. Fry, *J. Org. Chem.*, **29**, 1673 (1964).
123. H. Nozaki, R. Noyori and K. Sisido, *Tetrahedron*, **20**, 1125 (1964).
124. U. Schöllkopf and E. Wiskott, *Liebigs Ann. Chem.*, **694**, 44 (1966).
125. C. C. Leznott, *Can. J. Chem.*, **46**, 1152 (1968).
126. A. Dornow and W. Bartsch, *Angew. Chem.*, **67**, 209 (1955).
127. G. Kaufman, F. Cook, H. Shechter, J. Bayless and L. Friedman, *J. Amer. Chem. Soc.*, **89**, 5736 (1967).
128. R. H. Shapiro and M. J. Heath, *J. Amer. Chem. Soc.*, **89**, 5734 (1967).
129. R. H. Shapiro, *Tetrahedron Lett.*, 345 (1968).
130. A. P. Krapcho and J. Diamanti, *Chem. Ind. (London)*, 847 (1965).
131. J. Casanova and B. Waegell, *Bull. Soc. Chim. Fr.*, 1289 (1971).
132. J. Casanova and B. Waegell, *Bull. Soc. Chim. Fr.*, 1295 (1971).
133. K. Geibel and H. Mäder, *Chem. Ber.*, **103**, 1645 (1970).
134. T. Iwadare, I. Idachi, M. Hayashi, A. Matsunaga and T. Kitai, *Tetrahedron Lett.*, 4447 (1969).
135. A. Eschenmoser, D. Felix and G. Ohloff, *Helv. Chim. Acta*, **50**, 708 (1967).
136. M. Tanabe, D. F. Crowe, R. L. Dehn and G. Detre, *Tetrahedron Lett.*, 3739 (1967).
137. M. Tanabe, D. F. Crowe and R. L. Dehn, *Tetrahedron Lett.*, 3943 (1967).
138. J. W. Wilt, C. A. Schneider, H. F. Dabek, J. F. Kraemer and W. J. Wagner, *J. Org. Chem.*, **31**, 1543 (1966).
139. J. W. Wilt, R. G. Stein and W. J. Wagner, *J. Org. Chem.*, **32**, 2097 (1967).
140. J. W. Wilt, C. T. Parsons, C. A. Schneider, D. G. Schultenover and W. J. Wagner, *J. Org. Chem.*, **33**, 694 (1968).
141. J. W. Wilt and J. F. Kraemer, *J. Org. Chem.*, **33**, 4267 (1968).
142. J. G. Shelnut, S. Mataka and J.-P. Anselme, *J. Chem. Soc. Chem. Commun.*, 114 (1975).
143. D. E. Dana and J.-P. Anselme, *Tetrahedron Lett.*, 1565 (1975).
144. G. L. Closs and R. A. Moss, *J. Amer. Chem. Soc.*, **86**, 4042 (1964).
145. H. W. Davies and M. Schwarz, *J. Org. Chem.*, **30**, 1242 (1965).
146. D. G. Farnum, *J. Org. Chem.*, **28**, 870 (1963).

147. G. Diderich, *Helv. Chim. Acta*, **55**, 2103 (1972).
148. S. I. Murahashi, I. Moritani and M. Nishino, *Tetrahedron*, **27**, 5131 (1971).
149. I. Moritani, S. I. Murahashi, K. Yoshinaga and H. Ashitaka, *Bull. Chem. Soc. Jap.*, **40**, 1506 (1967).
150. W. Kirmse and H. Dietrich, *Chem. Ber.*, **100**, 2710 (1967).
151. T. Endo, K. Ikeda, Y. Kawamura and Y. Mizuno, *J. Chem. Soc. Chem. Commun.*, 673 (1973).
152. Y. Mizuno, T. Endo and T. Nakamura, *J. Org. Chem.*, **40**, 1391 (1975).
153. G. M. Kaufman, J. A. Smith, G. G. Vander Stouw and H. Shechter, *J. Amer. Chem. Soc.*, **87**, 935 (1965).
154. G. G. Vander Stouw, A. R. Kraska and H. Shechter, *J. Amer. Chem. Soc.*, **94**, 1655 (1972).
155. M. Rey, R. Begrich, W. Kirmse and A. S. Dreiding, *Helv. Chim. Acta*, **51**, 1001 (1968).
156. G. L. Closs and R. B. Larrabee, *Tetrahedron Lett.*, 287 (1965).
157. T. Sasaki, S. Eguchi, I. H. Ryu and Y. Hirako, *Tetrahedron Lett.*, 2011 (1974).
158. Z. Majerski, S. H. Liggero and P. v. R. Schleyer, *J. Chem. Soc. Chem. Commun.*, **949** (1970).
159. K. Kondo and J. Ojimo, *J. Chem. Soc. Chem. Commun.*, **949** (1970).
160. E. Piers, R. W. Britton, R. J. Keziere and R. D. Smillie, *Can. J. Chem.*, **49**, 2623 (1971).
161. A. G. Brook and P. F. Jones, *Can. J. Chem.*, **47**, 4353 (1969).
162. G. L. Closs and L. E. Closs, *J. Amer. Chem. Soc.*, **83**, 2015 (1961).
163. G. L. Closs, L. E. Closs and W. A. Böll, *J. Amer. Chem. Soc.*, **85**, 3796 (1963).
164. H. H. Stechl, *Chem. Ber.*, **97**, 2681 (1964).
165. *See* Reference 116, p. 328.
166. T. Sato and S. Watanabe, *Bull. Chem. Soc. Jap.*, **41**, 3017 (1968).
167. G. L. Closs and W. A. Böll, *Angew. Chem.*, **75**, 640 (1963); *Angew. Chem., Int. Ed.*, **2**, 399 (1963).
168. G. L. Closs, W. A. Böll, H. Heyn and V. Dev, *J. Amer. Chem. Soc.*, **90**, 173 (1968).
169. J. A. Pincock, R. Morchat and D. R. Arnold, *J. Amer. Chem. Soc.*, **95**, 7536 (1973).
170. R. H. Findlay and J. T. Sharp, *J. Chem. Soc. Chem. Commun.*, 909 (1970).
171. S. Sto, K. Takase, N. Kawabe and H. Sugiyama, *Bull. Chem. Soc. Jap.*, **39**, 253 (1966).
172. B. H. Freeman, J. M. F. Gagan and D. Lloyd, *Tetrahedron*, 4307 (1973).
173. M. Jones, A. M. Harrison and K. R. Rettig, *J. Amer. Chem. Soc.*, **91**, 7462 (1969).
174. E. Vedjs, *J. Chem. Soc. Chem. Commun.*, 536 (1971).
175. E. C. S. Jones and J. Kenner, *J. Chem. Soc.*, 363 (1933).
176. C. E. Redemann, F. O. Rice, R. Roberts and H. P. Ward, *Org. Syn. Coll.*, Vol. III, 244 (1955).
177. D. W. Adamson and J. Kenner, *J. Chem. Soc.*, 1551 (1937).
178. M. Berenbom and W. S. Fones, *J. Amer. Chem. Soc.*, **71**, 1629 (1949).
179. D. W. Adamson and J. Kenner, *J. Chem. Soc.*, 286 (1935).
180. P. D. Shevlin and A. P. Wolf, *J. Amer. Chem. Soc.*, **88**, 4735 (1966).
181. V. Hoȓàk and M. Prochazka, *Chem. Ind.* (*London*), 472 (1961).
182. V. Hoȓàk and M. Prochazka, *Chem.-Ztg.* **85**, 540 (1961).
183. H. v. Pechmann, *Ber. dt. chem. Ges.*, **27**, 1888 (1894).
184. H. v. Pechmann, *Ber. dt. chem. Ges.*, **28**, 855 (1895).
185. R. A. Moss, *Accounts Chem. Res.*, **7**, 421 (1974).
186. A. Hantzsch and M. Lehmann, *Ber. dt. chem. Ges.*, **35**, 897 (1902).
187. E. Müller, W. Hoppe, H. Hagenmaier, H. Haiss, R. Huber, W. Rundel and H. Suhr, *Chem. Ber.*, **96**, 1712 (1963).
188. E. Müller, W. Rundel, H. Haiss and H. Hagenmaier, *Z. Naturforsch. B*, **15**, 751 (1960).
189. H. Suhr, *Chem. Ber.*, **96**, 1720 (1963).
190. R. A. Moss, *J. Org. Chem.*, **31**, 1082 (1966).
191. R. A. Moss and F. C. Shulman, *Tetrahedron*, **24**, 2881 (1968).
192. H. Reimlinger, *Angew. Chem.*, **73**, 221 (1961).
193. T. K. Tandy and W. M. Jones, *J. Org. Chem.*, **30**, 4257 (1965).
194. R. Huisgen, *Liebigs Ann. Chem.*, **573**, 163 (1951).
195. C. D. Gutsche and H. E. Johnson, *J. Amer. Chem. Soc.*, **77**, 109 (1955).
196. W. M. Jones and D. L. Muck, *J. Amer. Chem. Soc.*, **88**, 3798 (1966).

197. W. Kirmse and G. Wächtershäuser, *Liebigs Ann. Chem.*, **707**, 44 (1967).
198. R. Huisgen and J. Reinertshofer, *Liebigs Ann. Chem.*, **575**, 174 (1952).
199. W. Pritzkow and P. Dietrich, *Liebigs Ann. Chem.*, **665**, 88 (1963).
200. W. Kirmse and U. Seipp, *Chem. Ber.*, **107**, 745 (1974); such as preceding papers.
201. A. E. Feiring and J. Ciabattoni, *J. Org. Chem.*, **37**, 3748 (1972).
202. H. Hart and J. L. Brewbaker, *J. Amer. Chem. Soc.*, **91**, 706 (1969).
203. S. M. Hecht and J. W. Kozarich, *Tetrahedron Lett.*, 5147 (1972).
204. S. M. Hecht and J. W. Kozarich, *J. Org. Chem.*, **38**, 1021 (1973).
205. E. A. Werner, *J. Chem. Soc.*, **115**, 1093 (1919).
206. W. M. Jones, D. L. Muck and T. K. Tandy, *J. Amer. Chem. Soc.*, **88**, 68 (1966).
207. W. W. Hartmann and M. R. Brethen, *Org. Syn. Coll.*, Vol. II, 278 (1943); W. W. Hartmann and R. Phillips, *Org. Syn. Coll.*, Vol. II, 464 (1943).
208. Schering A.G. (Inv. H. Meerwein), *DBP* 579 309 (1933); *Chem. Abstr.*, **27**, 4546 (1933).
209. C. D. Gutsche, *J. Amer. Chem. Soc.*, **71**, 3513 (1949).
210. C. D. Gutsche and H. E. Johnson, *Org. Syn. Coll.*, Vol. IV, 780 (1963).
211. B. Eistert, in *Neuere Methoden der präparativen organischen Chemie*, 3rd ed., Verlag Chemie, Weinheim, 1949, pp. 359, 398.
212. H. Meerwein and W. Burneleit, *Ber. dt. chem. Ges.*, **61**, 1840 (1928).
213. H. V. Pechmann, *Ber. dt. chem. Ges.*, **31**, 2640 (1898).
214. K. v. Auwers and E. Cauer, *Liebigs Ann. Chem.*, **470**, 284 (1929).
215. A. L. Wilds and A. L. Meader, *J. Org. Chem.*, **13**, 763 (1948).
216. H. E. Zimmerman and D. H. Paskovich, *J. Amer. Chem. Soc.*, **86**, 2149 (1964).
217. L. Hellerman and R. L. Garner, *J. Amer. Chem. Soc.*, **57**, 139 (1935).
218. D. Y. Curtin and S. M. Gerber, *J. Amer. Chem. Soc.*, **74**, 4052 (1952).
219. H. Philip and J. Keating, *Tetrahedron Lett.*, 523 (1961).
220. S. Hauptmann, F. Brandes, E. Brauer and W. Gabler, *J. Prakt. Chem.*, (4) **25**, 56 (1964).
221. V. D. Sheladyakov, G. D. Khatuntsev and V. F. Mironov, *Zh. Obsch. Khim.*, **39**, 2785 (1969); *Chem. Abstr.*, **72**, 111553p (1970).
222. M. S. Newman and T. B. Patrick, *J. Amer. Chem. Soc.*, **91**, 6461 (1969).
223. M. S. Newman and A. O. M. Akorodudu, *J. Org. Chem.*, **34**, 1220 (1969).
224. M. S. Newman and A. O. M. Akorodudu, *J. Amer. Chem. Soc.*, **90**, 4189 (1968).
225. D. J. Northington and W. M. Jones, *Tetrahedron Lett.*, 317 (1971).
226. D. J. Northington and W. M. Jones, *J. Org. Chem.*, **37**, 693 (1972).
227. H. Holter and H. Bretschneider, *Monatsh. Chem.*, **53/54**, 963 (1929).
228. E. Müller and S. Petersen, *Angew. Chem.*, **63**, 18 (1951).
229. C. M. Samour and J. P. Mason, *J. Amer. Chem. Soc.*, **76**, 441 (1954).
230. H. Hart and J. L. Brewbaker, *J. Amer. Chem. Soc.*, **91**, 706 (1969).
231. T. Lieser and G. Beck, *Chem. Ber.*, **83**, 137 (1950).
232. S. Nirdlinger and S. F. Acree, *Amer. Chem. J.*, **43**, 381 (1910).
233. C. D. Hurd and S. C. Lui, *J. Amer. Chem. Soc.*, **57**, 2656 (1935).
234. R. G. Salomon, M. F. Salomon and T. R. Heyne, *J. Org. Chem.*, **40**, 756 (1975).
235. J. L. Brewbaker and H. Hart, *J. Amer. Chem. Soc.*, **91**, 711 (1969).
236. D. M. Lemal, F. Menger and G. W. Clark, *J. Amer. Chem. Soc.*, **85**, 2529 (1963).
237. R. C. Atkins and B. M. Trost, *J. Org. Chem.*, **37**, 3133 (1972).
238. E. H. White, G. E. Maier, R. Graeve, U. Zirngibl and E. W. Friend, *J. Amer. Chem. Soc.*, **88**, 611 (1966).
239. D. M. Lemal and K. S. Shim, *Tetrahedron Lett.*, 3231 (1964).
240. S. Masamune, N. Nakamura, M. Suda and H. Ona, *J. Amer. Chem. Soc.*, **95**, 8481 (1973).
241. G. F. D'Alelio and E. E. Reid, *J. Amer. Chem. Soc.*, **59**, 109 (1937).
242. K. Heyns and O. F. Woyrisch, *Chem. Ber.*, **86**, 76 (1953).
243. K. Heyns and W. v. Bebenburg, *Chem. Ber.*, **86**, 278 (1953).
244. Du Pont (Inv. F. S. Fawcett), *U.S. Patent* 2675 378 (1954); *Chem. Abstr.*, **49**, 1777b (1955).
245. H. Reimlinger, *Chem. Ber.*, **94**, 2547 (1961).
246. BASF (Inv. H. Stummeyer and G. Hummel), *DBP*, 841 747 (1952); *Chem. Abstr.*, **52**, 10162c (1958).

247. J. A. Moore and D. E. Reed, *Org. Syn.*, **41**, 16 (1962).
248. G. A. Gladkovskii and S. S. Skorokhodov, *Zh. Org. Khim.*, **3**, 24 (1967); *Chem. Abstr.*, **66**, 85396u (1967).
249. A. Oppé, *Ber. dt. chem. Ges.*, **46**, 1095 (1919).
250. H. Reimlinger, *Angew. Chem.*, **74**, 252 (1962); *Angew. Chem., Int. Ed.*, **1**, 216 (1962).
251. H. Reimlinger, *Liebigs Ann. Chem.*, **713**, 113 (1968).
252. F. Arndt and J. Amende, *Angew. Chem.*, **43**, 444 (1930).
253. F. Arndt and H. Scholz, *Angew. Chem.*, **46**, 47 (1933).
254. F. Arndt, L. Loewe and S. Avan, *Ber. dt. chem. Ges.*, **73**, 606 (1940).
255. F. Arndt, *Org. Syn. Coll.*, Vol. II, 461 (1943).
256. *See* Reference 211, p. 395.
257. K. Clusius and F. Endtinger, *Helv. Chim. Acta*, **43**, 2063 (1960).
258. *See* Reference 1, p. 357.
259. H. Irie, T. Kishimoto and S. Uyeo, *J. Chem. Soc. C*, 1645 (1969).
260. F. Arndt, *Org. Syn. Coll.*, Vol. II, 165 (1943).
261. A. Roedig and K. Grohe, *Tetrahedron*, **21**, 397 (1965).
262. O. Dessaux and M. Durand, *Bull. Soc. Chim. Fr.*, 41 (1963).
263. W. H. Urry and N. Bilow, *J. Amer. Chem. Soc.*, **86**, 1815 (1964).
264. S. M. Hecht and J. W. Kozarich, *Tetrahedron Lett.*, 1397 (1973).
265. J. R. Dyer, R. B. Randall and H. M. Deutsch, *Liebigs Ann. Chem.*, **39**, 3423 (1964).
266. W. Kirmse and H. A. Rinkler, *Liebigs Ann. Chem.*, **707**, 57 (1967).
267. G. D. Buckley and N. H. Ray, *J. Chem. Soc.*, 3701 (1952).
268. D. E. Applequist and D. E. McGreer, *J. Amer. Chem. Soc.*, **82**, 1965 (1960).
269. W. M. Jones and M. H. Frasley, *Tetrahedron Lett.*, 927 (1962).
270. W. M. Jones, M. H. Grasley and W. S. Brey, *J. Amer. Chem. Soc.*, **85**, 2754 (1963).
271. H. Lettré and U. Brose, *Naturwissenschaften*, **36**, 57 (1949).
272. W. M. Jones and D. L. Muck, *J. Amer. Chem. Soc.*, **88**, 3804 (1966).
273. F. v. Bruchhausen and H. Hoffmann, *Chem. Ber.*, **74**, 1584 (1941).
274. W. Kirmse, H. J. Schladetsch and H. W. Bücking, *Chem. Ber.*, **99**, 2579 (1966).
275. H. Musso and H. Klasacek, *Chem. Ber.*, **103**, 3076 (1970).
276. W. Kirmse and K. Pöhlmann, *Chem. Ber.*, **100**, 3564 (1967).
277. D. Seyferth, A. W. Dow, M. Menzel and T. C. Flood, *J. Amer. Chem. Soc.*, **90**, 1080 (1968).
278. D. Seyferth, M. Menzel, A. W. Dow and T. C. Flood, *J. Organomet. Chem.*, **44**, 279 (1972).
279. J. M. Crossman, R. N. Haszeldine and A. E. Tipping, *J. Chem. Soc. Dalton*, 483 (1973).
280. J. Mars and L. Marx-Moll, *Chem. Ber.*, **87**, 1499 (1954).
281. J. Hooz and H. Kono, *Organic Preparations and Procedures*, **3**, 47 (1971).
282. P. S. Skell and J. Klebe, *J. Amer. Chem. Soc.*, **82**, 247 (1960).
283. A. F. McKay, W. L. Ott, G. W. Taylor, M. N. Buchanan and J. F. Crooker, *Can. J. Research*, **28B**, 683 (1950).
284. A. F. McKay and G. F. Wright, *J. Amer. Chem. Soc.*, **69**, 3028 (1947).
285. A. F. McKay and G. F. Wright, *J. Amer. Chem. Soc.*, **70**, 1974 (1948).
286. T. J. De Boer and H. J. Backer, *Org. Syn. Coll.* Vol. IV, 943 (1963).
287. T. J. De Boer and H. J. Backer, *Rec. Trav. Chim. Pays-Bas*, **73**, 229 (1954).
288. T. J. De Boer and H. J. Backer, *Org. Syn. Coll.* Vol. IV, 250 (1963).
289. C. G. Overberger and J.-P. Anselme, *J. Org. Chem.*, **28**, 592 (1963).
290. C. Groth, E. Pfeil, E. Einrich and O. Weissel, *Liebigs Ann. Chem.*, **679**, 42 (1964).
291. M. Regitz, *Angew. Chem.*, **79**, 786 (1967); *Angew. Chem., Int. Ed.*, **6**, 733 (1967).
292. M. Regitz, in *Neuere Methoden der Präparativen Organischen Chemie*, Vol. VI, 1st ed., Verlag Chemie, Weinheim, 1970, p. 76.
293. M. Regitz, *Synthesis*, 351 (1972).
294. *See* Reference 2, p. 237.
295. W. v. E. Doering and C. H. De Puy, *J. Amer. Chem. Soc.*, **75**, 5955 (1953).
296. M. Regitz and A. Liedhegener, *Tetrahedron*, **23**, 2701 (1967).
297. T. Weil and M. Cais, *J. Org. Chem.*, **28**, 2472 (1963).
298. D. Lloyd, M. I. C. Singer, M. Regitz and A. Liedhegener, *Chem. Ind.* (*London*), 324 (1967).

299. J. C. Martin and D. R. Bloch, *J. Amer. Chem. Soc.*, **93**, 451 (1971).
300. B. H. Freeman, J. M. F. Gagan and D. Lloyd, *Tetrahedron*, **29**, 4307 (1973).
301. D. Rewicki and C. Tuchscherer, *Angew. Chem.*, **84**, 31 (1972); *Angew. Chem., Int. Ed.*, **11**, 44 (1972).
302. M. Regitz, *Chem. Ber.*, **97**, 2742 (1964).
303. F. Klages and K. Bott, *Chem. Ber.*, **97**, 735 (1964).
304. Agfa AG (Inv. W. Pelz), *U.S. Patent* 2950 273 (1960); *Chem. Abstr.*, **55**, 2116l (1961).
305. G. Cauquis, G. Rverdy and M. Rastoldo, *C. R. Acad. Sci., Paris*, **260**, 2259 (1965).
306. M. Regitz and G. Himbert, *Liebigs Ann. Chem.*, **734**, 70 (1970).
307. R. Fusco, G. Bianchetti, D. Pocar and R. Ugo, *Chem. Ber.*, **96**, 802 (1963).
308. R. Huisgen, L. Möbius and G. Szeimies, *Chem. Ber.*, **98**, 1138 (1965).
309. D. Pocar, G. Bianchetti and P. Ferruti, *Gazz. Chim. Ital.*, **97**, 597 (1967); such as preceding papers of this series.
310. G. Bianchetti, D. Pocar, P. Dalla Croce and A. Vigevani, *Chem. Ber.*, **98**, 2715 (1965).
311. G. Bianchetti, D. Pocar and R. Stradi, *Gazz. Chim. Ital.*, **100**, 726 (1970).
312. K. D. Berlin and M. A. R. Khayat, *Tetrahedron*, **22**, 987 (1966).
313. R. Scarpati and M. L. Graziano, *Tetrahedron Lett.*, 2085 (1971).
314. R. Scarpati and M. L. Graziano, *Tetrahedron Lett.*, 4771 (1971).
315. R. Scarpati and M. L. Graziano, *J. Heterocyl. Chem.*, **9**, 1087 (1972).
316. G. Himbert, D. Frank and M. Regitz, *Chem. Ber.*, **109**, 370 (1976).
317. M. E. Hermes and F. D. Marsh, *J. Amer. Chem. Soc.*, **89**, 4760 (1967).
318. J. A. Pincock, R. Morchat and D. R. Arnold, *J. Amer. Chem. Soc.*, **95**, 7538 (1973).
319. E. Buchner, *Ber. dt. chem. Ges.*, **28**, 215 (1895).
320. F. Arndt, B. Eistert and W. Partale, *Ber. dt. chem. Ges.*, **60**, 1364 (1927); F. Arndt and J. Amende, *Ber. dt. chem. Ges.*, **61**, 1122 (1928).
321. E. L. Closs and J. J. Coyle, *J. Amer. Chem. Soc.*, **87**, 4270 (1965).
322. R. J. Bussey and R. C. Neuman, *J. Org. Chem.*, **34**, 1323 (1969).
323. D. J. Cram and R. P. Partos, *J. Amer. Chem. Soc.*, **85**, 1273 (1963).
324. E. Müller and D. Ludsteck, *Chem. Ber.*, **87**, 1887 (1954).
325. O. J. Scherer and M. Schmidt, *Z. Naturforsch. B*, **20**, 1009 (1965).
326. E. Müller and W. Rundel, *Chem. Ber.*, **90**, 1302 (1957).
327. E. Müller and H. Disselhoff, *Liebigs Ann. Chem.*, **512**, 250 (1934).
328. T. Dominh, O. P. Strausz and H. E. Gunning, *Tetrahedron Lett.*, 5237 (1968).
329. J. Lorberth, *J. Organomet. Chem.*, **27**, 303 (1971).
330. A. N. Wright, K. A. W. Kramer and G. Steel, *Nature*, **199**, 903 (1963).
331. E. T. Blues, D. Bryce-Smith, J. G. Irwin and I. W. Lawston, *J. Chem. Soc. Chem. Commun.*, 466 (1974).
332. M. F. Lappert and J. Lorberth, *J. Chem. Soc. Chem. Commun.*, 836 (1967).
333. M. F. Lappert, J. Lorberth and J. S. Poland, *J. Chem. Soc. A*, 2954 (1970).
334. R. Grüning and J. Lorberth, *J. Organomet. Chem.*, **78**, 221 (1974).
335. D. Seyferth and T. C. Flood, *J. Organomet. Chem.*, **29**, C25 (1971).
336. P. Krommes and J. Lorberth, *J. Organomet. Chem.*, **93**, 339 (1975).
337. B. Eistert and O. Ganster, *Chem. Ber.*, **104**, 78 (1971).
338. B. Eistert and P. Donath, *Chem. Ber.*, **103**, 993 (1970).
339. B. Eistert and H. Juraszyk, *Chem. Ber.*, **103**, 2707 (1970).
340. M. Dürr, *Thesis*, Technische Universität, München, 1971.
341. U. Schöllkopf and H.-U. Scholz, *Synthesis*, 271 (1976).
342. *See* Reference 2, p. 250.
343. P. Yates and B. L. Shapiro, *J. Org. Chem.*, **23**, 759 (1958).
344. A. Eitel and F. Wessely, *Monatsh. Chem.*, **95**, 1382 (1964).
345. C. Grundmann and E. Kober, *J. Amer. Chem. Soc.*, **79**, 944 (1957).
346. J. A. Hendry, F. L. Rose and A. L. Walpole, *J. Chem. Soc.*, 1138 (1958).
347. J. G. Krause and A. Wozniak, *Chem. Ind. (London)*, 326 (1973).
348. K. B. Tomer, N. Harrit, I. Rosenthal, O. Buchardt, P. L. Kumler and D. Creed, *J. Amer. Chem. Soc.*, **95**, 7402 (1973).
349. L. Horner, W. Kirmse and H. Fernekess, *Chem. Ber.*, **94**, 279 (1961).
350. M. Regitz, *Angew. Chem.*, **87**, 259 (1975); *Angew. Chem., Int. Ed.*, **14**, 222 (1975).
351. T. Tsuchiya, C. Kaneko and H. Igeta, *J. Chem. Soc. Chem. Commun.*, 528 (1975).

352. E. Bamberger and E. Renauld, *Ber. dt. chem. Ges.*, **28**, 1682 (1895).
353. H. Staudinger and O. Kupfer, *Ber. dt. chem. Ges.*, **45**, 501 (1912).
354. D. T. Sepp, K. V. Scherer and W. P. Weber, *Tetrahedron Lett.*, 2983 (1974).
355. H. E. Zimmerman and D. H. Paskovieh, *J. Amer. Chem. Soc.*, **86**, 2149 (1964).
356. W. Rundel and P. Kästner, *Liebigs Ann. Chem.*, **686**, 88 (1965).

CHAPTER **16**

Preparation and uses of isotopically labelled diazonium and diazo compounds

PETER J. SMITH
Department of Chemistry and Chemical Engineering, University of Saskatchewan, Saskatoon, Saskatchewan, Canada
and
KENNETH C. WESTAWAY
Department of Chemistry, Laurentian University, Sudbury, Ontario, Canada

I. SYNTHESES OF LABELLED DIAZONIUM AND DIAZO COMPOUNDS

A. Synthesis of Deuterium-labelled Diazonium Salts and Diazo Compounds

1. Synthesis of diazonium ions labelled with deuterium

Several deuterated diazonium salts have been prepared for use in mechanistic studies on diazonium salt reactions, but in almost every case the compounds have been substantially less than 100% deuterated at the desired position. The deuterated arenediazonium salts have been synthesized by treating the deuterated aniline precursors with nitrite ion and acid at 0 °C[1] and, therefore, a discussion of the synthesis of the deuterated anilines illustrates the important aspects of the preparation of deuterated diazonium ions.

$$ArNH_2 + NaNO_2 \xrightarrow[0\,^{\circ}C]{H^+} Ar\overset{+}{N}{\equiv}N \quad (1)$$

The first deuterated diazonium salt was reported by Lewis and coworkers[2] in 1956 (equation 2). He prepared the *p*-trideuteromethylbenzenediazonium fluoroborate (87% deuterated) (1) from the deuterated aniline. The aniline synthesis began with the conversion of trideuteromethylbenzene into *p*-trideuteromethylacetophenone which was subsequently converted into the amide in a Schmidt rearrangement[3]. Acid hydrolysis of the amide led to the required aniline.

$$CD_3C_6H_5 + CH_3COCl \xrightarrow{AlCl_3} p\text{-}CD_3C_6H_4COCH_3 \xrightarrow{HN_3} p\text{-}CD_3C_6H_4NHCOCH_3$$

$$\xrightarrow[H^+,\,\Delta]{H_2O} p\text{-}CD_3C_6H_4NH_2 \xrightarrow[HBF_4,\ 0\,^{\circ}C]{C_2H_5ONO} p\text{-}CD_3C_6H_4\overset{+}{N}{\equiv}N\ \overset{-}{B}F_4 \quad (2)$$

(1)

The syntheses of the ring-deuterated derivatives of benzenediazonium ion were first reported by Heaney[4] and Cadogan[5] who were attempting to detect the presence of benzyne intermediates in diazonium salt reactions in acidic medium. Heaney synthesized 2,6-dideuterobenzenediazonium ion (81% deuterated) from 2,6-dideuteroaniline that had been prepared by treating 2,6-dibromoaniline with *n*-butyllithium and then D_2O[6] (equation 3). This general method has also been used

$$2,6\text{-}Br_2C_6H_3NH_2 \xrightarrow[(2)\ D_2O]{(1)\ n\text{-}C_4H_9Li} 2,6\text{-}D_2C_6H_3NH_2 + C_4H_{10} + LiOD \quad (3)$$

by Swain and coworkers[7] in the preparation of both the 2- and the 4-deuterobenzenediazonium ions. These syntheses also gave incomplete incorporation of deuterium, i.e. the 2-deuteroaniline was only 82% d_1 and the 4-deuteroaniline was only 84% deuterated, even when care was taken to exchange the NH hydrogens for deuterium before adding the *n*-butyllithium. Swain also attempted to prepare the 4-deuterobenzenediazonium ion by converting 4-deuterotoluene into 4-deuterobenzoic acid and then into the amide. The 4-deuteroaniline was obtained from a Hofmann rearrangement of the amide (equation 4). Unfortunately, this was no more successful than the *n*-butyllithium reaction as the diazonium ion was only 83% deuterated.

$$\text{D}-\text{C}_6\text{H}_4-\text{CH}_3 + \text{KMnO}_4 \xrightarrow[(2)\ \text{H}^+]{(1)\ \Delta} \text{D}-\text{C}_6\text{H}_4-\text{C}(=\text{O})-\text{OH} \xrightarrow[(2)\ \text{NH}_4\text{Cl}]{(1)\ \text{SOCl}_2} \text{D}-\text{C}_6\text{H}_4-\text{C}(=\text{O})-\text{NH}_2 \tag{4}$$

$$\xrightarrow[\text{OH}^-]{\text{NaOBr}} \text{D}-\text{C}_6\text{H}_4-\text{NH}_2$$

Cadogan and coworkers[5] prepared the 2,4,6-trideuterobenzenediazonium acetate although the details of the preparation were not given. Swain[7] prepared the same diazonium ion by exchanging the *ortho* and *para* hydrogens of aniline three times in refluxing D_2O for 24 h. The final product, the 2,4,6-trideuterobenzenediazonium ion, was 98·6% trideuterated. This is the best synthesis of a deuterated diazonium salt that has been reported. Zollinger[8] also used this technique to prepare the same trideuterated aniline which was then converted into the nitrogen-15 labelled 2,4,6-trideuterobenzene [β-^{15}N]diazonium fluoroborate (**2**) by treatment with $Na^{15}NO_2$ (99·2% ^{15}N) and fluoroboric acid at 0 °C.

$$2,4,6\text{-D}_3\text{C}_6\text{H}_2-\overset{+}{\text{N}}_\alpha\equiv{}^{15}\text{N}_\beta\ \overline{\text{B}}\text{F}_4$$

(**2**)

Swain's group synthesized several other deuterated benzenediazonium ions[7]. These workers prepared perdeuteroaniline (d_7) by reacting the 2,4,6-trideuteroaniline with deuterium gas over a pretreated platinum black catalyst[9] at 130 °C for 72 h. The final diazonium ion prepared by diazotizing the perdeuteroaniline gave 2,3,4,5,6-pentadeuterobenzenediazonium ion which was found to be 87% d_5 and 9·4% d_4 by a mass spectrometric analysis of its decomposition products.

Swain's group also prepared the 3,5-dideuterobenzenediazonium ion[7]. The 3,5-dideuteroaniline required for this synthesis was obtained in an unusual reverse exchange reaction in which the 2, 4 and 6 deuteriums of the perdeuteroaniline were exchanged for hydrogen. This was accomplished simply by refluxing the perdeuteroaniline in water. The diazonium salt prepared by this route was more than 96% dideuterated.

Two other substituted deuterated benzenediazonium ions have been prepared. Franck and Yanagi[10] prepared the 2,5-di-*t*-butyl-4,6-dideuterobenzenediazonium ion from the corresponding deuterated aniline. The deuteriums were exchanged into 2,5-di-*t*-butylaniline by refluxing for 48 h with an excess of D_2O and a trace of D_2SO_4 in dioxane. Swain[7] prepared the 2,4,6-trideutero-3,5-dimethylbenzenediazonium fluoroborate. The aniline required for this synthesis, was obtained by the usual method, refluxing 3,5-dimethylaniline in D_2O. An n.m.r. analysis indicated that the aniline was 99% deuterated.

2. Synthesis of diazo compounds labelled with deuterium

Diazomethane-d_2 has been prepared in low yield by exchange of diazomethane with D_2O in basic solution[11–14] or in acidic medium[15] under heterogeneous conditions.

A recent method[16], utilizing a homogeneous solution, leads to the product in high yield and isotopic purity. Diazomethane generated from *N*-nitrosomethylurea by the action of potassium deuteroxide in a homogeneous solution of 1,2-dimethoxyethane–D_2O, subsequently undergoes deuteroxide-promoted exchange (equation 5). The water is removed by freezing the reaction solution and the resulting diazomethane-d_2 in 1,2-dimethoxyethane solution can be used directly.

$$CH_3-N(N{=}O)-C(=O)-NH_2 \xrightarrow[\text{1,2-dimethoxyethane–}D_2O]{\text{KOD}} CH_2{=}\overset{+}{N}{=}\overset{-}{N} \xrightarrow[D_2O]{OD^-} CD_2N_2 \qquad (5)$$

The acid-catalysed deuterium exchange of diazomethane can be carried out in dry dioxane[15]. An excess of deuterium oxide is added to the diazomethane–dioxane solution and the resulting homogeneous solution is treated with catalytic amounts of either benzoic acid, propionic acid, phenol or ammonium chloride.

A study of the base-catalysed deuterium exchange reaction of α-diazoketones, $R-\overset{O}{\overset{\|}{C}}-CHN_2$, has been carried out by investigating the reaction of 2-diazo-2′,4′,6′-trimethylacetophenone (3) with sodium methoxide in methanol-O-*d*[17]. Two

$$2,4,6\text{-}(H_3C)_3C_6H_2-\overset{O}{\overset{\|}{C}}-CH{=}\overset{+}{N}{=}\overset{-}{N}$$

(3)

mechanisms were considered for the formation of the 2-diazo-2′,4′,6′-trimethylacetophenone-2-*d*. The first involves 'initial proton transfer', (equation 6), while the

$$R-\overset{O}{\overset{\|}{C}}-CH-N_2 + CH_3O^- \rightleftharpoons R\overset{O}{\overset{\|}{C}}-\overset{..}{\bar{C}}-N_2 + CH_3OH$$
$$R\overset{O}{\overset{\|}{C}}-\overset{..}{\bar{C}}-N_2 + CH_3OD \rightleftharpoons R\overset{O}{\overset{\|}{C}}-CD-N_2 + CH_3O^- \qquad (6)$$

other involves a 'terminal addition' (equation 7). As illustrated, both mechanisms can lead to exchange after the initial step. The rates of reaction were much faster for strongly basic catalysts such as acetate, methoxide or azide ion than for weakly basic catalysts, halide ions. Moreover, the rates of the halide-ion catalysed reactions decreased from iodide to bromide to chloride ion. These results suggest that the reactions with the reasonably strong basic catalysts occur by the proton abstraction mechanism (equation 6), whereas the halide ion reactions proceed by way of the terminal addition pathway (equation 7).

Deuteration of the methyl and/or methine groups of diazoacetone, $CH_3\overset{O}{\overset{\|}{C}}CHN_2$, has been reported[18]. Exchange of the methyl group, under either acidic or basic conditions leading to exchange in acetone, led to a large extent of decomposition. A

$$\mathrm{R\overset{O}{\overset{\|}{C}}-CH=\overset{+}{N}=\ddot{\bar{N}}: \longleftrightarrow R\overset{O}{\overset{\|}{C}}-\ddot{\bar{C}}H-\overset{+}{N}\equiv N: + CH_3O^- \longrightarrow R\overset{O}{\overset{\|}{C}}-\ddot{\bar{C}}H-\ddot{N}=\ddot{N}-OCH_3}$$

$$\mathrm{R\overset{O}{\overset{\|}{C}}-\ddot{\bar{C}}H-\ddot{N}=\ddot{N}-OCH_3 + CH_3OD \rightleftharpoons R\overset{O}{\overset{\|}{C}}-CHD-\ddot{N}=\ddot{N}-OCH_3 + CH_3O^-}$$

$$\mathrm{R\overset{O}{\overset{\|}{C}}-CHD-\ddot{N}=\ddot{N}-OCH_3 + CH_3O^- \rightleftharpoons R\overset{O}{\overset{\|}{C}}-\ddot{\bar{C}}D-\ddot{N}=\ddot{N}-OCH_3 + CH_3OH} \quad (7)$$

$$\mathrm{R\overset{O}{\overset{\|}{C}}-\ddot{\bar{C}}D-\ddot{N}=\ddot{N}-OCH_3 \rightleftharpoons R\overset{O}{\overset{\|}{C}}-CD=\overset{+}{N}=\ddot{\bar{N}}: + CH_3O^-}$$

successful synthesis of $CD_3\overset{O}{\overset{\|}{C}}CHN_2$ with almost quantitative yield was accomplished by reacting acetyl chloride-d_3 with diazomethane (equation 8). Acetyl chloride-d_3 can be conveniently prepared from acetic acid-d_3 which is obtained by decarboxylating malonic acid which had undergone prior exchange in D_2O. Exchange of the

$$\mathrm{CD_3\overset{O}{\overset{\|}{C}}-Cl + \ddot{\bar{C}}H_2-\overset{+}{N}\equiv N \longrightarrow CD_3\overset{O}{\overset{\|}{C}}-CH_2-\overset{+}{N}\equiv N \xrightarrow{-H^+} CD_3\overset{O}{\overset{\|}{C}}CHN_2} \quad (8)$$

methine hydrogen is accomplished by reacting the diazoacetone with sodium methoxide in D_2O at room temperature (equation 6).

Primary α-diazoketones and esters, $R-\overset{O}{\overset{\|}{C}}CHN_2$ and $RO\overset{O}{\overset{\|}{C}}CHN_2$, also exchange the α-hydrogen for deuterium rapidly in acidic deuterium oxide[15]. The mechanism for acid-catalysed deuterium exchange of these substrates may be considered as a reversible carbon protonation (equation 9).

$$\mathrm{R-\overset{O}{\overset{\|}{C}}-\underset{H}{\underset{|}{C}}=\overset{+}{N}=\ddot{\bar{N}}: \longleftrightarrow R-\overset{O}{\overset{\|}{C}}-\underset{H}{\underset{|}{\ddot{\bar{C}}}}-\overset{+}{N}\equiv N: \xrightarrow{D^+} R-\overset{O}{\overset{\|}{C}}-\underset{H}{\underset{|}{\overset{D}{\overset{|}{C}}}}-\overset{+}{N}\equiv N: \xrightarrow{-H^+} R-\overset{O}{\overset{\|}{C}}-CD=\overset{+}{N}=\ddot{\bar{N}}:} \quad (9)$$

B. Synthesis of Diazonium Salts Labelled with 15-Nitrogen

The first use of a 15-nitrogen-labelled diazonium ion was reported by Holt and Bullock, in 1950[19]. These workers prepared benzenediazonium ion with 15-nitrogen (33% enrichment by 15-nitrogen) in the α-nitrogen (α to the benzene ring) of the diazonium salt. Since that report many 15-nitrogen-labelled compounds have been prepared and the enrichment by 15-nitrogen has constantly increased as the purity of 15-nitrogen-labelled precursors has improved. In recent times, 15-nitrogen-labelled diazonium salts have been prepared with more than 98% 15-nitrogen at the desired position[8, 20–23].

The synthesis of 15-nitrogen-labelled diazonium salts has followed two general pathways depending upon the specific site of the label. Diazonium ions with 15-nitrogen in the β or terminal position have been prepared by diazotizing substituted anilines with 15-nitrogen-labelled potassium[24] or sodium nitrite[8], ethyl nitrite[25] or isoamyl nitrite[26] under the usual conditions[1]. Since almost all of these diazonium salts have been synthesized for use in mechanistic studies, the most stable form of the diazonium ion, the fluoroborate salt, has been prepared either directly by using fluoroboric acid or by anion exchange (equation 10). Several substituted benzene-

$$ArNH_2 + Na^{15}NO_2 \xrightarrow[0\,°C]{HBF_4} Ar\overset{+}{N}\equiv{}^{15}N\ \overset{-}{B}F_4 \qquad (10)$$

diazonium salts labelled with 15-nitrogen in the β-position have been reported (Table 1).

TABLE 1. Diazonium salts labelled with 15-nitrogen at either the α- or the β-nitrogen

Labelled at the α-nitrogen		Labelled at the β-nitrogen	
Substituent	Reference	Substituent	Reference
p-CH_3	25	*p*-CH_3	8, 29
m-CH_3	25	*p*-OC_2H_5	26, 29
p-OCH_3	20, 25	*p*-OCH_3	8
p-Cl	25	*p*-NO_2	8, 29
p-H	21, 22, 25, 27	*p*-H	8, 24, 29, 30
		2,4-Dinitro	29
		p-Br	29
		m-NO_2	29
		m-Cl	29
		2,4,6-Tribromo	29
		o-NO_2	29

Diazonium salts with 15-nitrogen at the α-nitrogen have been prepared by diazotization of 15-nitrogen-labelled primary aromatic amines. These anilines have invariably been prepared by the Hofmann rearrangement of the 15-nitrogen-labelled primary amide obtained from the reaction of an acyl chloride with 15-nitrogen-labelled ammonium chloride[22, 25, 27, 28] (equation 11).

$$Ar\overset{O}{\overset{\|}{C}}{-}Cl + {}^{15}NH_4Cl \longrightarrow Ar\overset{O}{\overset{\|}{C}}{-}{}^{15}NH_2 \xrightarrow[OH^-]{NaOBr} Ar^{15}NH_2 \qquad (11)$$

$$Ar^{15}NH_2 + NaNO_2 \xrightarrow[0\,°C]{HBF_4} Ar^{15}\overset{+}{N}\equiv N\ \overset{-}{B}F_4$$

The diazonium salts synthesized by this route are listed in Table 1. Recently, Bubnov and coworkers[31] have prepared benzenediazonium fluoroborate labelled with 15-nitrogen at both the α- and the β-nitrogens and Zollinger and collaborators[8] have prepared 2,4,6-trideutero[β-^{15}N]benzenediazonium fluoroborate.

II. THE USE OF ISOTOPES TO DETERMINE BONDING

The ^{13}C-nuclear magnetic resonance spectrum of ketene has demonstrated[32] that the terminal carbon atom is shielded to a considerable degree. It was concluded, in agreement with other data, that the resonance structure, $[H_2\ddot{\bar{C}}-C\equiv\overset{+}{O}:]$ (4), is the major contributor to the ground state structure.

$$H_2C=C=\ddot{O}: \longleftrightarrow H_2\ddot{\bar{C}}-C\equiv\overset{+}{O}: \longleftrightarrow H_2\ddot{\bar{C}}-\overset{+}{C}=\ddot{O}:$$

(4)

The principal resonance structure in the valence bond description of the isoelectronic diazomethane is generally accepted as being $[H_2\ddot{\bar{C}}-\overset{+}{N}\equiv N:]$ (5). If this is the

$$H_2C=\overset{+}{N}=\ddot{\bar{N}}: \longleftrightarrow H_2\ddot{\bar{C}}-\overset{+}{N}\equiv N: \longleftrightarrow H_2\ddot{\bar{C}}-N=\overset{+}{N}:$$

(5)

case, the carbon would be as shielded as the terminal carbon in ketene. The position of the ^{13}C-n.m.r. absorption of the formally doubly bonded carbon of several diazoalkanes is shown in Table 2[33]. In comparison with analogously substituted

TABLE 2. ^{13}C-n.m.r. data for diazoalkanes $R^1R^2C=N_2$

R^1	R^2	$\delta_{C(TMS)}$, p.p.m. for ($>C=N_2$)
H	H	23·1
C_6H_5	H	47·2
C_6H_5	CH_3	51·2
C_6H_5	C_2H_5	57·2
C_6H_5	C_6H_5	62·5
H	CO_2CH_3	46·3

imines, the resonances of the formally sp² hybridized diazomethylene carbon is shifted 100–120 p.p.m. upfield. Since the theory of chemical shifts predicts that increased electron density will lead to increased shielding, then it must be concluded that there is high electron density on the sp² carbon for both the ketene and the diazoalkanes in accord with the resonance structures **4** and **5**.

$$H_2\ddot{\bar{C}}-C\equiv\overset{+}{O}: \quad \text{and} \quad R^1-\underset{R^2}{\underset{|}{\ddot{\bar{C}}}}-\overset{+}{N}\equiv N:$$

III. THE USE OF ISOTOPES AS TRACERS

A. Use of 15-Nitrogen as a Tracer in Diazonium Salt Reactions

Tracer studies utilizing 15-nitrogen-labelled diazonium ions began in 1950 with the work of Holt and Bullock[19] and are still being actively pursued today[8, 23]. 15-Nitrogen-labelled diazonium salts have been used extensively to unravel the details of the reactions with nucleophiles and the N_α–N_β exchange reaction during the decomposition of the salt itself. The reaction of 15-nitrogen-labelled diazonium ions with nucleophiles will be discussed first.

I. Reactions with nucleophiles

a. *Reaction with azide ion.* The reaction between azide ion and a diazonium ion was the first reaction investigated with a 15-nitrogen-labelled diazonium salt. Although the products of this reaction were the expected aryl azide and nitrogen gas, the reaction did display some unusual characteristics. If the diazonium ion was treated with azide ion at temperatures below −20 °C the yield of nitrogen gas (primary nitrogen) was less than quantitative. If the reaction mixture was heated above 20 °C then more nitrogen (secondary nitrogen) was produced. The total of the primary plus the secondary nitrogen represented a quantitative yield.

Clusius[34] reacted both α-15-nitrogen- and β-15-nitrogen-benzenediazonium ions with azide ion in order to determine if this process was a simple displacement reaction of nitrogen by azide ion. If this were the case, the nitrogen produced from either the α- or the β-labelled diazonium salt should contain all the 15-nitrogen label. In fact, the nitrogen contained only 15% of the 15-nitrogen label and Clusius concluded that the reaction must proceed by way of two intermediates, a benzenediazoazide (**6**) and another intermediate (**7**). This reaction is illustrated with benzene[β-^{15}N]diazonium ion (equation 12). The labelled nitrogen gas was thought

$$C_6H_5\overset{+}{N}{\equiv}{}^{15}N \xrightarrow{N_3^-} C_6H_5N{=}{}^{15}N{-}N{=}\overset{+}{N}{=}\overset{-}{N}\ (\mathbf{6}) \longrightarrow C_6H_5N{=}{}^{15}\overset{+}{N}{=}\overset{-}{N} + N{\equiv}N \tag{12}$$

$$C_6H_5\overset{+}{N}{\equiv}{}^{15}N \xrightarrow{N_3^-} C_6H_5\overset{+}{N}({=}{}^{15}N)({-}N{=}\overset{+}{N}{=}\overset{-}{N})\ (\mathbf{7}) \longrightarrow C_6H_5N{=}\overset{+}{N}{=}\overset{-}{N} + {}^{15}N{\equiv}N$$

to have been formed from the two terminal nitrogens (at the end of each branch) in intermediate **7**. Since the nitrogen released from intermediate **6** would not be labelled, Clusius concluded that 85% of the product was produced via **6** and that 15% was formed via **7**.

In a later study Clusius[29], on the basis of other 15-nitrogen studies, modified his ideas and proposed that the minor intermediate **7** was, in fact, an aryl pentazole (**8**) which decomposed to give nitrogen gas and phenyl azide by eliminating either nitrogens 4 and 5 (Cleavage A) or 2 and 3 (Cleavage B) (equation 13). Huisgen[35]

$$C_6H_5{-}N^1(N^2{=}N^3{-}N^4{=}N^5)\ (\mathbf{8}) \xrightarrow{A} C_6H_5N^1{=}\overset{+}{N}{}^2{=}\overset{-}{N}{}^3 + N^4{\equiv}N^5 \tag{13}$$

$$(\mathbf{8}) \xrightarrow{B} C_6H_5N^1{=}\overset{+}{N}{}^5{=}\overset{-}{N}{}^4 + N^2{\equiv}N^3$$

found that 83% of the 15-nitrogen label from a reaction of benzene[β-^{15}N]diazonium ion with azide ion was in the phenyl azide and concluded that 65% of the benzenediazoazide decomposes directly to nitrogen and phenyl azide, and that 35% decomposes by way of the pentazole (equation 14). This would mean that the nitrogen gas would contain 17·5% of the 15-nitrogen label and that the phenyl azide would contain 82·5% of the label as is observed. Still later, additional support for this scheme was found. Ugi and coworkers[26] were able to isolate the 15-nitrogen-labelled pentazole derivative from the reaction of *p*-ethoxybenzene[β-^{15}N]diazonium ion with azide ion. When this pentazole intermediate decomposed, 50% of the 15-nitrogen label was in the *p*-ethoxyphenyl azide and 50% was in the nitrogen gas, thus proving that the two decomposition pathways, A and B, of the pentazole **8** are equal (equations 13 and 14). More recent studies[36, 37] have not altered the mechanism shown in (equation 14) substantially, although it is now believed that the aryldiazo-azide is formed in a preequilibrium step of the reaction[36].

$$C_6H_5\overset{+}{N}\equiv{}^{15}N + N_3^- \xrightarrow{\text{fast}} C_6H_5N={}^{15}N-N=\overset{+}{N}=\bar{N} \xrightarrow{65\%} C_6H_5N={}^{15}\overset{+}{N}=\bar{N} + N\equiv N$$

$$C_6H_5N={}^{15}N-N=\overset{+}{N}=\bar{N} \rightleftharpoons (35\%) \text{ pentazole } C_6H_5N({}^{15}N=N-N=N) \begin{cases} \xrightarrow{17.5\%} C_6H_5N={}^{15}\overset{+}{N}=\bar{N} + N\equiv N \\ \xrightarrow{17.5\%} C_6H_5N=\overset{+}{N}=\bar{N} + {}^{15}N\equiv N \end{cases} \quad (14)$$

b. *Reaction with amines.* 15-Nitrogen-labelled diazonium salts have been used to show that other nucleophiles react with diazonium salts by way of an addition complex like **6**. For example, Clusius[38] has shown that primary aromatic amines add to the diazonium salt to form a triazene in a reversible reaction (equation 15).

$$ArNH_2 + Ar'\overset{+}{N}\equiv N \underset{}{\overset{-H^+}{\rightleftharpoons}} Ar'N=N-N(H)Ar \quad (15)$$

Clusius also found that when 15-nitrogen-labelled aniline was reacted with unlabelled benzenediazonium ion, some of the 15-nitrogen was incorporated into the diazonium ion. This result indicates that the triazene is involved in the tautomeric equilibrium **9** to **10** shown in equation (16). The labelled diazonium ion is produced when the triazene **10** decomposes (equation 16).

$$C_6H_5{}^{15}NH_2 + C_6H_5\overset{+}{N}\equiv N \overset{-H^+}{\rightleftharpoons} \underset{(\mathbf{9})}{C_6H_5{}^{15}N(H)-N=NC_6H_5} \rightleftharpoons \underset{(\mathbf{10})}{C_6H_5{}^{15}N=N-N(H)C_6H_5}$$

$$\overset{H^+}{\rightleftharpoons} C_6H_5{}^{15}\overset{+}{N}\equiv N + C_6H_5NH_2 \quad (16)$$

A similar reaction has been observed between diazonium ions and phenylhydrazine[39] (equation 17).

$$C_6H_5\overset{+}{N}\equiv N + C_6H_5NHNH_2 \xrightarrow{-H^+} C_6H_5N=N-N(H)-N(H)C_6H_5 \quad (17)$$

The presence of the tetrazine intermediate **(11)** was suggested by a 15-nitrogen tracer study. When benzene[α-^{15}N]diazonium ion was used as the reactant, the label was found in almost equal amounts at the α-nitrogen of the azide and the nitrogen of the aniline. This is also in accord with a tautomeric equilibrium, (equation 18).

$$C_6H_5{-}^{15}NH_2 + \bar{N}{=}\overset{+}{N}{=}NC_6H_5 \quad (\text{via } H^+)$$

$$C_6H_5{-}^{15}N{=}N{-}\underset{H\,H}{NN}C_6H_5 \rightleftarrows \underset{(11)}{C_6H_5{-}^{15}\underset{H}{N}{-}N{=}N{-}\underset{H}{N}C_6H_5} \quad (18)$$

$$\xrightarrow{H^+} C_6H_5{-}^{15}N{=}\overset{+}{N}{=}\bar{N} + C_6H_5NH_2$$

Clusius[40] has also investigated the reaction of the two specifically labelled diazonium ions with ammonia. The phenyl azide produced in this reaction is formed when the addition complex **(12)** is oxidized by the bromine (equation 19). In this

$$C_6H_5{}^{15}\overset{+}{N}{\equiv}N + NH_3 \longrightarrow \underset{(12)}{C_6H_5{}^{15}N{=}N{-}NH_2} \xrightarrow{Br_2} C_6H_5{}^{15}N{=}\overset{+}{N}{=}\bar{N} \quad (19)$$

case, however, there is no rearrangement of the addition product. For example, the phenyl azide formed when benzene[α-^{15}N]diazonium ion was used in this reaction had the label only in the nitrogen bonded to the benzene ring. These results are consistent with a linear structure for phenyl azide but cannot be rationalized on the basis of a cyclic structure **(13)**.

$$C_6H_5N\langle N{=}N \rangle \quad (13)$$

c. *Reaction with hydroxide ion.* The reaction of diazonium salts with hydroxide ion has been studied by groups of Swan[41] and of Clusius[42]. Both groups were interested in determining whether the phenyl group of a benzenediazonium salt migrated from one nitrogen to the other in the isodiazotate formed when hydroxide ion reacted with the diazonium ion (equation 20). Clusius concluded that the phenyl group did

$$C_6H_5\overset{+}{N}{\equiv}N + OH^- \longrightarrow C_6H_5N{=}N{-}OH \xrightarrow{OH^-} C_6H_5N{=}NO^- \quad (20)$$

not migrate since reduction of the isodiazotate salt isolated from the reactions of both the α- and the β-15-nitrogen-labelled benzenediazonium salts to phenylhydrazine and cleavage to aniline and ammonia gave aniline with no 15-nitrogen enrichment when benzene[β-^{15}N]diazonium ion was the reactant and ammonia with no 15-nitrogen enrichment when benzene[α-^{15}N]diazonium ion was the reactant. Swan[41] reached the same conclusion from an identical experiment using benzene-[α-^{15}N]diazonium ion as the reactant. The analysis for the 15-nitrogen label in Swan's experiment was done by another technique, however. The isodiazotate was converted back to the diazonium salt which was coupled to β-naphthol and then cleaved to aniline (equation (21).

$$C_6H_5{-}^{15}\overset{+}{N}{\equiv}N + OH^- \longrightarrow C_6H_5{-}^{15}N{=}N{-}O^- \xrightarrow{HCl} C_6H_5{-}^{15}\overset{+}{N}{\equiv}N \quad (21)$$

$$\xrightarrow{\beta\text{-naphthol (OH)}} \text{1-}(N{=}^{15}NC_6H_5)\text{-2-naphthol} \xrightarrow{\text{Sodium dithionite}} \text{1-}NH_2\text{-2-naphthol} + C_6H_5{-}^{15}NH_2$$

The possibility of the 15-nitrogen label rearranging during the coupling with β-naphthol and subsequent cleavage to aniline had been ruled out by Holt[19] who showed that benzene[α-^{15}N]diazonium ion only gave aniline 15-nitrogen in this reaction.

2. N_α, N_β exchange in diazonium ion decomposition

The second general use of 15-nitrogen as a tracer was to investigate the fate of the nitrogens in the diazonium ion both before and during decomposition in acidic medium. Holt and coworkers[24, 43] have shown that the 15-nitrogen in benzene[β-^{15}N]-diazonium ion was not decreased even after very long exposures with unlabelled nitrous acid. This important result demonstrates that the formation of the diazonium salt (equation 22) is not reversible and hence allows the use of β-15-nitrogen-labelled benzenediazonium ion in tracer experiments. The reverse experiment, reacting unlabelled benzene- and *p*-nitrobenzenediazonium ions with 15-nitrogen-enriched nitrous acid, also failed to lead to exchange of the β-nitrogen.

$$\mathrm{ArNH_2 + HNO_2 \rightleftharpoons Ar\overset{+}{N}{\equiv}N + H_2O} \quad \text{(reverse arrow crossed)} \tag{22}$$

Another problem associated with diazonium ion decomposition was to determine if internal α- and β-nitrogen exchange (equation 23) occurs during the reaction. This problem has been under active investigation for over 25 years. In 1950, Holt[19]

$$\mathrm{Ar\overset{+}{N}_\alpha{\equiv}N_\beta \rightleftharpoons Ar\overset{+}{N}_\beta{\equiv}N_\alpha} \tag{23}$$

reacted benzene[α-^{15}N]diazonium ion with β-naphthol and tested for exchange by cleaving the resulting azo compound with $SnCl_2$ to give aniline (equation 24). The

$$\mathrm{C_6H_5{}^{15}\overset{+}{N}{\equiv}N} + \beta\text{-naphthol} \longrightarrow \text{1-}(\mathrm{C_6H_5{}^{15}N{=}N})\text{-2-naphthol} \xrightarrow{SnCl_2} \mathrm{C_6H_5{-}^{15}NH_2} \tag{24}$$

aniline was converted to nitrogen gas and analysed by mass spectrometry. The results showed that all the 15-nitrogen label had remained in the α-nitrogen and thus Holt concluded that the N_α–N_β rearrangement, i.e. phenyl migration from N_α to N_β, did not occur. Swan[41], in an identical experiment, obtained the same results.

The next investigation of this problem was by Insole and Lewis in 1963[21]. Studies of diazonium salt reactions had suggested[44] that an intermediate existed along the reaction coordinate for diazonium salt decomposition and Lewis attempted to learn how nitrogen was involved in the intermediate. To this end, Lewis reacted benzene[α-^{15}N]diazonium fluoroborate (99% 15-nitrogen) to 80% of completion, recovered the unreacted diazonium salt, reacted it with azide ion (see previous section) and collected the secondary nitrogen. A mass spectrometric analysis showed that the nitrogen gas was 2·6% enriched in 15-nitrogen (natural abundance of 15-nitrogen is 0·36%), thus indicating that the α- and β-nitrogens had exchanged during the reaction. This small amount of exchange obviously eliminated the symmetrical intermediate **(14)** but was consistent with the phenyl cation intermediate **(15)** originally proposed by Waters[45] where the two nitrogens are not equivalent.

$$\mathrm{C_6H_5^{+}\!\!<\!\!N{=}N} \qquad\qquad \mathrm{C_6H_5^{+}} \;\; \mathrm{N_\alpha{\equiv}N_\beta}$$

(14) **(15)**

Lewis favoured an intermediate in which the carbon–α-nitrogen bond was mainly, but not completely, broken. He proposed this intermediate to account for the exchange because no evidence for the reaction between nitrogen gas and the phenyl cation in the reverse of the first step in Water's mechanism (equation 25) could be found.

$$C_6H_5\overset{+}{N}{\equiv}N \rightleftharpoons C_6H_5^+ \; N{\equiv}N \qquad (25)$$

Lewis extended these exchange studies to other *para*-substituted benzene[α-^{15}N]-diazonium ions[22, 46] and found that the ratio of exchange to solvolysis varied slightly, but in a non-regular fashion (Table 3). The k_{exch}/k_{solv} ratio was found to be independent of temperature[25]. The almost constant value of the exchange to solvolysis ratio for compounds whose rates varied by as much as 20,000 times, and the fact that this ratio is temperature independent, led Lewis to conclude[46] that similar intermediates must be involved in the exchange and the solvolysis reactions.

TABLE 3. Ratio of exchange to solvolysis for the decomposition of some *para*-substituted benzene[α-^{15}N]diazonium fluoroborates

Substituent	$\frac{k_{exch}}{k_{solv}}$	Reference
p-CH_3O	0·038	46
p-CH_3	0·029	22
p-CH_3	0·031	46
m-CH_3	0·018	46
p-H	0·014	22
p-Cl	0·023	46

Lewis and Holliday used the n.m.r. technique designed by Bose and Kugajevsky[27] to study the N_α–N_β exchange in the hydrolysis of *p*-methoxybenzene[α-^{15}N]-diazonium ion[20]. This compound was chosen because it had the largest k_{exch}/k_{solv} ratio. After the labelled diazonium ion had reacted part way to completion, the unreacted salt was reduced to *p*-methoxyphenylhydrazine and then converted into a phenylhydrazone (**16**, equation 26). If the labelled nitrogen had rearranged during

$$p\text{-}CH_3OC_6H_4{-}^{15}\overset{+}{N}{\equiv}N + SnCl_2 \xrightarrow[HCl]{C_2H_5OH} p\text{-}CH_3OC_6H_4{}^{15}NHNH_2 \xrightarrow[H^+]{C_6H_5CHO} p\text{-}CH_3OC_6H_4{-}^{15}N(H){-}N{=}CH(C_6H_5) \qquad (26)$$

(16)

the reaction the product would contain some **17**. Integration of the ^{15}N—H doublet (^{15}N has a nuclear spin quantum number of 1/2 and $J_{^{15}N-H} = 91$ Hz) and the broad ^{14}N—H multiplet indicated that N_α–N_β exchange had occurred and, more important, that the amount of exchange was within experimental error of that found mass spectrometrically by Lewis.

$$p\text{-}CH_3OC_6H_4NH\text{-}^{15}N{=}CH(C_6H_5)$$

(17)

This conclusion received support when Swain and coworkers[23] found almost the same amount of exchange. Swain reacted benzene[β-^{15}N]diazonium fluoroborate part way to completion and then coupled the unreacted diazonium ion with the disodium salt of 2-naphthol 3,6-disulphonic acid. Reduction of the resulting azo compound gave aniline (equation 27) which was subsequently oxidized to nitrogen gas (equation 28). The nitrogen was analysed mass spectrometrically. The results

$$C_6H_5\overset{+}{N}{\equiv}^{15}N + \text{2-naphthol-3,6-disulphonate} \longrightarrow \text{1-}(C_6H_5N{=}^{15}N)\text{-2-naphthol-3,6-disulphonate} \xrightarrow[OH^-]{Na_2S_2O_4} C_6H_5NH_2 + \text{1-}^{15}NH_2\text{-2-naphthol-3,6-disulphonate} \quad (27)$$

$$C_6H_5NH_2 \xrightarrow[\text{digestion}]{\text{Kjeldahl}} NH_3 \xrightarrow{\text{NaOBr}} N_2 \quad (28)$$

indicated that the k_{exch}/k_{solv} ratio was 0·016 which is within experimental error of the k_{exch}/k_{solv} ratio of 0·014 reported by Lewis (Table 3) for the reaction of benzene-diazonium ion labelled at the β-nitrogen and using a different method of analysis.

The observation of N_α–N_β exchange does not establish, however, the structure of the intermediate in these reactions. The results are consistent with a phenyl cation intermediate **(15)** although attempts to find the reverse reaction, the reaction of the phenyl cation with nitrogen gas, had failed. In fact, Lewis has attempted to detect this reverse reaction in two ways. In his first attempt[21], he carried out the decomposition of the diazonium salt under a pressure of 50 atm of carbon monoxide but was unable to detect any reverse reaction of the phenyl cation even though carbon monoxide is more nucleophilic than nitrogen. In a subsequent experiment, Lewis studied the reaction of 15-nitrogen-labelled benzenediazonium ion in a reaction vessel swept with unlabelled nitrogen gas in an effort to reduce the return of the labelled nitrogen to the phenyl cation. Even under these conditions, the ratio of k_{exch}/k_{solv} was within experimental error of the results obtained in the absence of the sweeping (excess) unlabelled nitrogen.

It is interesting to note that Lewis and Holliday found a slight increase in the N_α–N_β exchange in another experiment where the 15-nitrogen labelled diazonium ion was reacted in a closed system where the atmosphere becomes enriched in labelled nitrogen gas[25]. The authors, however, ignored this evidence and concluded that the reverse reaction did not occur. Consequently, Lewis proposed an unsymmetrical intermediate **(18)** with a very weak bond between the phenyl carbon and the β-nitrogen, and a weak (mainly broken) bond between the phenyl carbon and the α-nitrogen.

The final, definitive proof for the phenyl cation intermediate originally proposed by Waters in 1942[45] was provided by Zollinger's group[8, 30]. These workers decomposed

(18)

benzene[β-^{15}N]diazonium fluoroborate in trifluoroethanol rather than in water and observed a dramatic increase in the amount of N_α–N_β exchange. In fact, the k_{exch}/k_{solv} ratio increased to 0·072 at 30 °C and 0·079 at 5 °C in trifluoroethanol from the ratio of 0·014 found by Lewis in aqueous medium. More important, however, was the observation that the amount of 15-nitrogen in the diazonium salt recovered after benzene[β-^{15}N]diazonium ion had reacted part way to completion, decreased as the pressure of the external nitrogen gas was increased (Table 4). This relationship between the exchange and the pressure of the external nitrogen gas is consistent with a phenyl cation intermediate since the labelled nitrogen produced in the formation of the phenyl cation would become less and less concentrated as the amount of external, unlabelled, nitrogen is increased.

TABLE 4. The effect of external nitrogen pressure on the 15-nitrogen content of the unreacted benzene[β-^{15}N]diazonium fluoroborate

Reaction (%)	Nitrogen pressure (atm)	Unreacted diazonium ion containing a 15-nitrogen atom (%)[a]
0	—	99·2
70·4	1·0	98·6 ± 0·44
73·2	20	98·23
69·9	300	96·23, 96·99
62·5	1000	94·71 ± 0·43

[a] The analysis for the 15-nitrogen content of the unreacted diazonium ion was done by analysing the azo compound prepared by coupling the unreacted diazonium salt with β-naphthol mass spectrometrically.

Zollinger[8, 30] showed that the loss of the 15-nitrogen label from an arene diazonium salt is a general phenomenon. The label decreased from 98·9% to 97·55 ± 0·44% for *p*-methoxy[β-^{15}N]diazonium ion and from 99·33% to 97·88 ± 0·30% for *p*-nitrobenzene[β-^{15}N]diazonium ion when they were reacted to between 60 and 70% of completion at an external nitrogen pressure of 300 atm.

Further evidence for the reaction between the phenyl cation and nitrogen was obtained by Zollinger[8, 30] who showed that carbon monoxide reacted with the phenyl cation to yield the benzoyl cation, which was recovered as trifluoromethyl benzoate, when benzenediazonium ion was decomposed in trifluoroethanol under an external pressure of 320 atm of carbon monoxide (equation 29).

$$C_6H_5\overset{+}{N}\equiv N \xrightarrow{\text{TFE}} C_6H_5^+ + N\equiv N \xrightarrow{\text{CO}} C_6H_5\overset{+}{C}=O \xrightarrow{CF_3CH_2OH} C_6H_5CO_2CH_2CF_3 \quad (29)$$

Zollinger[8, 30] has also found that the ratio of k_{exch}/k_{solv} increased in more viscous solvents and with less nucleophilic solvents. The k_{exch}/k_{solv} ratio increased from 0·016 in 1 M-sulphuric acid to 0·087 in 85% phosphoric acid. These solvents have appoximately the same acidity but a very different viscosity. The increased N_α–N_β exchange would occur because the labelled nitrogen gas released in the ionization step cannot escape from the vicinity of the phenyl cation in the more viscous phosphoric acid.

The k_{exch}/k_{solv} ratio is also larger in the less nucleophilic solvent, trifluoroacetic acid. This result would be expected because the reaction between the solvent and the cation would be slower with a less nucleophilic solvent. This would obviously increase probability of reaction between the labelled nitrogen and the phenyl cation, and a larger k_{exch}/k_{solv} ratio would be found.

Finally, Zollinger showed that the exchange reaction between the labelled diazonium ion and external nitrogen was slower than the N_α–N_β exchange reaction. The relative rates in trifluoroethanol were $k_{\text{exchange with external nitrogen}}/k_{solv} = 0{\cdot}0246$ whereas the exchange resulting from reaction with the labelled nitrogen produced in the ionization step (k_{exch}/k_{solv}) was 0·072. These results suggest that the phenyl cation reacts primarily with the α-nitrogen atom of the released nitrogen with no N_α–N_β rearrangement. The second most probable reaction is between the phenyl cation and the β-nitrogen atom of the released nitrogen. Finally, the least probable reaction is with the external nitrogen.

Zollinger explains his results in terms of an ion–molecule pair (**19**) produced in the first step of the reaction (equation 30). This ion–molecule pair can dissociate into a free phenyl cation. It is tempting to suggest that the N_α–N_β exchange proceeds by way of the ion–molecule pair whereas the free phenyl cation is the reactant with the external nitrogen, the solvent or other nucleophiles.

$$\mathrm{Ar{-}\overset{+}{N}_\alpha{\equiv}N_\beta} \rightleftharpoons \underset{(\mathbf{19})}{[\mathrm{Ar^{+}\,N_\alpha{\equiv}N_\beta}]} \rightleftharpoons \mathrm{Ar{-}\overset{+}{N}_\beta{\equiv}N_\alpha} \tag{30}$$

$$(\mathbf{19}) \longrightarrow \mathrm{Ar^{+}}$$

$$\mathrm{Ar^{+}} \xrightarrow{H_2O} \mathrm{ArOH};\quad \mathrm{Ar^{+}} \xrightarrow{X^-} \mathrm{ArX};\quad \mathrm{Ar^{+}} \xrightarrow{\text{external } N_2} \mathrm{Ar{-}\overset{+}{N}{\equiv}N};\quad \mathrm{Ar^{+}} \xrightarrow{CO} \mathrm{Ar\overset{+}{C}{\equiv}O} \xrightarrow{HS} \mathrm{ArC({=}O)SH}$$

The results of 15-nitrogen tracer studies were also used by Zollinger and co-workers[8] to determine the relationship between the N_α–N_β exchange reaction and the solvolysis reaction. They found that the plot of k_{exch} versus k_{solv} had a slope of 1·00, i.e. the k_{exch}/k_{solv} ratio remains constant for reactions at different temperatures, pressures of external nitrogen, and even with different substituents on the phenyl ring (Table 5). This constant ratio of k_{exch}/k_{solv} can only be possible if the solvolysis and exchange proceed via the same rate-determining step, i.e. formation of the phenyl cation.

Finally, 15-nitrogen has been used in tracer experiments on the photochemical reaction of diazonium ions. Since Lewis[47] found that the photochemical decomposition of diazonium ions occurs much faster and gives different product ratios than the

TABLE 5. The k_{exch}/k_{solv} ratio for benzene[β-^{15}N]diazonium ions under different reaction conditions

Conditions[a]	*para* Substituent	k_{exch}/k_{solv}
85 wt-% TFE, 30 °C	H	0·069
97 wt-% TFE, 30 °C	H	0·074
TFE, 30 °C	H	0·072
TFE, 5 °C	H	0·079
TFE, 1000 atm N_2, 25 °C	H	0·072
TFE, 64 °C	MeO	0·075
TFE, 40 °C	Me	0·082

[a] TFE = trifluoroethanol.

thermal decomposition in water, he concluded that these reactions might involve different intermediates. The k_{exch}/k_{solv} ratio was determined for two *para*-substituted benzene[α-^{15}N]diazonium ions under thermal and photochemical conditions (Table 6).

TABLE 6. The k_{exch}/k_{solv} ratio[a] for the photochemical and thermal decomposition of *para*-substituted benzene[α-^{15}N]diazonium ions

Substituent	(k_{exch}/k_{solv}) Photochemical	(k_{exch}/k_{solv}) Thermal
p-Me	0·13	0·03
p-MeO	$\geqslant$0·066	0·038

[a] The N_α–N_β exchange was determined by examining the secondary nitrogen produced when the unreacted diazonium was treated with azide ion, in a mass spectroscopic analysis.

The ratio of k_{exch}/k_{solv} is much higher for the photochemical reaction than for the thermal reaction and indicates that a different intermediate is involved in the photochemical reaction. Lewis[47] and others[48] have proposed structure **20** which should lead to greater N_α–N_β rearrangement than the intermediate proposed for the thermal process.

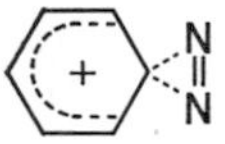

(20)

3. Use of a 15-nitrogen tracer to determine the decomposition mechanism of diazonium ions

Kirmse and coworkers[49] have used 15-nitrogen-labelled compounds to demonstrate that cyclopropanediazonium ions are more stable and thus react via a different mechanism than alkanediazonium ions. The authors found that the cyclopropyl azide obtained from the decomposition of cyclopropane[β-^{15}N]diazonium ion in the presence of azide ion was labelled with 15-nitrogen (68%) at the central nitrogen.

This indicated that the cyclopropyl azide was formed via the pentazene (**21**) and/or the pentazole (**22**) intermediates (equation 31). In contrast, the *n*-butyl azide

$$\triangleright\!-\overset{+}{N}\equiv{}^{15}N + N_3^- \longrightarrow \triangleright\!-N={}^{15}N-N=\overset{+}{N}=\bar{N}\ \textbf{(21)} \longrightarrow \triangleright\!-N={}^{15}\overset{+}{N}=\bar{N} \quad \text{(via pentazole } \textbf{(22)}\text{: } \triangleright\!-N\langle{}^{15}N{=}N{-}N{=}N\rangle) \tag{31}$$

produced from *n*-butyl[β-^{15}N]diazonium ion did not contain any of the 15-nitrogen label and must have been formed when azide ion reacted with the carbonium ion (equation 32). The fact that cyclopropanediazonium ion reacts in the same way as

$$\begin{aligned} CH_3(CH_2)_2\overset{+}{N}\equiv{}^{15}N &\xrightarrow{\text{fast}} CH_3(CH_2)_2\overset{+}{C}H_2 + N\equiv{}^{15}N \\ CH_3(CH_2)_2CH_2^+ + N_3^- &\longrightarrow CH_3(CH_2)_2CH_2N_3 \end{aligned} \tag{32}$$

arenediazonium ions indicates that the cyclopropanediazonium ions are considerably more stable than alkyldiazonium ions which decompose before they can react with azide ion to form the pentazene intermediate.

B. *Use of Deuterium as a Tracer*

1. Deuterium tracer studies on the reactions of diazo compounds

It has long been known[50] that the tosylhydrazones of aliphatic and aromatic aldehydes and ketones, when treated with base, undergo thermal decomposition to give diazoalkanes in the Bamford–Stevens reaction (equations 33 and 34). The

$$ArCH=\ddot{N}-\ddot{N}(H)-SO_2-p\text{-tolyl} \xrightarrow{\text{Base}} ArCH=\ddot{N}-\bar{\ddot{N}}-SO_2-p\text{-tolyl} \tag{33}$$

$$ArCH=\ddot{N}-\bar{\ddot{N}}-SO_2-p\text{-tolyl} \longrightarrow ArCH-\overset{+}{N}=\bar{\ddot{N}} + p\text{-Tolyl }SO_2^- \tag{34}$$

diazoalkanes thermally decompose[51] to give carbenes in aprotic media (equation 35). In protic solvents (SH) such as ethylene glycol on the other hand, competitive

$$ArCH=\overset{+}{N}=\bar{\ddot{N}} \longleftrightarrow Ar\bar{\ddot{C}}H-\overset{+}{N}\equiv\ddot{N} \longrightarrow Ar\dot{C}H\cdot + N_2 \tag{35}$$

protonation of the diazoalkanes occurs[52], and diazonium and/or carbonium ions are formed (equation 36).

$$Ar\bar{\ddot{C}}H-\overset{+}{N}\equiv N \xrightarrow{SH} ArCH_2-\overset{+}{N}\equiv N \longrightarrow Ar\overset{+}{C}H_2 + N_2 \tag{36}$$

In order to distinguish between the competitive carbenic and cationic pathways, the anion of 2-methylpropanal tosylhydrazone was decomposed in ethylene glycol-d_2 [53] (equation 37). It was found that the product, methylcyclopropane,

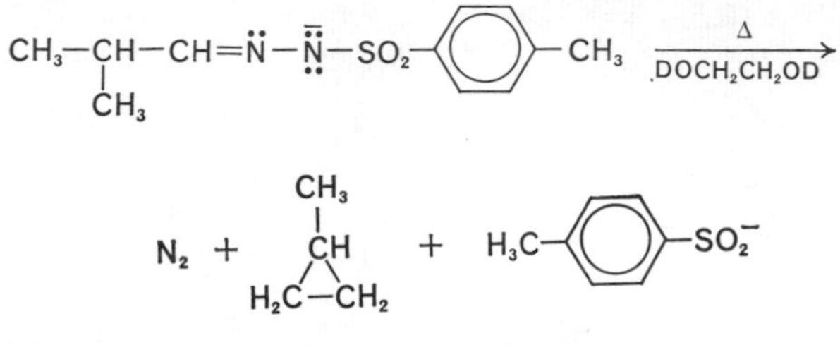

$$CH_3-\underset{\substack{|\\CH_3}}{CH}-CH=\ddot{N}-\underset{\cdot\cdot}{\overset{\cdot\cdot}{N}}-SO_2-C_6H_4-CH_3 \xrightarrow[DOCH_2CH_2OD]{\Delta} N_2 + \text{(methylcyclopropane)} + H_3C-C_6H_4-SO_2^- \tag{37}$$

contained d_0, d_1 and d_2 species in yields of 23, 55 and 22% respectively. The following reaction scheme, (equation 38) was proposed. The presence of the d_0 and d_1 products shows that the methylcyclopropane is formed by competitive carbenic and cationic pathways respectively. The d_2 product was accounted for by proposing a prior hydrogen–deuterium exchange to give $(CH_3)_2CH-CD=\overset{+}{N}=\overset{-}{N}$ which on reaction with D^+ and subsequent loss of N_2 and a proton gives methylcyclopropane-d_2.

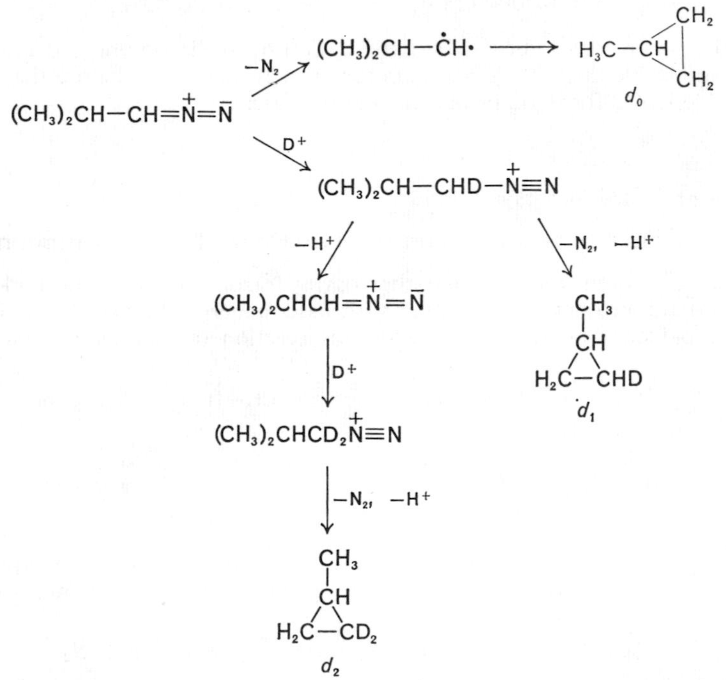

$$(CH_3)_2CH-CH=\overset{+}{N}=\overset{-}{N} \xrightarrow{-N_2} (CH_3)_2CH-\dot{C}H\cdot \longrightarrow H_3C-CH(CH_2)_2 \; (d_0)$$

$$(CH_3)_2CH-CH=\overset{+}{N}=\overset{-}{N} \xrightarrow{D^+} (CH_3)_2CH-CHD-\overset{+}{N}\equiv N \tag{38}$$

$$(CH_3)_2CH-CHD-\overset{+}{N}\equiv N \xrightarrow{-N_2,\ -H^+} CH_3-CH(CH_2)(CHD) \; (d_1)$$

$$(CH_3)_2CH-CHD-\overset{+}{N}\equiv N \xrightarrow{-H^+} (CH_3)_2CHCH=\overset{+}{N}=\overset{-}{N} \xrightarrow{D^+} (CH_3)_2CHCD_2\overset{+}{N}\equiv N \xrightarrow{-N_2,\ -H^+} CH_3-CH(CH_2)(CD_2) \; (d_2)$$

A further study of the effect of solvent on the decomposition pathway of the anion derived from a ketone tosylhydrazone was carried out by Nickon and Werstiuk[54]. They studied the thermal decomposition of the anion derived from norbornan-2-one **(23a)** and its 6-*exo*-deutero-**(23b)** and 6-*endo*-deutero-**(23c)** analogues in 'aprotic' medium (diglyme containing an excess of dissolved sodium methoxide), and in

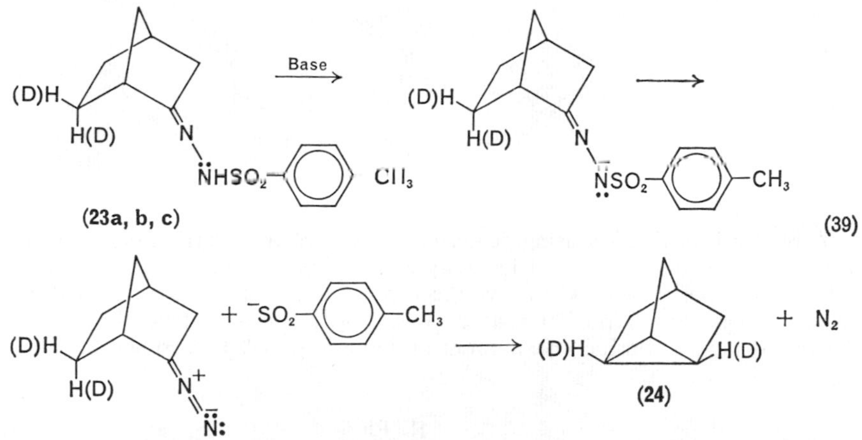

(23a, b, c) (39)

(24)

'protic' medium (ethylene glycol containing an excess of sodium) to give nortricyclene (24)[37] (equation 39). The nortricyclene (24) was isolated and analysed for deuterium mass spectroscopically (Table 7).

TABLE 7. Decomposition of norbornan-2-one tosylhydrazone

	Solvent	23a	23b	23c
24 in the hydrocarbon product (%)	Aprotic	>99	>99	>99
	Protic	93·2	92·9	92·3
Loss of original deuterium (%)	Aprotic	—	0	0
	Protic	—	19	52

In the 'aprotic' experiments it is noted that the nortricyclene contained the same amount of deuterium as its deuterated precursor. Therefore, nortricyclene (**24**) arises entirely by the insertion pathway with transfer of a hydrogen in the case of **23b** and a deuterium for **23c** (equation 40).

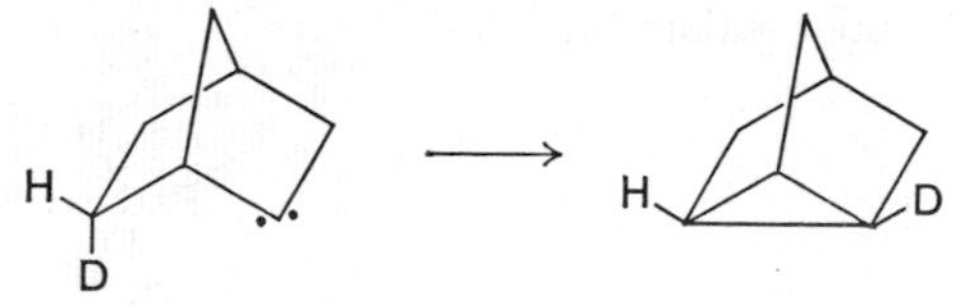

(40)

Carbene from **23c**

Each labelled substrate lost an appreciable amount of deuterium (loss of 19 and 52% from **23b** and **23c**, respectively) when the reaction was carried out in the protic solvent. This partial loss of the 6-hydrogen (deuterium) in the protic medium has to be balanced by a gain in hydrogen from the solvent. It was suggested that protonation of the diazoalkane produced from **23b** to give a diazonium ion or its equivalent (**25**) must occur before the nortricyclene is formed (equation 41). It was pointed out that *exo* protonation should be favoured sterically[55] and hence the C—N bond is shown in the *endo* position. 1,3-Elimination from **25** will give deuterated or non-deuterated nortricyclene depending upon whether H^+ or D^+ is lost from position 6.

from **23b** $\xrightarrow{H^+}$ **(25)**

A related investigation using deuterium as a tracer was carried out by Closs and coworkers[56] who studied the acid-catalysed reaction of phenyldiazomethane (**26**) in olefinic solvents at $-70\,^{\circ}C$ to give the products, (**27–31**) shown in equation (42). When **26** was reacted in trifluoroacetic acid-d_1 with *trans*-2-butene, the products **28**, **30** and **31** have a considerable amount of deuterium (80%). A somewhat surprising

$$PhCHN_2 + HA + >C{=}C< \longrightarrow \mathbf{(27)} + \mathbf{(28)} + \mathbf{(29)} + \mathbf{(30)} + PhCH_2A\ \mathbf{(31)} \qquad (42)$$

result was that only 21% of the phenylcyclopropane (**27**) contained a deuterium. One must conclude, therefore, that the products are not all derived from a common diazonium or carbonium ion since this would require that **27** be formed with at least half of the deuterium incorporated in the other products. Closs suggested that in the non-polar medium there is a hydrogen bond formation at the diazo carbon to give complex **32** (equation 43). Reaction of **32** with olefin leads to the formation of the cyclopropane via **33**, and loss of nitrogen and reaction with olefin leads to the other products via intermediate **34**.

$$Ph{-}C(=N_2){-}H + DA \longrightarrow PhC(=N_2)(H)\cdots DA\ \mathbf{(32)}$$

$$\mathbf{(32)} \xrightarrow{>C=C<} \mathbf{(34)}\ A^- \longrightarrow \mathbf{28, 30} \qquad (43)$$

$$\mathbf{(32)} \xrightarrow{>C=C<} \mathbf{(33)} \longrightarrow \text{Ph, H(D) cyclopropane}$$

The results of the previous studies on the Bamford–Stevens reaction, i.e. the thermal decomposition of the anion derived from an aldehyde or ketone tosylhydrazone, have been interpreted in terms of a carbene intermediate in 'aprotic' solvents while decomposition of the diazo intermediate (formed when the anion loses the tosyl group) proceeds through the diazonium and/or carbonium ion in 'protic' medium. A more recent study[57] using deuterated solvents has shown, however, that the decomposition of a diazo compound proceeds mainly via a carbene even in the protic solvent, acetic acid-d_1.

The decomposition of diethyl diazosuccinate (**35**) to give a mixture of diethyl maleate and fumarate (**36**) has been studied[57] in a series of deuterated hydroxylic solvents with varying pK_a. It was reasoned that olefin formation could follow path A

$$\underset{\textbf{(35)}}{\mathrm{EtO{-}\overset{O}{\overset{\|}{C}}{-}\underset{N_2}{\underset{\|}{C}}CH_2\overset{O}{\overset{\|}{C}}{-}OEt}} \longrightarrow \underset{\textbf{(36)}}{\mathrm{EtO{-}\overset{O}{\overset{\|}{C}}{-}CH{=}CH{-}\overset{O}{\overset{\|}{C}}{-}OEt}}$$

(insertion of the carbene into the C_β—H bond) or path B (β-elimination from the diazonium ion, equation 44). If the decomposition was carried out in deuterated hydroxylic solvent, SD, deuterium would not be incorporated into the olefin if path A were followed, while path B involves a prior addition of D^+ from solvent to give the α-deuterodiazonium ion which would subsequently undergo 1,2 elimination to give the deuterated olefin.

$$\mathrm{EtO{-}\overset{O}{\overset{\|}{C}}{-}\ddot{C}{-}CH_2{-}\overset{O}{\overset{\|}{C}}{-}OEt} \xrightarrow[\text{insertion}]{C_\beta\text{—H bond}} \mathrm{EtO{-}\overset{O}{\overset{\|}{C}}{-}CH{=}CH{-}\overset{O}{\overset{\|}{C}}{-}OEt}$$

$$\uparrow \text{A}$$

$$\mathrm{EtO{-}\overset{O}{\overset{\|}{C}}{-}\underset{N_2}{\underset{\|}{C}}{-}CH_2{-}\overset{O}{\overset{\|}{C}}{-}OEt} \qquad (44)$$

$$\downarrow \text{SD, } D^+ \text{ B}$$

$$\mathrm{EtO{-}\overset{O}{\overset{\|}{C}}{-}\underset{+N_2}{\underset{|}{C}D}{-}CH_2{-}\overset{O}{\overset{\|}{C}}{-}OEt} \xrightarrow[\text{1,2-elimination}]{} \mathrm{EtO{-}\overset{O}{\overset{\|}{C}}{-}CD{=}CH{-}\overset{O}{\overset{\|}{C}}{-}OEt}$$

The percentage of deuterium introduced into the olefinic positions for reaction of the substrate in several deuterated solvents is shown in Table 8 along with the calculated percentage from the carbenic process.

The results show that decomposition of **35** in strong mineral acid, DCl–D_2O, proceeds entirely via the diazonium ion–carbonium ion, pathway B. However, deuterium incorporation was considerably less for reaction in the other less acidic hydroxylic solvents, i.e. in cyclohexanol only 4% deuterium was incorporated and hence the elimination process proceeded approximately 92% via the carbene pathway. It was concluded, therefore, that the carbene process could proceed in protic solvents and that the percentage of the carbene pathway decreases as the acidity of the solvent increases. In fact, a plot of the pK_a of the solvent versus the amount of deuterium incorporation into the olefin (the amount of the ionic process) was linear.

TABLE 8. Percentage of deuterium incorporated into the olefin formed by thermal decomposition of diethyl diazosuccinate

Solvent	Percentage D incorporated into **36**	Percentage via carbenic process (calc.)
DCl–D_2O	51 ± 3	0
Acetic acid-d_1	17 ± 3	66
Ethanol-d_1	11 ± 5	78
Cyclohexanol-d_1	4 ± 2	92

Primary and secondary α-diazoketones react by a reversible carbon protonation in acidic media[58, 59]. Wentrup and Dahn[60] have recently studied the reaction of both primary and secondary diazoketones in super-strong acids. Contrary to the rapid exchange reaction of the α-hydrogens in acidic D_2O, primary α-diazoketones did not incorporate deuterium when treated with FSO_3D–SbF_5–SO_2 at −80 °C nor exchange deuterium for hydrogen when the α-deuterated diazoketone was reacted with HF–SbF_5 at −60 °C. N.m.r. spectroscopy was used to establish that the diazoketones were protonated entirely at oxygen in these acid solutions at low temperatures, i.e. where decomposition of the diazonium ion is not appreciable.

When solutions of protonated diazoketones were allowed to warm up to −25 °C, nitrogen gas was released and the deamination product, the fluorosulphate (**37**), was formed. These authors used a deuterium tracer study to investigate the mechanism

$$\mathrm{R{-}\overset{\overset{\displaystyle O}{\|}}{C}{-}CH_2OSO_2F}$$
(37)

for the formation of the fluorosulphate[60]. It was considered that the carbon-protonated form of the diazoketone could react with fluorosulphuric acid in an S_N2 reaction to yield the fluorosulphate.

When *p*-methoxydiazoactophenone-d_1 was reacted in fluorosulphuric acid, the fluorosulphate product had one deuterium in the methylene group (equation 45). This

$$p\text{-}\mathrm{CH_3OC_6H_4\overset{\overset{\displaystyle O}{\|}}{C}{-}CD{-}N_2} \underset{-\mathrm{H}^+\ k_{-1}}{\overset{\mathrm{H}^+\ k_1}{\rightleftarrows}} p\text{-}\mathrm{CH_3OC_6H_4\overset{\overset{\displaystyle O}{\|}}{C}{-}CHD{-}\overset{+}{N}_2} \xrightarrow[-\mathrm{N_2}]{\mathrm{FSO_3H},\ k_2} p\text{-}\mathrm{CH_3OC_6H_4\overset{\overset{\displaystyle O}{\|}}{C}{-}CHD{-}OSO_2F} \quad (45)$$

lack of exchange shows that, if protonation occurs on carbon, deamination must be faster than deprotonation, i.e. $k_2 \gg k_{-1}$. This behaviour is in contrast with that in aqueous acids where deprotonation by the base, water, is much faster than its attack on carbon. This is reasonable because water is a much stronger base than FSO_3^-.

2. Mechanism of alkylamine diazotization

A deuterium tracer study in both protic and aprotic solvents has been used to elucidate the mechanism of alkylamine diazotization. The aprotic diazotization of isobutylamine-d_2 in benzene produced a C_4 hydrocarbon mixture that contained deuterium to the extent of approximately 36% d_1 and approximately 12% d_2. However in protic solvents such as DOAc and D_2O–DOAc, deuterium uptake was diminished and only monodeuterated products were observed[61] (Table 9).

TABLE 9. Diazotization of isobutylamine

Deuterium source	Solvent	Deuterium content of product (mol.-%)		
		d_0	d_1	d_2
DOAc	C_6H_6	52	36	12
DOAc	DOAc	96	4	0

In order to explain the results obtained in the aprotic medium the authors proposed the reaction scheme in equation (46). The primary intermediate is considered to be

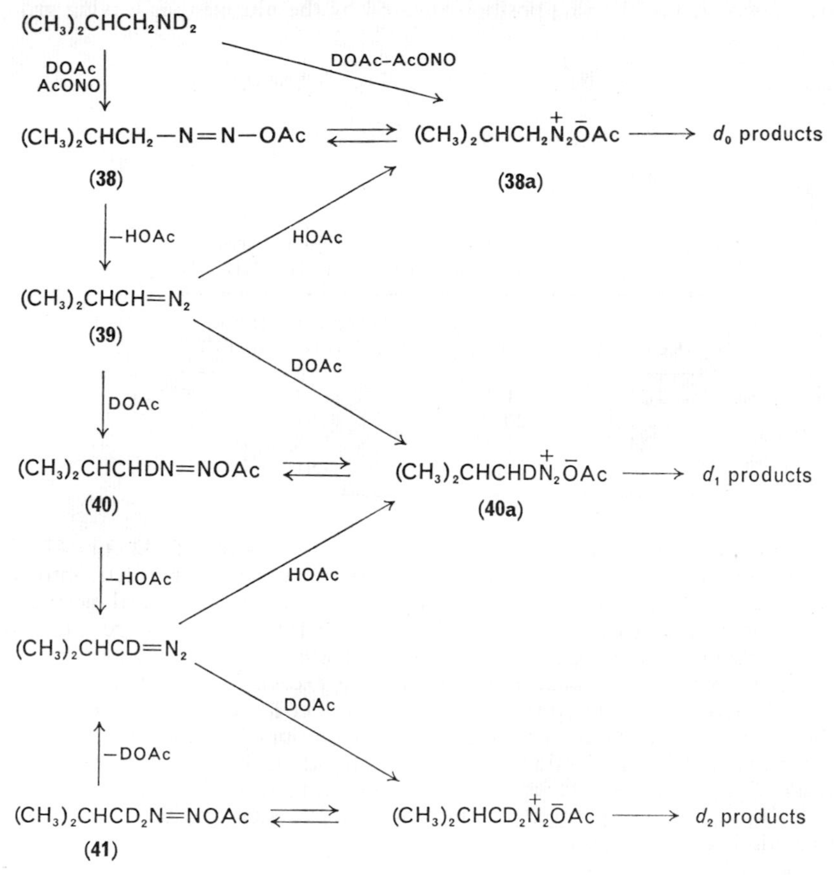

the covalent diazonium acetate (**38**) which is converted to the diazo compound (**39**) by the elimination of HOAc. Addition of DOAc to **39** leads to **40a** and thus to **40** with incorporation of one deuterium atom. Similarly, **41** and **41a** could be formed leading to incorporation of two deuterium atoms. The formation of non-, mono- and di-deuterated products will arise from **38a**, **40a** and **41a**, respectively.

In the aprotic solvent, benzene, the solvating power of the medium is low and the covalent diazonium acetate (**38**) would predominate over the ionic form (**38a**), and thus the amount of deuterated products would be greater than in protic solvents. An increase in acid concentration, on the other hand, would decrease the conversion of the covalent diazonium acetate (**38**) to the diazo compound (**39**) and the overall deuterium uptake would be lessened.

3. Decomposition of diazonium ions in alkaline medium

The decomposition of several arenediazonium salts and arylazo phenyl sulphones has been investigated[62–64] in CH_3OD solution with varying methoxide concentrations. The results observed for the reactions of 2-chlorobenzenediazonium ion (**42**) and the 2-chlorophenylazo phenyl sulphone (**43**) are shown in Table 10. In both instances, the principal product was chlorobenzene and the incorporation of the deuterium occurred at the ring position vacated by the nitrogenous leaving group.

N_2^+ Cl (**42**) $N{=}NSO_2C_6H_5$ Cl (**43**)

TABLE 10. Percentage monodeuteration of denitrogenation products obtained with $NaOCH_3$ in CH_3OD

Reactant	$[NaOCH_3]$ / [Substrate]	Monodeuteration of chlorobenzene (%)
42	1·0	3
	20	91
43	1·0	8
	20	91

When 1 mol of methoxide ion was reacted with 1 mol of **42** or **43**, i.e. $[NaOCH_3]/[substrate] = 1{\cdot}0$, very little deuterium was incorporated into the chlorobenzene product. This result was rationalized in terms of a radical mechanism in which the hydrogen atom added to the benzene ring during the reaction was abstracted from the methyl group of the solvent, methanol[65].

In solutions where methoxide ion was in large excess, i.e. 20 : 1, a significant amount of deuterium is incorporated into the product for the reaction of both **42** and **43** in CH_3OD. It has been suggested that deuterium is incorporated into the chlorobenzene when the aryl anion abstracts a deuterium from CH_3OD (equation 47). The methoxide anion first attacks the diazonium ion to form a covalent diazoether (**44**) and then a methyl hydrogen is removed by methoxide to give the aryl anion in an E2 elimination reaction.

$$\text{2-ClC}_6\text{H}_4\text{N}_2^+ \xrightarrow{CH_3O^-} \text{2-ClC}_6\text{H}_4\text{N=N-OCH}_3\ \mathbf{(44)} \xrightarrow{CH_3O^-} \text{2-ClC}_6\text{H}_4^{\bullet -} \xrightarrow{CH_3OD} \text{2-ClC}_6\text{H}_4\text{D} + CH_2O + N_2 \quad (47)$$

The results for compound **43** are very similar to those for the reaction of the diazonium salt (**42**). This is expected since the arylazo phenyl sulphones dissociate readily in methanol to diazonium and benzenesulphinate ions[66], equation (48).

$$\text{2-ClC}_6\text{H}_4\text{-N=N-SO}_2\text{C}_6\text{H}_5\ \mathbf{(43)} \underset{}{\overset{CH_3OH}{\rightleftharpoons}} \text{2-ClC}_6\text{H}_4\text{-N}_2^+ + C_6H_5SO_2^- \quad (48)$$

4. Aryne intermediates in diazonium salt reactions

Cadogan and Hibbert found that anthracene is converted into a triptycene derivative in reactions where diazonium ion intermediates are produced in dipolar aprotic solvents[67] (equation 49) and concluded that an aryne intermediate must be produced when diazonium ions decompose in aprotic solvents.

$$\text{2-}t\text{-BuC}_6\text{H}_4\text{N(NO)COCR}_3 \longrightarrow \text{2-}t\text{-BuC}_6\text{H}_4\text{N=NOCOCR}_3 \longrightarrow \text{2-}t\text{-BuC}_6\text{H}_4\overset{+}{\text{N}}\equiv\text{N} \xrightarrow{\text{anthracene}} t\text{-Bu-triptycene} \quad (49)$$

Franck and Yanagi[10] designed a deuterium tracer experiment to check this hypothesis. They prepared 2,5-di-*t*-butyl-4,6-dideuteroaniline and converted it into the diazonium salt by reacting it with *n*-butyl nitrite and one equivalent of a carboxylic acid in methylene chloride. When the deuterated diazonium ion decomposed in methylene chloride–furan solutions, three of the products were 1,4-di-*t*-butylbenzene (**45**), a 2,5-di-*t*-butylphenyl ester (**46**) and the Diels–Alder type cycloaddition adduct **47** (equation 50). If furan was not present in the solvent, product **47** was not observed, but the amount of **45** was increased by an amount equal to the yield of **47** in the methylene chloride–furan reaction (Table 11).

TABLE 11. Deuterium content and yield of some products from the decomposition of 2,5-di-*t*-butyl-4,6-dideuterobenzenediazonium ion

Solvent	Product	Yield (%)	Percentage[a] d_1	Percentage[a] d_2
Methylene chloride–furan	**45**	16	12·8	84·9
	46	25	6·2	92·0
	47	11	88·4	8·4
Methylene chloride	**45**	28	45·9	50·1
	46	26	17·6	81·1
	47	0	—	—

[a] The deuterium content of the products was determined mass spectrometrically.

If the products are produced from the aryne intermediate, L in equation (50) would be a hydrogen, and the products would be monodeuterated. If the products **45–47** are produced from either a phenyl cation or a phenyl radical however, they will contain two deuteriums per molecule. When the reaction was carried out in a furan–methylene chloride solution, **47** was 89% d_1 and the authors concluded that a major portion of this product is produced from the aryne. Only a small amount ($<13\%$ and $<6\%$) of **45** and **46**, respectively, were formed by the aryne pathway.

(51)

These results, which suggest that approximately 14% of the product is formed from an aryne intermediate, were confirmed when the reaction was carried out in the absence of furan. In this case, almost half of the 1,4-di-*t*-butylbenzene is monodeuterated and has, therefore, been produced by the aryne pathway. It is also noted that the amount of d_1 ester has increased over that formed in the presence of furan.

Cadogan and coworkers[5] used 2,4,6-trideuterobenzenediazonium ions generated from a deuterated aniline, pathway A, and from a deuterated nitrosourethane, pathway B, to investigate the mechanism of benzyne formation during diazonium salt decomposition. This study involved determining the deuterium content of the cycloaddition adduct formed when the benzyne intermediate reacts with tetraphenylcyclopentadiene (equation 51).

Analysis of the 1,2,3,4-tetraphenylnaphthalene recovered from these reactions showed that only one deuterium per molecule is lost during the reaction and thus an E1cB mechanism with $k_{-1} \gg k_2$, which would lead to a loss of more than one deuterium per molecule (equation 51), can be eliminated. Only one atom of deuterium would be lost if the benzyne intermediate were formed via an E1cB mechanism where proton abstraction (the k_1 step) is fully rate determining, i.e. $k_2 \gg k_{-1}$, or a concerted E2 mechanism where the proton abstraction and elimination of nitrogen occur simultaneously.

The possibility that a benzyne intermediate is involved in the diazonium salt decomposition in protic media has been investigated by Swain's group[68]. These workers determined the deuterium content of the phenol produced when benzenediazonium ion was decomposed in a D_2O–DCl solution. The benzyne intermediate, if formed, would react with D_2O to give phenol-d_1 as the final product (equation 52),

$$C_6H_5\overset{+}{N}{\equiv}N \longrightarrow N{\equiv}N + C_6H_4 \xrightarrow{D_2O} C_6H_4D{-}OH \qquad (52)$$

whereas phenol produced from the phenyl cation would not contain any deuterium. The phenol isolated from the reaction only contained 0·05% deuterium and the authors concluded that a benzyne intermediate does not form to any significant degree when diazonium ions are decomposed in protic media.

C. Use of 18-Oxygen as a Tracer

1. 18-Oxygen as a tracer in diazoester decompositions

18-Oxygen has been used as a tracer in experiments designed to elucidate the details of the thermal decomposition mechanism of diazoesters. White and coworkers[69, 70] prepared several diazoesters labelled with 18-oxygen at the carbonyl oxygen by the nitrosourethane route (equation 53). One of the products isolated from

$$RNH_2 \xrightarrow[(2)\ C_2H_5C(={}^{18}O)Cl]{(1)\ N_2O_4} RN(N{=}O){-}C({=}{}^{18}O)C_2H_5 \longrightarrow R{-}N{=}N{-}O{-}C({=}{}^{18}O)C_2H_5 \qquad (53)$$

the decomposition of the 18-oxygen-labelled diazoesters was an 18-oxygen-labelled alkyl propionate that formed when the ions in the carbonium ion–carboxylate ion pair **49** collapsed together in a fast step of the reaction (54).

$$R{-}N{=}N{-}O{-}\overset{^{18}O}{\overset{\|}{C}}C_2H_5 \longrightarrow \left[R{-}\overset{+}{N}{\equiv}N \quad \bar{O}{-}\overset{^{18}O}{\overset{\|}{C}}C_2H_5\right] \longrightarrow \left[R^{+} \quad N{\equiv}N \quad \bar{O}{-}\overset{^{18}O}{\overset{\|}{C}}C_2H_5\right]$$

(48) (49)

$$\xrightarrow{\text{fast}} R{-}O{-}\overset{^{18}O}{\overset{\|}{C}}C_2H_5 + R{-}^{18}O{-}\overset{O}{\overset{\|}{C}}C_2H_5 \tag{54}$$

Complete 18-oxygen scrambling is observed when the R group is 1-apocamphyl, i.e. decomposes via a bridgehead carbonium ion. The 18-oxygen label is mainly in the carbonyl group, however, when the R group is either a secondary (55–60% 18-oxygen in the carbonyl group when R = 1-phenylethyl-) or a tertiary (63–74% 18-oxygen in the carbonyl group when R = 2-phenyl-2-butyl-) alkyl group. The more extensive scrambling of the 18-oxygen label in the bridgehead diazoesters indicates that the lifetime of the diazonium ion–carboxylate ion pair (**48**) is longer in the bridgehead systems and almost certainly results from a difference in the rate of conversion of the diazonium ion–carboxylate ion pair (**48**) into the carbonium ion–carboxylate ion pair (**49**). The conversion of (**48**) to (**49**) would have a much larger free energy of activation and thus be more rate determining for the bridgehead system where a highly strained carbonium ion intermediate is produced.

IV. ISOTOPE EFFECTS IN DIAZONIUM SALT REACTIONS

A. Theory of Kinetic Isotope Effects

1. Heavy atom kinetic isotope effects

The Bigeleisen treatment[71–73], based on Eyring and coworkers' absolute rate theory[74], assumes that there is a single potential energy surface along which the reaction takes place, and that there is a potential energy barrier separating the reactants from the products of the reaction. The reaction occurs along the path corresponding to the lowest potential energy, i.e. it passes over the lowest part of the barrier. The transition state is located at the top of the barrier on the reaction path, i.e. it lies at the energy maximum for motion along the reaction coordinate but at an energy minimum in all other directions, and is assumed to have all the properties of a stable molecule for all degrees of freedom except that corresponding to the path of decomposition (motion along the reaction coordinate).

The expression for the rate constant (k) of the reaction according to these assumptions may be expressed by equation (55)

$$k = \frac{\bar{k}T\kappa K^{\ddagger}}{h} \tag{55}$$

where $\bar{k}$ is the Boltzmann constant, T is the absolute temperature, h is Planck's constant, κ is the so-called transmission coefficient and $K^{\ddagger}$ is the equilibrium constant between the activated complex (the molecule at the transition state) and the reactants. It is assumed that the transition state complex is in equilibrium with the reactants. The degree of freedom corresponding to the reaction path is not included for the activated complex, κ represents a factor which takes into account the non-classical correction required to allow molecules with insufficient classical energy to surmount the barrier to 'tunnel' through it[75]. Using equation (56) with a knowledge of the potential energy surface, $K^{\ddagger}$ may be calculated using the methods of statistical mechanics, since:

$$K^{\ddagger} = \frac{Q^{\ddagger}}{Q_{\mathrm{A}}\,Q_{\mathrm{B}}} \tag{56}$$

where Q's are the complete partition functions for reactants A, B, ..., etc., and $Q^{\ddagger}$ is the partition function for the transition state complex, omitting again the one vibrational energy level corresponding to the degree of freedom along the decomposition pathway.

The calculation of the potential energy surface from first principles is, at present, insufficiently accurate to allow this approach to yield reliable values of $Q^{\ddagger}$ and therefore of $K^{\ddagger}$. However, the effect of isotopes on these quantities can be predicted more accurately than can the quantities themselves and isotopic rate ratios may be calculated for fairly complex reactions with some confidence. For the reaction:

$$\mathrm{A+B+C+...} \longrightarrow [T.S.]^{\ddagger} \longrightarrow \mathrm{Products} \tag{57}$$

$$\frac{k_1}{k_2} = \frac{\kappa_1}{\kappa_2}.\frac{Q_1^{\ddagger}}{Q_2^{\ddagger}}.\frac{Q_{\mathrm{A}_2}}{Q_{\mathrm{A}_1}}.\frac{Q_{\mathrm{B}_2}}{Q_{\mathrm{B}_1}}.\frac{Q_{\mathrm{C}_2}}{Q_{\mathrm{C}_1}}$$

where the subscripts 1 and 2 refer to the molecules containing the lighter and heavier isotopes, respectively.

The assumption is that $\kappa_1 = \kappa_2$ initially, although these transmission coefficients are not known with certainty. To correct for any error introduced in this assumption, a 'tunnelling correction' factor is introduced. Bigeleisen and Goeppert–Mayer[76] expressed the partition functions in terms of the vibrational frequencies of the molecules in the gas phase. Hence, in the harmonic approximation, for all non-linear gas molecules except hydrogen, Q_2/Q_1 is given by equation (58) where S_1 and S_2 are the symmetry numbers of the respective molecules, the M's are the molecular weights, the I's are the moments of inertia about the three principal axes of the n-atom molecules and the ν's are the fundamental vibrational frequencies of the molecules in wave numbers.

$$\frac{Q_2}{Q_1} = \frac{S_1}{S_2}\left(\frac{I_{\mathrm{A}_2}\,I_{\mathrm{B}_2}\,I_{\mathrm{C}_2}}{I_{\mathrm{A}_1}\,I_{\mathrm{B}_1}\,I_{\mathrm{C}_1}}\right)^{\frac{1}{2}}\left(\frac{M_2}{M_1}\right)^{\frac{3}{2}}\pi_i^{3n-6}\exp\left(\frac{(\nu_{1i}-\nu_{2i})\,hc}{2kT}\right)\frac{(1-\exp(-hc\nu_{1i}/kT))}{(1-\exp(-hc\nu_{2i}/kT))} \tag{58}$$

Using various approximations, a solution to the isotopic rate ratio equation can be obtained. It is found that the isotope rate ratio, k_1/k_2, is dependent on the force constant changes which occur in passing to the transition state. Consequently, if C—X bond rupture, where X can be halogen, sulphur, nitrogen, etc., has not progressed at the transition state of the slow rate-determining step for the overall reaction, a rate ratio $k_{\mathrm{X}_1}/k_{\mathrm{X}_2}$ equal to one is expected. Accordingly, a value of the isotope rate ratio greater than one will be observed if there is a decrease in the force constants at the transition state of the slow step. The greater the decrease in the force constant the larger will be the magnitude of the isotope effect.

The observation of a heavy atom isotope effect, therefore, allows one to determine whether C—X bond weakening, decrease in force constant, has proceeded at the activated complex of the slow rate-determining step. The magnitude of the isotope effect provides information concerning the relative degree of C—X 'bond rupture' and hence provides information concerning the structure of the transition state.

Saunders[77] has recently calculated the dependence of the leaving group isotope effect on the extent of C—X rupture for trimethylamine and dimethyl sulphide as leaving groups. The calculations were performed for elimination processes where the degree of carbon–hydrogen cleavage was taken as 50%. A plot of the leaving group isotope effect versus the extent of C—X rupture is shown in Figure 1. It is noted that the heavy atom isotope effects are essentially linearly related to the extent of

C—X rupture. Sims and coworkers, in a similar calculation, found that the same relationship between the magnitude of the leaving group isotope effect and the extent of C—X bond rupture, existed for a nucleophilic substitution reaction[78].

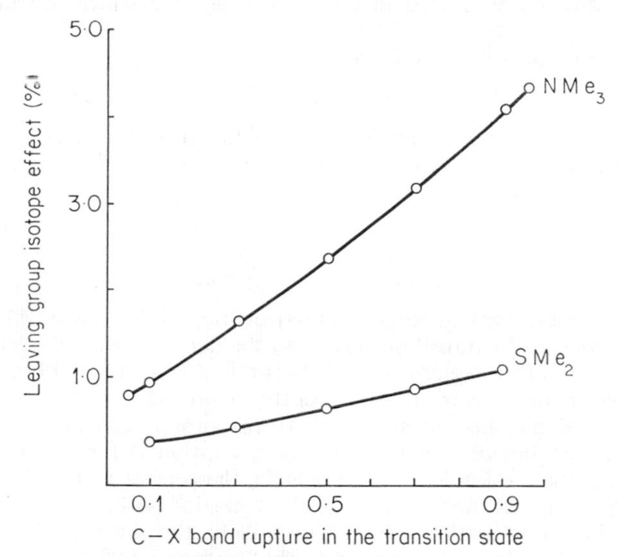

FIGURE 1. Relationship between the magnitude of the heavy atom kinetic isotope effect and the amount of C—X bond rupture in the transition state.

2. Primary hydrogen–deuterium kinetic isotope effects

It is apparent that the Bigeleisen formulation can be used to calculate transition state force constants with some confidence if a large computer is available. For some purposes, however, it is sufficient to have only a qualitative idea of the changes in force constants which have occurred at the transition state, and acceptable estimates of the isotope effect can be obtained without recourse to a complex calculation. It has been observed that zero-point energy differences between the isotopic molecule's vibrations, while not the only contributor to the isotope effect, are, however, often the dominant term. This is particularly true for the cases of hydrogen–deuterium where the zero-point energy difference is large, and also for large molecules where isotopic substitution does not affect the mass and moment of inertia term significantly. It is usual to assume that the stretching modes are the most important in determining the isotope effect. This is based on the assumptions that the bending vibrations are generally of a lower frequency and therefore have smaller zero-point energy differences for isotopic molecules, and further that the bending motions in the transition state will be largely similar to those in the substrates.

For a single C—H bond undergoing rupture in a unimolecular process

$$\frac{k^{\mathrm{H}}}{k^{\mathrm{D}}} = \exp\left(-\frac{hc}{2kT}(\nu_{\mathrm{H}}-\nu_{\mathrm{D}})\right) \qquad (59)$$

where ν_{H} and ν_{D} are the ground-state symmetric stretching frequencies for the C—H and C—D bonds, respectively. Substitution into equation (57) leads to an expected isotope effect of approximately seven at 25 °C.

For reactions involving a proton transfer from one molecule to another, however, the situation is more complex since bond formation and breaking are occurring concurrently and, as Westheimer[79] points out, it is essential to realize that new stretching vibrations are created in the transition state which are not present in the reactants.

Westheimer considers the reaction:

$$\text{AH+B} \longrightarrow [\text{A}\cdots\text{H}\cdots\text{B}] \longrightarrow \text{A+HB} \tag{60}$$

where [A···H···B] is a linear transition state. If this transition state is regarded as a linear molecule, there would be two independent stretching vibrational modes which may be illustrated as follows:

A···H···B	A···H···B
← ? →	← → ←
symmetric	antisymmetric

Neither of these vibrations corresponds to stretching vibrations of AH or BH. The translational mode in the transition state may be identified with the 'antisymmetric' vibrational mode, but the 'symmetric' mode is a real vibration, with a positive force constant. Westheimer[79], and more recently More O'Ferrall[80], show that the 'symmetric' vibration (transition state) may or may not involve motion of the central H(D) atom, depending on the relative 'force constants' for the A–H and H–B partial bonds. If the motion is truly symmetric, the central atom will be motionless in the vibration and thus the frequency of the vibration will not depend on the mass of this atom, i.e. the vibrational frequency will be the same for both isotopically substituted transition states. It is apparent that under such circumstances there will be no zero point energy differences between deuterium- and hydrogen-substituted compounds for the symmetric vibration in the transition state. Hence an isotope effect of $k^{\mathrm{H}}/k^{\mathrm{D}} = 7$ at room temperature is expected since the difference in activation energy $(E_{\mathrm{D}} - E_{\mathrm{H}})$ is the difference between the zero point energies of the symmetric stretching vibrations of the initial states.

For instances where bond breaking and bond making at the transition state are *more* or *less* advanced, the 'symmetric' vibration will be no longer truly symmetric, the frequency will have some dependence on the mass of the central atom, and there will be a zero point energy difference for the vibrations of the isotopically substituted molecules at the transition state. Hence:

$$\frac{k^{\mathrm{H}}}{k^{\mathrm{D}}} = \exp\left(\frac{-hc}{2kT}\,[(\nu_{\mathrm{H}} - \nu_{\mathrm{D}}) - \Delta\nu_{\mathrm{s}}]\right) \tag{61}$$

where $\Delta\nu_{\mathrm{s}}$ corresponds to the frequency difference of the symmetric mode of the transition state on isotopic substitution. For such situations, $k^{\mathrm{H}}/k^{\mathrm{D}}$ will have values smaller than 7.

It may be concluded that for reactions where the proton is less or more than one-half transferred in the transition state, i.e. the A–H and H–B force constants are unequal, the primary hydrogen–deuterium isotope effect will be less than the maximum of 7. The maximum isotope effect will be observed only when the proton is exactly half-way between A and B in the activated complex (Figure 2).

3. Secondary β-deuterium kinetic isotope effects

In the preceding sections the bond involving the isotopic atom was broken or formed in the rate-determining step of the reaction. In these cases, the change in rate is referred to as the primary kinetic isotope effect. Isotopic substitution at other sites

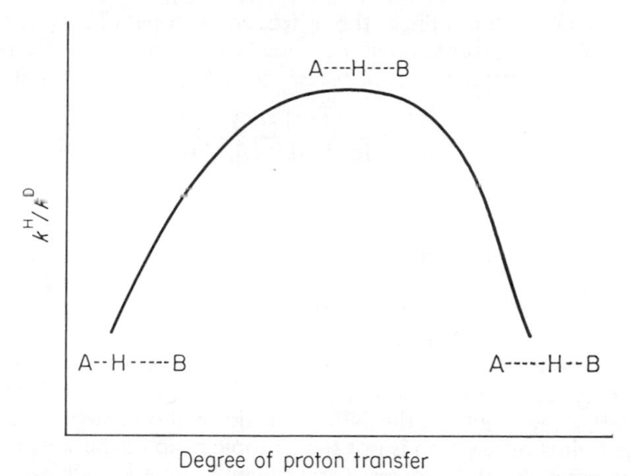

FIGURE 2. Plot of k^H/k^D versus degree of proton transfer.

in the molecule gives smaller rate effects and these are collectively referred to as secondary effects.

Secondary β-deuterium isotope effects arise when the hydrogen(s) on the β-carbon (adjacent to the carbon where the C—X bond rupture is progressing) are replaced by deuterium(s). These isotope effects (k_H/k_D) are greater than unity for solvolytic processes. In addition, the magnitude of the isotope effect increases as the amount of positive charge (carbonium ion character) on the α-carbon in the transition state **50** is increased. For example, the isotope effect per CD_3 group increases from about

$$\left[-\overset{|}{\underset{\substack{| \\ \text{H(D)}}}{\text{C}}}{}^{\beta}\text{—}{}^{\alpha}\overset{|}{\underset{|}{\text{C}}}{}^{\delta+}\cdots\cdots\text{X}^{\delta-} \right]^{\ddagger}$$

(50)

1·03 for ethyl compounds which undoubtedly react by an S_N2 mechanism to approximately 1·37 for a *t*-butyl compound which reacts by a limiting S_N1 mechanism[81]. A wealth of experimental evidence[82] indicates that these isotope effects are primarily, if not completely, a result of hyperconjugative electron release from the C_β—H bonds[83]. Other studies by Shiner[84, 85] have demonstrated that the magnitude of these isotope effects vary with the dihedral angle between the C_β—H orbital and the developing p orbital on the α-carbon. The maximum isotope effect in any system is observed when the dihedral angle is either 0° or 180°, i.e. where the overlap between the C_β—H and the p orbital on the α-carbon is maximized.

B. Nitrogen Kinetic Isotope Effects in Diazonium Salt Reactions

The first nitrogen kinetic isotope effect determined for a diazonium salt decomposition reaction was measured by Lewis[86]. Unlike later workers who used the normal competitive method of measuring the nitrogen isotope effect[87], Lewis determined an isotope effect, $k^{14}/k^{15} = 1{\cdot}019 \pm 0{\cdot}004$, by measuring the individual rate constants for the reactions of *p*-methylbenzenediazonium and *p*-methylbenzene[α-^{15}N]-diazonium ion at 49 °C.

Brown and Drury[28] determined the nitrogen isotope effects for the thermal decomposition of several substituted benzenediazonium ions. The isotopic ratios required for the calculation of the isotope effect (equation 62) were obtained by

$$\frac{k^{14}}{k^{15}} = \frac{\ln(1-f)}{\ln(1-[R_0/R_f]f)} \quad (62)$$

analysing the nitrogen gas released in the first few percentage points of reaction and samples of nitrogen gas obtained from a reaction taken to completion, in a 60° sector isotope ratio mass spectrometer. Equation (62), where f is the extent of reaction expressed as a fraction, R_0 is the N^{14}/N^{15} ratio in the starting material and R_f is the N^{14}/N^{15} ratio in the product after a few percentage points of reaction, is used to calculate the isotope effect[88].

If the isotope ratios determined in their experiments are used to calculate the nitrogen isotope effect, a value of 1·022 is obtained. However, since two nitrogen atoms are freed in the reaction, the N^{14}/N^{15} ratio in the nitrogen recovered after a few percentage points of reaction is not the isotopic ratio of the α-nitrogen, the atom involved in the bond rupture process. Consequently, an accurate α-nitrogen isotope effect could not be determined. These authors assumed that there would be no isotopic fractionation of the β-nitrogen, i.e. $k^{14}/k^{15} = 1{\cdot}000$ for the β-nitrogen, and concluded that the change in the isotopic ratio in the nitrogen recovered after a few percentage points of reaction was due to the change in the isotopic ratio of the α-nitrogen. When this assumption was incorporated into the calculation of the isotope effect, a value of approximately 1·044 was obtained.

The nitrogen kinetic isotope effects determined by Brown and Drury, using the method described above, for the thermal decomposition of several arenediazonium ions at 40·5 °C are shown in Table 12. It is seen that the nitrogen kinetic isotope effect is independent of both the counter ion in the diazonium salt and the substituent on the benzene ring.

TABLE 12. Nitrogen kinetic isotope effects for the thermal decomposition of arenediazonium ions at 40·5 °C

Substituent	Counter ion	k^{14}/k^{15}
H	Cl^-	1·044 ± 0·003
H	BF_4^-	1·043 ± 0·005
o-CH_3	BF_4^-	1·045 ± 0·001
m-CH_3	BF_4^-	1·047 ± 0·001
p-CH_3	BF_4^-	1·047 ± 0·001
m-Cl	BF_4^-	1·044 ± 0·001

The temperature dependence of these isotope effects was normal, i.e. the magnitude of the isotope effect increased from 1·043 at 68·5 °C to 1·053 at 6·9 °C. Finally, these values are as large as the maximum theoretical kinetic nitrogen isotope effect of 1·043 at 59 °C[89] and it was concluded that the C—N_α bond is almost completely broken in the transition state of the rate-determining step for the decomposition process for all the compounds studied.

Brown and Drury[28] measured the infrared stretching and bending frequencies associated with the diazonium group in benzene- and benzene-[α-^{15}N]diazonium fluoroborate (Table 13). The authors used these observed and other estimated frequencies to calculate the nitrogen kinetic isotope effects for these reactions. They were able to duplicate the experimental results (Table 12), including the temperature dependence, to within the limit of experimental error.

TABLE 13. Infrared frequencies of benzene- and benzene[α-^{15}N]-diazonium ion

Vibration	Frequency (cm^{-1})	
	$C_6H_5-\overset{+}{N}\equiv N\ \overset{-}{BF_4}$	$C_6H_5-{}^{15}\overset{+}{N}\equiv N\ \overset{-}{BF_4}$
C—N—N bend	455	451
C—N—N bend	533	526
N≡N stretch	2296	2251

In another study, Loudon, Maccoll and Smith[90] found nitrogen kinetic isotope effects of 1·043 for the decomposition of arenediazonium ions in water thus confirming the results of Brown and Drury. They found, however, that the nitrogen isotope effect did vary with the concentration of added salts when the substrate was *p*-nitrobenzenediazonium ion, and was considerably smaller for *p*-methoxy- and *p*-hydroxybenzenediazonium ion, which were not studied by Brown and Drury (Table 14).

TABLE 14. Nitrogen kinetic isotope effects[a] for the decomposition of *para*-substituted benzenediazonium fluoroborates at 40 and 69 °C

Substituent	Added salt	k^{14}/k^{15}	Temperature (°C)
p-H	—	1·043±0·005	40
p-H	5 M-KSCN	1·046±0·001	40
m-Cl	—	1·042±0·007	40
p-NO_2	—	1·044±0·004	69
p-NO_2	6 M–KBr	1·031±0·003	69
p-NO_2	5 M-$KHSO_4$	1·039±0·004	69
p-MeO	—	1·028±0·004	69
p-OH	—	1·025[b]	—

[a] Calculated on the assumption that the beta-nitrogen kinetic isotope effect is 1·000.
[b] Value is uncertain.

The authors stated that the reaction of the compounds with an isotope effect of approximately 1·044 are S_N1 processes where the formation of the phenyl cation occurs in the slow step of the reaction. Thus all of the compounds studied by Brown and Drury and by Maccoll and coworkers react by an S_N1 mechanism with the exception of the reaction of the *p*-nitrobenzenediazonium ion in the presence of added salt, and the reactions where the benzene ring is substituted with strongly electron-donating groups, i.e. the *p*-methoxy- and *p*-hydroxybenzenediazonium ions.

The kinetic expression for the decomposition of the *p*-nitrobenzenediazonium salt in the presence of bromide or other anions has a second-order term. This suggests that *p*-nitrobenzenediazonium ion reacts in the presence of bromide ion or hydrogen sulphate ion by concurrent S_N1 and S_N2 mechanisms.

The observed nitrogen isotope effects in these reactions are a composite of the nitrogen isotope effects for the S_N1 reaction of the diazonium ion and the S_N2 reaction of the diazonium ion with either bromide or hydrogen sulphate ion. Using the assumption that the S_N1 component had an isotope effect of 1·044, Maccoll was able to calculate the nitrogen isotope effects for the S_N2 component of these reactions and found much smaller isotope effects, i.e. k^{14}/k^{15} between 1·014 and 1·021 for the bromide ion assisted reaction and approximately 1·036 for the hydrogen sulphate ion reaction.

Maccoll concluded on the basis of the lower nitrogen kinetic isotope effects that the *p*-methoxybenzene- and the *p*-hydroxybenzenediazonium ions react entirely by an S_N2 mechanism. The S_N2 mechanism was preferred by Maccoll since the *p*-methoxy and *p*-hydroxy groups would stabilize the reactants (**51, 51a**) by resonance to a much greater degree than their respective transition states† (**52, 52a**) in the S_N1 reaction. The resonance stabilization of the transition state for the S_N1 process

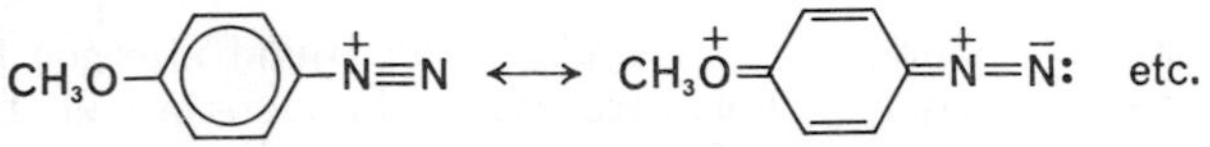

$CH_3O-C_6H_4-\overset{+}{N}{\equiv}N \longleftrightarrow CH_3\overset{+}{O}{=}C_6H_4{=}\overset{+}{N}{=}\overset{-}{N}:$ etc.

(51) (51a)

$CH_3O-C_6H_4^{+} \longleftrightarrow CH_3\overset{+}{O}{=}C_6H_4^{-}{}^{+}$ etc.

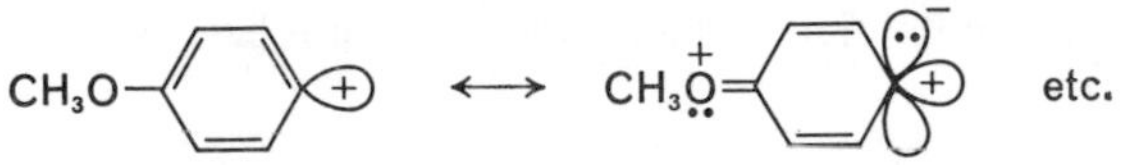

(52) (52a)

would be minimal because the developing positive charge in the empty sp^2 orbital is perpendicular to the π system of the benzene ring. As a result, the S_N1 mechanism would have a higher free energy of activation than the S_N2 mechanism.

Maccoll's interpretation[90] of the smaller kinetic isotope effects for the *p*-methoxybenzenediazonium ion seems doubtful, however, since other properties of the reaction of the *p*-methoxydiazonium ion require the formation of a *p*-methoxyphenyl cation. For example, it is impossible to explain the k_{exch}/k_{solv} ratio (0·038 in water and 0·075 in trifluoroethanol) observed for the decomposition of the *p*-methoxybenzenediazonium ion without assuming that an aryl cation is formed. In addition, the exchange between external nitrogen and the *p*-methoxybenzenediazonium ion is consistent with the formation of an aryl cation but is difficult to explain on the basis of an S_N2 mechanism. Finally, the much slower rate for the *p*-methoxy compound ($k_{p\text{-H}}/k_{p\text{-MeO}} \approx 15{,}000/1$) is consistent with an S_N1 mechanism (see above) although it does not rule out an S_N2 process.

An alternative explanation for the smaller isotope effect is that the *p*-methoxyphenyl cation intermediate is produced, but that it reverts to starting material *p*-methoxybenzenediazonium ion, to a much greater extent than the phenyl cation reverts to benzenediazonium ion. Thus, the k_{-1} and k_{exch} steps combined are larger than the k_{solv} step (equation 63) when Z is methoxy, whereas the k_{-1} and k_{exch} steps are smaller than the k_{solv} step for benzenediazonium ion, i.e. when Z = hydrogen. This means that the k_1 step is almost, or is completely, rate determining in the case where Z is hydrogen, the C—N_α bond is breaking in the slow step of the reaction, and the observed isotope effect is large. When Z is methoxy, the k_{solv} step is more

† The magnitude of the nitrogen isotope effects shows that the transition states are virtually identical to the intermediates.

rate determining (the k_{exch}/k_{solv} ratio is 0·038 rather than the 0·014 found for the unsubstituted compound and the $k_{-1_{p\text{-MeO}}}/k_{-1_{p\text{-H}}}$ would be large), the C—N$_\alpha$ bond is breaking in a step which is only partially rate determining, and the observed nitrogen isotope effect will be small. A precedent for this phenomenon, i.e. change in the magnitude of the observed isotope effect with a shift in the rate-determining step of a reaction, has been observed by Graczyk and Taylor[91] in the reaction of *p*-methoxybenzyl chloride with azide ion in water. Finally, although the aryl cation must exist, it is possible that concurrent S_N1 and S_N2 reactions are responsible for the smaller isotope effect.

$$\text{Z-C}_6\text{H}_4\text{-}\overset{+}{\text{N}}_\alpha\equiv\text{N}_\beta \underset{k_{-1}}{\overset{k_1}{\rightleftharpoons}} \text{Z-C}_6\text{H}_4^+ + \text{N}_\alpha\equiv\text{N}_\beta \xrightarrow{k_{exch}} \text{Z-C}_6\text{H}_4\text{-}\overset{+}{\text{N}}_\beta\equiv\text{N}_\alpha \qquad (63)$$

$$\text{Z-C}_6\text{H}_4^+ \xrightarrow{k_{solv},\ \text{H}_2\text{O}} \text{Z-C}_6\text{H}_4\text{-OH}$$

More recently, Swain has questioned Brown and Drury's, and Maccoll's assumption that the β-nitrogen kinetic isotope effect is 1·000, and has determined both the α- and the β-nitrogen kinetic isotope effects in an elegant set of experiments[23]. Swain was able to obtain nitrogen gas from the α-nitrogen and from the β-nitrogen separately. The nitrogen gas from the α- and the β-nitrogens was obtained separately by converting the unreacted diazonium ion from a partial reaction into an azo compound with the disodium salt of 2-naphthol-3,6-disulphonic acid, and then reducing the azo compound to aniline and disodium 1-amino-2-naphthol-3,6-disulphonate (equation 64). The aniline containing the α-nitrogen was separated by steam distillation, converted into ammonium ion in a Kjeldahl digestion and then oxidized to nitrogen gas with sodium hypobromite in a vacuum line (equation 65). The β-nitrogen could be recovered by treating the 1-amino-2-naphthol-3,6-disulphonate ion in the same way.

$$\text{Ar}\overset{+}{\text{N}}\equiv\text{N} + \text{2-HO-3,6-}(\text{SO}_3^-)_2\text{C}_{10}\text{H}_5 \longrightarrow \text{1-(Ar-N}_\alpha\text{=N}_\beta\text{)-2-HO-3,6-}(\text{SO}_3^-)_2\text{C}_{10}\text{H}_4 \xrightarrow{\text{Na}_2\text{S}_2\text{O}_4} \text{ArN}_\alpha\text{H}_2 + \text{1-NH}_2\text{-2-HO-3,6-}(\text{SO}_3^-)_2\text{C}_{10}\text{H}_4 \qquad (64)$$

$$\text{ArNH}_2+\text{H}_2\text{SO}_4+\text{K}_2\text{SO}_4 \xrightarrow[\Delta]{\text{Hg}^{2+}} (\text{NH}_4)_2\text{SO}_4 \xrightarrow{\text{NaOBr}} \text{N}_2 \qquad (65)$$

In addition, he recovered the nitrogen gas released during the reaction, i.e. both the α- and the β-nitrogens together. This nitrogen was recovered when the benzenediazonium ion slightly enriched with 15-nitrogen at the β-nitrogen, was reacted part way to completion. The isotopic composition of all the nitrogen atoms in the starting material was obtained from a mass spectrometric analysis of the nitrogen gas from a reaction taken to completion.

Swain used the isotopic ratios obtained by the above procedures to show that the β-nitrogen kinetic isotope was $1{\cdot}0106 \pm 0{\cdot}0003$ and the α-nitrogen kinetic isotope was $1{\cdot}0384 \pm 0{\cdot}0010$ at 25 °C.

It is worth noting that Swain's experiments were internally consistent. For example, the same value of the β-nitrogen isotope effect was obtained whether the isotopic content of the nitrogen gas from the β-nitrogen was measured directly or obtained from the difference between the total nitrogen and the gas from the α-nitrogen. Swain also showed that his experiments led to the same isotope effects as those found by other investigators if the data were treated in the same way.

Finally, the α-nitrogen kinetic isotope effect of 1·0384 is near the theoretical maximum nitrogen isotope effect of 1·043 [28] and indicates that the $C{-}N_\alpha$ bond is almost completely broken in the transition state leading to the phenyl cation. The smaller β-nitrogen kinetic isotope effect of 1·0106 is similar to that observed in bond formation reactions[92] and probably results from the tightening of the $N_\alpha{\equiv}N_\beta$ triple bond at the transition state.† In fact, the $N_\alpha{\equiv}N_\beta$ bond frequency increases from 2298 cm^{-1} in the diazonium ion to 2331 cm^{-1} in the nitrogen gas.

C. Secondary β-Deuterium Kinetic Isotope Effects in Diazonium Salt Reactions

Strong evidence supporting the existence of the phenyl cation intermediate in the decomposition of arenediazonium ions in acidic solutions has recently been published by Swain and coworkers[7]. These authors argued that large secondary β-deuterium isotope effects characteristic of those found for S_N1 solvolysis reactions, i.e. k_H/k_D of 1·1–1·3[93, 94], should be observed in the decomposition of deuterated benzenediazonium ions if the phenyl cation intermediate is produced during the reaction. These isotope effects should be large because the dihedral angle between the *ortho* C—H bonds and the empty p orbital of the phenyl cation is zero degrees and orbital overlap should be at a maximum (equation 66).‡

$$C_6H_5{-}\overset{+}{N}{\equiv}N \longrightarrow C_6H_4(H(D))^{+} \longleftrightarrow C_6H_4{=}H^{+} \qquad (66)$$

In fact, the secondary β-hydrogen–deuterium kinetic isotope effects measured in these systems are large. The actual isotope effects observed for 12 different deuterated benzenediazonium ions showed that the isotope effect was $1{\cdot}22 \pm 0{\cdot}01$ for each *ortho* hydrogen, $1{\cdot}08 \pm 0{\cdot}01$ for each *meta* hydrogen and $1{\cdot}02 \pm 0{\cdot}01$ for a *para* hydrogen. For example, the calculated isotope effect for the decomposition of 2,4,6-trideuterobenzenediazonium ion should be $(1{\cdot}22)^2\,(1{\cdot}08)^1 = 1{\cdot}52$. The observed value was $1{\cdot}52 \pm 0{\cdot}03$.

† The temperature-dependent factor (the terms related to the change in vibrational frequencies) can have a value less than unity in a reaction where a bond is strengthened in going to the transition state. The temperature-independent factor (the term representing the reduced mass effect), on the other hand, is always greater than unity and thus a small isotope effect is observed.

‡ The carbons of the benzene ring are in the plane of the paper with the cloud above and below this plane. The π bond formed by hyperconjugation is in the plane of the paper and perpendicular to the π cloud of the benzene ring.

These isotope effects remained constant under several different experimental conditions, i.e. in H_2SO_4–water mixtures of various concentrations, in acetic acid, in trifluoroethanol† and in methylene chloride. Consequently, the authors concluded that the diazonium ion must decompose by the same mechanism, i.e. through the phenyl cation intermediate, in both protic and aprotic solvents.

The isotope effects for the *ortho* hydrogens, i.e. on the carbon beta to the carbon bearing the positive charge, are the largest that have been observed in aromatic systems. Swain, on the basis of both theoretical calculations and experimental results, has concluded that these isotope effects are a result of hyperconjugation between the positive carbon and the *ortho* hydrogens, and to a lesser extent the *meta* and *para* hydrogens. Hyperconjugation is particularly important in these reactions because it is the only means of stabilizing the carbonium ion. This occurs because the positively charged sp^2 orbital of the phenyl cation is perpendicular to the π cloud of the benzene ring and cannot be delocalized by the ring in the usual manner.

V. ACKNOWLEDGEMENTS

The authors wish to thank the National Research Council of Canada and President E. J. Monahan of Laurentian University for the financial support required to complete this chapter. The authors also wish to thank Professor W. H. Saunders, Jr for granting permission to use Figure 1.

VI. REFERENCES

1. E. B. Starkey, *Organic Synthesis, Collective Vol.* 2 (Ed. A. H. Blatt), John Wiley, New York, 1943, p. 225.
2. E. S. Lewis, J. L. Kinsey and R. R. Johnson, *J. Amer. Chem. Soc.*, **78**, 4294 (1956).
3. J. B. Hendrickson, D. J. Cram and G. S. Hammond, *Organic Chemistry*, McGraw-Hill, New York, 1970, p. 707.
4. P. C. Buxton and H. Heaney, *J. Chem. Soc. Chem. Commun.*, 545 (1973).
5. J. I. G. Cadogan, C. D. Murray and J. T. Sharp, *J. Chem. Soc. Chem. Commun.*, 133 (1974).
6. R. Harrison, H. Heaney, J. M. Jablonski, K. G. Mason and J. M. Sketchley, *J. Chem. Soc. (C)*, 1684 (1969).
7. C. G. Swain, J. E. Sheats, D. G. Gorenstein and K. G. Harbison, *J. Amer. Chem. Soc.*, **97**, 791 (1975).
8. R. G. Bergstrom, R. G. M. Landells, G. H. Wahl, Jr and H. Zollinger, *J. Amer. Chem. Soc.*, **98**, 3301 (1976).
9. W. G. Brown and J. L. Garnett, *J. Amer. Chem. Soc.*, **80**, 5272 (1958).
10. R. W. Franck and K. Yanagi, *J. Amer. Chem. Soc.*, **90**, 5814 (1968).
11. L. C. Leitch, P. E. Gagnon and A. Cambron, *Can. J. Res.*, **28B**, 256 (1950).
12. W. B. DeMore, H. O. Pritchard and N. Davison, *J. Amer. Chem. Soc.*, **81**, 5874 (1959).
13. G. W. Robinson and M. MacCarty, Jr, *J. Amer. Chem. Soc.*, **82**, 1859 (1960).
14. T. D. Goldfarb and G. C. Pimentel, *J. Amer. Chem. Soc.*, **82**, 1865 (1960).
15. K. J. van der Merwe, P. S. Steyn and S. H. Eggers, *Tetrahedron Letters*, 3923 (1964).
16. S. M. Hecht and J. W. Kozarich, *Tetrahedron Letters*, 1501 (1972).
17. H. A. Morrison and P. Yates, *Chem. and Industry*, 931 (1962).
18. A. Santucci, A. Foffani and G. Piazza, *Chem. Commun.*, 1262 (1969).
19. P. F. Holt and B. I. Bullock, *J. Chem. Soc.*, 2310 (1950).
20. E. S. Lewis and P. G. Kotcher, *Tetrahedron*, **25**, 4873 (1969).
21. J. M. Insole and E. S. Lewis, *J. Amer. Chem. Soc.*, **85**, 122 (1963).
22. E. S. Lewis and J. M. Insole, *J. Amer. Chem. Soc.*, **86**, 32 (1964).

† The β-hydrogen–deuterium isotope effect observed by Zollinger[8] in trifluoroethanol is in good agreement with the value calculated from Swain's data.

23. C. G. Swain, J. E. Sheats and K. G. Harbison, *J. Amer. Chem. Soc.*, **97**, 796 (1975).
24. P. F. Holt, B. I. Hopson-Hill and C. J. McNae, *J. Chem. Soc.*, 2245 (1960).
25. E. S. Lewis and R. E. Holliday, *J. Amer. Chem. Soc.*, **91**, 426 (1969).
26. I. Ugi, H. Perlinger and L. Behringer, *Chem. Ber.*, **92**, 1864 (1959).
27. A. K. Bose and I. Kugajevsky, *J. Amer. Chem. Soc.*, **88**, 2325 (1966).
28. L. L. Brown and J. S. Drury, *J. Chem. Phys.*, **43**, 1688 (1965).
29. K. Clusius and M. Vecchi, *Helv. Chim. Acta*, **39**, 1469 (1956).
30. R. G. Bergstrom, G. H. Wahl, Jr and H. Zollinger, *Tetrahedron Letters*, 2975 (1974).
31. N. N. Bubnov, K. A. Bilevitch, L. A. Poljakova and O. Yu. Okhlobystin, *J. Chem. Soc. Chem. Commun.*, 1058 (1972).
32. J. Firl and W. Runge, *Angew. Chem. Int. Ed.*, **12**, 668 (1973).
33. J. Firl, W. Runge and W. Hartmann, *Angew. Chem. Int. Ed.*, **13**, 270 (1974).
34. K. Clusius and H. Hürzeler, *Helv. Chim. Acta*, **37**, 798 (1954).
35. R. Huisgen and I. Ugi, *Chem. Ber.*, **90**, 2914 (1957).
36. C. D. Ritchie and D. J. Wright, *J. Amer. Chem. Soc.*, **93**, 6574 (1971).
37. C. D. Ritchie, *J. Amer. Chem. Soc.*, **93**, 2429 (1971).
38. K. Clusius and H. R. Weisser, *Helv. Chim. Acta*, **35**, 1524 (1952).
39. K. Clusius and H. Craubner, *Helv. Chim. Acta*, **38**, 1060 (1955).
40. K. Clusius and H. Hürzeler, *Helv. Chim. Acta*, **37**, 383 (1954).
41. G. A. Swan and P. Kelly, *J. Chem. Soc.*, 416 (1954).
42. K. Clusius and H. Hürzeler, *Helv. Chim. Acta*, **38**, 1831 (1955).
43. P. F. Holt and C. J. McNae, *J. Chem. Soc.*, 1825 (1961).
44. E. S. Lewis and J. E. Cooper, *J. Amer. Chem. Soc.*, **84**, 3847 (1962).
45. W. A. Waters, *J. Chem. Soc.*, 266 (1942).
46. E. S. Lewis and R. E. Holliday, *J. Amer. Chem. Soc.*, **88**, 5043 (1966).
47. E. S. Lewis, R. E. Holliday and L. V. Hartung, *J. Amer. Chem. Soc.*, **91**, 430 (1969).
48. J. G. Calvert and J. N. Pitts, Jr, *Photochemistry*, John Wiley, New York, 1966, p. 471.
49. W. Kirmse, W. J. Baron and U. Seipp, *Angew. Chem. Int. Ed.*, **12**, 924 (1973).
50. W. R. Bamford and T. S. Stevens, *J. Chem. Soc.*, 4735 (1952).
51. L. Friedman and H. Shechter, *J. Amer. Chem. Soc.*, **81**, 5512 (1959).
52. J. W. Powell and M. C. Whiting, *Tetrahedron*, **7**, 305 (1959).
53. J. H. Bayless, L. Friedman, F. B. Cook and H. Shechter, *J. Amer. Chem. Soc.*, **90**, 531 (1968).
54. A. Nickon and N. H. Werstiuk, *J. Amer. Chem. Soc.*, **88**, 4543 (1966).
55. H. C. Brown and H. R. Deck, *J. Amer. Chem. Soc.*, **87**, 5620 (1965).
56. G. L. Closs, R. A. Moss and S. H. Goh, *J. Amer. Chem. Soc.*, **88**, 364 (1966).
57. Y. Yamamoto and I. Moritani, *Tetrahedron Letters*, 3087 (1969).
58. H. Dahn and M. Ballenegger, *Helv. Chim. Acta*, **52**, 2417 (1969).
59. R. A. More O'Ferrall, *Advan. Phys. Org. Chem.*, **5**, 331 (1967).
60. C. Wentrup and H. Dahn, *Helv. Chim. Acta*, **53**, 1637 (1970).
61. J. Bayless and L. Friedman, *J. Amer. Chem. Soc.*, **89**, 147 (1967).
62. J. F. Bunnett and H. Takayama, *J. Org. Chem.*, **33**, 1924 (1968).
63. J. F. Bunnett, D. A. R. Happer and H. Takayama, *Chem. Commun.*, 367 (1967).
64. J. F. Bunnett and H. Takayama, *J. Amer. Chem. Soc.*, **90**, 5173 (1968).
65. R. O. C. Norman and B. C. Gilbert, *Advan. Phys. Org. Chem.*, **5**, 74 (1967).
66. C. D. Ritchie, J. D. Saltiel and E. S. Lewis, *J. Amer. Chem. Soc.*, **83**, 4601 (1961).
67. J. I. G. Cadogan and P. G. Hibbert, *Proc. Chem. Soc.*, 338 (1964).
68. C. G. Swain, J. E. Sheats and K. G. Harbison, *J. Amer. Chem. Soc.*, **97**, 783 (1975).
69. E. H. White, R. H. McGirk, C. A. Aufdermarsh, Jr, H. P. Tiwari and M. J. Todd, *J. Amer. Chem. Soc.*, **95**, 8107 (1973).
70. E. H. White and J. E. Stuber, *J. Amer. Chem. Soc.*, **85**, 2168 (1963).
71. J. Bigeleisen, *Proc. Int. Symposium on Isotope Separation*, North Holland, Amsterdam (1958).
72. J. Bigeleisen and M. Wolfsberg, *Advan. Chem. Phys.*, **1**, 15 (1958).
73. J. Bigeleisen, *J. Chem. Phys.*, **17**, 675 (1949).
74. S. Glasstone, K. J. Laidler and H. Eyring, *The Theory of Rate Processes*, McGraw-Hill, New York, 1941.

75. H. Eyring, J. Walter and G. E. Kimbal, *Quantum Chemistry*, John Wiley, New York, 1944, Chapter XVI.
76. J. Bigeleisen and M. Goeppert-Mayer, *J. Chem. Phys.*, **15**, 261 (1947).
77. W. H. Saunders, Jr, *Chemica Scripta*, **8**, 27 (1975).
78. L. B. Sims, A. Fry, L. T. Netherton, J. C. Wilson, K. D. Reppond and W. S. Cook, *J. Amer. Chem. Soc.*, **94**, 1364 (1972).
79. F. H. Westheimer, *Chem. Rev.*, **61**, 265 (1961).
80. R. A. More O'Ferrall, *J. Chem. Soc.* (*B*), 785 (1970).
81. J. C. Evans and G. Y.-S. Lo, *J. Amer. Chem. Soc.*, **88**, 2118 (1966).
82. C. J. Collins, Jr and N. S. Bowman, *Isotope Effects in Chemical Reactions*, Van Nostrand–Reinhold, New York, 1970, pp. 122–150.
83. C. J. Collins, Jr and N. S. Bowman, *Isotope Effects in Chemical Reactions*, Van Nostrand–Reinhold, New York, 1970, p. 138.
84. V. J. Shiner, Jr and J. S. Humphrey, Jr, *J. Amer. Chem. Soc.*, **85**, 2416 (1963).
85. V. J. Shiner, Jr and J. G. Jewett, *J. Amer. Chem. Soc.*, **87**, 1382 (1965).
86. E. S. Lewis and J. M. Insole, *J. Amer. Chem. Soc.*, **86**, 34 (1964).
87. A. N. Bourns and E. Buncel, *Can. J. Chem.*, **38**, 2457 (1960).
88. K. C. Westaway and R. A. Poirier, *Can. J. Chem.*, **53**, 3216 (1975).
89. W. H. Saunders, Jr, *Chem. Ind.*, 1661 (1963).
90. A. G. Loudon, A. Maccoll and D. Smith, *J. Chem. Soc. Faraday*, **69**, 899 (1973).
91. D. G. Graczyk and J. W. Taylor, *J. Amer Chem. Soc.*, **96**, 3255 (1974).
92. E. R. Hayes, *Ph.D. Thesis*, McMaster University, Hamilton, Ontario, Canada, 1958.
93. E. S. Lewis and C. E. Boozer, *J. Amer. Chem. Soc.*, **76**, 791 (1954).
94. E. S. Lewis and C. E. Boozer, *J. Amer. Chem. Soc.*, **76**, 794 (1954).

CHAPTER **17**

Carbonyl, phosphoryl and sulphonyl diazo compounds

M. REGITZ

Department of Chemistry, University of Kaiserslautern, D-6750 Kaiserslautern, Federal Republic of Germany

---

I. INTRODUCTION

Ethyl diazoacetate is without doubt the best-known representative of diazoalkanes bearing electron-acceptor substituents, which are the subject of this survey. Its synthesis by amine diazotization[1] was the starting point of a continuing development[2–4], the end of which is not yet in sight, since not only the diazo group but also the substituents attached to the diazo carbon atom participate to a greater or lesser degree in the reactions of the compound. As electron-acceptor substituents they retard reactions based on the nucleophilic character of the diazo C atom and of course promote those in which the terminal electrophilic nitrogen plays a decisive role. Hence one would expect C-protonation to be slow in comparison to that of non-acylated diazoalkanes, or even not to occur at all, and indeed their stability to acids is greater than in the case of the non-acylated compounds. Disregarding nitrodiazoalkanes and cyanodiazoalkanes, the other CO-, PO- and SO_2-substituted diazo compounds considered below show little tendency to undergo spontaneous explosive decomposition.

II. SYNTHETIC METHODS

The following account is based on the general survey of synthetic approaches for diazoalkanes presented in another chapter in this volume and on the secondary and alternative reactions also reported therein; they will be supplemented in specific cases. In contrast to the general survey, however, the acyl cleavage of *N*-alkyl-*N*-nitroso-amides is practically without significance in the present context; the field is clearly dominated by the dehydrogenation of hydrazones, the Bamford–Stevens reaction and by diazo group transfer, a method which came to the forefront only 10 years ago, as well as by substitution reactions.

A. Carbonyl Diazo Compounds

The historical significance of ethyl diazoacetate has already been mentioned in the introduction. This class of compounds has also acquired considerable importance as synthetic aids; a correspondingly large number of carbonyl diazo compounds have been synthesized and we can only cover a representative selection in this section.

1. Diazotization of amines

Diazotization of glycine ester hydrochloride to give ethyl diazoacetate by Curtius[1] (equation 1) was subsequently optimized owing to its synthetic importance[5–10].

$$C_2H_5OOC{-}CH_2{-}NH_3^+\,X^- \xrightarrow[-NaX,\ -2\,H_2O]{NaNO_2} C_2H_5OOC{-}CH{=}N_2 \qquad (1)$$

Amine diazotization was already applied to the peptide series by Curtius himself[11–13], and recently also utilized in the production of D- and L-2-amino-3-diazoacetoxypropanoic acid (azaserine)[14, 15].

Only poor yields are obtained in the synthesis of homologues of ethyl diazoacetate[16], as well as diethyl diazosuccinate[17–19] and diazoglutaric diesters[19]. Diazotization of α-amino carboxylic esters with isoamyl nitrite in the presence of up to 30% of acetic acid fulfils a real synthetic need (equation 2)[20].

$$R^1{-}\underset{\displaystyle NH_2}{\underset{|}{C}H}{-}COOR^2 \xrightarrow[CH_3COOH]{i\text{-}C_5H_{11}ONO} R^1{-}\underset{\displaystyle N_2}{\underset{\|}{C}}{-}COOR^2 \qquad (2)$$

$R^1 = CH_3$, $R^2 = CH_2C_6H_5$ (61%)
$R^1 = (CH_3)_2CHCH_2$, $R^2 = C_2H_5$ (69%)
$R^1 = CH_3SCH_2CH_2$, $R^2 = CH_3$ (64%)

The successful double diazotization of aminoacetohydrazide is to be regarded as a curiosity: it affords the highly interesting compound diazoacetyl azide (**1**) which is very stable, in spite of the cumulation of the diazo and the azide group[21].

$$H_2N{-}NH{-}C(=O){-}CH_2{-}NH_2 \xrightarrow[CH_2Cl_2/H_2O,\ -5\,^{\circ}C]{NaNO_2/HCl} N_3{-}C(=O){-}CH{=}N_2$$

(**1**)
10–15%

During the production of α-diazo ketones, amine diazotization is sometimes followed by deamination and rearrangement[22, 23], as for example on attempted synthesis of 2-diazocyclohexanone by this method (equation 3)[22]. However, other

$$\text{2-oxocyclohexyl-}\overset{+}{N}H_3\ Cl^- \xrightarrow{NaNO_2} \text{2-oxocyclohexyl-}N_2^+\ Cl^- \xrightarrow{H_2O} \text{cyclopentyl-}COOH\ (57\%) \qquad (3)$$

representatives of the same class such as 3-diazocamphor[24], ω-diazoacetophenone[25] or benzyl 6-diazopenicillinate (**2**)[26] are accessible by diazotization of corresponding amines.

$$\text{benzyl 6-aminopenicillinate} \xrightarrow{NaNO_2,\ acetone/water} \text{benzyl 6-diazopenicillinate}$$

(**2**)

Closely related to α-diazo carbonyl compounds are *o*- and *p*-quinone diazides; according to the principle of vinylene homology the latter are comparable with the *ortho* isomers. Pertinent diazonium salts occur as intermediates according to equations (4) and (5)[27, 28].

Formation of quinone diazides from amino-substituted aromatic compounds is sometimes associated with secondary reactions such as subsitution and oxidation. The former type includes the transformation of 9-amino-10-nitrophenanthrene

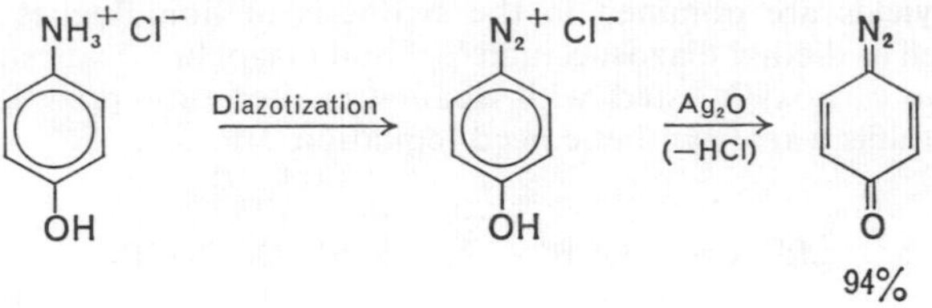

(4)

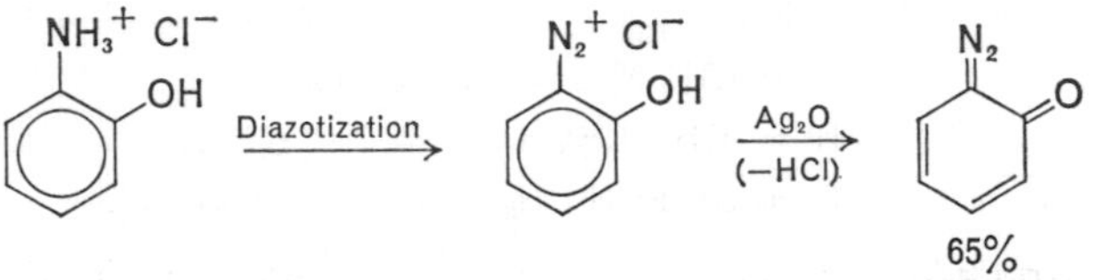

94%

(5)

65%

into phenanthrenequinone diazide (equation 6)[29] and also the formation of 3,4,5-trifluoro-6-nitro-1,2-benzoquinone from 2,3,4,5-tetrafluoro-6-nitroaniline (equation 7)[30].

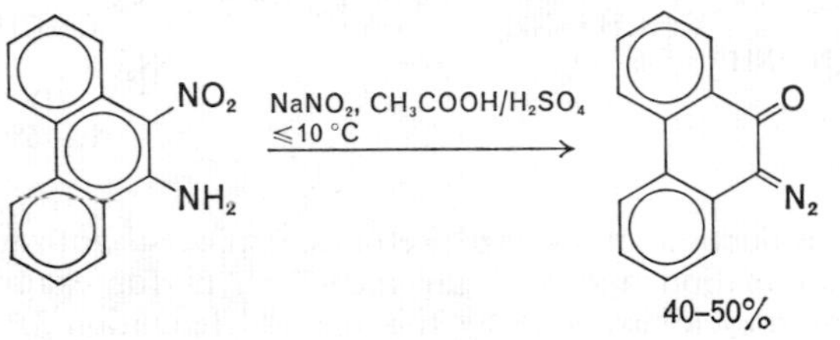

$NaNO_2$, CH_3COOH/H_2SO_4 ≤10 °C

(6)

40–50%

$NaNO_2$, H_2SO_4 −12 to −15 °C

−HF, −HX

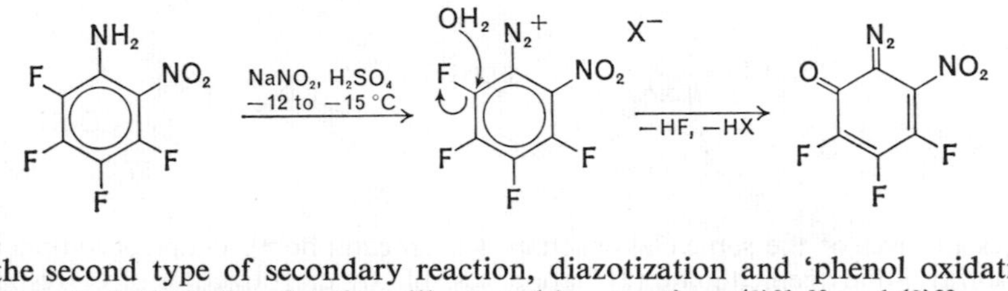

(7)

In the second type of secondary reaction, diazotization and 'phenol oxidation' are seen to be intimately related, as illustrated by equations (8)[31, 32] and (9)[33].

2-Diazo 1,3-dicarbonyl compounds (**4**) are in general readily accessible by amine diazotization because the deprotonation **3** → **4** is further facilitated by the second

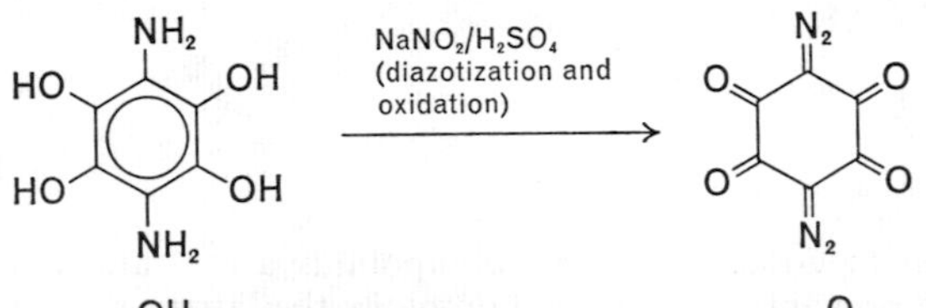

$NaNO_2/H_2SO_4$ (diazotization and oxidation)

(8)

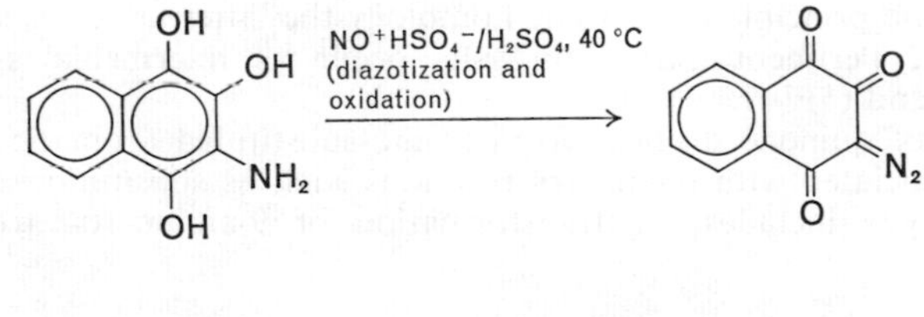

$NO^+HSO_4^-/H_2SO_4$, 40 °C (diazotization and oxidation)

(9)

acyl group, and their resistance to acids is further enhanced by electron delocalization according to **5**.

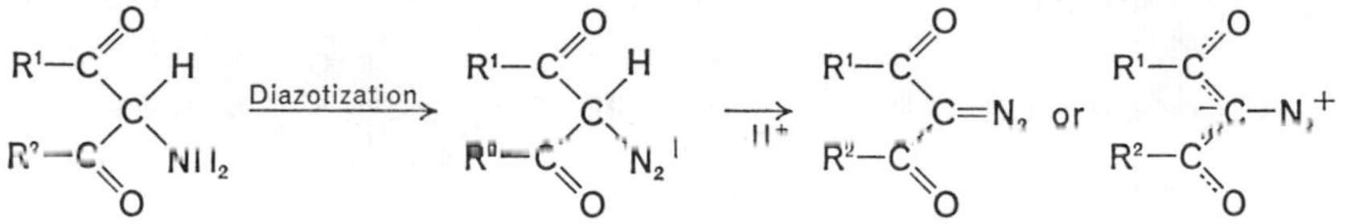

(3) (4) (5)

Diethyl diazomalonate[34, 35], ethyl diazoacetate[36] and 3-diazo-2,4-pentanedione[36, 37] can be viewed as historic examples; the parent substance, diazomalondialdehyde, has only recently become available[38]. Diazo derivatives of cyclic *trans*-fixed β-diketones such as 2-diazocyclohexane-1,3-dione[39], 2-diazo-5,5-dimethylcyclohexane-1,3-dione[40] and 3-diazospiro[5.5]undecane-2,4-dione[41] have likewise only been synthesized in the comparatively recent past.

Table 1 lists other α-diazo carbonyl derivatives, quinone diazides and α-diazo β-carbonyl derivatives which have been synthesized.

2. Forster reaction

In general, the production of carbonyl diazo compounds by the Forster reaction proceeds via initial oximation of the methylene components of the diazo products and subsequent reaction with chloramine solution. A typical reaction sequence is encountered in the conversion of 3,3-diphenyl-1-indanone via the oxime (**6**) into 2-diazo-3,3-diphenyl-1-indanone (**7**)[56]. A further aspect of this reaction is also of

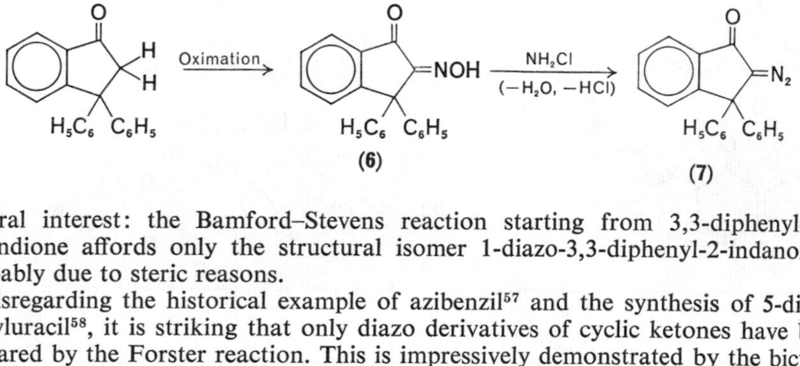

(6) (7)

general interest: the Bamford–Stevens reaction starting from 3,3-diphenyl-1,2-indandione affords only the structural isomer 1-diazo-3,3-diphenyl-2-indanone[56], probably due to steric reasons.

Disregarding the historical example of azibenzil[57] and the synthesis of 5-diazoacetyluracil[58], it is striking that only diazo derivatives of cyclic ketones have been prepared by the Forster reaction. This is impressively demonstrated by the bicyclic derivatives **8** (60%)[57], **9** (85%)[59] and **10** (67%)[60], and by the aromatic derivatives **11** (80%)[61], **12** (57%)[62], **13** (65%)[62] and **14** (66%)[63].

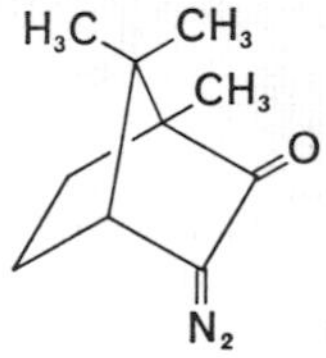

(8) 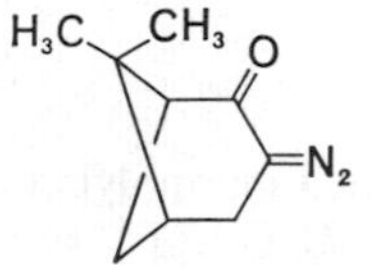 (9) 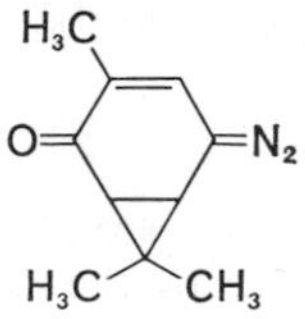 (10)

TABLE 1. Diazo carbonyl compounds by amine diazotization

Structure	Yield (%)	Reference	Structure	Yield (%)	Reference
$N_2=CH-COOCH_2-CH=CH_2$	72	42	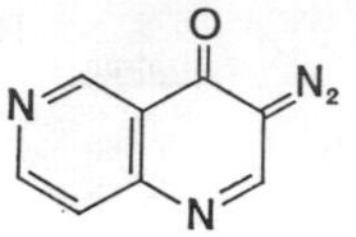	73	50
$N_2=CH-COOC(CH_3)_3$	78	43			
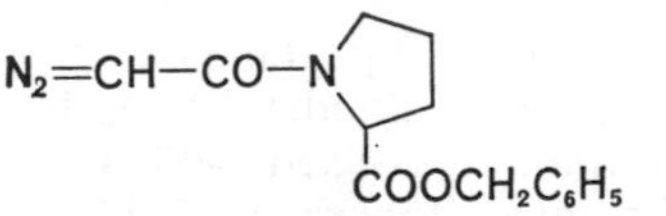	22	44	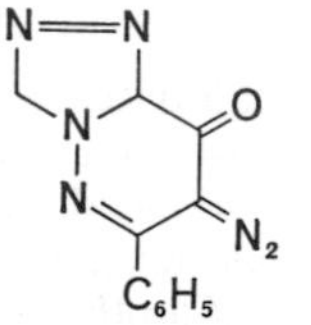	83	51
$CH_3-(CH_2)_3-C(=N_2)-COOC_2H_5$	30	45	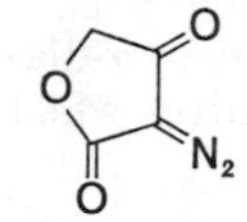	75	52
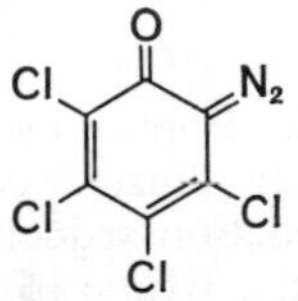	90	46	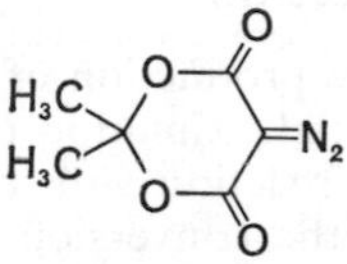	47	53
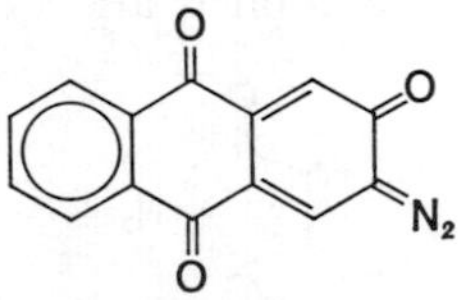	100	47	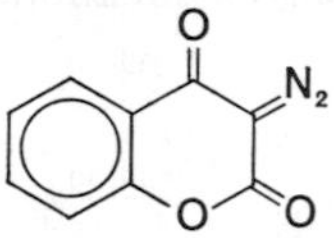	72	54
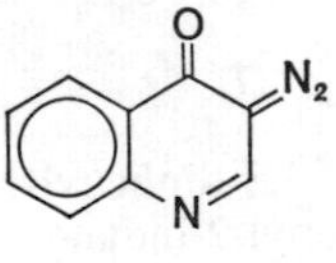	92	48, 49	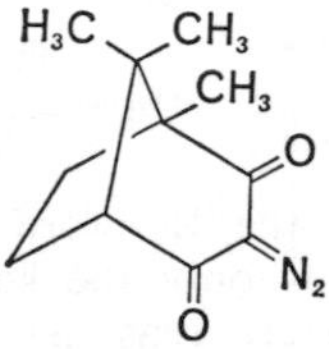	86	55

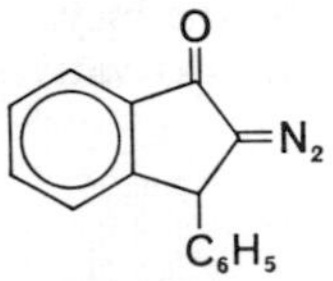

(11)

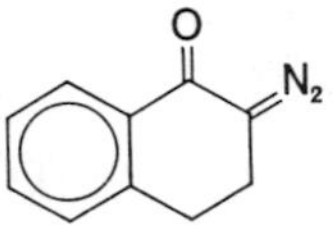

(12)

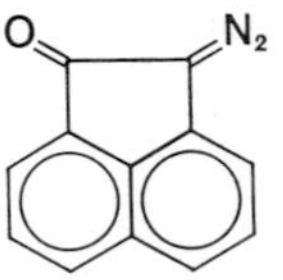

(13)

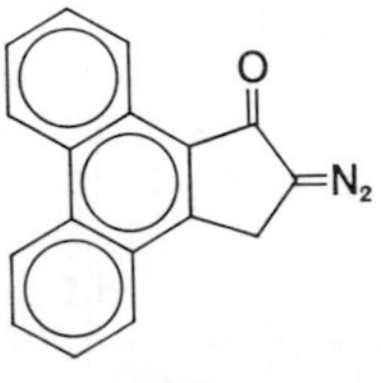

(14)

Excellent results are also obtained on applying the Forster reaction to the steroid series; as shown by the examples **15** (72%)[64], **16** (80%)[65], **17** (68%)[66], **18** (75%)[67, 68], **19** (77%)[69] and **20** (81%)[70], the diazo group can be incorporated into ring A or into the five-membered ring.

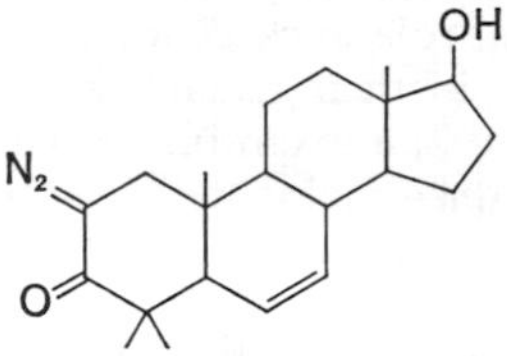

(15) 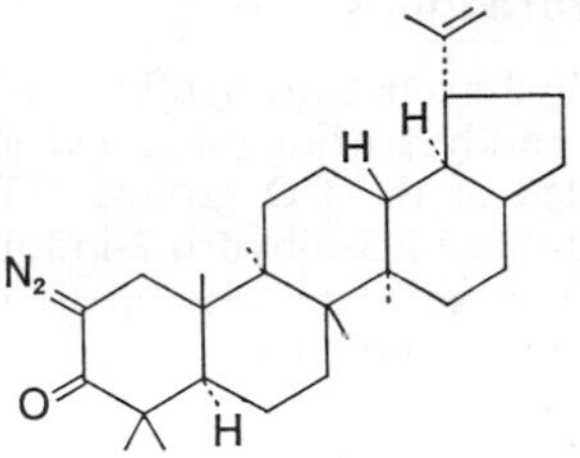 (16) 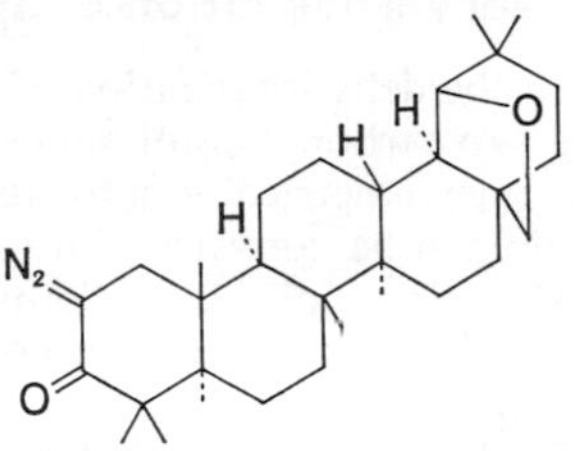 (17)

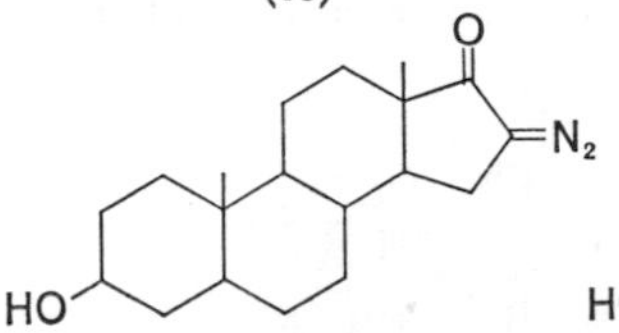

(18) 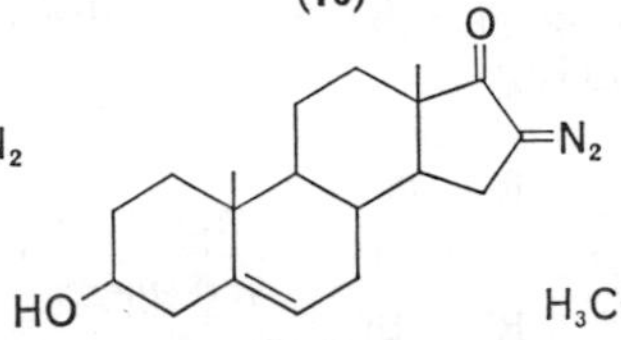 (19) 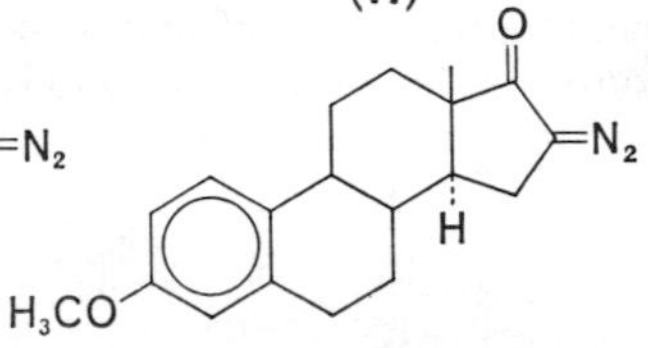 (20)

In the special case of the oxime **(21)**, condensation with phenylhydrazine gives 3-diazo-2,4-chromandione **(23)**[71]; assumption of the intermediacy of the triaza compound **(22)** does not appear unjustified; decomposition of **22** into aniline and the diazo compound would then display a certain analogy to diazo group transfer.

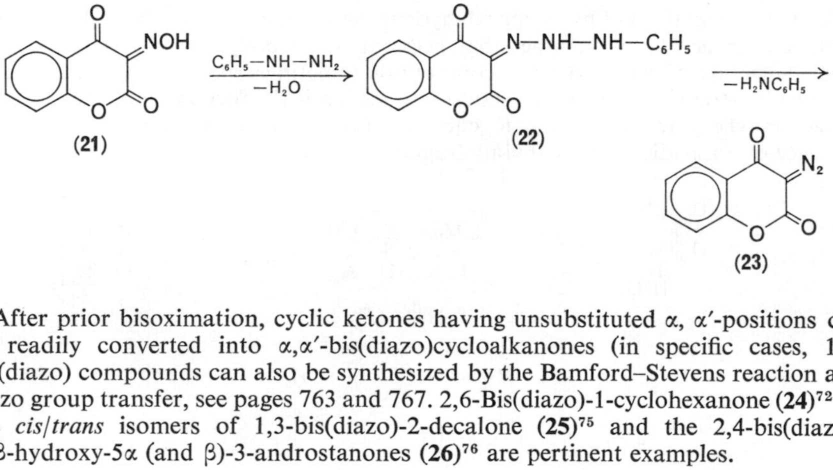

$$\textbf{(21)} \xrightarrow[-H_2O]{C_6H_5-NH-NH_2} \textbf{(22)} \xrightarrow{-H_2NC_6H_5} \textbf{(23)}$$

After prior bisoximation, cyclic ketones having unsubstituted α, α'-positions can be readily converted into α,α'-bis(diazo)cycloalkanones (in specific cases, 1,3-bis(diazo) compounds can also be synthesized by the Bamford–Stevens reaction and diazo group transfer, see pages 763 and 767. 2,6-Bis(diazo)-1-cyclohexanone **(24)**[72–74], the *cis*/*trans* isomers of 1,3-bis(diazo)-2-decalone **(25)**[75] and the 2,4-bis(diazo)-17β-hydroxy-5α (and β)-3-androstanones **(26)**[76] are pertinent examples.

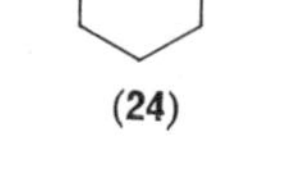

(24) 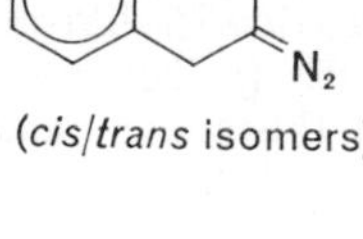 (25) (*cis*/*trans* isomers) 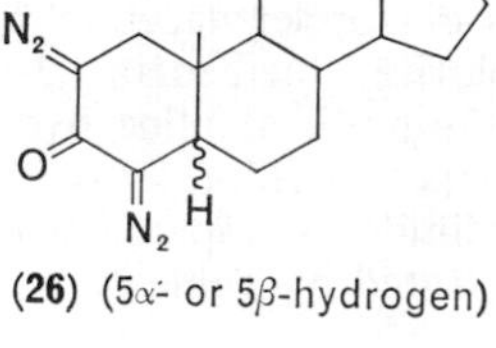 (26) (5α- or 5β-hydrogen)

3. Dehydrogenation of hydrazones

In the dehydrogenation of hydrazones the synthesis of the monohydrazones from α-diazo carbonyl compounds and hydrazine plays a crucial role in cases where they are unsymmetrically substituted or the CO groups differ in their reactivity. Thus isatin can be converted into 3-diazo-2,3-dihydro-2-indolone[77], alloxan into 5-diazobarbituric acid[78] and diethyl mesoxalate into diethyl diazomalonate[79], without formation of isomers. In contrast, both possible isomers are formed on reaction of phenyl-2-pyridyl-1,2-ethanedione with hydrazine[80]. Such problems are circumvented by use of α-methylene ketones as starting compounds; bromination and treatment with 3 moles of hydrazine give the desired hydrazones. The proposed reaction course is shown in equation (10)[81, 82].

$$R^1C(=O)CH_2R^2 \xrightarrow[-HBr]{Br_2} R^1C(=O)CHBrR^2 \xrightarrow[-H_2N{-}\overset{+}{N}H_3\,Br^-]{2\,H_2N{-}NH_2} R^1C(=O)CH(NH{-}NH_2)R^2 \rightleftharpoons R^1C(OH){=}C(R^2)NH{-}NH_2 \xrightarrow{-NH_3} R^1C(=O)C(=NH)R^2 \xrightarrow[-NH_3]{H_2N{-}NH_2} R^1C(=O)C(=N{-}NH_2)R^2 \qquad (10)$$

Secondary reactions of hydrazone dehydrogenation have already been described in another chapter of this volume; that account will merely be supplemented by a specific reaction of carbonyl diazo compounds sometimes observed on dehydrogenation with mercuric oxide. The reaction in question is the formation of mercuriobis-(diazomethylketones) according to equation (11)[83], which is dependent upon the presence of an acidic diazomethyl hydrogen.

$$2\,Mes{-}\underset{\displaystyle O}{\underset{\|}{C}}{-}\underset{\displaystyle N{-}NH_2}{\underset{\|}{CH}} \xrightarrow[-2\,Hg,\,-2\,H_2O]{2\,HgO} 2\,Mes{-}\underset{\displaystyle O}{\underset{\|}{C}}{-}\underset{\displaystyle N_2}{\underset{\|}{CH}} \xrightarrow[-H_2O]{HgO} \left[Mes{-}\underset{\displaystyle O}{\underset{\|}{C}}{-}\underset{\displaystyle N_2}{\underset{\|}{C}}\right]_2 Hg \qquad (11)$$

$$Mes = 2,4,6\text{-}(CH_3)_3C_6H_2{-}$$

As for non-carbonylated diazo compounds, the principal dehydrogenation reagents are mercuric oxide, manganese dioxide and silver oxide (see Table 2). Apart from these reagents, use has also been made in individual cases of lead tetraacetate (methyl diazophenylacetate)[84], nickel peroxide[85] (diethyl diazomalonate)[86], calcium hypochlorite (azibenzil, 3-diazocamphor, 2-diazo-1,5,5-trimethylbicyclo[2.2.1]-heptan-3-one)[87], alkaline hydrogen peroxide (5-diazobarbituric acid)[78] and oxygen with copper(II) chloride/pyridine catalysis (azibenzil)[88]. However, their significance is very limited compared with that of the above-mentioned metal oxides, as also demonstrated by Table 2.

TABLE 2. α-Diazo carbonyl compounds by hydrazone dehydrogenation

α-Diazo carbonyl compound	Dehydrogenation reagent/ Solvent	Yield (%)	Reference
$CH_3-C(=O)-C(=N_2)-CH_3$	MnO_2/chloroform	90–100	87
	Ag_2O/ether	76–79	89
$H_3C-C_6H_4-C(=O)-CH(=N_2)$	HgO/petroleum ether[a]	75	90
	MnO_2/chloroform	100	87
$C_6H_5-C(=O)-C(=N_2)-C_6H_5$	HgO/ether[a]	87–94	91
	MnO_2/chloroform	90–100	87
$C_5H_4N-C(=O)-C(=N_2)-C_6H_5$ (2-pyridyl)	HgO/ether[a]	80	80
	MnO_2/chloroform	80	80
2-diazoacenaphthen-1-one (N_2, O)	HgO/benzene	—	92
diazocamphor (H_3C, CH_3, CH_3, O, N_2)	HgO/benzene	81	93
	MnO_2/chloroform	90–100	87
C_6H_4 bridged by $-(CH_2)_4-C(=O)-C(=N_2)-(CH_2)_3-$	MnO_2/ether–dioxane[a]	—	94

[a] With addition of KOH as catalyst.

4. Bamford–Stevens reaction

The basis and possible secondary reactions of the Bamford–Stevens reaction have already been considered in another chapter; in the case of α-diazo carbonyl compounds only 1,5-cyclization is of importance. For instance, reaction of phenyl-2-pyridyl-1,2-ethanedione with tosyl hydrazide gives the triazolopyridine (**29**), there being no doubt as to the intermediacy of the diazo ketone (**27**) which preferentially aromatizes via the above-mentioned ring closure **27** → **29**[80].

The reaction also affords the tosylhydrazone (**28**) (as the hydrate) which is cleaved with aqueous sodium hydroxide in the usual way to give (**30**)[80]. Thus the same questions arise as in the transformation of α-diketones into α-diazo ketones by treatment with hydrazine and dehydrogenation of the hydrazones.

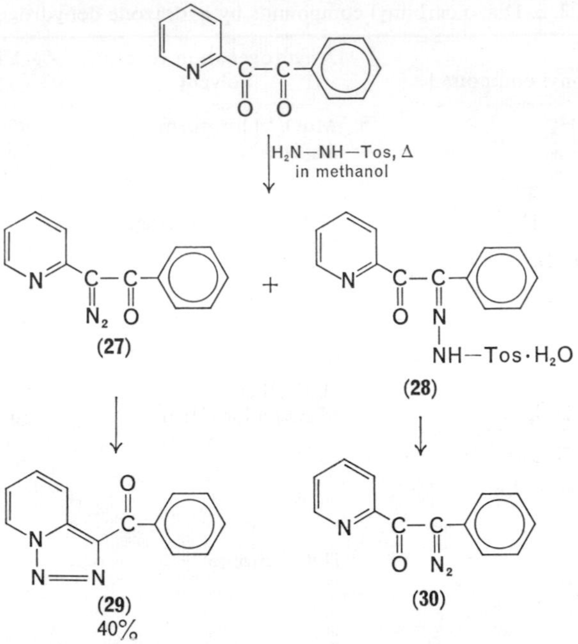

H_2N—NH—Tos, Δ
in methanol

(27) + (28)

N_2 O — NH—Tos·H_2O

(29) 40% (30)

The synthesis of α-diazo carbonyl compounds can be accomplished under much milder conditions than that of the non-acylated diazoalkanes. In particular, the cleavage of tosylhydrazones with basic alumina illustrates the difference in drastic manner[95]. Preparation of condensed quinone diazides by the title method requires no addition of base[56, 96, 97], as demonstrated by the transformation of phenanthrenequinone into phenanthrenequinone diazide with tosylhydrazine[56, 96]. Table 3 shows the Bamford–Stevens reaction to be a highly versatile method for synthesis of α-diazo carbonyl compounds and quinone diazides, whose only serious competitor at the present state of the art is diazo group transfer.

TABLE 3. α-Diazo carbonyl compounds by Bamford–Stevens reaction

α-Diazo carbonyl compound with yield, reaction conditions and reference	
R—C(=O)—C(=N_2)—R (R = CH_3, 86%) (R = C_6H_5, 84%) CH_2Cl_2/Al_2O_3/20 °C [95]	(65%) H_5C_6 C_6H_5 CH_2Cl_2/NaOH/20 °C [56]
$(CH_2)_{n-2}$ (n = 6–9, n = 6 : 75%) CH_2Cl_2 or petroleum ether/NaOH[39, 98]	(73%) CH_2Cl_2/NaOH/20 °C [56]

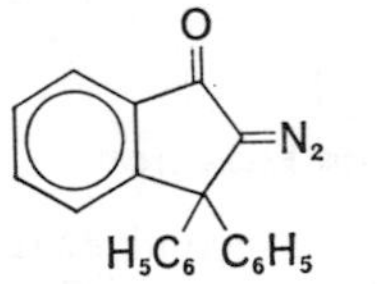

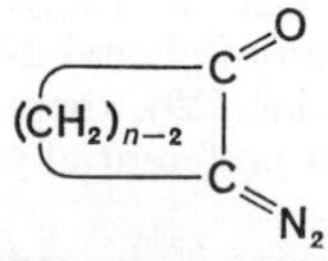

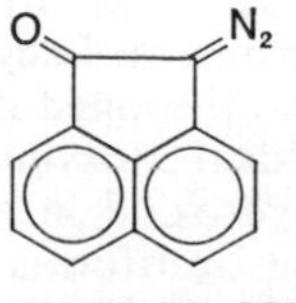

TABLE 3. (*cont.*)

α-Diazo carbonyl compound with yield, reaction conditions and reference
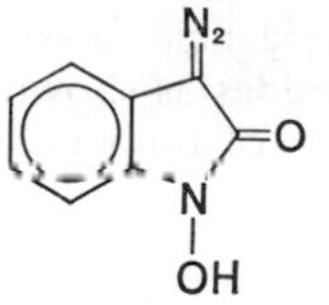 (79–86%) —/NaOH/50 °C [100]
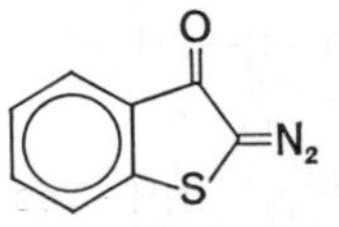 (75%) CH_2Cl_2/NaOH/20 °C [101]
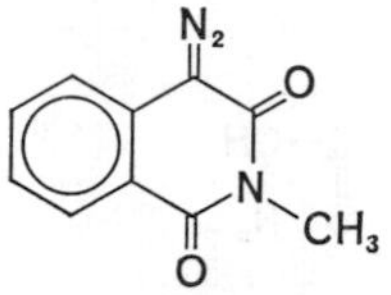 (90%) CH_2Cl_2/Al_2O_3/20 °C [95]
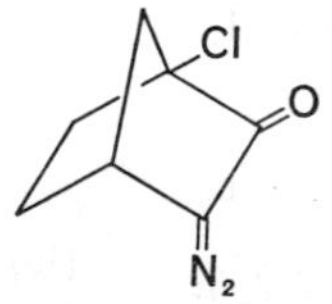 (59%) Pentane/NaOH/20 °C [102]
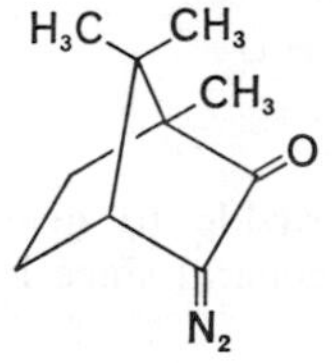 (76%[56], 99%[95]) Petroleum ether/NaOH/20 °C [56]; CH_2Cl_2/Al_2O_3/20 °C [95]
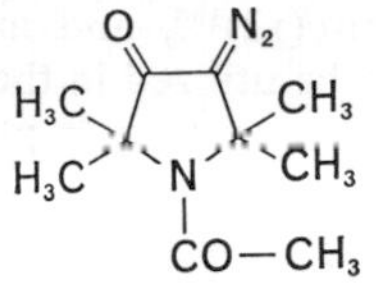 (87%) $CHCl_3/Al_2O_3$/20 °C [99]
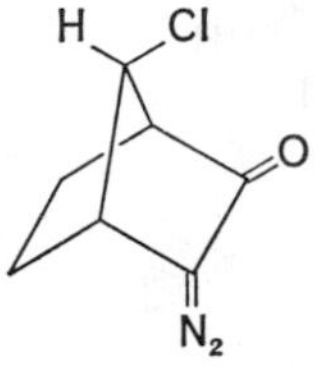 (94%) Pentane/NaOH/— [103]
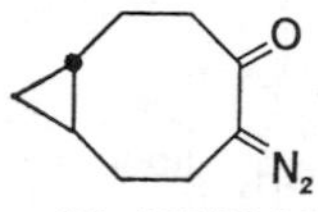 (—) —/NaOH/20 °C [104]
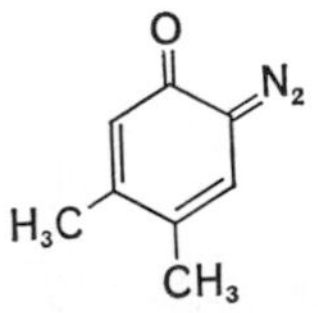 (57%) CH_2Cl_2/NaOH/20 °C [101]
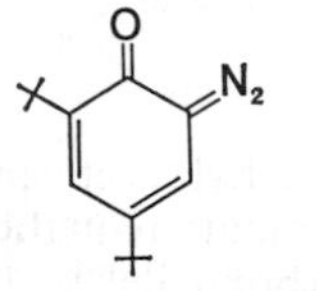 (100%) CH_2Cl_2/NaOH/20 °C [103]
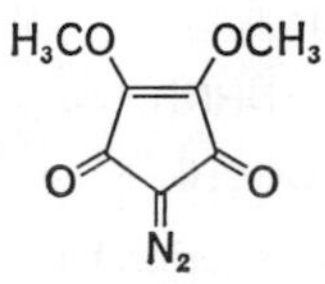 (68%) $CHCl_3/Al_2O_3$/20 °C [105]

Another method which should not go unmentioned is the formation of sometimes complex α-diazo carbonyl compounds from tosylhydrazones with a further group capable of condensation. One such suitable compound is tosylhydrazonoacetyl chloride, which is transformed into diazoacetic esters with unsaturated alcohols according to equation (12)[106], and another one is α-oxophenylacetic acid tosylhydrazone, which can be utilized in the synthesis of α-diazo carboxamides as shown in equation (13). These compounds play a key role in the production of 6-phenylpenicillanates[107].

$$\text{ROH } (n = 3, 4) \xrightarrow[CH_2Cl_2/N(C_2H_5)_3]{Cl(O=)C{-}CH{=}N{-}NH{-}Tos} RO{-}C(=O){-}CH{=}N{-}NH{-}Tos \xrightarrow{CH_2Cl_2/N(C_2H_5)_3} RO{-}C(=O){-}CH{=}N_2 \quad (12)$$

$$\text{thiazolidine-NH} \xrightarrow[-H_2O \text{ (in the presence of diphenylcarbodiimide)}]{HOOC{-}C(C_6H_5){=}N{-}NH{-}Tos} N{-}C(=O){-}C(=N{-}NH{-}Tos){-}C_6H_5 \xrightarrow{\text{NaH, monoglyme}} N{-}C(=O){-}C(=N_2){-}C_6H_5 \quad (13)$$

The reaction of dimethyl acetylenedicarboxylate with tosylhydrazide to give dimethyl 2-diazobutanedioate (equation 14) does not appear to be general since it fails with acetylenedicarbonitrile[108].

$$CH_3OOC{-}C{\equiv}C{-}COOCH_3 \xrightarrow{H_2NNH{-}Tos} CH_3OOC{-}CH{=}C(NHNH{-}Tos){-}COOCH_3 \xrightarrow{\sim H} CH_3OOC{-}CH_2{-}C(=NNH{-}Tos){-}COOCH_3 \xrightarrow[80\,°C]{K_2CO_3/H_2O} CH_3OOC{-}CH_2{-}C(=N_2){-}COOCH_3 \quad (14)$$

The Bamford–Stevens reaction has only limited synthetic utility for the preparation of α-diazo β-dicarbonyl compounds. While it is indeed the central CO group of alloxan, perinaphthindantrione or diethyl mesoxalate which is in each case converted smoothly into the diazo group[109] owing to its greater reactivity, the corresponding reaction in ninhydrin shows the formation of 2-diazo-1,3-dioxoindan to be accompanied by three further competing condensation reactions (equation 15)[110].

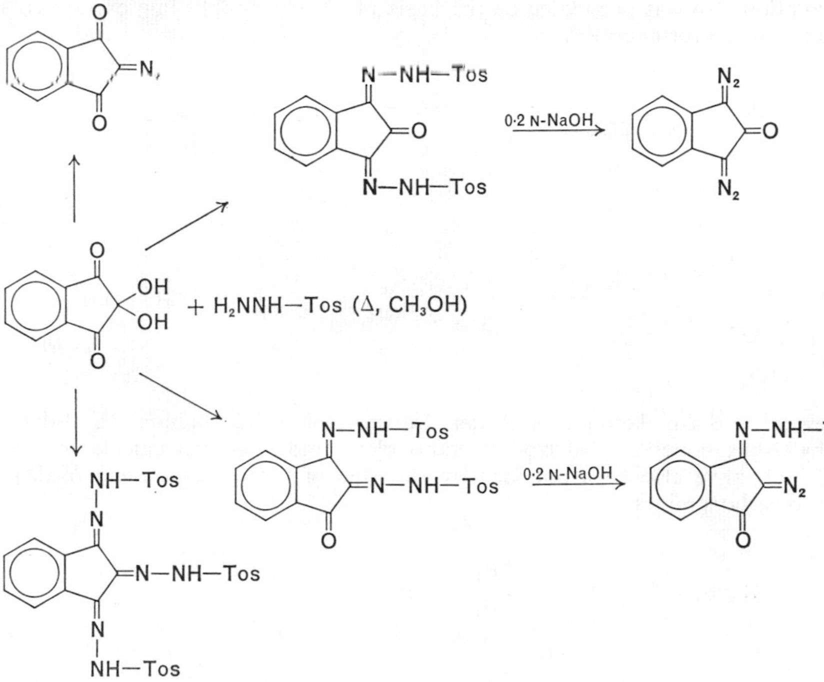

$$\text{Ninhydrin} + H_2NNH\text{—Tos} \ (\Delta, CH_3OH) \qquad (15)$$

N—NH—Tos

0·2 N-NaOH

N_2

N—NH—Tos

N—NH—Tos

0·2 N-NaOH

$=N_2$

NH—Tos

=N—NH—Tos

NH—Tos

5. Cleavage of *N*-alkyl-*N*-nitrosamides

The method given in the title is not only the oldest but still the most important preparation of non-acylated diazoalkanes. Although suitable in principle, it has nevertheless attracted hardly any attention for the preparation of α-diazo carbonyl compounds.

Thus the thermal cleavage of *N*-nitrosourethanes **(31)** presumably proceeds via the isomeric azo compounds **(32)** to give the α-diazo carboxylic ester **(33)**, with potassium carbonate acting as catalyst[111].

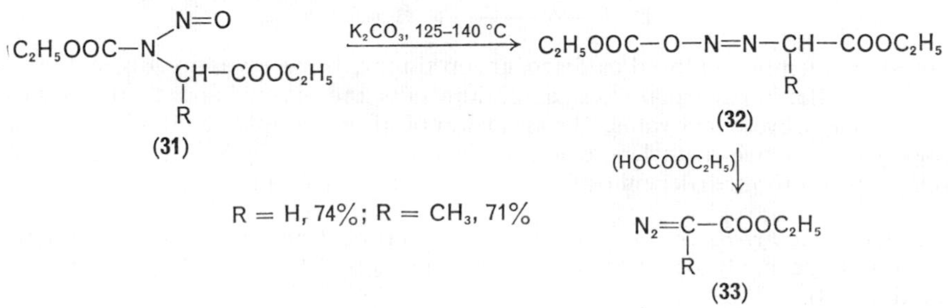

$$C_2H_5OOC\text{—}N(N{=}O)\text{—}CH(R)\text{—}COOC_2H_5 \ \textbf{(31)} \xrightarrow{K_2CO_3,\ 125\text{–}140\ ^\circ C} C_2H_5OOC\text{—}O\text{—}N{=}N\text{—}CH(R)\text{—}COOC_2H_5 \ \textbf{(32)}$$

$$\textbf{(32)} \xrightarrow{(HOCOOC_2H_5)} N_2{=}C(R)\text{—}COOC_2H_5 \ \textbf{(33)}$$

R = H, 74%; R = CH_3, 71%

Ethyl diazoacetate was also produced on acyl cleavage of ethyl *N*-acetyl-*N*-nitrosoaminoacetate when barium oxide/barium hydroxide was used in methanol as a base (equation 16)[112]; it could not be predicted that use of ammonia in methanol/ether as base would lead to generation of diazomethane. The reaction course shown in equation (16) was postulated on the basis of 18-oxygen-labelling experiments to account for its formation[113].

$$H_3C-C(=O)-N(NO)-CH_2COOC_2H_5 \xrightarrow{BaO,\ Ba(OH)_2,\ CH_3OH} N_2{=}CHCOOC_2H_5$$

$$\downarrow 2\,NH_3,\ -H_3CC(=O)NH_2 \qquad (16)$$

$$O^-{-}N{=}N{-}CH_2{-}C(=O){-}OC_2H_5\ \ NH_4^+ \longrightarrow \text{(oxadiazoline)}{-}OC_2H_5,\ O^-\ NH_4^+ \xrightarrow[-C_2H_5OH]{NH_3} N_2{=}CH_2 + H_2N{-}COO^-\ NH_4^+$$

Several α-diazo ketones, such as 2-diazo-1-phenyl-3-butanone[114], 3-diazo-5-methyl-2-hexanone[114], 3-diazo-2-heptanone[114] and 6-diazopenicillanic esters (**34**)[115, 116], have also been obtained by cleavage of the corresponding *N*-alkyl-*N*-nitroso carboxamides.

$$R^1{-}H_2C{-}N(NO){-}\text{(penam)}{-}COOR^2 \xrightarrow{\text{Cleavage}} N_2{=}\text{(penam)}{-}COOR^2 \quad (\mathbf{34})$$

$R^1 = C_6H_5$, $R^2 = CH_2{-}CCl_3$, (Δ, pyridine/chloroform)
$R^1 = OC_6H_5$, $R^2 = CH_2{-}C_6H_5$,
(chromatography on silicagel with chloroform)

6. Diazo group transfer

The principle of diazo group transfer consists in the transfer of a complete diazo group from a donor $D{=}N_2$ to an acceptor A according to reaction (17); the latter must of course possess replaceable substituents or be unsaturated[116-118]. Tosyl azide

$$D{=}N_2 + A \longrightarrow N_2{=}A + D \qquad (17)$$

is most frequently employed as diazo group donor, being exceptionally stable and easy to handle; in particular cases, azidinium salts and diazoalkanes themselves (see below) have proved their value. The spectrum of diazo group acceptors is very large, ranging from active methylene compounds via α-acyl aldehydes, alkenes, alkynes, cyclopropenes to methylenephosphoranes, and still expanding.

a. *Active methylene compounds.* Most diazo group transfers to active methylene compounds probably occur by a 'triazene mechanism'[116-118], such as is shown in equation (18).

$$R^1R^2CH_2 + N_2{=}N{-}Tos \longrightarrow R^1R^2C{=}N_2 + H_2N{-}Tos \quad (18)$$

(Base, $-H^+$: $R^1R^2\ddot{C}^-H$; $+\ :\overset{+}{N}{=}N{-}\ddot{N}^-{-}Tos$ → $R^1R^2C(H){-}N{=}N{-}\ddot{N}{-}Tos$; Cleavage, $+H^+$)

Diazo group transfers to active methylene compounds are therefore dependent upon the presence of a suitable base. In complete contrast, the highly reactive azidinium salts[119], which can be regarded as *N*-diazonium salts[120], transfer the N_2 group in neutral to acid media. The occurrence of triazene-type intermediates can again be assumed (equation 19)[121].

$$R^1R^2CH_2 + \text{3-ethylbenzothiazol-2-ylidene}{=}N{-}N_2^+\ BF_4^- \xrightarrow{-HBF_4} \text{3-ethylbenzothiazol-2-ylidene}{=}\dot{N}{-}N{=}N{-}C(H)R^1R^2 \quad (19)$$

$$\xrightarrow{HBF_4,\ \text{cleavage}} \text{3-ethylbenzothiazol-2-ylidene}{=}\overset{+}{N}H_2\ BF_4^- + N_2{=}CR^1R^2$$

A firm indication of the intermediacy of a triazene in diazo transfer is provided by the reaction of 1-aryl-2-phenyl-1-ethanones or their potassium salts with tosyl azide (aryl = mesityl)[122, 123]. In the case of aryl = phenyl, however, all available evidence suggests a triazoline structure of the intermediate, which can be isolated. Under the conditions of diazo group transfer it decomposes into azibenzil and potassium tosylamidate, while acid decomposition affords *N*-tosyldiphenylacetamide. Reaction (20) shows how its formation can be envisaged[122, 123].

$$H_5C_6{-}C(=O){-}\ddot{C}H^-{-}C_6H_5\ K^+ \leftrightarrow H_5C_6{-}C(O^-K^+){=}CH{-}C_6H_5$$

$$\xrightarrow{TosN_3} \text{triazoline: } H_5C_6C(O^-K^+){-}CH(C_6H_5){-}N{=}N{-}N(Tos) \xrightarrow{\text{Ring opening}} H_5C_6{-}C(O^-K^+)(\ddot{N}^-{-}Tos){-}CH(N_2^+){-}C_6H_5 \quad (20)$$

$$\xrightarrow{HX,\ -KX} H_5C_6{-}C(OH)(\ddot{N}^-{-}Tos){-}\overset{+}{C}H{-}C_6H_5 \xrightarrow{\sim C_6H_5,\ \sim H} Tos{-}HN{-}C(=O){-}CH(C_6H_5)_2$$

The decisive influence of the base upon the mechanism and product formation of diazo group transfer is illustrated, for instance, by diazo transfer onto diethyl malonate: while diethyl diazomalonate is expectedly formed in acetonitrile/triethylamine, ethyl *N*-tosyldiazomalonate is produced in ethanol/sodium ethoxide. A plausible explanation of these facts could assume the intermediacy of the triazene salt (**35**) in weakly basic media, but that of the triazoline salt (**36**) in the strongly alkaline range[124, 125].

$$C_2H_5OOC{-}CH_2{-}COOC_2H_5 \xrightarrow[-H^+]{TosN_3\ \text{in}\ CH_3CN/N(C_2H_5)_3} \mathbf{(35)} \xrightarrow{-Tos{-}\ddot{N}H^-} C_2H_5OOC{-}C(=N_2){-}COOC_2H_5$$

$$C_2H_5OOC{-}CH_2{-}COOC_2H_5 \xrightarrow{TosN_3\ \text{in}\ C_2H_5OH/C_2H_5ONa} \mathbf{(36)} \xrightarrow[-NaX]{(a)\ -C_2H_5OH;\ (b)\ \text{Ring opening};\ (c)\ HX} H_5C_2OOC{-}C(=N_2){-}CO{-}NH{-}Tos$$

The same applies to ethyl (4-nitrophenyl)acetate[123] and to some extent also to the reaction of diethyl malonate with 5-azido-2-methoxycycloheptatrienone in ethanol/sodium ethoxide[126].

In recent years the preparation of α-diazo β-dicarbonyl compounds by diazo transfer has completely ousted all the other methods. Organic amines, such as piperidine, di- or tri-ethylamine in solvents such as acetonitrile, methylene chloride or ethanol, normally suffice to transform the reactants into the reactive carbanions. Tosyl azide is used almost exclusively as diazo transfer agent[127]; azidinium salts are employed whenever a neutral or acid reaction medium prevents azo coupling, e.g. in the conversion of phloroglucinol into 1,3,5-tris(diazo)-2,4,6-cyclohexanetrione (95%)[128]. If diazo group transfer fails to go to completion then the use of excess 4-carboxybenzenesulphonyl azide is recommended since it can easily be removed together with the corresponding amide in an alkaline medium[129]. Polymer-bound sulphonyl azide has recently also been employed as diazo transfer agent (equation 21), without any obvious advantage[130].

Amberlite XE 305 (20% cross-linked polystyrene/divinylbenzene polymer)

$$\xrightarrow[(2)\ \Delta,\ CCl_4]{(1)\ ClSO_3H,\ 70\ ^\circ C} (P){-}SO_3H \xrightarrow[NaN_3]{C_2H_5OH/H_2O/\text{dioxane}}$$

$$(P){-}SO_2{-}N_3 \xrightarrow[C_2H_5OH/N(C_2H_5)_3]{R^1COCH_2COR^2} (P){-}SO_2{-}NH_2 + R^1{-}CO{-}C(=N_2){-}CO{-}R^2 \qquad (21)$$

Although of theoretical interest, diazo group transfers with diazo compounds themselves (ω-diazoacetophenone[131], ethyl diazonitroacetate[132]) are of no synthetic importance.

Examples involving acyclic α-diazo β-dicarbonyl compounds are so numerous that they cannot all be considered individually. Apart from symmetrical[130, 133–135] and unsymmetrical[136–138] diazomalonic diesters [(**37**) and (**38**), respectively], diazomalonamic esters (**39**)[139–141], diazoacetic esters (**40**)[127, 133, 135, 142], and diazoacetamides

(41)[141, 142], a large number of 2-diazo 1,3-diketones bearing alkyl, aryl or heteroaryl groups **(42)** have been synthesized[133, 143–145].

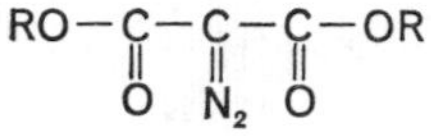

$$RO{-}C({=}O){-}C({=}N_2){-}C({=}O){-}OR \quad \textbf{(37)}$$

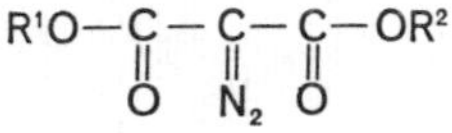

$$R^1O{-}C({=}O){-}C({=}N_2){-}C({=}O){-}OR^2 \quad \textbf{(38)}$$

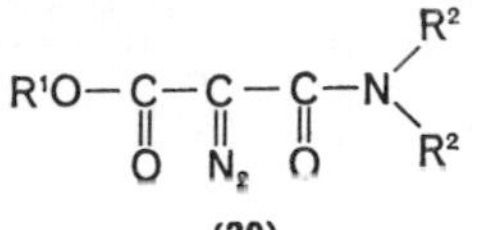

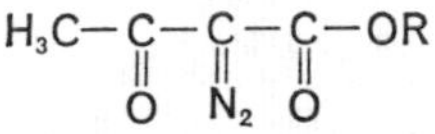

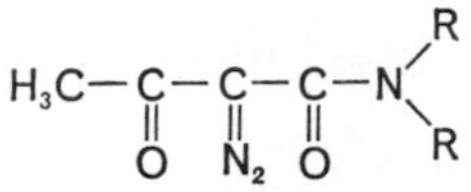

$$R^1O{-}C({=}O){-}C({=}N_2){-}C({=}O){-}NR^2_2 \quad \textbf{(39)}$$

$$H_3C{-}C({=}O){-}C({=}N_2){-}C({=}O){-}OR \quad \textbf{(40)}$$

$$H_3C{-}C({=}O){-}C({=}N_2){-}C({=}O){-}NR_2 \quad \textbf{(41)}$$

$$R^1{-}C({=}O){-}C({=}N_2){-}C({=}O){-}R^2 \quad \textbf{(42)}$$

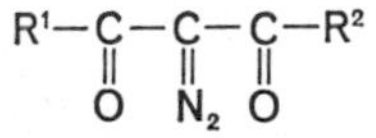

Diazo group transfer onto acyclic 1,3,5-tricarbonyl compounds undoubtedly first yields 2-diazo 1,3,5-tricarbonyl compounds **(43)**, which then rapidly undergo secondary reactions. Given the correct stoichiometry of the starting components and an appropriate choice of reaction conditions, reaction affords the 2,4-bis(diazo) 1,3,5-tricarbonyl compounds **(45)**[146]; otherwise the monodiazo compounds **(43)** are subject to intramolecular azo coupling to give 3,5-diacyl-4-pyrazolones **(44)**[147].

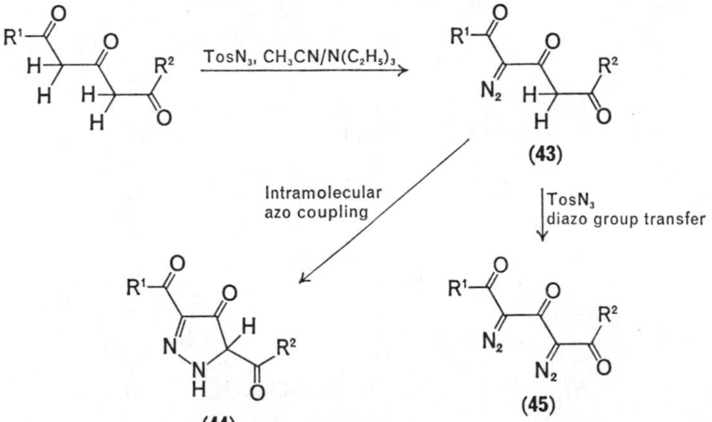

$TosN_3$, $CH_3CN/N(C_2H_5)_3$

(43)

Intramolecular azo coupling

$TosN_3$ diazo group transfer

(44)

(45)

Further secondary reactions following upon diazo transfer are observed when it is applied to α-methylene ketones with an additional C=N or C=S group. Thus β-oxo imines are converted into triazoles according to equation (22)[116, 148]. 2-Acylmethylpyridines bearing an *endo* azomethine group undergo a completely analogous reaction to give triazolopyridines (equation 23)[149]. In both cases, 1,5-cyclization occurs as a result of the electrophilicity of the terminal diazo nitrogen and the nucleophilicity of the imino N atom.

Acylthioacetamides also react with tosyl azide to give 4-acyl-5-amino-1,2,3-thiadiazoles via diazo group transfer and subsequent 1,5-cyclization (equation 24)[150]. Diazo transfers onto *trans*-fixed β-dicarbonyl compounds have also become so numerous that only a representative selection can be given (examples **46–57**).

In some cases, the formation of azo compounds, e.g. having structure **60**, is observed alongside or even instead of diazo transfer onto cyclic β-dicarbonyl compounds **(58)**. Their occurrence should be interpreted by assuming initial

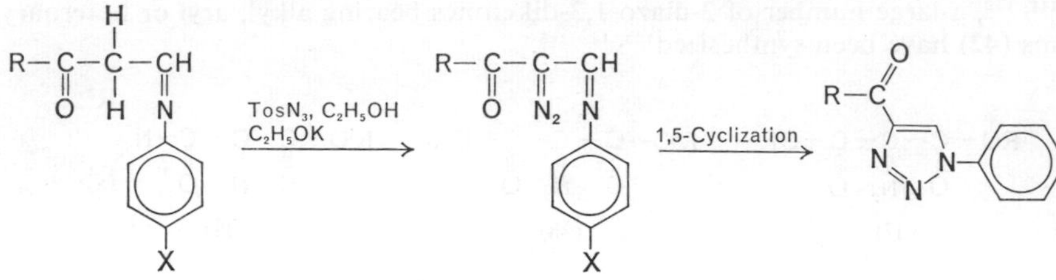

TosN$_3$, C$_2$H$_5$OH, C$_2$H$_5$OK → ; 1,5-Cyclization → (22)

$R = C_6H_5, CH_3; X = H, OCH_3, NO_2$

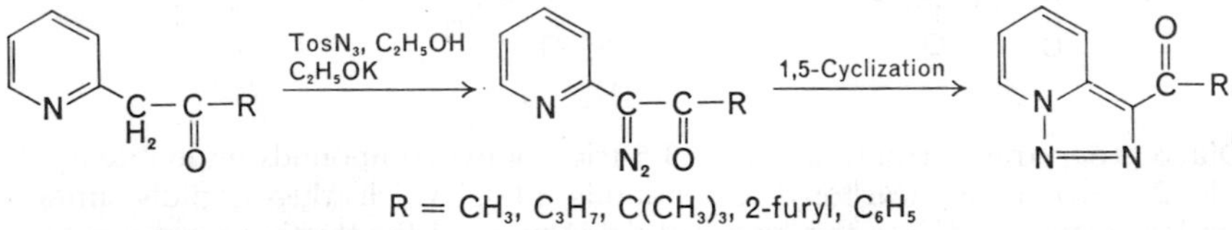

TosN$_3$, C$_2$H$_5$OH, C$_2$H$_5$OK → ; 1,5-Cyclization → (23)

$R = CH_3, C_3H_7, C(CH_3)_3$, 2-furyl, C_6H_5

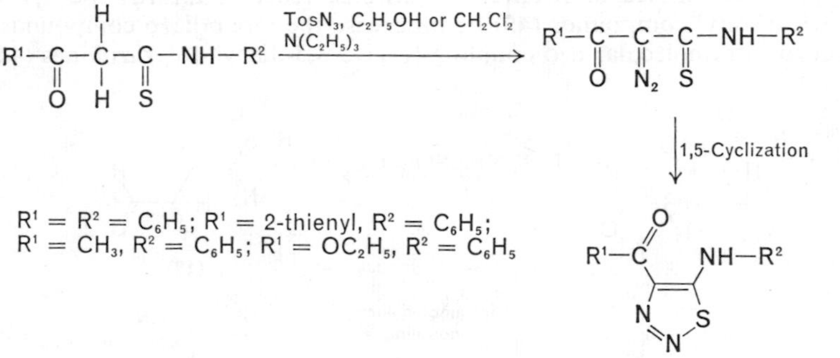

TosN$_3$, C$_2$H$_5$OH or CH$_2$Cl$_2$, N(C$_2$H$_5$)$_3$ → ; 1,5-Cyclization → (24)

$R^1 = R^2 = C_6H_5$; R^1 = 2-thienyl, $R^2 = C_6H_5$;
$R^1 = CH_3$, $R^2 = C_6H_5$; $R^1 = OC_2H_5$, $R^2 = C_6H_5$

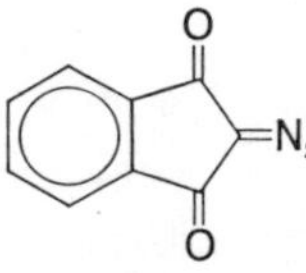

(46)[110,128,151] 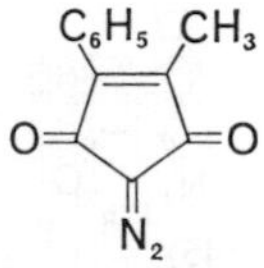 (47)[152] 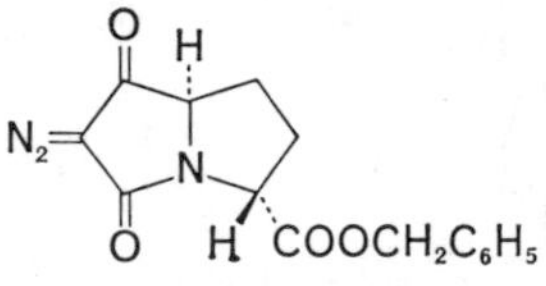 (48)[153] 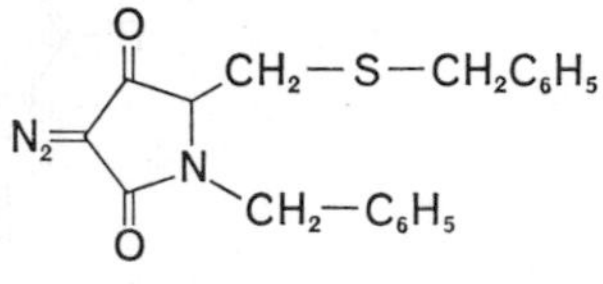 (49)[154]

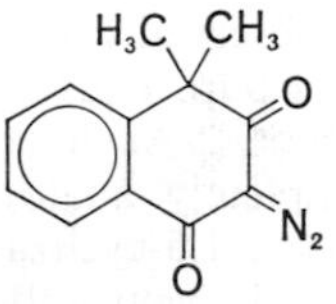

(50)[155] 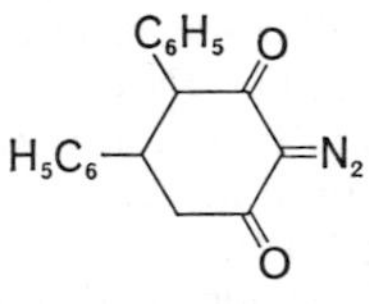 (51)[156] 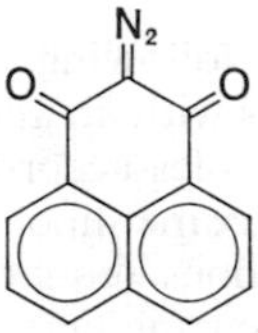 (52)[124] 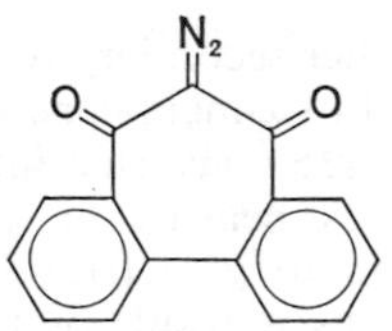 (53)[157]

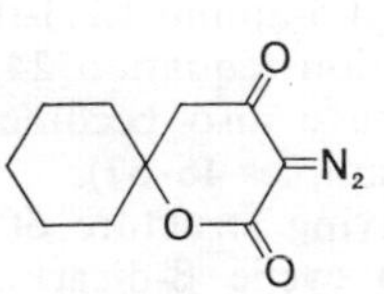

(54)[158] 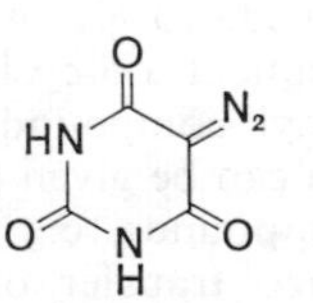 (55)[124,128] 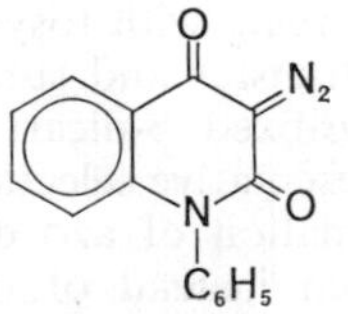 (56)[159] 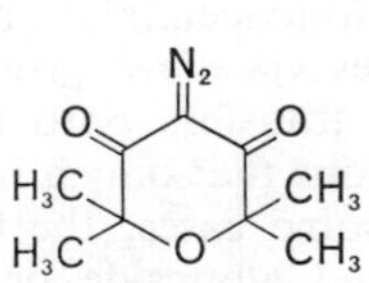 (57)[143]

formation—in the usual manner—of the diazo compounds **59** which possess a highly electrophilic CN_2 group[160]; this group undergoes the unusual aliphatic azo coupling reaction with unreacted **58** (**58**+**59** → **60**)[158, 160, 161]. Prior choice of the molar ratio methylene compound/tosyl azide readily permits the reaction course to be steered as a diazo or azo transfer (**59** or **60**).

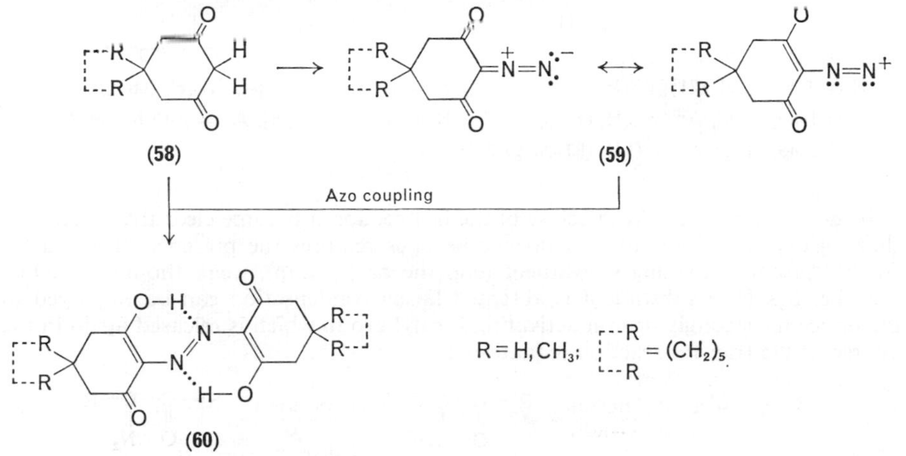

(58) (59)

Azo coupling

(60)

R = H, CH₃; R,R = $(CH_2)_5$

The triazole isomerization already observed in diazo-group transfer onto acyclic β-oxo imines also occurs in the cyclic series[116, 148, 162, 163] and has been studied particularly thoroughly in the 3-imino-1-indanone system[162, 163]. It was found that *para* donor substituents in the phenylimino group promote 1,5-ring closure (triazole formation in the case of X = OCH_3), while acceptor substituents (X = NO_2) or the unsubstituted group arrest the reaction at the diazo imine stage (equation 25).

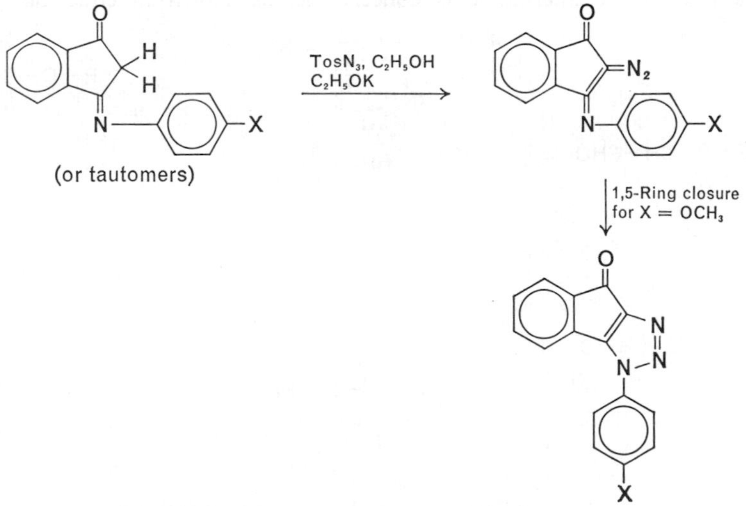

(or tautomers)

$TosN_3$, C_2H_5OH, C_2H_5OK

1,5-Ring closure for X = OCH_3

(25)

α-Methylene ketones exhibiting additional proton activation by aromatic groups are also accessible to diazo group transfer (equation 26); alkali alkoxides are necessary as bases[122, 123]. In the case of 4-nitrophenyl groups, formation of the α-diazo carbonyl compounds also occurs in the presence of organic amines[84, 164, 165].

$$R{-}\underset{\underset{O}{\|}}{C}{-}\underset{\underset{H}{|}}{\overset{\overset{H}{|}}{C}}{-}Ar \xrightarrow{TosN_3} R{-}\underset{\underset{O}{\|}}{C}{-}\underset{\underset{N_2}{\|}}{C}{-}Ar \qquad (26)$$

R = CH_3, Ar = C_6H_5 (72%)[122, 123]
R = Ar = C_6H_5 (71%)[122, 123]
R = 1-naphthyl, Ar = C_6H_5 (76%)[122, 123]
R = 2-naphthyl, Ar = C_6H_5 (87–90%)[122, 123]
R = C_6H_5, Ar = p-$O_2NC_6H_4$ (80%)[164]
R = OCH_3, Ar = p-$O_2NC_6H_4$ (60%)[84]
R = $N(CH_3){-}CO{-}C_6H_5$, Ar = p-$O_2NC_6H_4$ (—)[165]

b. *α-Acyl aldehydes.* At the close of the last section it became clear that successful diazo group transfer onto α-methylene ketones requires the presence of at least a further proton-activating substituent (e.g. the aryl group), weak though its effect may be. If such a substituent is absent, Claisen condensation can be employed to introduce the strongly proton-activating formyl group which is released again in the course of the transfer reaction (equation 27).

$$R^1{-}\underset{\underset{O}{\|}}{C}{-}CH_2{-}R^2 \xrightarrow[-ROH]{HCOOR} R^1{-}\underset{\underset{O}{\|}}{C}{-}\underset{\underset{CHO}{|}}{CH}{-}R^2 \xrightarrow[-H{-}C(=O){-}NH{-}Tos]{TosN_3\,(base)} R^1{-}\underset{\underset{O}{\|}}{C}{-}\underset{\underset{N_2}{\|}}{C}{-}R^2 \qquad (27)$$

R^2 = H, alkyl

This reaction, known as deformylating diazo transfer[116–118], has found wide acclaim within a few years. Two variants of equation (27) are commonly encountered in preparative work. Either the sodium salts resulting directly from the Claisen ester condensation (generally with R^2 = H), or the free formyl compounds or their tautomers (generally with R^2 = alkyl) are used as starting materials. In the former case a mechanistic alternative is conceivable, as shown in equation (28); the

$$R{-}\underset{\underset{O}{\|}}{C}{-}\ddot{C}H^{-}{-}CHO\ Na^+ \longleftrightarrow R{-}\underset{\underset{O}{\|}}{C}{-}CH{=}CH{-}O^-\ Na^+$$

(via TosN$_3$, C_2H_5OH)

Upper pathway: triazene intermediate $R{-}C(=O){-}C(H)(CHO){-}N{=}N{-}\ddot{N}^-{-}Tos\ Na^+$ $\xrightarrow{-Tos{-}\ddot{N}H^-\ Na^+}$ $R{-}\underset{\underset{O}{\|}}{C}{-}\underset{\underset{N_2}{\|}}{C}{-}CHO$ $\xrightarrow[-HCOOC_2H_5]{C_2H_5OH}$ $R{-}\underset{\underset{O}{\|}}{C}{-}\underset{\underset{N_2}{\|}}{CH}$ (28)

Lower pathway: triazoline intermediate $R{-}C(=O){-}CH{-}CH(O^-)$ ring with $N{=}N{-}N{-}Tos$, Na^+ $\xrightarrow{-HC(=O){-}\ddot{N}{-}Tos\ Na^+}$ $R{-}\underset{\underset{O}{\|}}{C}{-}\underset{\underset{N_2}{\|}}{CH}$

pathway leading via a triazene or α-diazo β-oxo aldehyde intermediate and subsequent solvolysis represents one possibility, and the triazoline mechanism another. A decision in favour of one pathway or the other on the basis of the product distribution has so far proved impossible because the sodium salt of *N*-tosylformamide formed by triazoline cleavage transforms into the sodium salt of tosyl amide under authentic conditions[166, 167]. That the lower pathway is followed at least when free formyl compounds or their tautomers (**27**, R^2 = alkyl) are subject to diazo transfer in the $CH_2Cl_2/N(C_2H_5)_3$ system follows from the isolation of *N*-tosylformamide[166, 167].

The cyclic transfer mechanism is even more significant in the conversion of α-formyl- or α-hydroxymethylenecycloalkanones into 1-diazo-2-cycloalkanones (equation 29)[166, 168]. On the one hand, the postulated *N*-tosylformamide is isolated in all cases and, on the other, formation of *N*-tosyl-2-oxocycloalkylcarboxamides is also observed, albeit to a lesser extent. The latter reaction, which is associated with elimination of N_2, is best understood in terms of a branching of the reaction at the triazoline stage. The homologous series of 1-diazo-2-cycloalkanones with 5 to 12 carbon atoms thus becomes readily accessible[166, 168, 169].

$TosN_3$, CH_2Cl_2, $N(C_2H_5)_3$; Decomposition; Ring opening; ~H⁻ (29)

n = 5 (98%), 6 (80–95%), 7 (83%), 8 (87%), 9 (73%), 10 (81%), 11 (79%), 12(57%)

A variant with respect to the second proton-activating group consists in replacement of the formyl by the alkoxyoxalyl group; the few examples reported so far do not permit any conclusion as to its value[170, 171].

The scope of deformylating diazo transfer, on the other hand, is so broad that only a rough survey can be given in Table 4. The method has been applied to the preparation of diazomethyl ketones, α-diazoalkyl ketones, α-diazo aldehydes, α-diazo carboxylic esters, α,β-unsaturated α′-diazo ketones and of 1-diazo-2-cycloalkanones of widely differing kinds.

c. *Alkenes*. Diazo group transfers onto alkenes occur primarily via 3+2 cycloadditions of azides, as demonstrated by the following examples; subsequent decomposition is spontaneous and isomerization with ring opening usually requires base catalysis.

Electron-rich enamines display a pronounced tendency to add azides[180]; addition is regiospecific and affords triazolines[181] which can be isolated in some cases[181, 182]. However, on preparation of α-diazo aldehydes according to equation (30) the triazoline intermediates prove just as impossible to isolate[183–185] as in the synthesis of ethyl diazoacetate by the same method (equation 31)[186].

TABLE 4. α-Diazo carbonyl compounds by deformylating diazo group transfer

α-Diazo carbonyl compound	Method/yield[a]	Reference	α-Diazo carbonyl compound	Method/yield[a]	Reference
$(CH_3)_2CH-C(=O)-CH(=N_2)$	A/52%	166, 167	$C_2H_5OOC-CH(=N_2)$	A/69%	166, 167
$(CH_3)_3C-C(=O)-CH(=N_2)$	A/83%	166, 167	$C_2H_5OOC-C(=N_2)-CH(CH_3)_2$	B/51%	167
$C_6H_5-C(=O)-CH(=N_2)$	A/73%	166, 167	$C_2H_5OOC-C(=N_2)-CH_2-C_6H_5$	B/62%	167
2-Thienyl$-C(=O)-CH(=N_2)$	A/69%	167	$HC(=O)-C(=N_2)-CH_3$	B/64%	167
3-Pyridyl$-C(=O)-CH(=N_2)$	A/34%	167	$HC(=O)-C(=N_2)-C_3H_7$	B/68%	167
Ferrocenyl$-C(=O)-CH(=N_2)$	A/84%	167	$HC(=O)-C(=N_2)-CH_2-C_6H_5$	B/94%	167
$CH_3-C(=O)-C(=N_2)-CH_3$	B/60	166, 167	$CH_3-C(CH_3)=CH-C(=O)-CH(=N_2)$	A/69%	172
			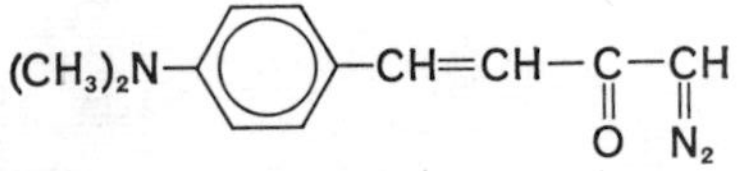 $(CH_3)_2N-C_6H_4-CH=CH-C(=O)-CH(=N_2)$	A/77%	172

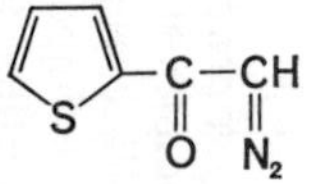

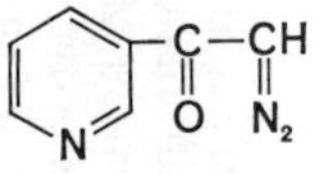

Compound	Method[a]/yield	Ref.	Compound	Method[a]/yield	Ref.
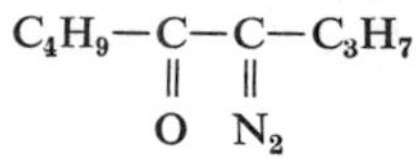 $C_4H_9-C(=O)-C(=N_2)-C_3H_7$	B/68	167	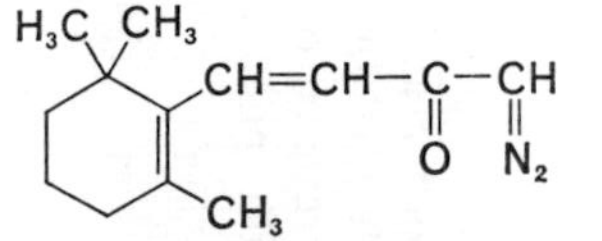 (2,6,6-trimethylcyclohexenyl)$-CH=CH-C(=O)-CH=N_2$	A/94%	172
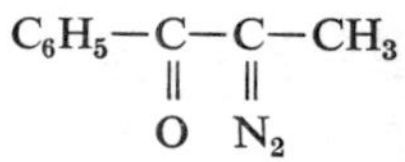 $C_6H_5-C(=O)-C(=N_2)-CH_3$	B/77	166, 167	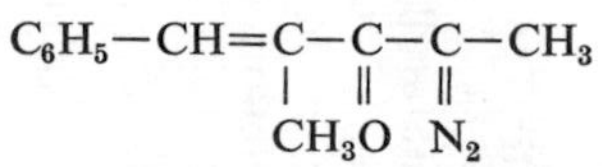 $C_6H_5-CH=C(CH_3)-C(=O)-C(=N_2)-CH_3$	A/81%	172
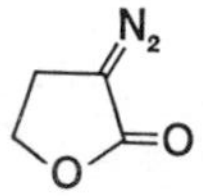	A/14%	173	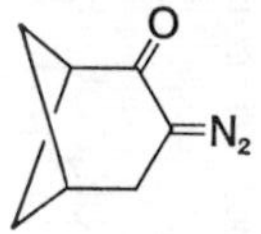	B/75%	176
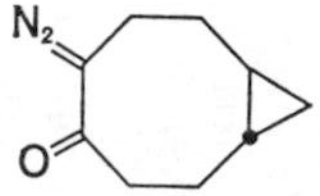	B/—	174	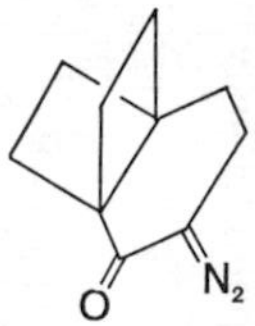	B/—	177
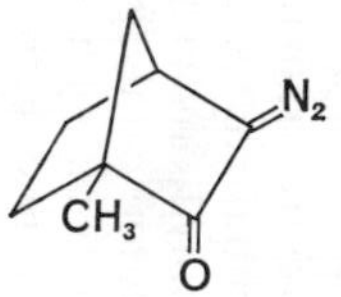	B/—	175	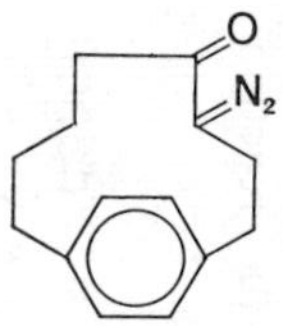	B/100	178, 179

[a] Method A: Deformylating diazo group transfer onto the alkali salts of β-oxo aldehydes.
Method B: Deformylating diazo group transfer onto β-oxo aldehydes or their tautomers.

Diacylated diazomethanes such as diethyl diazomalonate or 3-diazo-2,4-pentanedione are also accessible via suitable enamines[186].

$$R^1R^2N{-}CH{=}C(R^3){-}CH{=}O \xrightarrow{TosN_3} \text{[triazoline: } R^1R^2N, H, R^3, CH{=}O, \text{N-Tos]} \xrightarrow[-HC(NR^1R^2){=}N{-}Tos]{} R^3{-}C(=N_2){-}CH{=}O \quad (30)$$

$R^1 = R^3 = H$, $R^2 = C_6H_5$ (71%); $R^1 = R^2 = CH_3$, $R^3 = CH_3$ (68%); $R^1 = R^2 = CH_3$, $R^3 = C_2H_5$ (77%)

$$C_6H_5{-}NH{-}C(CH_3){=}CH{-}COOC_2H_5 \xrightarrow{TosN_3,\ CH_3CN,\ \Delta} N_2{=}CH{-}COOC_2H_5 + CH_3{-}C(NH{-}C_6H_5){=}N{-}Tos \quad (31)$$

Diazo group transfers onto electron-poor acylated alkenes naturally occur with relatively electron-rich azides. Thus dimethyl fumarate and benzyl azide give the triazoline (**61**), which isomerizes to the diazo ester (**62**) only in the presence of triethylamine[187]; the reaction spontaneously goes to completion with phenyl azide[187].

$$CH_3OOC{-}CH{=}CH{-}COOCH_3 \xrightarrow{C_6H_5CH_2{-}N_3} \text{(61)} \rightarrow CH_3OOC{-}CH(NH{-}CH_2{-}C_6H_5){-}C(=N_2){-}COOCH_3 \ \text{(62)}$$

Diazo transfers onto acrylic esters with aryl azides[187] or glycosyl azides[188] all require base-catalysed opening of the triazoline ring. Reaction between acrylonitrile and phenyl azide leads to an equilibrium mixture of the cycloadduct and diazo isomer[187].

Diazo transfers from diazoalkanes onto cyclopropenes have recently become topical. Thus diazomethane adds to dimethyl 3,3-dimethyl-1,2-cyclopropenedicarboxylate to form the bicyclo[3.1.0]diazahexene (**63**) which can be transformed into the diazo ester (**64**) by irradiation[189]; this reaction is to be interpreted as a cycloreversion. A potential source of interference is seen in the possible isomerization of **63** to the 1,4-dihydropyridazine (**65**); however, in the present case this reaction only occurs on proton catalysis[189].

Diazoalkane adducts with cyclopropenones are far less stable: The bicyclic intermediates (**66**) cannot be isolated, apparently undergoing spontaneous isomerization to the 2-diazo ketones (**67**)[189, 190].

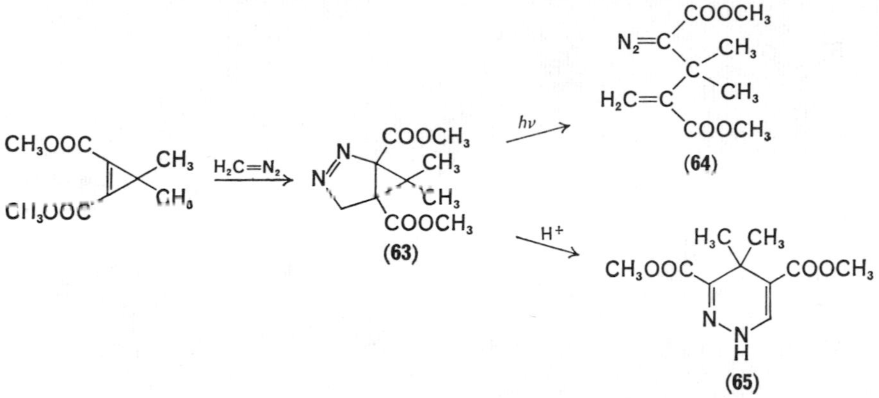

CH_3OOC, CH_3, CH_3, CH_3OOC $\xrightarrow{H_2C=N_2}$ **(63)** $\xrightarrow{h\nu}$ **(64)**; **(63)** $\xrightarrow{H^+}$ **(65)**

(63) (64) (65)

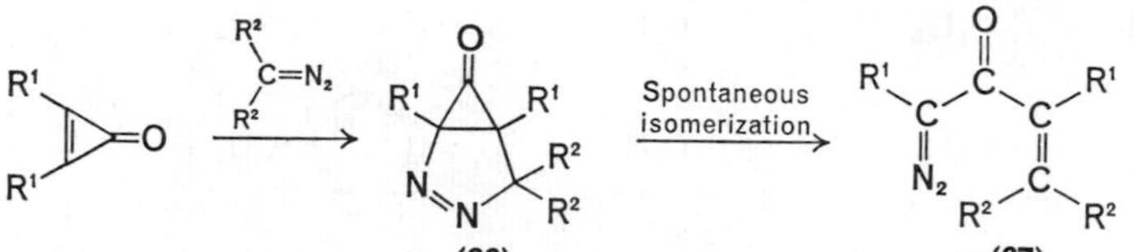

$\xrightarrow{R^2_2C=N_2}$ **(66)** $\xrightarrow{\text{Spontaneous isomerization}}$ **(67)**

(66) (67)

$R^1 = H, R^2 = C_6H_5; R^1 = C_6H_5, R^2 = CH_3$

d. *Methylenephosphoranes.* Carbonyl-substituted methylenephosphoranes react with azides either to form triazoles[191, 192], which is of no interest in the present context, or via diazo group transfer. Both the triazene (**68**) and the λ^5-phosphatriazole (**69**) are regarded as intermediates of the reaction by which ethyl diazoacetate[191] and

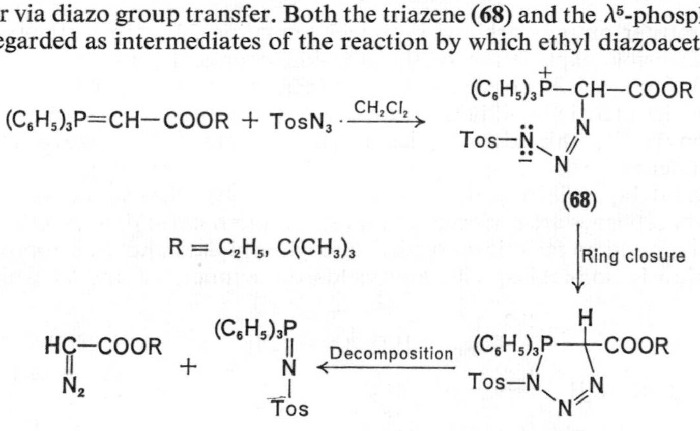

$$(C_6H_5)_3P{=}CH{-}COOR + TosN_3 \xrightarrow{CH_2Cl_2} (C_6H_5)_3\overset{+}{P}{-}CH{-}COOR\ (\text{Tos}{-}\bar{N}{-}N{=}N)$$

(68)

$R = C_2H_5, C(CH_3)_3$

Ring closure

$$HC{-}COOR\,(=N_2) + (C_6H_5)_3P{=}N{-}Tos \xleftarrow{\text{Decomposition}} (69)$$

(69)

t-butyl diazoacetate[193] have been synthesized. The preparation of *N,N*-diethyldiazo-acetamide and of ethyl 2-diazopropanoate by the same procedure[191] seems to be incorrect[194].

e. *Alkynes.* β-Carbonyl ynamines react with sulphonyl azides entirely in the sense of diazo group transfer; the 4-acyl-5-amino-1-arylsulphonyl-1,2,3-triazoles undoubtedly formed as intermediates transform quantitatively into diazo isomers (equation 32)[195].

A completely analogous reaction is observed between βH-ynamines and sulphonyl azides[196]. Even though no direct proof is available for the intermediacy of the triazole

$$\mathrm{R^1{-}C(=O){-}C{\equiv}C{-}NR^2R^3} + \mathrm{N_3{-}SO_2{-}R^4} \longrightarrow \text{triazole} \underset{?}{\overset{\text{Ring opening}}{\rightleftharpoons}} \mathrm{R^1{-}C(=O){-}C(=N_2){-}C(NR^2R^3){=}N{-}SO_2{-}R^4} \quad (32)$$

R^1	R^2	R^3	R^4	Yield (%)
OCH_3	C_2H_5	C_2H_5	$-C_6H_4-N(CH_3)_2$	85
OCH_3	C_2H_5	C_2H_5	$-C_6H_4-NO_2$	63
CH_3	C_2H_5	C_2H_5	$-C_6H_4-CH_3$	63
C_6H_5	CH_3	C_6H_5	$-C_6H_4-CH_3$	93
C_6H_5	$CH_2-C_6H_5$	C_6H_5	$-C_6H_4-CH_3$	96

in diazo transfer onto β-carbonyl ynamines (equation 32), it nevertheless provides the only plausible explanation of the reaction course. In numerous other cases, which will not be discussed in this context, 5-amino-1,2,3-triazoles could be isolated or detected in equilibrium with the α-diazo amidine isomers by ^{1}H-n.m.r. spectroscopy[197–201]; this also applies to the products of diazo transfer onto alkoxyacetylenes[202–204].

Understandably, diazo group transfers onto the electron-poorer, carbonyl-substituted acetylenes are performed with electron-rich diazo donors. One of the few examples reported so far utilizes methyl phenylpropiolate and 2-diazopropane. The cycloaddition is non-regiospecific and yields the adducts **70** and **71** which can be

$$\mathrm{CH_3OOC{-}C{\equiv}C{-}C_6H_5} \xrightarrow{\mathrm{(H_3C)_2C{=}N_2}} \textbf{(70)} + \textbf{(71)}$$

(70) $\xrightarrow{h\nu \text{ in benzene}}$ $\mathrm{(H_3C)_2C{=}C(COOCH_3){-}C(C_6H_5){=}N_2}$ **(72)**

(71) $\xrightarrow{h\nu \text{ in benzene}}$ $\mathrm{N_2{=}C(COOCH_3){-}C(C_6H_5){=}C(CH_3)_2}$ **(73)**

transformed photochemically into **72** and the structurally isomeric α-diazo carboxylic esters (**73**)[205, 206]. The reaction of dimethyl acetylenedicarboxylate with the same diazo transfer agent proceeds in a less clear-cut fashion since the normal reaction, shown in the above scheme, is masked by a double cycloaddition[207].

7. Substitution reactions

The following substitution reactions at the diazomethyl carbon atom bear witness to the stability of the diazo group, even under drastic chemical conditions. Although the acylation of diazomethane[208], which probably represents the most important reaction type, had long been known in numerous examples, the substitution approach has only come into its own during the past 10 years.

a. *Nitration.* Dinitrogen pentoxide in carbon tetrachloride is a suitable nitrating agent for ethyl diazoacetate[209, 210]; loss of 1 mole of ethyl diazoacetate, which reappears as nitric acid ester as shown in equation (33), simply has to be accepted and cannot be obviated by use of auxiliary bases[210].

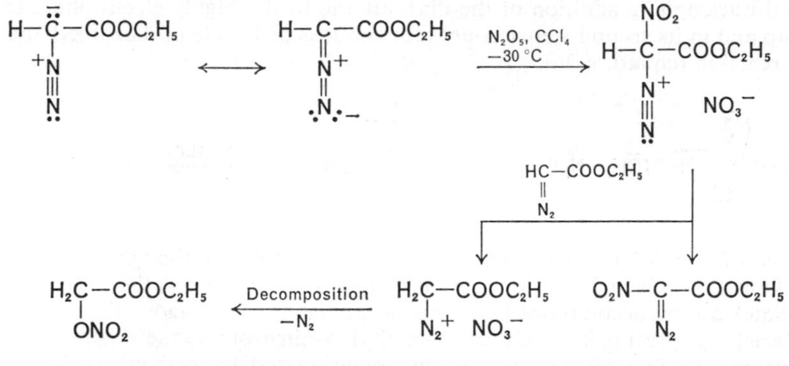

$$H\text{-}\ddot{C}(\text{-}N^{+}{\equiv}N{:})\text{-}COOC_2H_5 \longleftrightarrow H\text{-}C(=N^{+}=N^{-})\text{-}COOC_2H_5 \xrightarrow[-30\,^{\circ}C]{N_2O_5,\ CCl_4} H\text{-}C(NO_2)(N^{+}{\equiv}N)\text{-}COOC_2H_5\ \ NO_3^{-} \quad (33)$$

$$\xrightarrow{HC(=N_2)\text{-}COOC_2H_5} O_2N\text{-}C(=N_2)\text{-}COOC_2H_5 \ + \ H_2C(N_2^{+})\text{-}COOC_2H_5\ NO_3^{-} \xrightarrow[-N_2]{\text{Decomposition}} H_2C(ONO_2)\text{-}COOC_2H_5$$

t-Butyl diazonitroacetate prepared in the same way, and which is also accessible from di-*t*-butyl mercuriobis(diazoacetate)[210, 211], warrants further attention. Degradation of the ester group with trifluoroacetic acid in ether affords diazonitromethane[193, 211] whose acidic hydrogen can be nitrated afresh with dinitrogen pentoxide[193, 212] (equation 34).

$$O_2N\text{-}C(=N_2)\text{-}COOC(CH_3)_3 \xrightarrow[-H_2C=C(CH_3)_2-CO_2]{C_3F\text{-}COOH\ \text{in ether}} O_2N\text{-}C(=N_2)\text{-}H \xrightarrow{N_2O_5} O_2N\text{-}C(=N_2)\text{-}NO_2 \quad (34)$$

b. *Acylation.* Although acylation reactions with diazoacetic ester were known earlier[213], the work of Arndt and Eistert[208] provided the foundation for the ensuing rapid development of the method[214-216] when they conducted the acylation of diazomethane in a manner that suppressed the formation of α-halo ketones[217-221]. Possible acylation reagents are acyl halides, carboxylic anhydrides (or mixtures of carboxylic acids with carbodiimides) and acyl isocyanates. Sulphonyl chlorides do not react with diazomethane[222]; it remains to be seen how far this statement must be qualified in the light of the reported sulphonylation of diazophenylmethane by tosyl chloride[223]. In any case, equation (35) clearly shows that carbonylation takes precedence over sulphonylation when the two are in competition[224].

$$\text{o-}C_6H_4(COCl)(SO_2Cl) \xrightarrow[-CH_3Cl,\ -N_2]{2\ H_2C{=}N_2} \text{o-}C_6H_4(CO{-}CH{=}N_2)(SO_2Cl) \xrightarrow[C_6H_5NH_2]{\text{Identification as anilide}} \text{o-}C_6H_4(CO{-}CH{=}N_2)(SO_2{-}NH{-}C_6H_5) \quad (35)$$

Evidence has also recently become available in support of sulphoxylation of diazomethane, as indicated by its reaction with phenylsulphinyl chloride; formation of the α-diazo sulphoxide (**74**) is accompanied by production of the S/Cl insertion compound (**75**)[225].

$$C_6H_5{-}S(O){-}Cl \xrightarrow{CH_2{=}N_2,\ -78°\,C} \underset{(74)}{C_6H_5{-}S(O){-}CH{=}N_2} + \underset{(75)}{C_6H_5{-}S(O){-}CH_2Cl}$$

A mechanism was formulated for the reaction between diazomethane and acyl chlorides some time ago[226] and has remained unquestioned to this day. It invokes initial nucleophilic addition of the diazoalkane to the highly electrophilic carbonyl group and in its second step accounts for the loss of 1 mole of diazomethane during this reaction (equation 36).

$$R{-}C(=O){-}Cl + :\bar{C}H_2{-}N_2^+ \longrightarrow R{-}C(O^-)(Cl){-}CH_2{-}N_2^+ \xrightarrow[-CH_3Cl,\ -N_2]{H_2C{=}N_2} R{-}C(=O){-}CH{=}N_2 \quad (36)$$

Loss of a mole of diazomethane is generally accepted, although several examples are known in which its basic function (removal of HCl from the diazonium intermediate) can be accomplished by organic amines[227-230]. Diazomethane can also be replaced by precursors such as *N*-methyl-*N*-nitrosourethane[231] or *N*-methyl-*N*-nitrosourea[232]. Steric factors can prevent acylation of diazomethane, as demonstrated by the failure of 2,4,6-trimethylbenzoyl chloride to react[233], or strongly suppress it, as in 2-chloro-6-methoxybenzoyl chloride[234]. One of the rare cases in which diazomethane is not acylated in spite of favourable structural conditions is encountered in the reaction of hippuric acid. Contrary to the original reports[235], diazomethane acts only in its capacity as a base in the cyclization to the oxazolone (**76**) which involves elimination of HCl[236].

$$C_6H_5{-}C(=O){-}NH{-}CH_2{-}C(=O)Cl \underset{?}{\rightleftharpoons} C_6H_5{-}C(OH){=}N{-}CH_2{-}C(=O)Cl \xrightarrow[-N_2,\ -CH_3Cl]{H_2C{=}N_2} \underset{(76)}{\text{2-}C_6H_5\text{-oxazol-5(4H)-one}}$$

Secondary reactions are to be expected on acylation of diazomethane with α,β-unsaturated acyl chlorides. Hence in the reaction between diazomethane and cinnamoyl chloride, which has repeatedly been examined[172, 237-242], acylation is always found to be accompanied by 1,3-dipolar cycloaddition; whether Δ^1- or Δ^2-pyrazolines are formed will depend upon the duration of reaction, and whether further acylation of the cycloadducts occurs will depend upon the molar ratio of the reactants (equation 37)[172, 241, 242].

$$C_6H_5{-}CH{=}CH{-}C(=O)Cl \xrightarrow[-CH_3Cl,\ -N_2]{3\,H_2C{=}N_2} \text{4-}C_6H_5\text{-3-}(COCH{=}N_2)\text{-1-pyrazoline} \xrightarrow{\sim H} \text{4-}C_6H_5\text{-3-}(COCH{=}N_2)\text{-2-pyrazoline (NH)} \xrightarrow{C_6H_5-CH=CH-C(=O)Cl} \text{1-}(COCH{=}CH{-}C_6H_5)\text{-4-}C_6H_5\text{-3-}(COCH{=}N_2)\text{-2-pyrazoline} \quad (37)$$

Diazoacetylpyrazolines also appear during acylation of diazomethane with 3,4,5-tris(acetoxy)cyclohexene-1-carbonyl chloride[243], α-(4-chlorophenyl)cinnamoyl chloride[244], and α-bromocinnamoyl chloride[245]. In the last-named example the reaction can also be directed so as to largely suppress 3+2 cycloaddition. That steric factors are responsible for this behaviour follows from the corresponding reaction of α-methylcinnamoyl chloride, which initially leads to **77** and then only slowly to **78** [240].

$$C_6H_5{-}CH{=}C(CH_3){-}C(=O)Cl \xrightarrow[-CH_3Cl,\ -N_2]{2\,H_2C{=}N_2} \underset{(\mathbf{77})}{C_6H_5{-}CH{=}C(CH_3){-}C(=O){-}CH{=}N_2} \xrightarrow{H_2C{=}N_2} \underset{(\mathbf{78})}{\text{4-}H\text{-4-}C_6H_5\text{-3-}CH_3\text{-3-}(C(=O){-}CH{=}N_2)\text{-1-pyrazoline}}$$

Pyrazoline formation does not occur at all with chlorinated acryloyl chlorides[246, 247]. This also applies to acyl chlorides having carbon–carbon double bonds or triple bonds in the β,γ position or more remote from the carbonyl group (for examples, see Table 5).

Acylation reactions with carboxylic anhydrides[248] likewise require 2 moles of diazomethane, one for diazo ketone formation and one for carboxylic ester formation as shown in equation (38)[249, 250].

$$R^1{-}C(=O){-}O{-}C(=O){-}R^2 \xrightarrow[-N_2]{2\,H_2C{=}N_2} R^1{-}C(=O){-}CH{=}N_2 + R^2{-}COOCH_3 \quad (38)$$

$R^1 = C_6H_5$, $R^2 = OC_2H_5$; $R^1 = CH_2CH(CH_3)_2$, $R^2 = OC_2H_5$; $R^1 = H$, $R^2 = CH_3$

The uniform course of cleavage of unsymmetrical anhydrides can be attributed to preferential attack of the more electrophilic CO group by diazomethane (when R^1—CO is considerably more reactive than R^2—CO). A directed diazo ketone synthesis is therefore accomplished by subjecting carboxylic anhydrides with CO groups of differing reactivity to reaction with diazomethane. Unsymmetrical carboxylic anhydrides meeting these requirements can be synthesized from carboxylic acids and ethyl chloroformate in the presence of triethylamine. Selected examples illustrate the utility of this variant (**79** [251], **80** [252], **81** [253]).

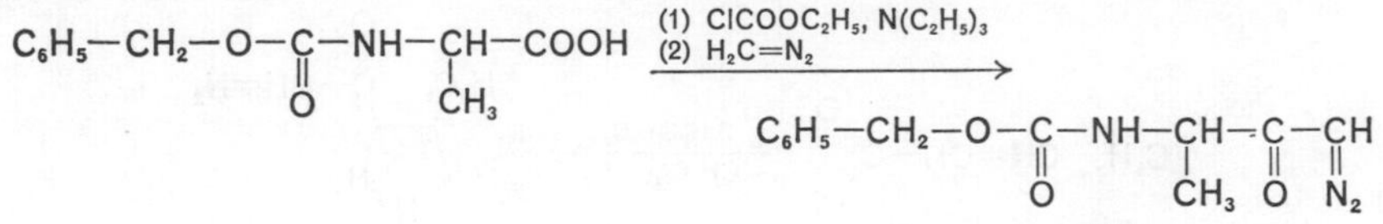

$$C_6H_5-CH_2-O-\underset{\underset{O}{\|}}{C}-NH-\underset{\underset{CH_3}{|}}{CH}-COOH \xrightarrow[(2)\ H_2C=N_2]{(1)\ ClCOOC_2H_5,\ N(C_2H_5)_3} C_6H_5-CH_2-O-\underset{\underset{O}{\|}}{C}-NH-\underset{\underset{CH_3}{|}}{CH}-\underset{\underset{O}{\|}}{C}-\underset{\underset{N_2}{\|}}{CH}$$

(79)

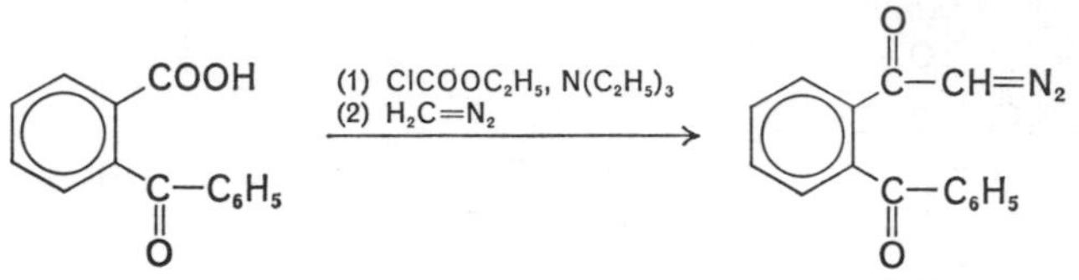

$$o\text{-}(C_6H_5CO)C_6H_4COOH \xrightarrow[(2)\ H_2C=N_2]{(1)\ ClCOOC_2H_5,\ N(C_2H_5)_3} o\text{-}(C_6H_5CO)C_6H_4-CO-CH=N_2$$

(80)

$$(H_5C_6)_2C\overset{O}{—}CH-COOH \xrightarrow[(2)\ H_2C=N_2]{(1)\ ClCOOC_2H_5,\ N(C_2H_5)_3} (H_5C_6)_2C\overset{O}{—}CH-CO-CH=N_2$$

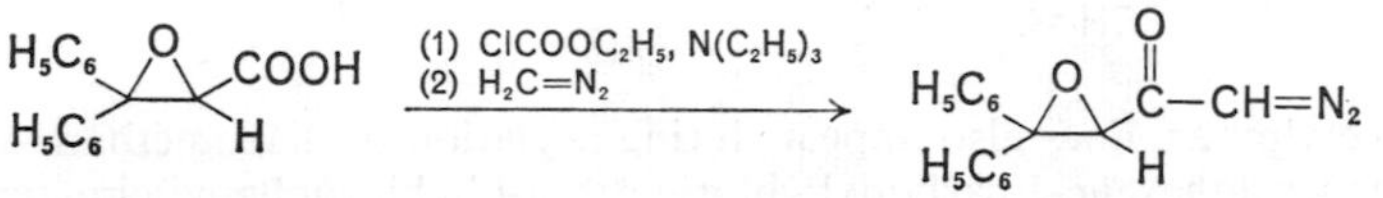

(81)

The acylation reactions can also be carried out with carboxylic acids after prior treatment with dicyclohexylcarbodiimide to form the corresponding anhydrides, which are then subjected without isolation to reaction with diazomethane; the diazo ketones **82a** [254] and **82b** [251] were obtained in this way.

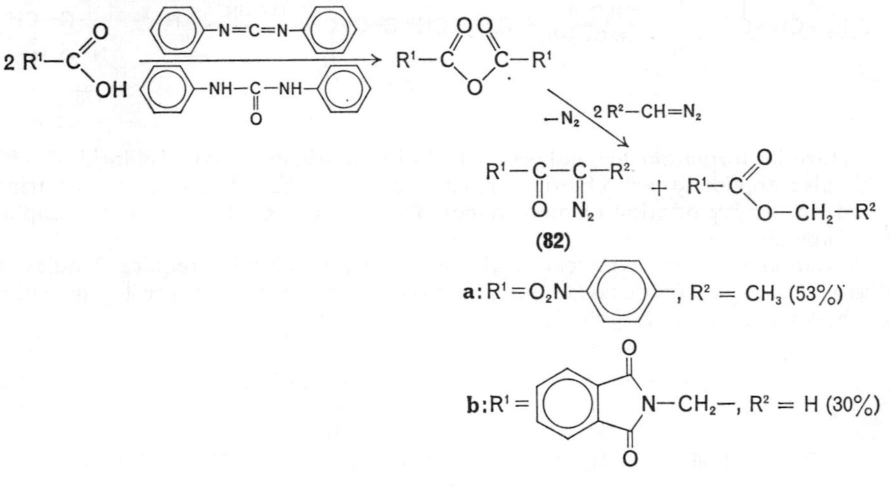

$$2\ R^1-C(=O)OH \xrightarrow[-\ C_6H_5-NH-CO-NH-C_6H_5]{C_6H_5-N=C=N-C_6H_5} R^1-CO-O-CO-R^1 \xrightarrow[-N_2]{2\ R^2-CH=N_2} R^1-\underset{\underset{O}{\|}}{C}-\underset{\underset{N_2}{\|}}{C}-R^2 + R^1-C(=O)-O-CH_2-R^2$$

(82)

a: $R^1 = O_2N-C_6H_4-$, $R^2 = CH_3$ (53%)

b: $R^1 =$ phthalimido$-N-CH_2-$, $R^2 = H$ (30%)

Acylations of α-diazo carbonyl compounds are restricted almost exclusively to reactions between acyl chlorides and diazoacetic ester. Once again moisture must be excluded because ester and anhydride formation have been observed as a consequence of partial hydrolysis of the acyl chloride to the carboxylic acid[255, 256]. In principle, ethyl diazoacetate can also be acylated with anhydrides, as demonstrated by the reaction with trifluoroacetic anhydride shown in equation (39); consumption of diazo ester can be reduced by neutralizing the trifluoroacetic acid inevitably formed with pyridine[257, 258].

$$F_3C{-}C(=O){-}O{-}C(=O){-}CF_3 + HC(=N_2){-}COOC_2H_5 \xrightarrow[-\,C_5H_5NH^+\;F_3C{-}COO^-]{C_5H_5N} F_3C{-}C(=O){-}C(=N_2){-}COOC_2H_5 \qquad (39)$$

Acyl isocyanates have recently been found to be suitable acylation reagents for diazomethyl carbonyl compounds (equation 40)[259, 260]. The success of this reaction, which has still not been exploited to the full, appears to be attributable to the enhanced electrophilic nature of the heterocumulene CO group due to the acyl group.

$$C_6H_5{-}C(=X){-}N{=}C{=}O + HC(=N_2){-}C(=O){-}R \xrightarrow{\Delta} C_6H_5{-}C(=X){-}NH{-}C(=O){-}C(=N_2){-}C(=O){-}R \qquad (40)$$

$X = O, R = OC_2H_5$; $X = O, R = C_6H_5$; $X = S, R = C_6H_5$

Since diazoacetic esters show less tendency to undergo cycloaddition with electron-poor olefins than diazomethane, they are smoothly acylated by cinnamoyl bromide[213]. In other cases 3+2 cycloaddition is actually observed, but acylation still dominates[261, 262].

The synthetic examples presented in Table 5 are intended to demonstrate the versatility of diazoalkane acylation.

c. *Metalation.* Systematic studies on the metalation of diazoalkanes were investigated only a few years ago, although the mercuration of ethyl diazoacetate has been known for 80 years[289]. The same diazo compound is metalated by butyllithium in ether or tetrahydrofuran/ether at −110 °C[290, 291]. Ethyl lithiodiazoacetate is also accessible by transmetalation from diethyl mercuriobis(diazoacetate) (equation 41)[290]. Since the lithium compound is unstable it is reacted directly with electrophiles at low temperatures[290, 291].

$$HC(=N_2){-}COOC_2H_5 \xrightarrow[-110\,°C]{C_4H_9Li} LiC(=N_2){-}COOC_2H_5 \xleftarrow[R{-}SLi,\ -70\,°C]{C_6H_5Li,\ CH_3Li\ \text{or}} Hg[C(=N_2){-}COOC_2H_5]_2 \qquad (41)$$

Metalation of ethyl diazoacetate with methylmagnesium iodide is likewise feasible at very low temperatures; it remains an open question whether the ethyl(iodomagnesio)diazoacetate (**83**) coexists with **84** and magnesium iodide in a conceivable 'Schlenck equilibrium'[291].

$$2\,HC(=N_2){-}COOC_2H_5 \xrightarrow[-2\,CH_4]{2\,H_3CMgI,\ \text{ether/THF},\ -65\,°C} 2\,I{-}Mg{-}C(=N_2){-}COOC_2H_5\ \mathbf{(83)} \qquad Mg[C(=N_2){-}COOC_2H_5]_2\ \mathbf{(84)} + MgI_2$$

While **83** is known only in solution, diethyl zincio- and cadmiobis(diazoacetate) can be isolated as tolerably stable oils[292]. They are prepared by metalating ethyl diazoacetate according to equation (42) with zinc or cadmium bis(trimethylsilyl) amide.

TABLE 5. Acylation of diazoalkanes and diazoesters

Diazo carbonyl compound	Yield (%)	Reference	Diazo carbonyl compound	Yield (%)	Reference
$HC(=O)-CH(=N_2)$	4	263	cyclopropyl(H)$-C(=O)-CH(=N_2)$	—	271
$O_2N-C_6H_4-O-C(=O)-CH(=N_2)$	97	264	decalin with $CH_2CH_2-C(=O)-CH(=N_2)$ and $COOCH_3$	—	272
$CH_3-S-C(=O)-CH(=N_2)$	25	265	adamantyl$-CH_2-C(=O)-CH(=N_2)$	100	273
$CH_3-C(=O)-CH(=N_2)$	—	208	steroid: $H_3C-C(=O)-O$~ (ring A), 5-H, 17-$CO-CH=N_2$	—	274
$C_6H_5-S-C(=O)-NH-CH_2-C(=O)-CH(=N_2)$	38	266	$HC(=N_2)-C(=O)-(CH_2)_n-C(=O)-CH(=N_2)$, $n = 2–8$	12–90	275, 276, 277
$CCl_3-CCl_2-C(=O)-CH(=N_2)$	48	267	$HC(CH_2-C(=O)-CH(=N_2))_3$	64	278
$C_6H_5OOC-CH_2-C(=O)-CH(=N_2)$	77	268			
$C_2H_5OOC-CH(NH-CO-CF_3)-CH_2-CH_2-C(=O)-CH(=N_2)$	95	269			
$CH_3-C(NO_2)_2-CH_2-CH_2-C(=O)-CH(=N_2)$	48	270			

Compound	Yield (%)	Ref.	Compound	Yield (%)	Ref.
$2\text{-}O_2NC_6H_4\text{-}C(=O)\text{-}CH=N_2$	87	208	$N_2=CH\text{-}C(=O)\text{-}C_6H_4\text{-}C(=O)\text{-}CH=N_2$	100	279
$2\text{-}(Tos\text{-}HN)C_6H_4\text{-}C(=O)\text{-}CH=N_2$	73	280	$[CH_2=C(CH_3)\text{-}CH_2]_2C[C(=O)\text{-}CH=N_2]_2$	—	285
$2\text{-}C_{10}H_7\text{-}C(=O)\text{-}C(=N_2)\text{-}CH_3$	78	281	$(\text{cyclohex-3-enyl})\text{-}CH_2\text{-}C(=O)\text{-}CH=N_2$	100	286
$C_6H_5\text{-}C(=O)\text{-}C(=N_2)\text{-}CH_2CH_2COOCH_3$	90	230	$(\text{methoxy-octahydrophenanthryl, } CH_3O, H)\text{-}C(=O)\text{-}CH=N_2$	100	287
$(\text{phthalimido})N\text{-}CH_2\text{-}C(=O)\text{-}C(=N_2)\text{-}CH_3$	58	282	$(\text{bicyclo, H})\text{-}C(=O)\text{-}CH=N_2$	100	288
$(2\text{-}H_5C_6\text{-}4\text{-}CH_3\text{-thiazol-5-yl})\text{-}C(=O)\text{-}CH=N_2$	66	283	$CH_3\text{-}C(=O)\text{-}C(=N_2)\text{-}COOC_2H_5$	—	213
$CH_2=C(CH_3)\text{-}CH(CH_3)\text{-}C(=O)\text{-}CH=N_2$	70–85	284	$Cl_2C=CCl\text{-}C(=O)\text{-}C(=N_2)\text{-}COOCH_3$	90	261
$CH_3\text{-}C\equiv C\text{-}(CH_2)_4\text{-}C(=O)\text{-}CH=N_2$	—	239	$(2\text{-furyl})\text{-}C(=O)\text{-}C(=N_2)\text{-}COOCH_3$	80	256
$N_2=CH\text{-}C(=O)\text{-}CH_2\text{-}CH=CH\text{-}CH_2\text{-}C(=O)\text{-}CH=N_2$	82	238	$CH_3OOC\text{-}C(=N_2)\text{-}C(=O)\text{-}(CH_2)_n\text{-}C(=O)\text{-}C(=N_2)\text{-}COOCH_3$, $n = 4–7$	50–65	265

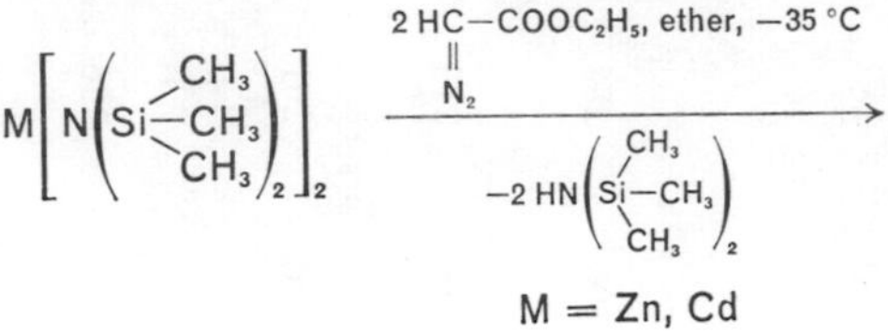

$$M\left[N\left(Si(CH_3)_3\right)_2\right]_2 \xrightarrow[-2\,HN(Si(CH_3)_3)_2]{2\,HC(=N_2)-COOC_2H_5,\ \text{ether},\ -35\,^\circ C} M\left(C(=N_2)-COOC_2H_5\right)_2 \qquad (42)$$

$M = Zn, Cd$

Mercuration of ethyl diazoacetate occurs on mere treatment with mercuric oxide[289], apparently due to its CH acidity. The reaction takes place in ether, light petroleum, or methylene chloride and can be applied generally to diazomethylcarbonyl compounds, as shown in equation (43).

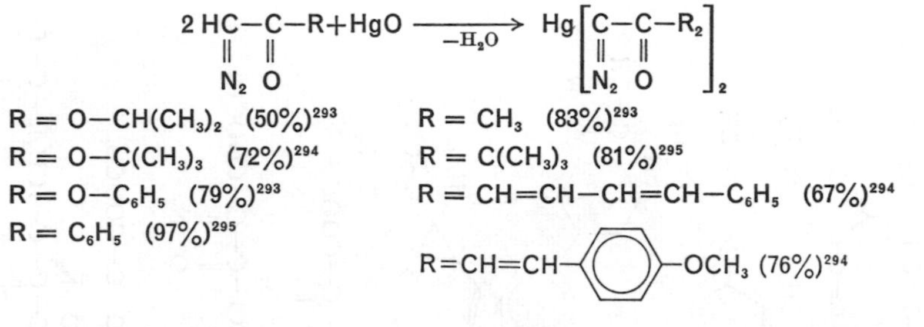

$$2\,HC(=N_2)-C(=O)-R + HgO \xrightarrow{-H_2O} Hg\left[C(=N_2)-C(=O)-R\right]_2 \qquad (43)$$

$R = O-CH(CH_3)_2$ (50%)[293]
$R = O-C(CH_3)_3$ (72%)[294]
$R = O-C_6H_5$ (79%)[293]
$R = C_6H_5$ (97%)[295]
$R = CH_3$ (83%)[293]
$R = C(CH_3)_3$ (81%)[295]
$R = CH=CH-CH=CH-C_6H_5$ (67%)[294]
$R = CH=CH-C_6H_4-OCH_3$ (76%)[294]

An alternative to mercuration of ethyl diazoacetate with mercuric oxide is provided by metalation with mercury bis(trimethylsilyl)amide[292]; mercuriobis(diazomethyl ketones) are also accessible via the same method[296]. Apart from the metal amide method[292, 296], a second possible approach to the preparation of alkyl- or aryl-mercuriodiazomethylcarbonyl compounds consists in mercuration with mercury alkoxides[297] (equation 44).

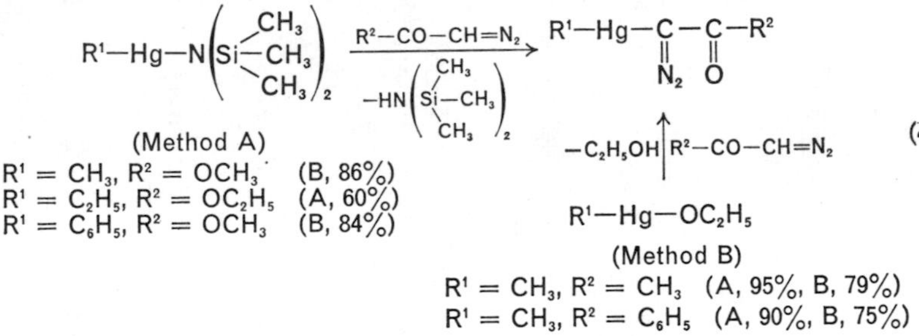

$$R^1-Hg-N\left(Si(CH_3)_3\right)_2 \xrightarrow[-HN(Si(CH_3)_3)_2]{R^2-CO-CH=N_2} R^1-Hg-C(=N_2)-C(=O)-R^2 \xleftarrow[-C_2H_5OH]{R^2-CO-CH=N_2} R^1-Hg-OC_2H_5 \qquad (44)$$

(Method A) — (Method B)

$R^1 = CH_3, R^2 = OCH_3$ (B, 86%)
$R^1 = C_2H_5, R^2 = OC_2H_5$ (A, 60%)
$R^1 = C_6H_5, R^2 = OCH_3$ (B, 84%)
$R^1 = CH_3, R^2 = CH_3$ (A, 95%, B, 79%)
$R^1 = CH_3, R^2 = C_6H_5$ (A, 90%, B, 75%)

The acidity of diazomethylcarbonyl compounds also permits the analogous direct introduction of silver with silver oxide[298, 299]. Unlike the corresponding Hg derivatives, argentiodiazomethylcarbonyl compounds (**85**) are thermally unstable and can be isolated only with difficulty.

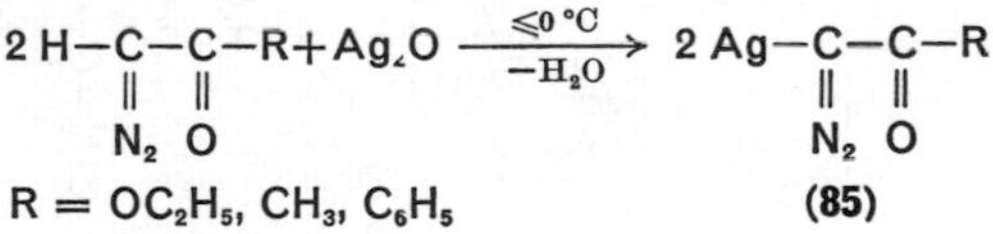

$$2\,H-C(=N_2)-C(=O)-R + Ag_2O \xrightarrow[-H_2O]{\leq 0\,^\circ C} 2\,Ag-C(=N_2)-C(=O)-R$$

$R = OC_2H_5, CH_3, C_6H_5$ — (**85**)

Methyl diazoacetate can only be silylated by a circuitous route proceeding via methyl diazotriethylstannylacetate, which exchanges its metal group on warming with bromotrimethylsilane (equation 45)[300].

$$(C_2H_5)_3Sn-\underset{\underset{N_2}{\|}}{C}-COOCH_3 \xrightarrow[-(H_5C_2)_3SnBr]{(CH_3)_3SiBr,\ 80\,^\circ C} (CH_3)_3Si-\underset{\underset{N_2}{\|}}{C}-COOCH_3 \qquad (45)$$

In contrast, ethyl diazoacetate can be germylated directly by the 'metal amide method' to give ethyl diazotrimethylgermylacetate[292, 301]. Ethyl diazotrimethylstannylacetate and other stannylated diazomethylcarbonyl compounds (**86**) are obtained by the same method[300, 302].

$$R^1_3Sn-N(CH_3)_2+H\underset{\underset{N_2}{\|}}{C}-\underset{\underset{O}{\|}}{C}-R^2 \xrightarrow[-HN(CH_3)_2]{ether} R^1_3Sn-\underset{\underset{N_2}{\|}}{C}-\underset{\underset{O}{\|}}{C}-R^2$$

(**86**)

$R^1 = CH_3, R^2 = OC_2H_5$ (100%); $R^1 = C_6H_5, R^2 = OC_2H_5$ (100%); $R^1 = C_2H_5, R^2 = CH_3$ (80%); $R^1 = C_2H_5, R^2 = C_6H_5$ (100%).

Bifunctional stannylation reagents such as bis(dimethylamino)dimethylstannane react with ethyl diazoacetate in the molar ratio 1 : 2, as demonstrated by the formation of **87**[302].

$$(CH_3)_2Sn[N(CH_3)_2]_2+2\,H\underset{\underset{N_2}{\|}}{C}-COOC_2H_5 \xrightarrow[-2\,HN(CH_3)_2]{} (CH_3)_2Sn\left[\underset{\underset{N_2}{\|}}{C}-COOC_2H_5\right]_2$$

(**87**)
60–70%

The 'metal amide procedure' is also the method of choice for introducing organolead groups into diazomethylcarbonyl compounds; some examples are shown in equation (46)[292, 303, 304].

$$(CH_3)_3Pb-N\left(Si\begin{matrix}CH_3\\CH_3\\CH_3\end{matrix}\right)_2 + H\underset{\underset{N_2}{\|}}{C}-\underset{\underset{O}{\|}}{C}-R \xrightarrow[-HN\left(Si\begin{matrix}CH_3\\CH_3\\CH_3\end{matrix}\right)_2]{Ether,\ -30\ to\ +20\,^\circ C} (CH_3)_3Pb-\underset{\underset{N_2}{\|}}{C}-\underset{\underset{O}{\|}}{C}-R \qquad (46)$$

$R = OC_2H_5, CH_3, C_6H_5$

The ethyl diazodimethylarsenio(antimonio and bismuthio)acetates are the first known representatives of diazomethylcarbonyl compounds bearing an As-, Sb-, or Bi-containing group on the diazo carbon atom; they were also synthesized by the 'metal amide method'[305].

d. *Substitution via metalated derivatives.* Since metal atoms such as lithium, silver or mercury are more readily replaced than hydrogen in substitution reactions at the CN_2 group, attention has recently been directed to performing transmetalations, halogenations, and also alkylations, via such derivatives of diazomethylcarbonyl compounds.

For instance, the highly reactive compound ethyl lithiodiazoacetate reacts at as low a temperature as −100 to −110 °C with trialkylmetal halides to form the correspondingly metalated diazoacetic esters (**88**)[291, 303].

$$R_3Me{-}\underset{\underset{N_2}{\|}}{C}{-}COOC_2H_5 \xleftarrow[-LiCl]{R_3MeCl} Li{-}\underset{\underset{N_2}{\|}}{C}{-}COOC_2H_5 \xrightarrow[-LiCl]{\text{2-chloro-1,3,2-benzodioxaborole}} (89)$$

(88) Me = Si, R = CH_3 [291]
Me = Sn, R = C_4H_9 [291]
Me = Pb, R = CH_3 [303]

(89): 1,3,2-benzodioxaborol-2-yl$-\underset{\underset{N_2}{\|}}{C}{-}COOC_2H_5$

The same method was used to prepare the first representative of the previously unknown boron-substituted diazoacetic esters, viz. **89** [291]; it can also be synthesized from diethyl mercuriobis(diazoacetate) and 2-chloro-1,3,2-benzodioxaborole[291]. Applying the same approach to the Hg derivative and trisubstituted iodosilanes affords silylated diazoacetic esters (80–90%)[306]. A variant of the procedure consists in the introduction of organosilicon, organotin or organogermanium groups by reaction of diethyl mercuriobis(diazoacetate) with corresponding organometal sulphides (equation 47)[303, 307, 308].

$$Hg\left[\underset{\underset{N_2}{\|}}{C}{-}COOC_2H_5\right]_2 + (R_3M)_2S \longrightarrow 2\,R_3M{-}\underset{\underset{N_2}{\|}}{C}{-}COOC_2H_5 + HgS \qquad (47)$$

R = CH_3, M = Si[307, 308], Ge[303], Sn[303]

Direct halogenation of diazomethylcarbonyl compounds has not yet been reported. Hence, the reaction of metal derivatives with halogens or halogenated reactants represents the only possible method of halogenating the diazomethyl carbon atom. Some details concerning the preparation of halogenated diazoacetic esters by this procedure are given in equation (48)[309, 310]; they are rather unstable oils. Bromination[309] and iodination[299] are also possible via ethyl argentiodiazoacetate.

$$Hg\left[\underset{\underset{N_2}{\|}}{C}{-}COOC_2H_5\right]_2 \xrightarrow{\text{Halogenation}} X{-}\underset{\underset{N_2}{\|}}{C}{-}COOC_2H_5 \qquad (48)$$

X = Cl (SO_2Cl_2 in $ClCF_3$ at −30 °C, 30%)[309]
X = Br (Br_2 in ether/THF at −100 °C, 80–90%)[309]
X = I (I_2 in ether at 0°C, 70–90%)[309, 310]

Diazomethylcarbonyl compounds can be C-alkylated by reaction of their lithium, mercury or silver derivatives with S_N1-active halides. Thus ethyl lithiodiazoacetate is smoothly alkylated with allyl iodide to give **90** [290]; the analogous reaction with tri-*t*-butylcyclopropenylium tetrafluoroborate plausibly affords **91** [311].

$$\underset{(90)}{H_2C{=}CH{-}CH_2{-}\underset{\underset{N_2}{\|}}{C}{-}COOC_2H_5} \xleftarrow[-LiI]{H_2C=CH-CH_2I} Li{-}\underset{\underset{N_2}{\|}}{C}{-}COOC_2H_5 \xrightarrow[-Li^+\ BF_4^-]{(t\text{-}Bu)_3C_3^+\ BF_4^-} (91)$$

(91): 1,2,3-tri-*t*-butylcycloprop-2-en-1-yl$-\underset{\underset{N_2}{\|}}{C}{-}COOC_2H_5$

Carbon-alkylations are most successful with silver compounds; several examples are shown in equations (49)[298, 299].

$$R^1{-}CH_2Hal + Ag{-}\underset{\underset{N_2}{\|}}{C}{-}\underset{\underset{O}{\|}}{C}{-}R^2 \longrightarrow R^1{-}CH_2{-}\underset{\underset{N_2}{\|}}{C}{-}\underset{\underset{O}{\|}}{C}{-}R^2 + AgHal \quad (49)$$

$R^1 = CH{=}CH_2$, $R^2 = OC_2H_5$ (66%); $R^1 = C_6H_5$, $R^2 = OC_2H_5$ (59%); $R^1 = CH{=}CH_2$, $R^2 = CH_3$ (25%); $R^1 = CH{=}CH_2$, $R^2 = C_6H_5$ (55%); $R^1 = CH{=}CH_2$, $R^2 =$ 2-thienyl (52%)

Particular mention should be made of the reaction between ethyl argentiodiazoacetate and crotyl bromide since it affords an isomeric mixture of **93** and **94** (51%, 85 : 15)[299]. Its interpretation assumes the intermediacy of the ion pair **92** which leads to branching of the reaction[299].

$$CH_3{-}CH{=}CH{-}CH_2I + Ag{-}\underset{\underset{N_2}{\|}}{C}{-}COOC_2H_5 \longrightarrow [\text{ion pair } \mathbf{(92)}:\ H_3C{-}\overset{+}{C}H{\cdots}CH{\cdots}CH_2 \ / \ I{-}\overset{-}{Ag}{-}\underset{\underset{N_2}{\|}}{C}{-}COOC_2H_5]$$

$$\mathbf{(92)} \xrightarrow{-AgI} CH_3{-}CH{=}CH{-}CH_2{-}\underset{\underset{N_2}{\|}}{C}{-}COOC_2H_5 \ \mathbf{(93)} \quad + \quad CH_2{=}CH{-}\underset{\underset{CH_3}{|}}{CH}{-}\underset{\underset{N_2}{\|}}{C}{-}COOC_2H_5 \ \mathbf{(94)}$$

Finally, the C-alkylation of diethyl mercuriobis(diazoacetate) with bromotriphenylmethane[299], which inexplicably fails when attempted with ethyl argentiodiazoacetate, also warrants attention.

e. *Addition reactions.* The CH acidity of the diazo component is of decisive importance for addition reactions of diazomethylcarbonyl compounds to C=O or C=C double bonds which leave the diazo group intact. An historical precursor of this kind of reaction is the aldol-type addition of ethyl diazoacetate to alloxan tetrahydrate which gives the diazo ester **95**[312].

$$\text{alloxan}\cdot 4H_2O \xrightarrow{HC(=N_2){-}COOC_2H_5,\ \text{without solvent}} \text{5-hydroxy-5-}[C(=N_2){-}COOC_2H_5]\text{-barbituric acid } \mathbf{(95)}$$

The base-catalysed reaction of 1-diazo-3-phenyl-2-propanone with benzaldehyde is of particular interest in that reaction occurs exclusively at the diazomethyl carbon to form **96** and not at the active CH_2 group (formation of **97**)[313].

$$C_6H_5{-}CH_2{-}\underset{\underset{O}{\|}}{C}{-}\underset{\underset{N_2}{\|}}{CH} + C_6H_5{-}CHO \xrightarrow{NaOH} C_6H_5{-}CH_2{-}\underset{\underset{O}{\|}}{C}{-}\underset{\underset{N_2}{\|}}{C}{-}\underset{\underset{OH}{|}}{CH}{-}C_6H_5 \quad \textbf{(96)}$$

$$\not\rightarrow C_6H_5{-}\underset{\underset{OH}{|}}{CH}{-}\overset{\overset{C_6H_5}{|}}{CH}{-}\underset{\underset{O}{\|}}{C}{-}\underset{\underset{N_2}{\|}}{CH} \quad \textbf{(97)}$$

Numerous cases of such base-catalysed aldol additions of ethyl diazoacetate and 1-diazo-2-propanone to aldehydes have been discovered; equation (50) shows some representative examples[314].

$$R^1{-}CHO + \underset{\underset{N_2}{\|}}{HC}{-}\underset{\underset{O}{\|}}{C}{-}R^2 \xrightarrow[\text{or methanol}]{\text{KOH in ethanol}} R^1{-}\underset{\underset{OH}{|}}{CH}{-}\underset{\underset{N_2}{\|}}{C}{-}\underset{\underset{O}{\|}}{C}{-}R^2 \qquad (50)$$

$R^1 = CH_3$, $R^2 = OC_2H_5$ (90%); $R^1 = CH(CH_3)_2$, $R^2 = OC_2H_5$ (80%); $R^1 = C(CH_3)_3$, $R^2 = OC_2H_5$ (70%); R^1 = Cyclohexyl, $R^2 = OC_2H_5$ (90%); $R^1 = C_6H_5$, $R^2 = OC_2H_5$ (60%); $R^1 = C_6H_5$, $R^2 = CH_3$ (68%);

$R^1 = O_2N{-}C_6H_4{-}$ (para), $R^2 = CH_3$ (86%); $R^1 = Cl{-}C_6H_3(Cl){-}$ (3,4-dichlorophenyl), $R^2 = C_6H_5$ (40%)

1,2-Additions of ethyl diazoacetate and 1-diazo-2-propanone to aldehydes occur at very low temperatures on use of organometallic bases such as lithium diisopropylamide or *n*-butyllithium[291]. The diazomethylcarbonyl compound may of course be metalated first and then reacted with aldehydes[290].

The result of the reaction between 1-chloro-3-diazo-2-propanone and benzaldehyde depends upon the molar ratio of the reactants; 1 : 1 reaction proceeds stereoselectively via Darzens condensation to give **98**, while working with an excess of benzaldehyde leads to both **98** and a diastereomeric mixture of **99**[314a].

$$ClH_2C{-}\underset{\underset{O}{\|}}{C}{-}\underset{\underset{N_2}{\|}}{CH} \xrightarrow[NaOH]{C_6H_5{-}CHO,\ CH_3OH} \text{(3-phenyloxiranyl)}{-}\overset{\overset{O}{\|}}{C}{-}\underset{\underset{N_2}{\|}}{CH}\ \textbf{(98)} \xrightarrow{C_6H_5CHO} \text{(3-phenyloxiranyl)}{-}\overset{\overset{O}{\|}}{C}{-}\underset{\underset{N_2}{\|}}{C}{-}\underset{\underset{OH}{|}}{CH}{-}C_6H_5\ \textbf{(99)}$$

The importance of the electron density at the carbonyl carbon atom for aldol additions becomes plain on going to the less reactive ketones. Hence ethyl diazoacetate still adds to cyclobutanone in a KOH-catalysed reaction[314]; however, other ketones require use of organometallic bases such as *n*-butyllithium which enhance the nucleophilic character of the diazo carbon atom (equation 51)[291].

$$(CH_2)_{n-1}\,C{=}O \xrightarrow{\underset{\underset{N_2}{\|}}{HC}{-}COOC_2H_5} (CH_2)_{n-1}\,C(OH){-}\underset{\underset{N_2}{\|}}{C}{-}COOC_2H_5 \qquad (51)$$

$n = 4$ (KOH); $n = 5$ (C_4H_9Li); $n = 6$ (C_4H_9Li)

Aldol additions lead to cyclization if the diazomethyl and the carbonyl group are components of the same molecule; this is demonstrated by equations (52) and (53)[315].

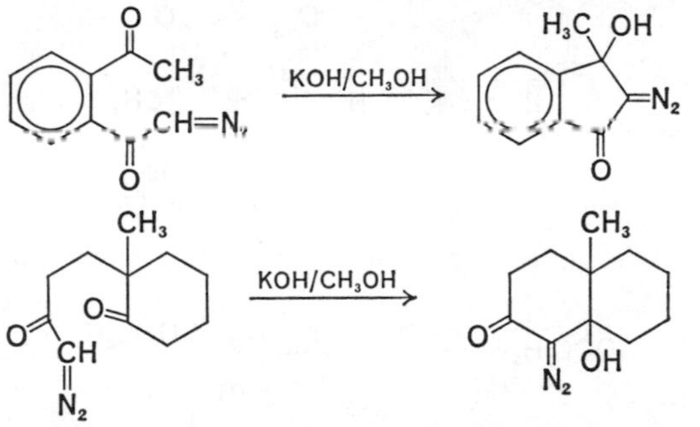
CH₃
CH=N₂
KOH/CH₃OH
H₃C OH
N₂
KOH/CH₃OH
CH
N₂
OH

(52)

(53)

There are only few examples of addition reactions with acyclic α-dicarbonyl compounds, such as that of ethyl phenylglyoxylate with ethyl diazoacetate leading to **100**[314]. Addition of the same diazo compound to dimethyl acetylphosphonate is also of interest in this context; it proceeds without catalysis and leads to **101**[316].

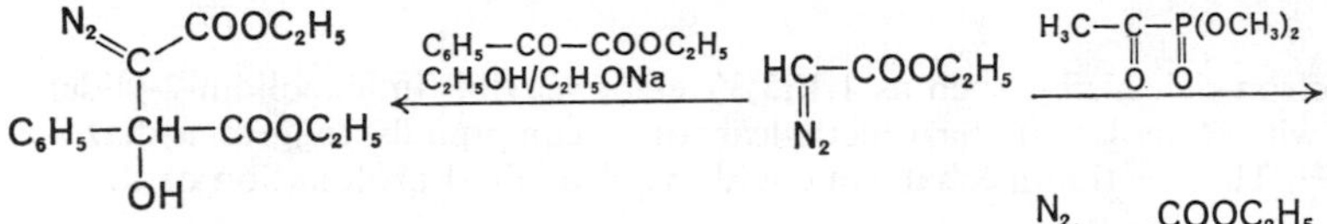
COOC₂H₅
C₆H₅—CH—COOC₂H₅
OH
C₆H₅—CO—COOC₂H₅
C₂H₅OH/C₂H₅ONa
HC—COOC₂H₅
H₃C—C—P(OCH₃)₂
H₃C—C—P(OCH₃)₂

(100)

(101)

By way of contrast, addition reactions to cyclic 1,2-di- and 1,2,3-tricarbonyl compounds are numerous. Owing to the pronounced reactivity of the central CO group, the latter compounds do not require base catalysis; neither do some of the 1,2-dicarbonyl compounds. Compounds **102–106** represent typical adducts.

Disregarding the addition of ethyl diazoacetate to *N*-cyclohexylidenebenzylamine[314], 1,2-addition has only been reported for azomethines bearing powerfully electron-withdrawing substituents, as illustrated by examples **107** and **108**[324].

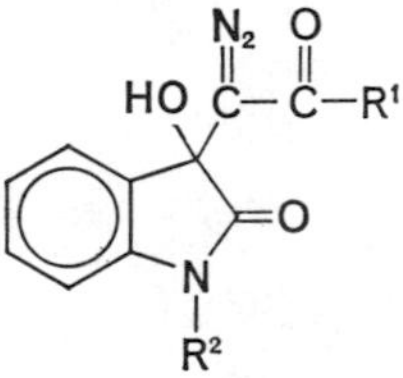
HO
R¹
R²

(102)

$R = OC_2H_5$, $R^2 = CH_3$ (86%)[317]
$R = C_6H_5$, $R^2 = CH_3$ (40%)[318]
$R = CH_3$, $R^2 = H$ (50%)[318]
$R = OC_2H_5$, $R^2 = OCOCH_3$ (74%)[319]

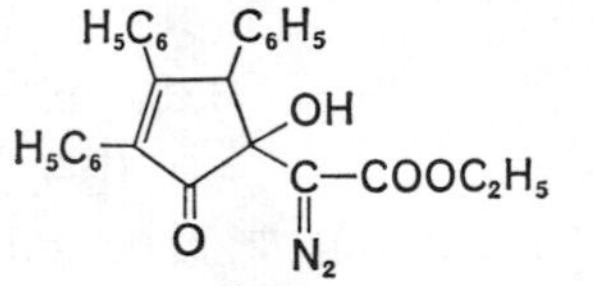
H₅C₆
C₆H₅
OH
C—COOC₂H₅
N₂

(103)
59%[320]

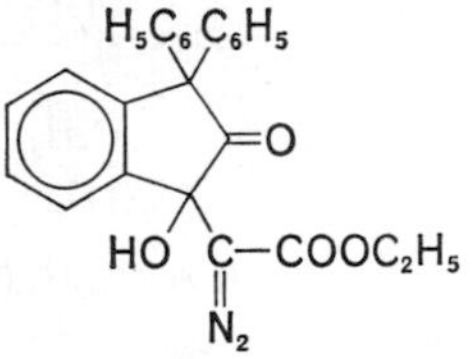
H₅C₆ C₆H₅
HO
C—COOC₂H₅
N₂

(104)
99%[321]

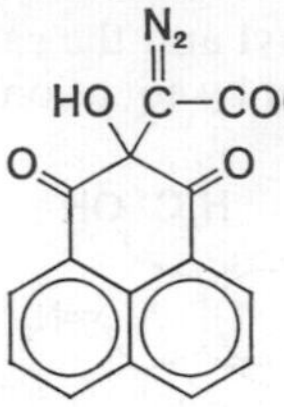

(105) 80%[322]

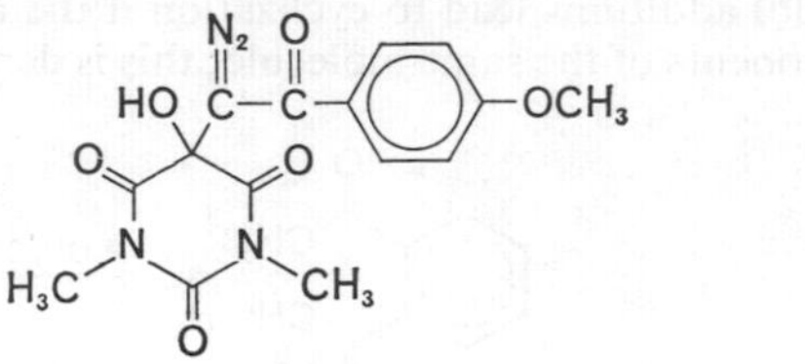

(106) 35%[323]

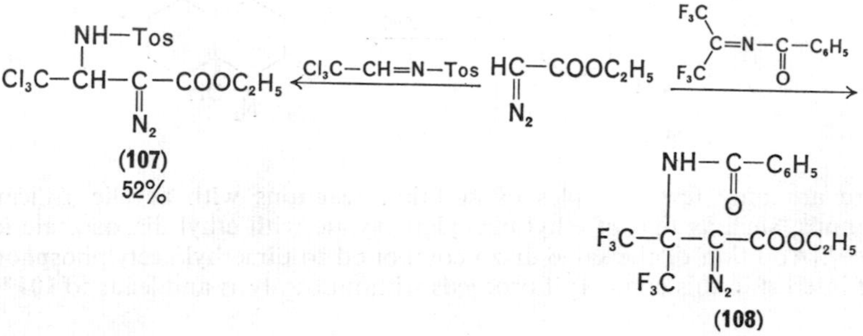

(107) 52%

(108) 78%

Electron-rich olefins such as 1,1′,3,3′-tetraphenylbis(imidazolidin-2-ylidene) (**109**) react with 2 moles of diazomethylcarbonyl compounds to give α-diazo aminals (**112**)[325]. The reaction mechanism could involve initial protonation of the olefin **109**

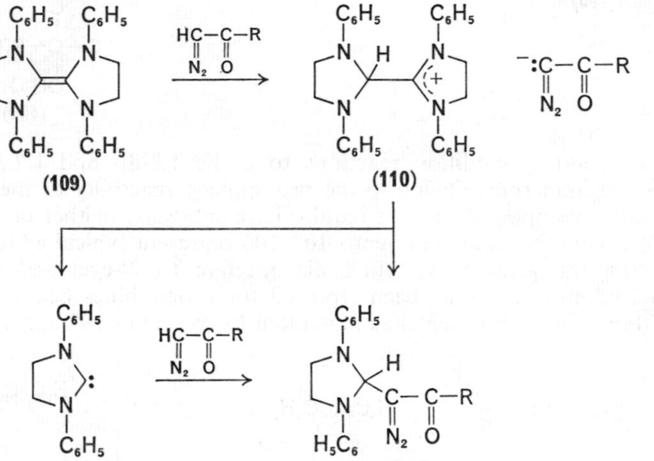

(109) (110)

(111) (112)

R = OC_2H_5 (78%); $OC(CH_3)_3$ (72%); $N(C_2H_5)_2$ (69%); N-piperidino (100%)

by the CH-acidic diazo compound to give **110**, which decomposes to form the diazo aminal **112** together with the nucleophilic carbene **111**. It has not yet been established whether the latter product undergoes direct CH insertion with the diazomethylcarbonyl compound to give **112** or re-enters the reaction after dimerization to **109**[325].

Diazomethylcarbonyl compounds react with enamines in the manner of a C-alkylation without cleavage; the possible 1,3-dipolar cycloaddition fails to occur (equation 54)[314]. This reaction is also to be interpreted as a consequence of the CH acidity of the diazo component.

$$R^1_2C{=}CH{-}N(CH_2)_4 + HC(=N_2){-}C(=O){-}R^2 \longrightarrow R^1_2CH{-}CH(N(CH_2)_4){-}C(=N_2){-}C(=O){-}R^2 \qquad (54)$$

$R^1 = CH_3$, $R^2 = OC_2H_5$ (81%); R^1,R^1 = cyclopentane ring, $R^2 = OC_2H_5$ (81%);

R^1,R^1 = cyclohexane ring, $R^2 = C_6H_5$ (36%)

f. *Acyl cleavage.* α-Diazo β-dicarbonyl compounds have long been known to undergo acyl cleavage in basic media (equation 55)[213, 326]. The first-mentioned

$$R^1{-}C(=O){-}C(=N_2){-}C(=O){-}R^2 \xrightarrow[\text{cleavage (X—OH)}]{\text{Alkaline acyl}} R^1{-}COOX + HC(=N_2){-}C(=O){-}R^2 \qquad (55)$$

$R^1 = C_6H_5$, $R^2 = OCH_3$, $X = CH_3$ [213]

$R^1 = R^2 = CH_3$, $X = H$ [326]

example and the following cleavages clearly show that in cases of unsymmetrical acyl substitution of the diazo carbon atom the group removed is always the one which is most susceptible to nucleophilic attack.

This reaction only acquired synthetic utility since α-diazo β-dicarbonyl compounds had become readily accessible by diazo group transfer. The most important examples are those involving α-diazo β-oxo carboxylic acid derivatives of type **113** from which the acetyl group is generally removed to give **114**. The reaction can also be executed

$$\underset{\textbf{(113)}}{CH_3{-}C(=O){-}C(=N_2){-}C(=O)X} \xrightarrow{(1)\ OH^-\ \text{or}\ OR^-;\ (2)\ H^+} CH_3{-}C(=O)OH + \underset{\textbf{(114)}}{HC(=N_2){-}C(=O)X}$$

$X = OC_2H_5$ [142,129]; OC_4H_9 [142]; $OC(CH_3)_3$ [127,142]; N-piperidino [142, 125]; $N(C_2H_5)_2$ [142]

$N(CH_3)(C_6H_5)$ [141]; CH_3 [129]; 3-O_2N-C_6H_4

by starting from suitable methylene compounds and choosing conditions that will permit diazo group transfer and acyl cleavage to occur in a one-pot reaction[142]. In some cases it is convenient to work in a two-phase system with addition of a quaternary ammonium salt[135].

The acyl cleavage of cyclic α-diazo β-dicarbonyl compounds is of course accompanied by ring opening, as demonstrated by the reaction of diazodimedone with caustic soda solution (equation 56)[129]. Ring cleavage may, however, also be followed by other reactions, as in the case of 2-diazo-1,3-dioxoindan (equation 57)[327].

$$\text{diazodimedone} \xrightarrow{NaOH,\ CH_3CN} (H_3C)_2C(CH_2COOH)CH_2-C(=O)-CH=N_2 \quad (56)$$

$$\text{2-diazo-1,3-dioxoindan} \xrightarrow{0{\cdot}5\ N\text{-}NaOH} o\text{-}(N_2=CH-C(=O))C_6H_4COO^-\ Na^+ \xrightarrow{2\ N\text{-}HCl} \text{isochroman-1,4-dione} \quad (57)$$

B. Phosphoryl Diazo Compounds

In contrast to carbonyl diazo compounds, our knowledge about phosphoryl diazo compounds is all very recent. Parallels and differences in syntheses and reactions are apparent from studies performed during the past 10 years; they were motivated in part by problems of organophosphorus chemistry.

1. Diazotization of amines

Although they are weaker proton activators than carbonyl groups, the influence of PO groups nevertheless suffices to promote deprotonation of the diazonium intermediates generated on diazotization of amines. Hence it is understandable that aminomethylphosphoryl compounds can be transformed into diazomethylphosphoryl compounds by reaction with nitrous acid (equation 58)[328–331].

$$R^1R^2P(=O)-CH_2NH_2 \xrightarrow{NaNO_2,\ H_3C-COOH} R^1R^2P(=O)-CH=N_2 \quad (58)$$

$R^1 = R^2 = OCH_3$[328]; $R^1 = R^2 = OC_2H_5$[329]; $R^1 = C_6H_5$, $R^2 = OCH_3$[330]; $R^1 = R^2 = C_6H_5$[329,331]

Apart from the above compounds, only (diazobenzyl)diphenylphosphine oxide ($R^1 = R^2 = C_6H_5$, C_6H_5 in place of H)[332] and (diazobenzyl)bis(4-methoxyphenyl)-phosphine oxide[333] have been synthesized by this method. Mineral acid media should be avoided to prevent acid-catalysed decomposition[334].

2. Bamford–Stevens reaction

The Bamford–Stevens reaction is an extremely good method for the synthesis of α-diazo phosphonic esters and α-diazo phosphinic esters; whether the procedure also leads to α-diazo phosphine oxides is not yet known. Its use requires ready availability of the α-oxo phosphoryl compounds, a condition generally satisfied by the Michaelis–Arbusov reaction which rarely fails to give the desired compound. The overall sequence is depicted in equation (59).

$$R^1(R^2O)P(OR^2) + X-C(=O)-R^3 \xrightarrow{-R^2X} R^1(R^2O)P(=O)-C(=O)-R^3 \xrightarrow{H_2N-NH-Tos} R^1(R^2O)P(=O)-C(=N-NH-Tos)-R^3 \quad (50)$$

$$\xrightarrow{OH^-} R^1(R^2O)P(=O)-C(=N_2)-R^3$$

R¹ = Aryl, OAlkyl

In numerous cases it is possible to isolate *syn/anti* isomeric tosylhydrazones and to establish the configuration at the C=N double bond by n.m.r. spectroscopy[330, 335, 336]; however, this is of no consequence for the ensuing alkaline cleavage (e.g. **115** → **117** and **116** → **117**)[335].

$$(C_2H_5O)_2P(=O)-C(=O)-C_6H_5 \xrightarrow{Tos-NH-NH_2,\ \Delta,\ C_2H_5OH} (C_2H_5O)_2P(=O)-C(C_6H_5)=N-NH-Tos \ \textbf{(115)} + (C_2H_5O)_2P(=O)-C(C_6H_5)=N-N(H\cdots O)-Tos \ \textbf{(116)}$$

$$\textbf{115}, \textbf{116} \xrightarrow{KOH} (C_2H_5O)_2P(=O)-C(=N_2)-C_6H_5 \ \textbf{(117)}$$

Dimethyl α-tosylhydrazonophosphonate has recently also been found to undergo cleavage to dimethyl α-diazophosphonate and toluenesulphinate on treatment with sodium borohydride in methanol[337]. However, the method does not appear to have any advantage over the usual alkaline cleavage. Surprisingly, when the same reaction is attempted in the aprotic solvent tetrahydrofuran, the tosylhydrazones are reduced to methylene compounds[337].

Table 6 conveys an impression of the scope and versatility of the Bamford–Stevens reaction as applied to the preparation of α-diazo phosphinic and α-diazo phosphonic esters. Many phosphoryl diazo compounds can be purified by distillation, being much more thermally stable than their carbonyl analogues[338, 339].

Two secondary reactions may attend the synthesis of unsaturated phosphoryl-diazoalkanes, viz. intramolecular 3+2 cycloaddition and 1,5-cyclization. The latter process is observed during the synthesis of dimethyl (1-diazo-2-alken-1-yl)phosphonate (**118**) and is extremely dependent upon the substituents attached to the double bond. Hence **118a** defies direct detection by rapid ring closure to give **119a**[347]; **118b** and **118c** can be isolated but cyclize slowly to **119b** and **119c**, respectively[344, 347]; while **118d** and **118e** do not display any tendency to undergo 1,5-ring closure[344, 346, 347]. The chemistry of the carbenes derived from **118** has become a topical field of study during recent years[346, 348, 349].

TABLE 6. α-Diazo phosphoryl compounds by the Bamford–Stevens reaction

$R^1(R^2O)P(=O)-C(=N_2)-R^3$

R^1	R^2	R^3	Yield (%)	Reference
CH_3O	CH_3	C_6H_5	93 51	328 336
C_2H_5O	C_2H_5	C_6H_5	70–87	335
$(CH_3)_2CHO$	$(CH_3)_2CH$	C_6H_5	54 95	336 340
CH_3O	CH_3	$p\text{-}XC_6H_4$		
		X = Br X = OCH_3	76 97	341 336
CH_3O	CH_3	1-naphthyl	53	342
C_2H_5O	C_2H_5	$p\text{-}XC_6H_4$		
		X = Cl X = CH_3	76 82	343 343

R^1	R^2	R^3	Yield (%)	Reference
CH_3O	CH_3	CH_3	44	344, 345
CH_3O	CH_3	$C_6H_5CH_2$	100	342
CH_3O	CH_3	$(CH_3)_2CH$	72	344
CH_3O	CH_3	$(CH_3)_3C$	64	344
CH_3O	CH_3	cyclopentyl	80	344
CH_3O	CH_3	cyclohexyl	82	344
C_6H_5	CH_3	C_6H_5	66	330
C_6H_5	CH_3	CH_3	59	330
C_6H_5	CH_3	$C_6H_5CH_2$	81	330
C_6H_5	CH_3	$H_2C{=}CH{-}CH_2$	68	330
C_6H_5	CH_3	$(CH_3)_2C{=}CH$	41	330

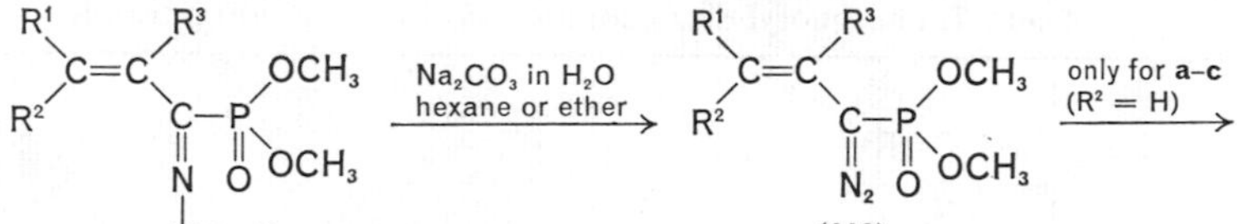

Na_2CO_3 in H_2O, hexane or ether →

only for **a–c** (R^2 = H) →

(118)

a: $R^1 = C_6H_{11}$, R^2 = H, $R^3 = CH_3$ (fast cyclization)
b: $R^1 = C_6H_5$, $R^2 = R^3$ = H (slow cyclization)
c: $R^1 = CH_3$, R^2 = H, $R^3 = CH_3$ (slow cyclization)
d: $R^1 = R^2 = CH_3$, R^3 = H (no cyclization)
e: $R^1 = C_6H_5$, $R^2 = CH_3$, R^3 = H (no cyclization)

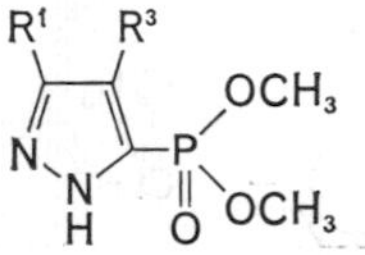

(119)

The second reaction type, viz. intramolecular 1,3-dipolar cycloaddition, is encountered with α-diazo phosphinic esters (**120**) bearing an unsaturated ether group. In all cases the diazo compounds can be detected at least by the appearance of the diazo stretching frequency in the i.r. spectrum; they transform at various rates into heterobicyclic compounds of type **121**[350]. A methyl substituent attached to one of the two double bond sites of the allyl group in the cases of R = C_6H_5 considerably retards the intramolecular cycloaddition[350].

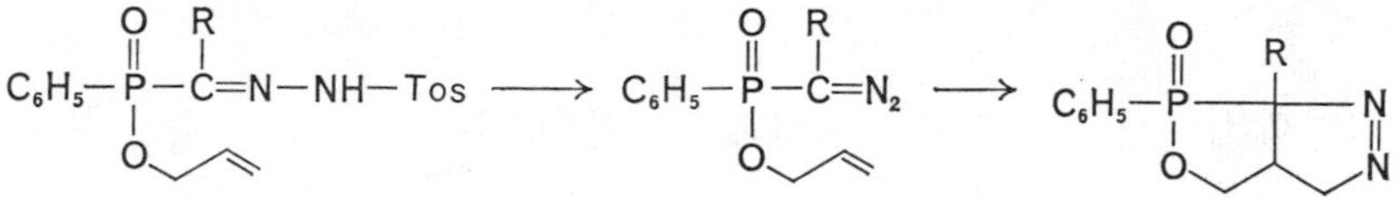

(120) **(121)**

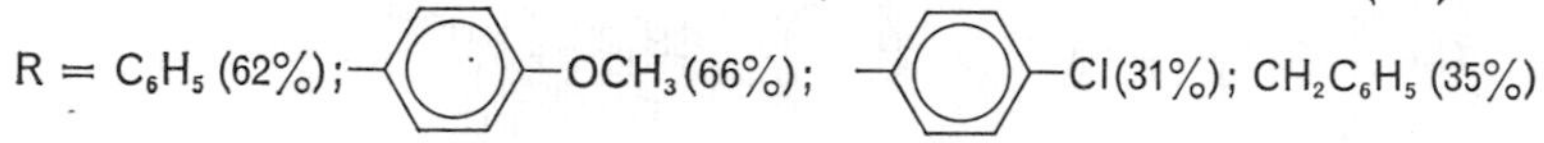

R = C_6H_5 (62%); C_6H_4–OCH_3 (66%); C_6H_4–Cl (31%); $CH_2C_6H_5$ (35%)

3. Diazo group transfer

Diazo group transfer has a very important part to play in the preparation of α-diazo phosphine oxides, α-diazo phosphonic esters and α-diazo phosphinic esters; this applies to both CH-acidic compounds and diazo acceptors containing partially unsaturated structures.

a. *Active methylene compounds*. As with the Bamford–Stevens reaction, the Michaelis–Arbusov reaction again proves invaluable for the synthesis of the starting materials. Deprotonation of the methylene compounds to give the reactive carbanions is effected with organic amines, alkali alkoxides, or organometallic bases as required; tosyl azide serves almost exclusively as diazo transfer reagent[127] (equation 60). Table 7 reveals the wide range of variation of substituents R^1 to R^3.

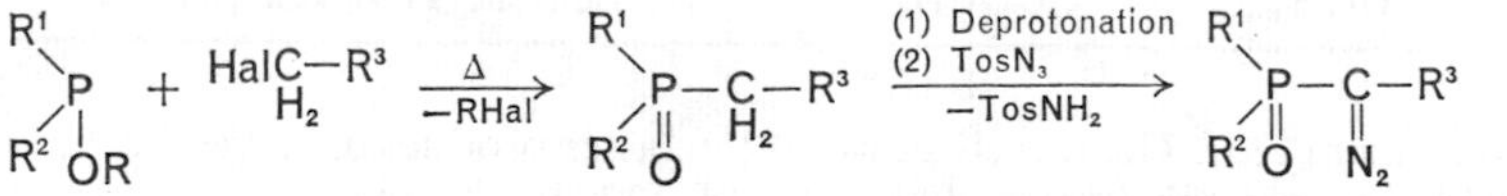

$$R^1R^2P{-}OR + HalCH_2{-}R^3 \xrightarrow[-RHal]{\Delta} R^1R^2P(O){-}CH_2{-}R^3 \xrightarrow[-TosNH_2]{(1)\ \text{Deprotonation}\ (2)\ TosN_3} R^1R^2P(O){-}C(N_2){-}R^3 \qquad (60)$$

An account of some secondary reactions associated with diazo transfer onto PO-activated methylene compounds completes this impression of the scope of the method. For instance, reaction of the phosphorylacetamide (**122**) with tosyl azide furnishes the triazole (**124**) which, however, can undergo thermal conversion into the

TABLE 7. Phosphoryl diazo compounds by diazo group transfer

General structure: $R^1R^2P(=O)-C(=N_2)-R^3$

R^1	R^2	R^3	Solvent/base	Yield (%)	Reference
C_6H_5	C_6H_5	C_6H_5	Ether/monoglyme, C_6H_5Li	30	338, 351
$4-ClC_6H_4$	$4-ClC_6H_4$	C_6H_5	Benzene/THF, C_4H_9Li	25	340
C_6H_5	C_6H_5	$COCH_3$	Benzene, $(CH_3)_3COK$	44	338, 351
C_6H_5	C_6H_5	COC_6H_5	CH_2Cl_2, piperidine	100	338, 351
C_6H_5	C_6H_5	$CO-C_6H_4-4-N(CH_3)_2$	CH_2Cl_2, piperidine	75	351
C_6H_5	C_6H_5	$CO-C_6H_4-4-NO_2$	$CH_2Cl_2,N(C_2H_5)_3$	72	352
C_6H_5	C_6H_5	$CO-C_6H_4-4-C_6H_5$	CH_2Cl_2, piperidine	71	353
C_6H_5	C_6H_5	$CO-(4-pyridyl)$	$CH_2Cl_2,N(C_2H_5)_3$	58	354
C_6H_5	C_6H_5	$COOC_2H_5$	Benzene/THF, $(CH_3)_3COK$	57–60	351, 355
C_2H_5O	C_2H_5O	C_6H_5	Benzene, C_6H_5Li	24	338, 335
CH_3O	CH_3O	$4-O_2NC_6H_4$	Benzene, $(CH_3)_3COK$	35	336
C_2H_5O	C_2H_5O	$COOC_2H_5$	Benzene, $(CH_3)_3COK$	37	335
C_2H_5O	C_2H_5O	$PO(OC_2H_5)_2$	Benzene, $(CH_3)_3COK$	35	335
C_6H_5	OCH_3	$COCH_3$	Benzene/THF, $(CH_3)_3COK$	31	330
C_6H_5	OCH_3	COC_6H_5	Benzene/THF, $(CH_3)_3COK$	42	330
C_6H_5	OCH_3	$COOCH_3$	Benzene/THF, $(CH_3)_3COK$	43	330

diazo isomer (**123**). The recyclization of the latter product observed in the presence of alkoxide suggests an intermediate role of **123** on the reaction pathway leading to **124**[351].

A diazoalkane intermediate again appears very likely in equation (61), although subsequent ring opening of the triazolopyridine is impossible[351]; an analogous reaction is known in the carbonyl series[149].

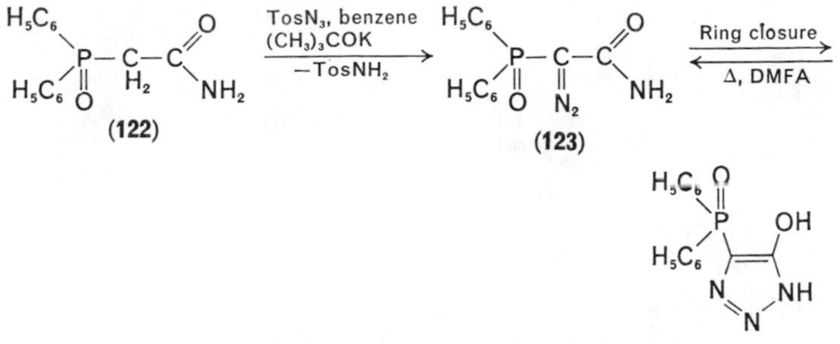

(122) —[TosN$_3$, benzene, $(CH_3)_3COK$, −TosNH$_2$]→ (123) ⇄ [Ring closure / Δ, DMFA] (124)

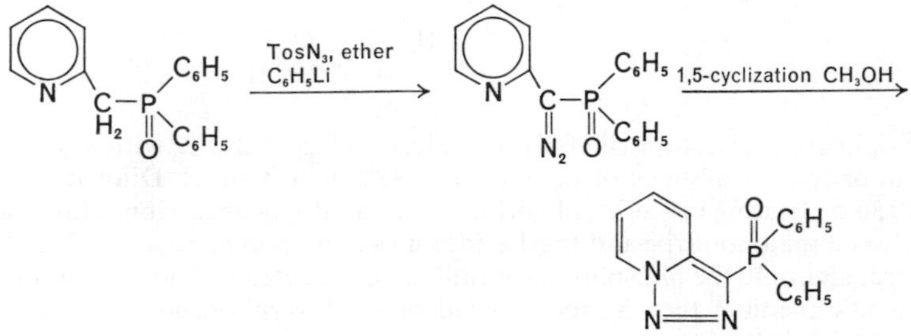

(61)

The first cyclic phosphoryl diazo compound has recently become accessible with the preparation of **125** by diazo transfer[354]; **125** is exceptionally sensitive to protic reagents and is cleaved, e.g. by methanol to the diazomethyl ketone **(126)**[354].

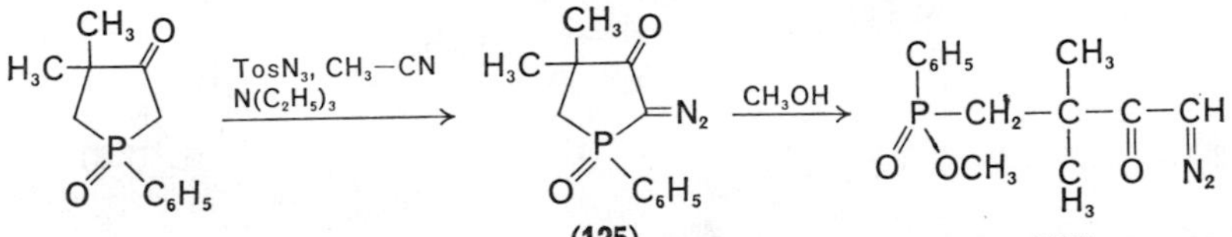

—[TosN$_3$, CH_3—CN, $N(C_2H_5)_3$]→ (125) —[CH_3OH]→ (126)

A surprising ligand exchange is observed on diazo group transfer onto diethyl phenacylphosphonate in the presence of phenyllithium as base. The expected diazo ester **(127)** is accompanied by the diazo phosphine oxide **(128)**: precisely at what stage substitution occurs remains an open question[335].

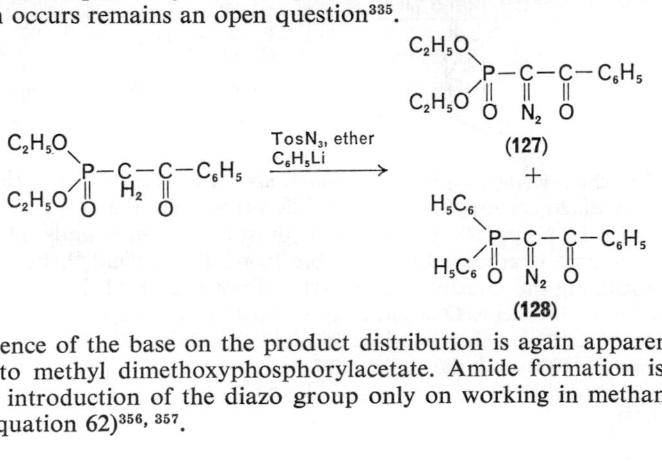

—[TosN$_3$, ether, C_6H_5Li]→ (127) + (128)

The influence of the base on the product distribution is again apparent in diazo transfer onto methyl dimethoxyphosphorylacetate. Amide formation is found to accompany introduction of the diazo group only on working in methanol/sodium ethoxide (equation 62)[356, 357].

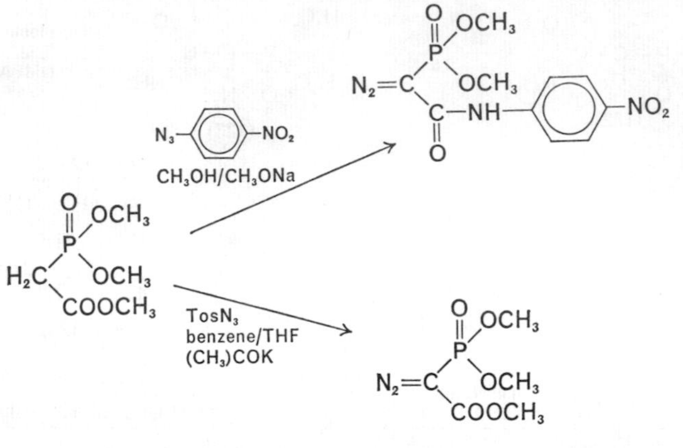

(62)

Diethylphosphorylacetonitrile fails to undergo diazo transfer with aryl azides but instead affords 1 : 1 adducts of composition **129**, which suffer Dimroth rearrangement to **130** on heating in acetic anhydride[357]. An analogous reaction is known to take place between malononitrile and tosyl azide; it occurs spontaneously[358–360]. However, if diazo transfer onto the phosphorylacetonitrile is executed with an azidinium salt in a weakly acidic medium then a modest yield of diethoxyphosphoryldiazoacetonitrile (**131**) is obtained[121].

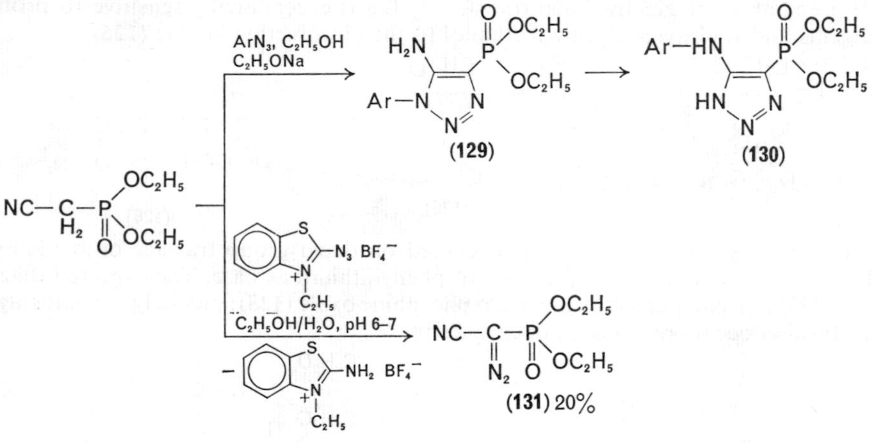

(**129**) (**130**)

(**131**) 20%

b. *α-Hydroxymethylenephosphoryl compounds*. In contrast to the situation described for α-diazo carbonyl compounds, deformylating diazo transfer plays only a minor role in the preparation of α-diazo phosphoryl compounds. Diethyl(diazomethyl)phosphonate is admittedly accessible in mediocre yield[361], but not so **135b** and **135c**. Assuming the intermediacy of triazolines **132** in all the reactions, that is where branching will occur. One pathway consists in cleavage after the manner of diazo transfer[361] and the other one in a kind of PO-activated olefination leading to the 1,2,3-triazole (**133**) and the anion **134**—it being immaterial at what stage the tosyl group is extruded[335, 351]. This reaction apparently prevents formation of **135b** and **135c**[335, 351].

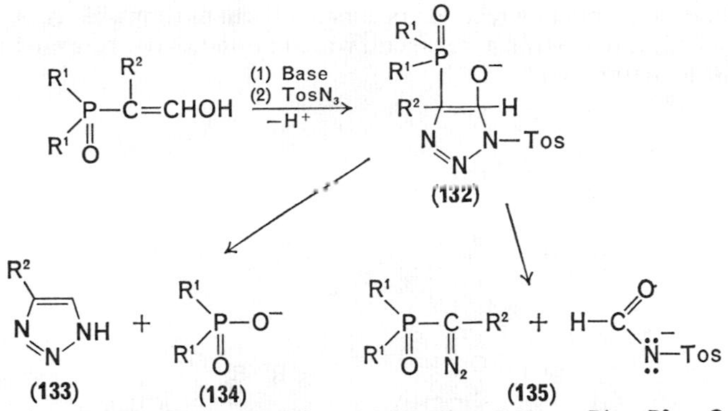

$(R^1)_2P(O)-C(R^2)=CHOH$ —(1) Base, (2) $TosN_3$, $-H^+$→ (132): triazoline bearing R^2, $P(O)(R^1)_2$, O^-, H and N—Tos

(132) → R^2-substituted 1,2,3-triazole (NH) (133) + $(R^1)_2P(O)-O^-$ (134)

(132) → $(R^1)_2P(O)-C(=N_2)-R^2$ (135) + $H-C(=O)-\ddot{N}^--Tos$

(133) (134) (135)

a: $R^1 = OC_2H_5$, $R^2 = H$; **b**: $R^1 = OC_2H_5$, $R^2 = C_6H_5$; **c**: $R^1 = R^2 = C_6H_5$

c. *Cyclopropenes.* Diazo group transfers onto cyclopropenes can be effected both with azides, which will be considered later, and with diazoalkanes. In the latter case the initial products are bicyclo[3.1.0]diazahexenes having the steric arrangement shown in structure **136**. They undergo thermal or photochemical isomerization to the α-diazo phosphonic esters (**137**)[346, 362].

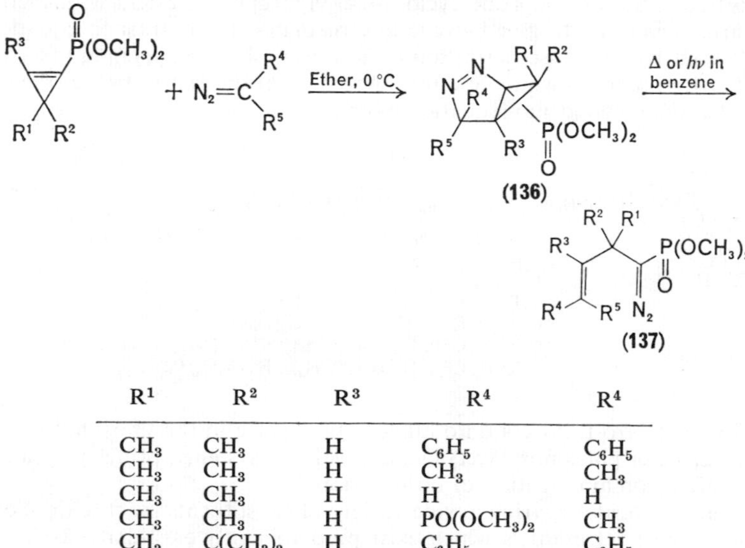

Cyclopropene (R^3, $P(O)(OCH_3)_2$, R^1, R^2) + $N_2=C(R^4)(R^5)$ —Ether, 0 °C→ bicyclic diazahexene (R^1, R^2, R^3, R^4, R^5, $P(OCH_3)_2$, O) (136) —Δ or $h\nu$ in benzene→ $R^4(R^5)C=C(R^3)-C(R^1)(R^2)-C(=N_2)-P(O)(OCH_3)_2$ (137)

(136)

(137)

R^1	R^2	R^3	R^4	R^4
CH_3	CH_3	H	C_6H_5	C_6H_5
CH_3	CH_3	H	CH_3	CH_3
CH_3	CH_3	H	H	H
CH_3	CH_3	H	$PO(OCH_3)_2$	CH_3
CH_3	$C(CH_3)_3$	H	C_6H_5	C_6H_5
CH_3	$C(CH_3)_3$	H	CH_3	CH_3
CH_3	$C(CH_3)_3$	H	H	H
H	CH_3	CH_3	C_6H_5	C_6H_5
CH_3	CH_3	CH_3	C_6H_5	C_6H_5

In the corresponding reactions of ethyl diazoacetate and dimethyl (diazomethyl)-phosphonate the cycloadduct of type **136**, which undoubtedly represents the primary product, isomerizes immediately. Without exception, the γ,δ-unsaturated α-diazo phosphonic esters predominate over the 1,4-dihydropyridazines in the approximate ratio 2 : 1. That the competing reaction occurs at all can most probably be ascribed to the CH-acidity of the diazo transfer reagent, which promotes rearrangement involving a proton shift (equation 63)[362].

$+ N_2{=}CH{-}R^3$

Ether, 25 °C

(63)

$R^1 = R^2 = CH_3$, $R^3 = COOC_2H_5$;
$R^1 = R^2 = CH_3$, $R^3 = PO(OCH_3)_2$;
$R^1 = CH_3$, $R^2 = C(CH_3)_3$, $R^3 = COOC_2H_5$;
$R^1 = CH_3$, $R^2 = C(CH_3)_3$, $R^3 = PO(OCH_3)_2$

Diazo transfers onto the same cyclopropenylphosphonic esters with methyl and phenyl azide again fail to give bicyclic intermediates. In contrast to equation (63), however, no heterocyclic isomerization occurs; exclusive ring opening takes place to give α-diazo γ-imino phosphonic esters. Apart from the last example, they readily hydrolyse at the imino group (equation 64)[362].

R^3N_3 → Hydrolysis → (64)

$R^1 = R^2 = CH_3$, $R^3 = CH_3$;
$R^1 = R^2 = CH_3$, $R^3 = C_6H_5$;
$R^3 = CH_3$, $R^2 = C(CH_3)_3$, $R^3 = CH_3$;
$R^1 = CH_3$, $R^2 = C(CH_3)_3$, $R^2 = C_6H_5$

d. *Alkynes*. Diazo transfers onto alkynes take place in the phosphoryl ynamine/sulphonyl azide or phosphorylacetylene/diazoalkane systems; primary cycloaddition reactions are a common feature of both variants.

Arylsulphonyl azides bearing a wide variety of substituents react with phosphoryl ynamines to give 1:1 adducts which exist predominantly as 2-diazo-2-phosphoryl-acetamidines (**139**), both in the crystalline state and in $CDCl_3$ solution. The examples appended to the reaction scheme reveal that electron-donating substituents in the aryl group of the sulphonyl azide promote formation of the triazole isomer (**138**). Its contribution to the solution equilibrium decreases along the series phosphoryl, thiophosphoryl, selenophosphoryl and phosphorimidoyl group[363].

Diazo transfers onto phosphorylacetylenes with 2-diazopropane and 1-diazo-1-phenylethane proceed initially via 3+2 cycloaddition to form 3*H*-pyrazoles (**140**)[364]; in some cases this primary step also encounters opposition from cycloaddition of non-specific orientation since isomeric cycloadducts and, as a result, isomeric diazo

(138) (139)

X	R^1	R^2	R^3	Isomer in the crystalline state	Equilibrium 138 ⇄ 139, 40 °C, $CDCl_3$ 138 (%)	139 (%)
O	CH_3	C_6H_5	Mesityl	**138**	30	70
O	CH_3	C_6H_5	*p*-Tolyl	**139**	8	92
O	CH_3	C_6H_5	C_6H_5	**139**	0	100
O	CH_3	C_6H_5	$C_6H_4NO_2$-*p*	**139**	0	100
S	CH_3	C_6H_5	Mesityl	**138**	60	40
S	CH_3	C_6H_5	*p*-Tolyl	**139**	38	62
S	CH_3	C_6H_5	C_6H_5	**139**	21	79
S	CH_3	C_6H_5	$C_6H_4NO_2$-*p*	**139**	0	100
Se	CH_3	C_6H_5	*p*-Tolyl	**139**	39	61
Se	CH_3	C_6H_5	C_6H_5	**139**	20	80
Se	CH_3	C_6H_5	$C_6H_4NO_2$-*m*	**139**	0	100
=N—Tos	CH_3	C_6H_5	Mesityl	**138**	65	35
≡N—Tos	CH_3	C_6H_5	*p*-Tolyl	**138**	32	68
=N—Tos	CH_3	C_6H_5	C_6H_5	**139**	15	85
=N—Tos	CH_3	C_6H_5	$C_6H_4NO_2$-*p*	**139**	0	100

Ether or chloroform/ether; $h\nu$ (350 nm) benzene; $h\nu$ (350 nm) benzene, $-N_2$

(140) (141) (142) (143)

R^1	R^2	R^3	140 (%)	141 (%)	142 (%)
H	$PO(C_6H_5)_2$	CH_3	100	38	39
$PO(C_6H_5)_2$	CH_3	CH_3	70	66	34
$PO(C_6H_5)_2$	C_6H_5	CH_3	87	46	47
C_6H_5	$PO(C_6H_5)_2$	C_6H_5	50	50	40
C_6H_5	$P(C_6H_5)_2$=N—Tos	CH_3	77	37	40
$PO(C_6H_5)_2$	$PO(C_6H_5)_2$	CH_3	77	63	37
$PO(C_6H_5)_2$	$PO(C_6H_5)_2$	C_6H_5	77	26	41

compounds can arise. Irradiation of the 3*H*-pyrazoles (**140**) in the 350 nm region (presumably $n \rightarrow \pi^*$ excitation) leads in all cases to the desired ring opening, i.e. to formation of diazo isomers (**141**), possibly having the opposite configuration at the C=C double bond[364]. Since not only the 3*H*-pyrazoles but also the diazo isomers (**141**) are subject to $n \rightarrow \pi^*$ excitation, photochemical decomposition of the latter to cyclopropenes (**142**) via the carbenes (**143**) cannot be avoided[364].

4. Substitution reactions

As with carbonyl diazo compounds, substitution at the diazomethyl group of phosphoryl diazo compounds has developed into a synthetic method in its own right whose potential is far from exhausted.

a. *Nitration.* The reaction scheme deduced for the nitration of diazoacetic esters with dinitrogen pentoxide is also valid without restriction for dimethyl (diazomethyl)-phosphonate, (diazomethyl)diphenylphosphine oxide, and methyl (diazomethyl)-phenylphosphinate. In all cases, their nitro derivatives are obtained together with the corresponding nitric esters (equation 65)[365]. Thus, once again, loss of 1 mole of (diazomethyl)phosphoryl compound cannot be avoided.

$$2\,HC(=N_2)-P(=O)R^1R^2 + N_2O_5 \xrightarrow{CCl_4,\ -20\,^{\circ}C} O_2N-C(=N_2)-P(=O)R^1R^2 + O_2N-O-CH_2-P(=O)R^1R^2 \quad (65)$$

$R^1 = R^2 = OCH_3$ (48%); $R^1 = R^2 = C_6H_5$ (40%); $R^1 = OCH_3$, $R^2 = C_6H_5$ (20%)

b. *Acylation.* Acylation of dimethyl(diazomethyl)phosphonate and (diazomethyl)diphenylphosphine oxide with acyl chlorides is always performed in the presence of triethylamine as auxiliary base so that no diazo compound is lost (equation

$$HC(=N_2)-P(=O)R^1R^1 + R^2-C(=O)Cl \xrightarrow[-HN(C_2H_5)_3^+\ Cl^-]{N(C_2H_5)_3} R^2-C(=O)-C(=N_2)-P(=O)R^1R^1 \quad (66)$$

$R^1 = C_6H_5$, $R^2 = OC_2H_5$ (57%); $R^1 = R^2 = C_6H_5$ (75%); $R^1 = OCH_3$, $R^2 = C_6H_5$ (61%); $R^1 = C_6H_5$, $R^2 = CH_3$ (47%); $R^1 = C_6H_5$, $R^2 = C(CH_3)_3$ (68%)

66)[356, 366]. Double reaction takes place between oxalyl chloride and (diazomethyl)-diphenylphosphine oxide to form 1,4-bis(diazo)bis(diphenylphosphoryl)-2,3-butanedione (71%)[366].

Acylation reactions with benzoyl isocyanate proceed under very mild conditions; yields are impaired by competing formation of oxazolinones. The point of branching of the reaction could be the primary adduct having betaine character (equation 67)[365].

The same reaction carried out with benzoyl isothiocyanate also involves initial acylation which is, however, followed by spontaneous 1,5-cyclization to give 1,2,3-thiadiazoles (equation 68)[365].

c. *Metalation.* Owing to its CH-acidity, diethyl (diazomethyl)phosphonate can be metalated directly with *n*-butyllithium in ether/tetrahydrofuran[365]. The same property makes for extremely facile metalation of diazomethylphosphoryl compounds with mercuric and silver oxide to give the metal derivatives **144** [329] and **145** [329, 330], respectively. The silver derivatives are considerably more stable than their carbonyl analogues.

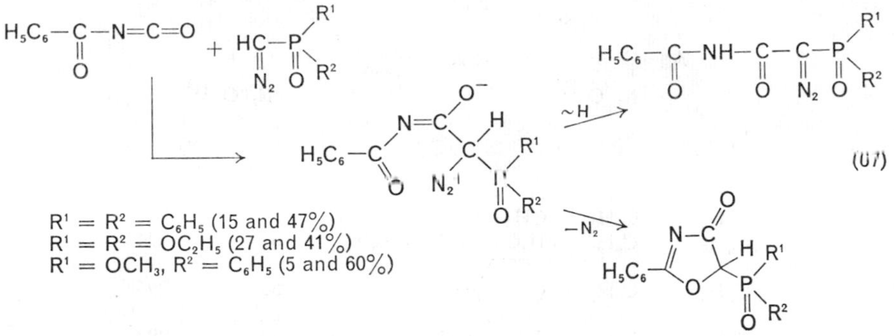

$$H_5C_6-C(=O)-N=C=O + HC(=N_2)-P(=O)R^1R^2 \longrightarrow H_5C_6-C(=O)-N=C(O^-)-C(H)(N_2^+)-P(=O)R^1R^2 \xrightarrow{\sim H} H_5C_6-C(=O)-NH-C(=O)-C(=N_2)-P(=O)R^1R^2 \tag{67}$$

$$\xrightarrow{-N_2} \text{2-}H_5C_6\text{-oxazol-4(5H)-one, 5-H, 5-}P(=O)R^1R^2$$

$R^1 = R^2 = C_6H_5$ (15 and 47%)
$R^1 = R^2 = OC_2H_5$ (27 and 41%)
$R^1 = OCH_3$, $R^2 = C_6H_5$ (5 and 60%)

$$H_5C_6-C(=O)-N=C=S + HC(=N_2)-P(=O)R^1R^2 \xrightarrow{\Delta,\ \text{benzene}} H_5C_6-C(=O)-NH-C(=S)-C(=N_2)-P(=O)R^1R^2 \xrightarrow{\text{Cyclization}} \text{5-}(H_5C_6C(=O)NH)\text{-4-}(P(=O)R^1R^2)\text{-1,2,3-thiadiazole} \tag{68}$$

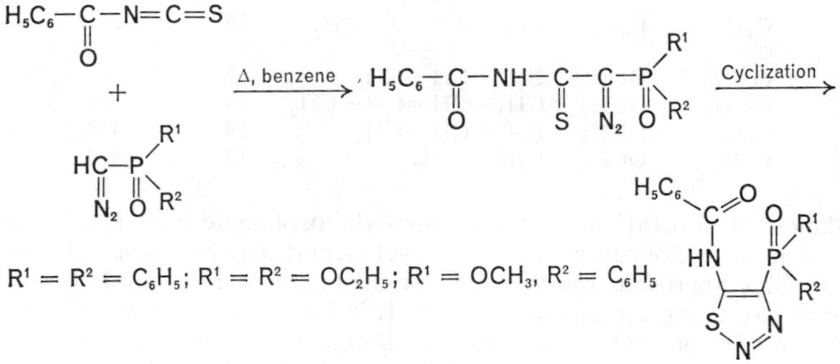

$R^1 = R^2 = C_6H_5$; $R^1 = R^2 = OC_2H_5$; $R^1 = OCH_3$, $R^2 = C_6H_5$

$$Hg\left[C(=N_2)-P(=O)R^1R^2\right]_2 \xleftarrow[-H_2O]{HgO} 2\,HC(=N_2)-P(=O)R^1R^2 \xrightarrow[-H_2O]{Ag_2O} 2\,Ag-C(=N_2)-P(=O)R^1R^2$$

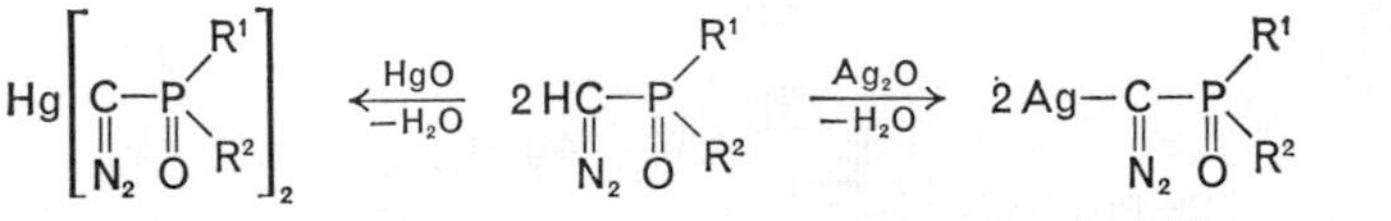

(144)
$R^1 = R^2 = C_6H_5$
$R^1 = R^2 = OC_2H_5$

(145)
$R^1 = R^2 = C_6H_5$
$R^1 = R^2 = OC_2H_5$
$R^1 = OCH_3$, $R^2 = C_6H_5$

d. *Substitution via metalated derivatives*. Diazomethylphosphoryl compounds can be halogenated via their silver derivatives; cyanogen bromide or iodine act as halogenation reagents, as seen in equation (69); the halogen compounds are very

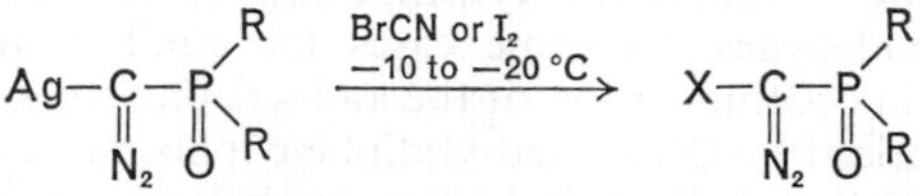

$$Ag-C(=N_2)-P(=O)R_2 \xrightarrow[-10\text{ to }-20\,^\circ C]{BrCN\text{ or }I_2} X-C(=N_2)-P(=O)R_2 \tag{69}$$

$R = C_6H_5$, $X = Br$; $R = C_6H_5$, $X = I$; $R = OCH_3$, $X = Br$; $R = OCH_3$, $X = I$

unstable. The same silver derivatives and also methyl argentiodiazomethylphenylphosphinate have been subjected to numerous alkylation reactions with S_N1 active halides such as allyl iodide, methallyl iodide, crotyl bromide, 3-bromocyclohexene, benzyl iodide and 4-substituted benzyl halides (equation 70)[294, 330, 367].

$$\text{Ag}-\underset{\underset{N_2}{\|}}{C}-\underset{\underset{O}{\|}}{P}(R^1)R^2 + R^3X \xrightarrow{-\text{AgX}} R^3-\underset{\underset{N_2}{\|}}{C}-\underset{\underset{O}{\|}}{P}(R^1)R^2 \tag{70}$$

R^1	R^2	R^3	%	Reference
C_6H_5	C_6H_5	$H_2C{=}CH{-}CH_2$	70	294
C_6H_5	C_6H_5	$CH_3{-}CH{=}CH{-}CH_2$	52	244
C_6H_5	C_6H_5	$H_2C{=}C(CH_3){-}CH_2$	67	294
C_6H_5	C_6H_5	cyclohex-2-enyl	53	294
C_6H_5	C_6H_5	$C_6H_5{-}CH_2$	54	294
C_6H_5	C_6H_5	$CH_3O{-}C_6H_4{-}CH_2$	35	294
OCH_3	OCH_3	$H_2C{=}CH{-}CH_2$	62	367
OCH_3	OCH_3	$CH_3{-}CH{=}CH{-}CH_2$	48	367
C_6H_5	OCH_3	$H_2C{=}CH{-}CH_2$	59	330
C_6H_5	OCH_3	$C_6H_5{-}CH_2$	72	330

Alkylation of dimethyl argentiodiazomethylphosphonate with Hückel aromatic species such as cyclopropenylium and cycloheptatrienylium salts also warrants attention. This approach was adopted for synthesis of **146** and **147** which hold promise of interesting carbene reactions[365]. It is hard to understand why attempted C-alkylation with bromotriphenylmethane should fail with the silver salt and yet proceed without difficulty with tetramethyl mercuriobis(diazomethylphosphonate)[367].

$$\text{(146)} \xleftarrow[-\text{AgBr}]{\text{triphenylcyclopropenylium } Br^-} \text{Ag}-C(=N_2)-P(O)(OCH_3)_2 \xrightarrow[-\text{AgBr}]{\text{tropylium } Br^-} \text{(147)}$$

(146): $(H_5C_6)_2C_3(C_6H_5)-C(=N_2)-P(O)(OCH_3)_2$

(147): $C_7H_7-C(=N_2)-P(O)(OCH_3)_2$

e. *Addition reactions.* Diazomethylphosphoryl compounds undergo base-catalysed aldol addition according to equation (71) with aliphatic, aromatic, heteroaromatic, and α,β-unsaturated aldehydes. In some cases the equilibrium character of the reaction is manifested in decomposition of the adducts into the starting materials on attempted recrystallization[368]. (Diazomethyl)diphenylphosphine oxide is seen to be more prone to addition than is dimethyl (diazomethyl)phosphonate. Therefore the phosphonic ester adds only benzaldehydes bearing electron-withdrawing substituents, whereas the phosphine oxide also reacts with the parent compound[368].

Simple ketones react only in exceptional cases; hence an adduct has so far only been synthesized from cyclohexanone and (diazomethyl)diphenylphosphine oxide (equation 72)[369]. No plausible explanation has yet been obtained for the failure of cyclobutanone, cyclopentanone, cycloheptanone and cyclooctanone to react in

$$R^1R^2P(=O)-CH(=N_2) + R^3-CHO \underset{}{\overset{KOH\ or\ N(C_2H_5)_3}{\rightleftharpoons}} R^1R^2P(=O)-C(=N_2)-CH(OH)-R^3 \quad (71)$$

$R^1 = R^2 = C_6H_5$; R^3: CH_3 (85%), C_6H_5 (63%), 2-naphthyl (68%), 9-phenanthryl (49%), 2-furyl (58%), 4-pyridyl (84%), $C_6H_5-CH=CH$ (55%), $C_6H_5-C\equiv C$ ([illegible]%)

$R^1 = R^2 = OCH_3$; R^3: $C_6H_4NO_2$-*p* (50%), 2-naphthyl (22%)

$R^1 = OCH_3$, $R^2 = C_6H_5$; R^3: $C_6H_5NO_2$-*p* (67%)

$$\text{cyclohexanone} + HC(=N_2)-P(=O)(C_6H_5)_2 \xrightarrow{CH_3OH,\ KOH} \text{1-hydroxycyclohexyl}-C(=N_2)-P(=O)(C_6H_5)_2 \quad (72)$$

this way. Numerous addition reactions to 1,2-dicarbonyl compounds have been reported. Reaction of phenylglyoxal with (diazomethyl)diphenylphosphine oxide takes an unusual course in that the diazo aldol doubtless formed as primary product yields a 'dioxane-like' dimer. The latter reaction can be reversed, however, in a weakly acidic medium (equation 73)[370].

$$C_6H_5-C(=O)-CHO + HC(=N_2)-P(=O)(C_6H_5)_2 \longrightarrow C_6H_5-C(=O)-CH(OH)-C(=N_2)-P(=O)(C_6H_5)_2$$

$$\xrightarrow{\text{Dimerization}} \quad \xleftarrow{H^+\text{-catalysis},\ CH_3OH/CHCl_3,\ 55\ ^\circ C} \quad (73)$$

Dimer: 1,4-dioxane ring bearing at each of two ring carbons HO and C_6H_5, and at each of the other two ring carbons H and $-C(=N_2)-P(=O)(C_6H_5)_2$.

Diacetyl undergoes smooth hydroxyl-ion-catalysed addition to the same diazo compound, yielding 2-(diazodiphenylphosphorylmethyl)-2-hydroxy-3-butanone[369]. α-Oxo phosphonic esters (**148**), of comparable reactivity to 1,2-dicarbonyl compounds, readily add (diazomethyl)diphenylphosphine oxide at the CO group to form **149** [369]. Examples of aldol-type additions of diazomethylphosphoryl compounds to cyclic 1,2-di- and 1,2,3-tricarbonyl compounds are numerous; the latter are so reactive that base catalysis becomes unnecessary.

$$\underset{\mathbf{(148)}}{R-C(=O)-P(=O)(OCH_3)_2} + HC(=N_2)-P(=O)(C_6H_5)_2 \xrightarrow{N(C_2H_5)_3} \underset{\mathbf{(149)}}{N_2=C[P(=O)(C_6H_5)_2]-C(R)(OH)-P(=O)(OCH_3)_2}$$

In the case of 1,2-indandiones, isatin, *N*-substituted isatins and coumarandione, reaction takes place at the benzoyl CO group. This is due to steric effects of the substituent in position 3 in the 1,2-indandiones, and to carboxamide and ester resonance in the heterosubstituted 1,2-dicarbonyl compounds[330, 370, 371]. Only in the case of thionaphthenequinone does addition occur at the 'thioester CO group', as shown in equation (74)[370].

$$\text{thionaphthenequinone} + HC(=N_2)-P(=O)R_2 \longrightarrow \text{3-oxo-2-hydroxy-2-}[C(=N_2)-P(=O)R_2]\text{-2,3-dihydrobenzothiophene} \qquad (74)$$

$$R = C_6H_5,\ OCH_3$$

In 1,2,3-tricarbonyl compounds reaction invariably ensues at the 'central' CO group, whose pronounced reactivity toward nucleophiles is general knowledge. Table 8 provides a representative survey of the aldol adducts hitherto synthesized.

Enamines are alkylated at the β-carbon atom by diazomethylphosphoryl compounds, no evidence being obtained for the seemingly likely 3 + 2 cycloaddition of the two reactants[372]. A hydrogen-bonded species is postulated as intermediate leading to formation of the β-amino α-diazo phosphoryl compound (equation 75)[372].

$$R^1_2C=CH-N\!\!\frown \xrightarrow{HC(=N_2)-P(=O)R^2_2} R^1_2\overset{-}{C}-CH=\overset{+}{N}\!\!\frown \cdots H-C(=N_2)-P(=O)R^2_2 \longrightarrow R^1_2CH-CH(N\!\!\frown)-C(=N_2)-P(=O)R^2_2 \qquad (75)$$

$R^1 = CH_3$, N⌒ = morpholino, $R^2 = C_6H_5$ or OCH_3 (89/89%)

$R^1 = CH_3$, N⌒ = piperidino, $R^2 = C_6H_5$ or OCH_3 (95/92%)

$R^1 = C_2H_5$, N⌒ = morpholino, $R^2 = C_6H_5$ (97%)

$R^1 = CH_3$, N⌒ = pyrrolidino, $R^2 = OCH_3$ (60%)

C. Sulphonyl Diazo Compounds

Although current interest in sulphonyl diazo compounds was awakened at about the same time as that in phosphoryl diazo compounds, synthetic exploitation and practical utilization have progressed much more slowly. The principal method of preparation has been diazo transfer[116–118]; considerably less significant are the cleavage of *N*-alkyl-*N*-nitrosourethanes and several other methods whose scope has not yet been ascertained.

TABLE 8. Aldol addition of diazomethyl phosphoryl compounds onto 1,2-Di- and 1,2,3-tricarbonyl compounds

Diazoaldol	Yield (%)	Reference	Diazoaldol	Yield (%)	Reference
H_3C CH_3; =O; HO C—P$(C_6H_5)_2$; N_2 O	65	370	N_2 O R^1; HO C—P; R^2; =O; N; R^3		
H_5C_6 C_6H_5; =O; HO C—P$(OCH_3)_2$; N_2 O	64	370	$R^1 = R^2 = C_6H_5$; $R^3 = H$	94	370, 371
			$R^1 = R^2 = C_6H_5$; $R^3 = OCOCH_3$	78	370, 371
			$R^1 = R^2 = OCH_3$; $R^3 = OH$	81	370, 371
			$R^1 = R^2 = OCH_3$; $R^3 = COCH_3$	56	370, 371
			$R^1 = OCH_3$, $R^2 = C_6H_5$, $R^3 = H$	95	330
N_2 O; C—P$(OCH_3)_2$; H_3C; OH; H_3C; =O; N	90	370	N_2 O; HO C—P$(C_6H_5)_2$; =O; O	38	370
O; OH R^1; C—P; O N_2 O R^2			N_2 O; HO C—P$(OC_2H_5)_2$; O O	71	371
$R^1 = R^2 = C_6H_5$	70	371			
$R^1 = R^2 = OCH_3$	60	371			
$R^1 = OCH_3$, $R^2 = C_6H_5$	71	330			
O OH; C—P$(OC_2H_5)_2$; N_2 O; N H O	86	371	N_2 O; HO C—P$(C_6H_5)_2$; O O; HN NH; O	78	366
			O; N_2 P$(C_6H_5)_2$; C; C_6H_5—C—C—C—C_6H_5; O OH O	83	366

I. Forster reaction

The reaction of the oxime **150** with chloroamine proceeds inhomogeneously; benzonitrile and benzyl *p*-tosyl sulphone are formed at the expense of the expected diazo (phenyl)tosylmethane (**151**), which is obtained in only modest yield. The intermediates involved in their formation are not yet known[373, 374].

$$C_6H_5-C(=N-OH)-SO_2-C_6H_4-CH_3 \ \textbf{(150)} \xrightarrow{NH_2Cl} C_6H_5-C(=N_2)-SO_2-C_6H_4-CH_3 \ \textbf{(151)}(6\%) + C_6H_5-CN + C_6H_5CH_2-SO_2-C_6H_4-CH_3$$

2. Dehydrogenation of hydrazones

Much to the detriment of sulphonyl diazo chemistry, α-oxo sulphones are still unknown. This has greatly restricted the utility of hydrazone dehydrogenation and completely ruled out use of the Bamford–Stevens reaction. There appears to be only one alternative route to suitable hydrazones. This is shown in equation (76) and affords bis(arylsulphonyl)formaldehyde hydrazones, which can be dehydrogenated with manganese dioxide[375, 376]. The transformation of the tetrasulphonylated azine into bis(benzenesulphonyl)diazomethane, mediated by benzenesulphinate (92%)[376], has no parallel in the literature.

$$Br_2C=N-N=CBr_2 \xrightarrow[-4\ NaBr]{4\ X-C_6H_4-SO_2^-\ Na^+,\ DMFA} (X-C_6H_4-SO_2)_2C=N-N=C(SO_2-C_6H_4-X)_2$$

$$(X-C_6H_4-SO_2)_2C=N-N=C(SO_2-C_6H_4-X)_2 \xrightarrow{\text{Hydrolysis}} (X-C_6H_4-SO_2)_2C=N-NH_2 \xrightarrow{MnO_2/THF} (X-C_6H_4-SO_2)_2C=N_2$$

$$\text{for } X = H: \ (C_6H_5SO_2)_2C=N-N=C(SO_2C_6H_5)_2 \xrightarrow[-(C_6H_5SO_2)_3C^-\ Na^+]{+C_6H_5SO_2^-\ Na^+} (C_6H_5SO_2)_2C=N_2 \qquad (76)$$

X = H, CH_3, $(CH_3)_2CHO$, CH_3CONH, NO_2

3. Cleavage of *N*-alkyl-*N*-nitrosourethanes

Alkylsulphonyl and arylsulphonyl diazo compounds first became accessible by cleavage of *N*-alkyl-*N*-nitrosourethanes[377, 378]. In most cases the acyl group is released by chromatography on alumina[378–380]. Secondary sulphonyldiazoalkanes have recently also been obtained by 'conventional' KOH cleavage[381], while both variants are employed in the preparation of β-sulphonyl diazo compounds[382]. Selected examples are appended to the general reaction scheme in equation (77).

$$R^1{-}SO_2{-}CH(R^2){-}N(NO){-}COOC_2H_5 \xrightarrow{Al_2O_3 \text{ or } KOH} R^1{-}SO_2{-}C(=N_2){-}R^2 \quad (77)$$

R^1	R^2	Base	Yield (%)	Reference
$CH_3O{-}C_6H_4{-}$ (para)	H	KOH	50	377
$o{-}NO_2C_6H_4{-}$	H	Al_2O_3	30–60	379
$(CH_3)_3C$	H	Al_2O_3	77	378
C_6H_5	H	Al_2O_3	≥81	378
$m{-}F_3CC_6H_4{-}$	H	Al_2O_3	51	380
$H_3C{-}C_6H_4{-}$ (para)	CH_3	KOH	55	381

4. Diazo group transfer

a. *β-Oxo sulphonyl compounds*. The conditions for successful diazo transfer with α-methylene sulphones are analogous to those valid for α-methylene carbonyl compounds: it proceeds smoothly only if the CH_2 group is activated by a second proton-activating substituent such as a carbonyl or a sulphonyl group. Remarkably, the first diazo transfer onto a β-oxo sulphonyl compound leads to a surprising result: on treatment with tosyl azide in alkaline medium, 3-oxo-2,3-dihydrothionaphthene 1,1-dioxide (**152**) afforded not the α-diazo β-oxo sulphone (**155**) but instead the dipotassium salt of the azo compound (**153**) or, after acidification, the hydrazone (**154**)[383]. Disregarding the problem of tautomerism, a diazo transfer has therefore actually taken place. A simple interpretation would assume that the primary product **155** undergoes fast azo coupling with **152** to give **154**[183]. Fortunately, **154** can be uncoupled to regenerate the methylene compound **152** and 2-diazo-3-oxo-2,3-dihydrothionaphthene (**155**) by heating in polar solvents. The anion **156** may occur as intermediate[383]. In contrast, diazo transfer with 1-ethyl-2-azidobenzothiazolium tetrafluoroborate in neutral to acidic media effects smooth transformation **152** → **155**[384].

The nucleophilic alkoxy displacement observed in the carbonyl and phosphoryl series during diazo transfers onto carboxylic esters in the alkanol/alkoxide system (see relevant sections) also occurs with sulphonyl compounds, as demonstrated by equation (78).

Although a number of smooth diazo transfers onto 1,3-disulphonyl compounds are known (see Table 9), some entirely unexpected secondary and side-reactions have also been reported[386]. Hence diazo transfer onto bis(mesitylsulphonyl)methane leads not only to the expected diazoalkane (**157**) but also to the thiosulphonate (**158**) and

(152) $\xrightarrow{TosN_3,\ C_2H_5OH,\ KOH}$ (153) $\xrightarrow{H^+}$ (154) $\xrightarrow{\Delta \text{ in DMFA}}$ (156) $\xrightarrow{-152 \text{ as carbanion}}$ (155)

(152) + 2-azido-3-ethylbenzothiazolium BF_4^- $\longrightarrow$ (155) − 2-amino-3-ethylbenzothiazolium BF_4^-

$$(CH_3)_3C-SO_2-CH_2-COOC_2H_5 \xrightarrow[C_2H_5ONa]{TosN_3,\ \text{ethanol/ether},} (CH_3)_3C-SO_2-C(=N_2)-C(=O)-\bar{N}-Tos\ Na^+ \quad (78)$$

the vinylhydrazone **(159)**[386]. A large excess of tosyl azide favours formation of the diazo compound; a reasonable interpretation of the complex reaction course is found in the literature[386].

$$Mes-SO_2-CH_2-SO_2-Mes \xrightarrow{TosN_3,\ THF,\ OH} Mes-SO_2-C(N_2)-SO_2-Mes\ \textbf{(157)} + Mes-SO_2-S-Mes\ \textbf{(158)} + (Mes-SO_2)_2C=N-NH-CH=C(SO_2-Mes)_2\ \textbf{(159)}$$

Table 9 lists some representative preparations of α-diazo β-oxo sulphonyl and α-diazo β-disulphonyl compounds. Triethylamine, potassium hydroxide and, in special cases, organolithium compounds served as bases.

b. *α-Acyl aldehydes.* Diazo transfers onto phenylsulphonylmethanes are not feasible directly, but only after formylation with formic ester[373, 374]; the diazo transfer reagent employed is 4-carboxybenzenesulphonyl azide whose acidic properties are of distinct advantage in work-up (equation 79).

c. *Thiirene* 1,1-*dioxides.* Compared with carbonylated and sulphonylated cyclopropenes, diazo transfers onto thiirene 1,1-dioxides such as **160** take a more complex course. Addition of diazoalkanes undoubtedly leads to the sulphur heterocycles **161**;

TABLE 9. α-Diazo β-oxo-sulphonyl- and α-diazo β-disulphonyl compounds by diazo group transfer

$R^1{-}SO_2{-}C(=N_2){-}R^2$

R^1	R^2	Yield (%)	Reference	R^1	R^2	Yield (%)	Reference
C_6H_5	COC_6H_5	91	387	C_2H_5	$SO_2C_2H_5$	46	389
C_6H_5	$COCH_3$	62	387	C_6H_5	$SO_2C_6H_5$	70	389
Tolyl	$COCH_3$	70	385	3,4-difluorophenyl (F–C_6H_3(F)–)	SO_2–$C_6H_3F_2$	84	389
$C_6H_4NO_2$-*p*	$COCH_3$	45	388	$(C_2H_5)(C_6H_5)N$	$SO_2{-}N(C_2H_5)(C_6H_5)$	50	389
C_6H_5	COC_6H_5	63	339				
C_6H_5	$COC_6H_4NO_2$-*p*	62	387				
Tolyl	$COOC_2H_5$	84	339				
Tolyl	$CON(CH_3)_2$	69	339				
C_6H_5	$COC(=N_2)COC_6H_5$	60	146				

$$R{-}SO_2{-}CH(CHO){-}C_6H_5 \xrightarrow{HOOC{-}C_6H_4{-}SO_2{-}N_3,\ CH_2Cl_2/H_2O,\ NH_3} R{-}SO_2{-}C(=N_2){-}C_6H_5 \quad (79)$$

R = $H_3C{-}C_6H_4{-}$, $C(CH_3)_3$, $CH_2C_6H_5$

although they are not isolable their intermediacy follows compellingly from the product distribution. Both the ring opening, characteristic of diazo group transfer, giving (**162**) and pyrazole formation (**163**) with extrusion of SO_2 are observed[390, 391].

(160) diphenylthiirene dioxide $\xrightarrow{R{-}CH{=}N_2}$ (161) → Isomerization → (162) $H_5C_6{-}C(=N_2){-}SO_2{-}C(C_6H_5){=}CHR$; (161) $\xrightarrow{\sim H,\ -SO_2}$ (163) 3(5)-R-4,5-diphenylpyrazole

a: R = C_6H_5 (44% **162**, 6% **163**)[390]
b: R = CH_3 (37% **162**, 52% **163**)[391]

5. Substitution reactions

In contrast to the other two classes of compounds already considered, substitution reactions of diazomethylsulphonyl compounds are still comparatively rare. The long-known and versatile acylation of diazoalkanes with acyl chlorides cannot be applied generally to sulphonyl chloride, even though an exception has been described[223]. The CH-acidity of diazomethylsulphonyl compounds has so far only been successfully exploited on a broader scale in the C-alkylation of enamines (equation 80)[392].

$$R^1R^2C{=}C(R^3)\text{-morpholino} + HC(=N_2)-SO_2-R^4 \xrightarrow{CH_3CN \text{ (or without solvent)}} R^1R^2CH-C(R^3)(\text{morpholino})-C(=N_2)-SO_2-R^4 \quad (80)$$

$R^1 = R^2 = CH_3$, $R^3 = H$, $R^4 = C_6H_4CH_3$-p (89%)
$R^1 = C_6H_5$, $R^2 = R^3 = H$, $R^4 = C_6H_4CH_3$-p (50%)
$R^1 = R^2 = H$, $R^3 = C_6H_5$, $R^4 = C_6H_4CH_3$-p (78%)
$R^1 = R^2 = CH_3$, $R^3 = H$, $R^4 = C(CH_3)_3$ (88%)
$R^1 = R^2 = CH_3$, $R^3 = H$, $R^4 = CH_2C_6H_5$ (82%)

Acyl cleavage of α-diazo β-oxo sulphonyl compounds are of limited synthetic utility; acyl cleavage of the carbonyl-containing group is effected by alumina in water or triethylamine in methanol, as shown by formation of **164a** and **164b**[393].

$$X-C_6H_4-SO_2-C(=N_2)-C(=O)-CH_3 \longrightarrow X-C_6H_4-SO_2-CH(=N_2)$$

a: $X = NO_2$, **b**: $X = CH_3$ **(164)**

III. REFERENCES

1. T. Curtius, *Ber. dt. chem. Ges.*, **16**, 2230 (1883).
2. B. Eistert, M. Regitz, G. Heck and H. Schwall, in *Methoden der organischen Chemie (Houben-Weyl-Müller)*, Vol. X/4, 4th ed., G. Thieme Verlag, Stuttgart, 1968, p. 557.
3. M. Regitz, in *Methodicum Chimicum*, Vol. 6, 1st ed., G. Thieme Verlag, Stuttgart, p. 211; Academic Press, New York, 1974, Engl. ed., Vol. 6, p. 206.
4. M. Regitz, *Aliphatic Diazo Compounds*, G. Thieme Verlag, Stuttgart, in the press.
5. T. Curtius, *Ber. dt. chem. Ges.*, **17**, 953 (1884).
6. T. Curtius, *J. Prakt. Chem.*, (2) **38**, 401 (1888).
7. O. Silberrad, *J. Chem. Soc.*, **81**, 600 (1902).
8. G. S. Skinner, *J. Amer. Chem. Soc.*, **46**, 731 (1924).
9. E. I. Du Pont (Inv. N. E. Searle), *U.S. Patent* 2 490 714 (6.12.1949); *Chem. Abstr.*, **44**, 3519d (1950); see also National Distillers Products (Inv. J. A. S. Hammond), *U.S. Patent* 2 691 649 (12.10.1954); *Chem. Abstr.*, **49**, 11690i (1955).
10. N. E. Searle, *Org. Syn. Coll.* **IV**, 424 (1963).
11. T. Curtius, *Ber. dt. chem. Ges.*, **37**, 1284 (1904).
12. T. Curtius and A. Darapsky, *Ber. dt. chem. Ges.*, **39**, 1373 (1906).
13. T. Curtius and J. Thompsen, *Ber. dt. chem. Ges.*, **39**, 1379 (1906).
14. E. D. Nicolaides, R. D. Westland and E. L. Wittle, *J. Amer. Chem. Soc.*, **76**, 2887 (1954).
15. J. A. Moore, J. R. Dice, E. D. Nicolaides, R. D. Westland and E. L. Wittle, *J. Amer. Chem. Soc.*, **76**, 2884 (1954).
16. T. Curtius, *Ber. dt. chem. Ges.*, **37**, 1261 (1904).
17. H. Lindemann, A. Walter and R. Groger, *Ber. dt. chem. Ges.*, **63**, 711 (1930).
18. A. Weissberger and H. Bach, *Ber. dt. chem. Ges.*, **65**, 265 (1932).
19. H. M. Chiles and W. A. Noyes, *J. Amer. Chem. Soc.*, **44**, 1798 (1922).
20. N. Takamura, T. Mizoguchi, K. Koya and S. Yamata, *Tetrahedron*, **31**, 227 (1975).
21. H. Neunhoeffer, G. Cuny and W. K. Franke, *Liebigs Ann. Chem.*, **713**, 96 (1968).
22. O. E. Edwards and M. Lesage, *J. Org. Chem.*, **24**, 2071 (1959).

23. H. E. Baumgarten and C. H. Andersen, *J. Amer. Chem. Soc.*, **83**, 399 (1961).
24. R. Schiff, *Ber. dt. chem. Ges.*, **14**, 1375 (1881); A. Angeli, *Gazz. Chim. Ital.*, **23** (II), 351 (1893) such as **24** (II), 318 (1894).
25. A. Angeli, *Ber. dt. chem. Ges.*, **26**, 1715 (1893).
26. D. Hauser and H. P. Sigg, *Helv. Chim. Acta*, **50**, 1327 (1967).
27. M. Puza and D. Doetschman, *Synthesis*, 481 (1971).
28. W. Ried and K. Wagner, *Liebigs Ann. Chem.*, **681**, 45 (1965).
29. J. W. Barton, A. R. Grinham and E. K. Whitaker, *J. Chem. Soc. C*, 1384 (1971).
30. M. Hudlicky and H. M. Bell, *J. Fluorine Chemistry*, **4**, 149 (1974).
31. G. B. Ansell, P. R. Hammond, S. V. Hering and P. Corradini, *Tetrahedron*, **25**, 2549 (1969).
32. F. Henle, *Liebigs Ann. Chem.*, **350**, 344 (1906).
33. W. L. Mosby and M. L. Silva, *J. Chem. Soc.*, 3990 (1964).
34. O. Piloty and J. Neresheimer, *Ber. dt. chem. Ges.*, **39**, 514 (1906).
35. H. Lindemann, A. Wolter and R. Groger, *Ber. dt. chem. Ges.*, **63**, 702 (1930).
36. L. Wolff, *Liebigs Ann. Chem.*, **325**, 129 (1902) (erroneously described as an oxadiazole).
37. L. Wolff, *Liebigs Ann. Chem.*, **394**, 23 (1912).
38. Z. Arnold and J. Šaulinová, *Coll. Czech. Chem. Commun.*, **38**, 2641 (1973).
39. H. Stetter and K. Kiehs, *Chem. Ber.*, **98**, 1181 (1965).
40. B. Eistert, H. Elias, E. Kosch and R. Wollheim, *Chem. Ber.*, **92**, 130 (1959).
41. B. Eistert, G. Bock, E. Kosch and F. Spalink, *Chem. Ber.*, **93**, 1451 (1960).
42. W. Kirmse and H. Dietrich, *Chem. Ber.*, **98**, 4027 (1965).
43. E. Müller and H. Huber-Emden, *Liebigs Ann. Chem.*, **660**, 54 (1962).
44. R. A. Franich, G. Lowe and J. Parker, *J. Chem. Soc. Perkin I*, 2034 (1972).
45. C. S. Marvel and W. A. Noyes, *J. Amer. Chem. Soc.*, **42**, 2259 (1920).
46. W. Ried and M. Butz, *Liebigs Ann. Chem.*, **716**, 190 (1968).
47. W. Ried and E. A. Baumbach, *Liebigs Ann. Chem.*, **713**, 139 (1968).
48. O. Süs and K. Möller, *Liebigs Ann. Chem.*, **593**, 91 (1955).
49. O. Süs, M. Glos, K. Möller and H. D. Eberhardt, *Liebigs Ann. Chem.*, **583**, 150 (1953).
50. K. Möller and O. Süs, *Liebigs Ann. Chem.*, **612**, 153 (1958).
51. H. G. O. Becker and H. Böttcher, *J. Prakt. Chem.*, **314**, 55 (1972).
52. L. Wolf and A. Lüttringhaus, *Liebigs Ann. Chem.*, **312**, 119 (1900); F. G. Fischer and E. Fahr, *Liebigs Ann. Chem.*, **651**, 64 (1962).
53. B. Eistert and F. Geiss, *Chem. Ber.*, **94**, 929 (1961).
54. F. Arndt, L. Loewe, R. Ün and E. Ayça, *Chem. Ber.*, **84**, 319 (1951); C. F. Huebner and K. P. Link, *J. Amer. Chem. Soc.*, **67**, 99 (1945).
55. B. Eistert, D. Greiber and I. Caspari, *Liebigs Ann. Chem.*, **659**, 64 (1962).
56. M. P. Cava, R. L. Litle and D. R. Napier, *J. Amer. Chem. Soc.*, **80**, 2257 (1958).
57. M. O. Forster, *J. Chem. Soc.*, **107**, 260 (1915).
58. L. O. Ross, *J. Org. Chem.*, **26**, 3395 (1961).
59. J. Meinwald and P. G. Gassman, *J. Amer. Chem. Soc.*, **82**, 2857 (1960).
60. I. W. J. Still and D. T. Wang, *Can. J. Chem.*, **46**, 1583 (1968).
61. A. T. Blomquist and C. G. Bottomley, *Liebigs Ann. Chem.*, **653**, 67 (1962).
62. L. Horner, W. Kirmse and K. Muth, *Chem. Ber.*, **91**, 430 (1958).
63. L. Horner, K. Muth and H. G. Schmelzer, *Chem. Ber.*, **92**, 2953 (1959).
64. M. P. Cava and P. M. Weintraub, *Steroids*, **4**, 41 (1964); *Chem. Abstr.*, **61**, 10738d (1964).
65. S. Huneck, *Chem. Ber.*, **98**, 2284 (1965).
66. S. Huneck, *Chem. Ber.*, **98**, 1837 (1965).
67. J. L. Mateos, O. Chao and H. Flores, *Tetrahedron*, **19**, 1051 (1963).
68. J. Meinwald, G. G. Curtis and P. G. Gassman, *J. Amer. Chem. Soc.*, **84**, 116 (1962).
69. G. Müller, C. Huynh and J. Mathieu, *Bull. Soc. Chim. Fr.*, 296 (1962).
70. M. P. Cava and E. Moroz, *J. Amer. Chem. Soc.*, **84**, 115 (1962).
71. G. Casini, F. Gualtieri and M. L. Stein, *Gazz. Chim. Ital.*, **95**, 983 (1965).
72. W. Kirmse, *Angew. Chem.*, **71**, 539 (1959).
73. R. Tasovac, M. Stefanović and A. Stojiljković, *Tetrahedron Lett.*, 2729 (1967).
74. J. M. Trost and P. J. Whitman, *J. Amer. Chem. Soc.*, **96**, 7421 (1974).

75. R. F. Borch and D. L. Fields, *J. Org. Chem.*, **34**, 1480 (1969).
76. M. P. Cava, E. J. Glamkovski and P. M. Weintraub, *J. Org. Chem.*, **31**, 2755 (1966).
77. T. Curtius and K. Thun, *J. Prakt. Chem.* (2), **44**, 551 (1891); T. Curtius and H. Lang, *J. Prakt. Chem.* (2), **44**, 544 (1891).
78. E. Fahr, *Liebigs Ann. Chem.*, **627**, 213 (1959).
79. E. Ciganek, *J. Org. Chem.*, **30**, 4366 (1965).
80. B. Eistert and E. Endres, *Liebigs Ann. Chem.*, **734**, 56 (1970).
81. S. Hauptmann, M. Kluge, K. D. Seidig and H. Wilde, *Angew. Chem.*, **77**, 678 (1965); *Angew. Chem. Int. Ed.*, **4**, 688 (1965).
82. S. Hauptmann and H. Wilde, *J. Prakt. Chem.*, **311**, 604 (1969).
83. P. Yates and F. X. Garneau, *Tetrahedron Lett.*, 71 (1967).
84. E. Ciganek, *J. Org. Chem.*, **35**, 862 (1970).
85. K. Nakagawa, R. Konaka and T. Nakata, *J. Org. Chem.*, **27**, 1597 (1962).
86. K. Nakagawa, H. Onoue and K. Minami, *J. Chem. Soc. Chem. Commun.*, 730 (1966).
87. H. Morrison, S. Danishefsky and P. Yates, *J. Org. Chem.*, **26**, 2617 (1961).
88. J. Tsuji, H. Takahashi and T. Kajimoto, *Tetrahedron Lett.*, 4573 (1973).
89. O. Diels and K. Pflaumer, *Ber. dt. chem. Ges.*, **48**, 223 (1915).
90. R. C. Fuson, L. J. Armstrong and W. J. Schenk, *J. Amer. Chem. Soc.*, **66**, 964 (1944).
91. C. D. Nenitzescu and E. Solomonica, *Org. Syn. Coll.*, Vol. **II**, p. 496 (1947).
92. L. Berend and J. Herms, *J. Prakt. Chem.*, (2), **60**, 16 (1899).
93. J. Bredt and W. Holz, *J. Prakt. Chem.*, (2), **95**, 133 (1917).
94. N. L. Allinger and L. A. Freiberg, *J. Org. Chem.*, **27**, 1490 (1962).
95. J. M. Muchovski, *Tetrahedron Lett.*, 1773 (1966).
96. O. Süs, H. Steppan and R. Dietrich, *Liebigs Ann. Chem.*, **617**, 20 (1958).
97. B. M. Trost and P. L. Kinson, *J. Amer. Chem. Soc.*, **92**, 2592 (1970).
98. A. T. Blomquist and F. W. Schlaeter, *J. Amer. Chem. Soc.*, **83**, 4547 (1961).
99. T. Chen, T. Sanjiki, H. Kato and M. Ohta, *Bull. Chem. Soc. Japan*, **40**, 2398 (1967).
100. L. Capuano and W. Ebner, *Chem. Ber.*, **104**, 2221 (1971).
101. W. Ried and R. Dietrich, *Chem. Ber.*, **94**, 387 (1961).
102. K. B. Wiberg, B. R. Lowry and T. H. Colby, *J. Amer. Chem. Soc.*, **83**, 3998 (1961).
103. J. Meinwald, C. Blomquist-Jensen, A. Lewis and A. Swithenbank, *J. Org. Chem.*, **29**, 3469 (1964).
104. A. J. Ashe, *Tetrahedron Lett.*, 523 (1969).
105. G. Seitz and W. Klein, *Tetrahedron*, **29**, 253 (1973).
106. H. O. House and C. J. Blankley, *J. Org. Chem.*, **33**, 53 (1968).
107. E. J. Corey and A. M. Felix, *J. Amer. Chem. Soc.*, **87**, 2518 (1965).
108. M. Franck-Neumann and G. Leclerc, *Bull. Soc. Chim. Fr.*, 247 (1975).
109. G. Heck, *unpublished results*, Universität Saarbrücken, 1964; see also Reference 2, p. 565.
110. M. Regitz and G. Heck, *Chem. Ber.*, **97**, 1482 (1964).
111. E. H. White and R. J. Baumgarten, *J. Org. Chem.*, **29**, 2070 (1964).
112. H. Reimlinger and L. Skatteböll, *Chem. Ber.*, **93**, 2162 (1960).
113. H. Reimlinger and L. Skatteböll, *Chem. Ber.*, **94**, 2429 (1961).
114. V. Franzen, *Liebigs Ann. Chem.*, **602**, 199 (1957).
115. J. C. Sheehan, Y. S. Lo, J. Löliger and C. C. Podewell, *J. Org. Chem.*, **39**, 1444 (1974).
116. M. Regitz, *Angew. Chem.*, **79**, 786 (1967); *Angew. Chem. Int. Ed.*, **6**, 733 (1967).
117. M. Regitz, in *Neuere Methoden der präparativen organischen Chemie*, Vol. VI, 1st ed., Verlag Chemie, Weinheim, 1976, p. 76.
118. M. Regitz, *Synthesis*, 351 (1972).
119. H. Balli and F. Kersting, *Liebigs Ann. Chem.*, **647**, 1 (1961).
120. H. Balli, *Liebigs Ann. Chem.*, **647**, 11 (1961).
121. H. Balli and H. Rempfler, *unpublished results*, Universität Basel, 1970.
122. M. Regitz, *Tetrahedron Lett.*, 1403 (1964).
123. M. Regitz, *Chem. Ber.*, **98**, 1210 (1965).
124. M. Regitz, *Liebigs Ann. Chem.*, **676**, 101 (1964).
125. M. Regitz and A. Liedhegener, *Chem. Ber.*, **99**, 3128 (1966).
126. H. Horino and T. Toda, *Bull. Chem. Soc. Jap.*, **46**, 1212 (1973).
127. M. Regitz, J. Hocker and A. Liedhegener, *Org. Syn.*, **48**, 36 (1968).

128. H. Balli and V. Müller, *Angew. Chem.*, **76**, 573 (1964); *Angew. Chem. Int. Ed.*, **3**, 644 (1964).
129. J. B. Hendrickson and A. Wolf, *J. Org. Chem.*, **33**, 3610 (1968).
130. W. R. Roush, D. Feitler and J. Rebeck, *Tetrahedron Lett.*, 1391 (1974).
131. D. G. Farnum and P. Yates, *Proc. Chem. Soc. (London)*, 224 (1960).
132. U. Schöllkopf, P. Tonne, H. Schäfer and P. Markusch, *Liebigs Ann. Chem.*, **722**, 45 (1969).
133. M. Regitz and A. Liedhegener, *Chem. Ber.*, **99**, 3128 (1966).
134. B. W. Peace, F. Carman and D. S. Wulfman, *Synthesis*, 658 (1971).
135. H. Ledon, *Synthesis*, 347 (1974).
136. H. Ledon, G. Linstrumelle and S. Julia, *Bull. Soc. Chim. Fr.*, 2071 (1973).
137. H. Ledon, G. Linstrumelle and S. Julia, *Tetrahedron*, **29**, 3609 (1973).
138. R. D. Clark and C. H. Heathcock, *Tetrahedron Lett.*, 529 (1975).
139. G. Lowe and J. Parker, *J. Chem. Soc. Chem. Commun.*, 577 (1971).
140. D. M. Brunwin, G. Lowe and J. Parker, *J. Chem. Soc. C*, 3756 (1971).
141. R. A. Franich, G. Lowe and J. Parker, *J. Chem. Soc. Perkin I*, 2034 (1972).
142. M. Regitz, J. Hocker and A. Liedhegener, *Organic Preparations and Procedures*, **1**, 99 (1969).
143. J. K. Korobitsyna and V. A. Nikolaev, *Zh. Org. Khim.*, **7**, 413 (1971); *Chem. Abstr.*, **74**, 111 859j (1971).
144. G. Heyes and G. Holt, *J. Chem. Soc. Perkin I*, 1206 (1973).
145. B. Eistert and J. Grammel, *Chem. Ber.*, **104**, 1942 (1971).
146. M. Regitz, H. J. Geelhaar and J. Hocker, *Chem. Ber.*, **102**, 1743 (1969).
147. M. Regitz and H. J. Geelhaar, *Chem. Ber.*, **101**, 1473 (1968).
148. M. Regitz and F. Menz, *unpublished results*, Universität Saarbrücken, 1966.
149. M. Regitz and A. Liedhegener, *Chem. Ber.*, **99**, 2918 (1966).
150. M. Regitz and A. Liedhegener, *Liebigs Ann. Chem.*, **710**, 118 (1967).
151. M. Regitz, H. Schwall, G. Heck, B. Eistert and G. Bock, *Liebigs Ann. Chem.*, **690**, 125 (1965).
152. F. B. Culp, K. Kurita and J. A. Moore, *J. Org. Chem.*, **38**, 2945 (1973).
153. G. Lowe and D. D. Ridley, *J. Chem. Soc. Perkin I*, 2024 (1973).
154. J. R. Hlubucek and G. Lowe, *J. Chem. Soc. Chem. Commun.*, 419 (1974).
155. B. Eistert, R. Müller, J. Mussler and H. Selzer, *Chem. Ber.*, **102**, 2429 (1969).
156. W. D. Barker, R. Gilbert, J.-P. Lapointe, H. Veschambre and D. Vocelle, *Can. J. Chem.*, **47**, 2853 (1969).
157. W. Ried and R. Conde, *Chem. Ber.*, **104**, 1573 (1971).
158. M. Regitz and D. Stadler, *Liebigs Ann. Chem.*, **687**, 214 (1965).
159. B. Eistert and P. Donath, *Chem. Ber.*, **106**, 1537 (1973).
160. M. Regitz, A. Liedhegener and D. Stadler, *Liebigs Ann. Chem.*, **713**, 101 (1968).
161. M. Regitz and D. Stadler, *Angew. Chem.*, **76**, 920 (1964); *Angew. Chem. Int. Ed.*, **3**, 748 (1964).
162. M. Regitz, *Tetrahedron Lett.*, 3287 (1965).
163. M. Regitz and H. Schwall, *Liebigs Ann. Chem.*, **728**, 99 (1969).
164. W. Jugelt, *Z. Chem.*, **5**, 456 (1965).
165. M. Hamaguchi and T. Ibata, *Tetrahedron Lett.*, 4475 (1974).
166. M. Regitz, F. Menz and J. Rüter, *Tetrahedron Lett.*, 739 (1967).
167. M. Regitz and F. Menz, *Chem. Ber.*, **101**, 2622 (1968).
168. M. Regitz and J. Rüter, *Chem. Ber.*, **101**, 1263 (1968).
169. M. Regitz, J. Rüter and A. Liedhegener, *Org. Syn.*, **51**, 86 (1971).
170. R. E. Harmon, V. K. Sood and S. K. Gupta, *Synthesis*, 577 (1974).
171. A. L. Fridman, Y. S. Andreichikov and L. F. Gein, *Zh. Org. Khim.*, **9**, 1754 (1973); *Chem. Abstr.*, **79**, 125 991m (1973).
172. M. Regitz, F. Menz and A. Liedhegener. *Liebigs Ann. Chem.*, **739**, 174 (1970).
173. A Schmitz, U. Kraatz and F. Korte, *Chem. Ber.*, **108**, 1010 (1975).
174. K. B. Wiberg and A. de Meijere, *Tetrahedron Lett.*, 519 (1969).
175. T. Gibson and W. F. Erman, *J. Org. Chem.*, **31**, 3028 (1966).
176. K. Grychtol, H. Musso and J. F. M. Oth, *Chem. Ber.*, **105**, 1798 (1972).
177. P. E. Eaton and G. H. Temme, *J. Amer. Chem. Soc.*, **95**, 7508 (1973).

178. N. L. Allinger and T. J. Walter, *J. Amer. Chem. Soc.*, **94**, 9267 (1972).
179. N. L. Allinger, T. J. Walter and M. G. Newton, *J. Amer. Chem. Soc.*, **96**, 4588 (1974).
180. R. Fusco, G. Bianchetti, D. Pocar and R. Ugo, *Chem. Ber.*, **96**, 802 (1963).
181. R. Huisgen, L. Möbius and G. Szeimies, *Chem. Ber.*, **98**, 1138 (1965).
182. D. Pocar, G. Bianchetti and P. Ferruti, *Gazz. Chim. Ital.*, **97**, 597 (1967); see also preceding papers in this series.
183. J. Kučera and Z. Arnold, *Tetrahedron Lett.*, 1109 (1966).
184. Z. Arnold, *J. Chem. Soc. Chem. Commun.*, 299 (1967).
185. J. Kučera, Z. Janoušek and Z. Arnold, *Coll. Czech. Chem. Commun.*, **35**, 3618 (1970).
186. M. Regitz and G. Himbert, *Liebigs Ann. Chem.*, **734**, 70 (1970).
187. R. Huisgen, G. Szeimies and L. Möbius, *Chem. Ber.*, **99**, 475 (1966).
188. M. T. Garcia-Lòpez, G. Garcia-Muñoz and R. Madroñero, *J. Heterocycl. Chem.*, **9**, 717 (1972).
189. M. Franck-Neumann and C. Buchecker, *Tetrahedron Lett.*, 2659 (1969).
190. R. Breslow and M. Oda, *J. Amer. Chem. Soc.*, **94**, 4787 (1972).
191. G. R. Harvey, *J. Org. Chem.*, **31**, 1587 (1966).
192. P. Ykman, G. L'abbé and G. Smets, *Tetrahedron Lett.*, 5225 (1970).
193. U. Schöllkopf and P. Markusch, *Liebigs Ann. Chem.*, **753**, 143 (1971).
194. M. B. Sohn, M. Jones, M. E. Hendrick, R. R. Rando and W. v. E. Doering, *Tetrahedron Lett.*, 53 (1972).
195. G. Himbert and M. Regitz, *Synthesis*, 571 (1972).
196. G. Himbert and M. Regitz, *Chem. Ber.*, **105**, 2963 (1972).
197. M. Regitz and G. Himbert, *Tetrahedron Lett.*, 2823 (1970).
198. G. Himbert and M. Regitz, *Liebigs Ann. Chem.*, 1505 (1973).
199. R. E. Harmon, F. Stanley, S. K. Gupta, R. A. Earl, J. Johnson and G. Slomp, *Chem. Ind. (London)*, 1021 (1970).
200. R. E. Harmon, F. Stanley, S. K. Gupta and J. Johnson, *J. Org. Chem.*, **35**, 3444 (1970).
201. G. Himbert, D. Frank and M. Regitz, *Chem. Ber.*, **109**, 370 (1976).
202. P. Grünanger and P. V. Finzi, *Tetrahedron Lett.*, 1839 (1963).
203. P. Grünanger, P. V. Finzi and C. Scotti, *Chem. Ber.*, **98**, 623 (1965).
204. G. Himbert and M. Regitz, *Chem. Ber.*, **105**, 2975 (1972).
205. G. E. Palmer, J. R. Bolton and D. R. Arnold, *J. Amer. Chem. Soc.*, **96**, 3708 (1974).
206. M. Franck-Neumann and C. Buchecker, *Tetrahedron Lett.*, 15 (1969).
207. M. Franck-Neumann and D. Martina, *Tetrahedron Lett.*, 1767 (1975).
208. F. Arndt, B. Eistert and W. Partale, *Ber. dt. chem. Ges.*, **60**, 1364 (1927); F. Arndt and J. Amende, *Ber. dt. chem. Ges.*, **61**, 1122 (1928).
209. U. Schöllkopf and H. Schäfer, *Angew. Chem.*, **77**, 379 (1965); *Angew. Chem. Int. Ed.*, **4**, 358 (1965).
210. U. Schöllkopf, P. Tonne, H. Schäfer and P. Markusch, *Liebigs Ann. Chem.*, **722**, 45 (1969).
211. U. Schöllkopf and P. Markusch, *Tetrahedron Lett.*, 6199 (1966).
212. U. Schöllkopf and P. Markusch, *Angew. Chem.*, **81**, 577 (1969); *Angew. Chem. Int. Ed.*, **8**, 612 (1969).
213. H. Staudinger, J. Becker and H. Hirzel, *Ber. dt. chem. Ges.*, **49**, 1978 (1916).
214. W. E. Bachmann and W. S. Struve, *Org. Reactions*, **1**, 38 (1942).
215. B. Eistert, in *Neuere Methoden der präparativen organischen Chemie*, 3rd ed., Verlag Chemie, Weinheim, 1949, p. 378.
216. See also Reference 2, p. 589.
217. D. A. Clibbens and M. Nierenstein, *J. Chem. Soc.*, 1491 (1915).
218. H. Staudinger and C. Mächling, *Ber. dt. chem. Ges.*, **49**, 1973 (1916).
219. M. Nierenstein, D. G. Wang and J. C. Warr, *J. Amer. Chem. Soc.*, **46**, 2551 (1924).
220. A. J. M. Kahil and M. Nierenstein, *J. Amer. Chem. Soc.*, **46**, 2556 (1924).
221. H. H. Lewis, M. Nierenstein and E. M. Rich, *J. Amer. Chem. Soc.*, **47**, 1728 (1925).
222. F. Arndt and H. Scholz, *Ber. dt. chem. Ges.*, **66**, 1012 (1933).
223. B. Michel, J. F. McGarrity and H. Dahn, *Chimia*, **27**, 320 (1973).
224. G. Heyes, G. Holt and A. Lewis, *J. Chem. Soc. Perkin I*, 2351 (1972).
225. C. G. Venier, H. J. Barager and M. A. Ward, *J. Amer. Chem. Soc.*, **97**, 3228 (1975).
226. B. Eistert, *Ber. dt. chem. Ges.*, **68**, 208 (1935).

227. M. S. Newman and P. Beal, *J. Amer. Chem. Soc.*, **71**, 1506 (1949).
228. M. Berenbom and W. S. Fones, *J. Amer. Chem. Soc.*, **71**, 1629 (1949).
229. V. Franzen, *Liebigs Ann. Chem.*, **602**, 199 (1957).
230. S. Hauptmann and K. Hirschberg, *J. Prakt. Chem.* (4), **34**, 262 (1966).
231. I. G. Farben (Inv. B. Eistert), *DRP* 724 757 (1940); *Chem. Abstr.*, **37**, 57337 (1943).
232. Schering AG (Inv. H. Eusenbach), *DBP* 875 659 (1953); *Chem. Abstr.*, **50**, 12124h (1956).
233. See Reference 214, p. 45
234. T. P. C. Mulholland, R. I. W. Honeywood, H. D. Preston and D. T. Rosevear, *J. Chem. Soc.*, 4939 (1965).
235. P. Karrer and R. Widmer, *Helv. Chim. Acta*, **8**, 203 (1925).
236. P. Karrer and G. Bussmann, *Helv. Chim. Acta*, **24**, 645 (1941).
237. W. Bradley and G. Schwarzenbach, *J. Chem. Soc.*, 2904 (1928).
238. C. Grundmann, *Liebigs Ann. Chem.*, **524**, 31 (1946).
239. J. H. Wotiz and S. N. Buco, *J. Org. Chem.*, **20**, 210 (1955).
240. J. A. Moore, *J. Org. Chem.*, **20**, 1607 (1955).
241. M. Itoh and A. Sugihara, *Chem. Pharm. Bull.* (*Tokyo*), **17**, 2105 (1969).
242. A. Nabeya and J. A. Moore, *J. Org. Chem.*, **35**, 2022 (1970).
243. R. Grewe and A. Bokranz, *Chem. Ber.*, **88**, 49 (1955).
244. J. F. Godington and E. Mosettig, *J. Org. Chem.*, **17**, 1027 (1952).
245. A. Nabeya, F. B. Culp and J. A. Moore, *J. Org. Chem.*, **35**, 2015 (1970).
246. A. Roedig and R. Maier, *Chem. Ber.*, **86**, 1467 (1953).
247. A. Roedig and R. Kloss, *Liebigs Ann. Chem.*, **612**, 1 (1958).
248. W. Bradley and R. Robinson, *J. Chem. Soc.*, 1310 (1928).
249. D. S. Tarbell and J. A. Price, *J. Org. Chem.*, **22**, 245 (1957).
250. J. Hooz and G. F. Morrison, *Org. Prep. and Proc. Int.*, **3**, 227 (1971).
251. B. Penke, J. Czombos, L. Baláspiri, J. Petres and K. Kovács, *Helv. Chim. Acta*, **53**, 1057 (1970).
252. L. Horner and H. Schwarz, *Liebigs Ann. Chem.*, **747**, 21 (1971).
253. B. Zwanenburg and L. Thijs, *Tetrahedron Lett.*, 2459 (1974).
254. D. Hodson, G. Holt and D. K. Wall, *J. Chem. Soc. C*, 971 (1970).
255. J. H. Looker and D. N. Thatcher, *J. Org. Chem.*, **23**, 403 (1958).
256. J. H. Looker and C. H. Hayes, *J. Org. Chem.*, **28**, 1342 (1963).
257. F. Weygand, W. Schwenke and H. J. Bestmann, *Angew. Chem.*, **70**, 506 (1958).
258. F. Weygand and H. J. Bestmann, *Angew. Chem.*, **72**, 535 (1960).
259. O. Tsuge, K. Sakai and M. Tashiro, *Tetrahedron*, **29**, 1883 (1973).
260. J. Goerdeler and R. Schimpf, *Chem. Ber.*, **106**, 1496 (1973).
261. A. Roedig, H. Aman and E. Fahr, *Liebigs Ann. Chem.*, **675**, 47 (1964).
262. J. H. Looker and J. W. Carpenter, *Can. J. Chem.*, **45**, 1727 (1967).
263. F. Kaplan and G. K. Meloy, *J. Amer. Chem. Soc.*, **88**, 950 (1966).
264. J. Schäfer, P. Boronowsky, R. Laursen, F. Finn and F. H. Westheimer, *J. Biol. Chem.*, **241**, 421 (1966).
265. S. S. Hixson, *J. Org. Chem.*, **37**, 1279 (1972).
266. J. Kollonitsch, A. Hajós and V. Gábor, *Chem. Ber.*, **89**, 2288 (1956).
267. A. Roedig and H. Lunk, *Chem. Ber.*, **87**, 971 (1954).
268. J. Ratuský and F. Sorm, *Chem. Listy*, **51**, 1091 (1957); *Chem. Abstr.*, **51**, 13 843a (1957).
269. F. Weygand, H. J. Bestmann and E. Klieger, *Chem. Ber.*, **91**, 1037 (1958).
270. A. L. Fridman and G. S. Ismagilova, *Z. Org. Khim.*, **8**, 1126 (1972); *Chem. Abstr.*, **77**, 125 863e (1972).
271. J. H. Turnbull and E. S. Wallis, *J. Org. Chem.*, **21**, 663 (1956).
272. G. Snatzke and G. Zanati, *Liebigs Ann. Chem.*, **684**, 62 (1965).
273. J. K. Chakrabarti, S. S. Szinai and A. Todd, *J. Chem. Soc. C*, 1303 (1970).
274. T. Reichstein and H. G. Fuchs, *Helv. Chim. Acta*, **23**, 658 (1940).
275. E. Fahr, *Liebigs Ann. Chem.*, **638**, 1 (1960).
276. I. Ernest and J. Hofman, *Chem. Listy*, **45**, 261 (1951); *Chem. Abstr.*, **46**, 7048h (1952).
277. J. Walker, *J. Chem. Soc.*, 1304 (1940).

278. H. Stetter and H. Stark, *Chem. Ber.*, **92**, 732 (1959); I. Font, F. López and F. Serratosa, *Tetrahedron Lett.*, 2589 (1972).
279. W. C. J. Ross, *J. Chem. Soc.*, 752 (1950).
280. W. Hampel, *J. Prakt. Chem.*, **311**, 78 (1969).
281. A. C. Wilds and A. L. Meader, *J. Org. Chem.*, **13**, 763 (1948).
282. W. Jugelt and P. Falck, *J. Prakt. Chem.* (4), **38**, 88 (1968).
283. G. Csávássy and Z. A. Györfi, *Liebigs Ann. Chem.*, 1195 (1974).
284. A. B. Smith, S. J. Branca and B. H. Toder, *Tetrahedron Lett.*, 4225 (1975).
285. S. Bien and D. Ovadia, *J. Chem. Soc. Perkin I*, 333 (1974).
286. H. Klusacek and H. Musso, *Chem. Ber.*, **103**, 3066 (1970).
287. T. R. Klose and L. N. Mander, *Aust. J. Chem.*, **27**, 1287 (1974).
288. R. Dran, B. Decock-Le Reverend and M. Polveche, *Bull. Soc. Chim. Fr.*, 2114 (1971).
289. E. Buchner, *Ber. dt. chem. Ges.*, **28**, 215 (1895).
290. U. Schöllkopf and H. Frasnelli, *Angew. Chem.*, **82**, 291 (1970); *Angew. Chem. Int. Ed.*, **9**, 301 (1970).
291. U. Schöllkopf, B. Banhidai, H. Frasnelli, R. Meyer and H. Beckhaus, *Liebigs Ann. Chem.*, 1767 (1974).
292. J. Lorberth, *J. Organomet. Chem.*, **27**, 303 (1971).
293. T. Dominh, O. P. Strausz and H. E. Gunning, *Tetrahedron Lett.*, 5237 (1968).
294. M. Regitz and U. Eckstein, *unpublished results*, Universität Saarbrücken, 1970.
295. P. Yates, F. X. Garneau and J. P. Lokensgard, *Tetrahedron*, **31**, 1979 (1975).
296. J. Lorberth, F. Schmock and G. Lange, *J. Organomet. Chem.*, **54**, 23 (1973).
297. S. J. Valenty and P. S. Skell, *J. Org. Chem.*, **38**, 3937 (1972).
298. U. Schöllkopf and N. Rieber, *Angew. Chem.*, **79**, 238 (1967); *Angew. Chem. Int. Ed.*, **6**, 261 (1967).
299. U. Schöllkopf and N. Rieber, *Chem. Ber.*, **102**, 488 (1969).
300. A. S. Kostyuk, J. B. Ruderfer, Y. I. Baukov and I. F. Lutsenko, *Zh. Obshch. Khim.*, **45**, 819 (1975); *Chem. Abstr.*, **83**, 79346 (1975).
301. M. F. Lappert, J. Lorberth and J. S. Poland, *J. Chem. Soc. A*, 2954 (1970).
302. J. Lorberth, *J. Organomet. Chem.*, **15**, 251 (1968).
303. U. Schöllkopf, B. Bánhidai and H.-U. Schulz, *Liebigs Ann. Chem.*, **761**, 137 (1972).
304. R. Grünig and J. Lorberth, *J. Organomet. Chem.*, **78**, 221 (1974).
305. P. Krommes and J. Lorberth, *J. Organomet. Chem.*, **93**, 339 (1975).
306. K. D. Kaufmann and K. Rühlmann, *Z. Chem.*, **8**, 262 (1968).
307. U. Schöllkopf and N. Rieber, *Angew. Chem.*, **79**, 906 (1967); *Angew. Chem. Int. Ed.*, **6**, 884 (1967).
308. U. Schöllkopf, D. Hoppe, N. Rieber and V. Jacobi, *Liebigs Ann. Chem.*, **730**, 1 (1969).
309. U. Schöllkopf, F. Gerhart, M. Reetz, H. Frasnelli and H. Schumacher, *Liebigs Ann. Chem.*, **716**, 204 (1968).
310. F. Gerhart, U. Schöllkopf and H. Schumacher, *Angew. Chem.*, **79**, 50 (1967); *Angew. Chem. Int. Ed.*, **6**, 74 (1967).
311. S. Masamune, N. Nakamura, M. Suda and H. Ona, *J. Amer. Chem. Soc.*, **95**, 8481 (1973).
312. H. Biltz and E. Kramer, *Liebigs Ann. Chem.*, **436**, 154 (1924).
313. N. F. Woolsey and M. H. Khalil, *J. Org. Chem.*, **37**, 2405 (1972).
314. E. Wenkert and C. A. McPherson, *J. Amer. Chem. Soc.*, **94**, 8084 (1972).
314a. N. F. Woolsey and M. H. Khalil, *J. Org. Chem.*, **38**, 4216 (1973).
315. T. L. Burkoth, *Tetrahedron Lett.*, 5049 (1969).
316. A. N. Pudovik, R. D. Gareev, A. B. Remizev, A. V. Aganov, G. I. Evstaf'er and S. E. Shtil'man, *Zh. Obshch. Khim.*, **43**, 559 (1973); *Chem. Abstr.*, **79**, 42 618u (1973).
317. B. Eistert and H. Selzer, *Chem. Ber.*, **96**, 1234 (1963).
318. B. Eistert and G. Borggrefe, *Liebigs Ann. Chem.*, **718**, 142 (1968).
319. B. Eistert, G. Borggrefe and H. Selzer, *Liebigs Ann. Chem.*, **725**, 37 (1969).
320. B. Eistert and E. A. Hackmann, *Liebigs Ann. Chem.*, **657**, 120 (1962).
321. B. Eistert, R. Müller, I. Mussler and H. Selzer, *Chem. Ber.*, **102**, 2429 (1969).
322. B. Eistert, W. Eifler and O. Ganster, *Chem. Ber.*, **102**, 1988 (1969).
323. B. Eistert and P. Donath, *Chem. Ber.*, **102**, 1725 (1968).
324. M. Dürr, *Thesis*, Technische Universität, München, 1971.

325. J. Hocker, M. Regitz and A. Liedhegener, *Chem. Ber.*, **103**, 1486 (1970).
326. L. Wolff, *Liebigs Ann. Chem.*, **394**, 23 (1912).
327. G. Holt and D. K. Wall, *J. Chem. Soc. C*, 857 (1966).
328. D. Seyferth and R. S. Marmor, *Tetrahedron Lett.*, 2493 (1970); D. Seyferth, R. S. Marmor and P. Hilbert, *J. Org. Chem.*, **36**, 1379 (1971).
329. M. Regitz, A. Liedhegener, U. Eckstein, M. Martin and W. Anschütz, *Liebigs Ann. Chem.*, **748**, 207 (1971).
330. U. Felcht and M. Regitz, *Chem. Ber.*, **108**, 2040 (1975).
331. N. Kreutzkamp, E. Schmidt-Samoa and K. Herberg, *Angew. Chem.*, **77**, 1138 (1965); *Angew. Chem. Int. Ed.*, **4**, 1078 (1965).
332. M. Regitz and H. Eckes, *unpublished results*, Kaiserslautern, 1972.
333. M. Hufnagel and M. Regitz, *unpublished results*, Kaiserslautern, 1975.
334. L. Horner, H. Hoffmann, H. Ertel and G. Klahre, *Tetrahedron Lett.*, 9 (1961).
335. M. Regitz, W. Anschütz and A. Liedhegener, *Chem. Ber.*, **101**, 3734 (1968).
336. H. Scherer, A. Hartmann, M. Regitz, B. D. Tunggal and H. Günther, *Chem. Ber.*, **105**, 3357 (1972).
337. G. Rosini and G. Baccolini, *Synthesis*, 44 (1975).
338. M. Regitz, W. Anschütz, W. Bartz and A. Liedhegener, *Tetrahedron Lett.*, 3171 (1968).
339. M. Regitz and W. Bartz, *Chem. Ber.*, **103**, 1477 (1970).
340. W. Jugelt, W. Lamm and F. Pragst, *J. Prakt. Chem.*, **314**, 193 (1972).
341. N. Gurudata, C. Benezra and H. Cohen, *Can. J. Chem.*, **51**, 1142 (1973).
342. H. Cohen and C. Benezra, *Can. J. Chem.*, **52**, 66 (1974).
343. W. Jugelt and D. Schmidt, *Tetrahedron*, **25**, 5569 (1969).
344. R. S. Marmor and D. Seyferth, *J. Org. Chem.*, **36**, 128 (1971).
345. D. Seyferth, P. Hilbert and R. S. Marmor, *J. Amer. Chem. Soc.*, **89**, 4811 (1967).
346. A. Hartmann, W. Welter and M. Regitz, *Tetrahedron Lett.*, 1825 (1974).
347. M. Regitz and A. Hartmann, *unpublished results*, Kaiserslautern, 1974.
348. M. Regitz, *Angew. Chem.*, **87**, 259 (1975); *Angew. Chem. Int. Ed.*, **14**, 222 (1975).
349. W. Welter and M. Regitz, *Tetrahedron Lett.*, 1476 (1976).
350. M. Regitz and M.-E. Jaeschke, *unpublished results*, Kaiserslautern, 1974.
351. M. Regitz and W. Anschütz, *Chem. Ber.*, **102**, 2216 (1969).
352. M. Regitz, A. Liedhegener, W. Anschütz and H. Eckes, *Chem. Ber.*, **104**, 2177 (1971).
353. M. Regitz and U. Förster, *unpublished results*, Kaiserslautern, 1974.
354. M. Regitz and M. Martin, *unpublished results*, Kaiserslautern, 1975.
355. G. Petzold and H. G. Henning, *Naturwissenschaften*, **54**, 469 (1967).
356. G. Maas and M. Regitz, *Chem. Ber.*, **109**, 2039 (1976).
357. U. Heep, *Liebigs Ann. Chem.*, 578 (1973).
358. R. Mertz, D. van Assche, J.-P. Fleury and M. Regitz, *Bull. Soc. Chim. Fr.*, 3442 (1973).
359. G. Keller, J.-P. Fleury, W. Anschütz and M. Regitz, *Bull. Soc. Chim. Fr.*, 1219 (1975).
360. D. Stadler, W. Anschütz, M. Regitz, G. Keller, D. van Assche and J.-P. Fleury, *Liebigs Ann. Chem.*, 2159 (1976).
361. M. Regitz and W. Anschütz, *Liebigs Ann. Chem.*, **730**, 194 (1969).
362. M. Regitz and W. Welter, *unpublished results*, Kaiserslautern, 1974.
363. G. Himbert and M. Regitz, *Chem. Ber.*, **107**, 2513 (1974).
364. M. Regitz and H. Heydt, *unpublished results*, Kaiserslautern, 1975.
365. M. Regitz and B. Weber, *unpublished results*, Kaiserslautern, 1975.
366. M. Regitz and U. Eckstein, *unpublished results*, Universität Saarbrücken, 1973.
367. M. Regitz and B. Weber, *unpublished results*, Universität Saarbrücken, 1970.
368. W. Disteldorf and M. Regitz, *Chem. Ber.*, **109**, 546 (1976).
369. M. Regitz and W. Disteldorf, *unpublished results*, Kaiserslautern, 1973.
370. W. Disteldorf and M. Regitz, *Liebigs Ann. Chem.*, 225 (1976).
371. M. Regitz, W. Disteldorf, U. Eckstein and B. Weber, *Tetrahedron Lett.*, 3979 (1972).
372. W. Welter and M. Regitz, *Tetrahedron Lett.*, 3799 (1972).
373. A. M. van Leusen, J. Strating and D. van Leusen, *Tetrahedron Lett.*, 5207 (1973).
374. A. M. van Leusen, B. A. Reith and D. van Leusen, *Tetrahedron*, **31**, 597 (1975).
375. E. I. du Pont de Nemours (Inv. J. Diekmann), *U.S. Patent* 3 332 936 (25.7.1967); *Chem. Abstr.*, **68**, 59 304j (1968).

376. J. Diekmann, *J. Org. Chem.*, **30**, 2272 (1965); **28**, 2933 (1963).
377. J. Strating and A. M. van Leusen, *Rec. Trav. Chim. Pays-Bas*, **81**, 966 (1962).
378. A. M. van Leusen and J. Strating, *Rec. Trav. Chim. Pays-Bas*, **84**, 151 (1965).
379. A. Wagenaar, G. Kransen and J. B. F. F. Engberts, *J. Org. Chem.*, **39**, 411 (1974).
380. J. B. F. N. Engberts, G. Zuidema, B. Zwanenburg and J. Strating, *Rec. Trav. Chim. Pay-Bas*, **88**, 641 (1969).
381. B. Michel, J. F. McGarrity and H. Dahn, *Chimia*, **27**, 320 (1973).
382. J. Strating, J. Heeres and A. M. van Leusen, *Rec. Trav. Chim. Pays-Bas*, **85**, 1061 (1966).
383. M. Regitz, *Chem. Ber.*, **98**, 36 (1965).
384. H. Balli, Universität Basel, *private communication* (9.6.1964).
385. A. M. van Leusen, P. M. Smid and J. Strating, *Tetrahedron Lett.*, 337 (1965).
386. G. Heyes and G. Holt, *J. Chem. Soc. Perkin I*, 189 (1973).
387. W. Illger, A. Liedhegener and M. Regitz, *Liebigs Ann. Chem.*, **760**, 1 (1972).
388. D. Hodson, G. Holt and D. K. Wall, *J. Chem. Soc. C*, 2201 (1968).
389. F. Klages and K. Bott, *Chem. Ber.*, **97**, 735 (1964).
390. L. A. Carpino, L. V. McAdams, R. H. Rynbrant and J. W. Spiewak, *J. Amer. Chem. Soc.*, **93**, 476 (1971).
391. M. Regitz and B. Mathieu, *unpublished results*, Kaiserslautern, 1972.
392. A. M. van Leusen, B. A. Reith, R. J. Multer and J. Strating, *Angew. Chem.*, **83**, 290 (1971); *Angew. Chem. Int. Ed.*, **10**, 271 (1971).
393. D. Hodson, G. Holt and D. K. Wall, *J. Chem. Soc. C*, 2201 (1968).

CHAPTER **18**

Synthetic applications of diazoalkanes, diazocyclopentadienes and diazoazacyclopentadienes

D. S. WULFMAN

Department of Chemistry, University of Missouri-Rolla, Rolla, Mo. 65401, USA and *Équipe de Recherche No. 12 du CNRS, Laboratoire de Chimie, École Normale Supérieure, Paris, France*

G. LINSTRUMELLE

Équipe de Recherche No. 12 du CNRS, Laboratoire de Chimie, École Normale Supérieure, Paris, France

and (in part)

C. F. COOPER

Department of Chemistry, University of Missouri-Rolla, Rolla, Mo. 65401, USA

INTRODUCTION

The role of diazoalkanes in synthesis has been surprisingly limited to a few very common reaction types although there are numerous examples where potentially difficult targets could in principle be attained efficiently using diazo chemistry. The first application to a major natural products synthesis dates back to the now classical synthesis of equilenin by Bachmann, Cole and Wilds[1469] † where an Arndt Eistert–Wolff Rearrangement sequence was employed to convert an acetic acid into a propionic acid. Very extensive studies by Gutsche[1831–1834] were directed towards the synthesis of colchicine. These studies ultimately proved unsuccessful because of the inherent nature of the target molecules. The use of diazomethane for ether formation played an important, though almost trivial, role in the first two successful syntheses of colchicine[2213, 2311] where a highly hindered hydroxyl group was involved. The synthetic role of diazoalkanes is well established in the area of valence isomerization processes[1824] where syntheses of tropilidenes, homotropilidenes, barbralenes, homobarbralenes and bullvalenes are well documented[2214]. In addition, syntheses in the tropone[1644], tropolone[1643], and azulene[2269] fields frequently employ diazoalkane chemistry. Numerous examples of diazoalkane application can be found in some recent treatises on the syntheses of natural products[1443] and on classical syntheses[1523, 1777] and include syntheses of agarospirol[2049], bokhenolide[1758], haemanthidine[1860, 1861], illuden-s[2024], lucidulene[2219], muscone[1757], patchouli alcohol[1548], sirenin[1526, 1527], sesquicarene[1526, 1527], trichodermin[1589], aristolone[1514], β-cubebene[2140], sabinene[2314] and cyclopropanone[1850].

The fact that diazoalkanes are not used more widely may well reflect their bad reputation as toxic substances and as explosives. Much of this is undoubtedly deserved but techniques for safe handling have removed many of the problems and some diazoalkanes are amazingly stable. Diazomalonic esters fall nicely into the 'safe' category, and even diazoacetates are moderately stable in non-polar solvents such as benzene and toluene at elevated temperatures. The fact that the C—N bond energy of diazomethane is estimated to be over 100 kcal[1979] (one theoretician states over 230 kcal[1838]) suggests that spontaneous generation of carbenes and nitrogen may not be the mode of decomposition, and that dilution leads to stabilization because appreciable activation by local hot spots resulting from chain reactions is prevented[1440, 1498, 1499, 1556, 1580, 1840, 1857, 1913, 1977–1979, 2078, 2113].

The reputation for carcinogenic activity is well documented in a few cases and surely warrants caution with any diazo compound just as it does with other untested chemicals.

I. CYCLOADDITION REACTIONS

In this section we are concerned with four basic cycloaddition reactions, three of which fall into the 1,3-dipolar addition category and one in the 1,4 category. Of these, two of the 1,3-dipolar processes carbonyl ylides (**2**) and ketocarbenes (**3**) result from the loss of nitrogen from a diazoalkane (**1**) and at best are normally difficult to study transient species. The most extensive literature lies in the area of 1,3-dipolar additions of diazoalkanes with an X=Y and X≡Y bond system[591]. Traditionally one associates such work with Huisgen and colleagues[397, 586, 588, 591, 1326, 1885, 1891]; however, a very large and important contribution has been made by the groups of Carrié, Vo-Quang, Franck-Neumann and Bastide. Since the work of the Munich school has

† References below 1400 occur at the end of Chapter 8. Those above 1400 appear at the end of this chapter.

been treated extensively in the English language literature as well as in the German, our current emphasis lies heavily upon work which has appeared almost exclusively in French (some unpublished) and upon the extensive pyrazoline investigations of the two Canadian schools of Crawford and McGreer.

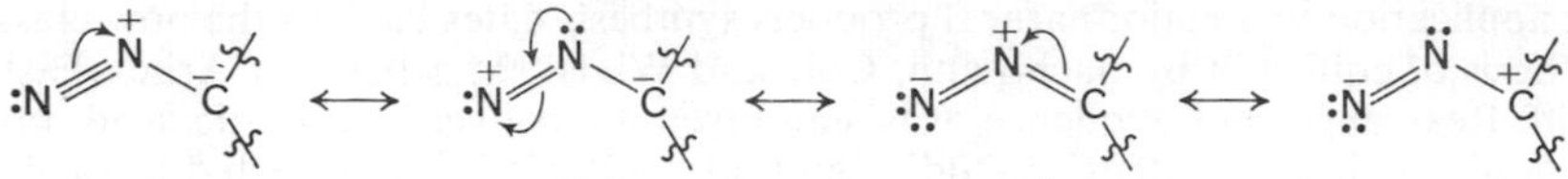

(1a) (1b) (1c) (1d)

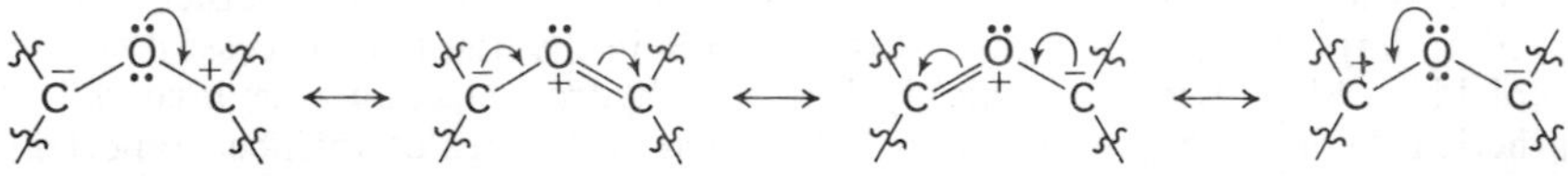

(2a) (2b) (2c) (2d) (1)

(3a) (3b)

The fourth category of reaction (1,4-additions of diazo functions) might better appear in Chapter 8 but the first correction appeared after the completion of that chapter and this paper is supplemented with recently obtained unpublished data furnished by Sheppard of DuPont. Since much of the chemistry of the DuPont workers involves heterocyclic analogues of diazocyclopentadiene it appears in this chapter. The basic features of cycloaddition reactions have been treated extensively by Huisgen[591], Woodward and Hoffmann[2354], Dewar[291, 292] and by others. The reactions of interest here all appear to proceed by *cis* additions but may not be fully concerted. The processes involving diazoalkanes clearly exhibit appreciable solvent effects in protic solvents; this is suggestive of sizeable charge separation in the transition state[1935]. One primary concern is that of regioselectivity. This question has been carefully examined in the diazoalkane case with a large variety of systems. Of equal importance for this chapter are the subsequent reactions of the cycloaddition products. In some cases the primary product is not isolable or even demonstrably an intermediate. Thus, some Δ^1-pyrazolines and 3 *H*-pyrazoles are not observed and only the Δ^2-pyrazoline or products resulting from loss of N_2 are found or the pyrazole is isolated directly.

A. 1,3-Dipolar Additions of Diazoalkanes[1406]

At first glance it would appear that diazoalkanes can add across almost any multiple bond whether it possesses polar character or not. This is perhaps an extreme view, but the probability is high although the rate and/or equilibrium constants may not be favourable. Both of these problems can in part be overcome by changing solvent polarity[1935] (perhaps the mechanism as well) and by employing very high pressures[2272]. The latter both speeds up the reaction and shifts the equilibrium (if any) to the right.

Huisgen[1888] and coworkers measured the relative rates of addition of diphenyldiazomethane to a number of olefins (Table 1).

The relative reactivity of cycloalkenes towards diazomethane and diazoethane has been examined by Paul and collaborators[2123] (Table 2).

TABLE 1. The relative rates of addition of diazomethane to some multiple carbon–carbon bonds[1888]

Dipolarophile	$10^7 k_2$ (1 mol^{-1} s^{-1})
Acetylenes	
Phenylacetylene	118
Prop-2-ynal di(*n*-propyl)acetal	222
Ethyl phenylpropiolate	333
Methyl propiolate	$106{\cdot}5 \times 10^3$
Diethyl acetylenedicarboxylate	968×10^3
Cycloalkenes	
Norbornene	286
Dicyclopentadiene	345
Dimethyl *endo*-bicyclo[2.2.1]hept-2-ene-5,6-dicarboxylate	528
Diethyl 2,3-diazobicyclo[2.2.1]hept-5-ene-2,3-dicarboxylate	2900
Conjugated olefins	
1,1-Diphenylethylene	27
Benzalacetone	72
Ethyl cinnamate	125
Styrene	140
Ethyl crotonate	246
Ethyl *p*-nitrocinnamate	8×10^2
Ethyl methacrylate	$5{\cdot}05 \times 10^3$
Dimethyl malonate	$6{\cdot}85 \times 10^3$
Acrylonitrile	$4{\cdot}34 \times 10^4$
Ethyl acrylate	$7{\cdot}07 \times 10^4$
trans-Dibenzoylethylene	$9{\cdot}79 \times 10^4$
Dimethyl fumarate	$2{\cdot}45 \times 10^5$
Phenyl acrylate	$2{\cdot}5 \times 10^5$
Maleic anhydride	$5{\cdot}83 \times 10^5$

TABLE 2. The effect of ring size on the addition of diazoalkanes to cyclic olefins[2123]

Ring size	Reagent	Reaction time (days)	Yield (%)
4	CH_2N_2	10	42
5	CH_2N_2	20	31
6	CH_2N_2	25	<1
4	CH_3CHN_2	3	61
5	CH_3CHN_2	6	16
6	CH_3CHN_2	10	2·2 *syn* 3·3 *anti*

Basingdale and Brook[1493] have examined the effect of structure upon the addition of silyl diazomethanes to diethyl fumarate and found relative rates $Me_3SiCN_2Me(1) > Me_3SiCN_2C_6H_5\ (2 \times 10^{-2}) > (C_6H_5)_3SiCN_2Me\ (2{\cdot}5 \times 10^{-3}) > (C_6H_5)_3SiCN_2C_6H_5\ (2{\cdot}5 \times 10^{-6}) > Me_2SiCN_2CO_2Et\ (2{\cdot}5 \times 10^{-7})$. All of the silyldiazomethanes and pyrazolines have enhanced thermal stability. The trimethylplumbyl pyrazolines from trimethylplumbyl diazoalkanes also appear to possess high stability[1886].

Kadaba and Colturi[1935] compared the reactivity of diazomethane towards substituted styrenes using anhydrous ether and wet dioxane as solvents. The wet dioxane was clearly the better solvent (Table 3). The reactions involving equilibria can, on occasion, be further enhanced by subsequent conversion of the primary product into a secondary product by an irreversible process.

TABLE 3. The addition of diazoalkanes to substituted styrenes[1935]

Substituent	Reaction time (h)	Yield wet dioxane (%)	Ether
p-OCH_3	91	42	
p-Me	96	36	
H	168	60	28
p-Cl	168	76	57
m-Cl	95	48	
m-NO_2	90	58	50

I. Additions to X=Y Systems

By far the most thoroughly examined class of substrate is the olefinic link with or without activation by substituents in conjugation with the C=C bond. A common activation process is the introduction of strain into a cycloalkene[1468, 1557, 1587, 1786, 1787, 2123, 2388, 2346, 2347, 2361, 2373, 2188], but this problem in part can be overcome by operating at pressures in the area of 5000 atm. Thus, de Suray and coworkers[2272] were able to prepare the Δ^1-pyrazolines from *trans*-stilbene in 100% yield in 170 h at ambient temperature whereas no product resulted after 72 h at 1 atm (Table 4).

TABLE 4. The effect of pressure upon the 1,3-dipolar additions of diazomethane[2272]

Dipolarophile	Product	Reaction time (h)	Pressure (atm)	Yield (%)
9-Carbomethoxy phenanthrene	Δ^1-Pyrazoline	49	1	7
			5000	45
Methyl *cis*-α-methyl cinnamate	Δ^1-Pyrazoline	49	1	15
			5000	100
Methyl *trans*-α-phenyl cinnamate	Δ^1-Pyrazoline	49	1	0
			5000	66
Methyl-*cis*-α-phenyl cinnamate	Δ^1-pyrazoline	49	1	33
			5000	100
trans-Stilbene	Δ^1-Pyrazoline	72	1	0
		170	5000	100
cis-Stilbene	Δ^2-Pyrazoline	72	1	0
		199	5000	40
p-Nitro-*trans*-stilbene	Δ^2-Pyrazoline	144	1	5
		144	5000	70
Benzylideneaniline	Triazoline	70	1	5
			5000	650
p-Nitrobenzylideneaniline	Triazoline	70	1	34
			5000	72

The direction of addition to an unsymmetrical olefin can in principle occur in two senses, and too frequently does so even when the double bond is strongly polarized by substituents. It is attractive to assign such behaviour to steric effects, since when this phenomenon occurs, it usually involves disubstituted diazomethanes. This assumes of course that the carbon of the diazoalkane possesses excess electron density and the terminal nitrogen is partially positive. (A point of some dispute amongst theoreticians[1440, 1498, 1499, 1556, 1580, 1838, 1857, 1912, 1977, 1978, 1979, 2113, 2216] but generally felt to be the case by most theoreticians and by experimentalists, see References 1440 and 2113 for alternative conclusions.)

(4) → (5) Normal sense

(6) → (7) Abnormal or inverse sense

(2)

The early studies of Parham[2114–2119] upon the addition of diazoalkanes to nitro-olefins revealed that diazomethane and diazoethane add normally whereas diphenyldiazomethane gives the reverse orientation. With allenic ketones and esters this effect is not observed[1504]. However, the anomalous addition was observed by Franck-Neumann in the addition of 2-diazopropane to a Δ^3-β-octalone[1514]. Anomalous orientation was not observed with alkylidene and benzylidene malonates, cyanoacetates and acetylacetones. Anomalously oriented products are observed with arylacetylenes and diazomethane[1946] while disubstituted diazomethanes furnish only the anomalous product[1652, 1946, 1921]. Other acetylenes furnishing at least some anomalous products include alcohols[1921, 2058, 2061], ethers[1494, 2185], esters[1496], phenyl propiolates[1496, 1525, 1965, 2288, 2322], ketones[1496, 1631, 1652], aldehydes and alkoxyacetylenes[1827].

Huisgen has compared the relative reactivity of diazoalkanes as 1,3-dipoles. The general trend is 2-diazo-1,3-dicarbonyl compounds $<$ diazoketones $<$ diazoacetic ester $<$ diphenyldiazomethane $<$ diazomethane[2217].

a. *Simple olefins and substituted olefins other than ketones and aldehydes.* A wide variety of diazoalkanes has been found to add across double bonds to furnish pyrazolines. These include diazomethane, diazoethane, diazocyclopropane, 2-diazopropane, diazoacetic esters, dimethyl diazomalonate, aryldiazomethanes and diaryldiazomethanes as well as diazocyclopentadienes. The reactions often tend to be slow or non-existent unless the double bond is (i) activated by being part of a strained ring system (cyclopropenes, cyclobutenes and cyclopentenes show decreasing reactivity), (ii) in conjugation with an electron-withdrawing or -donating substituent, (iii) elevated temperatures or pressures are employed or (iv) polar solvents are employed. Since many diazo compounds and Δ^1-pyrazolines are quite temperature sensitive, the temperature variable is normally not attractive.

Strained five-membered rings are more reactive if they are part of a bicyclic system such as norbornene or dicyclopentadiene. In the last case, the relatively unreactive dimethyl diazomalonate adds quantitatively in ~60 days at 30 °C to the 8,9 double bond while failing to react with cyclopentene under identical conditions or with the 4,5 double bond of dicyclopentadiene[2361].

(8) $\xrightarrow[\sim 30\ ^\circ C]{N_2C(COOMe)_2}$ (9) ~100% $\xrightarrow[15\ min]{\Delta,\ C_6H_6,\ 0{\cdot}01\ mmol\ Cu(BF_4)_2}$ (10) ~100% (3)

The addition of diazomethane or diazoethane to *cis*-3,4-dichlorocyclobutene[1781, 1786, 1787] is illustrative of one of the problems of stereochemistry; it furnishes the *syn*-pyrazoline quantitatively with the alkyl residue *exo* (equation 4).

(11) $\xrightarrow{RCHN_2}$ (12) R = H, CH_3

(13) $\xrightarrow{RCHN_2}$ (14) $\xrightarrow{RCHN_2}$ (15) (4)

X = Cl, I, Br, OH, OAc

When, however, 7*X*-substituted norbornadienes are employed, *endo-anti* addition occurs for the first diazoalkane molecule and the second enters *exo* (X = Cl, Br, I, OAc or OH). With dimethyl tetracyclo[3.3.2.$0^{2,4}$.$0^{6,8}$]dec-9-ene-2,4-dicarboxylate **(14)**, addition is on the least hindered and least unsaturated face **(15)**[2293]. With diazoalkanes and *trans* dichloronorbornene or *cis-endo*-dichloronorbornene, only *exo* adducts are obtained[1488, 1785]. With several diaryldiazomethanes the addition to 7-*anti-t*-butoxynorbornene occurs exclusively on the *exo* face while the results with the related norbornadiene exhibit little selectivity except that the *t*-butoxy group considerably reduces attack on that face.

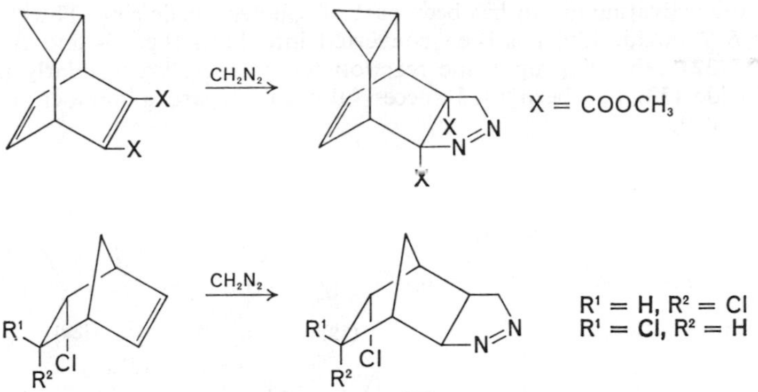

CH_2N_2 — X = $COOCH_3$ (5)

CH_2N_2 — $R^1 = H, R^2 = Cl$; $R^1 = Cl, R^2 = H$

(16) **(17)**

The work of Callot, Cohen and Benezra[1557, 1587] with aryldiazoalkanephosphonates and alkyldiazoalkanephosphonates employing norbornene and norbornadiene as substrates furnished only the *exo* adducts **19, 20, 21,** and **22.** The relative amounts of the two possible orientations of the substituents R^1 and R^2 were found to be a function of steric and electronic factors (Table 5).

TABLE 5. The reactions of some diazophosphonates with norbornadiene and norbornene[1557, 1587]

R^1	19 (%)	20 (%)	22 (%)	23 (%)
Ph	100			
CH_3	81	19		
Ph			100	
CH_3			78	22
p-$CH_3OC_6H_4$			95	5
p-$CH_3C_6H_4$			90	10
p-BrC_6H_4			90	10
C_6H_4			95	5
p-$NO_2C_6H_4$			93	7
p-$CH_3OCOC_6H_4$			91	9
o-$CH_3OC_6H_4$			100	
α-Naphthyl			100	
β-Naphthyl			90	10
CH_3			75	25
C_2H_5			60	40
i-C_3H_7			50	50
t-C_4H_9			0	100
$PhCH_2$			55	45

With open-chain olefins and unsymmetrical diazoalkanes, two possible orientations exist in the resulting pyrazolines. The reactions proceed stereoselectively in a *cisoid* fashion with steric factors dominating[1442, 1482, 2110] (equation 7).

A variety of functional groups can serve as activating substituents and these can be electron-withdrawing groups such as nitro, cyano, ester, aldehyde, keto or sulphonyl, or electron-donating groups such as thioether, amines and ethers. In some cases, the activating group has been part of a heterocyclic ring. Thus the benzthiophene *S,S*-dioxide (**30**) has been converted into both the Δ^1- and Δ^2-pyrazolines (**31** and **32**) depending upon the reaction temperature[2190]. Similarly the thietane-1,1-dioxide (**33**) has been used successfully to prepare a number of substituted pyrazoles[1641].

(**18**) $\xrightarrow{R^1R^2CN_2}$ (**19**) + (**20**)

(**21**) $\xrightarrow{R^1R^2CN_2}$ (**22**) + (**23**) (6)

R^1 = Me, C_6H_5, *p*-$MeOC_6H_4$, *p*-MeC_6H_4, *p*-BrC_6H_4,
R^1 = R^2, *p*-$NO_2C_6H_4$, *p*-$AcOC_6H_4$, *o*-$MeOC_6H_4$,
α-naphthyl, β-naphthyl, Et, *i*-Pr, *t*-Bu, $C_6H_4CH_2$
R^2 = $P(O)(OCH_3)_2$

(**24**) + (**25**) ⟶ (**26**) + (**27**) (7)

(**25**): X = H; X = CH_3; X = OCH_3; X = Cl

(**26**): + ; + ; + ; +

(**27**): — ; + some ; + some ; —

Δ ⟶ (**28**) + (**29**)

(**28**) X = H, 50 °C, 6%; 80 °C, 8·8%

(**29**) 94%; 91·2%

Ojima and Kondo[2102] have compared the relative reactivities towards diazomethane of vinyl and allyl ethers with the related thioethers. They concluded that there was pπ–dπ interaction between sulphur and the double bond. With benzyl allyl sulphide they obtained only a single pyrazoline, whereas the related ether furnished both possible orientations. With the related vinyl systems, the reverse situation occurs (equation 9). It is perhaps noteworthy that acrolein acetals are claimed to furnish a single (normal) orientation. Subsequent removal of N_2 furnishes the cyclopropylcarboxaldehyde acetals in good yield[2268].

R = H, CH_3
R¹ = H, CH_3

(8)

R	R¹	Yield (%)	
		34	**35**
Ph	H		55
CH_3	H		60
Ph	Ph		50
Ph	CH_3		83
CH_3	CH_3		60
H	H		73
C_6H_6	heat	50–57%	**38**
	hν	45%	**38**
H	heat	52%	**39**

An amusing application of the cycloaddition of a diazomethane to an enamide has been developed by Barton's group[1490, 1491]. They treated **50** with diazomethane and subsequently reacted the product **51** with Zn and acetic acid or KOBu-*t*/*t*-BuOH to furnish the β-lactam (**52**) (equation 10).

Unlike the Diels–Alder addition to cycloheptatriene[1506] (which proceeds via the norcaradiene form), the reaction of the azepine (**53**) with diazomethane occurs with both valence isomers[2144]. This point has been employed as an argument against the presence of a dynamic equilibrium in the diazepine (**58**). The product with diazomethane furnishes a tetra-aza-azulene (**60**) which upon treatment with $Pb(OAc)_4$ gave a diazanorcaradiene (**63**). When 2-diazopropane is employed, thermolysis yields the diazahomotropilidene (**61**) which does not undergo the expected Cope rearrangement (**61** ⇄ **62**)[1943].

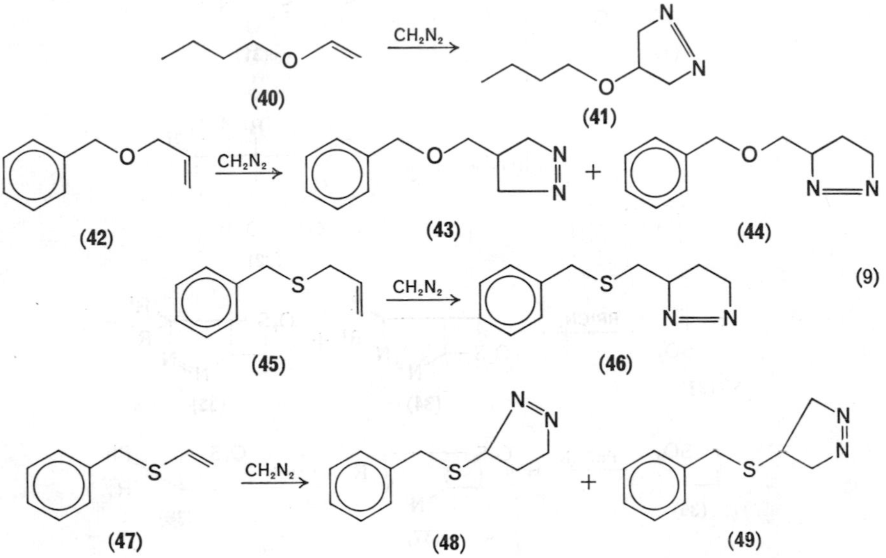

CH₂N₂ (40) → (41)

CH₂N₂ (42) → (43) + (44)

CH₂N₂ (45) → (46) (9)

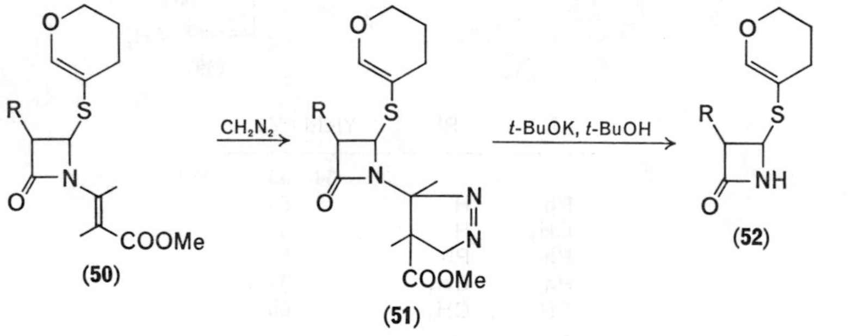

CH₂N₂ (47) → (48) + (49)

(50) CH₂N₂ → (51) t-BuOK, t-BuOH → (52) (10)

Sharp and coworkers[1413, 1640, 2256] have examined the intramolecular cyclization of vinyl diazomethanes to furnish 3*H*-pyrazoles. Some of these are able to undergo isomerization to 3*H*-1,2-diazepines. In some cases the 3*H*-diazepines are formed directly.

The parent 3*H*-pyrazol synthesis was first observed by Hund and Liu in 1935[1898] and the bimolecular cycloaddition of vinyl diazomethane has been examined by Tabuski and coworkers[2273] and by Salomon[2188]. The transient participation of an intramolecular cycloaddition nicely accounts for the conversion of **73** into **75**, **76** and **77**[1968].

The presence of two geminal activating groups on an olefin apparently does not lead to abnormal addition. Thus, Carrié and collaborators have carried out extensive studies upon alkylidenecyanoacetates[256, 1845–1847, 1849], alkylidenemalonates[1619], alkylideneacetoacetates[1581, 1613–1617, 1621, 1624, 1625, 2023], alkylidenedibenzoylmethanes[1625],

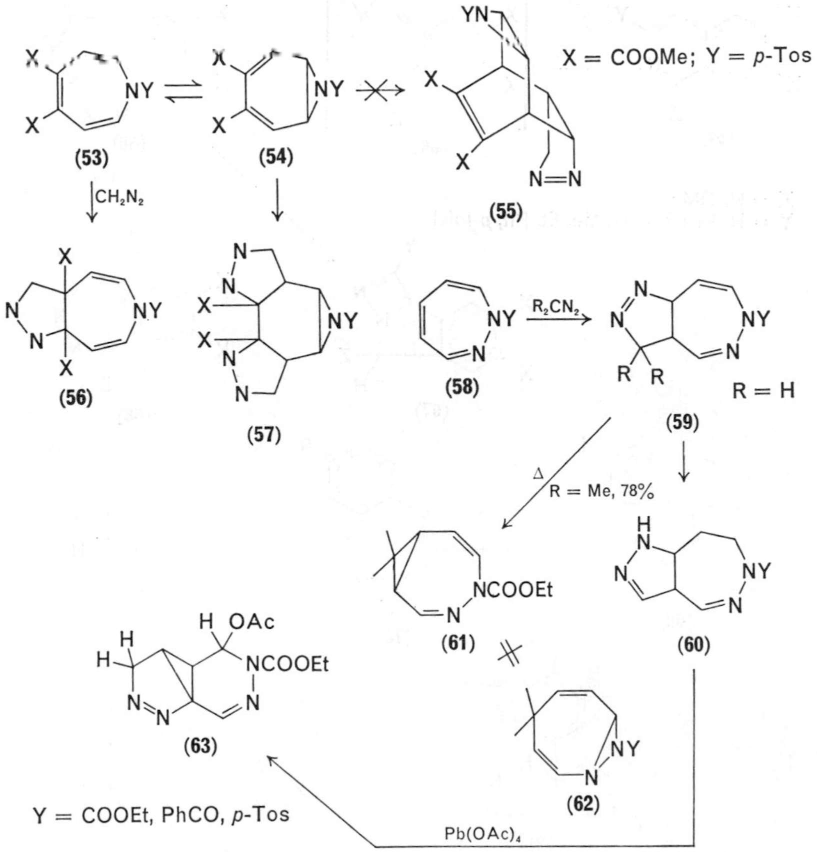

X = COOMe; Y = *p*-Tos

(53) (54) (55) (56) (57) (58) (59) (60) (61) (62) (63)

CH_2N_2 R_2CN_2 R = H Δ R = Me, 78% $Pb(OAc)_4$

Y = COOEt, PhCO, *p*-Tos (11)

cinnamylidenecyanoacetates[2022], cinnamylidenemalononitriles[2022], cinnamylidenemalonic esters[2021], benzylidenemalonic esters[256, 1618, 1848, 1849], alkylidenecyanoacetamides[1849], alkylidenenitroacetates[988], alkylidenenitroacetonitriles[988], nitroolefins[219], benzylideneacetoacetic esters[256], benzylidenecyanoacetates[256], and benzylidenenitroacetates[219]. The diazoalkanes primarily employed were diazomethane, diazoethane and diazoacetic ester. The reactions studied are summarized in equations (14)–(33) and the products arising initially and after thermolysis are given there and in Tables 6–17. The basic processes observed are summarized in equation (14). The large variety of products arise in part because of the conformational equilibria between the two possible conformers, e.g. **81a** and **81b** (see also equation 30).

What is surprising, are the reaction conditions successfully employed when diazoacetic ester was the 1,3-dipole. Some of the reactions (e.g. **143** → **142**, equation 23, Table 8) were run in refluxing toluene or benzene and the cycloaddition reaction occurred preferentially to simple loss of N_2 to furnish 'carbene'-derived products.

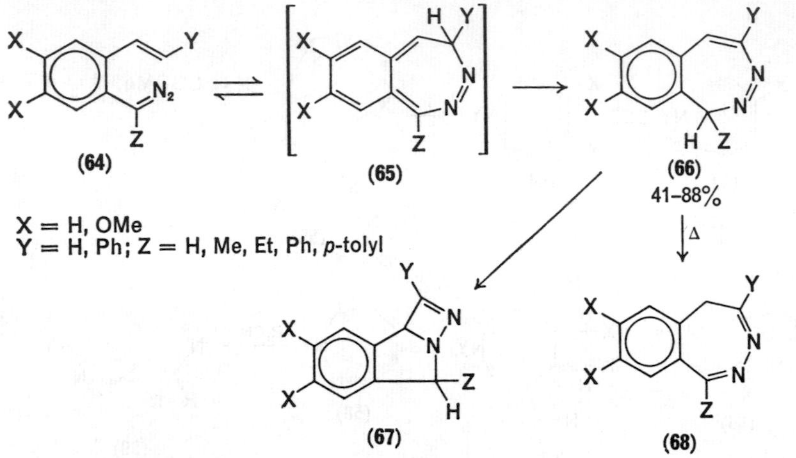

(64) (65) (66) 41–88% (67) (68)

X = H, OMe
Y = H, Ph; Z = H, Me, Et, Ph, *p*-tolyl

(12)

(69) (70) (71) (72)

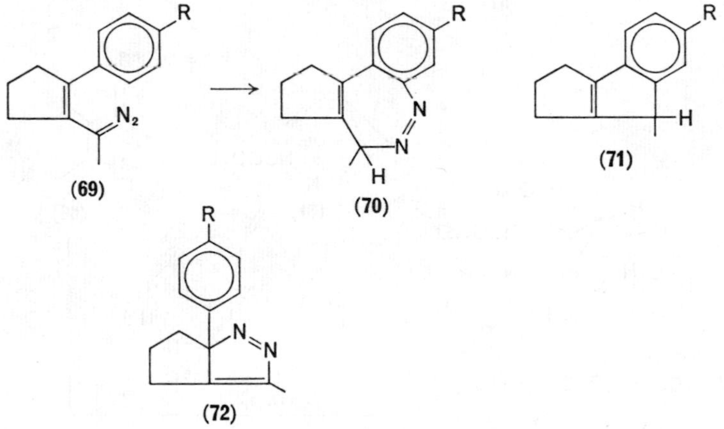

R	70 (%)	71 (%)	72 (%)
H	32	20	46
F	38	26	36
CH_3	37	20	43

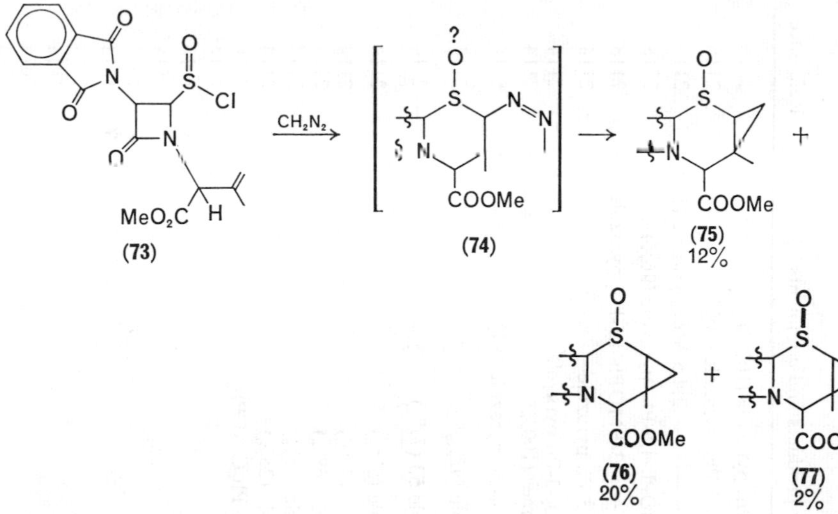

(73) $\xrightarrow{CH_2N_2}$ [(74)] → (75) 12% + (76) 20% + (77) 2% (13)

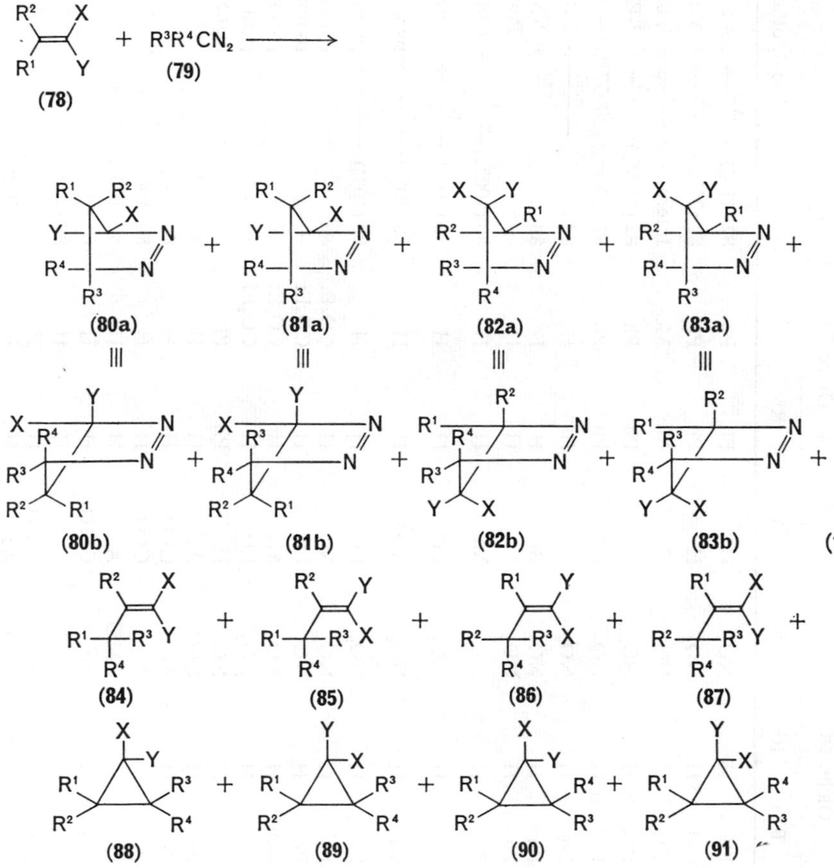

(78) + $R^3R^4CN_2$ (79) ⟶

(80a) + (81a) + (82a) + (83a) +

(80b) + (81b) + (82b) + (83b) + (14)

(84) + (85) + (86) + (87) +

(88) + (89) + (90) + (91)

TABLE 6. Additions of some diazoalkanes to olefins

Olefin **78**				Diazo **79**			
R^1	R^2	X	Y	R^3	R^4	Yield, other products and comments	Reference
Ph	H	NO_2	H	Ph	Ph	**82** (41%) $\xrightarrow{\Delta\Sigma}$ 3,4,5-Ph_3 pyrazole (100%)	2116
CH_3	H	NO_2	H	Ph	Ph	**82** (27%) $\xrightarrow{\Delta}$ 3-Me-4,5-Ph_2 pyrazole (90%)	2116
CH_3	H	NO_2	H	Me	Me	Intermediate observed, **82** $\xrightarrow{\Delta}$ 3,4,5-Me_3 pyrazole (10·5%)	2116
PBOPh	H	NO_2	H	Ph	Ph	**82** (41·5%) $\xrightarrow{\Delta}$ 3-PBOPh-4,5-Ph_2 pyrazole (96%)	2118
PBOPh	H	NO_2	H	Ph	Ph	**82** (19·7%) (20–37%) $\xrightarrow{\Delta}$ 3-PMOPh-4,5-Ph_2 pyrazole (100%)	2118
Ph	H	NO_2	Br	H	H	**80** (42·3%) $\xrightarrow{\text{acid}}$ 2-Br-3-Ph pyrazole	2114
Ph	H	NO_2	Br	H	H	**80** (42·3%) $\xrightarrow{\text{base}}$ 2-NO_2-3-Ph pyrazole	2114
Ph	H	NO_2	H	H	H	— Polymer (100%)	2114
Ph	H	NO_2	Me[a]	H	H	**80** (100%) $\xrightarrow[\Delta]{H^+}$ 3-Me-4-Ph pyrazole (84%)	2114
Ph	H	NO_2	Et[a]	H	H	**80** (10%) $\xrightarrow[\Delta]{H^+}$ pyrazole (92%)[c]	2114
n-Pr	H	NO_2	Me[a]	H	H	**80** (10%) $\xrightarrow[\Delta]{H^+}$ pyrazole **53** (77%)	2114
Et	H	NO_2	Et[b]	H	H	**80** (10%) $\xrightarrow[\Delta]{H^+}$ pyrazole (57%)	2114
Ph	H	NO_2	Me	H	CO_2Et	— pyrazole (92%)	2114
Ph	H	NO_2	Et	H	CO_2Et	— pyrazole (68%)	2114
n-Pr	H	NO_2	Me[b]	H	CO_2Et	— pyrazole (48%)	2114
Et	H	NO_2	Et[b]	H	CO_2Et	— pyrazole (25%)	2114
Ph	Me	NO_2	H	Ph	Ph	— azine + $Ph_2C{=}CPh_2$	2116
Ph	H	CEO	Ac	H	H	**80** (100%)	1614
Ph	H	Ac	CEO	H	H	**80** (100%)	1614
Ph	H	Ac	CEO	H	H	**80** (100%)	1614
CH_3	H	CMO	Ac	H	H	**80** (100%)	1614
CH_3	H	Ac	CMO	H	H	**80** (100%)	1614
n-Pr	H	CMO	Ac	H	H	**80** (100%)	1614
n-Pr	H	Ac	CMO	H	H	**80** (100%)	1614
n-Pr	H	CEO	Ac	H	H	**80** (100%)	1614
n-Pr	H	Ac	CEO	H	H	**80** (100%)	1614
Ph	H	CMO	Ac	H	H	**80** (100%)	1614
Ph	H	Ac	CMO	H	H		1614

PNPh	H	CEO	Ac	H	H	**80** (100%)	1614
PNPh	H	Ac	CEO	H	H		1614
Et	H	CMO	CMO	H	H	**80** (100%) $\xrightarrow{\Delta}$ **84** (55–60%) + **86** (15–20%) + **88** (25%)	1618
n-Pr	H	CMO	CMO	H	H	**80** (100%) $\xrightarrow{\Delta}$ **84** (55–60%) + **86** (15–20%) + **88** (25%)	1618
n-Pr	H	CMO	CMO	H	H	**80** (100%) $\xrightarrow{\Delta}$ **84** (55–60%) + **86** (15–20%) + **88** (25%)	1618
n-Bu	H	CMO	CMO	H	H	**80** (100%)	1618
n-Bu	H	CEO	CEO	H	H	**80** (100%)	1618
CH_3	CH_3	CMO	CMO	H	H	**80** (100%) $\xrightarrow{\Delta}$ **84** (33%) + **88** (67%)	1618
PNPh	Me	CN	CEO	H	H	**80** (82%)	1845
PCPh	Me	CN	CEO	H	H	**80** (83%) $\xrightarrow{\Delta}$ **84** (36%)[d] + **86** (42%)[d] + **88** (22%)[d]	1846
Me	Ph	CN	CEO	H	H	**80** (84%) $\xrightarrow{\Delta}$ **84** (29%)[d] + **86** (41%)[d]	1846
Me	PTL	CN	CEO	H	H	**80** (87%) $\xrightarrow{\Delta}$ **84** (74%)[d] + **86** (6%)[d] + **88** (7%)[d]	1846
Me	PNPh	CN	CEO	H	H	**80** (85%)	1846
Ph	Ph	CN	CEO	H	H	**80** (~100%) $\xrightarrow{\Delta}$ **84** (92%)[d] + **88** (8%)[d]	1846
Me	Me	CN	CEO	H	H	**80** (~100%)	1846
Me	Et	CN	CEO	H	H	**80** (~100%)	1846
Bz	Bz	CN	CEO	H	H	**80** (~100%)	1846
Bz	Ph	CN	CEO	H	H	**80** (~100%), **84** (89%)[d] + **86** (8%)[d] + **88** (3%)[d]	1846
PMPh	Me	CN	CEO	H	H	s.m. **80**, **84** (48%)[d] + **86** (18%)[d] + **88** (18%)[d]	1846
Me	PMOPh	CN	CEO	H	H	s.m. **80**, **84** (78%)[d] + **86** (5%)[d] + **88** (4%)[d]	1846
PCPh	Me	CEO	CN	H	H	s.m. **80**, **84** (88%)[d] + **86** (2%)[d] + **88** (7%)[d]	1846
PTL	Me	CN	CEO	H	H	s.m. **80**, **84** (35%)[d], **86** (35%)[d], **88** (19%)[d]	1846
Me	Me	CN	CEO	H	H	**80** (100%), **84** (65%)[d], **88** (35%)[d]	1846
Me	Et	CN	CEO	H	H	**80** (100%), **84** (52%)[d], **86** (10%)[d], **88** (38%)[d]	1846
Et	Me	CN	CEO	H	H	**80** (100%), **84** (24%)[d], **86** (40%)[d], **88** (36%)[d]	1846
Me	BPR	CN	CEO	H	H	**80** (100%), **84** (42%)[d], **86** (24%)[d], **88** (34%)[d]	1846
BPr	Me	CN	CEO	H	H	**80** (100%), **84** (12%)[d], **86** (62%)[d], **88** (26%)[d]	1846
H	Me	Ac	CMO	H	H	s.m. **80**, (62%) Et_2NH, Δ′-pyrazoline, 15 days at 0 °C in Et_2O	1617
H	Et	Ac	CMO	H	H	s.m. **80**, (58%) as above	1617
H	*n*-Pr	Ac	CMO	H	H	s.m. **80**, (55%) as above	1617
H	*n*-Pr	Ac	CEO	H	H	s.m. **80**, (48%) as above	1617
H	Ph	Ac	CMO	H	H	s.m. **80**, (75%) as above	1617
H	Ph	Ac	CEO	H	H	s.m. **80**, (95%) as above	1617
Me	PTL	CN	CEO	H	H	s.m. **80**, **84** (50%), **86** (50%); toluene solvent	1847
Me	PTL	CN	CEO	H	H	s.m. **80**, **84** (27%), **86** (73%); methanol solvent	1847
Me	PTL	CN	CEO	H	H	s.m. **80**, **84** (46%), **86** (54%); acetonitrile solvent	1847

TABLE 6 (*cont.*)

Olefin **78**				Diazo **79**			
R^1	R^2	X	Y	R^3	R^4	Yield, other products and comments	Reference
Me	PCPh	CN	CEO	H	H	s.m. **80**, **84** (46%), **86** (32%); toluene solvent	1847
Me	PCPh	CN	CEO	H	H	s.m. **80**, **84** (34%), **86** (66%)	1847
Me	PCPh	CN	CEO	H	H	s.m. **80** (~100%), **84** (40%), **86** (60%)	1847
Me	PTL	CEO	CN	H	H	s.m. **80** (~100%), **84** (2%), **86** (98%)	1847
Me	PTL	CEO	CN	H	H	s.m. **80** (~100%), 84 (4%), **86** (96%)	1847
Me	PTL	CEO	CN	H	H	s.m. 80 (~100%), **84** (4%), **86** (96%)	1847
Me	PCPh	CEO	CN	H	H	s.m. **80** (~100%), **84** (3%), **86** (97%)	1847
Me	PCPh	CEO	CN	H	H	s.m. **80** (~100%), **84** (5%), **86** (95%)	1847
Me	PCPh	CEO	CN	H	H	s.m. **80** (~100%), **84** (4%), **86** (96%)	1847
H	Ph	Ac	CEO	H	H	**80** (85%)	1616
H	PCPh	Ac	CMO	H	H	**80** (94%); 53% Et_2NH, Δ'-pyrazoline, 15 days at 0 °C in Et_2O	1616
H	PNOPh	Ac	CMO	H	H	**80** (98%); 95% *N*-acetyl-Δ^2-pyrazoline, 15 days at 0 °C in Et_2O	1616
H	PCPh	Ac	CEO	H	H	**80** (94%); 97% as above	1616
H	PNPh	Ac	CEO	H	H	**80** (93%)	1848
H	PCPh	CMO	Ac	H	H	**80** (75%)	1848
H	Ph	CN	CEO	H	H	No intermediate, **84** (100%)	1848
H	PNPh	CN	CEO	H	H	No intermediate, **84** (100%)	1848
H	PCPh	CN	CEO	H	H	No intermediate, **84** (100%)	1848
H	Ph	CN	CA	H	H	**80** (100%) $\xrightarrow{\Delta}$ **84** (100%)	1848
H	PCPh	CN	CA	H	H	**80** (100%) $\xrightarrow{\Delta}$ **84** (100%)	1848
H	Ph	CEO	CEO	H	H	**80** (100%) $\xrightarrow{\Delta}$ **84** (100%)	1848
H	PNPh	CEO	CEO	H	H	**80** (100%) $\xrightarrow{\Delta}$ **84** (100%)	1848
H	PMOPh	Ac	CEO	H	H	**80** (95%); 95% *N*-acetyl-Δ^2-pyrazoline, 15 days at 0 °C in Et_2O	1617
H	PCPh	CMO	Ac	H	H	**80** (~100%); 87% Et_2NH, Δ'-pyrazoline, 15 days at 0 °C in Et_2O	1617
H	H	*p*-MOPh	H	*p*-MOPh	H	s.m. **80** $\xrightarrow{\Delta}$ **88** (43%) + **89** (57%)	2112
H	H	H	*p*-MOPh	*p*-MOPh	H	s.m. **80** $\xrightarrow{\Delta}$ **88** (7%) + **89** (93%)	2112
H	H	Me	H	Me	H	s.m. **80** $\xrightarrow{\Delta}$ **88** (33%) + **89** (66%)	1604
H	H	Me	H	Me	H	s.m. **80** $\xrightarrow{h\nu}$ **88** (48%) + **89** (41%)	2051
H	H	H	Me	Me	H	s.m. **80** $\xrightarrow{\Delta}$ **88** (73%) + **89** (25%)	1604
H	H	H	Me	Me	H	s.m. **80** $\xrightarrow{h\nu}$ **88** (60%) + **89** (26%)	2051

H	H	CMO	Me	Me	H	s.m. **80** $\xrightarrow{\Delta}$ **88** (25%) + **89** (56%)	Stereochemistry defined relative to CH_3	2007
H	H	CMO	Me	Me	H	s.m. **80** $\xrightarrow{h\nu}$ **88** (61%) + **89** (23%)		2007
H	H	CMO	H	Me	H	s.m. **80** $\xrightarrow{\Delta}$ **88** (66%) + **89** (18%)		2007
H	H	CMO	H	Me	H	s.m. **80** $\xrightarrow{h\nu}$ **88** (22%) + **89** (65%)		2007
H	D	Me	H	H	H	s.m. **80** $\xrightarrow{\Delta}$ **88** (43%) + **89** (43%)		2007
D	H	Me	H	H	H	s.m. **80** $\xrightarrow{\Delta}$ **88** (42%) + **89** (42%)		2007
H	Me	Me	H	H	H	s.m. **80** $\xrightarrow{\Delta}$ **88** (44%) + **89** (35%)		1605a
H	Me	Me	H	H	H	s.m. **80** $\xrightarrow{h\nu}$ **88** (43%) + **89** (38%)		2051
H	Me	H	Me	H	H	s.m. **80** $\xrightarrow{\Delta}$ **88** (46%) + **89** (22%)		1605a
H	Me	H	Me	H	H	s.m. **80** $\xrightarrow{h\nu}$ **88** (25%) + **89** (41%)		2051
Me	Me	CMO	CMO	H	H	**80** (90%) $\xrightarrow{\Delta}$ **83** (33%) + **88** (67%)		1619
Et	H	CMO	CMO	H	H	**80** (100%) $\xrightarrow{\Delta}$ **83** (61%) + **85** (15%) + **88** (24%)		1619
n-Pr	H	CMO	CMO	H	H	**80** (100%) $\xrightarrow{\Delta}$ **83** (52%) + **85** (22%) + **88** (26%)		1619
n-Pr	H	CEO	CEO	H	H	**80** (100%) $\xrightarrow{\Delta}$ **83** (58%) + **85** (13%) + **88** (29%)		1619
Me	H	Ac	Ac	H	H	**80** (100%)		1620
Me	H	PhCO	Ac	H	H	**80** (100%)		1620
Ph	H	Ac	Ac	H	H	**80** (100%)		1620
Ph	H	PhCO	Ac	H	H	**80** (100%)		1620
Ph	H	PhCO	PhCO	H	H	**80** (100%)		1620
Me	Me	Ac	CMO	H	H	**80** (100%)		1621
Ph	Ph	Ac	CMO	H	H			1621
H	Me	CN	PhCO	H	H	No intermediate, **83**		1621
H	Ph	CN	PhCO	H	H	No intermediate, **83**		1621
H	Ph	CN	POE	H	H	**80** (75%) $\xrightarrow{\Delta}$ **83** (39·8%) + **85** (16·2%); + **89** (100%, 25 °C, 2 days)		1622

Abbreviations: PBOPh = *p*-benzyloxyphenyl; PMOPh = *p*-methoxyphenyl; CEO = carboethoxy; PNPh = *p*-nitrophenyl; CMO = carbomethoxy; PCPh = *p*-chlorophenyl; PTL = *p*-tolyl; Bz = benzyl; CA = $-CONH_2$; PhCO = benzoyl; POE = $P(O)(OEt)_2$; s.m. = starting material.

[a] Stereochemistry not specified.
[b] Assume *cis* alkyl groups.
[c] Pyrazole substituted as in supposed or real pyrazoline.
[d] Stereochemistry not determined.

In cases where two possible stereochemistries existed for the olefin, the thermodynamically most stable has been assumed where stereochemistry has not been supplied. This was based on the synthetic approaches applied in the olefin syntheses which might be expected to furnish such products.

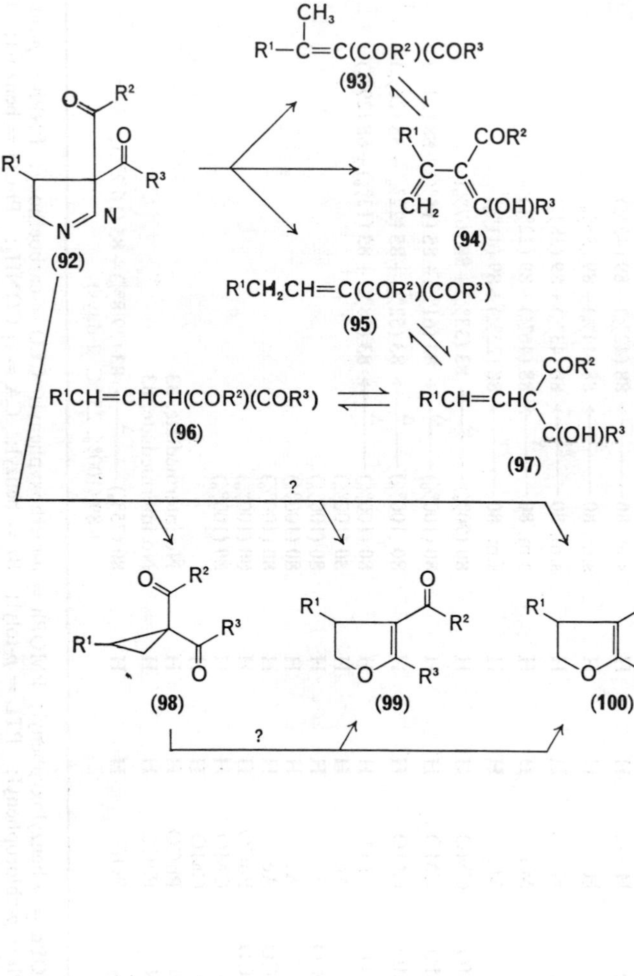

(92) → (93) R^1—C(CH_3)=C(COR^2)(COR^3) ⇌ (94) CH_2=C(R^1)—C(COR^2)=C(OH)R^3

(92) → (95) R^1CH_2CH=C(COR^2)(COR^3) ⇌ (97) R^1CH=CHC(COR^2)=C(OH)R^3

(96) R^1CH=CHCH(COR^2)(COR^3) ⇌ (97)

(92) → ? (98), (99), (100)

(98) → ? (99), (100)

(15)

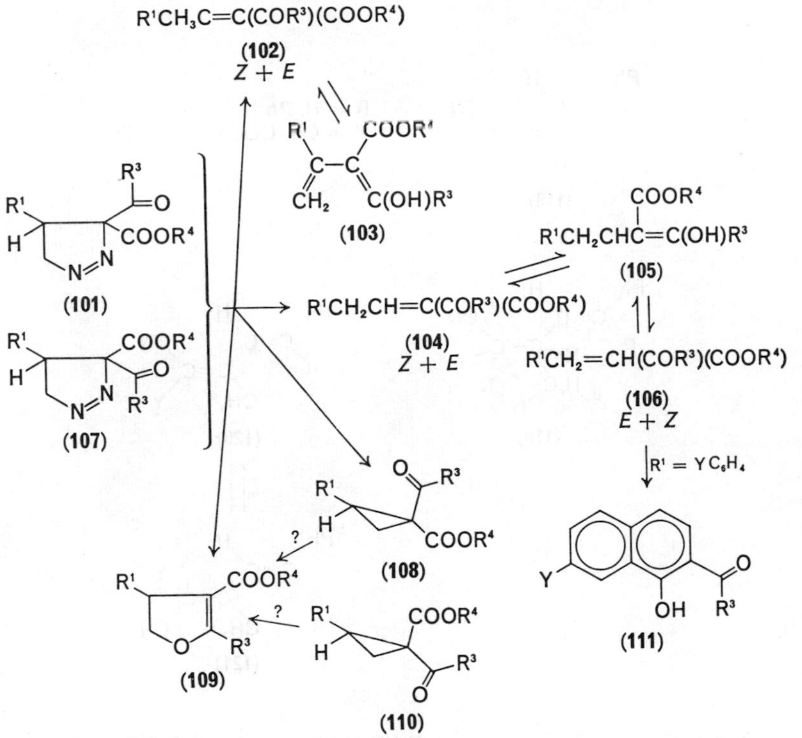

$R^1CH_3C{=}C(COR^3)(COOR^4)$ **(102)** $Z + E$

(103) $R^1(CH_2{=})C{-}C(COOR^4){=}C(OH)R^3$

(101), **(107)** → $R^1CH_2CH{=}C(COR^3)(COOR^4)$ **(104)** $Z + E$

$R^1CH_2CHC(COOR^4){=}C(OH)R^3$ **(105)**

$R^1CH_2{=}CH(COR^3)(COOR^4)$ **(106)** $E + Z$

$R^1 = YC_6H_4$

(111)

(108) ? **(109)** ? **(110)**

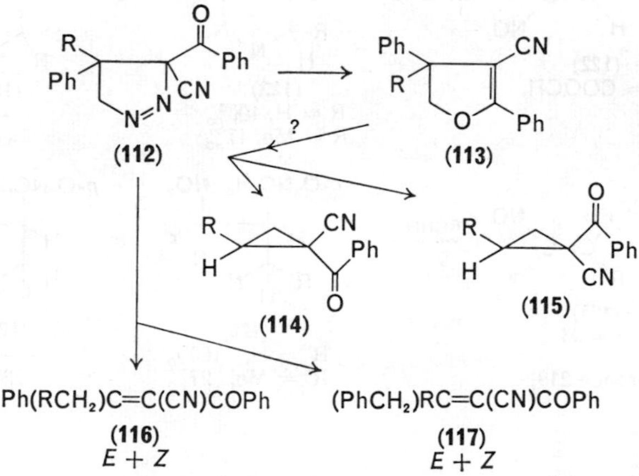

(16)

(112) → **(113)**

?

(114) **(115)**

$Ph(RCH_2)C{=}C(CN)COPh$ **(116)** $E + Z$

$(PhCH_2)RC{=}C(CN)COPh$ **(117)** $E + Z$

(17)

R = H, Ph
Y = CN, COOMe, COOEt

(118) $\xrightarrow{CH_2N_2}$ (119) $\xrightarrow{\Delta}$ (120) $\underset{h\nu}{\rightleftarrows}$ (121) (18)

(122) X = $COOCH_3$ $\xrightarrow{RCHN_2}$ (123) + (124)

(123): R = H, 100%; R = Me, 17%
(124): —; 83%

(125) R = H $\xrightarrow{RCHN_2}$ (126) + (127) (19)

(126): R′ = H, 100%; R′ = Me, 27%
(127): —; 73%

(Reference 219)

126
R = H; X = COOMe $\xrightarrow{\Delta}$ p-$O_2NC_6H_4$, H, COOMe, N→O, O (**128**) 100%

123
R = H; X = COOMe $\xrightarrow{\Delta}$ p-$O_2NC_6H_4$, H, NO_2, O, OMe (**130**) 48% + p-$O_2NC_6H_4$, CO_2Me, H, NO_2 (**129**) 43% + **128** 4% (20)

124
R = Me; X = CO_2Me $\xrightarrow{\Delta}$ p-$O_2NC_6H_4$(CH$_3$)C=CHCHXNO$_2$ (**131**) 21% + p-$O_2NC_6H_4$(CH$_3$)CHCH=C(NO_2)X (**132**) 79%
(Reference 219)

123
R = Me; X = COOMe $\xrightarrow{\Delta}$ p-$O_2NC_6H_4$(Et)C=C(X)NO_2 (**133**) 100% $\xrightarrow{h\nu}$ p-$O_2NC_6H_4$(Et)C=C(NO_2)X (**134**)

126
R = Me; X = COOMe $\xrightarrow{\Delta}$ **134** 100%

127
R = Me; X = COOMe $\xrightarrow{\Delta}$ p-$O_2NC_6H_4$, H, X, H, CH_3, O, N→O (**135**) 11% + p-$O_2NC_6H_4$, NO_2, R, X, CH_3 (**136**) 6% + **131** 83% (21)
(Reference 219)

123, 124, 126 or **127** $\xrightarrow[RCHN_2]{H_2O\ or}$ p-$O_2NC_6H_4$, X, R, N, NH (**137**)

125 R = H, Me; X = Ac $\xrightarrow{CH_2N_2}$ **126** $\xrightarrow{\Delta}$ **128** (R = H)

126 $\xrightarrow{R = Me}$ **131** 23%

126 $\xrightarrow{CH_2N_2}$ (**138**) R = Me 29%

(**138**): p-$O_2NC_6H_4$, NO_2, R, N–N, acetyl

(Reference 219)

(22)

$RCH{=}CXY + N_2CHCO_2R' \longrightarrow$ (**141**) $\longrightarrow$ (**143**)

(**139**) (**140**) R′ = Me, Et

(Reference 256)

(**141**) $\longrightarrow$ (**142**) + $R(CH_2CO_2R')C{=}CXY$ (**144**)

(**143h**) $\xrightarrow{N_2CHCO_2Et}$ (**146**)/(**145**)

X = Ac; Y = CO_2Me (**146**)
X = CO_2Me; Y = Ac (**145**)

(23)

(**142**) (Reference 256)

R^1	R^2	R^3	R^4	R^5	R^6	
*COOEt	CH_3	CO_2Me	*H	CO_2Me	H	**142b**
COOEt	C_6H_5	CN	H	H	COOMe	**142c**
*COOEt	C_6H_5	COOMe	*H	H	COOMe	**142e**
*COOEt	p-ClC_6H_4	COOMe	*H	H	Ac	**142h**
*COOEt	p-ClC_6H_4	Ac	*H	H	COOMe	**142i**

* Assignment not given.

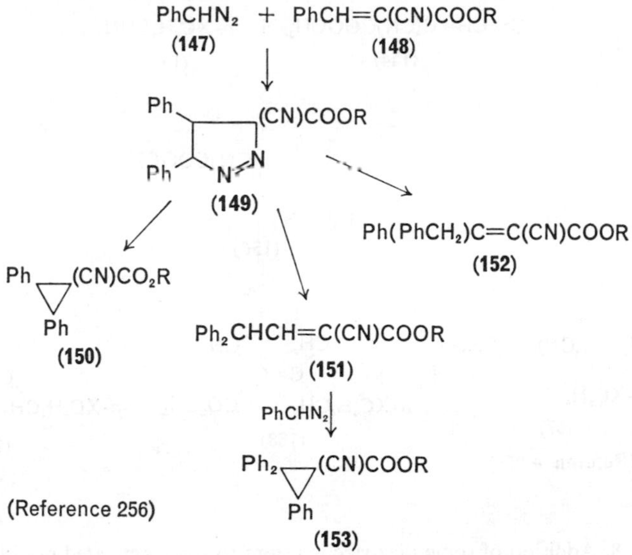

$$PhCHN_2 + PhCH{=}C(CN)COOR$$
(147) (148)

↓

Ph, Ph, (CN)COOR, N=N

(149)

$Ph(PhCH_2)C{=}C(CN)COOR$
(152)

Ph, (CN)CO_2R, Ph
(150)

$Ph_2CHCH{=}C(CN)COOR$
(151)

$PHCHN_2$ ↓

Ph_2, (CN)COOR, Ph
(153)

(Reference 256) (24)

TABLE 7. Additions of some diazoacetates to alkylidene cyanoacetates[256]

No.	Compound **139** R	X	Y	**140** R′	Conditions[a]	Time (days)	A (%)	B (%)
139a	CH_3	CN	CO_2CH_3	C_2H_5	1	15	50	—
139b	CH_3	CO_2CH_3	CO_2CH_3	C_2H_5	1	15	100	—
139c	C_6H_5	CN	CO_2CH_3	C_2H_5	3	3	50	30
139d	C_6H_5	CN	$CO_2C_2H_5$	C_2H_5	{2 {3	30 3	— 50	35 30
139d	C_6H_5	CN	$CO_2C_2H_5$	CH_3	3	3	45	30
139e[b]	C_6H_5	CO_2CH_3	CO_2CH_3	C_2H_5	3	10	100	92
139f[b]	C_6H_5	$CO_2C_2H_5$	$CO_2C_2H_5$	CH_3	3	13	100	85
139g	C_6H_5	CN	CN	C_2H_5	{1 {3	3 2	0 0	
139h (*cis*)	*p*-ClC_6H_4	$COCH_3$	CO_2CH_3	C_2H_5	3	6	100	
139i[b] (*trans*)	*p*-ClC_6H_4	$COCH_3$	CO_2CH_3	C_2H_5	3	10	100[c]	65
139h (*cis*)	*p*-ClC_6H_4	$COCH_3$	CO_2CH_3	CH_3	3	6	100	90
139i[b] (*trans*)	*p*-ClC_6H_4	$COCH_3$	OO_2CH_3	CH_3	3	13	100[c]	—

[a] 1, Room temperature in solvent; 2, at reflux in ether; 3, at 50 °C in solvent.
[b] Excess of diazo compound relative to olefin **139** was 15%, except for the compounds **139e**, **139f** and **139i** where the excess was 25%.
[c] Olefin **139i** isomerizes partially during the reaction and the pyrazoline **143i** (or **143′i**) obtained is contaminated with a small quantity of isomer **143h** (or **143′h**).
A indicates the percentage determined by n.m.r. of the pyrazoline existing in the product.
B indicates the yield of pyrazoline-2 relative to starting material.

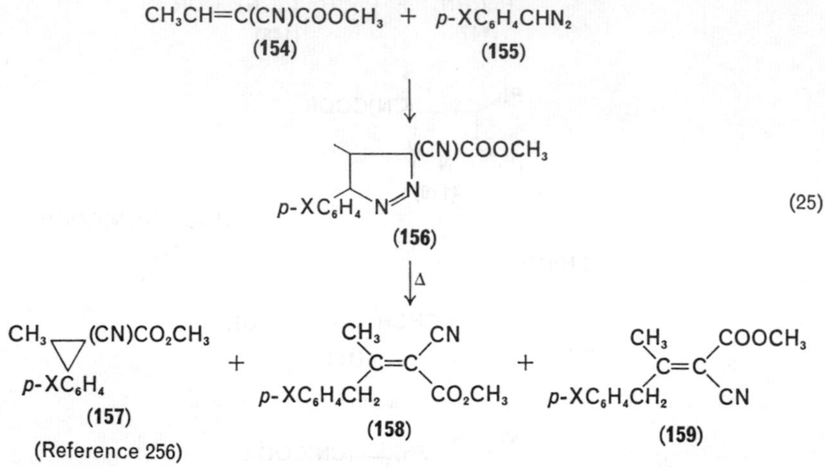

$$CH_3CH{=}C(CN)COOCH_3 + p\text{-}XC_6H_4CHN_2$$
(154) (155)

↓

(CN)COOCH$_3$
p-XC$_6$H$_4$ N=N
(156) (25)

↓ Δ

CH$_3$ (CN)CO$_2$CH$_3$ / p-XC$_6$H$_4$ (157) (Reference 256) + CH$_3$ CN C=C p-XC$_6$H$_4$CH$_2$ CO$_2$CH$_3$ (158) + CH$_3$ COOCH$_3$ C=C p-XC$_6$H$_4$CH$_2$ CN (159)

TABLE 8. Addition of some diazoacetic esters to some activated double bonds[256]

	Compound **141**							
	R	X	Y	R′	Conditions[a]	143 (%)	142 (%)	144 (%)
a	CH_3	CN	CO_2CH_3	C_2H_5	1	50	30	20
b	CH_3	CO_2CH_3	CO_2CH_3	C_2H_5	1	100	0	0
					3	0	100	0
c	C_6H_5	CN	CO_2CH_3	C_2H_5	2	50	30	20
					3	0	60	40
d	C_6H_5	CN	$CO_2C_2H_5$	C_2H_5	2	50	30	20
					3	0	60	40
d	C_6H_5	CN	$CO_2C_2H_5$	CH_3	2	45	30	25
e	C_6H_5	CO_2CH_3	CO_2CH_3	C_2H_5	2	100	0	0
					3	*b*	100	0
f	C_6H_5	$CO_2C_2H_5$	$CO_2C_2H_5$	CH_3	2	100	0	0
g	C_6H_5	CN	CN	C_2H_5	2	0	50	50
h (*cis*)	p-ClC_6H_4	$COCH_3$	CO_2CH_3	C_2H_5	2	100	0	0
					3	40	60	0
					4	10[c]	90	0
i (*trans*)	p-ClC_6H_4	$COCH_3$	CO_2CH_3	C_2H_5	2	100	0	0
					3	10[c]	90	0
					4	10[c]	90	0
h (*cis*)	p-ClC_6H_4	$COCH_3$	CO_2CH_3	CH_3	2	100	0	0
i (*trans*)	p-ClC_6H_4	$COCH_3$	CO_2CH_3	CH_3	2	100	0	0

[a] 1, Without solvent at room temperature; 2, without solvent at 50 °C; 3, in solution (benzene) at reflux; 4, in solution (toluene) at reflux.
[b] Traces.
[c] Compounds **145** and **146** result from reaction with the Δ^2 in the reaction mixture.

p-$XC_6H_4(CH_3)C{=}C(CN)CO_2Et$ + $PhCHN_2$ ⟶ No appreciable reaction (2 months, room temperature; 12 days, reflux C_6H_6)
(**160**) *cis* (**161**) *trans* X = Cl

(**161**, X = H) = **162** $\xrightarrow{CH_3CHN_2}$ $PhCHCH_3(CN)C{=}C(H)(COOEt)$ (**163**) $\xrightarrow{CH_2N_2}$ $PhCHCH_3(CN)C{=}C(CH_3)(COOEt)$ (**164**) *cis* (96)

(Reference 256)

TABLE 9. Reactions of some benzylidene cyanoacetates with ethyl diazoacetate[256]

Olefin 139			A	B	C	D
R	X	Y	(%)	(%)	(%)	(%)
C_6H_5	CN	CO_2CH_3	0	20	20	*a*
C_6H_5	CN	$CO_2C_2H_5$	0	20	20	*a*
C_6H_5	CN	CN	0	10	10	*c*
CH_3	CO_2CH_3	CO_2CH_3	26	20	100	*a*
p-ClC_6H_4	$COCH_3$	CO_2CH_3	10	30	100	*a*
(*cis*)			0	100	100	*b*
p-ClC_6H_4	$COCH_3$	CO_2CH_3	60	50	500	*a*
(*trans*)			15	115	200	*b*
C_6H_5	CO_2CH_3	CO_2CH_3	58	0	500	*a*
			67	0	*x*	*b*

A Olefin **139** (in %) not transformed after 12 days' reaction.
B Excess (in %) of diazo compound utilized in the preceding reaction.
C Excess (in %) of diazo compound necessary for complete transformation of the olefin **139**.
D Conditions employed for complete transformation.
x The reactions were not worked up until all olefin had been consumed.
Ethyl diazoacetate employed as diazo compound.
[a] Refluxing benzene, 12 days.
[b] Refluxing toluene, 12 days.
[c] 50 °C, 45 h, no solvent.

TABLE 10. The reaction of methyl-α-cyanocrotonate with *para*-substituted phenyldiazomethanes[256]

X	157 (%)	159 (%) *cis*	159 (%) *trans*
H	80	10	10
CH_3O	79·5	14	6·5
NO_2	84·5	15·5	0

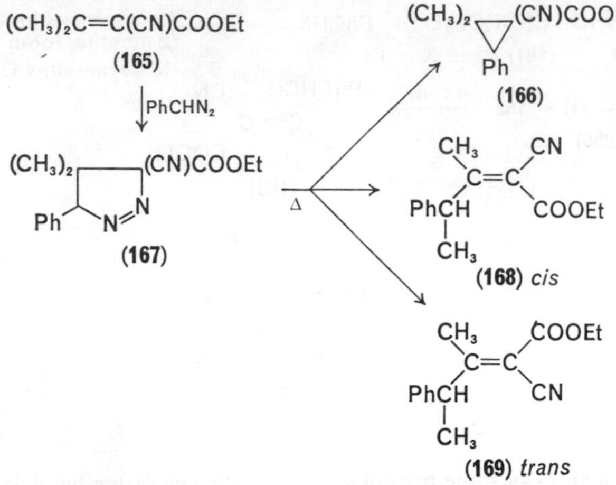

$(CH_3)_2C{=}C(CN)COOEt$ (165) $\xrightarrow{PhCHN_2}$ (167) $\xrightarrow{\Delta}$ (166) + (168) *cis* + (169) *trans* (27)

Temperature	166 (%)	168 (%)	169 (%)
10 °C	24	40·5	35·5
Room temp.	41·5	24·5	34

(Reference 256)

TABLE 11. Decomposition of some Δ^1-pyrazolines derived from some monocyclic olefins

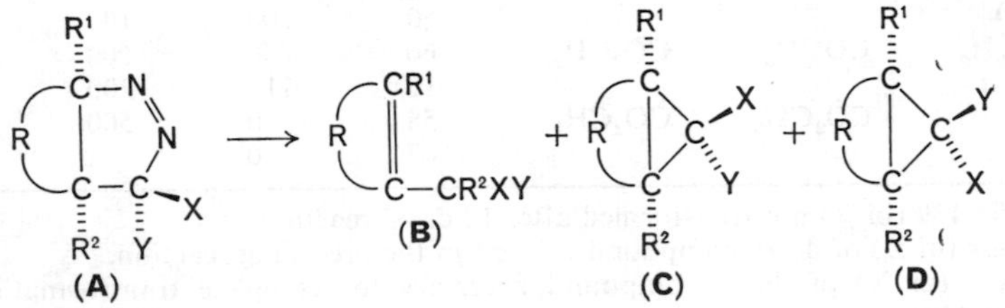

(A) → (B) + (C) + (D)

Conditions[a]	R	R¹	R²	X	Y	B (%)	C (%)	D (%)	Reference
Δ	$-(CH_2)_3-$	H	H	H	$CH_2{=}CH-$		100		1415
Δ	$-(CH_2)_2-$	H	H	H	H	2·7	15		1416
$h\nu$ (T)	$-(CH_2)_2-$	H	H	H	H	90			1416
$h\nu$ (S)	$-(CH_2)_2-$	H	H	H	H	59	1		1416
Δ	$-(CH_2)_3-$	H	H	H	CH_3	11	8·1	72·8	*b*
Δ	$-(CH_2)_3-$	H	H	CH_3	H	5·3	21·2	67·2	*b*
$h\nu$ (S)	$-(CH_2)_3-$	H	H	H	CH_3	9·0	62·2	21·3	*b*
$h\nu$ (S)	$-(CH_2)_3-$	H	H	CH_3	H	5·8	61·0	29·8	*b*
$h\nu$ (T)	$-(CH_2)_3-$	H	H	H	CH_3		68·9	31·1	*b*
$h\nu$ (T)	$-(CH_2)_3-$	H	H	CH_3	H		40·6	59·4	*b*
$h\nu$	$-(CH_2)_2-$	CO_2R	CO_2R	H	H		+		*c*

[a] $h\nu$ (T)≡photolytic, triplet state; $h\nu$ (S)≡photolytic, singlet state.
[b] P. B. Condit and R. G. Bergman, *J. Chem. Soc. Chem. Commun.*, 4 (1971).
[c] H. Prinzbach and H.-P. Martin, *Chimia*, **23**, 37 (1969).

(170) $\xrightarrow{Ph_2CN_2}$ (171) $\xrightarrow{\Delta}$ (172) + (173) (28)

(Reference 256)

	(172)		(173)
	89%	$R = CH_3$	11%
	94%	$R = Ph$	6%

TABLE 12. Decompositions of some Δ^1-pyrazolines derived from bicyclic olefins

(E) → (F) + (G) + (H) + (I) + (J)

X	Y	R	R′	G (%)	H (%)	Conditions	Reference
C_6H_5	$P(O)(OCH_3)_2$	$-CH_2-$	$-(CH_2)_2-$	96		$h\nu$	1557
C_6H_5	$P(O)(OCH_3)_2$	$-CH_2-$	$-(CH_2)_2-$	47	47	Sensitized	1557
CH_3	$P(O)(OCH_3)_2$	$-CH_2-$	$-(CH_2)_2-$	96		$h\nu$	1557
$P(O)(OCH_3)_2$	CH_3	$-(CH_2)_2-$	$-(CH_2)_2-$	76	19	$h\nu$	1557
C_6H_5	$P(O)(OCH_3)_2$	$-CH_2-$	$-CH{=}CH-$	54		$h\nu$	1557
C_6H_5	$P(O)(OCH_3)_2$	$-CH_2-$	$-CH{=}CH-$	15·4	28·6	Sensitized	1557
CH_3	$P(O)(OCH_3)_2$	$-CH_2-$	$-CH{=}CH-$	75		$h\nu$	1557
$P(O)(OCH_3)_2$	CH_3	$-CH_2-$	$-CH{=}CH-$	56·7	6·3	$h\nu$	1557
C_6H_5	C_6H_5	$-$CHOBu-γ (*anti*)	$-(CH_2)_2-$	~90		Δ	2346
p-Anisyl	*p*-Anisyl	CHOBu-γ (*anti*)	$-(CH_2)_2-$	~90		Δ	2346
p-Tolyl	*p*-Tolyl	$-$CHOBu-γ (*anti*)	$-(CH_2)_2-$	~90		Δ	2346
p-ClC_6H_4	*p*-ClC_6H_4	$-$CHOBu-γ (*anti*)	$-(CH_2)_2-$	~90		Δ	2346
C_6H_5	C_6H_5	$-$CHOBu-γ (*anti*)	$-CH{=}CH-$	~96		Δ	2346
C_6H_5	C_6H_5	$-$CHOBu-γ (*syn*)	$-CH{=}CH-$	~90		Δ	2235
C_6H_5	C_6H_5	$-CH{=}CH-$	=CHOBu-γ (*syn*)	~90		Δ	2235
C_6H_5	C_6H_5	$-CH{=}CH-$	=CHOBu-γ (*anti*)	~90		Δ	2235

(170) $\xrightarrow{CH_2N_2}$ (174) ⇌ (175)

(174) $\xrightarrow{\Delta}$ (176) + (178) + (179)

(175) $\xrightarrow{\Delta}$ (177) + ... (Reference 988) (29)

TABLE 13. Reaction conditions for the 1,3-cycloaddition of some diazoacetic acid esters to some benzylidene cyanoacetates, malonates and acetoacetates[256]

Olefin	R	R′	X	Y	Conditions[a]	Reaction time[b]
139a [c]	CH_3	H	CN	CO_2CH_3	1	30 min
					2	10 min
139c	C_6H_5	H	CN	CO_2CH_3	1	4 h
139d	C_6H_5	H	CN	$CO_2C_2H_5$	1	4 h
139e	C_6H_5	H	CO_2CH_3	CO_2CH_3	2	36 h
139g	C_6H_5	H	CN	CN	1	30 min
139h (*trans*)	*p*-ClC_6H_4	H	$COCH_3$	CO_2CH_3	1	12 h
139i (*cis*)	*p*-ClC_6H_4	H	$COCH_3$	CO_2CH_3	1	12 h
139j	CH_3	CH_3	CN	$CO_2C_2H_5$	1	12 days
					2	3 days
139e	*p*-ClC_6H_4	CH_3	CN	$CO_2C_2H_5$	2	60 days
					3	10 days
139m (*trans*)	*p*-ClC_6H_4	CH_3	CN	CO_2H_5	2	60 days
					3	10 days
139d	C_6H_5	$C_6H_5CH_2$	CN	$CO_2C_2H_5$	2	60 days

[a] 1, Solution (ether) at 10 °C; 2, solution (ether) at 25 °C; 3, solution (benzene) at 40 °C.
[b] Time necessary for reaction to go to completion.
[c] With *p*-methoxy- and *p*-nitro-phenyldiazomethanes, the reaction is complete after 4 days and 5 minutes, respectively.

TABLE 14. Conversion of some alkylidene cyanoacetates to cyclopropanes[256]

Olefin	Compound				Cyclopropane obtained	Yield (%)
	R	R′	X	Y		
139a	CH_3	H	CN	CO_2CH_3	**157** (X = H)	50
					157 (X = O—Me)	55
139c	C_6H_5	H	CN	CO_2CH_3	**153**	15
139d	C_6H_5	H	CN	$CO_2C_2H_5$	**150**	50
					153	15
139g	C_6H_5	H	CN	CN	**150**	55

TABLE 15. Addition of diphenyl diazomethane to some benzylidene and alkylidene cyanoacetates[256]

	RR′C=CXY				Conditions[a]	Reaction time (days)	A (%)	B (%)	Yield (%)
	R	R′	X	Y					
139	CH_3	H	CN	CO_2CH_3	1	21	8	5	55
139	CH_3	H	CO_2CH_3	CO_2CH_3	1	20	9	20	80
139	C_6H_5	H	CN	CO_2CH_3	2	4	17	15	62
139	C_6H_5	H	CN	$CO_2C_2H_5$	2	4	17	15	64
139	C_6H_5	H	CO_2CH_3	CO_2CH_3	3	20	78	400	
139	C_6H_5	H	CN	CN	2	3	10	10	66
139 (*cis*)	*p*-ClC_6H_4	H	$COCH_3$	CO_2CH_3	3	15	33	200	
139 (*trans*)	*p*-ClC_6H_4	H	$COCH_3$	CO_2CH_3	3	15	49	200	
139	CH_3	CH_3	CN	$CO_2C_2H_5$	1	—[b]	81	5	

[a] 1, Solution (ether) at 25 °C; 2, solution (benzene) at 50 °C; 3, solution (benzene) at 40 °C.
[b] The solution was left at 25 °C for 14 months.
A (%) Olefin unreacted.
B (%) Excess of diphenyl diazomethane added. The last column gives the yield of cyclopropane isolated.

TABLE 16. Reaction of isopropylidene norboacetate with diazomethane[988]

Temperature (°C)	Solvent	**176**+**177** (%)	**178** (%)	**179** (%)
0	Without solvent	17	33	50
20	Without solvent	20[a]	35	45
70	Benzene	40	35	25
90	Toluene	42	39	19
110	Toluene	50	37	13
140	Xylene	56	40	4

[a] Relative percentage **176** 42% and **177** 58%.

TABLE 17. Pyrolyses of some 4-aryl-3-cyano-3-nitro-1-pyrazolines[988]

No.	X	192 (%)	193 (%)	194 (%)
191a	H	85	10	5
191b	p-CH_3O	34	16	50
191c	p-NO_2	88	12	—

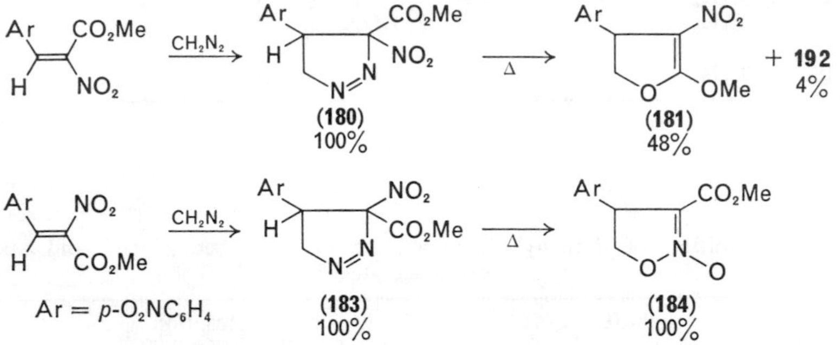

Ar(H)C=C(CO_2Me)(NO_2) $\xrightarrow{CH_2N_2}$ **(180)** 100% $\xrightarrow{\Delta}$ **(181)** 48% + **192** 4% (30)

Ar(H)C=C(NO_2)(CO_2Me) $\xrightarrow{CH_2N_2}$ **(183)** 100% $\xrightarrow{\Delta}$ **(184)** 100%

Ar = p-$O_2NC_6H_4$

(Reference 988)

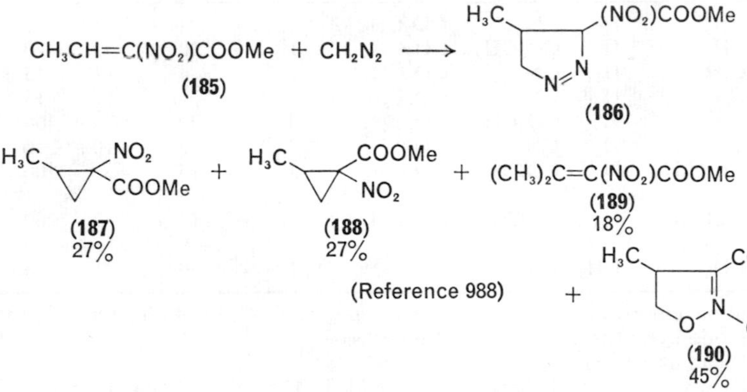

$CH_3CH{=}C(NO_2)COOMe$ + CH_2N_2 ⟶ **(186)**
(185)

(187) 27% + **(188)** 27% + $(CH_3)_2C{=}C(NO_2)COOMe$ **(189)** 18% + **(190)** 45% (31)

(Reference 988)

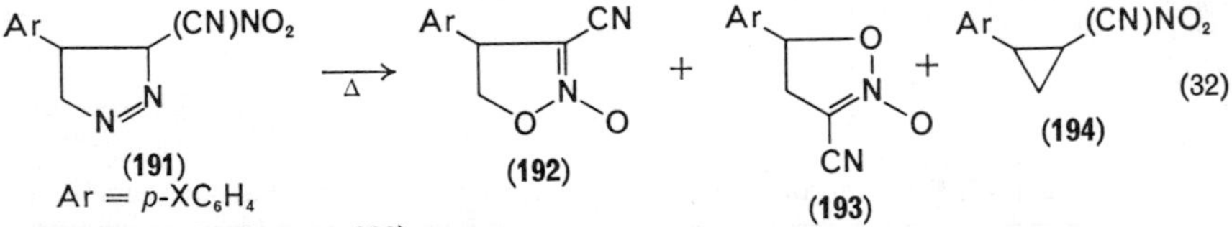

(191) $\xrightarrow{\Delta}$ **(192)** + **(193)** + **(194)** (32)

Ar = p-XC_6H_4

(Reference 988)

In a number of studies, initially formed Δ^1-pyrazolines undergo prototropy to the Δ^2-pyrazoline. Related processes occur with 3-acyl and 3-silylated Δ^1-pyrazolines to furnish *N*-silyl-Δ^2-pyrazolines and *N*-acyl-Δ^2-pyrazolines. It is clear that many Δ^1-pyrazolines have very appreciable stabilities whereas others are only transitory in nature, either undergoing loss of N_2 or rearrangement to the Δ^2 systems.

The latter phenomenon is clearly facilitated by the presence of a hydrogen on C_3 or C_5 which is accompanied by a strong electron-withdrawing group such as

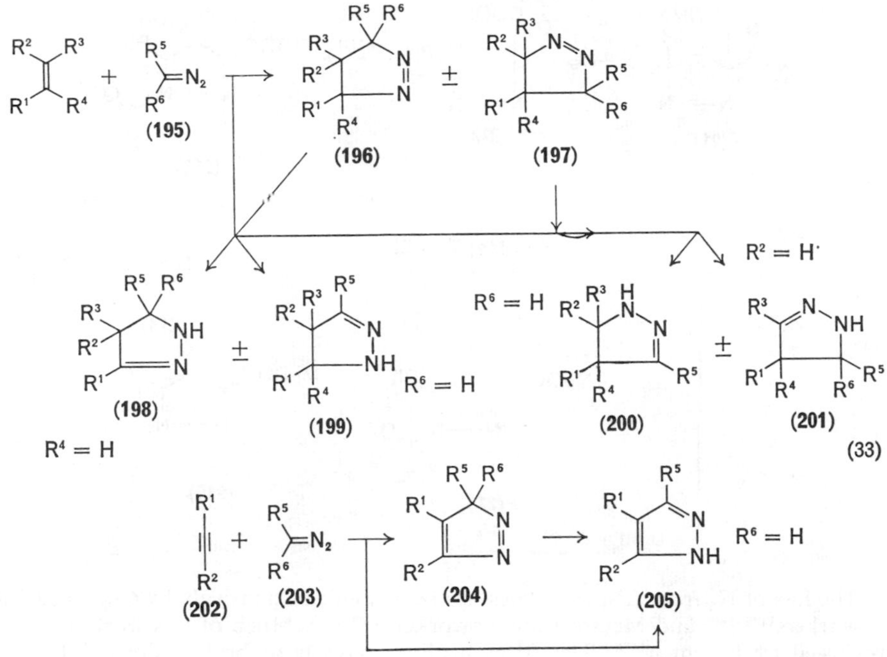

(33)

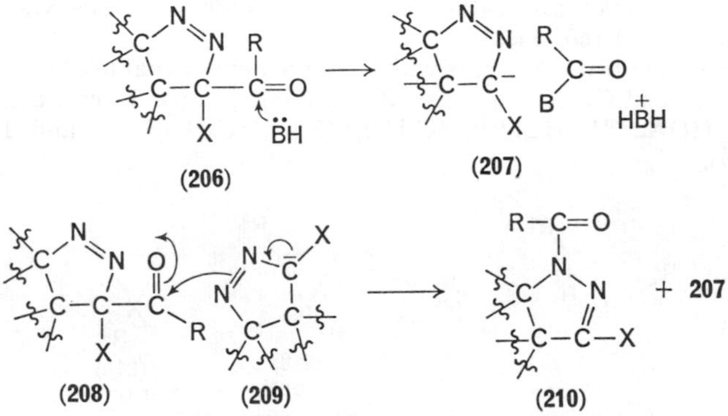

(34)

X = CN, COR′, COOR′; BH = CH_2N_2, $EtNH_2$, Et_2NH

acyl or carboalkoxy. The migration of an acyl group was shown by Danion-Bougot and Carrié to result from base catalysis and was rationalized by equation (34), and supported by the data in equation (35)[1617].

The thermolyses of diacyl Δ^1-pyrazolines lead to a number of products (equation 15). Some of these represent various tautomers **95**, **96** and **97** and **93** and **94**, whereas the two possible dihydrofurans, **99** and **100**, may well be secondary products arising from rearrangements of the cyclopropane **98**. Similar but less complex results are observed with monoacyl systems (equation 16).

The possibility must be considered that the dihydrofuran **109** is a secondary product.

(211) + (212) $\xrightarrow{Et_2NH}$ (213) + (214) + (215) + (216); (211) → $Et_2NH + EtCONEt_2$ (35)

X = Me; Y = Et

The loss of N_2 from Δ^1-pyrazolines has been studied extensively by Crawford and coworkers[1599–1609] and McGreer and coworkers[2005–2013]. Much of this work has been reviewed by Bergman[1515]. The Δ^1-pyrazolines have been broken down into four basic categories by McGreer. There appears to be only one reported and verified example of the copper salt catalysed decomposition of a Δ^1-pyrazoline[2361] (see equation 3, page 3 of reference).

A not uncommon process occurs with Δ^1-pyrazolines capable of losing a molecule HX between C_3 and C_4 (X is a reasonable leaving group). Several common combinations are HOAc [1981], HCl [1981], $(C_6H_5)_3PO$ [2374], $(C_6H_5)_3P$ [2217] and HNO_2 [2114–2119] (equation 39).

(217) → Type 1 → (218) rr or ii; (219) ir or ri

(220) → Type 2 → (221) rn; (222) in (36)

r = retention
i = inversion
n = no information

(223) → Type 4 → (224) rr; (225) ri; (226) ir; (227) ii

(228) → (229) rr + (230) ii Type 3 (37)

(231) + (232) → (233), (234) → (235), (236) (38)

X = Halogen, NO_2, OAcyl, OR, $\overset{+}{P}PH_3$, SiR_3, $OP(O)(OR)_2$

R_3	R^4	R^5
X	H	H
H	X	H
H	H	X

It is possible to convert Δ^2-pyrazolines into Δ^1-pyrazolines by the action of $Pb(OAc)_4$. The resulting 3-acetoxy-Δ^1-pyrazolines can be converted into pyrazoles by loss of acetic acid or into cyclopropyl acetates by pyrolysis.

(237) $\xrightarrow{Pb(OAc)_4}$ (238) $\xrightarrow{\Delta}$ (239); (238) $\xrightarrow{H^+}$ (240) (39)

b. *Imines*. The formation of triazolines via the 1,3-dipolar cycloaddition of diazoalkanes to imines has been examined extensively by Kadaba[1932–1938]. The reaction is facilitated by the employment of very high pressures (see Table 4)[2272]. Kadaba has investigated the effects of changes in solvent dielectric constant, the use of protic and dipolar aprotic solvents, and the use of wet solvents. This work expanded upon an earlier observation of Mustafa that diazomethane formed stable addition complexes with anils. With the substituted anils, the rate of addition is modestly altered by substituents with *p*-NO_2 causing a 10-fold rate increase in DMF and only a two-fold change when *p*-dioxane is the solvent. The additions are accelerated by

$$\underset{\mathbf{(241)}}{R^1N{=}CR^2R^3} + \underset{\mathbf{(242)}}{R^4R^5CN_2} \longrightarrow \underset{\mathbf{(243)}}{\text{triazoline}} \longleftarrow \underset{\mathbf{(244)}}{R^1N_3} + \underset{\mathbf{(245)}}{R^2R^3C{=}CR^4R^5} \qquad (40)$$

the presence of *ortho* substituents, presumably as a consequence of relief of steric inhibition of resonance. The use of aqueous dioxane greatly improves the yields of triazolines and, in fact, facilitated the formation in very modest yield of 3,4-diphenyl-1-pyrazoline from diazomethane and *cis*-stilbene; this latter reaction goes somewhat better under 5000 atm pressure in ether. Apparently the combination of wet dioxane and high pressure has not been examined.

The introduction of strain into the imine substrate such as is encountered in a 3*H*-azirene leads to enhanced rates of cycloaddition accompanied by subsequent ring opening to furnish azides (equation 41).

$$\underset{\mathbf{(246)}}{\text{Me-azirene(H)(Ar)}} \xrightarrow{CH_2N_2} \underset{\mathbf{(247)}}{ArCH{=}C(CH_2N_3)CH_3} \qquad Ar = 2,4\text{-}(NO_2)_2C_6H_3$$

$$\underset{\mathbf{(248)}}{\text{Ph-azirene(H)(Ph)}} \xrightarrow{PhCHN_2} \underset{\mathbf{(249)}}{(PhCH_2)(Ph)C{=}C(N_3)Ph} \qquad (41)$$

This behaviour of triazolines can under some conditions be reversible in a modified form where triazolines are in equilibrium with diazo compounds. Himbert and Regitz[1869] have carried out extensive investigations in this area which are summarized in equation (42) and Tables 18, 19 and 20. The reaction is not new. Dimroth[1639] observed similar phenomena both with the reaction of phenyl azide with malonamide and the related half methyl ester half amide in 1910. A somewhat similar reaction is observed when phenyl azide reacts with ethyl acrylate[1989].

A slightly different phenomenon occurs with the aryl hydrazones of 1-nitro-aldehydes (equation 43). With diazomethane these hydrazones furnish methyl nitronate esters with concomitant conversion of the hydrazone into an azo function[1475–1481].

c. *Isonitriles*. The reactivity of diazoalkanes towards isonitriles has apparently attracted little attention, although isonitriles can be looked upon in one sense as imino carbenes and should be active in the formation of azine-type products as well

$$\text{(250)} + \text{(251)} \longrightarrow \text{(252)} \rightleftharpoons \text{(253)}$$

R^1 = A CH_3, B C_2H_5, C C_4H_9, D Ph

$$\text{(254)} + \text{(255)} \longrightarrow \text{(256)} \rightleftharpoons \text{(257)} \tag{42}$$

R^1 = A CH_3, B C_3H_7, C $CH_2{=}CHCH_2$, D C_4H_9, E Ph

$$ArNHN{=}C(NO_2)R \; \textbf{(258)} \xrightarrow{CH_2N_2} ArN{=}N{-}C(R){=}N(\rightarrow O){-}O{-}CH_3 \; \textbf{(259)} \tag{43}$$

TABLE 18. Equilibria between triazolines and diazoimines[1869]

R^1	Equilibrium 252 ⇄ 253 R^2	252 (%)	253 (%)
$2\text{-}NO_2C_6H_4$	CH_3		100
$3\text{-}NO_2C_6H_4$	CH_3		100
$4\text{-}NO_2C_6H_4$	CH_3		100
3-[*N*-(1-Diethylamino-2-diazopropylidene)amino sulphonyl phenyl]	CH_3	22	78
$4\text{-}IC_6H_4$	CH_3	33	67
$4\text{-}BrC_6H_4$	CH_3	32	68
$4\text{-}ClC_6H_4$	CH_3	30	70
$4\text{-}FC_6H_4$	CH_3	49	51
α-Naphthyl	CH_3	48	52
β-Naphthyl	CH_3	59	41
C_6H_5	CH_3	62	38
CH_3	CH_3	63	37
$4\text{-}CH_3C_6H_4$	CH_3	80	20
$4\text{-}CH_3OC_6H_4$	CH_3	84	16
$2,4,6\text{-}(CH_3)_3C_6H_2$	CH_3	90	10
$4\text{-}(CH_3)_2N{-}C_6H_4$	CH_3	100	

TABLE 19. Equilibria between triazolines and diazoimines[1869]

	Equilibrium **252** ⇄ **253**		
R^1	R^2	**252** (%)	**253** (%)
C_6H_5	3-$NO_2C_6H_4$	0	100
C_6H_5	4-$NO_2C_6H_4$	0	100
C_6H_5	α-Naphthyl	0	100
C_6H_5	4-$CH_3C_6H_4$	10	90
C_6H_5	4-$CH_3OC_6H_4$	13	87
C_6H_5	2,4,6-$(CH_3)_3C_6H_2$	14	86
C_6H_5	4-$(CH_3)_2NC_6H_4$	50	50
n-C_4H_9	4-$NO_2C_6H_4$	0	100
n-C_4H_9	α-Naphthyl	28	72
n-C_4H_9	2,4,6-$(CH_3)_3C_6H_2$	77	23
n-C_4H_9	4-$(CH_3)_2NC_6H_4$	100	0
C_2H_5	4-$NO_2C_6H_4$	0	100
C_2H_5	2,4,6-$(CH_3)_3C_6H_2$	80	20

TABLE 20. Equilibria between triazolines and diazoimines[1869]

	Equilibrium **256** ⇄ **257**		
R^1	R^2	**256** (%)	**257** (%)
CH_3	4-$CH_3OC_6H_4$	100	
CH_3	4-$CH_3C_6H_4$	100	
CH_3	C_6H_5	100	
CH_3	4-$NO_2C_6H_4$	80	20
CH_3	3-$NO_2C_6H_4$	88	12
CH_3	2-$NO_2C_6H_4$	30	70
n-C_3H_7	2-$NO_2C_6H_4$	22	78
$CH_2-CH=CH_2$	4-$CH_3C_6H_4$	100	0
$CH_2-CH=CH_2$	4-BrC_6H_4	93	7
$CH_2-CH=CH_2$	3-$NO_2C_6H_4$	88	12
$CH_2-CH=CH_2$	2-$NO_2C_6H_4$	18	82
n-C_4H_9	4-$CH_3C_6H_4$	100	
n-C_4H_9	4-$NO_2C_6H_4$	75	25
n-C_4H_9	3-$NO_2C_6H_4$	87	13
n-C_4H_9	2-$NO_2C_6H_4$	20	80
C_6H_5	4-$CH_3C_6H_4$	100	
C_6H_5	4-BrC_6H_4	100	
C_6H_5	4-$NO_2C_6H_4$	100	
C_6H_5	4-$NO_2C_6H_4$	76	24
C_6H_5	3-$NO_3C_6H_4$	79	21
C_6H_5	2-$NO_2C_6H_4$	17	83

as ketenimines. Staudinger[2250] found aryl isonitriles very reactive towards diphenyldiazomethane and obtained the ketenimines. The reaction can be photocatalysed. Diphenyldiazomethane does not react with cyclohexyl isonitrile in the dark, only upon photolysis[1538, 1821]. Thermolysis of methyl phenyldiazoacetate in the presence of *t*-butyl isonitrile also furnishes a ketenimine. More detailed studies of Staudinger's systems indicate that the reaction furnishes many products of which the ketenimine is only a minor product. Other diazo compounds and isonitriles were examined as well[2079].

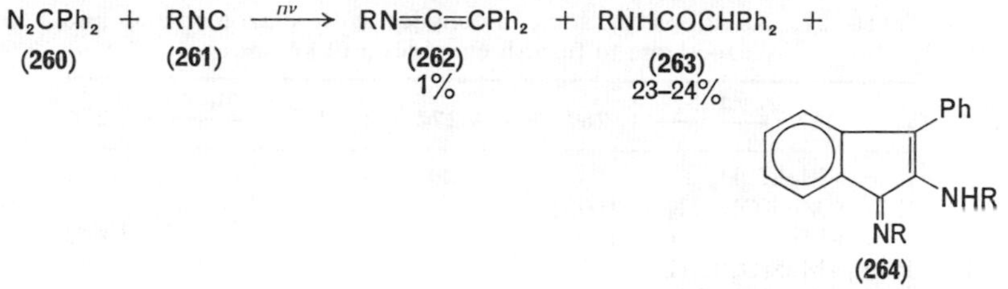

$$N_2CPh_2 + RNC \xrightarrow{h\nu} RN{=}C{=}CPh_2 + RNHCOCHPh_2 + \text{(264)} \quad (44)$$

(260) (261) (262) 1% (263) 23–24%

(264): indenone imine bearing Ph, NHR and =NR substituents

$$N_2CHCOR' + RNC \xrightarrow{Cu} \text{(266)}$$

(265) (261) R = 2,6-Xylyl, Ph; (266) 10–12%, R′ = OEt, Ph

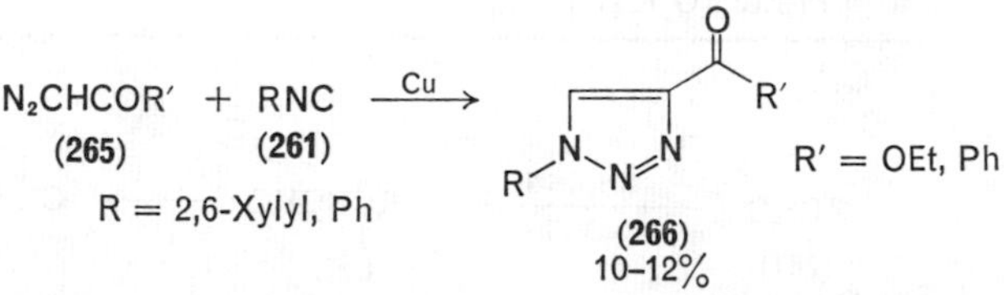

d. *Carbonyl groups.* The reactions of aldehydes, ketones and a few special categories of esters with diazoalkanes encompass two basic cycloaddition processes: (i) those involving addition to the carbonyl group, and (ii) those involving addition to double and triple bonds in conjugation with the carbonyl groups. However, unlike the α, β unsaturated esters, the reactions of this latter category are not limited to the C=C bond and both processes often occur simultaneously. One can therefore expect to encounter complex mixtures, and often does with members of this last category.

i. Unconjugated systems. Considerable space has been devoted in the literature to the question of homologating aldehydes and ketones by employing diazoalkanes. Two very thorough reviews exist[477, 478]. Eistert and coworkers have published a series of papers on the transformations of substituted aldehydes with diazoalkanes.[1735, 1737, 1742, 1750]. Their results are summarized in equations 45 and 46.

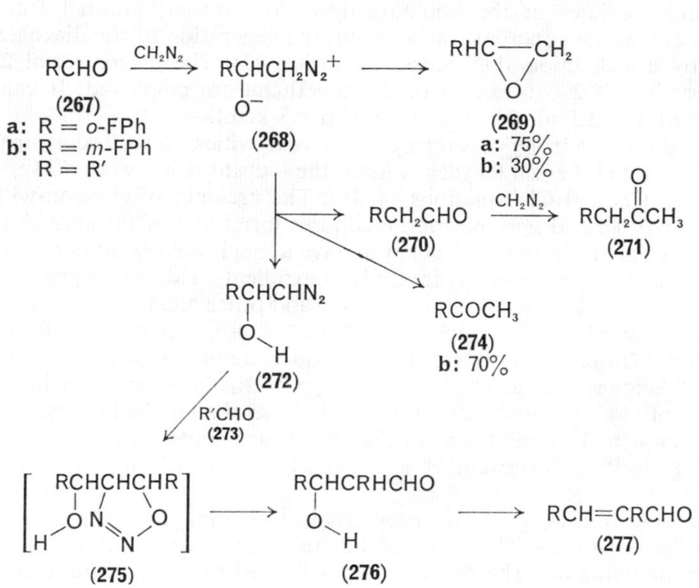

$$RCHO \xrightarrow{CH_2N_2} RCH(O^-)CH_2N_2^+ \longrightarrow RHC{-}CH_2 \text{ (epoxide, O bridge)}$$

(267) a: R = *o*-FPh; b: R = *m*-FPh; R = R′ — (268) — (269) a: 75%, b: 30%

$$\longrightarrow RCH_2CHO \xrightarrow{CH_2N_2} RCH_2COCH_3$$

(270) (271)

$$\longrightarrow RCOCH_3 \quad \text{(274) b: 70\%}$$

$$\longrightarrow RCH(OH)CHN_2 \quad (272) \xrightarrow{R'CHO\ (273)} \quad (45)$$

$$\left[\text{RCHCHCHR (O–H; N=N–O ring)}\right] \longrightarrow RCH(OH)CRHCHO \longrightarrow RCH{=}CRCHO$$

(275) (276) (277)

TABLE 21. Reactions of some aldehydes with diazomethane and diazoethane to furnish epoxides and ketones

	267 $\xrightarrow{CH_2N_2}$ 269	+ 274	: 267 $\xrightarrow{CH_3CHN_2}$ 280
R = *o*-$MeSC_6H_4$		100%	
R = *o*-$MeS(O)C_6H_4$	100%		
R = *o*-$MeS(O_2)C_6H_4$	100%		100%
R = *p*-$MeS(O_2)C_6H_4$	35%	65%	100%
R = *m*-$MeS(O_2)C_6H_4$		100%	

$$\text{RCHO (267)} \xrightarrow[\text{MeOH free}]{CH_3CHN_2} \text{RCH(CH}_3\text{)CHO (278)} \longrightarrow CH_3CHC(O)CH(CH_3)R \text{ (279)} \tag{46}$$

$$\text{RCHO (267)} \xrightarrow[\text{MeOH free}]{CH_3CHN_2} RC(O)CH_2CH_3 \text{ (280)}$$

a: R = *o*-FPh; **b**: R = *m*-FPh

(278) **a**: 95%; (280) **b**: 100%

The principal reactions with diazomethane arc formal insertion into the aldehyde C—H bond and epoxide formation. These products can be nicely rationalized as resulting from loss of nitrogen from an intermediate oxadiazoline. The remaining products represent aldol-type condensations and/or further homologations. An interesting feature of this latter group of reactions is the ability to bring about cross condensations between highly electron-deficient aldehydes such as chloral and the intermediates in the homologation. An unusual product from such a reaction is **286** which apparently arises from the generation of the diazoalcohol **288** followed by attack upon the chloral (equation 47). The diazoalcohol **288** is the primary product if 2 equivalents of diazomethane are employed. It can then be condensed with a second aldehyde to furnish a β-ketol.

Eistert's group[1735, 1737, 1750] compared the reactivities toward diazomethane of several substituted benzaldehydes where the substituents were SMe, S(O)Me, SO_2Me, CF_3, CN and Cl (equations 44, 46). The presence of electron-withdrawing groups in the *ortho* and *para* position facilitates formation of the epoxides, whereas the same substituents in the *meta* position have a much weaker influence. The use of diazoethane with the same aldehydes leads to excellent yields of the propiophenones.

The reaction of diazomethane with chloral and other aldehydes was examined by Arndt and Eistert[1451–1453, 1458, 1738, 1742, 1750] and more recently by Bowman and coworkers[1536]. Gutsche[477] points out that proper credit for discovering the homologation of ketones and aldehydes should go to Buchner and Curtius (1885)[1549], von Pechmann (1895)[2323] and Meyer (1905)[2041], rather than to Schlotterbeck (1907)[2196]. However, the actual development of the processes should be credited to work by Arndt[1450], Eistert[1728], Meerwein[2030] and Mosettig[2057] in the late 1920's and early 1930's.

The homologation of aldehydes proceeds in yields ranging from 20 to 100% for the formation of ketones. The principal side reactions are the formation of epoxides and higher homologues. The reaction is not limited to diazomethane and good to

$$\mathbf{267} \xrightarrow[R'CHO]{CH_2N_2} \mathbf{269} + \mathbf{271} + \mathbf{272} + R'CH(OH)CH_2C(=O){-}R\ \mathbf{(281)} + R'CH{=}CHC(=O){-}R\ \mathbf{(282)} + RCH_2CH(OH)CH_2C(=O)R\ \mathbf{(283)} + \mathbf{274} \quad (47)$$

$$\mathbf{(283)} + \mathbf{274} \xrightarrow{CH_2N_2} \mathbf{(281)}$$

R	R′	269 (%)	271 (%)	272 (%)	281 (%)	282 (%)	283 (%)	274 (%)
o-NCC$_6$H$_4$	R				33			
o-ClC$_6$H$_4$	R				13·1		5·3	36
o-NCC$_6$H$_4$	CCl$_3$				38, R = R′ +4·2			
o-NCC$_6$H$_4$	p-NO$_2$				4·3	32		
m-NCC$_6$H$_4$	R	10			9·6			44
m-ClC$_6$H$_4$	R	8·7			13		3	42
p-NCC$_6$H$_4$	R	6	2		24			27 (5% epoxide[2812])
p-ClC$_6$H$_4$	CCl$_3$				6 11, R = R′		2	35
p-ClC$_6$H$_4$	R	14			11		2	35

excellent yields are obtained with diazoethane[1402, 2058, 2250], diazopropane[1402] and diazobutane[1402]. The reaction with diazoacetic ester apparently goes beyond the formation of β-keto esters and furnishes dioxolanes instead[1636]. The presence of alcohols such as methanol appears on occasion to have a deleterious effect upon the yield and even the course of reaction. Methanol in some circumstances tends to favour aryl migration with aryl aldehydes to furnish an aralkyl aldehyde and subsequent products whereas the methanol-free reaction leads to H migration and formation of a methyl ketone[2057]. The choice of diazoalkane also influences the course of the reaction with an aryl aldehyde. When nitropiperonal is treated with diazomethane, 90% of the product is the oxide and only 10% is the methyl ketone. When diazoethane is employed, the ethyl ketone is obtained in 73% yield[2058]. With piperonal an 84% yield of ketone results[2058].

$$\underset{(284)}{Cl_3CCHO} + CH_2N_2 \longrightarrow \underset{(285)}{Cl_3CCH(O^-)CH_2N_2^+}$$

$$\mathbf{(285)} \longrightarrow \underset{(287)}{Cl_3CCH{-}CH_2\ (\text{epoxide, O})}$$

$$\mathbf{(285)} \rightleftharpoons \underset{(288)}{Cl_3CCH(O{-}H)CHN_2} \xrightarrow{284} \underset{(286)}{Cl_3CCH(OH)CH{-}CHCCl_3\ (\text{epoxide, O})} \quad (48)$$

The homologation of ketones has been extensively reviewed by Gutsche[477, 478] and occurs with acyclic and cyclic ketones. The mechanism is probably the same as that with aldehydes. The reactions are slower than with aldehydes and the use of a polar solvent such as methanol is frequently advantageous[2029, 2030]. The acyclic ketones give only fair yields of homologues with the yields decreasing with chain length[477]. The major products are the related oxides and can often be obtained in very high yields. When enolization is favourable (e.g. acetoacetic ester) one may encounter O and C alkylated products as well as the oxide[1454, 2029].

With 1,2-diketones, methylenedioxy derivatives may result. This holds both with simple 1,2-diones and *o*-quinones[2212] although exceptions exist[1529, 1737, 1831] (equation 49).

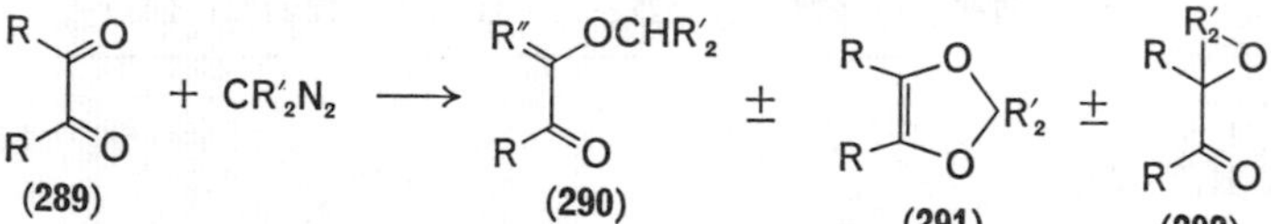

$$\textbf{(289)} + CR'_2N_2 \longrightarrow \textbf{(290)} \pm \textbf{(291)} \pm \textbf{(292)}$$

R = Me; R′ = H ⟶ **290**
R = Me; R′ = C_6H_5 or *p*-$MeOC_6H_4$ ⟶ **291**
R = Me,OMe; R′ = H ⟶ **292**

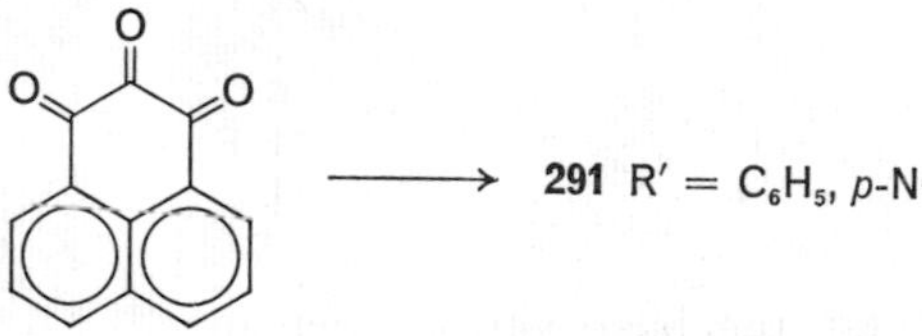

⟶ **291** R′ = C_6H_5, *p*-$NO_2C_6H_4$

(49)

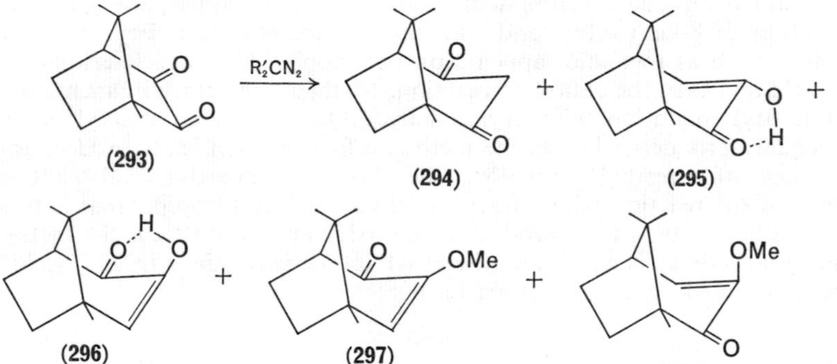

(293) $\xrightarrow{R'_2CN_2}$ **(294)** + **(295)** + **(296)** + **(297)** + **(298)**

With cyclic ketones the smallest ring might be considered as two, that in ketene, but it will be treated in Section I.A.2.a along with the product, cyclopropanone.

The carbocyclic ketones up through at least cyclopentadecanone undergo homologations along with some oxide formation. The relative rates have been measured[2124, 2126] and are quoted by Gutsche[477] as: cyclopentanone, 1·00; cyclohexanone, 1·80–2·65; cycloheptanone, 1·25; cyclooctanone, 0·62; cyclopentadecanone, 1·70 and cycloheptadecanone, 2·42. A frequent side reaction is *bis* homologation accompanied by epoxide formation. When cyclohexanone is treated with 2 equivalents of diazomethane, cyclooctanone is obtained in ~60% yield[1956, 2126]. The presence of substituents on a cyclic ketone frequently leads to the formation of both possible isomers if the starting ketone does not possess a C_2 symmetry axis.

This holds whether the substituent is α or far from the reaction site. Thus Tchoubar[2284] obtained both possible cycloheptanones from 3,5,5-trimethyl cyclohexanone.

The reaction with α-tetralone is extremely slow and yields oxides. Benzophenone is unreactive. In contrast, moderate yields of phenanthrones are obtained with fluorenone and tetramethoxyfluorenone[1492, 1832].

The use of higher diazoalkanes has been examined. Gutsche and coworkers have inserted ethyl ω-diazocaproate into cyclohexanone and 4-(3′,4′,5′-trimethoxyphenyl)-1-diazobutane into the same substrate as part of studies directed towards the synthesis of colchicine[1831, 1832].

ii. α,β-Unsaturated ketones and aldehydes. Although many α,β-unsaturated ketones and aldehydes undergo pyrazoline formation faster than homologation, examples do exist where this is not the case. Johnson and coworkers[2085] have homologated Δ^4-3-ones in steroid systems to furnish A-homosteroids by employing Lewis-acid catalysts and diazomethane (equations 50 and 51).

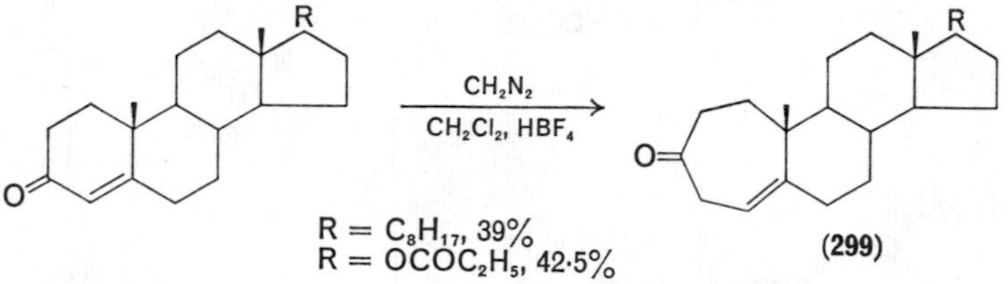

$$\xrightarrow[CH_2Cl_2,\ HBF_4]{CH_2N_2} \quad (50)$$

(299)

$R = C_8H_{17}$, 39%
$R = OCOC_2H_5$, 42·5%

$$PhCH{=}CHCOCH_3 \ \textbf{(300)} \xrightarrow[(2)\ H_2,\ Pd/C]{(1)\ CH_2N_2,\ HBF_4} PhCH_2CH_2CH_2COCH_3 \ \textbf{(301)}\ 60\%$$

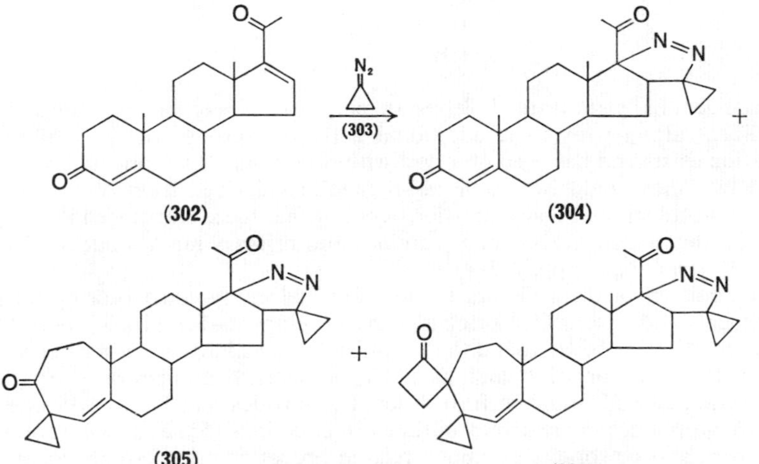

(302) $\xrightarrow{N_2 \text{ (303)}}$ (304) + (305) + (306) (51)

An uncatalysed homologation of the steroidal A ring has been accomplished using diazocyclopropane and a Δ^4-3-one. The reaction is accompanied by pyrazoline formation at the Δ^{16}-20-one function as well as by subsequent attack of the saturated keto group in the homologated A-ring. This secondary product presumably arose from the transitory generation of a cyclopropylidene epoxide or a cyclopropyl carbonium ion[1529].

The addition of diazoalkanes to α,β-unsaturated ketones occurs in the normal sense if the diazoalkane is monosubstituted or diazomethane. The intermediate products are 3-acyl-Δ^1-pyrazolines and therefore possess a hydrogen acidified by both the acyl group and the —N=N—. Since diazoalkanes are reasonably basic, it is not surprising that the product actually isolated from such a reaction is not the Δ^1-pyrazoline but rather the Δ^2 isomer which possesses the conjugated chromophore H—N̈—N=C—C=O. If, however, the diazomethane is disubstituted, simple prototropy will not furnish the Δ^2 system unless the substituent on $C_{(5)}$ is electron withdrawing. In such a case, the identical arguments apply. It is not surprising that dimethyl maleate and dimethyl fumarate both react with diazomethane and with diazoacetic ester to furnish identical products irrespective of the stereochemistry of the dipolarophile[283, 2004].

(307) + N_2CHCO_2Et ⟶ [(308)]

(309) + N_2CHCO_2Et ⟶ [(310)] ⟶ (311) (52)

When geminally substituted olefins are employed where an acyl group serves as a substituent along with some additional activating group, the possibility of acyl migration exists and has been observed with one system. Danion-Bougot and Carrié found that they could isolate the Δ^1-pyrazolines derived from diazomethane and some monosubstituted α-acyl acrylic esters but that these rearranged in the presence of excess diazomethane, secondary amines and primary amines but not by tertiary amines[1617] (*vide supra*; pp. 853–4).

The addition of diazoalkanes to polycarbonyl compounds bearing unsaturated substituents and unsaturated ketones such as cyclopentadienones has been examined by Eistert[1741, 1745–1750]. He found that both isatin and ninhydrin add β- and γ-pyridyl diazomethanes to furnish homologated ring systems. A number of other quinisatins were examined and found to furnish ketals, epoxides and homologues.

With *p*-quinones the reactions outlined in equations (55) and (56) were observed. Schönberg and coworkers examined related processes using *o*-quinones and diaryl diazomethanes. They obtained only epoxides with isatin and ninhydrin[2212].

With cyclopentadienones in the presence of alcohols, diazoalkanes lead to the addition of alcohol across a double bond[1744]. In aprotic solvents such compounds can form pyrazolines which on expulsion of nitrogen furnish bicyclo[3.1.0] systems, but if they are isomerized first to Δ^2 pyrazolines, furnish phenols on expulsion of the nitrogen. If the reactions are run in methanol rather than benzene, epoxides, hydroxy-fulvenes and phenols can result.

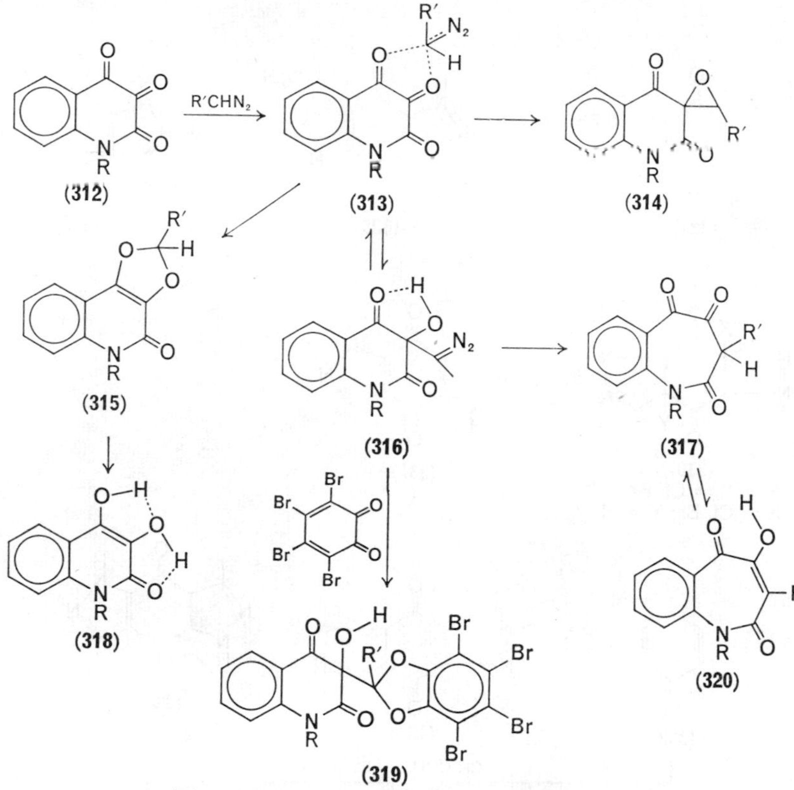

(312) (313) (314) (315) (316) (317) (318) (319) (320)

(R′ = Alkyl; R = H, CH_3, C_6H_5, 1,8-$(CH_2)_3$—

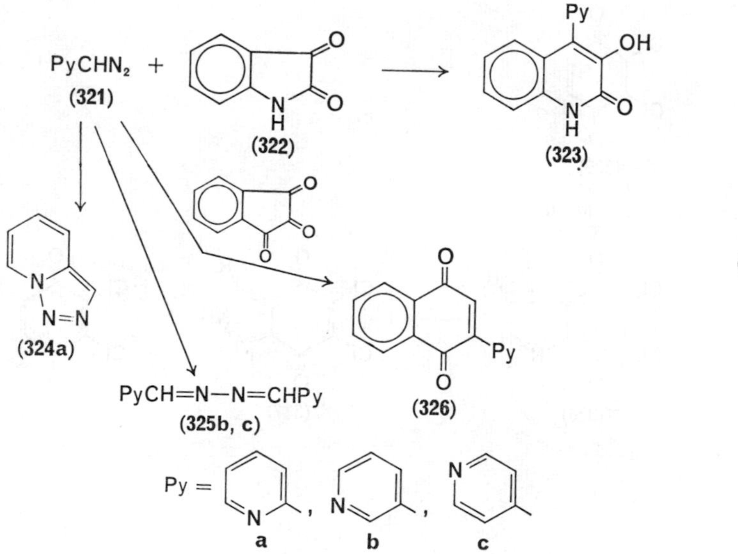

(53)

$PyCHN_2$ + (321) (322) (323) (324a)

PyCH=N—N=CHPy (325b, c) (326)

Py = a, b, c

(54)

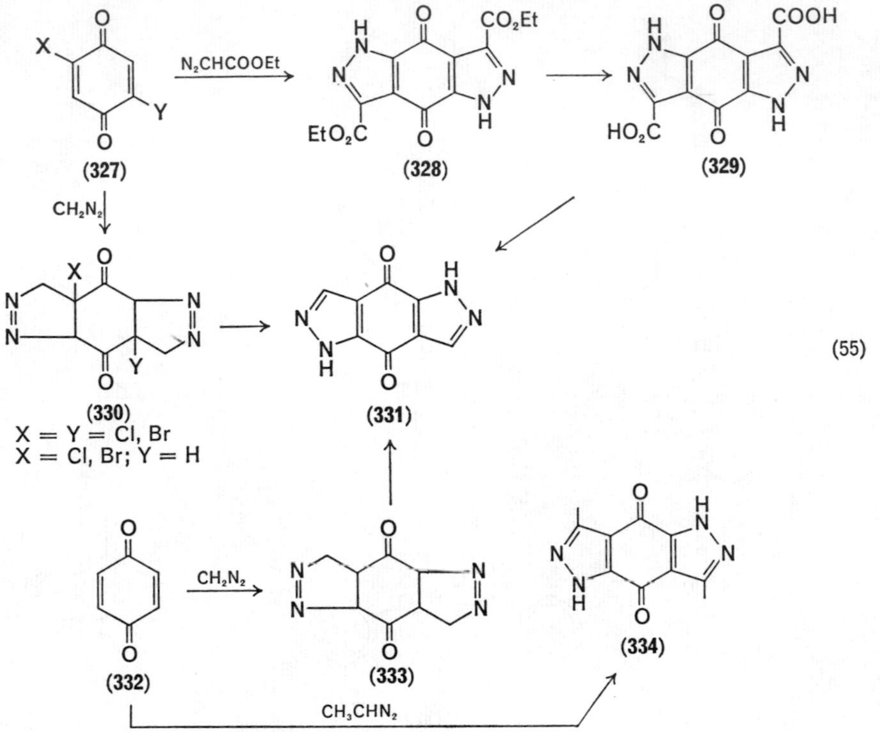

X, Y, O, O
$N_2CHCOOEt$
H, N, N, O, CO_2Et, N, N, H, EtO_2C, O
H, N, N, O, COOH, N, N, H, HO_2C, O

(327) (328) (329)

CH_2N_2

N, N, X, O, N, N, N, O, Y
O, H, N, N, N, N, H, O

(330) (331)

X = Y = Cl, Br
X = Cl, Br; Y = H

(55)

O, O
CH_2N_2
N, N, O, N, N, O
O, H, N, N, N, N, H, O

(332) (333) (334)

CH_3CHN_2

O, Cl, Cl, O

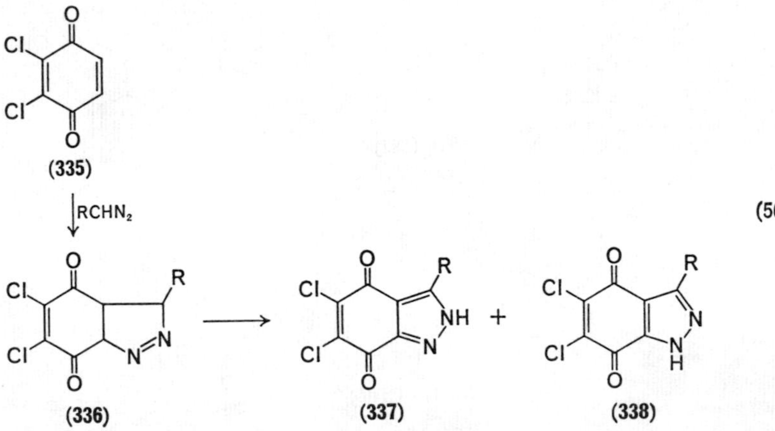

(335)

$RCHN_2$

(56)

O, R, Cl, Cl, N=N, O
O, R, Cl, Cl, NH, N, O
+
O, R, Cl, Cl, N, N, H, O

(336) (337) (338)

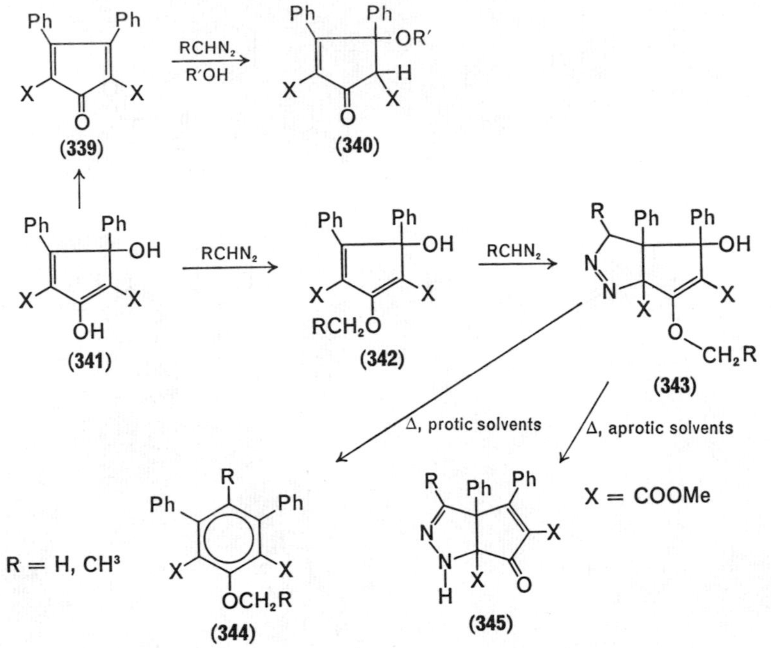

(339) (340) (341) (342) (343) (344) (345)

Δ, protic solvents

Δ, aprotic solvents

R = H, CH³ X = COOMe (57)

Although esters are supposed to be inert to attack on the carbonyl group by diazoalkanes, Dean has reported the isolation of an epoxide from the reaction of diazoethane and 3-nitro coumarin[1632a] (equation 58). This observation strengthens the earlier claim of Fleming and coworkers that attack on an ester carbonyl group was a part of the reaction sequence responsible for the conversion of **348** into **350**[1521, 1776].

The reactions of quinoidomethanes[1486] and imines[2092] with diazomethane have been examined and representative examples appear in equations (59) and (60).

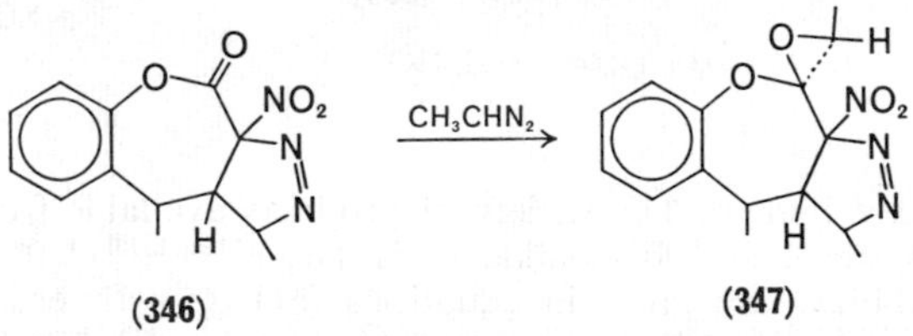

(346) (347) (58)

(348) (349) (350)

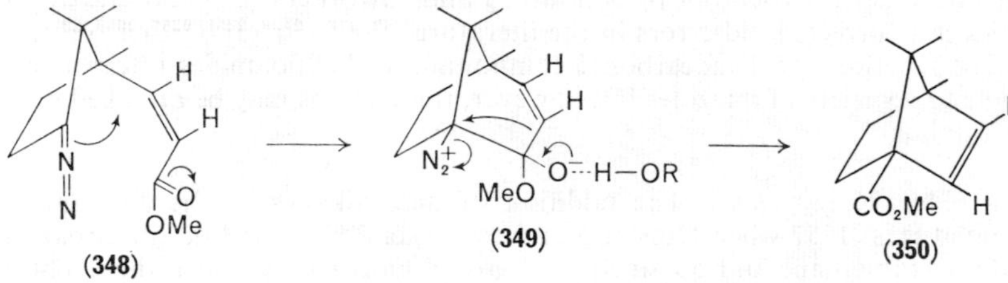

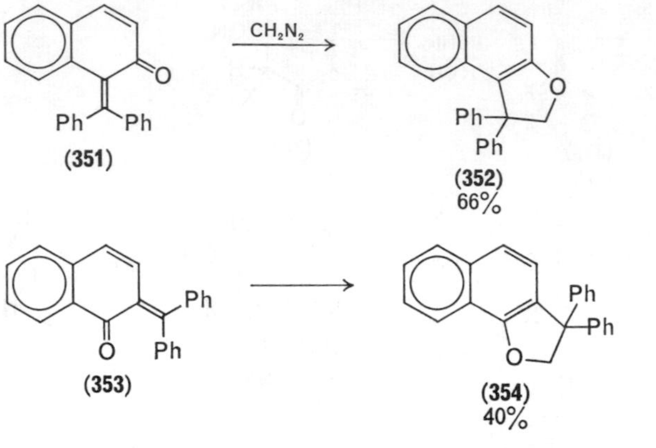

(351) $\xrightarrow{CH_2N_2}$ (352) 66% (59)

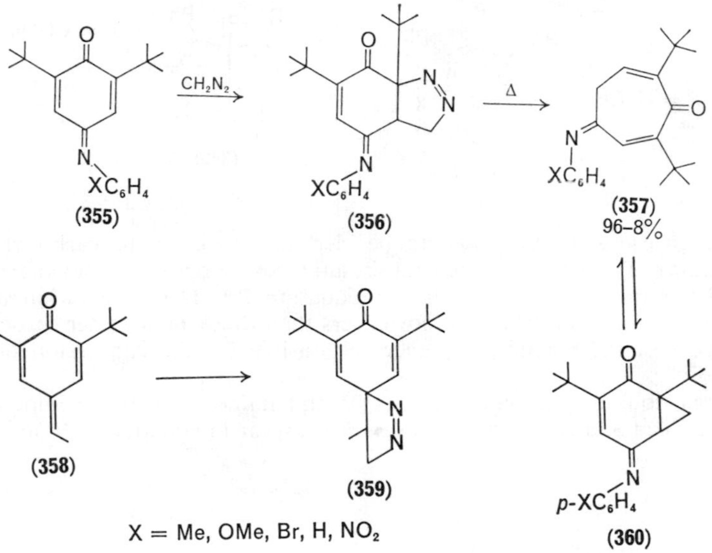

(353) ⟶ (354) 40%

(355) $\xrightarrow{CH_2N_2}$ (356) $\xrightarrow{\Delta}$ (357) 96–8% (60)

(358) ⟶ (359) (360)

X = Me, OMe, Br, H, NO_2

e. *Thiocarbonyl systems*. The variety of products available from the reaction of thiocarbonyl systems and diazoalkanes is large[1513a, 1513b, 1963a, 1968a, 2014a]. A few observed possibilities are given in equations (61)–(65). Extensive studies in this area have been carried out by Schönberg and coworkers[2199–2206]. Some of the newer work has corrected old errors in the literature[2043, 2188, 2199, 2251, 2257, 2258, 2371].

The reactions of diazocarbonyl compounds with thioamides offers a reasonable route to a variety of thiazoles[1945]; however, thiolactams may be attacked on nitrogen or sulphur[1637, 1815].

f. —N=N— *systems*. The addition of diazoalkanes to the azo linkage dates from at least 1952 when Ginsburg and coworkers[1804] treated hexafluoroazomethane with diazomethane and obtained *N,N′-bis*(trifluoromethyl)diaziridine. The reaction does not, however, appear to be general since the examples reported in the literature all possess electron-withdrawing substituents on both nitrogens. Extensive studies have been performed since 1962 by Fahr and coworkers[1759–1764, 2019], using a variety of diacyl derivatives. All of the above work was apparently overlooked in a recently published diaziridine synthesis[1910]. Some representative reactions of diazocompounds with the —N=N— bond are given in equation (66). In all of the examples, nitrogen is lost[1759–1764, 1804, 1910, 1940, 2019], but the processes can be rationalized as passing through a tetrazole. Only in the Ginsburg study has a tetrazole been isolated.

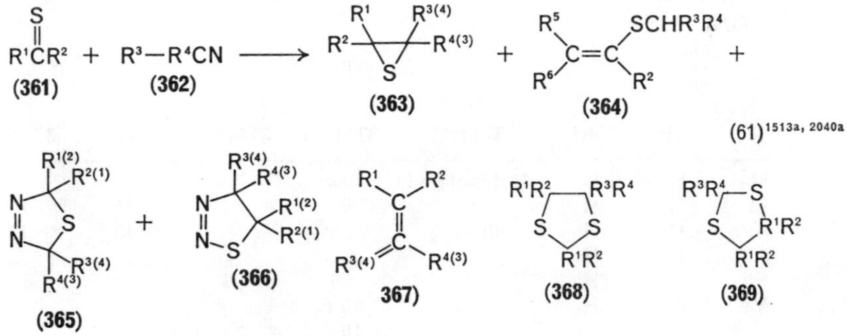

$$R^1C(=S)R^2\ \textbf{(361)} + R^3{-}R^4CN\ \textbf{(362)} \longrightarrow \textbf{(363)} + \textbf{(364)} + \textbf{(365)} + \textbf{(366)} + \textbf{(367)} + \textbf{(368)} + \textbf{(369)} \qquad (61)^{1513a,\ 2040a}$$

R^1	R^2	R^3	R^4	R^5	R^6	363 (%)	364 (%)	367 (%)	368 (%)	369 (%)
Me	*i*-Pr	H	H	H	H	70	25			
Me	*t*-Bu	H	H	H	H	60	35			
Et	*i*-Pr	H	H	Me	H	85	5 *E* 5 *Z*			
Me	*t*-Bu	Me	H	H	H	75	25			
Ph	*t*-Bu	Me	H	H	H	70		15 *Z* 15 *E*		
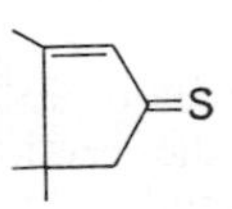		H	H				20			85
		H	Me					75		45
		Me	Me					85		
		Ph	H					55		
		Ph	Ph					40		
		EtO_2C	H				60			
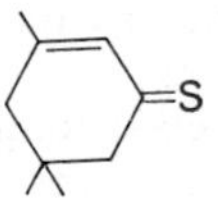		H	H			45		20		85
		H	Me					95		90
		Me	Me					90		
		Ph	H					45		
		Ph	Ph			75				
		EtO_2C	H				70	20		
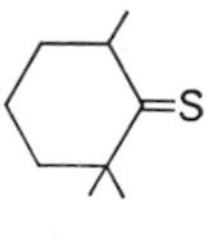		H	H				75	10		
		H	Me					90		
		Me	Me					90		
		Ph	H			50				
		Ph	Ph			70				
		EtO_2C	H				25	35		

The yields for the last three thioketones are the maximum values reported and are a function of temperature and order of addition.

$R^1C(=S)SCH_3$ **(370)** $\xrightarrow{R^2R^3CN_2}$ **(371)** + $CH_3S{-}C(R^1){=}CR^2R^3$ **(372)** + **(373)** + **(374)** (62)[1513a]

(371) $\xrightarrow[-OMe]{CS_2}$ **(375)** $\longrightarrow$ **(376)**

R^1	R^2	R^3	372 (%)	373 (%)	374 (%)	375 (%)	376 (%)
Me	H	H	Not isolated				
Et	H	H				42	11
i-Pr	H	H	40–60%			40	13
t-Bu	H	H					
Ph	H	H		66	33		
Me	Me	H	40 *E*	10 *E*			
			40 *Z*	10 *Z*			
Et	Me	H	30 *E*	20 *E*			
			40 *Z*	10 *Z*			
i-Pr	Me	H	50 *E*	4 *E*			
			40 *Z*	6 *Z*			
t-Bu	Me	H	3 *E*	32 *E*			
			60 *Z*	5 *Z*			
Ph	Me	H			60		
Me	Me	Me	Yields of **371** and **372** not given				
t-Bu	Me	Me	Suggestion that GC converted **365** and **366** to **371** and **372**				

g. —N=O *systems*. The nitroso[1527] and nitro groups[2091] have both been topics in this series[1769], and both undergo reactions with diazoalkanes. Whether the reactions proceed via cycloaddition to the N=O group is a matter of conjecture. It is a reasonable hypothesis[2243] since the N=O group is isoelectronic with —CHO, and cycloaddition has been demonstrated as being involved in attack on carbonyl groups. An adduct has been observed between trifluoronitrosomethane and diazomethane which subsequently loses N_2 [2015].

The nitroso group reacts with diazoalkanes to furnish nitrones or their oxidation products to furnish bis-nitrones[1474, 1485, 1537, 1974–1978, 2015, 2243, 2327]. The nitrones are 1,3-dipoles in their own right and undergo cycloaddition reactions with suitable substrates.

$$CF_3CHN_2 + (CH_3)_3CN{=}O \longrightarrow CF_3CH{=}N(O)C(CH_3)_3$$

$$2\ C_6F_5NO + 2\ CH_2N_2 \xrightarrow{[O]} C_6H_5N(O)CHCHN(O)C_6H_5$$

The reaction of diazoalkanes with the nitro group furnishes nitronic esters when there is a sufficiently acidic proton on the carbon atom bearing the nitro group[2091]. It is not clear whether nitro groups on non-enolizable carbon are attacked to furnish

$$R^1C(=S)OR^2 \ (\mathbf{377}) \xrightarrow{R^3R^4CN_2} R^1CR^3R^4C(=S)OR^2 \ (\mathbf{378}) + \mathbf{379} + \mathbf{380} + \mathbf{381} + \mathbf{382} + R^1CR^3R^4CR^3R^4C(=S)OR^2 \ (\mathbf{383}) + R^1R^4R^3CC(=S)OR^2 \ (\mathbf{384}) \quad (63)^{1513b}$$

(379): thiirane, R^2O, R^1 on one C; $R^{3(4)}$, $R^{4(3)}$ on the other C
(380): $(R^2O)(R^1)C{=}C(R^{3(4)})(R^{4(3)})$
(381): 1,2,3-thiadiazoline, R^1O, R^2 on C5; $R^{3(4)}$, $R^{4(3)}$ on C4
(382): 1,2,3-thiadiazole, $R^1R^3R^4C$ on C5; R^3 on C4

R^1	R^2	R^3	R^4	378 (%)	379 (%)	380 (%)	381 (%)	382 (%)	383 (%)	384 (%)
H	Et	H	H					50	10	30
CH_3	CH_3	H	H	30			40			
CH_3	Et	H	H	30			60			
Et	Et	H	H	30			50			
Ph	Et	H	H				30			
Me	Et	H	H					+ } in presence of CH_3OH		
PhH	Et	H	H					+ } in presence of CH_3OH		
Me	Me	Me	Me	+	+		+ (not isolated)			
i-Pr	Me	Me	Me	+	+		+ (not isolated)			
Me	Me	Me	H	20	80					
Me	Me	Me	H	50 *E* 50 *Z*						
Et	Et			45 *E* 55 *Z*						
Ph	Me			20 *E* 30 *Z*		20 *E* 30 *Z*				

$$2\,PhC(=O)C(=N_2)Ph \ (\mathbf{385}) + CS_2 \longrightarrow \mathbf{386} \quad \text{(Reference 2043)}$$

(386): Ph–C(Ph)–S / O=C–S–C=C(Ph)–C(=O)Ph ring

$$\mathbf{385} + S{=}CCl_2 \ (\mathbf{387}) \longrightarrow \mathbf{388} \quad \text{(Reference 2251)} \qquad (64)$$

(388): PhC=CPh–O–CCl$_2$–S ring

$$\mathbf{387} + XC_6H_4C(=O)CHN_2 \longrightarrow \mathbf{389} + XC_6H_4C(=O){-}CHClC(=S)Cl \ (\mathbf{390})$$

(389): $XC_6H_4C(=O)$–C=C(Cl)–S–N=N ring; (Reference 2168)

$$PhC(=O)CHN_2\ (391) + H_2NC(=S)Ph\ (392) \xrightarrow{R = Ph} (393)\ \text{(Reference 1945)}$$

$$H_2NC(=S)NH_2\ (394) + 391 \xrightarrow{R = NH_2} (393)\ \text{(Reference 2334a)} \qquad (65)$$

$$(395) \xrightarrow{CH_2N_2} (396) \qquad (397) \xrightarrow{CH_2N_2} (398)$$

$$F_3CN{=}NCF_3\ (399) + CH_2N_2 \longrightarrow (400) \longrightarrow (401)\ \text{(Reference 1804)}$$

$$EtO_2CN{=}NCO_2Et\ (402) + N_2CHCO_2Et \xrightarrow{<80\ ^\circ C} (403)$$

$$\xrightarrow{>100\ ^\circ C} (EtO_2C)_2N{-}N{=}CH{-}COEt\ (404)\ \text{(Reference 1760)} \qquad (66)$$

homologated nitro compounds, oxatriazolines or any other products. Nitronic acids react directly with diazomethane to furnish methyl nitronic esters[1460, 1816, 1960, 2224]. The related nitro compounds also furnish the esters[1460, 1960].

$$p\text{-}NO_2C_6H_4CH{=}NO_2H + CH_2N_2 \longrightarrow p\text{-}NO_2C_6H_4CH{=}NO_2CH_3 + N_2$$

$$p\text{-}NO_2C_6H_4CH_2NO_2 + CH_2N_2 \longrightarrow p\text{-}NO_2C_6H_4CH{=}NO_2CH_3 + N_2$$

When there is an electron-withdrawing group present on the same carbon as the nitro group there exist two possible geometries for the nitronic esters[1460, 1960]. Both isomers are formed. Their relative amounts for some nitronic esters derived from diazomethane and diazoethane and four monosubstituted nitromethanes were determined by Grée and Carrié[1817] and are given in equation (67). They have examined the cycloaddition reaction of these 1,3-dipoles[1819–1821].

The formation of α-arylazonitronic esters via the reaction of diazomethane with aldehyde 1-nitrohydrazones is discussed in Section A.1.b.

The NO molecule reacts readily with phenyldiazomethane to furnish the dimer of the nitrile oxide (**416**). This on treatment with excess NO converts it to benzaldehyde and benzalnitramine in ~25% yield, presumably via **418** [1882].

$$RCH_2NO_2 + R'CHN_2 \longrightarrow \underset{(406)}{R(H)C{=}N(\to O)OR'} + \underset{(407)}{R(H)C{=}N(\to O)OR'} \qquad (67)^{1817}$$

(405)

R	R′	406 (%)	407 (%)
CN	CH_3	57	43
$COCH_3$	CH_3	45	55
CO_2CH_3	CH_3	60	40
p-$NO_2C_6H_4$	CH_3	40	60
CN	C_2H_5	60	40
$COCH_3$	C_2H_5	61	39
CO_2CH_3	C_2H_5	70	30

Oximes are tautomers of nitroso compounds and undergo reactions characteristic of both tautomers with *ortho* alkylation and nitrone formation often occurring in the same reaction[2242, 2264, 2280, 2281, 2291, 2295].

$$PhCHN_2 + NO \longrightarrow \left\{ \underset{(408)}{PhCH(N{=}O){-}N{=}N} \right\}$$

$$\underset{(409)}{PhCH{=}N{\to}O} \longrightarrow \underset{(410)}{PhCH{=}N(\to O){-}N(\to O){=}CHPh} \xrightarrow{NO} \underset{(411)}{PhCH{=}N(\to O){-}N{=}O} \qquad (68)^{1882}$$

$$\underset{\substack{syn\\(412)}}{p\text{-}NO_2C_6H_4CH{=}NOH} + CH_2N_2 \longrightarrow \underset{(413)}{p\text{-}NO_2C_6H_4CH{=}NOCH_3} + \underset{(414)}{p\text{-}NO_2C_6H_4CH{=}N(\to O)CH_3}$$

The reaction with **412** and diazomethane indicates the range of products available[1470, 2261].

Engbersen and Engberts[1754] have found the reaction of diazoalkanes with N_2O_3 a convenient route to oxadiazole oxides.

h. *Oxygen*. Diphenyldiazomethane reacts photochemically with oxygen, in the presence of a singlet sensitizer and benzaldehyde, to furnish benzophenone (67%)

the triphenylethylene ozonide (26%) and N_2O. Somewhat similar observations were made with phenyl diazomethane and benzaldehyde where the ozonide is formed in 50% yield[1867].

$$RCHN_2 + ONONO \longrightarrow \text{RC—CR, N—O—N}\rightarrow\text{O ring}\ \textbf{(415)} \qquad (69)^{1754}$$

R	Yield (%)
CO_2Et	89
$m\text{-}O_2NC_6H_4SO_2$	100
$m\text{-}ClC_6H_4SO_2$	100
$PhCH_2SO_2$	97
$p\text{-}O_2NC_6H_4CO$	71
$PhCO$	62
$p\text{-}CH_3C_6H_4SO_2$	75

$$Ph_2CN_2 \xrightarrow[\text{CH}_3\text{CN, C}_6\text{H}_5\text{CHO, methylene blue}]{h\nu,\ O_2} \left\{ \textbf{(416)} \right\} \longrightarrow Ph_2\overset{+}{C}O\overset{-}{O} \xrightarrow{PhCHO} \textbf{(417)} + Ph_2CO\ \textbf{(418)} \qquad (70)^{1867}$$

Diphenyldiazomethane reacts under photolysis conditions without a triplet sensitizer to furnish benzophenone[2279]. Whether this involves a cycloaddition reaction as proposed for the singlet-sensitized case has not been demonstrated, nor has a free carbene as suggested by Nagai[2279] been proved. Similar results are encountered with diazobenzil[2115]. The reaction requires photolysis. Simple treatment with oxygen at room temperature does not lead to any reaction; however, in refluxing benzene, diazobenzil does react[2279]. In a related experiment, Hillhouse[1868] measured the rate of disappearance of the chromophore with dimethyl diazomalonate dissolved in extremely pure and oxygen-free cyclohexene using a medium-pressure Hg lamp and a Pyrex filter. When air was admitted to the system the rate of reaction doubled and then slowly fell back to the original value. Such behaviour suggests that the diazo compound in a photoexcited state is involved and not the carbene; hence the intermediate **416** (equation 70) seems a probable candidate to explain the oxidation[1867].

2. Additions to X=Y=Z cumulene systems

The reactions of diazoalkanes with various cumulene systems have received very little attention compared with X=Y and X≡Y systems. There exist a number of examples for most of the cumulenes and no attempt at exhaustive coverage has been made.

a. *Allenes.* The addition of diazomethane to allene furnishes 4-methylene-Δ^2-pyrazoline in good yield[1718] if the product is allowed to stand for a prolonged period, or the Δ^1-pyrazoline if it is isolated carefully[1602]. When the allene bears electron-attacking groups, the double bond bearing the substituent becomes part of the pyrazoline ring. With electron-donating groups such as acyloxy and alkoxy, addition occurs on the β—C=C bond[1503, 1629]. Thermolysis of the product **420** leads to the methylenecyclopropane (**422**) and not to **421** [1503] which arises as a consequence of prolonged reaction times (equation 71). The addition can occur in two ways and does

$$H_2C{=}C{=}CH_2\ (\mathbf{419}) + CH_2N_2 \longrightarrow (\mathbf{420}) \longrightarrow (\mathbf{421});\quad (\mathbf{420}) \xrightarrow{\Delta \text{ or } h\nu} (\mathbf{422}) \qquad (71)$$[1503, 1602, 1718]

so with diphenyldiazomethane with both types of substituted allenes[1503, 1506] (Table 22). The relative rates of addition of diazomethane to a number of substituted allenes have been measured and are summarized in Table 23 [1506]. Kinetics have also been run on substituted 1,2,3-trienes[1507]. With allenic ketones, one obtains Δ^1- or Δ^2-pyrazolines and pyrazoles if diazomethane, diphenyldiazomethane or diazoacetic ester are employed[1503, 1504, 1506].

TABLE 22. Additions of diazoalkanes to substituted allenes[1503, 1504, 1506, 1529]

					Yield (%)					
R^1	R^2	R^3	R^4	R^5	428	429	430	431	Methyl pyrazole derived from	Δ^2-Pyrazoline derived from
I	CH_3O	H	H	H			61			
I	C_6H_5O	H	H	H			69			
:H_3	C_6H_5O	H	H	H			60			
I	CH_3O	H	C_6H_5	C_6H_5				73		
I	C_6H_5	H	C_6H_5	C_6H_5				75		
:H_3	C_6H_5O	H	C_6H_5	C_6H_5				43		
:tO_2C	H	H	C_6H_5	C_6H_5	75	25				
:tO_2C	CH_3	CH_3	H	H		80 (*anti*) 20 (*syn*)				
:H_3CO	H	H	H	H					49 (*E*+*Z*)	**429**
:H_3CO	H	H	H	CO_2Et					7	**429**
:H_3CO	H	H	C_6H_5	C_6H_5						(51%) **429**
$_3H_7CO$	H	H	H	H					34 (*E*+*Z*)	**429**
$_3H_7CO$	H	H	H	CO_2Et					58	**429**
$_3H_7CO$	H	H	C_6H_5	C_6H_5						(74%) **429**
IeO_2C	H	H	H	H					84 (*E*+*Z*)	**429**
IeO_2C	H	H	H	CO_3Et					40	**429**
IeO_2C	H	H	C_6H_5	C_6H_5	44					(25%) **429**
tO_2C	H	H	H	H					48	**429**

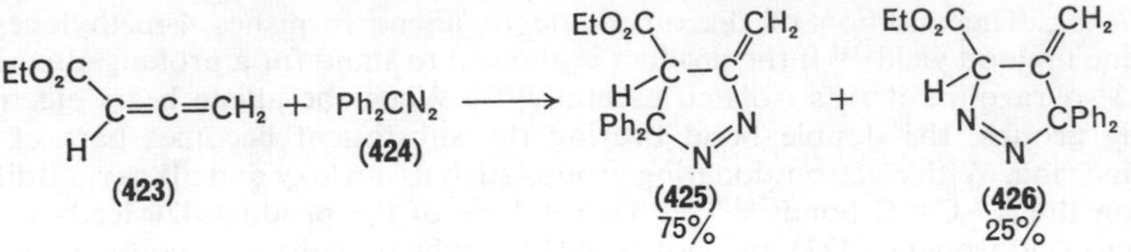

$EtO_2C(H)C{=}C{=}CH_2$ (423) + Ph_2CN_2 (424) ⟶ (425) 75% + (426) 25% (72)[1503]

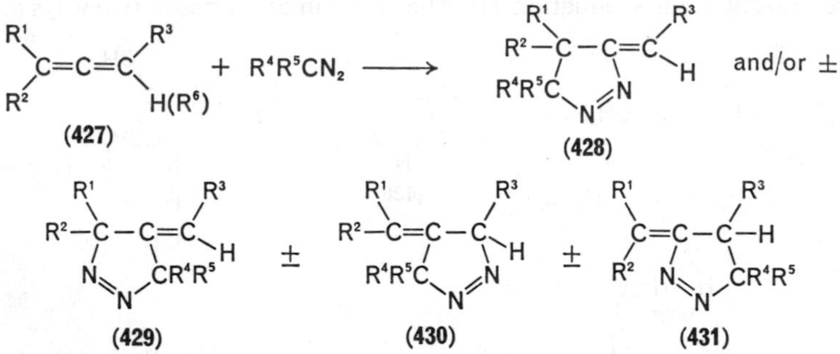

(427) + $R^4R^5CN_2$ ⟶ (428) and/or ± (429) ± (430) ± (431) (73)[1506]

$RCH{=}C{=}CH_2$ (432) + N_2CXY (433) ⟶ (434) + (435) + (436)

(434) ⟶ (437) ⟶ (440); (435) ⟶ (438); (435) ⟶ (439); (436) ⟶ (439) (74)[1503–1505, 1507]

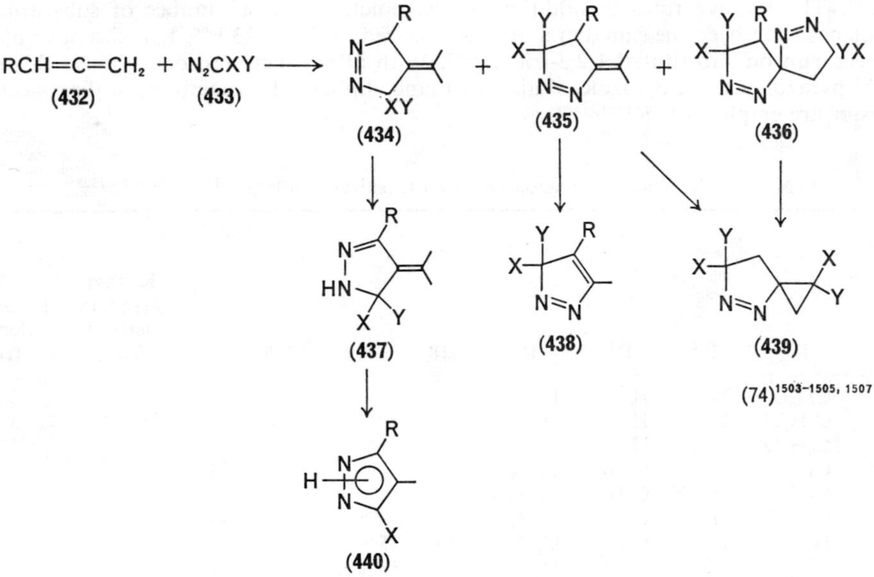

R	Product	X	Y	%
Ac	**438+439**	H	H	49
PrCO	**438+439**	H	H	34
MeOCO	**438+439**	H	H	84
EtOCO	**439**	H	H	48
	Prolonged reaction time			
Ac	**439**	EtO_2C	H	7
PrCO	**439**	EtO_2C	H	58
MeOCO	**439**	EtO_2C	H	40

TABLE 23. Rates of addition of CH_2N_2 to substituted allenes[1506]

Allene 427					
R^6	R^1	R^2	R^3	Relative rate	Absolute rate, k^2 ($l\ mol^{-1}\ s^{-1}$)
H	H	H	H	1	6×10^{-5}
H	CH_3O	H	H	30	$1{\cdot}8 \times 10^{-3}$
H	C_6H_5O	H	H	148	$8{\cdot}9 \times 10^{-3}$
H	EtO_2C	H	H	~1200 (34)	$7{\cdot}7 \times 10^{-2}$
CH_3	EtO_2C	Me	Me	(1)	$2{\cdot}2 \times 10^{-3}$
H	EtO_2C	Me	Me	(~5)	$1{\cdot}0 \times 10^{-2}$
H	EtO_2C	Me	H	(~5)	$1{\cdot}16 \times 10^{-2}$

b. *Ketenimines.* Ketenimines can be prepared from isonitriles and diazoalkanes (*vide supra*, p. 856), and it is clear from the work of Muramatsu and coworkers[2079] that they are reactive towards excess diazoalkane. Diazoalkanes also react with carbodiimides to furnish triazoles[2180].

$$C_6H_5{-}N{=}C{=}N{-}C_6H_5 \xrightarrow{CH_2N_2} \text{(442)} \quad (75)^{2180}$$

(441) **(442)**

c. *Ketens.* Keten is relatively reactive towards diazomethane even at −15 °C. The reaction leads to cyclobutanone via cyclopropanone which is very reactive[1947, 1987, 1988, 2192, 2302]. Turro isolated cyclopropanone by operating at −78 °C in methylene chloride with an appreciable excess of keten[2302]. DeBoer obtained the hydrate by operating in wet ether at 0 °C[2192].

$$H_2C{=}C{=}O + CH_2N_2 \xrightarrow[-78\,°C]{\text{Anhyd. ether}} \text{(444)} \quad \text{(Reference 2302)}$$

$$\xrightarrow[0\,°C]{\text{Wet ether}} \text{(445)} \quad \text{(Reference 2192)} \quad (76)$$

(443) **(444)** **(445)**

Diphenyl diazomethane reacts with diphenylketen in the abnormal sense to furnish a methylene oxadiazole (**447**)[1947]. With diazoacetic ester, an indene and a furanone result[1941]. When dimethylketen and ethyldiazoacetate react, a furanone and an acrylic ester result[1941]. The latter product involves loss of CO and N_2 probably from an intermediate Δ^1-3-pyrazolone which could also furnish the furanone. The reversed sense of addition will not account for either product.

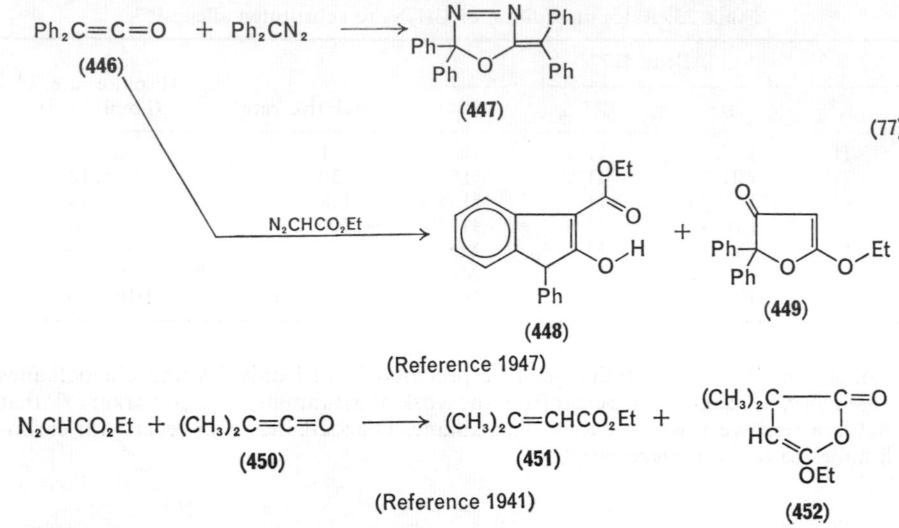

$$Ph_2C{=}C{=}O\ (446) + Ph_2CN_2 \longrightarrow (447)$$

$$(446) \xrightarrow{N_2CHCO_2Et} (448) + (449) \quad (77)$$

(Reference 1947)

$$N_2CHCO_2Et + (CH_3)_2C{=}C{=}O\ (450) \longrightarrow (CH_3)_2C{=}CHCO_2Et\ (451) + (452)$$

(Reference 1941)

Ried[2164, 2165, 2171, 2172] has examined the relative usefulness of the addition of diazoketones to ketens for synthesizing butenolides and found alkyl diazoketones gave better yields than their aromatic counterparts.

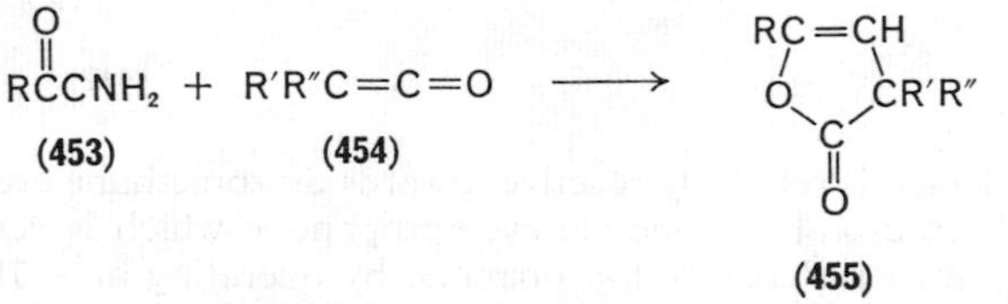

$$RCONH_2\ (453) + R'R''C{=}C{=}O\ (454) \longrightarrow (455) \quad (78)^{2172}$$

Carbon suboxide reacts with diazomethane in a rather messy fashion to furnish a large number of products. In the presence of methanol one obtains amongst other things, dimethyl succinate, dimethyl glutarate and methyl-3-oxocyclobutane carboxylate[2199].

d. *Sulphines and sulphenes.* The reactions of sulphenes with diazoalkanes on cursory examination appears very well studied[1774, 2105]. However, almost all of the examples involve alkyl and aryl diazomethanes, and other substituted diazomethanes would appear to have been neglected. In Section A.2.j we discuss the reaction of diazo compounds with sulphur dioxide. This offers a route to sulphenes known as the Staudinger–Pfenninger method[1775, 2106, 2254]. Sulphenes can also be prepared by the action of bases upon alkylsulphonyl chlorides possessing an α-hydrogen. The normal technique is to employ a tertiary amine such as triethylamine to take up the hydrogen chloride and to operate in the presence of the diazo compound[1534, 1773, 2106, 2164, 2254]. Several products can arise from the reactions. Episulphones, olefins, 1,3,4-thiadiazoline-1,1-dioxides and ketazines are the major products. Thermolysis of the episulphones or the thiadiazoline dioxides furnishes a route to olefins. Similarly, treatment of the episulphones with bases furnishes olefins. Unlike the Staudinger–Pfenninger method, the use of this second method offers a route to unsymmetrical olefins. Representative episulphones available via this second

$$R^1R^2CHSO_2Cl + R_3N \;\; (456)\;(457) \longrightarrow \{R^1R^2C{=}SO_2\}\;(458) \xrightarrow{R^3R^4CN_2} R^1R^2C{-}CR^3R^4 \text{ (episulphone, } SO_2\text{) } (459) \xrightarrow{\Delta \text{ or base}} R^1R^2C{=}CR^3R^4 \;(460)$$

$$458 + R^3R^4CN_2 \longrightarrow R^1R^2C(N{=}N)(SO_2)CR^3R^4 \;(461) \xrightarrow{\Delta} 460 \qquad (79)$$

approach are given in Table 24 and olefins available by either method in Table 25. The tendency to form thiadiazoline dioxides is greatest with alkyl diazomethanes and their pyrolysis furnishes at best only modest yields of olefins[1863] (equation 79, Table 26). Thiocarbonamide-*S*-oxides are α-amino sulphines. Methylation of nitrogen occurs when they are treated with diazomethane[2331] (equation 80).

$$RC({=}SO)NR'H \;(462) \xrightarrow{CH_2N_2} RC({=}SO)NR'CH_3 \;(463) \qquad (80)^{2331}$$

TABLE 24.

$$R^1R^2CHSO_2C + R^3R^4CN_2 \longrightarrow R^1R^2C{-}CR^3R^4 \; (SO_2 \text{ bridge})$$

R^1	R^2	R^3	R^4	Temperature (°C)	Yield (%)	Reference
H	H	H	H	0	64	2106
C_2H_5	H	H	H	10	95	2106
C_6H_5	H	H	H	−20	90	1534
8-Camphoryl	H	β-Pr	H	−10	64	2106
$C_6H_5CH_2$	H	β-Pr	H	−5	75	2106
Cl	H	H	H	−10	83	2106
8-Camphoryl	H	H	H	0	92	2106
$C_6H_5CH_2$	H	H	H	0	99	2106
CH_3	Br	H	H	8	64	1561
C_6H_5	H	CH_3	H	0	70	2149
C_6H_5	H	C_6H_5	H	−2	44 : 56 (*cis* : *trans*)	2149

TABLE 25

$$R^1R^2C\overset{SO_2}{—}CR^3R^4 \longrightarrow R^1R^2C{=}CR^3R^4$$

R^1	R^2	R^3	R^4	Configuration	Yield (%)	Reference
$C_6H_5CH_2$	H	CH_3	H	(*cis* : *trans*) 2 : 1	70	2149
C_2H_5	H	CH_3	H	(*cis* : *trans*) 2 : 3	20	2149
CH_3	H	CH_3	H	(*cis* : *trans*) 48 : 52	11	2149
CH_3	H	CH_3	H	(*cis* : *trans*) 49 : 51	55	2087
C_6H_5	H	H	H		93	2106
$C_6H_5CH_2$	H	H	H		97	2106
8-Camphoryl	H	H	H		71	2106
8-Camphoryl	H	CH_3	H	(*cis* : *trans*) 1 : 9	65	1510
8-Camphoryl	H	β-Pr	H		54	1510
$C_6H_5CH_2$	H	CH_3	H	(*cis* : *trans*) 2 : 1	70	1774
C_6H_5	H	C_6H_5	H	(*cis* : *trans*) 45 : 55	49	2294
p-BrC_6H_4	C_2H_5	*p*-BrC_6H_4	C_2H_5	(*cis* : *trans*) 1 : 4	80	2312
C_6H_5	C_6H_5	C_6H_5	C_6H_5		48	2026

TABLE 26[1863, 1907]

$$R^1R^2C\overset{SO_2}{\diagup\diagdown}CR^1R^2 \ (\text{N=N}) \ \textbf{(461)} \longrightarrow R^1R^2C{=}CR^1R^2 \ \textbf{(460)}$$

R^1	R^2	Yield **461** (%)	Yield **460** from **461** (%)
C_2H_5	4-Oxocyclohexyl	19·5	26·3
C_2H_5	Cyclohexyl	36	9·6
C_2H_5	4-Cyclohexenyl	12·3	25·8
CH_3	4-Oxo-1-methylcyclohexyl	63	10·7
CH_3	1-Methyl-3-cyclohexenyl	17·2	2·1

Zwanenburg and collaborators have investigated the reactions of 2-diazopropane and diazomethane with arylsulphonyl sulphines and found that they furnish a regio-selective and stereospecific route to Δ^3-1,3,4-thiadiazolene-1-oxides. The products from diazomethane undergo rearrangement to furnish 2-sulphonyl-1,3,4-thiadiazoles[1533, 2290, 2380, 2381].

$$R^1(SO_2R^2)C{=}S{=}O \ \textbf{(464)} + CH_2N_2 \longrightarrow \textbf{(465)} \longrightarrow \textbf{(466)} \quad (81)$$

$R^1 = R^2 = Ph$; $R^1 = p\text{-}MeOC_6H_4$, $R^2 = p\text{-}MeC_6H_4$

e. *Isocyanates*. Isocyanates react with diazoalkanes[2228–2230]. Several sulphonyl isocyanates have been treated with hexafluoro-2-diazopropane and 1,2,3-oxathiazole-4-one-2-oxides have resulted.

$$RSO_2NCO + (CF_3)_2CN_2 \longrightarrow$$ (469) — (82)[2228-2230]

(467) (468)

R = F, Cl, CF_3

f. *Nitrile oxides.* Benzonitrile oxide reacts with two equivalents of diazomethane to furnish 1-nitroso-3-phenyl-Δ^2-pyrazoline as the major product[1989]. This represented the first example of such a process.

$$PhC{\equiv}N{-}O + CH_2N_2 \longrightarrow$$ (470) — (83)[1989]

(469)

g. *Isothiocyanates.* The reaction of diazomethane with phenyl isothiocyanate to furnish 5-anilino-1,2,3-thiadiazole was discovered by von Pechmann in 1895[2324, 2326] and later reinvestigated by Sheehan[2228]. More recently Martin and Mucke have extended the reaction[2022a,b] where others had previously failed[1981, 2292]. With alkyl isothiocyanates the resulting initial product is an alkyl aminothiodiazole and these can react further to furnish the related thiourea derivatives[2022a,b]. Diazomethyl ketones react with isothiocyanato formates and sulphonyl isothiocyanates to furnish similar systems[1814, 2169].

$$RNCS + R'CNH_2 \longrightarrow$$ (473) — (84)[2022a, b]

(471) (472)

R	R′	Yield (%)
o-ClC_6H_4	H	47·8
C_6H_5CO	H	59·6
C_2H_5OCO	H	59·4
$C_6H_5O_2C$	H	27·6
C_6H_5	C_2H_5OCO	3·0
α-$C_{10}H_7$	C_2H_5OCO	3·0
o-ClC_6H_4	C_2H_5OCO	10·4
m-ClC_6H_4	C_2H_5OCO	23·3
p-ClC_6H_4	C_2H_5OCO	25·3
p-$NO_2C_6H_4$	C_2H_5OCO	48·5
C_6H_5CO	C_2H_5OCO	62·8
C_2H_5OCO	C_2H_5OCO	82·4
$C_6H_5O_2C$	C_2H_5OCO	34·6
$CH_2{=}CHCH_2$	C_2H_5OCO	25·5

h. *Isoselenocyanates*. Aryl isoselenocyanates react with diazomethane in a manner completely analogous to arylisothiocyanates[2270].

i. *Azasulphines*. Musser[2081, 2082] has examined the reaction of Ar—N=S=O systems with diazoalkanes and found a variety of products, Table 18. The authors[2081, 2082] rationalized their observations on the basis that a carbene intermediate was involved. The argument was that although the reaction can proceed in the dark with several diazoalkanes, with diphenyldiazomethane the reaction is so slow that photolysis was required. However, no carbene-trapping experiments were performed and all of the products can be reasonably rationalized as resulting from a series of cycloaddition reactions (equation 86).

$$\underset{(474)}{R{-}N{=}S{=}O} + \underset{(475)}{Ph_2CN_2} \xrightarrow[\text{hexane}]{h\nu} N_2 + \underset{(476)}{\text{[R–N–S(→O)–C(Ph)(Ph) ring]}} \longrightarrow \underset{(477)}{R{-}N{=}CPh_2} + \underset{(478)}{SO} \quad (85)$$

$$R = C_6H_5,\ p\text{-}MeOC_6H_4,\ p\text{-}NO_2C_6H_4,\ C_6H_{11}$$

$$R'NSO + R_2CN_2 \longrightarrow \underset{(479)}{\text{[R'–N–SO–CR}_2\text{–N=N ring]}} + \underset{(480)}{\text{[R'–N–SO–N=N–CR}_2\text{ ring]}} \longrightarrow \underset{(482)}{R'N{=}CR_2} + N_2 + SO \quad (86)$$

$$R'N_3 + R_2CSO \longrightarrow \underset{(483)}{\text{[R}_2\text{C–SO–N=N–CR}_2\text{ ring]}} \longrightarrow \underset{(484)}{R_2C{=}CR_2} + N_2 + SO$$

$$SO + R_2CN_2 \longrightarrow \underset{(485)}{\text{[S–O–N=N–CR}_2\text{ ring]}} + \underset{(486)}{\text{[O–S–N=N–CR}_2\text{ ring]}}$$

$$(485) \longrightarrow \underset{(487)}{R_2C{=}S} + N_2O \qquad (486) \longrightarrow \underset{(488)}{R_2C{=}O} + N_2 + S$$

The scheme suggested involves two modes of cycloaddition of the diazo compound.

The collapsings of the intermediates **479** and **480** are symmetry allowed. The SO produced will be in the singlet state and, like singlet oxygen (Section I.A.h), could add to the diazo compound in two senses which on collapse (symmetry allowed) will furnish the ketone and the thioketone. Conversion of the intermediate **479** to the

TABLE 27. Reaction of sulphines with 0·1 mol diphenyl diazomethane

N-Sulphinyl compound	Yield of **477** (%)	Yield of $Ph_2C{=}O$ (%)
N-Sulphinylaniline, 0·1 mol	36·9	23·3
N-Sulphinylaniline, 0·05 mol	54·5	19·7
N-Sulphinylaniline, 0·025 mol	86·8	*a*
N-Sulphinyl-*p*-anisidine, 0·1 mol	43·7	17·9
N-Sulphinyl-*p*-nitroaniline, 0·1 mol	25·1	*a*
N-Sulphinylcyclohexylamine, 0·09 mol	47·1	*a*

[a] Not determined.

sulphine **481** will lead to the formation of the intermediate **483** which can collapse (symmetry allowed) to olefin. The azide can decompose and react with diazo compound to furnish imine **477**. The method of analysis employed would destroy any azide formed and fail to detect any N_2O. Theoretical calculation by Halevi and collaborators[1838] and LeRoy[1979] suggest that a photoexcited diazoalkane would be a hotter 1,3-dipole than the ground state. This contention is further bolstered by recent kinetic evidence and was proposed previously to account for the photochemical behaviour of dimethyl diazomalonate[1759, 2360, 2362].

j. *Sulphur dioxide*. Sulphur dioxide reacts with diazoalkanes to furnish olefins, azines, episulphones and thiadiazoline 1,1-dioxides (Section I.2.A.c) in what is known as the Staudinger–Pfenninger method[1774, 2106, 2254]. Representative examples of results employing the method are given in Table 28. If the diazoalkane is in the

TABLE 28

$$R^1R^2CN_2 + SO_2 \longrightarrow R^1R^2C\overset{SO_2}{\diagup\diagdown}CR^1R^2$$

R^1	R^2	Yield (%)	Solvent	Reference
H	C_6H_5	SO_2 (*l*), 20 °C; 23 (*cis*)	Ether	2312
H	C_6H_5	SO_2 (*l*), 20 °C; 60 (*cis*)	Hexane	2312
H	C_6H_5	SO_2 (aq), 0 °C; 55 (*cis*)	H_2O/ether	2312
H	C_6H_5	SO_2 (*l*), 60 °C; 6 (45/55; *c*/*t*)	Pentane	1534
C_6H_5	C_6H_5	48	CS_2	2026
C_2H_5	C_6H_5	SO_2 (g), 0 °C; 33 (*trans*)	Ligroin	2312
C_2H_5	*p*-BrC_6H_4	SO_2 (g), 0 °C; 80 (20/80; *c*/*t*)	Ligroin	2312
C_2H_5	*p*-$MeOC_6H_4$	SO_2 (g), 0 °C; 25 (*trans*)	Ligroin	2312

presence of excess sulphur dioxide it can be converted into the related carbonyl compound. Presumably this involves interception of the intermediate sulphene by SO_2 [1774]. If traces of water are present, sulphite esters can arise[1864] and if alcohols are present, mixed sulphite esters are generated[1865]. Similarly, the presence of amines leads to sulphonamides. With diazoacetic ester and aniline, followed by heating at 100 °C, ethyl *N*-phenyl-glycinate results[1865].

On occasion pyrolysis of the episulphones does not lead to olefins[1955] (equation 87).

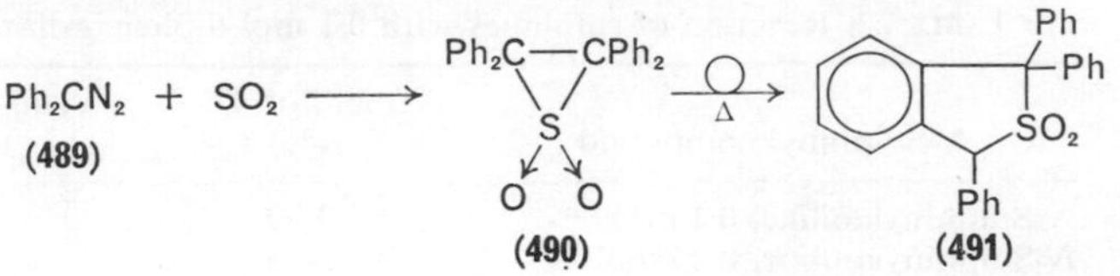

Ph_2CN_2 + SO_2 ⟶ $Ph_2C{-}CPh_2$ (episulfone, S with O O) ⟶(Δ) (benzo-fused cyclic sulfone, Ph, Ph, SO_2, Ph)
(489) (490) (491)

(References 1955a and 2254)

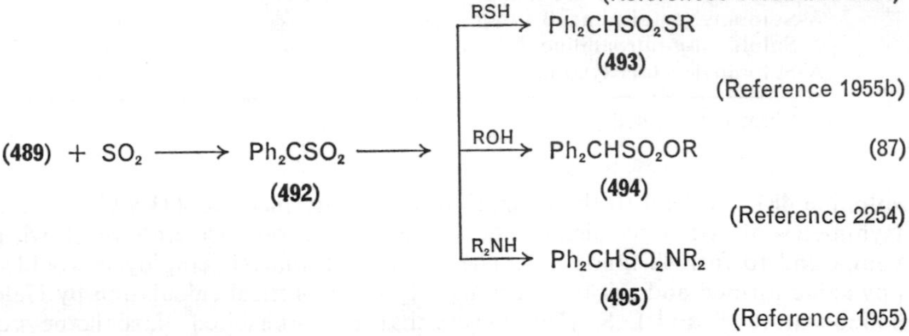

(489) + SO_2 ⟶ Ph_2CSO_2 (492) ⟶
- RSH ⟶ Ph_2CHSO_2SR (493) (Reference 1955b)
- ROH ⟶ Ph_2CHSO_2OR (494) (Reference 2254)
- R_2NH ⟶ $Ph_2CHSO_2NR_2$ (495) (Reference 1955)

(87)

3. Addition to X≡Y systems

a. *Acetylenes*

i. Simple acetylenes. The addition of diazoalkanes to acetylenes has been reviewed[1497]. Thc products are pyrazoles and 3*H*-pyrazoles (pyrazolenines). The latter products only arise if the diazoalkane is disubstituted. With substituted acetylenes lacking a C_{2v} symmetry axis, two possible modes of addition will exist. With terminal acetylenes the tendency is to attack the terminal carbon with the diazo-carbon and extend the carbon chain[1497, 1518, 1859, 1897, 2158–2160, 2162] (equation 88).

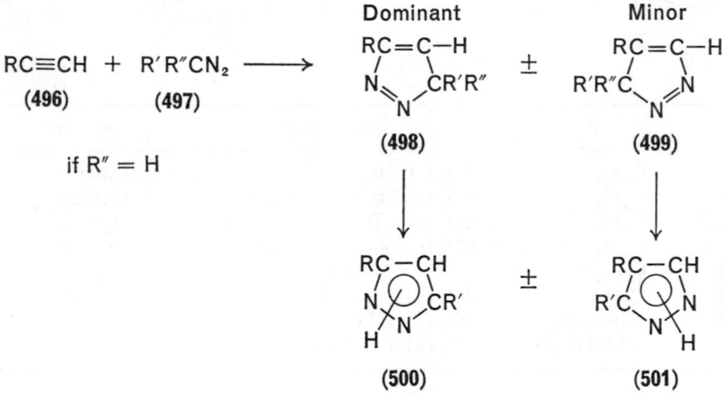

RC≡CH + R′R″CN_2 ⟶ Dominant: RC=C–H / N=N–CR′R″ ring (498) ± Minor: RC=C–H / R′R″C–N=N ring (499)
(496) (497)

if R″ = H

(498) ⟶ (500) RC–CH / N(H)–N=CR′ (pyrazole) ± (499) ⟶ (501) RC–CH / R′C–N(H)–N (pyrazole)

(88)

The only apparent exception to this rule is the reaction of diazomethane with phenylacetylene where the ratio is 10 : 1 in favour of the ‘normal’ orientation[1950]. With functionalized acetylenes the reactions are not as clear. Hence with a series of ethynyl carbinols both orientations occur even with diazomethane[1916, 2061, 2110]. With a series of ethynyl carbinyl ethers similar behaviour is observed[1494, 1517]. Surprisingly, the additions of ethynyl carbinols by substituted diazoalkanes occurs in a single sense[1517, 1518, 1630, 1652, 2061, 2167].

$$HC\equiv CC(OH)R^1R^2 + CH_2N_2 \longrightarrow (503) + (504)$$ (89)[1921, 2061, 218]

(502) → (503) 4-($HOCR^1R^2$)pyrazole; (504) 3(5)-($HOCR^1R^2$)pyrazole

R^1 =	H	H	H	Me	$(CH_2)_4$	$(CH_2)_5$
R^2 =	H	Me	Ph	Me		
503 (%)	2	20	60	35	46	46
504 (%)	53	80	40	65	54	54

$$RC\equiv CH + CH_2N_2 \longrightarrow (506) + (507)$$ (90)[1494, 2185]

(505)

R =	CH_2O-(2-tetrahydropyranyl)	$CH(Me)O$-(2-tetrahydropyranyl)	$C(Me)_2O$-(2-tetrahydropyranyl)	CH_2OMe	$CH(Me)OMe$
506 (%)	67	79	80	67	70
507 (%)	33	21	20	33	30

R =	$C(Me)_2OMe$	$C(CH_2)_5OMe$	CHC_6H_4OMe
506 (%)	68	50	33
507 (%)	32	50	67

$$HC\equiv CCHOHR' + RCHN_2 \longrightarrow (510)$$ (91)[2061, 2167]

(508) (509) → (510) 3-R-5-(CHOHR')pyrazole

R′	R	**510** (%)
H	EtOCO	
H	Ph—CO	
Me	p-$O_2NC_6H_4CO$	37
Me	p-$MeOC_6H_4CO$	36
Et	p-$O_2NC_6H_4CO$	23
Et	p-ClC_6H_4CO	27
Pr	p-$O_2NC_6H_4CO$	16·5
Pr	Naphthoyl	20

$$HC\equiv CCOHR^1R^2 + RR'CN_2 \longrightarrow (513)$$ (92)[1517, 1518, 1630, 1652]

(511) (512) → (513) 3,3-RR′-4-($COHR^1R^2$)-3H-pyrazole

R^1	R^2	R	R′	**513** (%)
Ph	H	Ph	Ph	30
Me	Me	Me	Me	20

With acetylenic esters the additions go well and the orientation is normally controlled by electronic effects[1551, 1552, 1768, 1779, 1839, 1854, 1981, 2321].

However, 2-diazopropane gives predominantly reverse orientation with methyl buty-2-ynoate (70 : 12)[1631]. With aryl propiolates and diazomethane, both senses of addition are observed[1496, 1524, 1965, 2288, 2321]. With alkynyl ketones and alkynyl

$$R^1C{\equiv}CCO_2R^2 \; (514) + RCHN_2 \longrightarrow (515) \qquad (93)^{1554, 1854, 1901, 2273a}$$

R^1	R^2	R	515 (%)
H	R	H	82
H	Et	$CH_2{=}CH$	—
H	Et	$EtOCO(CH_2)_n$ ($n = 1, 2, 3, 4$)	70
Me	Me	H	—
Me	Me	EtOCO	—
MeOCO	Me	$CH_2{=}CH$	81
MeOCO	Me	EtOCO	—
MeOCO	Me	$EtOCO(CH_2)_n$ ($n = 1, 2, 3, 4$)	—

$$R^1C{\equiv}CCO_2R^2 \; (514) + RR'CN_2 \longrightarrow (515) \qquad (94)^{1652, 1901}$$

R^1	R^2	R	R′	515 (%)
H	Et	Me	Me	80
H	Me	Ph	Ph	100
MeOCO	Me	Ph	Ph	100
H	Et	p-BrC_6H_4	Ph	—
H	Et	p-$MeOC_6H_4$	Ph	—
MeOCO	Me	Me	Me	65
MeOCO	Me	9-Fluorenyl		100

$$RC{\equiv}CCOOMe + Me_2CN_2 \longrightarrow (516) + (517) \qquad (95)^{1552, 1782}$$

	(516)	(517)
R = Me	12%	70%
R = Ph	30%	70%

aldehydes the problem of regioselectivity also occurs and can present problems even with diazomethane. Some results are summarized in Table 29[1495, 1496, 1520, 1631, 1652, 2016]. With aldehydes and their acetals, regioselectively is strongly influenced by steric factors. The reactions with diazoalkanes are difficult and furnish poor yields in much the same manner as simple acetylenes[1652, 1854, 1862, 1899–1901, 1935, 2016].

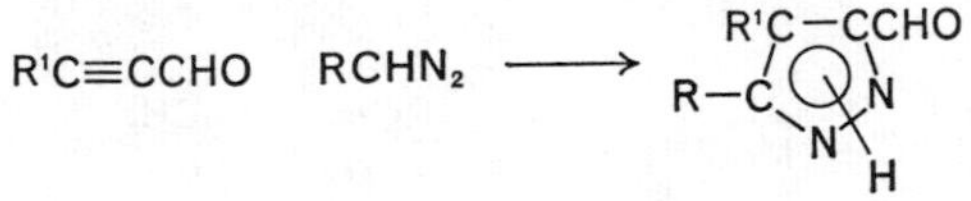

$$R^1C{\equiv}CCHO \quad RCHN_2 \longrightarrow \text{(pyrazole: } R^1C{-}CCHO\text{, } R{-}C\text{, N, N–H)} \qquad (96)^{1859, 1899, 2016, 2016a}$$

R^1	R	518 (%)
H	Allyl	02
C_6H_5	Allyl	7
CH_3	Allyl	20

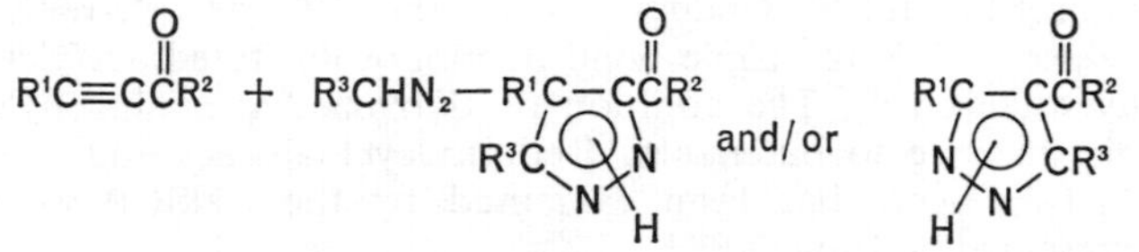

$$R^1C{\equiv}C\overset{O}{\overset{\|}{C}}R^2 + R^3CHN_2 \longrightarrow \text{(519)} \text{ and/or } \text{(520)} \qquad (97)$$

(519): $R^1C{-}C(=O)R^2$, R^3C, N, N–H; (520): $R^1C{-}C(=O)R^2$, N, CR^3, H–N

TABLE 29. Addition of diazoalkanes to acetylenic ketones[1494, 1495, 2016a]

R^1	R^2	R^3	519 (%)	520 (%)
H	CH_3	H	93	
H	CH_3	$CH_2{=}CH$	94	
H	C_6H_5	H	95	
H	C_6H_5	H	85	
H	p-$MeOC_6H_4$	H	90	
H	$C{\equiv}CH$	H	57	
C_6H_5	CH_3	H	65	35
C_6H_5	C_6H_5	H	100	
$(CH_3)_3C_6H_2$	C_6H_5	H	50	
C_6H_5CO	C_6H_5	H	60	
CO_2CH_3	C_6H_5	H	95	

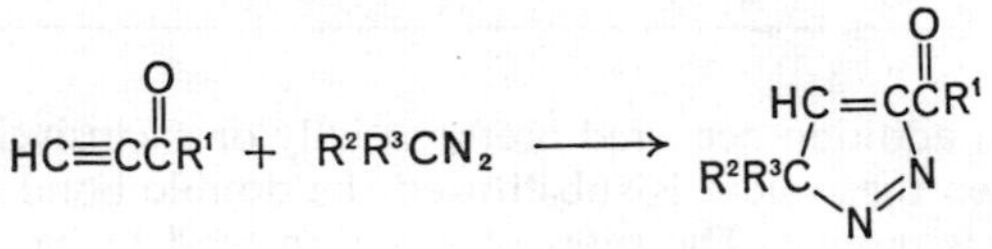

$$HC{\equiv}C\overset{O}{\overset{\|}{C}}R^1 + R^2R^3CN_2 \longrightarrow \text{(521)} \qquad (98)$$

(521): $HC{=}C(=O)R^1$, R^2R^3C, N=N

(References 1517, 1518, 1652)

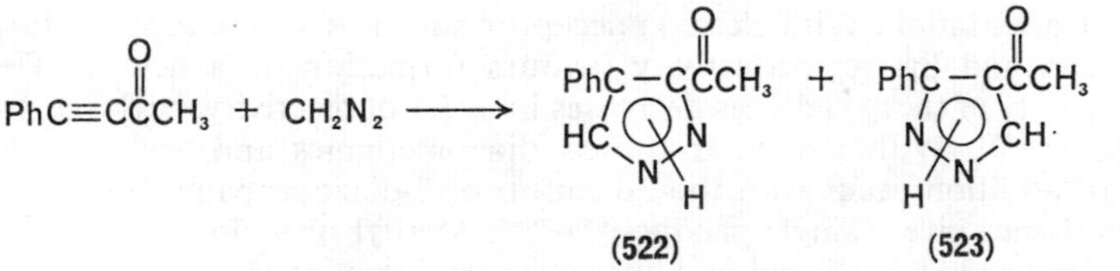

$$PhC{\equiv}C\overset{O}{\overset{\|}{C}}CH_3 + CH_2N_2 \longrightarrow \text{(522)} + \text{(523)}$$

(522): $PhC{-}C(=O)CH_3$, HC, N, N–H — 65%

(523): $PhC{-}C(=O)CH_3$, N, CH, H–N — 35%

(Reference 1496)

$$CH_3C{\equiv}CC(=O)CH_2R + Me_2CN_2 \longrightarrow \textbf{(524)} + \textbf{(525)} \qquad (99)^{1524, 1631, 1652}$$

R = H, Me

ii. Enynes. The question of the relative reactivity of enynes toward diazoalkanes has been examined by the Vo-Quangs[2319] and others[1518, 2093, 2298]. With the exception of vinylacetylene[2093, 2298], the triple bond in monosubstituted acetylenes (1-ynes) is preferentially attacked[1525]. The addition of diazomethane, diphenyldiazomethane and ethyl diazoacetate to butenynes, 1-ethynylcycloalkenes and *cis*- and *trans*-1-methoxybuta-1-en-3-ynes has been examined by the ENSCP workers and the results are presented in Table 30[2329].

TABLE 30. Addition of diazoalkanes to conjugated enynes[1518, 2094, 2325, 2328]

$$R^1CH{=}CR^2C{\equiv}CH + R^3CHN_2 \longrightarrow R^1CH{=}CR^2{-}C{-}CH \text{ (pyrazole ring with N, NH, CR}^3)$$

R^1	R^2	R^3	Yield (%)
H	CH_3	C_6H_5	37
C_6H_5	H	C_6H_5	65
$-(CH_2)_4-$		C_6H_5	65
$HOCH_2$	H	C_6H_5	76
HOCO	H	C_6H_5	65
MeO(*Z*)	H	Me	50
MeO(*E*)	H	Me	50
MeO(*E*)	H	C_6H_5	50

They found that addition occurred preferentially or exclusively to the triple bond of alkenynes. When the system is substituted the double bond is attacked, and the sense of addition is unique. The rates of reaction tend to be slow[1518, 1566, 1567, 2094, 2161, 2298, 2308, 2328].

iii. Diynes. A variety of diynes have been studied[1518, 2163, 2262–2264] and are summarized in equations (100), (101) and Table 31. Not surprisingly, one encounters both senses of orientation. With electron-acceptor substituents the α,β-unsaturated bond is attacked and the regioselectivity favours formation of a new β—C—C bond. The reactivity of the acyl diynes decreases in order of reactivity, acyl > carboalkoxy > benzoyl[2253, 2254]. With mono-substituted diazomethanes and arylpropiolates, both senses of addition occur with the exception of 2-diazopropane while di-substituted diazomethanes give a single product[1652, 1854]. Methyl phenylpropiolate and 2-diazopropane furnish 30% normal and 70% abnormal products[1551, 1782].

Numerous examples exist of the addition of diazoalkanes to acetylenic ketones[1497]. The nuances are similar to those treated above for the acetylenic esters.

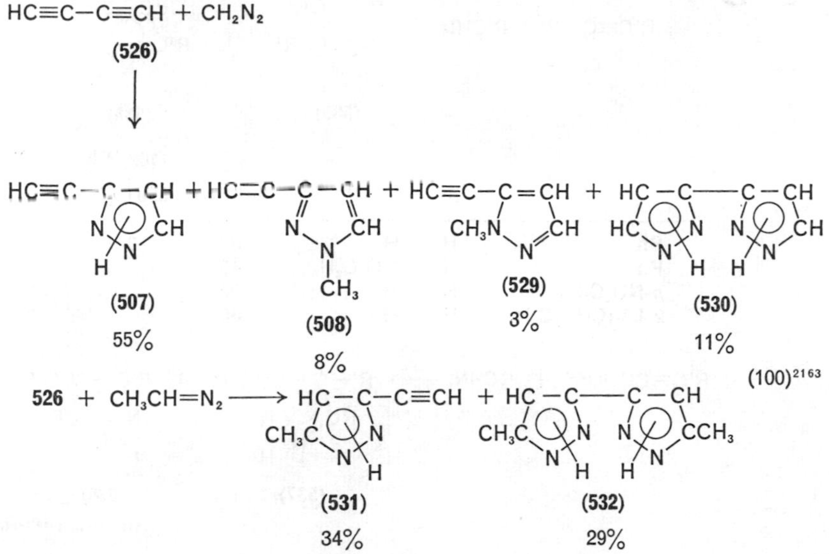

$$HC\equiv C-C\equiv CH + CH_2N_2 \; (\mathbf{526}) \longrightarrow \mathbf{507}\ (55\%) + \mathbf{508}\ (8\%) + \mathbf{529}\ (3\%) + \mathbf{530}\ (11\%)$$

(507) 55%, (508) 8%, (529) 3%, (530) 11% (100)[2163]

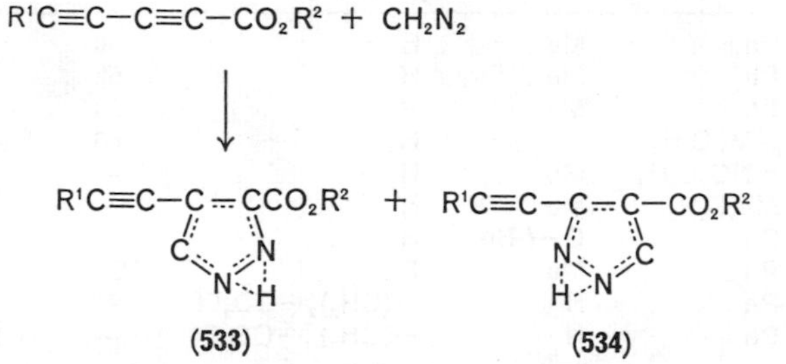

$$\mathbf{526} + CH_3CH{=}N_2 \longrightarrow \mathbf{531}\ (34\%) + \mathbf{532}\ (29\%)$$

(531) 34%, (532) 29%

$$R^1C\equiv C-C\equiv C-CO_2R^2 + CH_2N_2 \longrightarrow \mathbf{533} + \mathbf{534}$$

(533), (534) (101)[2263]

TABLE 31. Addition of diazomethane to some diyne ketones and carboxylic esters[2263]

R^1	R^2	Yield of **533** (%)	Yield of **534** (%)
p-$MeOC_6H_4$	Me	85	
p-MeC_6H_4	Me	80	5
C_6H_5	Me	78	7
p-ClC_6H_4	Me	91	
p-BrC_6H_4	Me	95	
p-$NO_2C_6H_4$	Me	100	
C_6H_5	OMe	72	6
C_6H_5	OC_6H_5	73	7
C_6H_5	p-$MeOC_6H_4$	63	5

iv. Hetero-substituted acetylenes. Diazomethane does not add to hetero-substituted acetylenes of the type $Ph_2M-C\equiv C-$ where M = P, P(O), As, As(O)[1829, 1980] but does react with silicon and tin derivatives[2246a, b] as well as sulphones[1829, 2313], sulphoxides[1829, 2313] and M = $(EtO)_2P(O)$[2146].

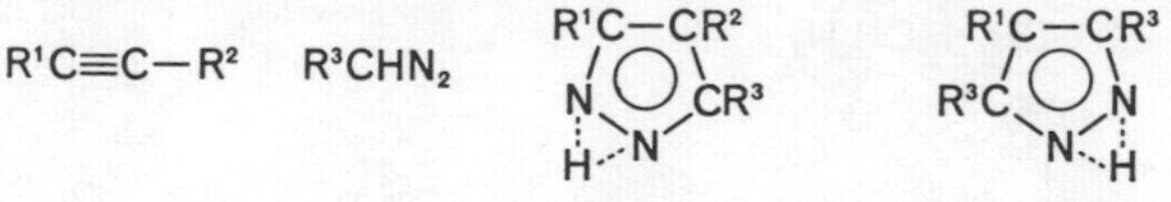

$R^1C{\equiv}C{-}R^2$ R^3CHN_2

(535) (536)

(102)[1496, 1946, 2016a, 2164a, 2288]

R^1	R^2	R^3	535 (%)	536 (%)
Ph	H	H	30	3
Ph	H	$CO_2C_2H_5$	45	
p-$NO_2C_6H_4$	H	H	93	
2,4,6-$(CH_3)_3C_6H_2$	H	H	30	

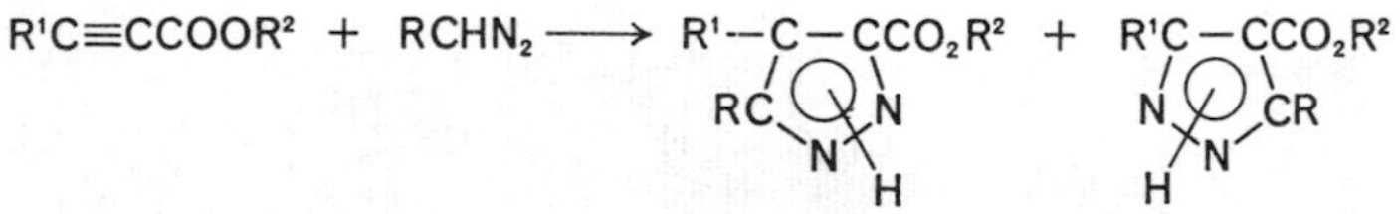

$R^1C{\equiv}CCOOR^2 + RCHN_2 \longrightarrow$ (537) + (538)

(103)[1551, 1552, 1768 1854, 2016, 2321]

R^1	R^2	R	537 (%)	538 (%)
Ph	Me	H	50	50
Ph	Me	H	52	48
Ph	Me	H	23	65
p-MeC_6H_4	Me	H	46	44
p-$NO_2C_6H_4$	Me	H	—	70
$Me_3C_6H_2$	Me	H	—	70
Ph	O—t-Bu	H	—	30
Ph	Me	Ph	25–30	70–75
Ph	H	$-(CH_2)_2-CO_2Et$	24	18
Ph	H	$-(CH_2)_n-CO_2Et$ $n = 3, 4$	—	27–30

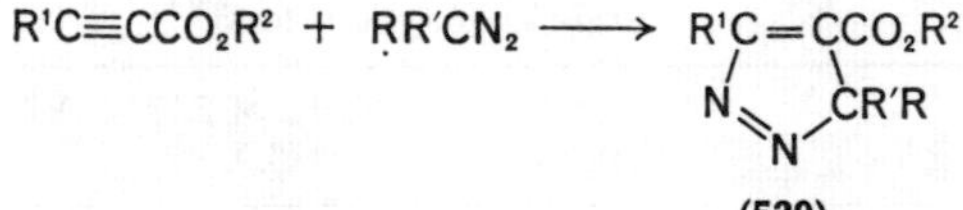

$R^1C{\equiv}CCO_2R^2 + RR'CN_2 \longrightarrow$ (539) (104)[1401a, 1631, 1732]

R^1	R^2	R	R'	539 (%)
Ph	Et	Ph	Ph	—
Ph	Et	Ph	Me	48
Ph	Et	p-BrC_6H_4	Ph	—
Ph	Et	p-$Me_2NC_6H_4$	Ph	—
p-MeC_6H_4	Me	Ph	Ph	24

b. *Nitriles.* Except under special conditions, the cyano group is inert to attack by diazo compounds. Photolytic and copper-catalysed decomposition of diazo carbonyl compounds leads to the generation of 'ketocarbenes' which can add to nitriles. This is discussed below. With highly acidic nitriles such as tricyanomethane and acetylmalononitrile, attack by diazomethane leads to methylation of nitrogen and in the latter, oxygen as well[1461, 1463]. A similar phenomenon occurs with carboalkoxy cyano sulphones but some C-alkylation also occurs[1638]. When the cyano group is activated by a strong electron-withdrawing group attached to the cyano carbon, cycloadditions occur to form 1,2,3-triazoles[2023, 2104, 2134, 2277] (equation 107).

$$HC(CN)_3 + CH_2N_2 \longrightarrow \underset{(540)}{(NC)_2C{=}C{=}N{-}CH_3} \qquad (105)^{1461,\,1463}$$

$$\underset{(541)}{CH_3C(=O)CH(CN)_2} + CH_2N_2 \longrightarrow \underset{(542)}{(NC)_2C{=}C(OCH_3)CH_3} + \underset{(543)}{CH_3N{=}C{=}C(COCH_3)CN}$$

$$RCHN_2 + R'CN \longrightarrow \underset{(544)}{\text{RC}{-}\text{CR}' \text{ (1,2,3-triazole, N-H)}} \qquad (106)^{2023,\,2104,\,2134,\,2277}$$

R = H, COOEt R′ = CN, Cl, Br, $-COCO_2CH_3$, $C_6H_5OC(=O)-$

$$2\,Ph_2C{=}N_2 + \underset{(545)}{MeOCOCN} \longrightarrow \underset{(546)}{Ph_2C{=}C(CO_2CH_3){-}N{=}CPh_2} \qquad \text{(Reference 1519)}$$

$$Ph_2CN_2 + \underset{(547)}{Cl_3CCN} \longrightarrow \underset{(548)}{Ph_2C(Cl){-}CCl_2CN} \qquad \text{(Reference 1519)}$$

$$\longrightarrow \underset{(549)}{\text{4,5-bis}(CCl_3)\text{-2,2-diphenyl-2H-imidazole}} \qquad (107)^{1929}$$

$$\text{fluorenylidene}{=}N_2 + \underset{(550)}{p\text{-}ClC_6H_4N{=}N{-}CN} \rightarrow\rightarrow \underset{(551)}{\text{fluorenylidene}{=}\overset{+}{N}(C{\equiv}N){-}\overline{N}{-}C_6H_4Cl\text{-}p} \qquad \text{(Reference 1886)}$$

The reactions of methyl cyanoformate[1519] trichloroacetonitrile[1519], benzoyl cyanide[1519] and *p*-chlorophenyldiazocyanide[1886, 1887] are exceptions[1894] and their reactions are summarized in equation (107).

Hoberg[1872] discovered that it was possible to activate C=N and C≡N bonds by employing organoaluminium compounds. His results are summarized in Table 32 and equation (108).

$$C_6H_5CN + CH_2N_2 \xrightarrow{R''R'RAl} \underset{\textbf{(552)}}{C_6H_5C{=}N\text{–}NH\text{–}N{=}CH\ (\text{ring})} \quad \underset{\textbf{(553)}}{C_6H_5C{=}N\text{–}NCH_3\text{–}N{=}CH\ (\text{ring})} \quad \underset{\textbf{(554)}}{C_6H_5C{=}N\text{–}N{=}\ldots N(CH_3)\text{–}CH\ (\text{ring})} \qquad (108)$$

TABLE 32. Catalysed addition of diazomethane to benzonitrile[1872]

Catalyst	Ratio (mmol) Catalyst/CH_2N_2/PhCN	Time (h)	Temperature (°C)	Triazole (%) 552	553	554
Et_3Al	100/100/200	2	0[a]	0	0	0
Et_3Al	100/100/200	80	−78	4·9	30	10
Et_2AlCl	100/100/200	0·5	0	30	23	7
Et_2AlCl	100/100/200	10	−78	28	12·6	15·7
Et_2AlI	100/100/200	12	−78	6	32	7
Et_2AlCH_2I	100/100/200	5	−78	9	33	12
Et_2AlOEt	100/100/200	150	−78	0	0	0
Et_2AlCH_2I	100/100/100	5	−78	9·5	42·7	2·5
Et_2AlCH_2I	100/100/100	5	−78	8·2	31·4	145
Et_2AlCH_2I	100/100/100	50	−78	39·4	7·4	9·9
$AlCl_3$	100/100/200	0·5	0	22	6·3	3·8
$AlCl_3$	100/100/200	5	−78	28	9	6·3
Et_2AlCH_2I	50/200/200	20	−78	4·9	19	7

[a] Polymethylene accounts for most if not all CH_2N_2 not furnishing triazoles.

B. Carbonyl Ylid Cycloadditions

If a photoexcited ketone (singlet state) were to interact with a diazoalkane, a carbene were to interact with a carbonyl oxygen, a 1,3,4-oxadiazole were to lose nitrogen or a photoexcited diazo compound interact with a carbonyl group, carbonyl ylids might result (equation 109).

Until fairly recently the existence of carbonyl ylids has been inferred[591, 2233], but recently Hamaguchi and Ibata have isolated stable examples[1840–1843, 1904–1906]. There is, however, good reason to suspect their intermediacy and biological importance. A large number of papers have appeared on the biological activity of diazo dipeptides and related compounds[1471, 1472, 1484, 1525, 1540–1542, 1558, 1806–1814, 1852–1854, 1909, 2050–2053, 2080, 2100, 2101, 2120, 2149–2151, 2275, 2300, 2301, 2363]. In all cases it seems highly probable that the activity is either due to the presence of carbonyl ylids or their isomerization to other heterocycles. Under the biological conditions employed, bimolecular carbene or carbonium ion processes seem highly unlikely (equation 109). The enzymic cleavage of diazoacetyl units to furnish diazomethane may possibly occur but this is inconsistent with the observance of copper catalysis being effective prior to treatment of biological systems. Shermer[2234] has prepared **570** and **571** by the schemes shown but, as he noted, this did not establish the existence of the carbonyl ylide intermediates because reasonable alternatives exist. Hamaguchi and Ibata[1840–1843, 1903–1906]

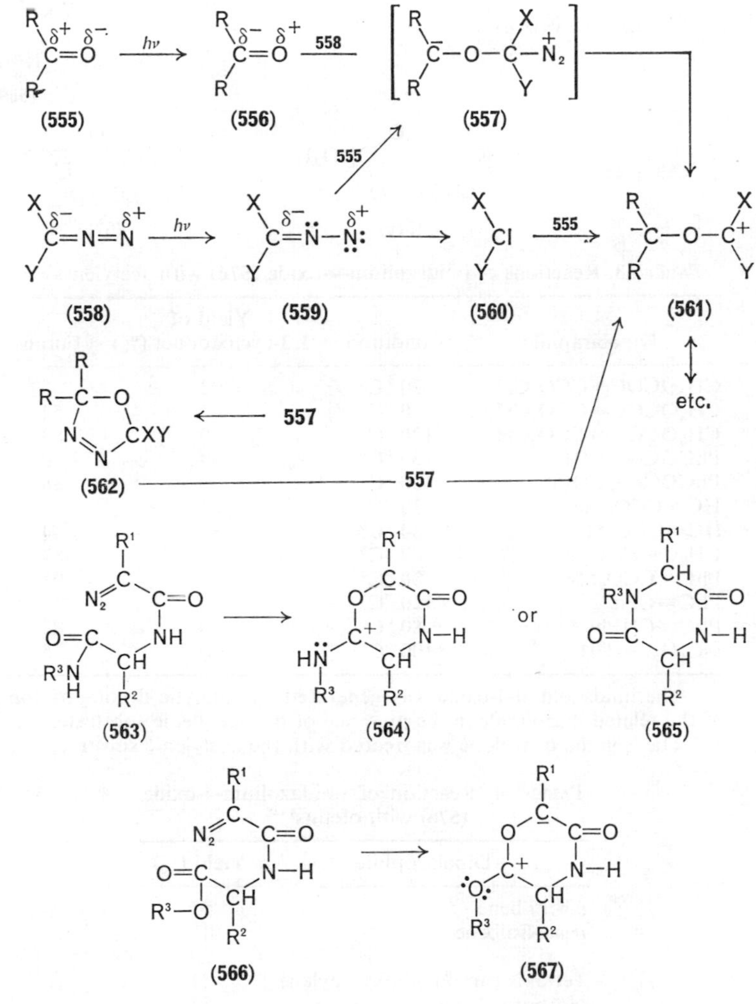

(555) (556) (557) (558) (559) (560) (561) (562) (109)

(563) (564) or (565)

(566) (567)

chose to trap their carbonyl ylids by forming them intramolecularly. This approach proved successful and they have been able to isolate the products which they call iso-münchons and employ them in cycloaddition reactions. They also ran similar reactions on 8-diazoacetyl methyl-1-naphthoate and *o*-carbomethoxyphenyldiazomethanes. These compounds also decompose to carbonyl ylids[1843, 1906]. A summary of their results occurs in equations (111–114) and Tables 33 and 34.

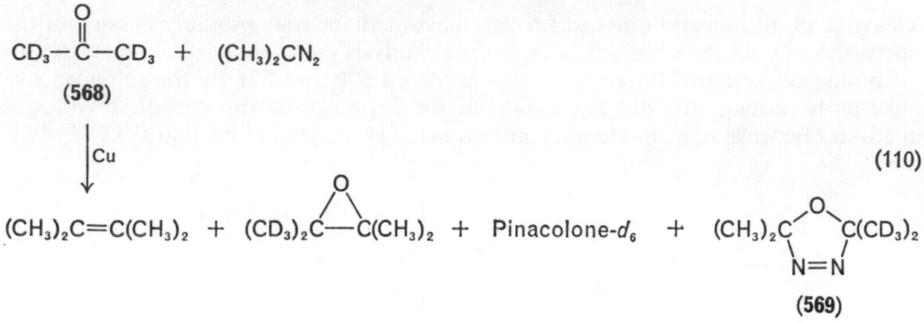

$$CD_3{-}C({=}O){-}CD_3 + (CH_3)_2CN_2 \xrightarrow{Cu} (CH_3)_2C{=}C(CH_3)_2 + (CD_3)_2\overset{O}{C{-}C}(CH_3)_2 + \text{Pinacolone-}d_6 + (CH_3)_2C\overset{O}{\;}C(CD_3)_2\,(N{=}N) \tag{110}$$

(568) (569)

$$\mathbf{569} + \mathbf{568} \longrightarrow (CH_3)_2C\overset{O}{\;}C(CD_3)_2\,/\,(CD_3)_2C{-}O \quad (CD_3)_2C\overset{O}{\;}C(CD_3)_2\,/\,(CH_3)_2C{-}O$$

(570) (571)

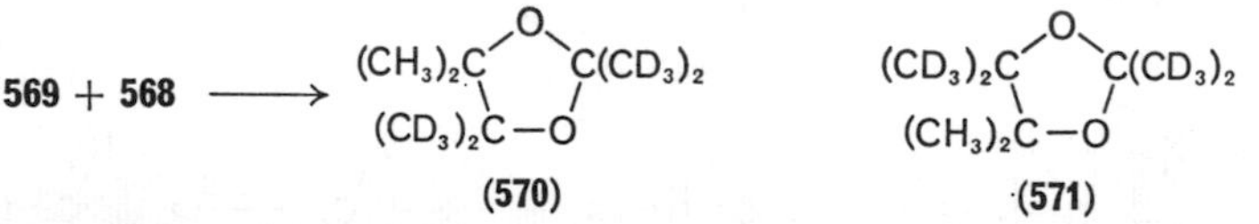

TABLE 33. Reactions of imidazolium-4-oxide (**576**) with acetylenes[1842]

Dipolaraphile	Condition	Yield of 1,3-cycloadduct (%)	Furan
$CH_3OCOC{\equiv}CCO_2CH_3$	30 °C[a]	92	5
$CH_3OCOC{\equiv}CCO_2CH_3$	80 °C[b]	42	54
$CH_3OCOC{\equiv}CCO_2CH_3$	120 °C[b]	0	83
$PhCOC{\equiv}CCOPh$	30 °C[a]	69	0
$PhCOC{\equiv}CCOPh$	120 °C[b]	0	88
$HC{\equiv}CCO_2Me$	30 °C[a]	82	1
$HC{\equiv}CCO_2Me$	80 °C[b]	35	44
$CH_3C{\equiv}CCO_2Me$	80 °C[b]	0	81
$PhC{\equiv}CCO_2Me$	80 °C[b]	0	91
$PhC{\equiv}CPh$	120 °C[b]	0	27
$PhC{\equiv}CHPh$	80 °C[b]	0	82
$n\text{-}C_4H_9C{\equiv}CH$	100 °C[b]	0	75

[a] The imidazolium-4-oxide was generated by catalytic decomposition of the related diazoimide in the presence of the acetylenic substrate.
[b] The isolated oxazalone was treated with the acetylenic substrate.

TABLE 34. Reaction of imidazolium-4-oxide (**576**) with olefins[1842]

Dipolarophile	Yield (%)
cis-Stilbene	59
trans-Stilbene	17
	29
Tetrakis carbomethoxyethylene	31
Norbornadiene	55
Acenaphthylene	100
N-Phenylmaleimide	87
Dimethyl maleate	61
	34
Dimethyl fumarate	100
trans-1,2-Dibenzoyl ethylene	23
	77

(572) → (573) ↔ (574)

(573) + MeO_2C–CH=CH–CO_2Me → (575)

X = H, OMe, Br (111)[1840–1843, 1903–1906]

(576) $\xrightarrow{XC\equiv CX}$ (577) → (578)

X = COOMe, R = Me (112)[1840–1843, 1903–1906]

(579) $\xrightarrow[\Delta]{200\ °C}$ **576** $\xrightarrow[in\ situ]{R^1C\equiv CR^2}$ **577** + **578** + MeNCO

R = Ph (113)[1840–1843, 1903–1906]

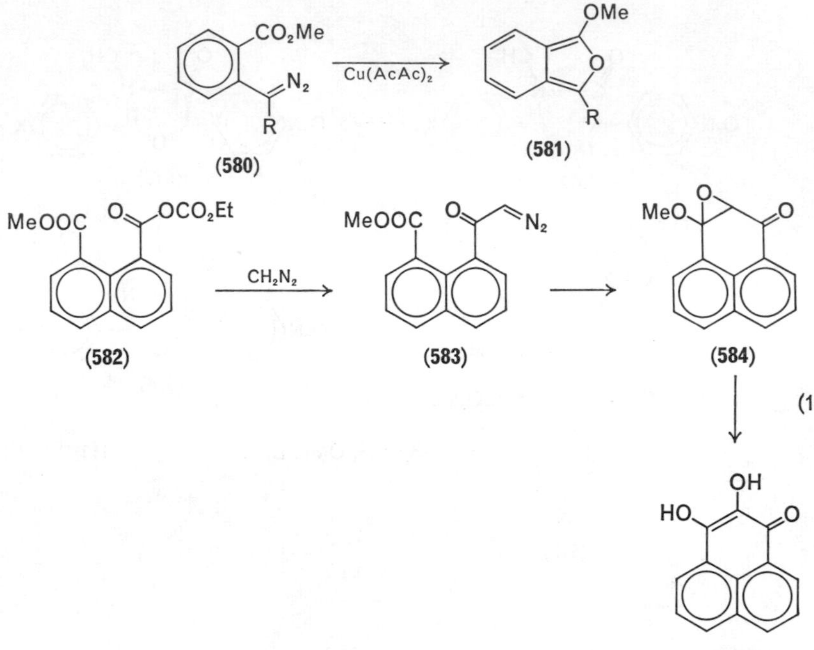

(580) $\xrightarrow{Cu(AcAc)_2}$ (581)

(582) $\xrightarrow{CH_2N_2}$ (583) $\longrightarrow$ (584) $\longrightarrow$ (585) (114)

C. Ketocarbene Cycloadditions

The loss of nitrogen from a diazocarbonyl compound is not a particularly facile process but it does occur. The resultant intermediate 'ketocarbene' **(586)** can be classified as a 1,3-dipole. With metal catalysis such a claim is probably spurious since there is reasonable evidence that the species is a carbenoid **(587)** and not a free carbene. Alternatively, one can argue that the species from photolysis is a photoexcited 1,5-dipole **(588)** which reacts in two steps. Whatever the mechanism,

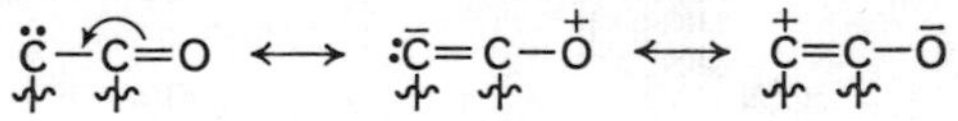

(586) (115)

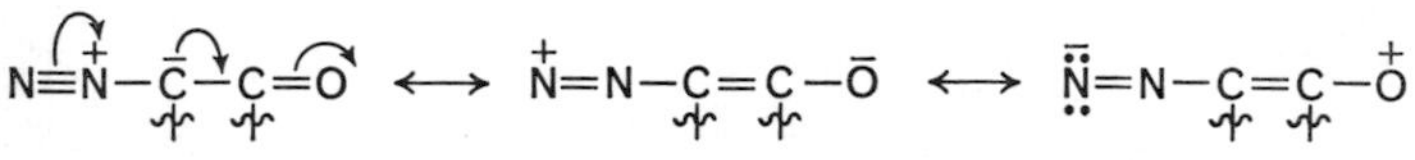

(587)

$L_4Cu{=}CXY$

(588)

it is possible to add formal ketocarbenes to various substrates via formal 1,3-cycloadditions. However, it should be remembered that acyl cyclopropanes undergo thermal rearrangement to furnish the same products as would arise from the 1,3-dipole[1545, 1722]. D'yakonov[1722] has shown that the claim[1545, 1889] that the furans (**590**) arising from reaction between ethyl diazoacetate and dialkyl and diaryl acetylenes come from 1,3-cylcoadditions is false and that the furans result from rearrangement of the related cyclopropenes (equation 116). Bien and Gillon[1520] generated a furanone by the intramolecular cyclization of a ketocarbenoid, presumably via a carbonyl-ylid intermediate (equation 117). The addition of keto carbenoids (using copper catalysis) to nitriles forms oxazoles[1889].

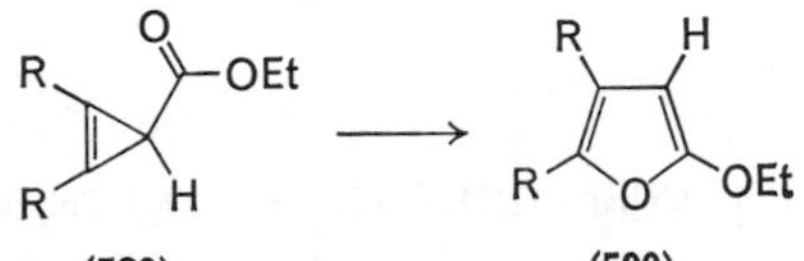

(589) → (590) (116)[1722,]

R = C_3H_7, C_4H_9, Ph

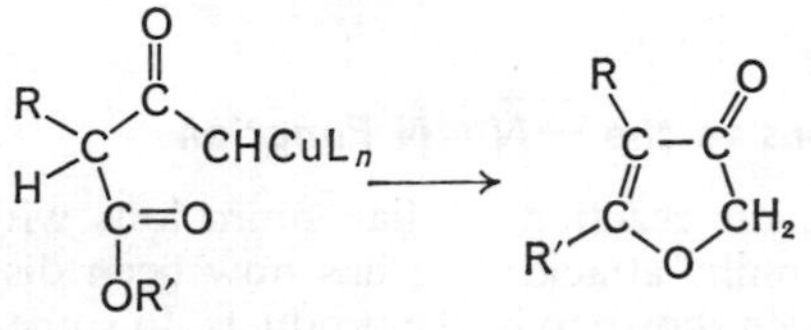

(591) → (592) (117)[1520]

R = Ph, R′ = Et
R = $PhCH_2$, R′ = Me
R = R′ = Et

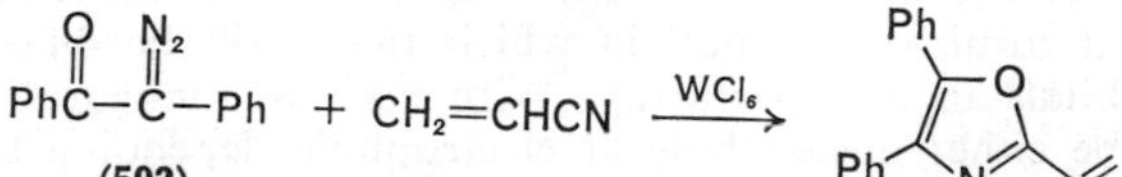

$$PhC(=O)-C(=N_2)-Ph + CH_2{=}CHCN \xrightarrow{WCl_6}$$ (118)[1951]

(593) (594)

Diazobenzil adds in the form of the ketocarbenoid to acrylonitrile in the presence of tungsten hexachloride to furnish the 2-vinyloxazole in 50% yield[1951].

A rather novel olefin-forming reaction has been studied by Marchand and Brockway[2017] in which photolysis of ethyl diazoacetate in the presence of an alkyl bromide furnishes the alkene and ethyl bromoacetate. The desired process, insertion into the C—Br bond, only occurs if there is no possibility of β elimination. Hence with neopentyl bromide, no elimination or rearrangement occurs. The mechanistic data for the elimination are totally consistent with the Wulfman–Poling hypothesis[2143, 2360, 2362] that many photochemical processes with diazoalkanes do not involve 'carbenes' (equation 119).

$$CH_3CH_2Br + N_2CHCO_2Et \xrightarrow{h\nu} CH_2{=}CH_2 + BrCH_2CO_2Et + N_2$$

$$(CH_3)_3CCH_2Br \xrightarrow[DAA]{h\nu} (CH_3)_3CH_2CH(Br)CO_2Et$$

The potential of the C—Br insertion for synthetic chemistry is obvious and appears to offer a ready route to many, otherwise difficult to synthesize, compounds.

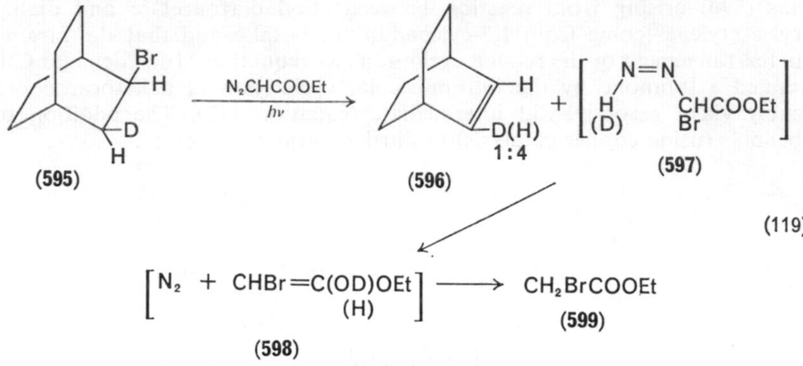

$N_2CHCOOEt$, $h\nu$

(595) → (596) 1:4 + (597)

$$\left[N_2 + CHBr{=}C(OD)OEt \; (H) \right] \longrightarrow CH_2BrCOOEt$$

(598) (599)

(119)[2107]

D. 1,4-Cycloadditions to the $-\overset{+}{N}{\equiv}N$ Function

Until very recently, the reaction of diazonium ions with 1,3-dienes was felt to involve simple electrophilic attack. This has now been disproved and 1,2-diazines and derivatives have been shown to be the products. In some cases these processes are observed with neutral species such as diazocyclopentadienes and its heterocyclic analogues. These derivatives can be envisaged as internal diazonium ion salts of cyclopentadienylide systems. They serve as a formal bridge (along with diazoxides) between diazoalkane and diazonium ion chemistry.

The recent interest in diazocycloalkenes has arisen from the recognition that such compounds can furnish 'carbenes' in which there will be $4n+2$ π electrons in contiguous orbitals in a cyclic array. With such a configuration, the resulting carbenes will be either nucleophilic or electrophilic depending upon whether the two carbene electrons are not required for aromaticity (nucleophilic carbene) or required (electrophilic carbene). The subject of unsaturated carbenes has been reviewed extensively by Dürr[316].

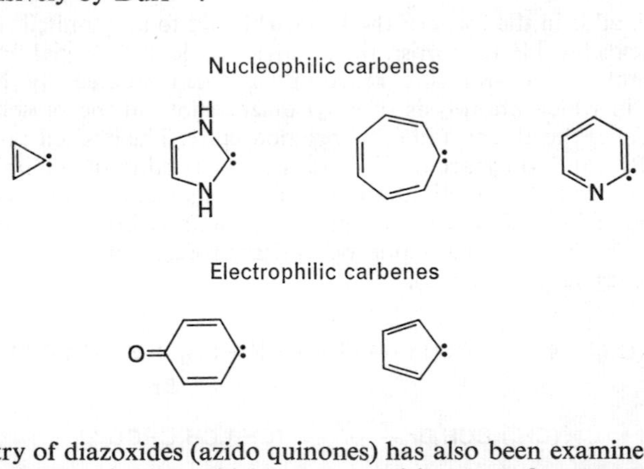

Nucleophilic carbenes

Electrophilic carbenes

(120)

The chemistry of diazoxides (azido quinones) has also been examined by Dürr[316]. Although the keyword 'diazoxide' appears with some frequency in *Chemical Abstracts*, it refers to an anaesthetic agent which is not of interest to the chemist pursuing diazoalkane or diazonium ion chemistry.

$\xrightarrow[\text{base}]{TosN_3}$ (600) $\xrightarrow{h\nu \text{ or } \Delta}$ (601)

(602) $\xrightarrow{HONO}$ (603) $\xrightarrow{h\nu \text{ or } \Delta}$ (604) (121)

(605) $\longrightarrow$ (606) $\xrightarrow[\Delta]{h\nu}$ (607)

(608) $\xrightarrow{h\nu \text{ or } \Delta}$ (609)

(610) $\xrightarrow{h\nu}$ (611) $\xrightarrow{PhCH_3}$ (612) + (613) + (614) (122)

Ph Ph Ph (615) + Olefin $\xrightarrow{h\nu}$ Cyclopropane + CH insertion (ratios statistically corrected)

(616)	40 : 1
(617)	34 : 1
(618)	228 : 1
(619)	14·8 : 1
(620)	17 : 1

(123)[316]

(607) + (621) → (622) + (623) + (624)

(607) → (625) → (626) ⇌ (627) → (628) (124)

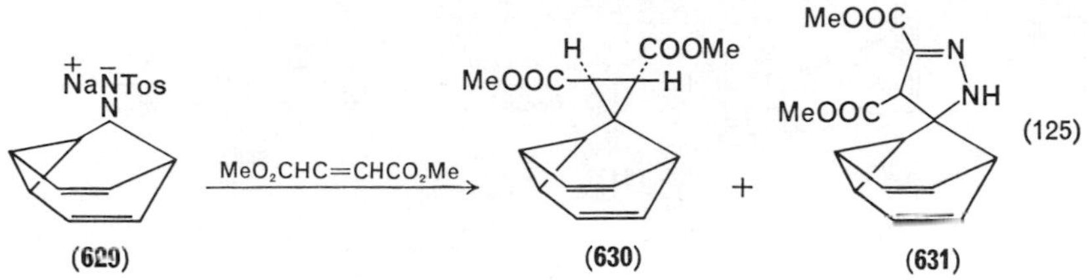

$$\text{(629)} \xrightarrow{MeO_2CHC{=}CHCO_2Me} \text{(630)} + \text{(631)} \qquad (125)$$

I. Arenediazonium ions

Recently Carlson, Sheppard and Webster[1560] have reinvestigated the reactions of aryl diazonium ions with butadienes and found that, contrary to the earlier reports in the literature, linear structures did not result[981, 1447, 2042, 2286].

The study was a direct outgrowth of earlier work with 2-diazo-4,5-dicyanoimidazole and butadiene[2232] where the DuPont works obtained a 1,6-dihydropyridazene.

Sheppard has pointed out[1560, 2232, 2233] that there are at least two other general precedents where the N=N chromophore serves as a dienophile (equation 126).

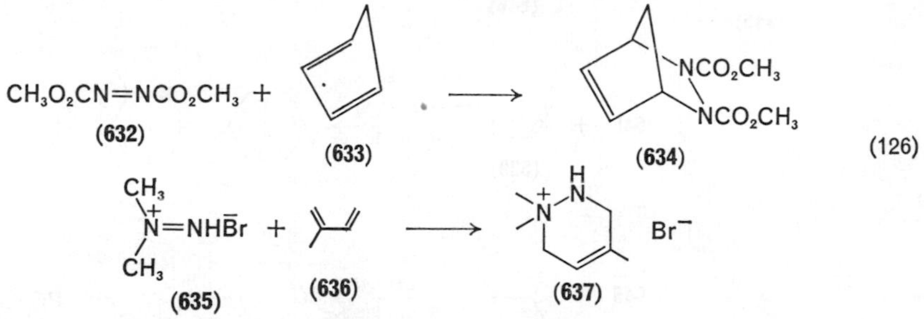

$$CH_3O_2CN{=}NCO_2CH_3 \text{ (632)} + \text{(633)} \longrightarrow \text{(634)}$$

$$\text{(635)} + \text{(636)} \longrightarrow \text{(637)} \; Br^- \qquad (126)$$

Some examples of the cycloaddition reactions of arenediazonium ions are summarized in equations (127–129). A number of other additions to diazonium ions have been observed by DuPont workers which will probably be published in the late 1970s[2233].

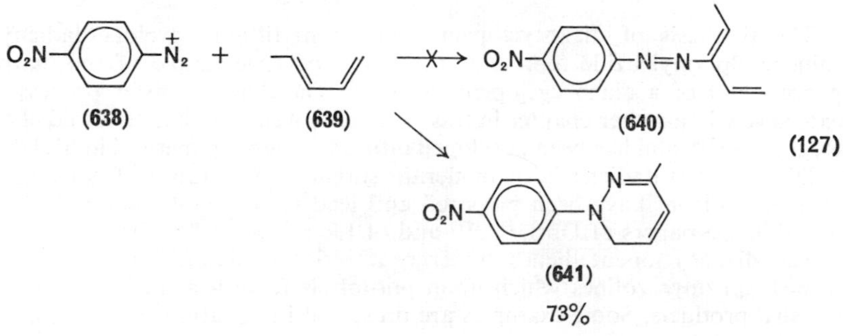

$$O_2N\text{–}C_6H_4\text{–}\overset{+}{N}_2 \text{ (638)} + \text{(639)} \not\longrightarrow \text{(640)}$$

$$\searrow \text{(641) } 73\% \qquad (127)$$

(641) + (642) $\xrightarrow[\text{Room temp.}]{CH_3CN}$ (643) or (644) $\bar{P}F_6$ (128)

X	Yield (%)
H	42
p-Cl	60
m-F	52
p-F	72
p-O_2N	79

(645) $-N_2^+ \bar{P}F_6$ + (646) $\xrightarrow[\text{Room temp.}]{CH_3CN}$ No reaction.

645 + (639) $\xrightarrow[\text{Room temp.}]{CH_3CN}$ (647) 3% $\bar{P}F_6$ (129)

645 + (642) $\xrightarrow[\text{Room temp.}]{CH_3CN}$ (648) 42% $\bar{P}F_6$

2. Diazocyclopentadiene and its aza analogues

The synthesis of diazocyclopentadiene from lithium cyclopentadienyl and *p*-toluenesulphonyl azide represents the first modern example of the diazo transfer process and of a diazo cyclopentadiene[1645]. The diazo transfer process is treated extensively in another chapter in this volume as well as being reviewed elsewhere by Regitz[1045, 1046], and has been developed into a very general method in his laboratories.

Diazocyclopentadiene is of moderate stability. A number of substituted diazocyclopentadienes have been prepared and lead references to other workers can be found in the papers of Dürr[1653–1717] and of Lloyd[1794–1797, 1990–2001].

The diazocyclopentadienes undergo a series of interesting cycloadditions to furnish spiropyrazolines which upon photolysis furnish a number of strained and unusual products. Some examples are presented in equations (130–133).

(649) (650) (651) (652) 0–0·1% (130)

R[1]	R[2]	R[3]	R[4]	R[5]	R[6]
H	H	H	H	Me	Me
Cl	Cl	Cl	Cl	Et	Et
Ph	Ph	Ph	Ph	COOMe	COOMe
Ph	Ph	CH=CH−CH=CH		H	COOMe
CH=CH−CH=CH		CH=CH−CH=CH		H	COOMe
Ph	Ph	Ph	H	Ph	COOMe
Ph	H	H	Ph	Ph	COOMe

(653) $\xrightarrow{h\nu}$ (654) X = COOMe (131)

R[1]	R[2]	R[3]	R[4]	Yield (%)
Ph	Ph	Ph	Ph	85
Ph	H	H	Ph	86
CH=CH−CH=CH		Ph	Ph	30

(132)

The 'carbenes' available from diazo cycloalkenes fall into two categories, the $4n$ such as cycloheptatrienylidene and the $4n+2$ such as cyclopentadienylidene or that derived from *p*-benzenediazoxide[1697] (see equation 120).

Such 'carbenes' can be generated in a variety of ways including the now classical routes to diazoalkanes such as basic decomposition of nitrosoamides, tosyl hydrazones, thermolysis and photolyses of the diazo compounds, and even decarboxylation of pyridinium carboxylates.

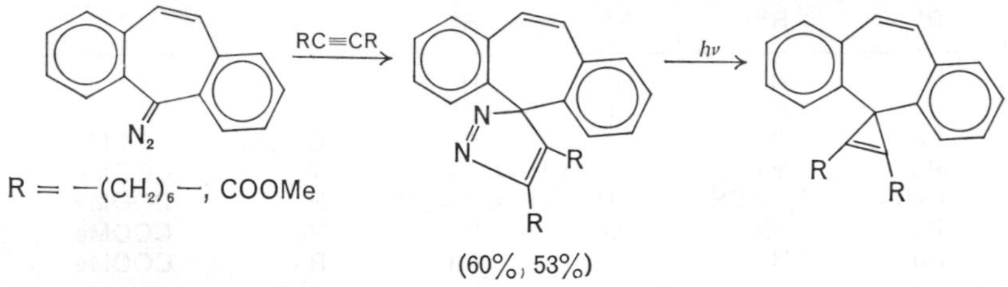

RC≡CR; $h\nu$ (133)

R = $-(CH_2)_6-$, COOMe

(60%, 53%)

An important reaction of such 'carbenes' is insertion into C—H bonds. This reaction can proceed either directly or via abstraction processes which may lead to several products, including reduced dimers and oxidative coupling of the hydrocarbon RH. The presence of more than one type of C—H bond (e.g. primary, secondary, tertiary or allylic will lead to mixtures of products). The degree of selectivity is a function of the diazo precursor and some examples are summarized by Dürr[1697] along with comparisons between C—H insertion and addition reactions.

Hammett plots of the addition of cycloheptatrienylidene to *para*-substituted styrenes clearly reveal that the 'carbene' is nucleophilic and chemical evidence indicates that cyclopropenylidene is also nucleophilic. The electrophilic nature of the cyclohexadienylidene system is not as clearly established although it appears evident. The electrophilic nature of cyclopentadienylidene has been established by employing Hammett plots[1702]. Hence the overall trend seems to be a preference for electron-rich olefins. The reaction of diazocyclopentadiene with tetramethylethylene is illustrative[2060].

With the diazo azacyclopentadienes there is less information available. Webster and Sheppard[2232] have examined a number of reactions of 2-diazo-4,5-dicyano-imidazole (**655**) and these are summarized in equations (134–137). Diazo diaza- and triaza-cyclopentadienes undergo typical coupling reactions with β-naphthol[2038] in moderate to quantitative yields, and undergo conversion to the related azides under the influence of hydrazine or dimethyl hydrazine[2226]. Hence their reactions strongly resemble aromatic diazonium salts.

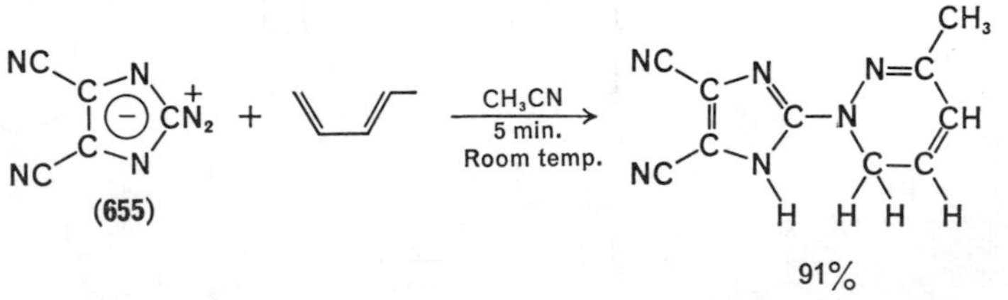

(**655**) + diene $\xrightarrow[\text{5 min. Room temp.}]{CH_3CN}$ product 91% (134)

655 + diene $\xrightarrow{\text{Room temp.}}$ No reaction

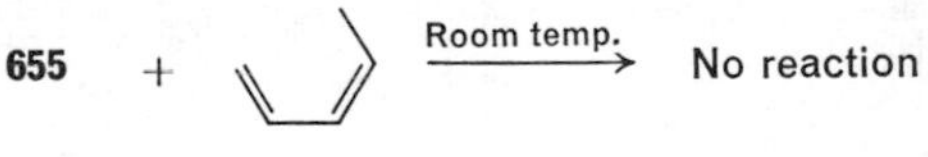

655 + [diene]; Room temp. standing, CH_3CN → ~40% + ~40% + ~20% (135)

655 + MeO–CH=CH–OMe → ... (136)

PhH, $PhCF_3$, Cyclohexane, Ethylene chloride, PhX; 655 (137)

Mukai and coworkers[2062–2064] examined some nucleophilic carbenes, in particular cycloheptatrienylidene and barbaralylidene (equation 125), whereas Jones and coworkers examined cyclopropenylidene, cycloheptatrienylidene and 11-annulenylidene[1802, 1916, 1923–1926, 2047].

Lloyd and coworkers[1794–1797, 1990–2001] have performed extensive studies on the formation of cyclopentadienyl ylids and related compounds. These include the reactions of diazocyclopentadienes with 2,6-dimethyl-4-thiopyrone[1990, 1993], 2,6-dimethyl-1-thiapyran-4-thione[1990] (equation 138), triphenylstibine[1999], triphenylbismuth[1995], diphenyltelluride[2000], pyridine[2001], triphenylphosphine[1794, 1997], triphenylarsine[1993, 1998], diphenylselenide[1994] and diphenyl sulphide[1996]. Ylid formation failed with triphenyl and diphenylamines[2001]. The reactions with triphenylphosphine also

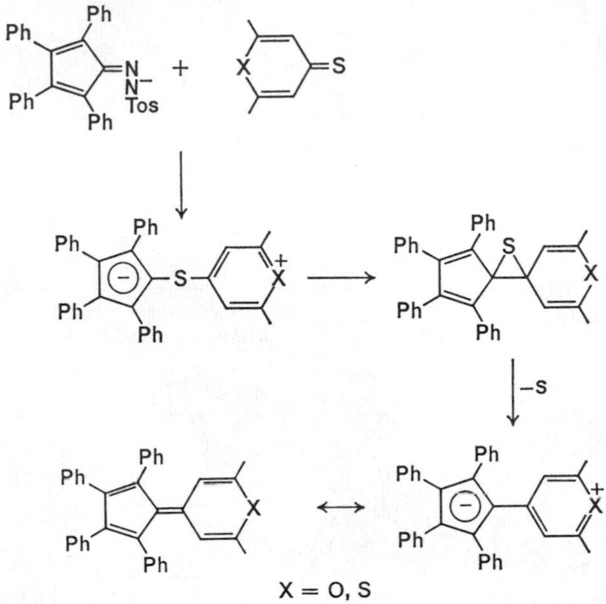

(138)[1990, 1993]

X = O, S

lead to phosphinazenes[1794, 1795, 1997]. This process is sufficiently general that it is possible to employ the reaction for preparing derivatives of a large number of diazo compounds. In the cases of interest here, heating at 100 °C in the melt or in benzene led to the phosphinazine, whereas heating at 140 °C for 1 h furnished the ylids.

Both diazo tetraphenylcyclopentadiene and diazocyclopentadiene insert into the halogen bridge in halogen-bridged dirhodium species with loss of nitrogen to furnish *pentahapto*rhodium complexes[1632].

II. REARRANGEMENTS

Rearrangements involving electron-deficient species are well known and have been extensively studied and reviewed[2027, 2028, 2034]. By far the most common process employing diazoalkanes is the Wolff rearrangement[2034] in which a α-diazoketone furnishes a keten or a keten-derived product. The Wolff rearrangement can be used for the purpose of ring contraction[1040] or for chain homologation. Since both classes of diazo compounds required are readily obtained it is not surprising that their use is widespread. The syntheses for homologation usually proceed via the Arndt–Eistert synthesis[1726–1728] where an acid chloride reacts with an excess of diazoalkane (normally diazomethane) to furnish an acyl diazoalkane and an alkyl chloride. The formation of α-diazocyclanones has been greatly advanced by the work of Regitz and coworkers[2153, 2154], who initially form the hydroxymethylene ketone and then subject it to diazo transfer conditions. This reaction also works with 1,3-diones[2003] and β-keto esters[1473].

$$RC(=O)X + CH_2N_2 \longrightarrow RC(=O)CHN_2 + HX \xrightarrow{\text{excess } CH_2N_2} CH_3X$$

X = Halogen, OCOOR′ X = Halogen

$$\begin{array}{c} \text{C=CHOH} \\ | \\ \text{C=O} \end{array} \xrightarrow[\text{Base}]{TosN_3} \begin{array}{c} \text{C=N}_2 \\ | \\ \text{C=O} \end{array} \quad (139)$$

$$\text{H}-\overset{|}{\text{C}}(\text{C(=O)CH}_3)-\text{C(=O)R} \xrightarrow[\text{Base}]{TosN_3} \text{C(=N}_2\text{)}-\text{C(=O)}-\text{R} \qquad R = OEt, CH_3$$

The other rearrangements of diazoalkanes primarily involve hydrogen or alkyl migrations to furnish olefins.

A. *Wolff Rearrangements*

Wolff rearrangements of α-diazo carbonyl compounds can be induced to occur by thermolysis, catalysis and photolysis. The processes observed are not always equivalent and in any particular instance one method may be preferable to the others. The photochemical process apparently involves a singlet carbene for Ando[1917] has successfully suppressed the rearrangement by employing triplet sensitizers. In general, copper-based catalysts also suppress the reaction whereas with silver catalysts the reactions proceed smoothly.

1. Thermolytic processes

Thermolysis occurs over a wide range of temperatures (room temperature to 750 °C)[1569]. Stability is primarily a function of electronic effects, either as a consequence of substitution altering the electron density of the $COCN_2$ unit or causing a twisting of the C—C bond and thereby altering the extent of overlap between the C=O and CN_2 chromophores[1825, 1927, 2039]. Since a number of processes can occur in competition with the Wolff rearrangement, it may be necessary to experiment over a range of conditions to obtain optimized yields. Meier has noted that initially increases in temperature tend to favour the Wolff rearrangement over side reactions[2034]. The reactions are frequently carried out in boiling aniline[1731, 1732, 1734, 1736] or benzyl alcohol to furnish the anilides or benzyl esters[2344]. Some examples are presented in equation (140)[1543, 1544, 1568, 1581, 1731, 1734, 1736, 1805, 1855, 1879, 2157, 2299, 2344, 2375].

2. Photolytic processes

Photolysis in the absence of sensitizers provides a convenient technique for Wolff rearrangements and in some cases (e.g. diazocamphor) succeeds[1879] where thermolysis fails. This has the advantage that one is capable of employing low-boiling solvents such as methanol and can operate at very low temperatures if the products are heat sensitive. Meier[2034] has concluded that one should operate at as long a wavelength as possible, but notes that the lowest singlet state is only moderately active. The photolytic approach also permits the incorporation of sensitive reagents capable of

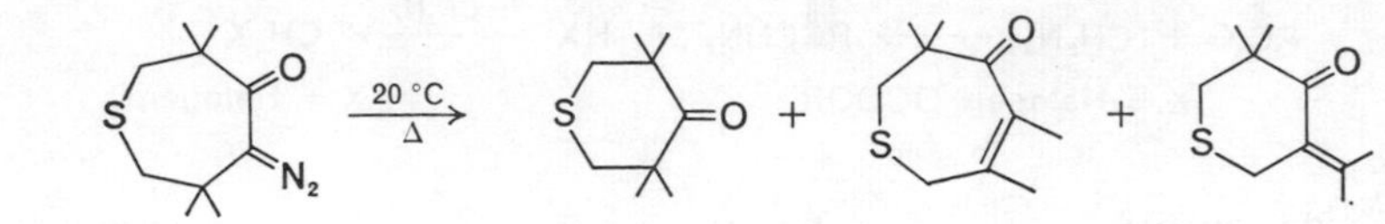

20 °C, Δ

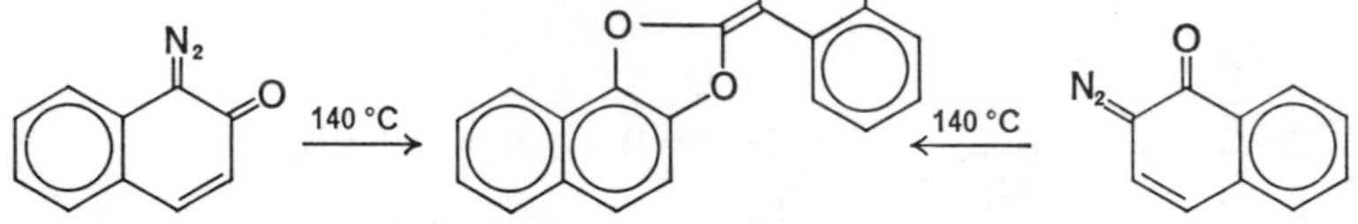

140 °C

140 °C

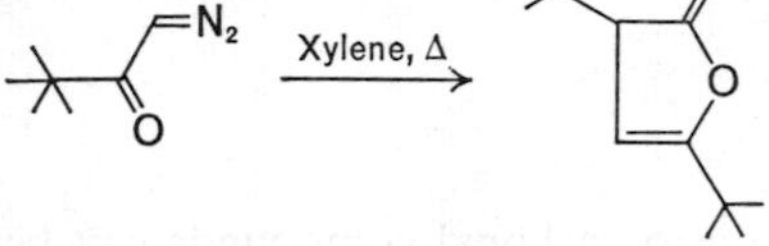

Xylene, Δ

(140)

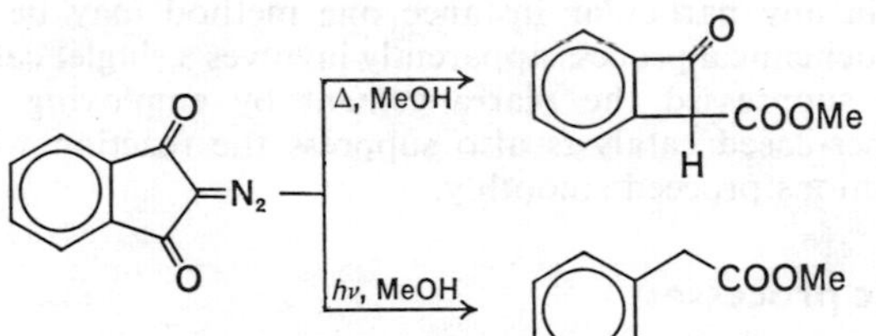

Δ, MeOH

COOMe

hν, MeOH

COOMe

COOMe

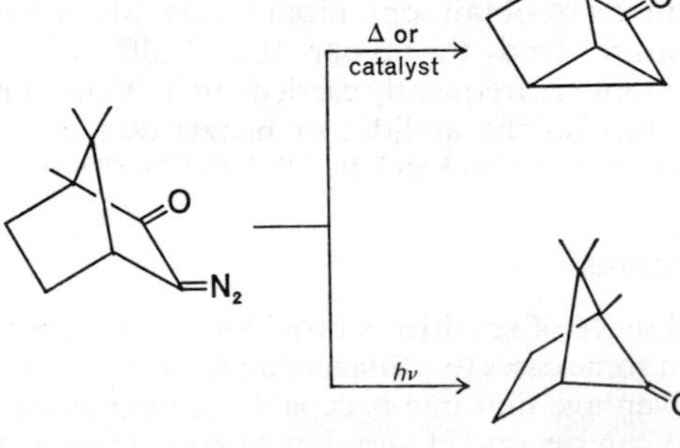

Δ or catalyst

hν

(141)

subsequent reaction with the keten and may actually furnish different products than those available from catalytic reactions[1798, 2034] (equation 141). Some typical photolytic Wolff rearrangements are summarized in equation (142)[1628, 1801, 1878, 1881, 1917, 1959, 2035, 2037, 2336, 2337, 2342]. On occasion the Wolff rearrangement will fail as a consequence of interaction with the nucleophile either by reduction (e.g. trifluoroacetyl diazoacetic ester[2335]) or by nucleophilic attack upon the diazo carbon (e.g. the Cu powder CH_3CN decomposition of diazoacetophenone in excess methanol[2372]).

3. Catalytic processes

The most common catalysts employed in Wolff rearrangements are based upon Ag^0 and Ag^I. Some Cu^{II}, Cu^I and Cu^0 systems and platinum systems have also been used; however, by far the best general systems involve silver. Copper and its salts tend to form relatively stable transient copper carbenoids which do not rearrange.

$$PhC(O)CHN_2 \xrightarrow{h\nu} \begin{cases} \xrightarrow[\text{Dioxane}]{HOH} PhCH_2COOH \\ \xrightarrow{CH_3OH} PhCH_2COOCH_3 \\ \xrightarrow[\text{Dioxane}]{PhNH_2} PhCH_2CONHPh \\ \xrightarrow{EtSH} PhCH_2COSC_2H_5 \end{cases}$$

$\xrightarrow[\text{THF, } h\nu]{HOH}$ COOH

N_2 O $\xrightarrow{\text{Methanol, } h\nu}$ CO_2CH_3 O

$\xrightarrow[h\nu]{\text{HOH, THF}}$ =O N_2 —COOH 47% (142)

COOCH$_3$ $\xrightarrow{CH_3OH}$ O N_2 $h\nu$ $\xrightarrow{\text{Sensitized}}$ O

O N_2 $\xrightarrow[CH_3OH]{h\nu}$ CH_3OOC—

Some rather commonly employed silver systems include $Ag_2O/Na_2S_2O_3$ [2352], $Ag_2O/Na_2S_2O_3/Na_2CO_3$ [1451], $C_6H_5CO_2Ag$/*t*-amine[2089] and CF_3CO_2Ag[1727]. The Newman amine system employing a tertiary amine has found wide usage. Surprisingly, one also encounters the use of catalysts in a number of photolytic Wolff rearrangements. It is not clear whether the catalyst plays an active role or represents a retreat to the chemistry of von Hohenheim. The use of $Ag_2O/Na_2S_2O_3/Na_2CO_3$ is the classic condition employed in the Arndt–Eistert homologation syntheses[1459, 1462]. This approach was of sufficient importance in the early days of steroidal synthesis that it was chosen as one of the topics for review in *Organic Reactions*, Volume 1[31], and was employed by Bachmann, Cole and Wilds in the first successful steroidal total synthesis[1469]. An examination of Bachmann's work from that period reveals a prodigous amount of work on the reaction[31]. Some typical examples employing catalysis are given in equation (143)[1564, 2045, 2089, 2348, 2372].

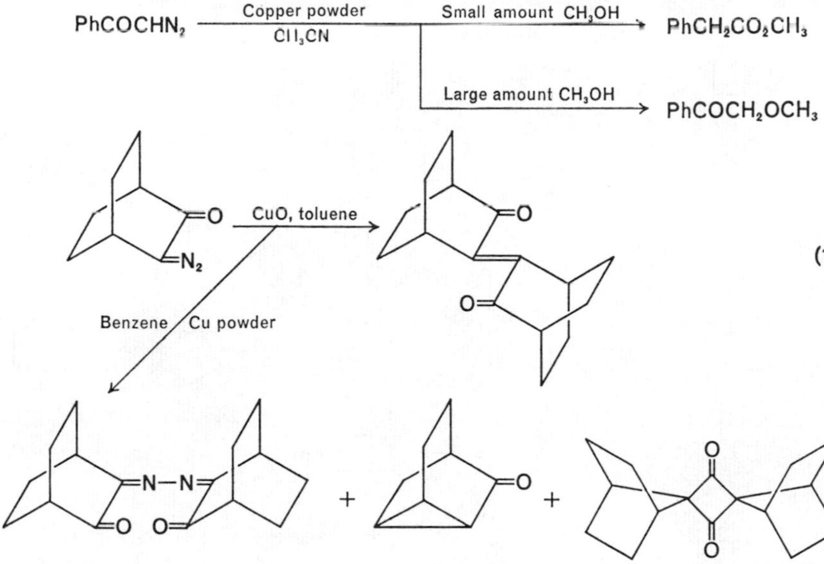

$$PhCOCHN_2 \xrightarrow[CH_3CN]{\text{Copper powder}} \begin{cases} \xrightarrow{\text{Small amount } CH_3OH} PhCH_2CO_2CH_3 \\ \xrightarrow{\text{Large amount } CH_3OH} PhCOCH_2OCH_3 \end{cases}$$

CuO, toluene

Benzene Cu powder

(143)

B. Other Rearrangements

A variety of processes can compete with the Wolff rearrangement. The processes are not unique to α-diazocarbonyl compounds. These rearrangements involve 1,2-hydrogen and 1,2-alkyl migrations. Insertion reactions are discussed in Section V and in a strict sense do not involve skeletal rearrangements of the type concerned here. A surprisingly large number of rearrangements are observed under Bamford–Stevens reaction conditions. These have been treated by Gutsche and Redmore[478, 1040]. Representative examples are presented in equations (144), (145) and (146).

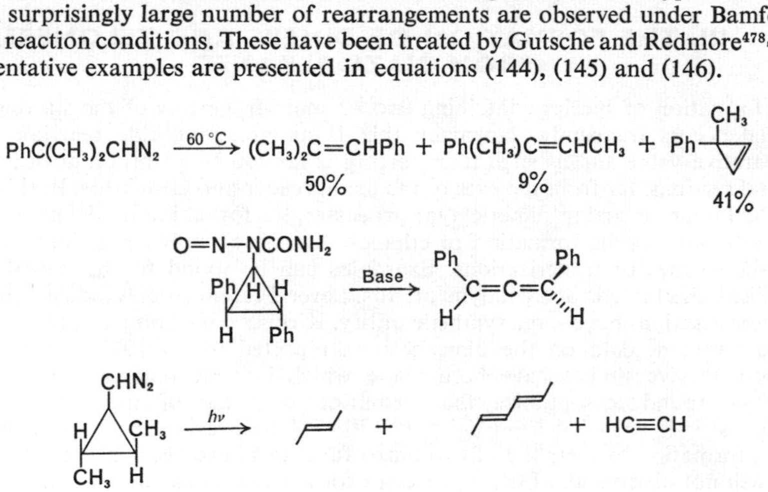

$$PhC(CH_3)_2CHN_2 \xrightarrow{60\,^\circ C} \underset{50\%}{(CH_3)_2C{=}CHPh} + \underset{9\%}{Ph(CH_3)C{=}CHCH_3} + \underset{41\%}{Ph\text{-}C(CH_3)(\text{cyclopropyl})}$$

$$O{=}N{-}NCONH_2 \text{ (Ph, H, H, Ph cyclopropane)} \xrightarrow{\text{Base}} \underset{H}{\overset{Ph}{}}C{=}C{=}C\underset{H}{\overset{Ph}{}} \qquad (144)$$

$$CHN_2 \text{ (H, } CH_3\text{, } CH_3\text{, H cyclopropane)} \xrightarrow{h\nu} \text{propene} + \text{pentadiene} + HC{\equiv}CH$$

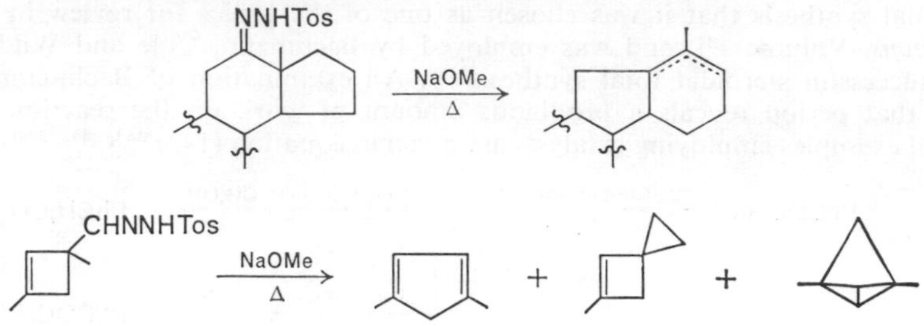

NNHTos

NaOMe, Δ

CHNNHTos

NaOMe, Δ

(145)

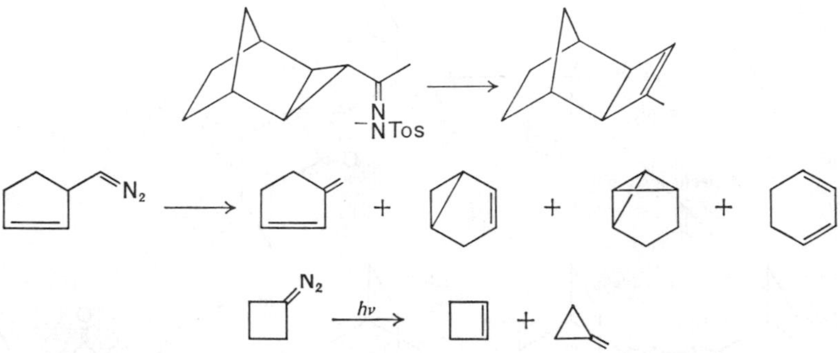

$^{-}$NTos

N_2

N_2

$h\nu$

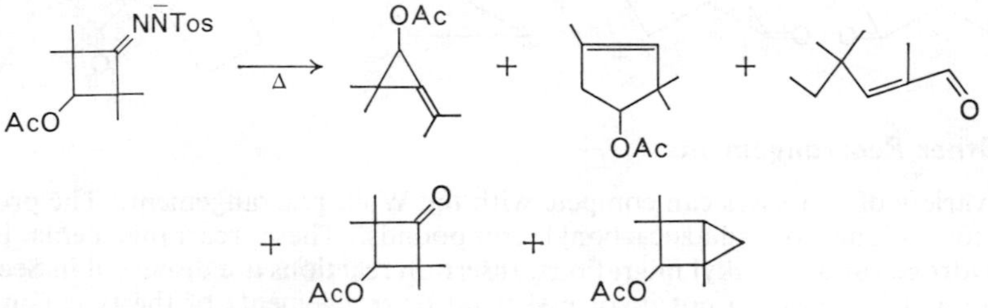

$N\bar{N}Tos$

AcO

Δ

OAc

OAc

O

(146)

O

AcO

AcO

III. THE FORMATION OF DIMERS AND TELOMERS FROM DIAZOALKANES

The formation of species containing two or more fragments of the starting diazo-alkane occurs frequently. Normally this is an annoying side reaction of little preparative value and as such its reporting is far too frequently hidden in experimental sections, far from the eyes of the usual reader and abstractor. In this section we shall treat the major 'dimeric' type processes, the formation of olefins, azines and pinacols but not the formation of ethanes, tetrazenes, polymerizations to furnish polymethylenes or trimerizations. Examples can be found for most with photochemical, thermolytic and catalytic origins. Several recent studies indicate that some of these reactions have some synthetic utility. Kirmse[654] has summarized some of the more pertinent data on the dimerizations reported up to 1970. More recently, fairly extensive studies have been made which indicate that suitable conditions may be found to optimize the formation of some of the major 'dimeric' products[1778, 1779, 2099, 2148, 2221, 2222, 2236, 2237, 2365, 2368], and several syntheses based upon olefin formation by coupling of two diazo functions have been realized. Frequently one will not observe all of these processes for a given diazo compound. There have

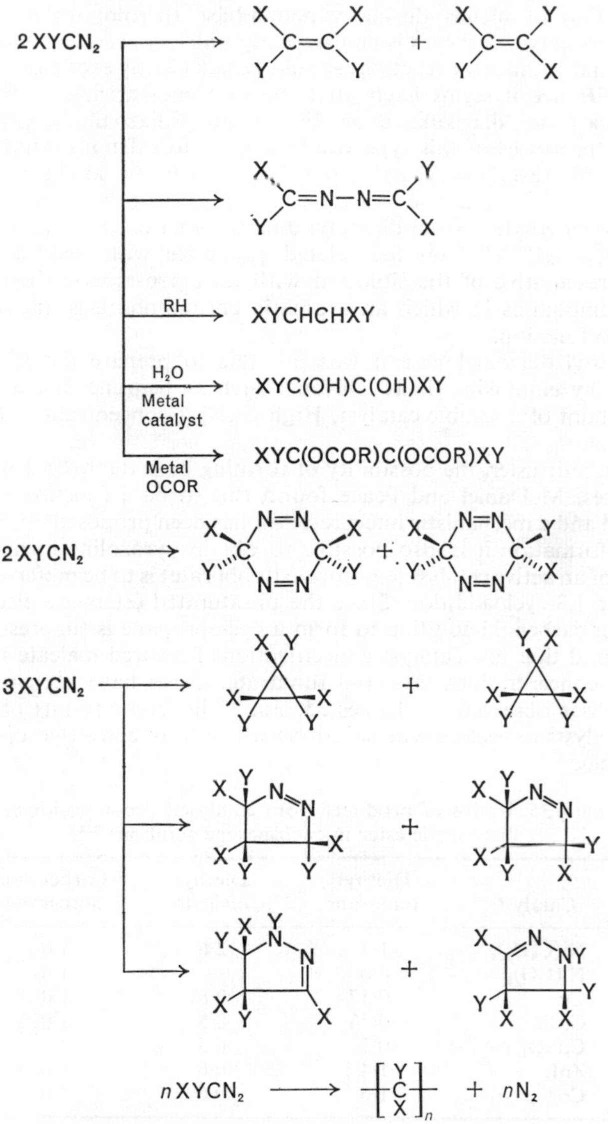

$$2\,XYCN_2 \longrightarrow XYC{=}CXY\ (cis) + XYC{=}CXY\ (trans)$$

$$\longrightarrow XYC{=}N{-}N{=}CXY$$

$$\xrightarrow{RH} XYCHCHXY$$

$$\xrightarrow[\text{Metal catalyst}]{H_2O} XYC(OH)C(OH)XY$$

$$\xrightarrow{\text{Metal OCOR}} XYC(OCOR)C(OCOR)XY$$

$$2\,XYCN_2 \longrightarrow \text{tetrazine dimers (cis + trans)} \qquad (147)$$

$$3\,XYCN_2 \longrightarrow \text{cyclopropanes} + \text{pyrazolines} + \text{pyrazolines}$$

$$n\,XYCN_2 \longrightarrow {-}[CXY]_n{-} + n\,N_2$$

been few mechanistic studies regarding how these products arise and almost all studies simply report the isolation of a particular set of products. Since the isolation and recognition of such products are not usually considered of prime importance, the absence of a particular product type in any given report should not be given great weight. Similarly, care must be exercised in accepting structural assignments. In at least one case an initially incorrect assignment made in 1893[1553] has been repeated in the most recent review (1970)[283], even though the error had been noted previously (1901)[1555, 1651, 2238]. In this instance the error was committed with diazoacetic ester, probably the most widely studied diazoalkane after diazomethane.

The formation of olefins during vapour-phase thermolyses or photolysis of diazoalkanes from two carbenes is an extremely unlikely event, both on the basis of probabilities and because the reactions should be sufficiently exothermic to favour the free carbene. Hence it seems likely that the ethylenes arising from diazoalkanes result from carbene + diazoalkane or diazoalkane + diazoalkane processes. Since vapour-phase processes of this type would appear to offer no synthetic utility we will consider only reactions occurring in solution or in the solid state.

Peace[2127, 2130, 2131, 2365] examined the affect of catalyst on the formation of tetrakis-methoxycarbonyl ethylene from dimethyl diazomalonate. These studies, along with those of McDaniel[2004, 2362] on the related processes with diazoacetic ester, are probably representative of the situation with all diazo-monocarbonyl and diazo-dicarbonyl compounds in which an aromatic chromophore is not in conjugation with the diazo function.

With dimethyl diazomalonate it was possible to prepare the ethylene in over 80% yield[2132] by employing 'inert' solvents such as benzene in the presence of a moderate amount of a soluble catalyst. High catalyst concentrations lead to poorer yields.

With diazoacetic ester, the possibility of forming both diethylmaleate and diethyl fumarate arises. McDaniel and Peace found this to be a function of the catalyst concentration and a mechanistic interpretation has been proposed[2365]. With fumarate and maleate formation it is also possible to obtain pyrazoline formation, and the employment of an active catalyst (e.g. cupric fluoborate) is to be preferred to minimize the competing 1,3-cycloaddition. Since the unsaturated esters are electron-deficient olefins, copper carbenoid addition to form a cyclopropane is suppressed. Peace and McDaniel found that low catalyst concentrations favoured maleate formation and high catalyst concentrations favoured fumarate. Cases have been reported where only maleate was observed[2189]. Tables 35, 36, 37 list some results obtained with a variety of catalysts in cyclohexene solutions which have converted diazoacetic ester to the ethylenes.

TABLE 35. Ratios of products from catalysed decompositions of diazoacetic ester in cyclohexene solutions[2334]

Catalyst	Diethyl fumarate	Diethyl maleate	7-Carboethoxy norcarane
$Ni(C_5H_5)_2$	1·4	2·6	1·0
$Ni(CO)_4$	6·0	—	1·0
Cu	0·57	0·67	1·0
CuBr	0·50	0·5	1·0
$CuSO_4$	0·7	0·5	1·0
ZnI_2	1·4	0·6	1·0
$Cr(C_5H_5)_2$	1·5	0·6	1·0

Seratosa and coworkers[1778, 1779, 2147, 2221, 2222] have decomposed diazoketones to furnish sulphur ylid intermediates which then form ethylenes, and in one case bullvaltrione. These processes are not of great efficiency but may be capable of improvement by optimization of catalyst concentration. In the bullvaltrione synthesis it is noteworthy that a cyclopropane (a trimer) is formed. This occurs because the second step involves addition of the nucleophilic ylid to the electron-poor 1,4-dioxo-2-butene (equation 148). The process operates in competition with pyrazoline formation which is essentially unimolecular whereas the desired process is bimolecular (e.g. involving substrate and catalyst).

TABLE 36. Product distribution in the reaction of cyclohexene with ethyl diazoacetate using $[(CH_3O)_3P]_n.CuI$ as catalyst[2365]

Catalyst (mmol)	Norcarane				Dimer		
	n	*exo*	*endo*	Ratio	*cis*	*trans*	Ratio
0·140	1	7·80	0·65	12·0	0·088	0·123	0·715
5·00	1	2·91	1·16	2·51	0·218	0·710	0·307
0·140	2	7·87	0·684	11·5	0·100	0·150	0·667
5·00	2	0·965	0·946	1·02	0·077	0·254	0·304
0·140	3	8·05	0·533	15·1	0·067	0·111	0·608
5·00	3	0·333	0·832	0·40	0·00	0·00	—

TABLE 37. The effect of $Cu(AcAc)^2$ concentration upon the ratio of diethyl fumarate : diethyl maleate from the decomposition of diazoacetic ester in cyclohexene solutions[2365]

Catalyst concentration (mg/50 ml)	Diethyl fumarate : diethyl maleate
0	0·53
1	0·69
4	0·69
16	0·89
32	1·03
64	1·39
256	2·18

COCHN₂ / COCHN₂ / COCHN₂ $\xrightarrow{R_2SCuX}$ COCHN₂ (with C=O, C–H, C–H, C=O ring) ⟶ CO, C–H, CH–$\overset{+}{S}R_2$, H, Cu ⟶ (148)

Although it would appear that no simple rule exists for predicting when an azine will be formed photochemically or when an olefin will result, there is ample reason to suspect that azines will be formed when the 'carbene' will be in the 'triplet' state. The formation of the intermediate diradical $Ar_2C\uparrow N_2C\uparrow Ar_2$ will be extensively delocalized throughout the π systems of the Ar grouping, whereas the 'singlet' species can collapse directly to olefin plus nitrogen. The apparent exception is dimesityldiazomethane[2377]. A reexamination of the published data in this case is equally consistent with *o,o'*-disubstituted phenyldiazomethanes being unable to conjugate fully the diazo chromophore with the aromatic rings and thus leading initially to the 'singlet' carbene rather than the 'triplet'. The formation of glycols (or their esters) when metal salts are employed in aqueous media can be easily accommodated by equation (149).

Peace and McDaniel have presented evidence that the formation of olefins proceeds by two mechanisms of which equation (150) (path A and path C) most closely accounts for the observed results[2133, 2359, 2364, 2365]. A summary of some azine, olefin and glycol (glycol ester) studies appears in Tables 35–41[2099, 2236]. In those cases

$$R_2CN_2 + ML_n \rightleftarrows R_2\overset{N_2^+}{C}\bar{M}L_n \longrightarrow R_2C(L)ML_{n-1} \overset{\pm R_2CN_2}{\rightleftarrows} \quad (149)$$

$$R_2C(L)-\bar{M}(L_{n-1})-CR_2(N_2^+) \longrightarrow R_2C(L)-M(L_{n-2})-CR_2(L) \longrightarrow R_2CLCLR_2$$

$$L = O-\overset{O}{\overset{\|}{C}}R'$$

A: $XYC^{-}-N^{+}{\equiv}N$ + $L_2Cu{=}CXY$ ⟶ $L_2\bar{C}u-CXY-CXY(N_2^+)$ ⟶ $X-C(Y)(N_2^+)-C(X)(Y)-\bar{C}uL_2$ ⟶ $XYC{=}CXY$

B: $L_2Cu{=}CXY$ + $XYC{=}CuL_2$ ⟶ [$L_2Cu-CXY$ / $CXY-CuL_2$ four-membered ring] ⟶ $XYC{=}CXY$

C: $L_2Cu{=}CXY$ + $XYC{=}CuL_2$ ⟶ $L_2\bar{C}u-CXY-CXY-\overset{+}{C}uL_2$ ⟶ $XYC{=}CXY$

$$(150)^{2365}$$

$$2\ (\text{fluorenylidene})=N_2 \xrightarrow{Cu} \text{bifluorenylidene} \quad (151)$$

$$2\ (\text{2-diazo-1-indanone, } =N_2) \xrightarrow{CuO} \text{dimer}$$

$$2\ CH_3O_2C(CH_2)_8COCHN_2 \xrightarrow[h\nu]{CuO} CH_3O_2C(CH_2)_8COCH{=}CHCO(CH_2)_8CO_2CH_3$$

TABLE 38. The catalytic decomposition of 9-diazofluorene by copper(II) carboxylates in aqueous dimethyl formamide[2237]

Cu(II) carboxylate	$Ar_2C\dashv_2$ (with AcO on C)	$Ar_2C_2\ddagger_2$	Ar_2CO	$Ar_2C{=}N\dashv_2$
Cu(II) carboxylate	(%)	(%)	(%)	(%)
Acetate	55	24	15	
Proprionate	47	17	18	
n-Butyrate	51	16	23	
Isobutyrate	36	36	19	
Tartrate		85		12

TABLE 39. Decomposition of diphenyldiazomethane by copper(II) carboxylates in aqueous dimethyl formamide[2236]

Cu(II) carboxylate	$Ph_2C\dashv_2$ (with AcO on C)	$Ph_2C\ddagger_2$	$Ph_2C{=}N\dashv_2$
Cu(II) carboxylate	(%)	(%)	(%)
Acetate	70	—	—
Proprionate	61	—	—
n-Butyrate	54	—	—
Isobutyrate	48	—	—
Benzoate	14	—	63
Tartrate	—	58	4
Glycinate	—	5	48
Salicylate	—	7	45

TABLE 40. Decomposition of diazoalkanes by metal acetates[2236, 2237]

Metal acetate	Solvent	Ar_2CHOAc	$Ar_2C\ddagger_2$	$Ar_2C\dashv_2$ (with AcO on C)	$ArC(OAc)_2$	Ar_2CO	Azine
		Products from diazofluorene (% yield)					
$Cr(OAc)_3$	DMF (aq)	9				55	
$Cu(OAc)_2$	DMF (aq)		24	55		15	
$Tl(OAc)_3$	CH_2Cl_2				42	56	
$Pb(OAc)_4$	CH_2Cl_2				61	30	
		Products from diphenyl diazomethane (% yield)					
$Cr(OAc)_3$	DMF	35				30	25
$Cr(OAc)_3$	DMF (aq)	40				27	
$Cu(OAc)_2$	DMF			52		40	
$Cu(OAc)_2$	EtOH			47		40	
$Cu(OAc)_2$	H			52		32	
AgOAc	DMF			17		20	
$Hg(OAc)_2$	DMF			5	27	50	
$Hg(OAc)_2$	Et_2O			50	3	30	
$Tl(OAc)_3$	DMF				21	50	
$Tl(OAc)_3$	Et_2O				53	Trace	

where metal salts are employed as catalysts, we suspect the intermediacy of carbenoids in the olefin-forming step and thermal processes or MT-1 process are responsible for any azines obtained. It is noteworthy that the processes examined by Nozaki and coworkers employ very large quantities of catalyst whereas those examined by Wulfman and collaborators operated at very low catalyst concentrations.

TABLE 41. Decomposition of Ar_2CN_2 by copper(II) acetylacetonates in benzene

Ar	Acetyl acetone	Yield (%)	
		$Ar_2C{=}CAr_2$	$Ar_2C{=}N{-}N{=}CAr_2$
C_6H_5	2,4-Pentadione	60	30
C_6H_5	1-Phenyl-1,3-butadione	47	46
C_6H_5	1,3-Diphenyl-1,3-propadione	41	50
C_6H_5	Acetoacetic ester	74	0
p-$MeOC_6H_4$	Acetoacetic ester	48	0
C_6H_5	1,1,1-Trifluoro-2,4-pentadione	84	15

Some additional examples of dimer formation and azine formation appear in equation (151)[1597, 1756, 1883, 2249, 2377].

Internal dimers (acetylenes) can be prepared by decomposition of 1,2-bisdiazoalkanes. The process has been reviewed by Meier[2033] and selected examples can be found there.

IV. CYCLOPROPANATION REACTIONS

A. Choosing Reaction Conditions

Perhaps the most common technique used by chemists to establish reaction conditions is that of precedent. One examines a limited selection of the literature to find a reported reaction similar to the one under consideration, and adapts it to fit one's needs or, more commonly, uses it with a minimum of change. The precedent may well have arisen in the same fashion. Consequently the conditions employed for performing a reaction may be far from optimum. In the extreme, precedent can become a case of the blind leading the blind. Syntheses with diazo alkanes are not unique in suffering from this situation. There is very little data available to permit answering the questions—Should photolysis, thermolysis or catalysis be employed? Should a pyrazoline be generated and then be decomposed? What catalyst is best for the reaction under consideration? Will solvents and/or temperature changes improve yields? Is it advantageous to purify in order to remove adventitious materials? If so, how? How might one avoid the problem?

We know of two relatively thorough analyses of the question whether photolysis or catalysis is preferred for decomposing diazoalkanes to furnish cyclopropanoid products. They both indicate that the catalytic approach is preferred. However, it should be remembered that this is not an all-encompassing rule. Thus, with dimethyl diazomalonate, photochemical decomposition leads to favoured generation of the most highly substituted cyclopropane while catalysis strongly favours the least-substituted cyclopropane. It is not possible to generate the adduct **690** catalytically but it is possible to do so photolytically[2368]. With 1-methylcyclohexene, the methods appear comparable, but with cyclohexene homogeneous catalysis is better than heterogeneous catalysis which is equal to or superior to photolysis. With cyclohexadienes, heterogeneous catalysis appears preferable for adding diazomalonates.

$X = COOCH_3 = n_a$
$X = COOC(CH_3)_3 = n_b$

( )$_n$ $\xrightarrow[(Cu)]{N_2CX_2}$ CHX_2 ( )$_n$ + ( )$_n$ X_2 + $X_2C{=}CX_2$

(656) $n = 1$ **(657a, b)** **(660a, b)** **(663a, b)**
$n = 2$ **(658)** **(661)** **(663a)**
$n = 3$ **(659)** **(662)** **(663a)**

(152)[2187, 2368]

(664) $\xrightarrow[(Cu)]{N_2CX_2}$ CHX_2 **(665)** + CHX_2 **(666)** + CHX_2 **(667)** + X_2 **(668)** + **663a**

(669) $\xrightarrow[(Cu)]{N_2CX_2}$ (666) (667) + X_2 **(670)** + **663a**

(671) $\xrightarrow[(Cu)]{N_2CX_2}$ CHX_2 **(672)** + CHX_2 **(673)** + **663a**

(674) ⟶ X_2 **(675)** $\xrightarrow[Rh/Al_2O_3]{H_2}$ X_2 **(662)**

(676) ⟶ X_2 **(677)** + X_2 **(678)** +

(679) ⟶ **(680)** + **663a** + X_2 **(681)** $\xrightarrow{H_2, Rh/Al_2O_3}$ **657a**

(682) ⟶ **663a** + **680** + X_2 **(683)** $\xrightarrow{H_2, Rh/Al_2O_3}$ **657a**

$\xrightarrow{N_2CHCOOEt}$ COOEt **(684)** + COOEt **(685)** + COOEt, COOEt **(686)** + EtOOC, COOEt **(687)**

(688) $\xrightarrow[\text{Catalyst}]{N_2CX_2}$ (689) $\xrightarrow[N_2CX_2]{h\nu}$ (690) + 689

(691) + (692) + N_2CHCO_2Et (693) ⟶ (694) + (695)

Catalyst	694+695 (%)	694 (mol.-%)	695 (mol.-%)
$CuOSO_2CF_3$	84	20	80
$Cu(OSO_2CF_3)_2$	98	21	79
$Cu(BF_4)_2$	98	19	81
$CuI.P(OCH_3)_3$	36	64	36
$CuCl.P(OCH_3)_3$	60	58	42
$Cu(AcAc)_2$	96	64	36
$CuSO_4$		64	36

CH_2N_2 | $Cu(OSO_2CF_3)_2$

1 : 2·5 : 0·4

(with $Cu(AcAc)_2$ 1 : 1 : 0·6)

With diazomethane, catalysis was found superior to photolysis. When two possible sites are involved with diazoacetate ester, the type of catalyst employed is of considerable importance and the same holds for diazomethane. Thus when copper(I) fluoroborate or copper(I) triflate were employed in the sequence **691**+**692**→**694**+**695**, **694** and **695** were generated 20:80 whereas with copper(I) iodide-trimethylphosphite, the order is reversed (64 : 36)[2187].

The work of Peace and McDaniel clearly reveals that the concentration of the catalyst is important for the formation of all the products from dimethyl diazomalonate and work with vinyl diazomethane also exhibits a yield maxima for cyclopropanation as a function of catalyst concentration. One must therefore find the optimum catalyst concentration and the optimum catalyst. With the diazomalonate system, copper(II) acetylacetonate proved to be the best catalyst. Copper(II) fluoroborate was comparable, but less convenient. The studies of Wulfman, Peace and McDaniel[2004, 2127–2133, 2355–2368] with diazomalonates and diazoacetates may be suitable for extrapolation to nearly all diazo carbonyl compounds and diazosulphones. (The only justification for including the sulphones is a strong similarity in their mass spectral behaviour with the esters. Both classes of molecules fail to fragment to a P-28 ion when glass inlet systems are employed.) This classification system may prove to be of use if sufficient data are forthcoming on other diazo compounds.

It is clear from tabulations of relative reactivities that it is possible to find conditions which frequently favour one unsaturated substrate over another. The data in Table 42 compare the selectivity of 'bis(methoxycarbonyl) carbene' generated by

TABLE 42[2004]

Cyclohexene + 1-methylcyclohexene $\xrightarrow{N_2CX_2}$ A + B

X = CO_2Me

Means	Solvent	Temperature (°C)	Product ratio (B : A)
$CuCl.P(OCH_3)_3$	Neat	Reflux	1 : 4·90
$Cu(BF_4)_2$	Neat	Reflux	1 : 13·82
$Cu(BF_4)_2$	C_6F_6	Reflux	1 : 2·29
$Cu(AcAc)_2$	Neat	Reflux	1 : 2·73
$Cu(AcAc\text{-}f_3)_2$	Neat	Reflux	1 : 2·96
$Cu(AcAc\text{-}f_6)_2$	Neat	Reflux	1 : 2·58
Photolysis	Neat	Reflux	1·1 : 1
Photolysis (Ph_2CO)	Neat	35	2 : 1
Photolysis	Neat	35	3 : 1

several methods. It also contains information relative to the effect of temperature upon the processes. This information and that in Table 43 is almost non-existent for other systems. This may be a consequence of the view which has been put forth to the effect that the activation energy for adding a carbene to a double bond is zero[654] and there should be no temperature effect on product distribution for thermally and photochemically generated carbenes. There appears to be no *a priori* justification for making the same assumption for catalysed processes and the data in Table 42 demonstrate the importance of temperature.

It is equally clear that the actual catalyst employed for catalytic processes is of importance. Wulfman, Peace and McDaniel examined a large number of catalysts in their studies (Tables 43–47) and found that the yields and partial rate data were

TABLE 43[2368]. Effect of catalyst upon yield in the reaction of dimethyl diazomalonate with cyclohexenes

Olefin:	Cyclohexene			1-Methylcyclohexene					1,2-Dimethylcyclohexene		
Catalyst Products:	660a[b]	657a[b]	663a[b]	668[b]	665[b]	666[b]	667[b]	663a[b]	673[b]	672[b]	663a[b]
Cu[a]	38·0	1·71	8·05	22·9	1·23	4·13	18·5	5·64	27·5	7·88	16·6
CuCl[a]	42·8	2·18	8·38	19·1	1·45	3·67	17·5	4·83	27·5	7·44	22·4
$CuSO_4$[a]	45·3	2·08	9·07	24·8	2·49	6·07	24·0	7·36	16·0	5·55	21·9
$(CH_3O)_3P\text{–}CuCl$[c]	63·7	4·67	18·4	29·6	2·30	10·7	26·0	5·15	32·8	10·4	19·5
$(CH_3O)_3P\text{–}CuI$[c]	73·5	5·98	11·5								
$Cu(AcAc)_2$[c]	78·1	5·92	12·4								
$(CH_3O)_3PCuCN$	41·5	0·94	1·0								
$(CH_3O)_3P\text{–}CuNCS$	28·1	1·80	1·8								
$(CH_3O)_3P\text{–}CuBr$[c]	68·1	6·63	20·6								
$(CH_3O)_3P\text{–}CuBF_4$	59·2	7·99	10·0								
$[(CH_3O)_3P]_2CuI$[c]	65·1	6·52	22·4								
$[(CH_3O)_3P]_3CuI$[c]	63·0	6·56	27·0								
$(C_6H_5O)_3PCuBr$[c]	72·7	7·39	19·9								
$[(CH_3)_2CHO]_3PCuCl$[c]	66·0	5·39	27·4								
$AgBF_4$[a]	19·6	3·40	8·20								

[a] Heterogeneous systems.
[b] Yield (%) based on VPC analysis and available dimethyl diazomalonate.
[c] Optimized yield for cyclopropane formation.

TABLE 44. Copper(II) catalyst versus products in the reaction of diazomalonate in cyclohexene[2368]

Catalyst ligand	Product yield (relative to norcarane)			
	660a	**657a**	**663a**	Tetrakis-methoxy-carbonyl ethane
Dipivaloylmethane	0·89 (1·00)	0·07 (0·09)	6·75 (7·52)	3·99 (4·43)
Acetylacetone	79·45 (1·00)	2·54 (0·03)	8·07 (0·10)	0·93 (0·01)
Acetylacetone-f_3	8·47 (1·00)	0·68 (0·08)	0·28 (0·03)	0·40 (0·05)
Acetylacetone-f_6	18·44 (1·00)	0·94 (0·05)	17·25 (0·94)	0·72 (0·04)
Thenoyltrifluoroacetylmethane	21·42 (1·00)	1·36 (0·05)	2·01 (0·09)	0·59 (0·03)
Benzoylacetylmethane	9·83 (1·00)	0·26 (0·03)	7·37 (0·75)	2·47 (0·25)
Acetate.(H_2O)	22·59 (1·00)	1·96 (0·09)	7·14 (0·32)	1·84 (0·08)
Octoate	34·475 (1·00)	1·57 (0·05)	4·20 (0·12)	3·50 (0·10)
Stearate	32·53 (1·00)	2·13 (0·07)	4·04 (0·12)	3·74 (0·11)
Ethyl acetoacetate	27·67 (1·00)	1·70 (0·06)	8·75 (0·32)	2·24 (0·08)

Numbers in parentheses are relative yields based on **660a** ≡ 1.

TABLE 45. Products and yields from the reaction of dimethyl diazomalonate with 2-heptenes[2368]

0·14 mmol of $(CH_3O)_3P.CuX$	2-Heptene (purity of isomer)	Temperature (°C)	Cyclopropane		C—H insertion products	**663a**
			cis	*trans*		
X = I	*cis* (96%)	98	95·6	Trace	0·52	3·36
I	*cis* (96%)	85	92·1	Trace	3·30	4·64
Br	*cis* (96%)	85	87·6	Trace	7·08	5·32
Cl	*cis* (96%)	85	81·3	Trace	12·8	6·00
I	*trans* (99%)	98	0·00	76·0	6·60	17·4
I	*trans* (99%)	85	0·00	69·8	18·6	11·6
Br	*trans* (99%)	85	0·00	72·5	18·4	9·20
Cl	*trans* (99%)	85	0·00	41·7	35·2	23·1

TABLE 46. Effect of additives upon yields and product distribution in the reaction of cyclohexene with dimethyl diazomalonate using $(CH_3O)_3P.CuCl$ as catalyst[2368]

Additive (30 mmol)	**660a**	**663a**	**657a**
None	63·92 (1·00)	13·25 (0·207)	4·69 (0·074)
CuCl	33·46 (1·00)	33·44 (1·00)	2·29 (0·071)
$(CH_3O)_3P$	46·07 (1·00)	21·18 (0·460)	3·75 (0·082)
$(CH_3O)_3PO$	68·13 (1·00)	16·23 (0·238)	5·92 (0·087)
$(CH_3O)_2CH_3PO$	60·58 (1·00)	19·20 (0·317)	5·08 (0·084)
$[(CH_3)_2N]_3P$	49·94 (1·00)	24·28 (0·486)	4·61 (0·092)
$[(CH_3)_2N]_3PO$	45·82 (1·00)	17·34 (0·378)	3·72 (0·081)
$CuCl_2$	55·00 (1·00)	17·96 (0·326)	5·34 (0·097)
$(CH_3O)_2HPO$	59·54 (1·00)	25·14 (0·422)	4·48 (0·077)

Numbers in parentheses are relative yields based on **660a** ≡ 1.

TABLE 47. Product distribution as a function of amount of catalyst or means of carbenoid species generation with 1-methylcyclohexene as substrate at 110 °C[2127]

Amount of catalyst (mg)	Product and yield[a]									
	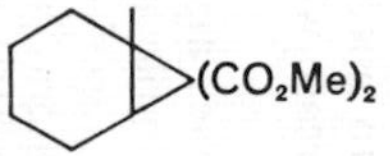 $>(CO_2Me)_2$		 $CH(COOMe)_2$		 $CH(COOMe)_2$		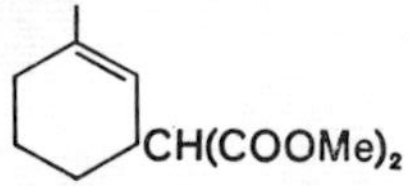 $CH(COOMe)_2$		Dimer	
	A (%)	R	A(%)	R	A(%)	R	A(%)	R	A(%)	R
1	21·53	1·0	1·51	0·070	7·78	0·361	19·54	0·904	4·03	0·187
2	23·07	1·0	1·68	0·073	8·72	0·378	20·12	0·874	3·84	0·166
4	22·34	1·0	1·63	0·073	8·45	0·378	19·59	0·877	2·48	0·111
8	21·80	1·0	1·68	0·077	8·05	0·369	18·91	0·868	3·56	0·163
16	23·21	1·0	1·80	0·077	8·38	0·361	20·39	0·878	3·69	0·159
32	20·39	1·0	1·45	0·071	7·11	0·349	16·90	0·830	4·78	0·234
64	20·12	1·0	1·54	0·077	7·98	0·396	19·45	0·965	3·84	0·191
128	19·38	1·0	1·49	0·077	7·44	0·384	18·71	0·966	5·08	0·262
256	18·51	1·0	1·01	0·055	7·63	0·358	14·89	0·805	7·77	0·420
512	8·72	1·0	0·40	0·046	3·76	0·431	9·05	1·040	22·67	2·60
1024	6·77	1·0	No appreciable amount		2·07	0·307	4·81	0·715	19·32	2·87
Thermal	12·54	1·0	1·09	0·087	0·84	0·067	8·65	0·686	None	
Photolytic 35 °C	6·61	1·0	0·80	0·121	No appreciable amount		9·05	1·37	No appreciable amount	

[a] A = absolute yield; R = relative yield.

a function of the associated ligands. These can be assumed to alter the hardness or softness of the catalyst, and likewise it is clearly possible to alter the hardness of the various species by changing the valence state of the catalyst, the metal ion, the solvent, the substrate and the diazo compound. With the majority of the diazo compounds which have been catalytically decomposed, copper salts were almost uniformly better than salts of other metals when one is considering cyclopropanation, dimeric olefin formation, pinacol formation and azine formation. For simple insertion into C—X bonds (X≠H) the picture is not as clear as to which salts (Lewis acids) should be employed.

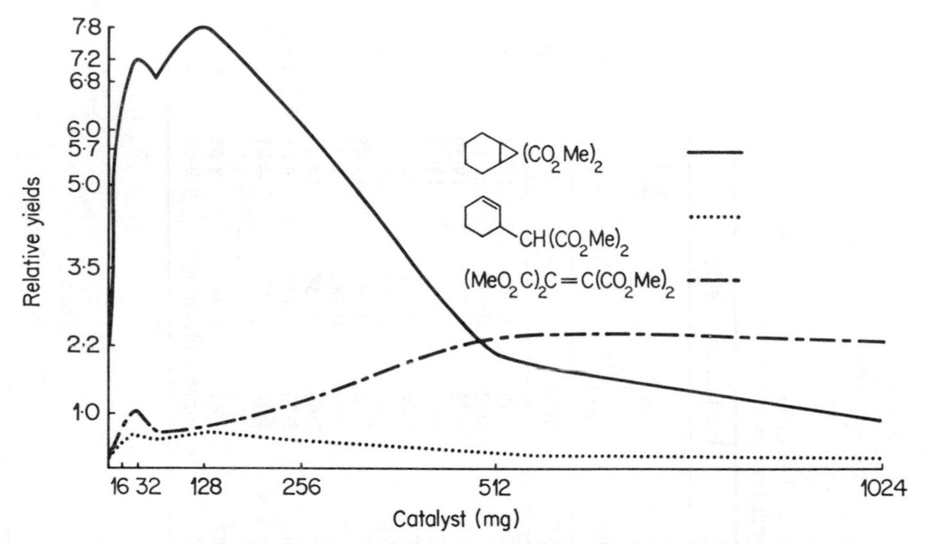

FIGURE 1. Product distribution as a function of $[(CH_3)_2CHO]_3P.CuCl$ concentration.

The choice between photochemical methods (sensitized or unsensitized) is even less clear. Moss[832] has summarized the relative rates of cyclopropanation for a number of 'carbenes' at 25 °C. It is not unusual to find sufficient selectivity to justify a choice. Whether the discrimination will be further enhanced by operating at lower temperatures should probably be examined on a case-by-case basis. However, with even a simple circulating system it is possible to operate at ~ −10 °C using brine cooling and one would expect selectivity to increase with decreasing temperatures if there is an observable temperature effect.

The use of triplet sensitizers has been used by Jones and Ando[1917] to suppress Wolff rearrangements of diazoketones and favour bimolecular cyclopropanations. In addition Ando and coworkers have successfully discriminated between C—X insertion and cyclopropanations by using catalytic systems or photolysis, when they employed an allyl—X type system (X = O, S, halogen)[1415, 1416, 1418, 1423–1427, 1432, 1438].

With diazoacetate and diazomalonate systems, homogeneous catalysis appears to be the preferred method when all methods give the same cyclopropane. However, the yield is a function of catalyst concentration[2127, 2130, 2132, 2148, 2368] and it is desirable where possible to optimize against this variable (Figure 1). With diazoacetates the *syn–anti* ratio is a function of catalyst concentration. Similarly, for dimerizations, the *cis–trans* olefin ratio is a function of catalyst concentration (see Tables 35–37).

Cowan and collaborators[1597] have examined the decomposition of diazoacetophenone in the presence of catalysts as well as photochemically and thermally. They concluded that better yields of cyclopropanes and greater stereoselectivity were obtained using copper catalysis.

The question of which copper catalyst system should be employed for cyclopropanation is fairly evident from the data in Tables 43, 44 and equation (152) and the best catalyst would appear to be copper(II) fluoroborate if a fairly accurate knowledge of the amount of water present can be made. Copper(II) acetylacetonate is much more convenient, is soluble at optimum concentration and furnishes essentially the same results. However, the fluoroborate has the advantage that when selectivity is desired between two differently substituted double bonds, one can in some cases obtain complementary results by using it instead of the acetylacetonate. Of course this proviso only applies when a tendency to give the most substituted cyclopropane exists, such as occurs with diazomethane or diazoacetic esters[2187].

The reaction of ethyl diazoacetate with $Cu(AcAc)_2$ has been studied by Sato[2189]. The main product was found to result from a reaction between one 'carbene' species and two molecules of acetylacetonate.

$$Cu(AcAc)_2 + N_2CHCOOEt \longrightarrow \begin{array}{c} CH(COMe)_2 \\ | \\ HC{-}COOEt \\ | \\ CH(COMe)_2 \end{array}$$

Diethyl maleate, a frequent product from diazoacetic ester reactions, was isolated in 45% yield. It is perhaps mechanistically significant that the thermodynamically more stable diethyl fumarate was not reported to be among the products.

Peace, McDaniel and Wulfman have examined the dependence of the dimerization reaction upon catalyst concentration (Tables 35–37) and they found that maleate formation is favoured by low catalyst concentrations whereas fumarate is favoured by high catalyst concentrations when operating in cyclohexene solutions. They ascribe this phenomenon to dipole–dipole repulsions in the transition states and to increased steric requirements at high catalyst concentrations[2365]. The rationale employed is that at low catalyst concentrations a carbenoid reacts with diazo compound to form dimer, but at high catalyst concentrations two carbenoids also react to furnish product.

One of the few early studies which has compared the catalytic activities of copper chelates to copper salts was performed by Hammond and coworkers[1597]. The copper(II) dipivaloylmethide complex was shown to be inferior to other copper salts in the addition of α-diazoacetophenone to cyclohexene. However, all the reactions were carried out at 28 °C and no effort was made to optimize reaction conditions or catalyst concentrations. It is most probably dangerous to draw conclusions from any study which fails to examine the obvious variables of temperature, concentration and olefin structure.

There does seem to be some question as to the actual catalytic species present when the phosphite catalysts are used. Moser[2055, 2056] noted that the catalyst solution turned brown early in the addition of the diazo compound. House and Blankley noted the appearance of an apparently insoluble material in all room temperature reactions utilizing the trialkylphosphite copper(I) halide catalysts[1883].

The observations of Peace and Wulfman cast doubt as to the actual catalyst present in Moser's studies. They found Moser's concentration conditions highly unsatisfactory for generation of bis(methoxycarbonyl) carbenoid. By operating at much lower catalyst concentrations they obtained far better yields (Table 44) and were able to demonstrate that copper(II) species are probably the active catalysts. Since copper(I) salts are almost invariably contaminated with copper(II) salts, the use by Moser of approximately 400 times as much formal copper(I) catalyst as Wulfman and Peace required under oxidative conditions is highly suspect. Peace[2359] found that the amount of copper(II) initially present significantly altered the melting points of copper(I) salt–phosphite complexes. Indeed, three different preparations of copper(I) chloride–trimethyl phosphite had three different melting points, exhibited different ^{31}P magnetic resonance spectra but had essentially identical infrared spectra, proton n.m.r. and total elemental analyses. The presence of the additional ligands when copper iodide catalysts are employed most probably alters the redox potentials of the Cu(II), Cu(I), I^- and I_2 systems as well and further precludes using the non-existence of copper(II) iodide in aqueous media as evidence of the presence of copper(I) iodide systems [copper(II) iodide is a known compound].

The presence of phosphite is inherently bad from the standpoint of carbenoid generation, for phosphites trap carbenes[1880] and generate phosphinazenes[1611] from diazo compounds. All of these factors were examined and Peace found that several additives grossly depressed yields. In the case of copper(I) and copper(II) chlorides, the effects may well result from the common-ion effect. With phosphite esters they found that the resulting phosphinazenes do not generate carbenes upon catalytic decomposition as is the case with phosphinazenes derived for diphenyldiazomethanes[2359].

Phosphorus derivatives do not always cause troubles. Regitz[1046] prepared a number of phosphorus-containing diazo compounds containing the phosphoro and phosphenyl groups which undergo 'carbene' reactions. This is clearly a consequence of the phosphorus being pentavalent.

Peace and Wulfman found the catalytic activity of $Cu(AcAc)_2$ is completely independent of the presence or absence of peroxides (Tables 48–50)[2358]. Hence they

TABLE 48. Copper(0) catalysed reactions[2359]

Condition	Yield (%) A (bicyclo[4.1.0]heptane, 7,7-X_2)	B (cyclohex-2-enyl-CHX_2)	C ($X_2C{=}CX_2$)
Thermal	12·7	Trace	0·00
Cu metal	38·0	8·05	1·71
Metal-free filtrate	36·0	7·20	2·78

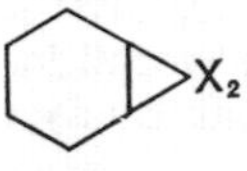

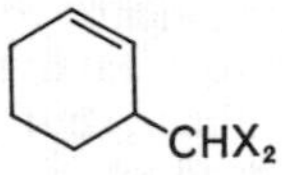

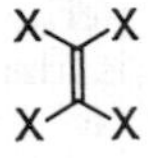

TABLE 49. The effect of peroxide upon the $Cu(AcAc)_2$-catalysed decomposition of dimethyl diazomalonate in cyclohexene[2359]

Condition	Yield (%) A (bicyclo[4.1.0]heptane, 7,7-X_2)	B (cyclohex-2-enyl-CHX_2)	C ($X_2C{=}CX_2$)
Peroxide free	78·5	12·4	5·92
Peroxide present	78·1	12·4	5·81

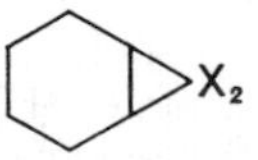

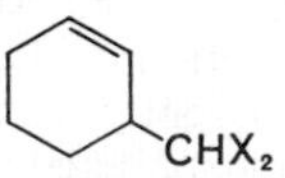

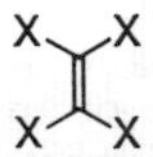

believe that copper(II) is the active catalyst species or is at least far superior to copper(0) and copper(I) for decomposing diazoesters and, by implication, diazoketones.

A somewhat surprising result was obtained when AR grade copper powder was employed as catalyst[2358]. The copper powder can be removed by filtration from the olefin after 4 h reflux and the supernatant solution will exhibit catalytic activity towards dimethyl diazomalonate (Table 48).

TABLE 50. Product distribution and yields in the reaction of cyclohexene and dimethyl diazomalonate as a function of catalyst and peroxide content of the olefin[2359] (A, B, C as in Tables 48 and 49)

$(CH_3O)_3P.CuZ$ 0·14 mmol	Yield (%)			Ratios
	A	B	C	
Z = Br[a]	68·7	15·1	6·66	1·00 : 0·219 : 0·097
I[a]	74·3	12·9	7·43	1·00 : 0·174 : 0·100
Br[b]	22·0	1·88	2·08	1·00 : 0·085 : 0·095
I[b]	19·9	2·02	1·79	1·00 : 0·101 : 0·090
Br[c]	71·6	22·9	5·72	1·00 : 0·320 : 0·080
I[c]	78·8	15·0	5·92	1·00 : 0·191 : 0·075
No catalyst[c]	9·45	1·82	0·00	1·00 : 0·192

[a] Commercial cyclohexene.
[b] Commercial cyclohexene filtered through alumina.
[c] Commercial cyclohexene filtered through alumina, then 0·07 mmol benzoyl peroxide added.

The effect of peroxide, hydroperoxide and molecular oxygen impurities is also of importance for the photochemical reactions of diazomalonate systems. Poling and Wulfman[2143, 2360, 2362] have found that the photolysis of dimethyl diazomalonate in cyclohexene in the absence of sensitizers has $\Delta E_a \approx 0$ kcal/mol. They also found that the rate of cyclopropanation did not relate directly to the loss of diazomalonate and that the cyclopropane was being formed even after all diazomalonate was consumed and therefore must arise from an intermediate other than the carbene. They found that the presence of small quantities of peroxide could double the rate of photolysis, that filtration through alumina and working under an argon atmosphere did not remove and prevent peroxides from persisting (it only removed hydroperoxides), and that addition of air to the system leads to a very impressive increase in rate of loss of diazo compound, perhaps by oxidation of the 'carbene'. In addition, Wulfman has shown that there is a symmetry-allowed pathway for stereospecific cyclopropanation with radical catalysis[2366]. Hughes[1884] has recommended purifying olefins by filtration through a column of calcined magnesia. The magnesia is calcined at ~500 °C and cooled in an inert and dry atmosphere. This furnishes olefins suitable for metatheses, processes extremely sensitive to oxygen, peroxides and hydroperoxides. Filtration is of course performed under an inert atmosphere.

The catalyst species most probably contributing to Peace and McDaniel's results with Moser's catalysts is an undefined copper(II) alkoxy halide. Fortunately, sufficient peroxy impurities were always present to ensure attaining optimum conditions. This conclusion is not particularly unreasonable because the amount of catalyst (0·14 mmol) is present in about 500 mmol of olefin. Thus, a peroxide content of only 0·03% will furnish sufficient alkoxy radicals to convert all catalyst present to the copper(II) species.

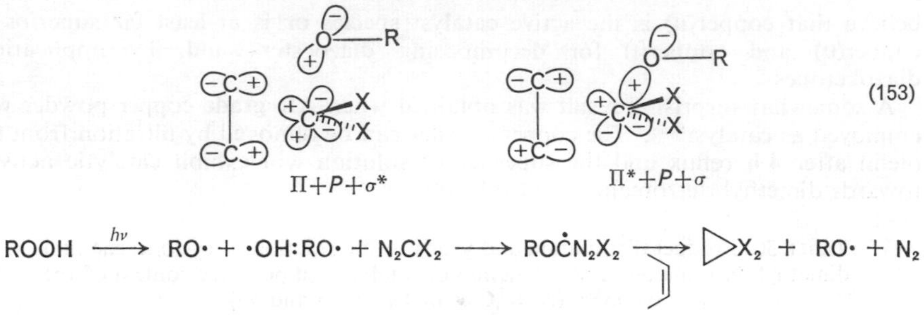

$\Pi + P + \sigma^*$ $\Pi^* + P + \sigma$ (153)

$$\text{ROOH} \xrightarrow{h\nu} \text{RO}\cdot + \cdot\text{OH}:\text{RO}\cdot + \text{N}_2\text{CX}_2 \longrightarrow \text{RO}\dot{\text{C}}\text{N}_2\text{X}_2 \longrightarrow \triangleright\text{X}_2 + \text{RO}\cdot + \text{N}_2$$

Symmetry-allowed path for peroxide-catalysed cyclopropanation of olefins by diazoalkanes

A comparison of the various catalysts generally employed for 'carbenoid' generation was made with cyclohexene and 1-methylcyclohexene (Table 43). In order to perform these studies it was necessary to examine the variable, concentration, when operating in the homogeneous systems. The fact that there should be optimum catalyst concentrations for the various products is not surprising.

Employing catalysts of the type $(CH_3O)_3P.CuX$, Peace and Wulfman found the amount of allylic C—H insertion (the triploid behaviour) relative to cyclopropanation increases as the leaving group ability of X increases (Table 43). When common anions were added to reactions involving trialkylphosphite copper(I) chloride and iodide, the decomposition of diazo compound was severely depressed even though there was insufficient anion added to saturate the ligancy[2359, 2367]. On the other hand, addition of fluoroborate ion to the same reaction mixtures had only a modest effect (Table 51).

TABLE 51. Effect of common ion on the reaction of cyclohexene with dimethyl diazomalonate (A, B, C as in Table 48)[2359]

Catalyst	Salt	Yield (%)		
		A	B	C
$(CH_3O)_3P$–CuI	None	74·23 (1·00)	12·88 (0·174)	7·42 (0·101)
$(CH_3O)_3P$–CuI	$(CH_3)_4NI$	3·33 (1·00)	0·00	7·08 (2·12)
$(CH_3O)_3P$–CuI	$(CH_3)_4NBF_4$	63·66 (1·00)	11·36 (0·178)	5·00 (0·078)
$(CH_3O)_3P$–CuCl	None	63·92 (1·00)	13·25 (0·207)	4·69 (0·074)
$(Ch_3O)_3P$–CuCl	$(CH_3)_4NCl$	8·58 (1·00)	1·57 (0·183)	0·00
$(CH_3O)_3P$–CuCl	$(CH_3)_4NBF_4$	49·33 (1·00)	5·26 (0·107)	4·08 (0·083)

The question of what solvent to use when excess olefin cannot conveniently be employed was examined by Peace (Table 52) and when this is considered in conjunction with Marchand's[2017] and Ando's studies[1415–1439], it becomes clear that heteroatom-containing solvents such as ethers and halocarbons are generally to be avoided because of side reactions.

Conversely, Regitz[1046] has decomposed a number of phosphorus-substituted diazoalkanes in CH_2Cl_2 and obtained excellent yields. Apparently the phosphorus substituents moderate the reactivity sufficiently so that C—Cl bonds are not affected.

When the degree of selectivity is sufficiently high, one might employ benzene as solvent. Peace[2129] found that with diazomalonates, hexafluorobenzene was an inert solvent which furnished highly stereoselective reactions for photolyses of diazomalonate and stereospecific addition when copper catalysts were involved. Jones[1920] found similar behaviour for the photolysis, but with the photolysis of 9-diazofluorene in hexafluorobenzene, he found the 'triplet' processes were favoured.

TABLE 52. Solvent effects in the reaction of dimethyl diazomalonate with cyclohexene using $(CH_3O)_3P{-}CuCl$ (0·14 mmol) as catalyst (**A**, **B**, **C** as in Table 48)[2127]

Solvent (90 mol.-%)	Yield (%)		
	A	**B**	**C**
Cyclohexene	63·92 (1·00)	13·25 (0·207)	4·69 (0·074)
Cyclohexane	31·23 (1·00)	52·52 (1·681)	3·90 (0·125)
Benzene	57·91 (1·00)	39·64 (0·684)	3·34 (0·058)
Hexafluorobenzene	81·00 (1·00)	15·87 (0·196)	4·24 (0·052)
Carbon tetrachloride	28·31 (1·00)	5·71 (0·202)	31·66 (1·13)
Dimethoxyethane	0·00	0·00	0·00

The question of stereospecificity is exceedingly complex and much has been written on whether a singlet or triplet carbene is responsible for observed behaviour, whether heavy atoms favour intersystem crossing from the singlet state to the triplet state, whether oxygen removes triplet carbenes, and whether photolyses even involve carbenes as a general rule.

The copper-catalysed decompositions of diazoalkylcarbonyl compounds appear to occur in a stereospecific *cis* fashion. However, when an aralkyldiazo ketone was employed[1597] (diazoacetophenone), the reaction was only stereoselective and favoured the *cis* pathway. This can be nicely accounted for using the systematology of Wulfman[2363] that classes methylene transfers (**MT**) as being of the **MT-1** or **MT-2** type (equation 154) and the various possible **MT-2** paths (*vide infra*). Komendantov found that stereospecific addition of alkoxycarbonyl copper carbenoids does occur[1957].

$$\text{alkene} + CX_2{=}CuL_n \xrightarrow{\textbf{MT-2sync}} CuL_n + \text{cyclopropane}{>}X_2$$

$$CX_2{=}CuL_n \underset{}{\overset{\textbf{MT-1}}{\rightleftarrows}} CX_2 + CuL_n;\ CX_2 + \text{alkene} \longrightarrow \text{cyclopropane}{>}X_2 \qquad (154)$$

$$CuL_n{=}CX_2 + \text{alkene} \xrightarrow{2+2} \underset{\textbf{(696)}}{\text{metallacyclobutane }(CuL_n, X_2)} \xrightarrow{\textbf{ia}} CuL_n + \text{cyclopropane}{>}X_2$$

In the case where the process is non-stereospecific and the evidence available indicates that photolysis furnishes ultimately 'triplet carbenes', the failure of copper catalysis to lead to a stereospecific process may be indicative of mixed **MT-1, MT-2** processes. Alternatively it may be evidence that the **MT-2** process is not concerted and is occurring in a stepwise fashion such as is shown in equations (155), (156) and (157).

Existing data indicate a greater degree of stereoselectivity when a copper catalyst system is employed. The question of *syn*/*anti* addition of unsymmetrical groups is even less clearly defined and the original literature needs to be consulted. However,

TABLE 53. Dependence of yield of 7-ethoxycarbonyl norcarane, diethyl maleate and diethyl fumarate upon cyclohexane concentration at 40 °C with diazoacetic ester concentration at 0·22 M

Concentration		Yield of reaction products (%, based on diazoacetic ester)									
Cyclo-hexene (M)	Catalyst (10^{-4} M)	*endo*-Norcarane	*exo*-Norcarane	Total norcarane	*endo*/*exo*[a]	Fumarate (F)	Maleate (M)	Total Dimer	F/M[a]	Total yield	Norcarane/dimer
—	2·36	—	—	—	—	49·0	48·0	97·0	1·02	97·0	—
0·20	2·36	0·3	5·5	5·8	0·054	40·5	47·5	88·0	0·85	93·8	0·07
0·31	2·36	0·5	8·7	9·2	0·057	37·0	44·0	81·0	0·84	90·2	0·11
0·68	2·36	1·1	17·2	18·3	0·064	34·2	42·5	76·7	0·80	95·0	0·24
1·23	2·36	1·6	25·5	27·1	0·063	28·3	37·0	65·3	0·76	92·4	0·42
3·74	4·15	2·6	42·5	45·1	0·062	21·5	29·8	51·3	0·72	96·4	0·88
6·24	4·15	3·2	51·0	54·2	0·063	16·8	21·3	39·9	0·79	94·1	1·36
9·96	4·15	3·5	56·0	59·5	0·063	14·2	19·5	33·7	0·73	93·2	1·77

[a] Calculated from data presented in Reference 2316a.

TABLE 54. Reaction of *cis*- and *trans*-4-octenes with methoxycarbonylcarbene from photolytic decomposition of methyl diazoacetate[1957]

4-Octene		Methyl	Hexafluoro-	Reaction	Total	Relative yield of esters (%)		
Isomer	Amount (g)	diazoacetate (g)	benzene (g)	time (min)	yield of esters (%)	*trans*	*cis anti*	*cis syn*
trans	26·5	3·0	—	90	23	98	2	—
	22·0	2·0	18·6	70	19	84	16	—
	19·3	4·3	32·4	150	24	78	22	—
	13·2	3·0	67·0	150	22	76	24	—
	9·2	2·25	83·7	100	23	75	25	—
	5·5	1·25	93·0	60	22	76	24	—
cis	51·0	5·0	—	180	21	~3	62	35
	29·7	1·0	24·2	60	24	16	60	24
	19·3	2·0	32·4	90	28	30	57	13
	7·7	1·0	39·1	60	21	77	23	—
	7·7	1·0	65·1	60	25	76	24	—
	5·5	1·2	93·0	60	20	76	24	—

TABLE 55. Reaction of *cis*- and *trans*-4-octenes with methoxycarbonylcarbene from catalytic decomposition of methyl diazoacetate[1957]

4-Octene		Methyl diazo-	Catalyst		Hexafluoro-	Total	Relative yield of esters (%)		
Isomer	Amount (g)	acetate (g)	Name	Amount (g)	benzene (g)	yield of esters (%)	*trans*	*cis anti*	*cis syn*
trans	1·0	0·25	$CuSO_4$	0·008	—	25	100	—	—
	1·1	0·25		0·005	9·3	27	100	—	—
	1·1	0·25	$Cu(C_{18}H_{36}O_2)_2$	0·004	—	24	100	—	—
	0·5	0·0625		0·004	4·65	21	100	—	—
	1·1	0·25	Cu	0·004	—	22	100	—	—
	1·1	0·25		0·004	9·3	19	100	—	—
cis	1·0	0·25	$CuSO_4$	0·003	—	24	—	65	35
	1·1	0·25		0·003	9·3	23	—	64	36
	1·0	0·25	$Cu(C_{18}H_{36}O_2)_2$	0·0035	—	28	—	66	34
	1·1	0·25		0·003	9·3	22	—	64	36
	1·1	0·25	Cu	0·003	9·3	20	—	62	38

since Wulfman, Peace and McDaniel have shown that this is affected by catalyst concentration, this feature should be examined on a case-by-case basis and it may be possible to optimize this variable.

Wulfman† has generated a systematology for the purpose of classification of carbene transfer processes (MT). (The initial **M** for methylene was used in preference to C. CT is the standard abbreviation of Charge Transfer.) Since the reactions

† This section is presented with the permission of Pergamon Press and is taken from Reference 2363.

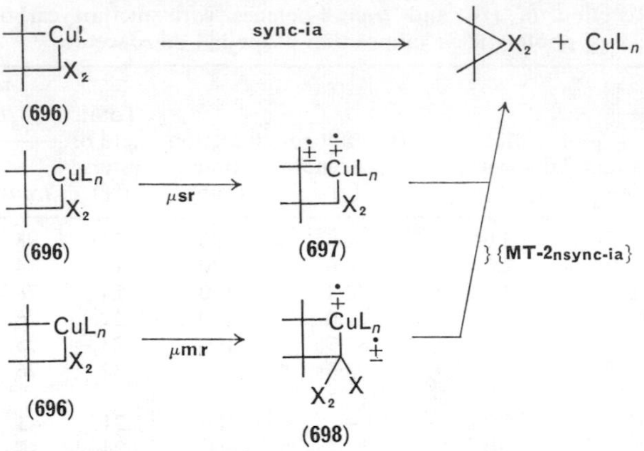

(155)

(For explanation of symbols, see text.)

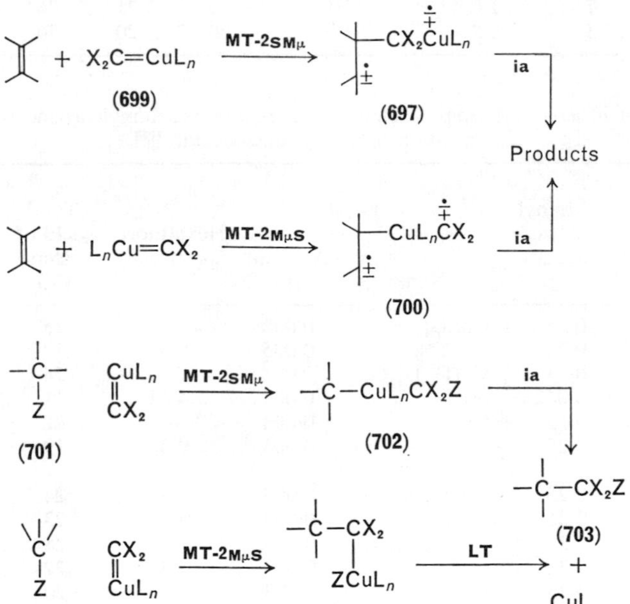

(156)

(For explanation of symbols, see text.)

involved are often exceedingly complex and it is difficult to identify the rate-determining step (r.d.s.), and because 'readily' identifiable features of reactions are the products, he based his analyses on the product-determining step. Since he was concerned with the group CWZ which is capable of forming a maximum of two bonds, he divided the processes into two basic categories for carbene transfer: (a) those in which only a single bond is made or broken, and (b) those in which two bonds are made or broken.

Accordingly, **MT** processes are either synchronous (**MT$_{sync}$**) or non-synchronous (**MT$_{nsync}$**). For the sake of simplicity, he assumes that most processes are of the **MT$_{nsync}$** type and writes these as **MT**, and only employs **MT$_{sync}$** for the less common process.

$CuL_n{=}CX_2$ MT-2MμS (705) → (706) → 1,3 → (707) $\bar{C}uL_nCX_2H$

MT-2MμS (705) → (708) $H\underline{C}uL_n{=}CX_2$

(705) MT-2SMμ $CX_2{=}CuL_n$

(709) CuL_n (710) (157)

N Processes

(711) $CuL_n{=}\bar{C}X_2$ MμS (712)

(711) $CX_2{=}\bar{C}uL_n$ SMμ (713) → Products

(For explanation of symbols, see text.)

In terms of molecularities, the group CWZ always arises from species **Q** to which it is bonded and is going to become bonded to a substrate **S**. For convenience he designated the carbene as **M** and any metal as **μ**. If the intermediate **Q** contains **M** and **μ**, he writes the species **Mμ**. If **M** is not formally doubly bonded to **μ**, but rather is attached to some other group (resideo) as well, e.g. N_2 as in a diazoalkane metal complex, he designates the species as **RMμ**. He writes a diazoalkane as **RM**.

There are two basic ways in which **Mμ** or **RMμ** can transfer **M** or **RM** to **S** to furnish a product **RMS** or **MS**: (a) dissociation to furnish **M** or **RM** which then reacts with **S** to give the product, an **MT-1** process, or (b) a bimolecular process

MT-2. The **MT-1** or **MT-2** process may however involve a substrate activated by the catalyst **μ**. These processes he designates **MTE** for carbene transfer with substrate enhancement.

A well-defined example of the **MT-1** process exists in the work of Seyferth[2224a] with compounds of the type $PhHgCX_3$ where warming liberates CX_2 and PhHgX.

MT-2 Processes: Wulfman believes that the possibilities for **MT-2** processes are numerous, but not unmanageable. For **MT-2$_{sync}$** there is the possibility of forming an intermediate possessing all of the atoms of the substrate or intermediates in which a portion of the carbenoid or *proto*-carbenoid is lost (equation 154). The intermediate **696** loses the CuL_n system by an internal alkylation (**ia**) and this process can be synchronous or non-synchronous where rupture occurs between metal and substrate (**μsr**) or metal (**μ**) and methylene (**μmr**), e.g. equation (155).

The possibilities for **MT-2$_{nsyn}$** fall into two basic categories **MT-2$_{nsync}$MμS** and **MT-2$_{nsyn}$SMμ** (equation 156).

In the last case there has been a ligand transfer (**LT**) from metal to carbene carbon. Applying this to the various C—H insertion processes known for bis-carboalkoxy carbenes, he writes the mechanistic schemes shown in equation (157).

The third and fourth examples in equation (157) are simply 'ene' reactions followed in one case by an **LT** process leading to a common intermediate which then participates in an **MT-2$_{ia}$** process to furnish products.

The replacement of Z by H in equation (157) will offer two routes to CH insertions. He also invisages C—Z insertions by yet another pathway in which non-bonded electron pairs on the atom Z are attacked by the carbenoid carbon or copper. These last two possibilities are especially attractive for accounting for the work of Ando with allylic ethers and sulphides. For purposes of designation he calls these the **N** processes to distinguish them from the **B** processes involving direct attack on the bonding electrons.

MTE-2 Reactions: The author thinks there is a real possibility that some substances will be activated by complexation to the catalyst and that these need not lead to enhancement towards all **MT** product formations but may instead lead to a favouring of one process over another. Since termolecular processes are of very low probability, it seemed preferable to class such reactions as proceeding by enhancement of substrate, where the enhancement is the result of the formation of substrate–catalyst complex which then reacts with carbenoid. Since the substrate–catalyst complex is simply a new substrate (**S′**), the analyses remain the same as those employed for the **MT-2** processes. A similar situation arises in the case of an **MTE-1** reaction.

If one assumes that this analysis is fundamentally correct, then it is possible to understand how some copper-catalysed cyclopropanations are stereospecific whereas others are only stereoselective; intermediates of the type **696** will undergo **μSr** or **Mμr** rupture in the **MT-2$_{nsync\text{-}ia}$** processes and permit free rotation. Any combination of substituents on the metal, the olefin and the methylene being transferred capable of stabilizing **697** or **698** will probably lead to nonstereospecific carbene transfer.

The intermediate **696**, if formed reversibly, can do so in two fashions. One will lead to the olefin and carbenoid but the other will lead to a new olefin and a new carbenoid. This latter process is probably the means by which olefin metatheses occur, and nicely accounts for the generation of high molecular weight products when cycloalkenes are subjected to metathesis conditions.

The efficacy of catalysis versus photolysis has been examined for the formation of tropilidenes (cycloheptatrienes) from benzenes and diazomethane. Catalysis furnished better yields.

The basic conclusions one reaches are: (1) Homogeneous catalysis is to be preferred over heterogeneous catalysis if one optimizes the catalyst structure, the concentration of catalyst, the addition rate of diazoalkane and the temperature; (2) catalysis is better than photolysis if the steric requirements of the substrate and/or 'carbene' are not excessive; (3) when two or more potential sites for attack are present, the photochemical and catalytic processes may be either similar or complementary (in some cases, the proper selection of catalysts may furnish the complementary conditions); (4) there is great need for a systematic study of sufficient generality to permit establishing clear guidelines so that future chemists can know what conditions should be employed as a function of (i) the product type desired, (ii) the nature of the olefin, and (iii) the nature of the diazo compound. Although such a series of studies is pedantic in nature, it would remove the choice of conditions from black art to science and greatly benefit everyone.

Some day it may prove preferable to prepare the pyrazoline and to decompose this catalytically if the cyclopropane is desired. Thus Wulfman and McDaniel[2361] were able to prepare **10** only via this route and only if they employed copper(II) fluoroborate as catalyst (equation 3). If that process is truly general, then coupled with the use of super-pressure conditions to facilitate the formation of the pyrazoline, the pyrazoline route would be preferred over other methods.

B. Cyclopropanation of aromatic substrates

The addition of 'carbenes' to benzenes affords cycloheptatrienes and/or bicyclo-[4.1.0]heptadienes (norcaradienes). They are 'valence isomers' which rapidly equilibrate even at low temperatures. The position of the equilibrium is strongly affected by the substituents on $C_{(7)}$ (the original 'carbene' carbon) and by incorporation of part of the triene system into additional rings[654]. Facile H shifts, rearrangements and cycloadditions are frequent secondary reactions.

1. Diazomethane and diazoalkanes

Photolysis of a benzene solution of diazomethane[1643, 1646, 1976, 2018], furnishes 32% of cycloheptatriene, accompanied by some toluene. If the diazomethane decomposition is catalysed by cuprous salts, the toluene by-product can be avoided[2074, 2077]. This is the method of choice for the ring expansion of numerous aromatic systems, including heterocycles. The reaction is strongly influenced by the steric and electronic effects[2075]

CH_2N_2, $h\nu$; CH_2N_2 CuBr

(158)[1643, 1650]

of various substituents on the aromatic ring. Many of the resultant cycloheptatrienes have been employed for the preparation of tropolones[478], azulenes[1837], and hydroazulenes.

Intramolecular reactions[478] of aromatic diazoalkanes were studied in attempts to synthesize colchicine. Eschenmoser's synthesis proceeded through a norcaradiene which was, however, formed by an alternative method[2213].

A number of substituted diazoalkanes have been examined[654, 1739, 1919]. These include aryl, vinyl[1919] (particularly diazocyclopentadienes[1696]), trifluoromethyl[1799, 1800], cyano[1547, 1573, 1575, 1577, 1579, 1784], phosphoryl[1045] and sulphonyl[2310] diazoalkanes. The product from benzene and dicyanodiazomethane exists predominantly as the norcaradiene.

2. Diazocarbonyl compounds

a. *Diazoesters, diazomalonates and derivatives.* The action of ethyl diazoacetate on benzene is a classic example[1548]. Cycloheptatriene esters are obtained[1409, 1647, 2285, 1983–1985, 2192] photochemically or thermally.

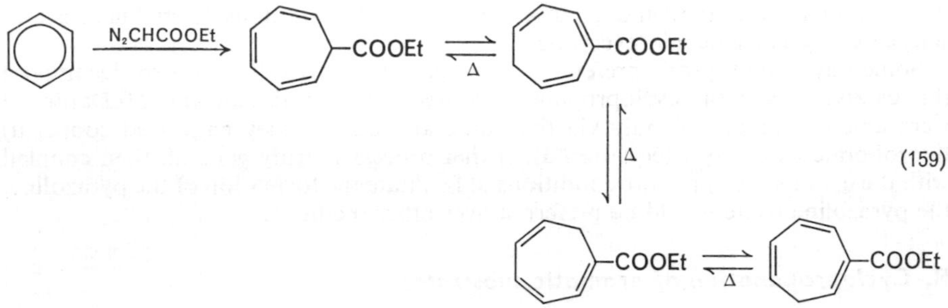

(159)

In contrast with the benzene series, addition to naphthalene[654, 1890], anthracene and phenanthrene gives norcaradiene compounds[283, 654, 757, 1919]. In the polycyclic cases the structures of the products normally are those which least alter the aromatic character of the remaining unsaturated rings.

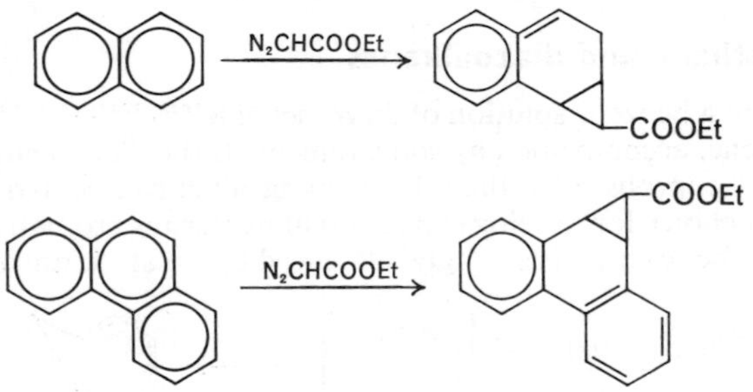

(160)

The additions of diazoketones and diazoesters to aromatic and heteroaromatic compounds have been reviewed[283, 654, 757, 1919, 2018].

Ring expansion of indanes has been applied in the synthesis of azulenes and is known as the Pfau–Plattner synthesis[1837, 2136]. With alkyl substitution of the aromatic ring, insertion into the C—H bond of the chain becomes increasingly important at the expense of the ring expansion[2244, 2245].

Cyanoesters[1579] and diazomalonates[2132] have been examined. The photochemical[1920, 1971, 2132] or copper-catalysed[1971] decomposition of dimethyl diazomalonate in benzene or substituted benzenes gives phenylmalonates (perhaps via the cycloheptatriene esters) and insertions in the side chain. The isomerization of these cycloheptatrienes is extremely facile and can occur on work-up. One group of workers has isolated the cycloheptatriene[1920].

The copper-catalysed reactions of a benzyl malonate[1970, 1974] lead only to intramolecular ring enlargement along with some insertion into the hydrocarbon solvent; the presence of a *para*-methoxy group furnished higher yields and gave rise to the formation of a spirodienone.[1974]

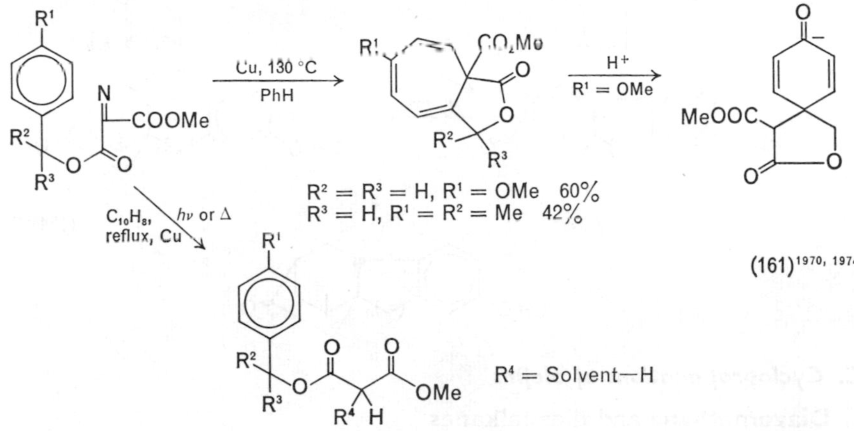

$R^2 = R^3 = H, R^1 = OMe$ 60%
$R^3 = H, R^1 = R^2 = Me$ 42%

R^4 = Solvent—H

(161)[1970, 1974]

b. *Diazoketones*

i. Synthesis of cycloheptatriene compounds. The intermolecular reaction of a diazoketone with aromatic systems[1757, 2094, 2297] has been used to produce azulenic compounds[2297]. The intramolecular process has been more successful and good yields have been obtained[1590, 1928, 2318, 2319]. This procedure has also been applied to a synthesis of azulenes[2218].

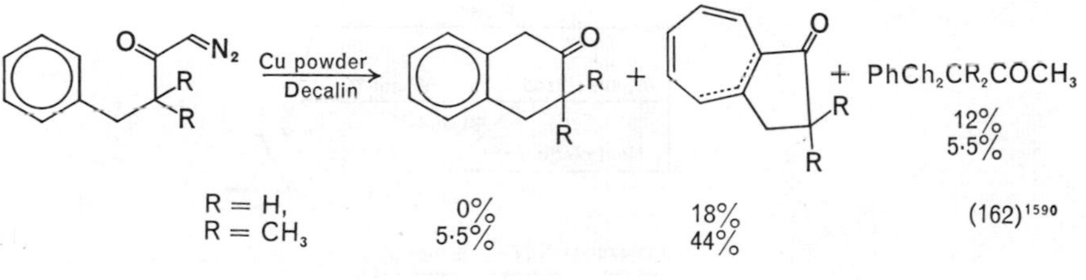

			$PhCH_2CR_2COCH_3$
R = H,	0%	18%	12%
$R = CH_3$	5·5%	44%	5·5%

(162)[1590]

ii. Intramolecular C alkylation by aromatic diazoketones. Although aromatic ketocarbenoids attack the aromatic ring and lead mainly to ring expansion, treatment of these diazoketones with acids affords intramolecular C alkylations[1590, 2090].

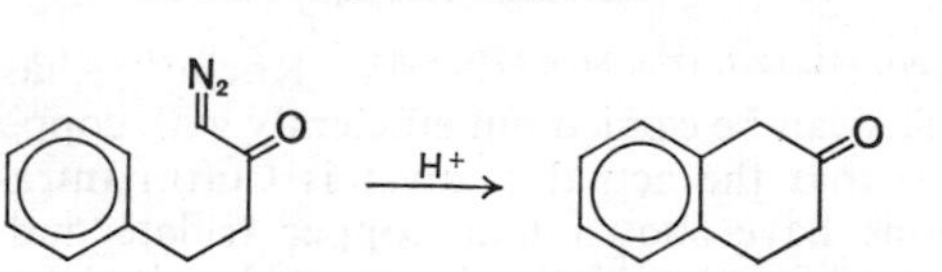

(163)[1590]

If the aromatic ring contains a hydroxy or methoxy group, a useful annelation and the formation of spirodienones result[1511–1513]. Some of these products are key intermediates in the synthesis of gibberellin.

Cu–CuO

$BF_3 \cdot Et_2O$, CH_2Cl_2

$CHCl_3$, HCl

H_2, 10% Pd/C

H_2, 10% Pd/C

(164)[1511–1513]

C. Cyclopropanations of olefins

1. Diazomethane and diazoalkanes

a. *Additions to simple olefins.* Diazomethane can be added under photochemical, thermal or catalytic conditions to olefins to furnish cyclopropanes. When performed photochemically or thermally, the formation of side products (insertion products) is important and this approach is seldom employed for synthesizing three-membered rings[654, 1487]. Catalytic decomposition, particularly with copper salts, avoids insertion products and diazomethane leads to stereospecific cyclopropanation of olefins[1409, 1508, 1649, 1650, 1739, 1774, 1949, 2077, 2078, 2317]. Numerous examples can be found

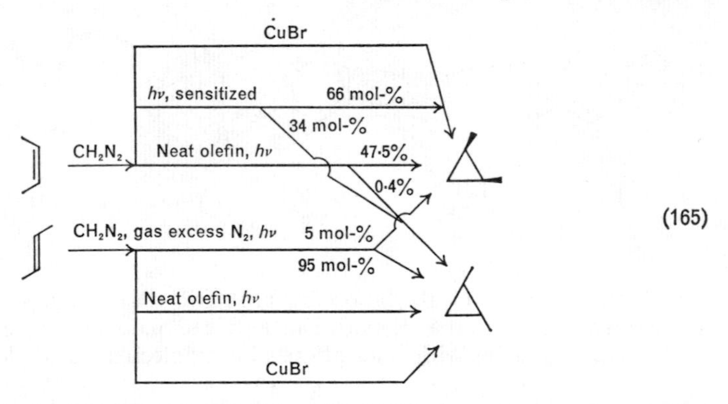

(165)

in recent reviews[264, 654, 1487, 1775, 2018, 2078, 2333]. Kochi[2187] has shown that cyclopropanation of olefins can be carried out efficiently with copper (II) triflate as formal catalyst. (He argues that the actual catalyst is Cu(I).) Intra- and inter-molecular competition reactions have shown that copper triflate and copper fluoroborate promote the cyclopropanation of the least alkylated olefins; in contrast, other catalysts possessing softer anions or ligands favour the most highly substituted olefins (see equation 152).

With styrene[2125] or with strained double bonds[1962], the Paulissen reagent (palladium acetate) gives good yields of the cyclopropanes.

An asymmetric synthesis with an optically active catalyst has been reported[2099]. For synthetic purposes, the cyclopropanations of olefins by diazomethane is often avoided by employing the Simmons–Smith procedure[2239, 2240].

Substituted diazoalkanes have been used to prepare three-membered rings.

The generation of alkyl 'carbenes' by decomposition of salts of tosyl hydrazones (Bamford–Stevens reaction) has recently been reviewed[1563, 1653]. This reaction appears to proceed by the generation of transient diazoalkanes and in some cases the diazoalkanes can be isolated[1563]. Addition to double bonds may give cyclopropanes but their chemistry is dominated by intramolecular H shifts[1885, 1958].

Bicyclobutanes can be obtained by irradiation[1975] or via the cuprous cyanide-catalysed[1549] decomposition of vinyldiazoalkanes. These decompositions are distinguished from those with more remote unsaturation by the absence of H shifts.

$$CH{=}CHCH_2CHN_2 \xrightarrow[-78\,°C]{h\nu} CH_2{=}CH{-}CH{=}CH_2 + \text{(bicyclobutane)} \qquad (166)$$

$$(CH_3)(R^2)C{=}C(R^1){-}C(R^3){=}NNHTos \xrightarrow{NaOMe} \text{cyclopropene } C(R^1)(CH_3),\ C(R^2){=}C(R^3) \qquad (167)$$

R^1	R^2	R^3	Yield (%)
CH_3	CH_3	H	72
CH_3	H	CH_3	39
CH_3	H	H	50
H	CH_3	H	4
H	H	H	3

Vinylcyclopropanes can be prepared[2188]. Closs, Closs and Böll[1583], Stechl[2260] and Dürr[1677] prepared many cyclopropanes by the pyrolysis or photolysis of basic salts of tosylhydrazones of α,β-unsaturated carbonyl compounds. The decompositions of vinyldiazoalkanes were key steps in a synthesis of thujopsene[1547] and sesquicarene[264, 1592].

$h\nu$; X = $Tos\bar{N}N{=}$ $\overset{+}{Na}$

HgI_2, THF; N_2; CO_2Me → CO_2Me (168)

N_2, Cu powder → [ ] →

Cyclopropanes have been obtained from olefins and a variety of diazoalkanes[1653] including aryl- and diaryl-diazomethanes, cyclic diazoalkenes, and sulphonyl[2310], and phosphoryl[2155, 2195] diazoalkanes.

b. *Additions to activated double bonds.* Frequently diazomethane adds to double bonds to furnish pyrazolines whose decomposition gives cyclopropanes, β-methyl derivatives and derivatives in which CH_2 has been inserted into a β C—C bond.

The combination[2126], $CH_2N_2/Pd(OAc)_2$, in some cases allows the direct cyclopropanation of activated double bonds[2040, 2148]. It adds stereospecifically *cis* to α,α or α,β-disubstituted, α,β-unsaturated ketones or esters in excellent yields. Trisubstituted α,β-unsaturated carbonyl compounds do not react. This method is applicable for the preparation of annelated alicyclic ketones although certain α,β unsaturated steroidal ketones do not react.

$$PhCH{=}CH_2 + N_2CHCOOEt \xrightarrow{Pd(OAc)_2} \text{Ph, COOEt (cis) cyclopropane} + \text{Ph, COOEt (trans) cyclopropane} \quad 96\%$$

$$R^1R^2C{=}CR^3R \xrightarrow[Pd(OAc)_2]{CH_2N_2} \text{cyclopropane}(R^1, R^2, R^3, R) \qquad (169)^{1962, 2040, 2125, 2148}$$

R^1	R^2	R^3	R^4	Yield (%)
Ph	H	H	CH_3	85
Ph	H	H	COOEt	90
Ph	H	H	Ph	98
H	Ph	H	COOEt	85
Ph	H	H	H	90
CH_3	CH_3	H	CH_3	0
CH_3	H	H	COOEt	85

2. Diazocarbonyl compounds

The diazo carbonyl compounds are of considerable importance synthetically because of the large number of transformations which can be realized with cyclopropyl carbonyl compounds which do not affect the three-membered ring[2369]. The cyclopropanes are also capable of serving as Michael acceptors[1626, 2356]. The 'carbenes' are stabilized by the electron-withdrawing resonance effect of the carbonyl group and they have high electrophilicity.

a. *Diazoesters, diazomalonates and derivatives.* Carbalkoxy diazomethanes, particularly ethyl or methyl diazoacetate, have been extensively studied and their additions to olefins has long been known. Several recent complete review articles have appeared[283, 757, 975, 1628, 1653, 1919, 2018]. General and new aspects will be given here. The corresponding diazoesters can be added thermally† at ~150 °C. This method is rarely employed synthetically and may involve pyrazoline intermediates. Photolytic additions are common (either via direct photolysis or photolysis in the presence of known sensitizers such as benzophenone).

Catalytic processes are by far the most important way of decomposing diazoesters. This permits the use of lower reaction temperatures and favours reactions with olefins

† The term 'thermal' is understood to include any reaction not carried out with irradiation or added catalyst.

to furnish cyclopropanes. C—H insertions are usually avoided except with diazomalonates.

The role and importance of the catalysts are discussed above. Copper and copper salts are the most frequently used catalysts. The catalysts 'copper triflate' and 'copper fluoroborate' furnish preferentially the cyclopropane from the least-substituted olefin when diazoacetates are employed[2187]. The addition is stereospecific. Both *syn* and *anti* isomers are formed. The *anti* isomer is the major one. This stereoselectivity has been discussed[2059]. Often the *anti* isomer is desired and, being the more stable isomer, can be obtained by epimerization with the base of the *syn*/*anti* mixture. The *syn*/*anti* ratio is frequently a function of catalyst concentration[975] and most probably of reaction temperature as well. Jones[2246] studied the reaction of alkyldiazoesters. Whereas the direct photolysis in an olefin is virtually useless for preparing cyclopropanes because of intramolecular reactions (especially insertions into C—H bonds), the photosensitized reaction, by generation of the triplet state, reduces the intramolecular reactions and allows cyclopropanation of olefins. Jones and Ando[1917] have used the same approach to overcome the tendency of diazoketones to undergo Wolff rearrangements. Halogeno[2157] and mercuri[2242, 2309] diazoesters have recently been studied.

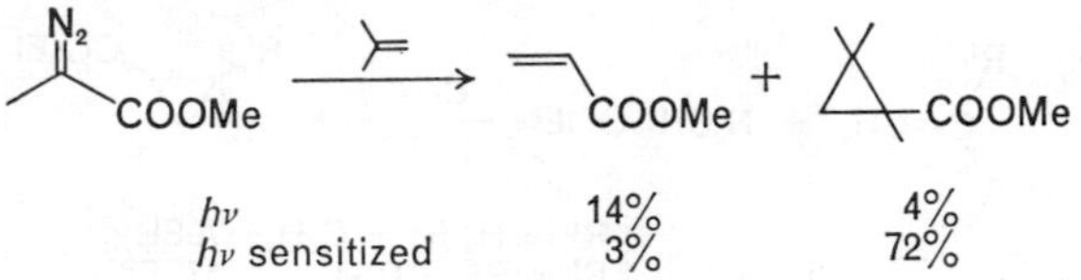

N_2 COOMe → COOMe + COOMe

$h\nu$	14%	4%
$h\nu$ sensitized	3%	72%

(170)[1917, 2246]

N_2 CO_2Me → ≡—COOMe + COOMe + COOMe

$h\nu$	3%	20%	8·5%
$h\nu$ sensitized	3%	11%	33%

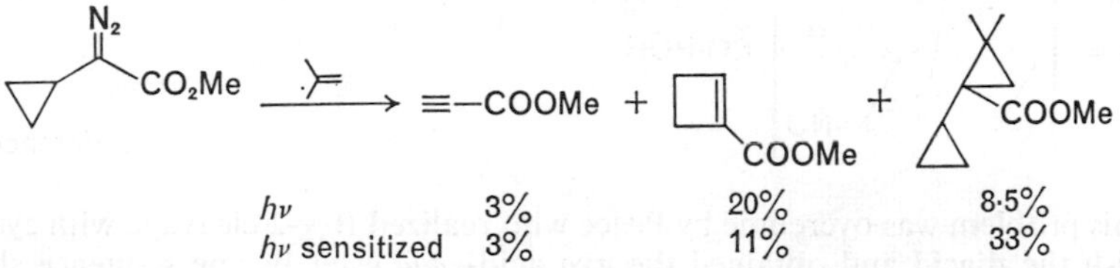

Asymmetric inductions: Optically active cyclopropanecarboxylates have been obtained from olefins by decomposition of ethyl diazoacetate with optically active copper complexes[1445, 2096]. However, in the early work, the stereoselectivity achieved was less than 10%. Otsuka[2283] found that optically active cyclopropane carboxylates can be prepared in 90% yield with enantioselectivity as high as 70% by use of a chiral cobalt chelate complex.

An asymmetric synthesis of chrysanthemic acid has been realized by using an asymmetric copper catalyst in 54% yield, the optical yield being 60–70%, and thus opens a route to synthetic pyrythrenoid insecticides[1446].

The chemistry of diazomalonic esters have been studied extensively[1415, 1422–1426, 1437, 1920, 1928, 1970–1974, 1983–1985, 2004, 2082, 2127–2132, 2355–2365, 2368]. Definitive papers on the photochemical[1920] and catalytic[975, 2359, 2364, 2365, 2368] behaviour have appeared. Much of the chemistry involving catalysis is presented in Sections III and IV.A.

The use of diazomalonates for cyclopropanation offers a route to *endo* cyclopropyl carboxylates. Thus Musso[2083] prepared **719** from 1,4-cyclohexadiene whereas for all practical purposes this adduct is not available from diazoacetic ester. The catalyst employed was Cu–bronze and good yields of the cyclopropane were obtained. The one troublesome aspect of the Musso sequence is hydrolysis to **717** or **721**.

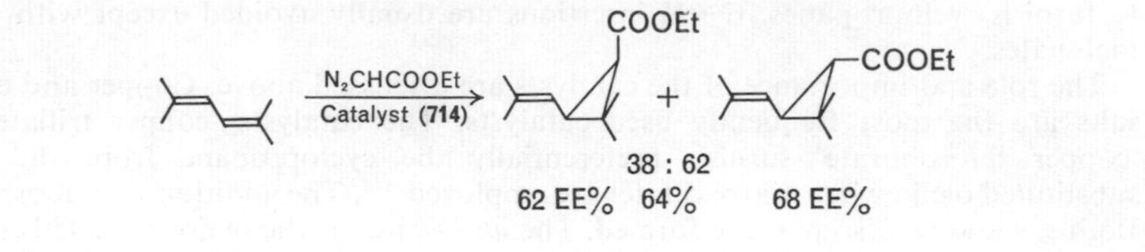

N₂CHCOOEt / Catalyst (714)

38 : 62
62 EE% 64% 68 EE%

(171)[1446]

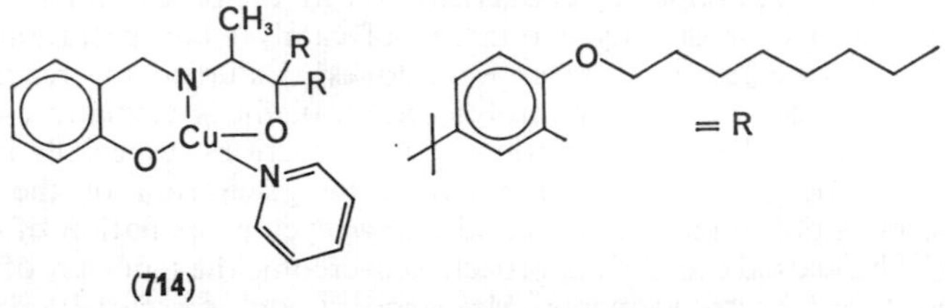

(714)

EE% = enantiomeric excess

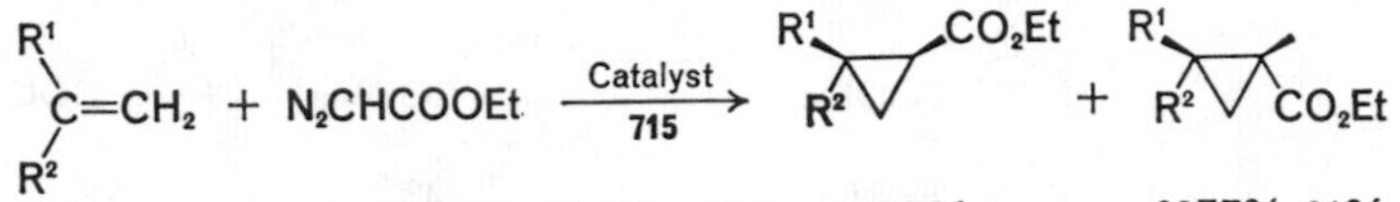

$R^1R^2C{=}CH_2 + N_2CHCOOEt \xrightarrow{\text{Catalyst } 715}$

$R^1 = H, R^2 = C_6H_5$ 70EE% 68EE% 91%
$R^1 = R^2 = C_6H_5$ 37EE% 70%

715 = [... NOH ... =NO]$_2$ CO·HOH

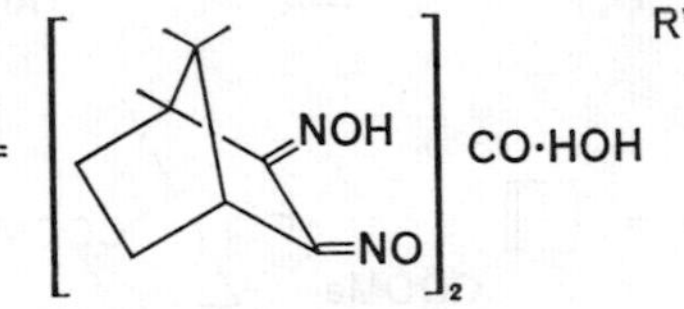

Reference 2283

This problem was overcome by Peace who realized B_{AL}-2 cleavage with cyanide to furnish the diacid and obtained the *exo* acid-*endo* ester by the sequence shown in equation (173)[2133, 2357].

The photochemical and catalytic additions to 1,4-dihydrobenzenes were employed by Berson to synthesize the norcaradiene **723** and examine its valence isomerization to the tropilidene[1516]. The sensitivity of these compounds to heat readily accounts

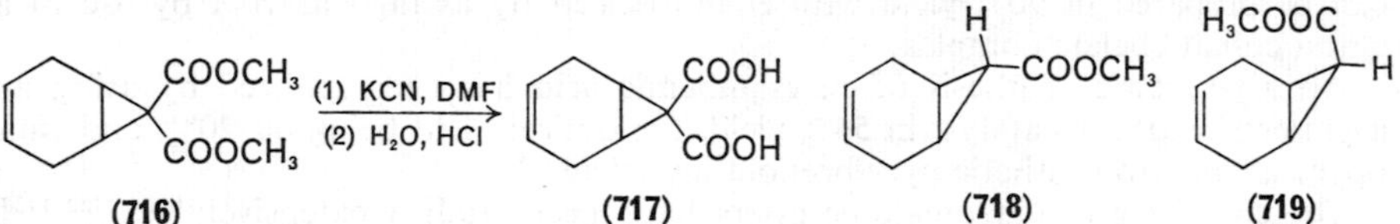

(1) KCN, DMF
(2) H_2O, HCl

(716) (717) 65% (718) 30% (719) 3%

(172)[2132]

for the failure of earlier workers to observe the tropilidene as a product from the photolysis of diazomalonates in benzenes[1574, 1970]. Compound **723**, for example, rearranges rapidly to **725** at ~100 °C and none of these norcaradienes survive GC analyses[1516].

The related studies of Julia and Linstrumelle are summarized in equations (177, 178, 188)[1898–1902, 1928, 1970–1974, 1984–1986].

Extensive photochemical studies on dimethyl diazomalonate were performed by the Princeton school[1920] after preliminary work was performed by Karustis at Yale[1939]. The noteworthy aspects of that study have been treated in Section IV.A. However,

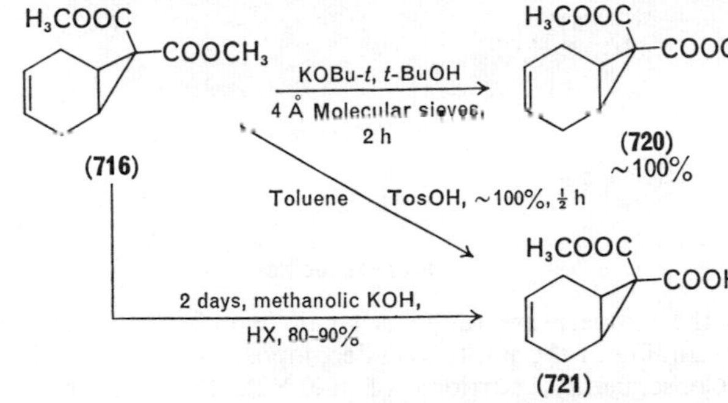

H_3COOC / $COOCH_3$ (716) —KOBu-*t*, *t*-BuOH, 4 Å Molecular sieves, 2 h→ H_3COOC / $COOC(CH_3)_3$ (720) ~100%

(716) —Toluene TosOH, ~100%, ½ h→ H_3COOC / $COOH$ (721)

(716) —2 days, methanolic KOH, HX, 80–90%→ (721) —Δ→ 719

(173)[2132]

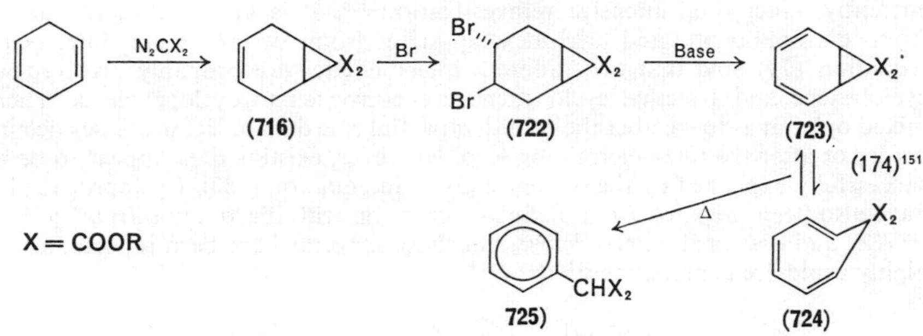

—N_2CX_2→ (716) —Br→ (722) —Base→ (723) ⇌ (724) —Δ→ (725) CHX_2

$X = COOR$

(174)[1516]

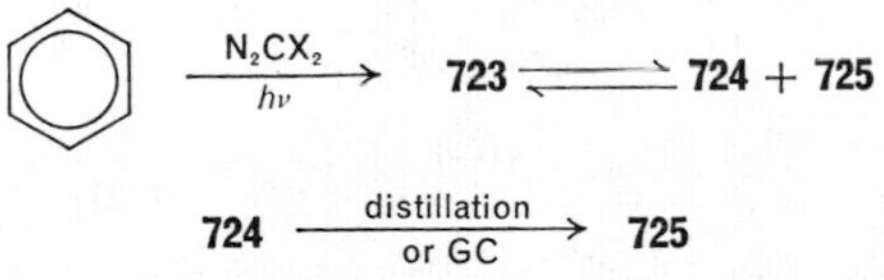

—N_2CX_2, $h\nu$→ 723 ⇌ 724 + 725

724 —distillation or GC→ 725

(175)

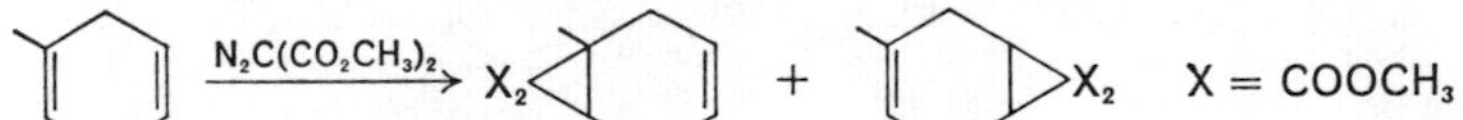

—$N_2C(CO_2CH_3)_2$→ X_2 + X_2 $X = COOCH_3$

(176)[1516, 2127]

Cu Catalysis	minor	major
$h\nu$	major	minor
$h\nu$	major	minor

nowhere in the Princeton or Yale studies is there evidence that great care was taken to avoid oxygen, hydroperoxides and peroxides. This may prove to be a minor point; however, rate studies indicate that these impurities grossly affect the rate of disappearance of the diazo compound. Hence the Hillhouse–Poling[1868] investigations have shown that it is necessary to distil cyclohexene from benzophenone sodium ketyl under an inert atmosphere (argon or nitrogen) to obtain reproducible rates; opening of the sampling port of the photolysis apparatus to the atmosphere for a few seconds led to a doubling of the rate of photolysis; addition of *t*-butyl peroxide or *t*-butyl hydroperoxide greatly enhanced the rate of reaction; peroxide impurities are only slowly consumed and appear to perform a catalytic role, and the rate of photolysis is temperature independent but product partitioning is temperature

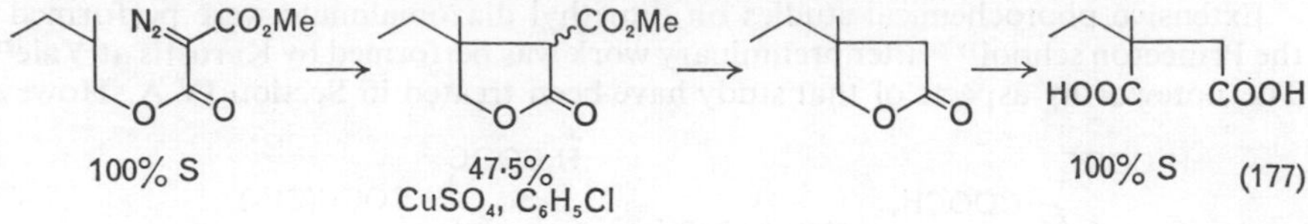

100% S → 47·5% ($CuSO_4$, C_6H_5Cl) → → 100% S (177)

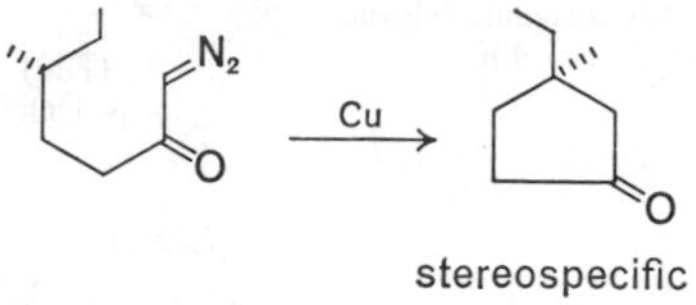

Cu → stereospecific (178)

dependent, even when the temperature range is small (e.g. 5–10 °C). All of this suggests that the understanding of the photolytic decompositions of diazoalkanes is at best fragmentary. Consequently, portions of the Yale–Princeton studies are currently undergoing intensive reinvestigation[2366]. It is known that the products from diazomalonate and cyclohexene arise from two different intermediates (equation 179), and that two different intermediates are probably involved when cyclohexene and 1-methyl cyclohexene are converted to cyclopropanes. There is mixed opinion as to whether the resulting partial rate data reflect processes occurring before or after the rate-determining step. However, existing data appear to be most successfully explained by the sequence given in equation (180). Cyclopropyl nitriles have also been prepared from olefins by reaction with diazoacetonitrile[1635, 1851, 1784, 2138, 2241] and diazomalonitrile[1573–1579], but these reagents have been reported as being highly explosive compounds[1574, 1635, 2138].

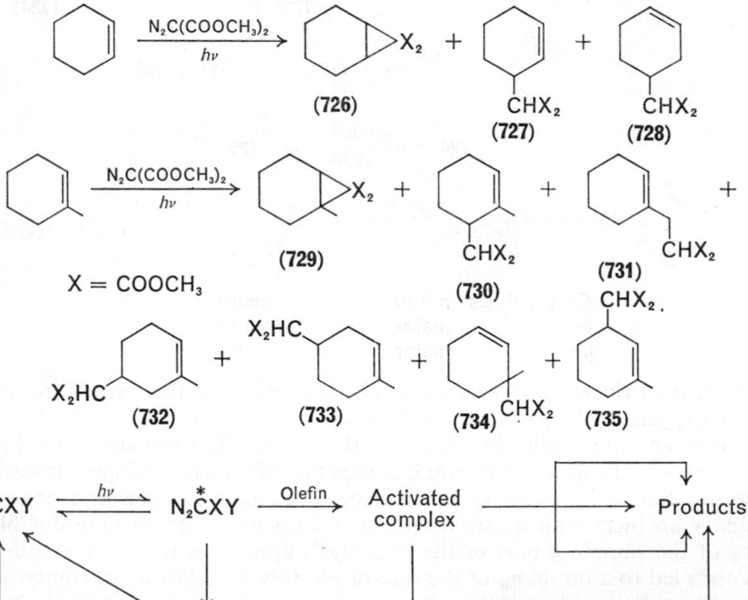

$N_2C(COOCH_3)_2$, $h\nu$ → (726) + (727) CHX_2 + (728) CHX_2

$N_2C(COOCH_3)_2$, $h\nu$ → (729) + (730) CHX_2 + (731) CHX_2 + (179)

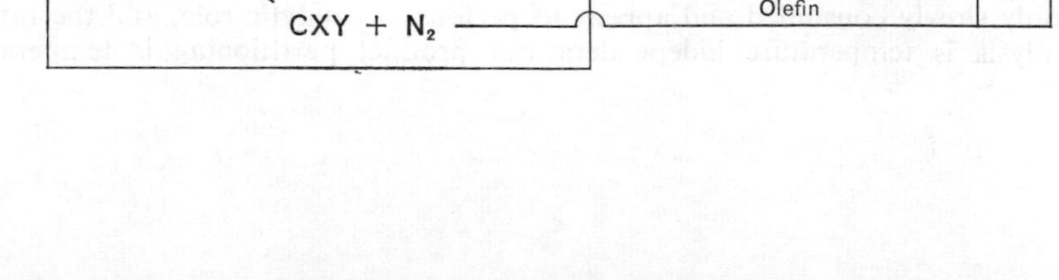

$X = COOCH_3$

(732) X_2HC + (733) X_2HC + (734) CHX_2 + (735) CHX_2.

$N_2CXY \underset{}{\overset{h\nu}{\rightleftharpoons}} N_2\overset{*}{C}XY \xrightarrow{\text{Olefin}}$ Activated complex → Products

$CXY + N_2 \xrightarrow{\text{Olefin}}$ Products (180)

V. INSERTIONS INTO X—Y BONDS

The vast majority of insertions into X—Y bonds involve loss of N_2 from the diazoalkanes; therefore, in a formal sense they are carbene and carbenoid reactions. The reality is somewhat more complex and a variety of intermediates involving radicals, carbonium ions and their complexes as well as ylids are also implicated. The insertion of 'carbenes' into C—H bonds has received considerable attention as regards the mechanism, but in general the processes are not employed synthetically. The subject has been treated extensively by Kirmse[654].

The most common insertion reactions involve X—H bonds where X is nucleophilic and H is somewhat acidic. The processes can occur either by catalysis (e.g. formation of ethers from alcohols) or by protonation by the acidic protons (e.g. conversion of a carboxylic acid into its esters).

The processes may involve clearly identifiable intermediates such as a diazoketone from the reaction of an acid chloride with diazomethane to furnish a chloromethyl ketone, or proceed directly as in the C—H insertion shown in equation (181)[1969, 1986].

N_2CHCO_2Et → CO_2Et + CO_2Et, CO_2Et (181)[1911]

The reactions of diazoalkane with M—X bonds have been reviewed by Lappert and Poland[712]. A number of reactions involving metals actually lead to metallo-diazoalkanes which in some instances are very stable. Frequently nitrogen is lost and M—C bonds are generated.

A. C—Y Bonds

1. C—H bonds

a. *Aromatic* C—H *bonds*. Insertions into aromatic C—H bonds by carbenes are normally of importance only in the absence of benzylic hydrogens. However, by employing Lewis acid catalysts the nature of the reactions is changed from carbene–carbenoid processes to Friedel–Crafts alkylations.

The α-hydrogen of pyrroles and *N*-substituted pyrroles is attacked by diazoacetic ester and diazoketones in the presence of Cu[2086]. Indole, under somewhat similar conditions, furnishes both N—H insertion and C—H(3) insertion[1751, 1911]. With diazosuccinic ester, attack at C—H(3) offers a ready route to 3-indolyl propionic acid[1751].

Anderson[1414] found that azulene could be alkylated at $C_{(1)}$ using either HBF_4 or light to decompose the diazo compound. Whitlock[2340] obtained only a modest yield of the corresponding 4-pyranylidene derivative from the reaction of 2,6-diphenylpyrylium perchlorate with diazoacetic ester.

N_2CHCO_2Et (182)[2340]

Ph, Ph, $\bar{C}lO_4$ → CO_2Et, Ph, O, Ph

The decomposition of aromatic diazoketones by strong acids (e.g. H_2SO_4) can lead to cyclization[2215]. The presence of hetero-atom-containing substituents *ortho* to the diazoketone branch frequently leads to attack of the substituent and is discussed below.

N_2 2 N-H_2SO_4 O Ph Ph O Ph Ph (183)

The reaction of benzenoids with diazomalonic esters furnishes appreciable quantities of arylmalonic esters[2133]. When the processes are performed at low or moderate temperatures, the reaction will lead to the norcaradiene–tropilidene system. However, these products are negligible and readily rearrange to the arylmalonic esters. When copper catalysis is used, the necessary reaction temperatures are sufficiently high to cause rearrangement. Whether this actually occurs is unclear. In unpublished studies, Peace and coworkers[2004, 2132, 2366] found 57 products from the reaction of prehnitene with dimethyl diazomalonate in the presence of $Cu(BF_4)_2$. Of these products *c*. 30 derived from the diazo compound but none involved addition to a norcaradiene or tropilidene intermediate. The major products are shown in equation (184).

$N_2C(COOMe)_2$ $Cu(BF_4)_2$ 140 °C CO_2Me CO_2Me + CO_2Me CO_2Me + CO_2Me CO_2Me

(184)[2004, 2127, 2366]

+ MeOOC MeOOC OMe CO CO OMe + Me O O CO_2Me OC O O OMe Me + X Me_3 Me_3

(736) X = C≡C
(737) X = CH=CH
(738) X = CH_2CH_2

The dimethyl tetramethylphenylmalonate could be easily isolated in ~30% yield since it crystallized from the reaction mixture.

With durene and catalysis by copper powder, diazoacetic ester forms the product from insertion into a benzylic hydrogen (30%).

The use of naphthalenes as substrate leads to insertion into the α-C—H bond when diazoacetic ester is pyrolysed at ~140 °C. Thus 2-methoxynaphthalene furnished ~80% of the related ethyl naphthalene 1-acetate[1633]. There is reason to suspect that the reaction proceeded via an intermediate benzonorcaradiene which was thermolysed under the reaction conditions (equation 185).

b. *Allylic* C—H *bonds*. Allylic and benzylic C—H bonds are normally the weakest hydrocarbon C—H bonds and as such can be expected to undergo attack by carbenes with some facility. The photolysis of CH_2N_2 inserts into isobutylene without rearrangement[1648]. Photolysis of diazoacetic ester also leads to appreciable quantities

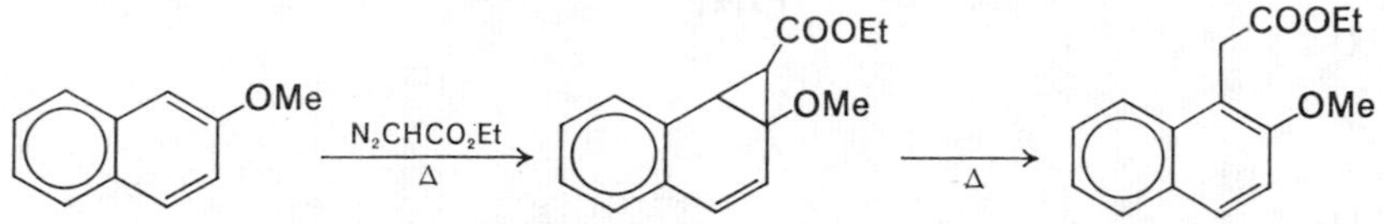

(185)[1633]

of allylic C—H insertion products as do the photolyses of diazocyclopentadienes. Normally this is not a problem with copper catalysed-reactions. However, Peace found this to be an important process with dimethyl diazomalonate[2368] and cycloalkenes. In this instance, the process has a lower activation energy than cyclopropanation (equation 186). Diazoacetic ester inserts into the benzylic C—H bond of durene in the presence of copper[2364].

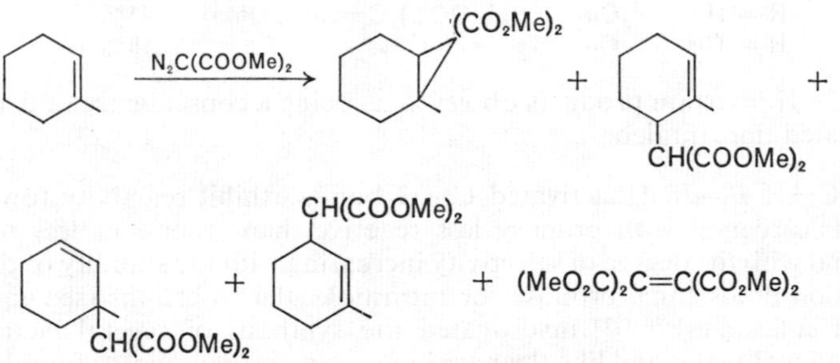

(186)[2368]

As is evident from equation (186), the insertion products involved some rearrangement. An explanation has been advanced[2364]. An earlier claim[2132] that cyclohexenone furnishes the 4-dimethyl malonyl derivative either by thermolysis or by using copper catalysts may well be in error. The product was insufficiently characterized to distinguish between the 4-derivative **(739)** and the 3-derivative **(742)** with an equilibrium between the 3-*exo*, α, and the α,β unsaturated isomers. In light of the work of Carrié and coworkers, this situation might seem more probable[1612–1625, 1817–1820, 1844–1849, 2021, 2022].

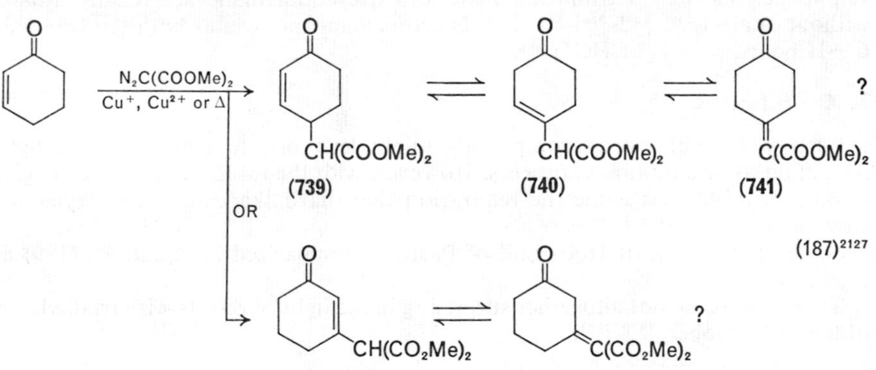

(187)[2127]

Ledon and coworkers[1970–1974] examined a number of C—H insertion reactions with mixed diazomalonic esters. The results with toluene and *p*-xylene are summarized in equation (188). In view of the results of Berson[1516] it seems reasonable to assign

$N_2=C(COOMe)_2$ + C_6H_5R ⟶ $C_6H_4R(CH_2CHX_2)$ + ... + ... + X_2RC... (188)[170–174]

Yield (%)	R	Conditions	M, yield (%)			
20	H	$h\nu$	33	40	4	23
78	H	Cu, 160 °C	24	27	8	39
53	CH_3	$h\nu$	37		63	
75	CH_3	Cu, 135 °C	38		62	
R = H	Cu	$(MeOOC)_2C=C(COOMe)_2$			45%	
R = CH_3	Cu				15%	

the aryl C—H insertion products observed as being a consequence of decomposition of the related norcaradienes.

c. C—C—H *Bonds*. Unactivated C—H bonds exhibit reactivity towards photo-generated 'carbenes' with primary less reactive than secondary less reactive than tertiary and with the degree of selectivity increasing with the stability of the 'carbene'. The reaction holds some promise for intramolecular insertions (see equation 177). Julia and colleagues[1970–1973] investigated the synthesis of several lactones derived from mixed malonates and like the reaction with a diazoketone[1986] found the reaction to be stereospecific. The processes proceeded with much better yields when copper-based catalysts were employed (see equations 178, 179, 188).

d. *Other* C—H *bonds*. The presence of a variety of substituents on C will render a C—H bond more readily susceptible to attack by 'carbenes' and 'carbenoids'. Hence dimethyl malonate reacts smoothly with dimethyl diazomalonate in the presence of Cu catalyst systems to furnish *syn*-tetrakiscarbomethoxy ethane[2132]. More acidic hydrogens such as those involved in 1,1-*bis*-sulphones and 1,1-*bis*-sulphonates as well as trinitromethane and tricyanomethane are readily attacked without catalysis[1455, 1456, 1461, 1464, 1530]. Diazomethane methylates both the O—H and C—H bonds in $(CF_3)_2CHOH$[1963].

2. C—B bonds

Early work with boron compounds might lead one to expect polymethylene formation to be a dominant process. However, with the rapid expansion of organo-borane chemistry has come the realization that diazoalkanes can be alkylated by alkyl boranes[1874–1877, 2121, 2122].

Some of the work of Hooz and of Pasto is summarized in equations (189) and (190).

These results are not altogether surprising in the light of results with trialkylalanes obtained by Hoberg[1870, 1871].

3. C—C bonds

Insertions into C—C bonds, like aliphatic C—H bonds, are highly unfavoured as a consequence of the bond strength. However, there are examples where a single C—C bond is formed which do not involve actual insertions but rather involve

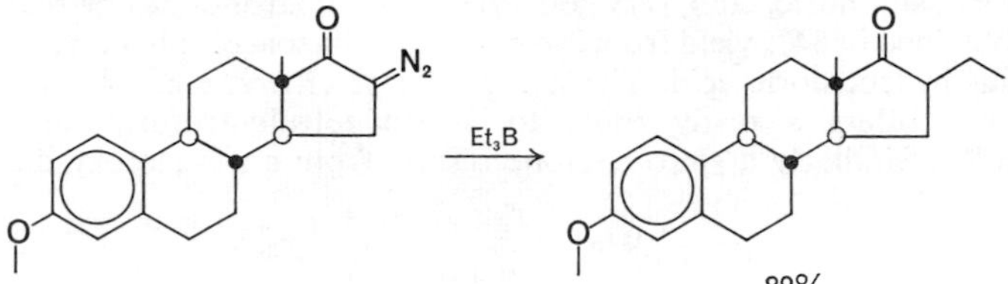

$\xrightarrow{Et_3B}$ 82% (189)

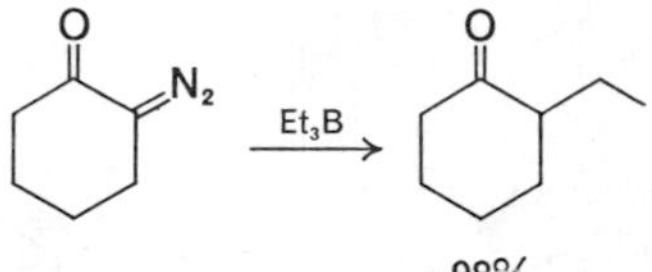

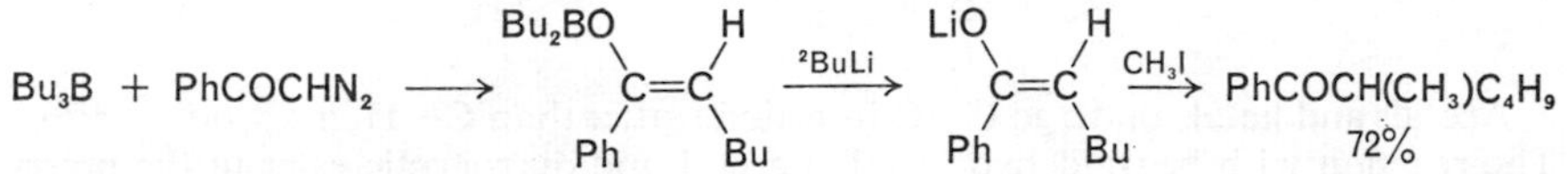

$\xrightarrow{Et_3B}$ 98%

$$Bu_3B + PhCOCHN_2 \longrightarrow Bu_2BO(Ph)C{=}CH(Bu) \xrightarrow{^{s}BuLi} LiO(Ph)C{=}CH(Bu) \xrightarrow{CH_3I} PhCOCH(CH_3)C_4H_9 \quad 72\% \qquad (190)$$

$\xrightarrow{Et_3B}$ $\longrightarrow$

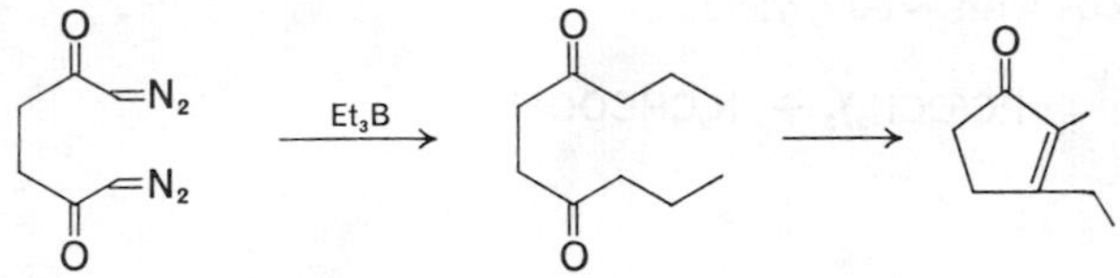

formally allylic hydrogen abstraction and bond formation at $C_{(1)}$ of the allylic system; essentially a formal 'ene' reaction. The most noteworthy examples are an outgrowth of the synthetic work by Ghatak on diterpenoids (see equation 164)[1803]. The process can be realized using the strong Lewis acid $BF_3{\cdot}OEt_2$ [1569, 1627, 1803] and thus differs mechanistically from the reactions of Peace which also generated a quaternary C atom[2364, 2368]. Since the same products can also be derived from the related cyclopropane by treatment with acid, the possibility exists that these reactions proceeded via a transient cyclopropane.

4. C—N bonds

The C—N bond is fairly resistant to attack by 'carbenes'; however, the 9-C—N bond in 9-dimethylaminofluorene is very susceptible to attack and thermolyses of phenyl diazomethane[1481] and diazofluorene[1481] lead to C—N insertions in ~40–45% yield. Similarly diazoacetic ester reacts either thermolytically or photolytically with 9-dimethylaminofluorene and benzyldimethyl amine[1782].

5. C—O bonds

Diazoacetic ester inserts into C—O bonds of 2-phenyloxetane and styrene oxide[5341, 5342] in the presence of copper catalysts; with simple unsymmetrical ethers (e.g. methyl heptyl ether) insertion occurs in both C—O bonds[1633, 1634].

The reactions of anisoles are of particular interest[1535, 1546, 1633, 1725, 1835, 1913, 1964, 2065, 2066, 2135, 2220, 2231]. With simple alkoxybenzenes, reaction with diazoacetic ester furnishes the related aryloxyacetic ester (anisole, 7%; anethole, 20%; 1,2-dimethoxybenzene, 38%; 1,3-dimethoxybenzene, 14%; 1,4-dimethoxybenzene, 40%). With intramolecular systems employing diazoketones and acid catalysis (acetic, formic, hydrochloric, sulphuric, etc.), very good yields of cumarones can be realized. Cumar-3-one is obtained in 84% yield from 2-methoxy-ω-diazoacetophenone in the presence of aqueous hydrochloric acid. The use of benzyl ethers, e.g. 4-benzyloxy-2-oxo-1-diazobutane, offers a ready route to 3-keto tetrahydrofuran and the related thiophene[2248]. Similarly a spiro oxetone results from an α-acetoxy diazoketone[2020].

OMe; HCl, HOH → ; N_2; O; O (191)

OAc; N_2; O; (1) ^-OH (2) HOAc → ; O; O (192)

Acetals and ketals undergo C—O insertion rather than C—H insertion[1833, 2205, 2207]. The reaction with benzaldehyde diethyl acetal and diazoacetic ester in the presence of BF_3 proceeds with ~80% yield.

$$RC(OCH_3)_3 + N_2CHCOOEt$$

$$\downarrow$$

$$RC(OCH_3)_2CH(OCH_3)CO_2Et \longrightarrow (R)(CH_3O)C{=}C(OCH_3)(CO_2Et) \quad (193)^{2207}$$

With ortho esters a route to 1,2-dialkoxy ethylenes results[2207]. Schonberg[2211] has compared the relative ease of insertion of the C—O and C—N bonds by fluorenylidene by the room temperature reaction (without catalyst) of acetoxymethyl diethylamine and 9-diazofluorene. The product involves C—O insertion (equation 194).

N_2; $(C_2H_5)_2NCH_2OAc$, Room temp. → ; $(C_2H_5)_2NCH_2$ OAc (194)[2211]

6. C—S, C—Se, C—Te bonds

There appears to be little relevant information on C—Se and C—Te bonds. However, considerable work has been carried out on C—S systems. Trithioortho-formate reacts with diazoacetic ester in the presence of BF_3 by insertion into the formyl C—S bond. Subsequent heating with $KHSO_4$ offers a route to the 1,2-bis-thiomethyl acrylates (equation 195)[2209]. The analogous reaction with diazoketones is well documented[2210]. More recently Ando[1416–1419, 1423] and coworkers have examined the reactions of dimethyl diazomalonate with allylic sulphoxides and sulphides using light (sensitized and unsensitized conditions), heat and copper catalysts. The sulphoxide reactions not involving triplet sensitizers undergo rearrangements which appear to involve sulphoxonium ylides. Thus the results strongly parallel those

$$HC(SCH_3)_3 + N_2CHCO_2C_2H_5 \xrightarrow{BF_3} HC(SCH_3)_2CH(SCH_3)CO_2C_2H_5 \quad \textbf{(744)}\ 90\% \qquad (195)^{2209}$$

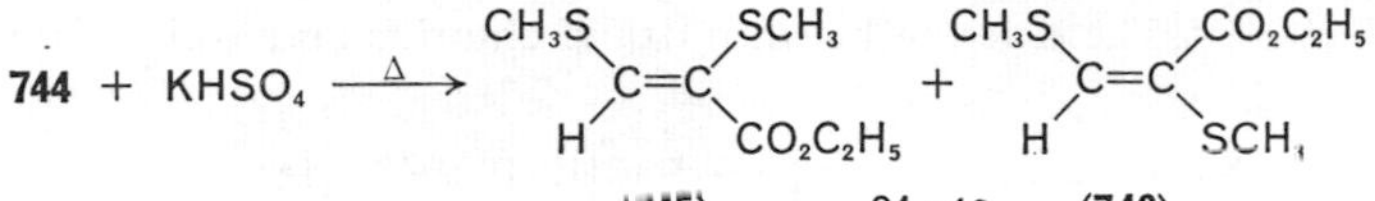

$$\textbf{744} + KHSO_4 \xrightarrow{\Delta} (CH_3S)(H)C{=}C(SCH_3)(CO_2C_2H_5)\ \textbf{(745)} + (CH_3S)(H)C{=}C(CO_2C_2H_5)(SCH_3)\ \textbf{(746)}$$

84 : 16, 99%

reported by Julia for such ylid systems[1902, 1929, 1930, 2046]. The process with biscarbomethoxycarbonyl 'carbene' has been extended to phenyl propargyl sulphides by Grieco[1823].

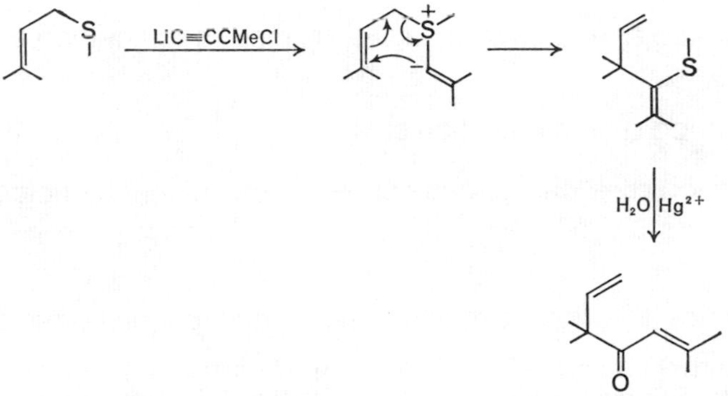

LiC≡CCMeCl

H_2O | Hg^{2+}

(196)

$$PhSCH_2C{\equiv}CPh + :C(CO_2CH_3)_2 \longrightarrow (CH_3O_2C)_2C(SPh)C(=CH_2)Ph$$

$$N_2C(CO_2CH_3)_2 \xrightarrow{h\nu} :C(CO_2CH_3)_2$$

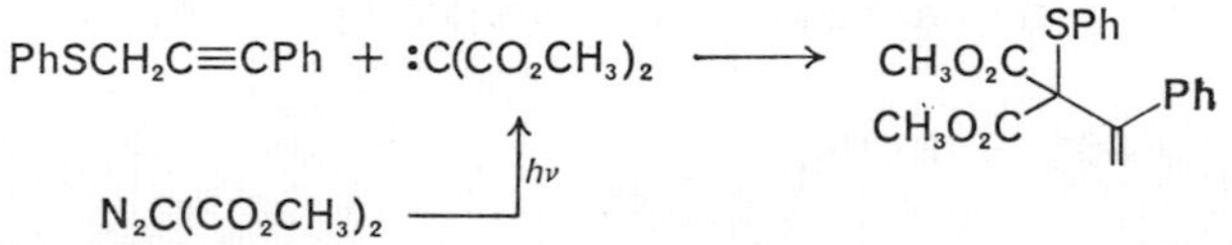

7. C–halogen bonds

The reactions of C–halogen bonds with diazoalkanes under photolytic conditions can be exceedingly messy and have been the subject of extensive investigations by Rothe[2175–2179]. The reaction of diazomethane with CCl_4 under photolytic conditions furnishes pentaerythritol tetrachloride in 33% yield operating with a large excess of CCl_4 [2303]. Other C—Cl systems behave somewhat similarly[1539, 1793, 2002, 2304].

The reactions with bromoform and iodoform are rather complex. Bromoform and diazomethane furnish dibromomethane and high molecular weight products; tetrabromomethane gives dibromomethane and 1,1-dibromoethylene; iodoform yields diiodomethane and vinyl iodide[2305]. As mentioned in Section I.C, alkyl bromides possessing an α-hydrogen undergo elimination of HBr upon photolysis in the presence of diazoacetic ester. When there is no α-hydrogen, insertion normally occurs[1531, 1719–1721, 2137]; however, trityl bromide reacts with diazoacetic ester with formation of ethyl triphenyl acrylate and ethyl bromoacetate[1718a] and benzhydryl bromide furnishes ethyl 3-bromo-2,3-diphenylpropionate. The two reactions involve copper catalysis. With allylic halides one expects rearrangements to play some role and both the chlorides[2137] and bromides[1415–1423] exhibit rearrangements. Phillips'[2137] results are summarized in equation (197). Insertions into acyl halides are part of the

Arndt–Eistert synthesis of diazo compounds and are treated elsewhere. Diazo compounds bearing no hydrogen on the diazo carbon cannot furnish diazoketones but do furnish α-chloroketones. Thus both 9-diazofluorene and diphenyldiazomethane react with phosgene and oxalyl chloride to furnish α-chloro compounds[2208, 2253, 2255]. The reaction between oxalyl chloride and 9-diazofluorene leads

$$Ph_2CHBr + N_2CHCO_2Et \longrightarrow PhCHBrPhCHCO_2Et$$

$$N_2CHCO_2Et + PhCH_2Br \longrightarrow PhCH_2CHBrCO_2Et$$

$$N_2CHCO_2Et + CH_2{=}CHCH_2Cl_2 \longrightarrow ClCH{=}CHCH_2CHClCO_2Et$$

$$N_2CHCO_2Et + ClCH{=}CHCH_2Cl \longrightarrow ClCH{=}CHCH_2CHClCO_2Et \qquad (197)^{1719-1721, 2137}$$

$$N_2CHCO_2Et + CH_3CH{=}CHCH_2Cl \longrightarrow CH_3CH{=}CHCH_2CHClCO_2Et$$

$$N_2CHCO_2Et + CH_2{=}CHCHClCH_3 \longrightarrow CH_2{=}CHCH(CH_3)CHClCO_2Et \text{ (and } CH_3CH{=}CHCH_2CHClCO_2Et)$$

to insertion into both acyl—Cl bonds even at 0 °C whereas it is possible to isolate the acid chloride with diphenyldiazomethane[2208]. Formyl fluoride furnishes the fluoroacetaldehyde with diazomethane or diazoacetaldehyde depending upon reaction conditions[2206].

$$RCOCl + R'CHN_2 \longrightarrow RCOR'CHCl + \text{excess } R'CHN \longrightarrow RCOR'CN_2 + R'CH_2Cl \qquad (198)$$

8. C—M (non-transition metal)

Diazomethane inserts into R_3Al systems to furnish homologated systems which distribute the CH_2 units in a statistical fashion[1870, 1871].

9. C—M (transition metal)

Fischer has observed the insertion of diazomethane into M—CO systems to furnish M—C(OCH_3)CH_3 systems (e.g. $W(CO)_6 \rightarrow (CO)_5WC(OCH_3)CH_3$ [1770–1773]).

B. X—H Bonds

1. O—H bonds

a. *Alcohols.* The ability to alkylate alcohols with diazo compounds extends back at least to 1930[2031]. The reaction was studied in detail by Johnson at Wisconsin and Stanford, and Caserio and Roberts at Pasadena, as well as by Müller in the late

50s[1565, 1914, 1915, 2070, 2076, 2072, 2085]. Kuhn extended the reactions to sugars[1966]. Müller employed $AlCl_3$ as a catalyst while Johnson's group used BF_3 and HBF_4. The yields of the respective methyl ethers from diazomethane are good to excellent: [cyclohexanol (84%); cholesterol (73%); glycerin (73% trimethoxypropane); L-menthol (78%); hydroxymalonic acid (100% dimethyl methoxymalonate); *N*-acetyl glucosamine (40% β-1-OCH_3)]. An anomalous reaction occurred with 2-hydroxyethyl trichloroacetate to furnish 2-methoxy-2-trichloromethyl-1,3-dioxolane in 78% yield[2032, 2070]. The reaction does, however, have its limitations in that ketone and unsaturated ketone functions elsewhere in the molecule will still undergo homologation reactions[1915, 1916, 2084].

b. *Enols*. Enols tend to be more acidic than simple alcohols and would therefore be expected to be more readily alkylated by diazomethane. This presents some minor problems in the homologation of 1,2-diones for the initial product, a 1,3-dione is frequently methylated on oxygen[1708, 1766, 1908, 2183, 2184]. Unsymmetrical 1,3-diones

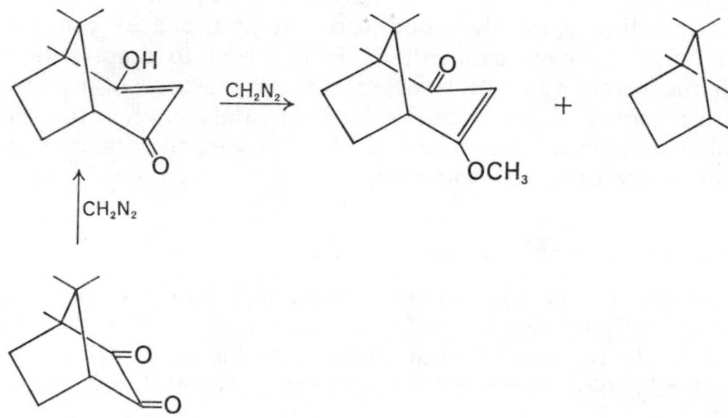

OH, CH_2N_2, O, OCH_3, +, OCH_3, O, CH_2N_2, O, O

(199)[1734]

furnish mixtures of both possible ethers[1730, 1768, 1826]; with β-keto esters both *cis*- and *trans*-β-methoxyacrylate systems arise. The possibility of C alkylation exists and is suppressed by using methanol–ether solvent systems rather than ether alone[1401, 1466, 2325].

The hydroxyl functions of phenols tend to exhibit appreciable degrees of acidity and diazomethane offers a clean and efficient alternative to Williamson's ether synthesis. In addition, the process does not involve large solvent cages and is successful when applied to hindered phenols such as the 1-hydroxy group in purpurogallin and its derivatives.

Occasionally problems are encountered; hence, 8-hydroxy quinoline undergoes both C and N alkylation[2139, 2194]. However, such problems are unusual.

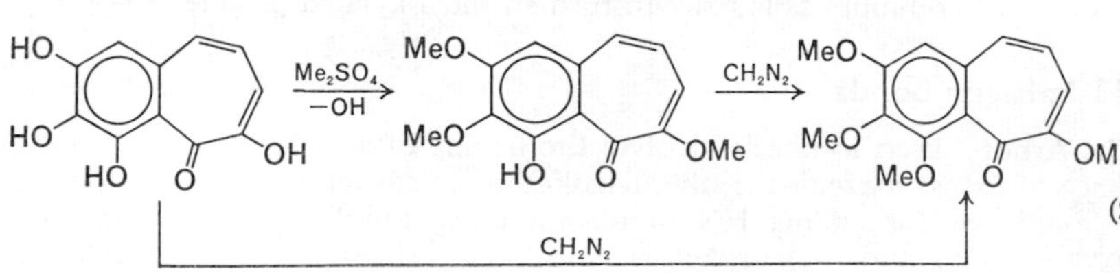

HO, HO, HO, O, OH, Me_2SO_4, −OH, MeO, MeO, HO, O, OMe, CH_2N_2, MeO, MeO, MeO, O, OMe, CH_2N_2

(200)

c. *Carboxylic acids*. The formation of esters from diazoalkanes and carboxylic acids must surely be the most common application of diazoalkanes. The reaction is not limited to diazomethane. The reaction apparently proceeds via a carbonium ion intermediate. Diazo compounds which would furnish carbonium ions exhibiting a tendency towards rearrangement do lead to both rearranged and unrearranged esters[1610, 1858, 1950, 2109]. The reaction of 4-methoxypyridine-*N*-oxide-2-carboxylic acid and diazomethane is unusual in that the products are 4-methoxypyridine, CO_2 and formaldehyde[2145].

d. *Other* O—H *bonds*. Hydroperoxides react with diazomethane to furnish good yields of the related $ROOCH_3$ systems[1873]. The low temperature reactions of dialkyloxonium hexachloroantimonates in sulphur dioxide with diazomethane furnish the related methyloxonium salt[1952–1954].

Sulphonic acids, like carboxylic acids, furnish methyl esters upon treatment with diazomethane[1834, 2235]. (The reaction with alcoholic solutions of sulphur dioxide is treated in Section I.A.2.j.) Similarly, hypophosphoric acid furnishes the tetramethyl ester[1509]. Phosphorous acid furnishes the diester[1449] and can also undergo P—H insertions[1449]. Depending upon the conditions, sulphuric acid can furnish half esters[2038] or diesters[2088]. Tosyldiazomethane is reported to react with perchloric acid to furnish the perchlorate[1755]; however, in the presence of water sulphonyl diazomethanes are converted into alcohols by acid catalysis[2379]. In the presence of other nucleophiles or even a large excess of chloride ion, the apparent carbonium ion is trapped in competition with water[2269].

2. S—H, Se—H and Te—H bonds

Hydrogen sulphide reacts with diphenyldiazomethane to furnish benzhydryl mercaptan. However, hydrogen sulphide will frequently reduce diazoalkanes to hydrazones[1639, 2352]. The reaction of ammonium bisulphide with diazoketones offers a route to 1,2,3-thiadiazoles[2251, 2343, 2350, 2351]. Thiophenols and thiols are alkylated[1642, 1723, 1931, 2068, 2252].

3. N—H bonds

The N—H bond in amines, ammonium salts, hydroxylamines and some amides can be alkylated by diazo compounds. Glycine and hydroxylamine-*o*-sulphonic acid undergo tris methylation with diazomethane[1522, 2332]. In order to obtain the methyl ester using a diazo compound one must first block the amine function and subsequently remove the blocking group[2271]. The problem of nucleosides has been examined by Todd[1836]. The simple mixing of a primary amine such as methyl amine and diazoacetic ester at room temperature leads to the formation of dihydro-1,2,4,5-tetrazene-3,6-carboxyamides. Secondary amines behave similarly; however, the use of a CuCN catalyst leads to alkylation[2067, 2069, 2071, 2186]. A direct contrast is the preparation of *N*-phenylglycine ethyl ester from aniline which does not require a catalyst[2073]. Pyridinium salts react to furnish the alkylated products[1528, 1944].

4. H-halogen bonds

The Arndt–Eistert synthesis involves the freeing of a molecule of HX and in the absence of excess diazoalkane, the diazoketone is converted into an α-haloketone. The conditions for adding HX to diazoketones can however be important for α-bromoketones and α-chloroketones undergo rearrangements in acidic media.

(201)[2174]

$R = H, CH_3$

Thus the unexpected product in equation (201) may well be the consequence of secondary processes[2174]. The reaction of methane-tris-acetyldiazomethane with hydrogen chloride furnishes a tris-chloromethyl-trioxaadamantane[2267]. The same compounds in the presence of a Cu–Bu_2S catalyst system furnished bullvaltrione[2222].

C. Insertions into other X—Y bonds

Diazomethane reacts smoothly with *N*-chlorosuccinimide to furnish the related chloromethyl derivative. This compound possesses a very labile chlorine and is thus potentially a useful synthetic intermediate[1595].

Many other X–halogen bonds are also susceptible to attack by diazoalkanes. Representative examples can be found listed in Houben–Weyl[1738]. These include both metal and non-metal halides including S—Cl and P—Cl bonds. Insertion into alkyl hypochlorites furnish routes to ketals and acetals, while insertions into S—S bonds furnish dithioketals. Halogens lead to R_2CX_2 functionality, while peroxides lead to a variety of products (equation 202).

(202)

$R^1 = R^2 = F, R^3 = p\text{-MeO}$
$R^1 = C_6H_5, R^2 = H, R^3 = p\text{-Me}$
$R^1 = R^2 = C_6H_5, R^3 = H$
$R^1 = C_6H_5CO, R^2 = H, R^3 = p\text{-Cl}$
$R^1 = C_2H_5OCO, R^2 = H, R^3 = p\text{-Br}$
$R^1 = R^2 = \text{fluorenyl}, R^3 = p\text{-NO}_2$
$R^1 = R^2 = \text{fluorenyl}, R^3 = o\text{-NO}_2$
$R^1 = Ph, R^2 = R^3 = H$
$R^1 = R^2 = Ph, R^3 = H$

VI. ACKNOWLEDGEMENTS

The writing of this and Chapter 8 was supported in part by salary grants to D. S. Wulfman by the CNRS and National Science Foundation Grant NSF–ENG–76–01321. This support is gratefully acknowledged along with support for C. F. Cooper from the Office of the Dean of the Graduate School, University of Missouri-Rolla.

VII. REFERENCES

References below 1400 occur at the end of Chapter 8.
Starred references are reviews or contain short reviews.

1400. L. J. Aarons, J. A. Connor, I. H. Hillier, M. Schwarz and D. R. Lloyd, *J. Chem. Soc. Faraday Trans. II*, **70**, 1106 (1974).
1401. R. Adams, R. S. Voris and L. N. Whitehill, *J. Amer. Chem. Soc.*, **74**, 5588 (1952).
1402. D. Adamson and J. Kenner, *J. Chem. Soc.*, 181 (1939).
1403.* A. A. Akhrem, D. I. Metelitsa and M. E. Skurko, *Uspekhi Khim.*, **43**, 868 (1975).
1404. K. Alder and G. Stein, *Ann. Chem.*, **501**, 1 (1933).
1405. K. Alder and G. Stein, *Ann. Chem.*, **515**, 165 (1934).
1406. K. Alder and G. Stein, *Ann. Chem.*, **515**, 185 (1935).
1407. K. Alder and G. Stein, *Ann. Chem.*, **525**, 221 (1936).
1408. K. Alder and H. F. Rickert, *Ann. Chem.*, **543**, 1 (1940).
1409. K. Alder and F. H. Flock, *Chem. Ber.*, **87**, 1916 (1954).
1410. K. Alder, H. Jungen and K. Rust, *Ann. Chem.*, **602**, 94 (1957).
1410a. J. V. Alphen, *Rec. Trav. Chim.*, **62**, 485 (1943).
1411. B. H. Al-Sader and R. J. Crawford, *Can. J. Chem.*, **46**, 3301 (1968).
1412. B. H. Al-Sader and R. J. Crawford, *Can. J. Chem.*, **48**, 2745 (1970).
1413. C. D. Anderson, J. T. Sharp, E. Stefanivk and R. S. Strathdcc, *Tetrahedron Letters*, 305 (1976).
1414. G. Anderson and R. C. Rhodes, *J. Org. Chem.*, **30**, 1616 (1965).
1415. W. Ando, T. Hagihara, S. Tozune and T. Migita, *J. Amer. Chem. Soc.*, **91**, 2786 (1969).
1416. W. Ando, K. Nahayama, K. Ichibori and T. Migita, *J. Amer. Chem. Soc.*, **91**, 5164 (1969).
1417. W. Ando, S. Kondo and T. Migita, *J. Amer. Chem. Soc.*, **91**, 6516 (1969).
1418. W. Ando, T. Hagihara, S. Tozune, S. Nakadio and T. Migita, *Tetrahedron Letters*, 1979 (1969).
1419. W. Ando, T. Hagihara and T. Migita, *Tetrahedron Letters*, 1983 (1969).
1420. T. Migita, W. Ando, S. Kondo, H. Matsuyama and M. Kosugi, *Nippon Kagaku Zasshi*, **91**, 374 (1970).
1421. W. Ando, S. Kordo and T. Migita, *Bull. Chem. Soc. Japan*, **44**, 571 (1971).
1422. W. Ando, N. Ogimo and T. Migita, *Bull. Chem. Soc. Japan*, **44**, 2278 (1971).
1423. W. Ando, T. Hagihara, S. Kondo, K. Nakayama, K. Yamato, S. Nakaido and T. Migita, *J. Org. Chem.*, **36**, 1732 (1971).
1424. W. Ando, *Int. J. Sulfur Chem.* **B7**, 189 (1972).
1425. W. Ando, I. Imai and T. Migita, *J. Chem. Soc. Chem. Commun.*, 822 (1972).
1426. W. Ando, T. Hagihara, S. Tozune, I. Imai, J. Suzuki, T. Toyama, S. Nakaido and T. Migita, *J. Org. Chem.*, **37**, 1721 (1972).
1427. W. Ando, I. Imai and T. Migita, *J. Org. Chem.*, **37**, 3596 (1972).
1428. W. Ando, M. Yamada, E. Matsuzaki and T. Migita, *J. Org. Chem.*, **37**, 3791 (1972).
1429. W. Ando, T. Hagiwara and T. Migita, *J. Amer. Chem. Soc.*, **95**, 7518 (1973).
1430. W. Ando, J. Suzuki, Y. Saiki and T. Migita, *J. Chem. Soc. Chem. Commun.*, 366 (1973).
1431. W. Ando, Y. Saiki and T. Migita, *Tetrahedron*, **29**, 3511 (1973).
1432. W. Ando, H. Fujii, T. Takeuchi, H. Higuchi, Y. Saiki and T. Migita, *Tetrahedron Letters*, 2117 (1973).
1433. W. Ando, K. Konishi, T. Hagiwara and T. Migita, *J. Amer. Chem. Soc.*, **96**, 1601 (1974).

1434. W. Ando, A. Sekiguchi, T. Hagiwara and T. Migita, *J. Chem. Soc. Chem. Commun.*, 372 (1974).
1435. W. Ando, H. Higuchi and T. Migita, *J. Chem. Soc. Chem. Commun.*, 523 (1974).
1436. W. Ando, K. Komishi and T. Migita, *J. Organometallic Chem.*, **67**, C7 (1974).
1437. W. Ando, T. Hagiwara and T. Migita, *Tetrahedron Letters*, 1425 (1974).
1438. W. Ando, A. Sekiguchi, T. Migita, S. Kammula, M. Green and M. Jones, *J. Amer. Chem. Soc.*, **97**, 3818 (1975).
1439. W. Ando, S. Kordo, K. Nakayama, K. Ichibovi, H. Kahoda, I. Imai, S. Nakaido and T. Migita, *J. Amer. Chem. Soc.*, **94**, 3870 (1972).
1440. J. M. André, M. Cl. André, G. Leroy and J. Weiler, *Int. J. Quant. Chem.*, **111**, 1013 (1969).
1441. S. D. Andrews and A. C. Day, *J. Chem. Soc. Chem. Commun.*, 902 (1967).
1442. J.-P. Anselme, *Org. Prep. Proc.*, **1**, 73 (1969).
1443.* J. Ap Simon (Ed.), *The Total Synthesis of Natural Products*, Wiley–Interscience, London, 1973.
1444. T. J. Arackal and B. Eistert, *Chem. Ber.*, **108**, 2660 (1975).
1445. T. Aratani, Y. Nakanisi and H. Nozaki, *Tetrahedron*, **26**, 1765 (1970).
1446. T. Aratani, Y. Yoneyoshi and T. Nagase, *Tetrahedron Letters*, 1707 (1975).
1447. B. Arbuzov and S. Rafikov, *Zh. Obshch. Khim.*, **7**, 2195 (1937).
1448. B. Arbuzov and S. Rafikov, *Zh. Obshch. Khim.*, **7**, 2195 (1937).
1449. B. A. Arbuzov and A. O. Vizel, *Izvest. Akad. Nauk S.S.S.R.*, 749 (1963).
1450. F. Arndt, *Angew. Chem.*, **40**, 1099 (1927).
1451. F. Arndt, B. Eistert and W. Partale, *Chem. Ber.*, **61**, 1107 (1928).
1452. F. Arndt and B. Eistert, *Chem. Ber.*, **61**, 1118 (1928).
1453. F. Arndt, B. Eistert and W. Ender, *Chem. Ber.*, **62**, 44 (1929).
1454. F. Arndt and C. Martius, *Ann. Chem.*, **499**, 228 (1932).
1455. F. Arndt and C. Martius, *Ann. Chem.*, **499**, 246 (1932).
1456. F. Arndt and C. Martius, *Ann. Chem.*, **499**, 265 (1932).
1457. F. Arndt, J. Amende and W. Ender, *Monatsh. Chem.*, **59**, 202 (1932).
1458. F. Arndt and B. Eistert, *Chem. Ber.*, **68**, 196 (1935).
1459. F. Arndt and B. Eistert, *Chem. Ber.*, **68**, 200 (1935).
1460. F. Arndt and J. D. Rose, *J. Chem. Soc.*, 1 (1935).
1461. F. Arndt, H. Scholz and E. Frobel, *Ann. Chem.*, **521**, 95 (1936).
1462. F. Arndt and B. Eistert, *Chem. Ber.*, **69**, 1805 (1936).
1463. F. Arndt and V. Loewe, *Chem. Ber.*, **71**, 1627 (1938).
1464. F. Arndt, L. Loewe, E. Özsöy, M. Ögüt, A. Arslan and L.Bage vi, *Chem. Ber.*, **71**, 1637 (1938).
1465. F. Arndt, L. Locwe, T. Senerge and I. Turegun, *Chem. Ber.*, **71**, 1640 (1938).
1466. F. Arndt, L. Loewe and B. Beyer, *Chem. Ber.*, **74**, 1460 (1941).
1467. A. J. Ashe, III, *Tetrahedron Letters*, 523 (1969).
1468. D. H. Aue and G. S. Helwig, *Tetrahedron Letters*, 721 (1974).
1469. W. E. Bachmann, W. Cole and A. L. Wilds, *J. Amer. Chem. Soc.*, **62**, 824 (1940).
1470. H. J. Backer, *Rec. Trav. chim.*, **69**, 1223 (1950).
1471. L. Baldini and G. Brambilla, *Cancer Res.*, **26**, 1754 (1966).
1472. L. Baldini, G. Brambilla, M. Cavanna, C. E. Caraceni and S. Parodi, *Transplantation*, **13**, 224 (1972).
1473. J. Baldwin, unpublished work.
1474. E. Bamberger, *Chem. Ber.*, **33**, 941 (1900).
1475. E. Bamberger, O. Schmidt and H. Levenstein, *Chem. Ber.*, **33**, 2043 (1900).
1476. E. Bamberger and O. Schmidt, *Chem. Ber.*, **34**, 574 (1901).
1477. E. Bamberger, *Chem. Ber.*, **35**, 54 (1902).
1478. E. Bamberger and J. Grob, *Chem. Ber.*, **35**, 67 (1902).
1479. E. Bamberger and J. Frei, *Chem. Ber.*, **35**, 82 (1902).
1480. E. Bamberger, *Chem. Ber.*, **36**, 90 (1903).
1481. W. R. Bamford and T. S. Stevens, *J. Chem. Soc.*, 4675 (1952).
1482. B. K. Bandlish, A. W. Garner, M. L. Hodges and J. W. Timberlake, *J. Amer. Chem. Soc.*, **97**, 5856 (1975).
1483. P. K. Banerjee, D. Mukhopdhyay and N. D. Choudhury, *J. Indian Chem. Soc.*, **42**, 115 (1965).

1484. E. Banfi, M. Tamaro, B. Pani and C. Monti-Bragadin, *Boll. Inst. Steroter. M. Ianese*, **53**, 5 (1974).
1485. R. E. Banks, W. T. Flowers, R. N. Hazeldine and P. E. Jackson, *J. Chem. Soc. Chem. Commun.*, 210 (1965).
1486. G. Bannikor, G. Nikiforov and V. Ershov, *Izvest. Akad. Nauk S.S.S.R. Ser. Khim.*, 2541 (1974).
1487. W. J. Baron, M. R. Decamp, M. E. Hedrick, M. Jones, Jr, R. H. Levin and M. E. Sohn, in *Carbenes,* Vol. 1 (Ed. M. Jones, Jr and R. A. Moss), Wiley–Interscience, 1972.
1489. P. D. Bartlett, R. Helgesson and O. A. Wersel, *Pure Appl. Chem.*, **16**, 187 (1968).
1490. D. H. R. Barton, D. G. T. Greig, P. G. Sammes and M. V. Taylor, *J. Chem. Soc. Chem. Commun.*, 845 (1971).
1491. D. H. R. Barton, P. G. Sammes, M. V. Taylor, C. M. Cooper, G. Hewitt, B. E. Looker and W. G. E. Underwood, *J. Chem. Soc. Chem. Commun.*, 1137 (1971).
1492. N. Barton, J. Cook, J. Loridon and J. MacMillan, *J. Chem. Soc.*, 1079 (1949).
1493. A. R. Bassindale and A. G. Brook, *Can. J. Chem.*, **52**, 3474 (1974).
1494. J. Bastide and J. Lematre, *C. R. Acad. Sci. Paris, Sér. C*, **268**, 532 (1969).
1495. J. Bastide, J. Lematre and J. Soulier, *C. R. Acad. Sci. Paris, Sér. C*, **269**, 358 (1969).
1496. J. Bastide and J. Lematre, *Bull. Soc. Chim. France*, 3543 (1970).
1497.* J. Bastide, J. Hamelin, F. Texier and Y. Vo-Quang, *Bull. Soc. Chim. France*, 2555 (1973).
1498. J. Bastide, O. Henri-Rousseau and E. Stephan, *C. R. Acad. Sci. Paris, Sér. C*, 278, 195 (1974).
1499. J. Bastide, O. Henri-Rousseau and L. Asport-Pascot, *Tetrahedron*, **30**, 3355 (1974).
1500. J. Bastide, J. Hamelin, F. Texier and Y. Vo Quang, *Bull. Soc. Chim. France*, 2555 (1973).
1501. J. Bastide, O. Henri-Rousseau and E. Stephen, *C. R. Acad. Sci. Paris, Sér. C*, **278**, 195 (1974).
1502. J. Bastide, O. Henri-Rousseau and L. Aspart-Pascot, *Tetrahedron*, **30**, 3355 (1974).
1503. P. Battioni and Y. Vo-Quang, *C. R. Acad. Sci. Paris, Sér. C*, **266**, 1310 (1968).
1504. P. Battioni, A. Aspect, L. Vo-Quang and Y. Vo-Quang, *C. R. Acad. Sci. Paris, Sér. C*, **268**, 1263 (1969).
1504a. P. Battioni, L. Vo-Quang and Y. Vo-Quang, *C. R. Acad. Sci. Paris, Sér. C*, **269**, 1063 (1969).
1505. P. Battioni, L. Vo-Quang and Y. Vo-Quang, *Bull. Soc. Chim. France*, 3938 (1970).
1505a. P. Battioni, L. Vo-Quang and Y. Vo-Quang, *C. R. Acad. Sci. Paris, Sér. C*, **271**, 1468 (1970).
1506. P. Battioni, L. Vo-Quang and Y. Vo-Quang, *C. R. Acad. Sci. Paris, Sér. C*, **275**, 1109 (1972).
1507. P. Battioni, L. Vo-Quang and Y. Vo-Quang, *Tetrahedron Letters*, 4803 (1972).
1508. M. A. Battiste and M. E. Brennan, *Tetrahedron Letters*, 5857 (1966).
1509. M. Baudler, *Z. Naturforsch.*, **8b**, 326 (1953).
1510. B. C. Baumann and N. H. Fischer, unpublished, cited in Reference 1729.
1511. D. J. Beames and L. N. Mander, *Austr. J. Chem.*, **24**, 343 (1971).
1512. D. J. Beames, T. R. Klase and L. N. Mander, *J. Chem. Soc. Chem. Commun.*, 773 (1971).
1513. D. J. Beames, L. N. Mander and J. V. Turner, *Austr. J. Chem.*, **27**, 1977 (1974).
1513a. J. M. Beiner, D. Lecadet, D. Paguer, A. Thuiller and J. Vialle, *Bull. Soc. Chim. France*, 1979 (1973).
1513b. J. M. Beiner, D. Lecadet, D. Paguer and A. Thuiller, *Bull. Soc. Chim. France*, 1983 (1973).
1514. G. Berger, M. Franck-Neumann and G. Ourisson, *Tetrahedron Letters*, 3451 (1968).
1515.* R. G. Bergman, in *Free Radicals*, Vol. 1 (Ed. J. Kochi), Wiley, London, 1973, p. 191.
1516. J. A. Berson, D. R. Hartter, H. Klinger and R. W. Grubb, *J. Org. Chem.*, **33**, 1669 (1968).
1517. G. P. Bettinetti and G. Desimoni, *Gazzetta*, **93**, 658 (1963).
1518. G. P. Bettinetti, G. Desimoni and P. Grünanger, *Gazzetta*, **94**, 92 (1964).

1519. G. P. Bettinetti, A. Donetti and G. Grünanger, *Tetrahedron Letters*, 2933 (1966).
1520. S. Bien and A. Gillon, *Tetrahedron Letters*, 3073 (1974).
1521. E. H. Billett and I. Fleming, *J. Chem. Soc. Perkin Trans. I*, 1658 (1973).
1522. H. Biltz and H. Pacetzold, *Chem. Ber.*, **55**, 1066 (1922).
1523. J. S. Bindra and R. Bindra, *Creativity in Organic Synthesis*, Academic Press, London, 1975.
1524. C. Bischoff and K. H. Platz, *J. prakt. Chem.*, **312**, 2 (1970).
1525. L. Bispink and H. Matthaei, *FEBS Letters*, **37**, 291 (1973).
1526. U. Blalero, J. J. Plattner and H. Rapoport, *J. Amer. Chem. Soc.*, **91**, 4933 (1969).
1527. U. Blalero, J. J. Plattner and H. Rapoport, *J. Amer. Chem. Soc.*, **92**, 3429 (1970).
1528. C. E. Blades and A. L. Wilds, *J. Org. Chem.*, **21**, 1013 (1956).
1529. P. Bladon and D. R. Rae, *J. Chem. Soc. Perkin I*, 2240 (1974).
1530. H. Böhme and R. Marx, *Chem. Ber.*, **74**, 1667 (1941).
1531. H. Böhme, E. Mundlos, W. Lehners and O.-E. Herboth, *Chem. Ber.*, **90**, 2008 (1957).
1532. I. B. M. Bond, D. Lloyd, M. I. C. Singer and F. I. Wasson, *J. Chem. Soc. Chem. Commun.*, 544 (1966).
1533. B. F. Bonini, G. Maccagni, A. Wagenaar, L. Thijs and B. Zwanenburg, *J. Chem. Soc. Perkin Trans. I*, 2490 (1972).
1534. F. G. Bordwell, J. M. Williams, E. B. Hoyt and B. B. Jarvis, *J. Amer. Chem. Soc.*, **90**, 429 (1968).
1535. A. K. Bose and P. Yates, *J. Amer. Chem. Soc.*, **74**, 4703 (1952).
1536. R. E. Bowman, A. Campbell and W. R. N. Williamson, *J. Chem. Soc.*, 3846 (1964).
1537. J. H. Boyer, in Reference 1769, p. 215.
1538. J. H. Boyer and W. Beverung, *J. Chem. Soc. Chem. Commun.*, 1377 (1969).
1539. J. N. Bradley and A. Ledwith, *J. Chem. Soc.*, 1495 (1961).
1540. G. Brambilla, S. Parodi, M. Cavanna and L. Baldini, *Transplantation*, **10**, 100 (1970).
1541. G. Brambilla, M. Cavanna, S. Parodi and L. Baldini, *Europ. J. Cancer.*, **8**, 127 (1972).
1542. G. Brambilla, M. Cavanna, A. Maura, S. Parodi, A. Furlani, V. Scarcia and R. Della-Loggia, *Arzneim. Forsch.*, **23**, 690 (1973).
1543. J. Bredt and W. Holz, *J. prakt. Chem.*, [2], **95**, 133 (1917).
1544. R. Breslow, R. Winter and M. Battiste, *J. Org. Chem.*, **24**, 415 (1959).
1545. R. Breslow and D. Chipman, *Chem. and Ind.*, 1105 (1960).
1546. F. V. Bruchhausen and H. Hoffman, *Chem. Ber.*, **74**, 1584 (1941).
1547. G. Buchi and J. D. White, *J. Amer. Chem. Soc.*, **86**, 2884 (1964).
1548. G. Buchi, W. McCleod and J. Padilla, *J. Amer. Chem. Soc.*, **86**, 4438 (1964).
1549. E. Buchner and T. Curtius, *Chem. Ber.*, **18**, 2371 (1885).
1550. E. Buchner and T. Curtius, *Chem. Ber.*, **18**, 2377 (1885).
1551. E. Buchner, *Chem. Ber.*, **22**, 842 (1889).
1552. E. Buchner, *Chem. Ber.*, **22**, 2165 (1889).
1553. E. Buchner and H. Witter, *Ann. Chem.*, **273**, 239 (1893).
1554. E. Buchner and M. Fritsch, *Chem. Ber.*, **26**, 256 (1893).
1555. E. Buchner and C. von der Heide, *Chem. Ber.*, **34**, 345 (1901).
1556. R. Caballol, R. Carbo and M. Martin, *Chem. Phys. Letters*, **28**, 422 (1974).
1557. H. J. Callot and C. Benezra, *Can. J. Chem.*, **50**, 1078 (1972).
1558. L. E. Cannon, D. K. Woodard, M. E. Woehler and R. E. Lovins, *Immunology*, **26**, 1183 (1974).
1559. L. Capuano, H. Dürr and R. Zander, *Ann. Chem.*, **721**, 75 (1969).
1560. B. A. Carlson, W. A. Sheppard and O. W. Webster, *J. Amer. Chem. Soc.*, **97**, 5291 (1975).
1561. L. A. Carpino and R. H. Rynbrandt, *J. Amer. Chem. Soc.*, **88**, 5682 (1966).
1562. E. Carstensen-Oeser, B. Müller and H. Dürr, *Angew. Chem.*, **84**, 434 (1972).
1563. J. Casanova and B. Waegell, *Bull. Soc. Chim. France*, 922 (1975).
1564. R. Casanova and T. Reichstein, *Helv. Chim. Acta*, **32**, 649 (1949).
1565. M. C. Caserio, J. D. Roberts, M. Neeman and W. S. Johnson, *J. Amer. Chem. Soc.*, **80**, 2584 (1958).
1566. J. Castaner, J. Castells and J. Pascual, *Anales real Soc. españ. Fís. Quim.*, **55B**, 739 (1959).

1567. J. Castells, R. Mestres and J. Pascual, *Anales real Soc. españ. Fis. Quim.*, **60B**, 803 (1964).
1568.* M. P. Cava and M. V. Lakshmikanthan, *Accounts Chem. Res.*, **8**, 139 (1975).
1569. P. N. Chakrabortty, R. Dasgupta, S. K. Dasgupta, S. R. Ghosh and U. R. Ghatak, *Tetrahedron*, **28**, 4653 (1972).
1570. J. N. Chatterjea, S. N. P. Gupta and V. N. Mehrotra, *J. Indian Chem. Soc.*, **42**, 208 (1965).
1571. F. Chi and G. Leroi, *Spectrochim. Acta*, **31A**, 1759 (1975).
1572. L. M. Christen, L. W. Waale and W. M. Jones, *J. Amer. Chem. Soc.*, **94**, 2118 (1972).
1573. E. Ciganek, *J. Org. Chem.*, **30**, 4198 (1965).
1574. E. Ciganek, *J. Org. Chem.*, **30**, 4366 (1965).
1575. E. Ciganek, *J. Amer. Chem. Soc.*, **87**, 652 (1965).
1576. E. Ciganek, *J. Amer. Chem. Soc.*, **88**, 1979 (1966).
1577. E. Ciganek, *J. Amer. Chem. Soc.*, **89**, 1454 (1967).
1578. E. Ciganek, *J. Org. Chem.*, **35**, 862 (1970).
1579. E. Ciganek, *J. Amer. Chem. Soc.*, **93**, 2207 (1971).
1580. D. T. Clark, D. B. Adams, I. W. Scanlan and I. S. Woolsey, *Chem. Phys. Letter*, **25**, 263 (1974).
1581. T. C. Clarke, L. A. Wendling and R. G. Bergman, *J. Amer. Chem. Soc.*, **97**, 5638 (1975).
1581a. R. Clinging, F. M. Dean and G. H. Mitchell, *Tetrahedron*, **30**, 4065 (1974).
1582. G. L. Closs and L. E. Closs, *J. Amer. Chem. Soc.*, **83**, 2015 (1961).
1583. G. L. Closs, L. E. Closs and W. A. Böll, *J. Amer. Chem. Soc.*, **85**, 3796 (1963).
1584. G. Closs ahd W. Böll, *J. Amer. Chem. Soc.*, **85**, 3904 (1963).
1585. G. Closs, W. Böll, H. Heyn and V. Dev, *J. Amer. Chem. Soc.*, **90**, 173 (1968).
1586. G. L. Closs and P. E. Pfeffer, *J. Amer. Chem. Soc.*, **90**, 2452 (1968).
1587. H. Cohen and C. Benezra, *Can. J. Chem.*, **52**, 66 (1974).
1588. L. A. Cohen and W. M. Jones, *J. Amer. Chem. Soc.*, **85**, 3397 (1963).
1589. E. W. Colvin, R. A. Raphael and J. S. Roberts, *J. Chem. Soc. Chem. Commun.*, 858 (1971).
1590. A. Constantino, G. Linstrumelle and S. Julia, *Bull. Soc. Chim. France*, 907 (1970).
1591. J. W. Cook and R. Schoendal, *J. Chem. Soc.*, 288 (1945).
1592. E. J. Corey and K. Achiwa, *Tetrahedron Letters*, 3257 (1969).
1593. E. J. Corey, K. Achiwa and J. Katzenellenbogen, *J. Amer. Chem. Soc.*, **91**, 4318 (1969).
1594. E. J. Corey and K. Achiwa, *Tetrahedron Letters*, 2245 (1970).
1595. R. A. Corral and O. O. Orazi, *Tetrahedron Letters*, 1693 (1964).
1596. S. Carsano, L. Capito and M. Bonamico, *Ann. Chem. Ital.*, **48**, 140 (1958).
1597. D. D. Cowan, M. M. Couch, K. R. Kopecky and G. S. Hammond, *J. Org. Chem.*, **29**, 1922 (1964).
1598. D. J. Cram and R. D. Partas, *J. Amer. Chem. Soc.*, **85**, 3397 (1963).
1599. R. J. Crawford, R. J. Dummel and A. Mishra, *J. Amer. Chem. Soc.*, **87**, 3023 (1965).
1600. R. J. Crawford and A. Mishra, *J. Amer. Chem. Soc.*, **87**, 3768 (1965).
1601. R. J. Crawford and A. Mishra, *J. Amer. Chem. Soc.*, **87**, 3907 (1965).
1602. R. J. Crawford and D. M. Cameron, *J. Amer. Chem. Soc.*, **88**, 2589 (1966).
1603. R. J. Crawford, A. Mishra and R. J. Dummel, *J. Amer. Chem. Soc.*, **88**, 3959 (1966).
1604. R. J. Crawford and A. Mishra, *J. Amer. Chem. Soc.*, **88**, 3963 (1966).
1605. R. J. Crawford and D. M. Cameron, *Can. J. Chem.*, **45**, 691 (1967).
1605a. R. J. Crawford and G. L. Erikson, *J. Amer. Chem. Soc.*, **89**, 3907 (1967).
1606. R. J. Crawford and L. H. Ali, *J. Amer. Chem. Soc.*, **89**, 3908 (1967).
1607. R. J. Crawford and T. R. Lynch, *Can. J. Chem.*, **46**, 1457 (1968).
1608. R. J. Crawford, J. Hamelin and B. Strehlke, *J. Amer. Chem. Soc.*, **93**, 3810 (1971).
1609. R. J. Crawford and M. Ohno, *Canad. J. Chem.*, **52**, 3134 (1974).
1610. D. Y. Curtin and S. M. Gerber, *J. Amer. Chem. Soc.*, **74**, 4052 (1952).
1611. D. R. Dalton and S. A. Liebman, *Tetrahedron*, **25**, 3321 (1969).
1612. D. Danion and R. Carrié, *Bull. Soc. Chim. France*, 1130 (1972).
1613. R. Danion-Bougot and R. Carrié, *C. R. Acad. Sci. Paris, Sér. C*, **264**, 1141 (1967).
1614. R. Danion-Bougot and R. Carrié, *C. R. Acad. Sci. Paris, Sér. C*, **264**, 1457 (1967).

1615. R. Danion-Bougot and R. Carrié, *Tetrahedron Letters*, 5285 (1967).
1616. R. Danion-Bougot and R. Carrié, *Bull. Soc. Chim. France*, 2526 (1968).
1617. R. Danion-Bougot and R. Carrié, *Bull. Soc. Chim. France*, 4241 (1968).
1618. R. Danion-Bougot and R. Carrié, *C. R. Acad. Sci. Paris, Sér. C*, **266**, 645 (1968).
1619. R. Danion-Bougot and R. Carrié, *Bull. Soc. Chim. France*, 313 (1969).
1620. R. Danion-Bougot and R. Carrié, *C. R. Acad. Sci. Paris, Sér. C*, **270**, 1135 (1970).
1621. R. Danion-Bougot and R. Carrié, *Bull. Soc. Chim. France*, 263 (1972),
1622. D. Danion and R. Carrié, *Bull. Soc. Chim. France*, 1130 (1972).
1623. R. Danion-Bougot and R. Carrié, *Bull. Soc. Chim. France*, 3511 (1972).
1624. R. Danion-Bougot and R. Carrié, *Bull. Soc. Chim. France*, 3521 (1972).
1625. R. Danion-Bougot and R. Carrié, *Org. Mag. Res.*, **5**, 453 (1973).
1626. S. Danishefsky and G. Rovnyak, *J. Org. Chem.*, **39**, 2924 (1974).
1627. S. K. Dasgupta, R. Dasgupta, S. R. Ghosh and U. R. Ghatak, *J. Chem. Soc. Chem. Commun.*, 1253 (1969).
1628. W. G. Dauben and D. L. Whalen, *Tetrahedron Letters*, 3743 (1966).
1629. A. C. Day and M. C. Whiting, *J. Chem. Soc., C*, 464 (1966).
1630. A. C. Day and M. C. Whiting, *J. Chem. Soc., C*, 1719 (1966).
1631. A. C. Day and R. N. Inwood, *J. Chem. Soc., C*, 1065 (1969).
1632. V. W. Day, B. R. Stults, K. J. Reimer and A. Shaver, *J. Amer. Chem. Soc.*, **96**, 1227 (1974).
1632a. F. M. Dean and B. K. Park, *J. Chem. Soc. Chem. Commun.*, 162 (1974).
1633. G. B. R. DeGraaf, J. H. van Dijck-Rothius and G. van de Kolk, *Rec. Trav. Chim.* **74**, 143 (1955).
1634. G. B. R. DeGraaf and G. van der Kolk, *Rec. Trav. Chim.*, **77**, 224 (1958).
1635. M. J. S. Dewar and R. Pettit, *J. Chem. Soc.*, 2026 (1956).
1636. W. Dieckmann, *Chem. Ber.*, **43**, 1024 (1910).
1637. D. J. Dijksman and G. T. Newbold, *J. Chem. Soc.*, 1216 (1951).
1638. A. Dijkstra and H. J. Backer, *Rec. Trav. chim.*, **73**, 575 (1954).
1639. O. Dimroth, *Ann. Chem.*, **373**, 343 (1910).
1640. J. Dingwall and J. T. Sharp, *J. Chem. Soc. Chem. Commun.*, 128 (1975).
1641. D. C. Dittmer and R. Glassman, *J. Org. Chem.*, **35**, 999 (1970).
1642. C. Djerassi and A. L. Nussbaum, *J. Amer. Chem. Soc.*, **75**, 3700 (1953).
1643. W. von E. Doering and L. Knox, *J. Amer. Chem. Soc.*, **72**, 2305 (1950).
1644. W. von E. Doering and F. Detert, *J. Amer. Chem. Soc.*, **73**, 876 (1951).
1645. W. von E. Doering and C. H. DePuy, *J. Amer. Chem. Soc.*, **75**, 5955 (1953).
1646. W. von E. Doering, L. H. Knox and F. Detert, *J. Amer. Chem. Soc.*, **75**, 297 (1953).
1647. W. von E. Doering, G. Laber, R. Vanderwahl, N. F. Chamberlain, and R. B. Williams, *J. Amer. Chem. Soc.*, **78**, 5448 (1956).
1648. W. von E. Doering and H. Prinzbach, *Tetrahedron Letters*, 27 (1959).
1649.* W. von E. Doering and W. R. Roth, *Angew. Chem.*, **75**, 27 (1963).
1650. W. von E. Doering and W. R. Roth, *Tetrahedron*, **19**, 715 (1963).
1651. A. Drapsky, *Chem. Ber.*, **43**, 1112 (1910).
1652. C. DuMont, J. Naire, M. Vidal and P. Arnaud, *C. R. Acad. Sci. Paris, Sér. C*, **268**, 348 (1969).
1653.* H. Dürr, *Methoden der Organischen Chemie* (Houben–Weyl–Müller), Vol. IV, Georg Thieme, Stuttgart, 1945, p. 1158.
1654. H. Dürr, *Ber. Bunsenges. Phys. Chem.*, **69**, 641 (1965).
1655. H. Dürr, G. Ourisson and B. Waegell, *Chem. Ber.*, **98**, 1858 (1965).
1656. H. Dürr, *Tetrahedron Letters*, 5829 (1966).
1657. H. Dürr, *Angew. Chem.*, **79**, 1104 (1967).
1658. H. Dürr and G. Scheppers, *Chem. Ber.*, **100**, 3236 (1967).
1659. H. Dürr, *Ann. Chem.*, **703**, 109 (1967).
1660. H. Dürr, *Tetrahedron Letters*, 1649 (1967).
1661. H. Dürr, *Z. Naturforsch.*, **22b**, 786 (1967).
1662. H. Dürr and G. Scheppers, *Angew. Chem.*, **80**, 359 (1968).
1663. H. Dürr, *Chem. Ber.*, **101**, 3047 (1968).
1664. H. Dürr, *Ann. Chem.*, **711**, 115 (1968).
1665. H. Dürr and P. Heitkämper, *Ann. Chem.*, **716**, 212 (1968).

1666. H. Dürr and G. Scheppers, *Tetrahedron*, **24**, 6059 (1968).
1667. H. Dürr and W. Benz, *Tetrahedron*, **24**, 6503 (1968).
1668. H. Dürr and L. Schrader, *Angew. Chem.*, **81**, 426 (1969).
1669. H. Dürr and L. Schrader, *Angew. Chem. Int. Ed.*, **8**, 446 (1969).
1670. H. Dürr and L. Schrader, *Chem. Ber.*, **102**, 2026 (1969).
1671. H. Dürr, G. Scheppers and L. Schrader, *J. Chem. Soc. Chem. Commun.*, 257 (1969).
1672. H. Dürr, *Ann. Chem.*, **723**, 102 (1969).
1673. H. Dürr and L. Schrader, *Z. Naturforsch.*, **24b**, 536 (1969).
1674. H. Dürr and P. Heitkämper, *Z. Naturforsch.*, **24b**, 779 (1969).
1675. H. Dürr, *Z. Naturforsch.*, **24b**, 1490 (1969).
1676. H. Dürr, *Chem. Ber.*, **103**, 369 (1970).
1677. H. Dürr and G. Scheppers, *Chem. Ber.*, **103**, 380 (1970).
1678. H. Dürr and L. Schrader, *Chem. Ber.*, **103**, 1331 (1970).
1679. H. Dürr and G. Scheppers, *Ann. Chem.*, **734**, 141 (1970).
1680. H. Dürr, R. Sergio and G. Scheppers, *Ann. Chem.*, **740**, 63 (1970).
1681. H. Dürr and H. Kober, *Ann. Chem.*, **740**, 74 (1970).
1682. H. Dürr, P. Heitkämper and P. Herbst, *Tetrahedron Letters*, 1599 (1970).
1683. H. Dürr and H. Kober, *Angew. Chem.*, **83**, 362 (1971).
1684. H. Dürr and L. Schrader, *Chem. Ber.*, **104**, 391 (1971).
1685. H. Dürr and H. Kober, *Allg. prakt. Chem.*, **23**, 73 (1972).
1686. H. Dürr and B. Ruge, *Angew. Chem.*, **84**, 215 (1972).
1687. H. Dürr, R. Sergio and W. Gombler, *Angew. Chem.*, **84**, 215 (1972).
1688. H. Dürr, B. Heu and G. Scheppers, *J. Chem. Soc. Chem. Commun.*, 1257 (1972).
1689. H. Dürr, H. Kober, V. Fuchs and P. Orth, *J. Chem. Soc. Chem. Commun.*, 973 (1972).
1690. H. Dürr, P. Herbst and K. E. Rozumek, *J. Chromatog.*, **73**, 287 (1972).
1691. H. Dürr, P. Heitkämper and P. Herbst, *Synthesis*, 261 (1972).
1692. H. Dürr and H. Kober, *Tetrahedron Letters*, 1255 (1972).
1693. H. Dürr and H. Kober, *Tetrahedron Letters*, 1259 (1972).
1694. H. Dürr and R. Sergio, *Tetrahedron Letters*, 3479 (1972).
1695. H. Dürr, B. Ruge and H. Schmitz, *Angew. Chem.*, **85**, 616 (1973).
1696. H. Dürr and H. Kober, *Chem. Ber.*, **106**, 1565 (1973).
1697.* H. Dürr, *Fortschr. Chem. Forsch.*, **40**, 103 (1973).
1698. H. Dürr, H. Kober, I. Halberstadt, U. Neu, W. M. Jones and T. T. Coburn, *J. Amer. Chem. Soc.*, **95**, 3818 (1973).
1699. H. Dürr, B. Ruge and T. Ehrhardt, *Ann. Chem.*, 214 (1973).
1700. H. Dürr and W. Bujnoch, *Tetrahedron Letters*, 1433 (1973).
1701. H. Dürr and W. Bujnoch, *Ann. Chem.*, 1691 (1973).
1702. H. Dürr and F. Werndorff, *Angew. Chem.*, **86**, 413 (1974).
1703. H. Dürr, M. Kausch and H. Kober, *Angew. Chem.*, **86**, 739 (1974).
1704. H. Dürr, B. Ruge and B. Weiss, *Ann. Chem.*, 1150 (1974).
1705. H. Dürr, P. Herbst, P. Heitkämper and H. Leismann, *Chem. Ber.*, **107**, 1835 (1974).
1706. H. Dürr and R. Sergio, *Chem. Ber.*, **107**, 2027 (1974).
1707. H. Dürr, H. Kober, R. Sergio, W. Schmidt and V. Formacek, *Chem. Ber.*, **107**, 2037 (1974).
1708. H. Dürr, H. Kober and M. Kausch, *Chem. Ber.*, **107**, 3415 (1974).
1709. H. Dürr, W. Schmidt and R. Sergio, *Ann. Chem.*, 1132 (1974).
1710. H. Dürr and W. Schmidt, *Ann. Chem.*, 1140 (1974).
1711. H. Dürr, A. C. Ranade and I. Halberstadt, *Synthesis*, 878 (1974).
1712. H. Dürr, A. C. Ranade and I. Halberstadt, *Tetrahedron Letters*, 3041 (1974).
1713. H. Dürr, D. Barth and M. Schlosser, *Tetrahedron Letters*, 3045 (1974).
1714. H. Dürr and B. Weiss, *Angew. Chem. Int. Ed.*, **14**, 646 (1975).
1715. H. Dürr and H. Schmitz, *Angew. Chem. Int. Ed.*, **14**, 647 (1975).
1716. H. Dürr and H. Kober, *Tetrahedron Letters*, 1941 (1975).
1717. H. Dürr, H. Kober and M. Kausch, *Tetrahedron Letters*, 1975 (1975).
1718. I. A. D'yakonov, *Zh. Obshch. Khim. S.S.R.*, **15**, 473 (1945).
1719. I. A. D'yakonov and N. B. Vinogradova, *Zh. Obshch. Khim.*, **23**, 244 (1953).
1720. I. A. D'yakonov and T. V. Domareva, *Zh. Obshch. Khim.*, **25**, 934 (1955).

1721. I. A. D'yakonov and T. V. Domareva, *Zh. Obshch. Khim.*, **25**, 1486 (1955).
1722. I. A. D'yakonov, M. I. Komendantov and T. S. Smirnova, *Zh. Org. Khim.*, **5**, 1742 (1969).
1723. B. L. Dyatkin and E. P. Mochalina, *Izvest. Akad. Nauk S.S.S.R.*, 1225 (1964).
1724. W. G. H. Edwards, *Chem. and Ind.*, 112 (1951).
1725. B. Eistert, *Chem. Ber.*, **69**, 1074 (1936).
1726. B. Eistert, *Angew. Chem.*, **54**, 99 (1941).
1727. B. Eistert, *Angew. Chem.*, **54**, 124 (1941).
1728. B. Eistert, *Angew. Chem.*, **54**, 193 (1941).
1729.* B. Eistert, in *Newer Methods of Preparative Organic Chemistry*, Vol. 1, Interscience, New York, 1948, p. 513.
1730. B. Eistert, and W. Reiss, *Chem. Ber.*, **87**, 112 (1954).
1731. B. Eistert, H. Elias, E. Kosch and R. Wollheim, *Chem. Ber.*, **92**, 130 (1959).
1732. B. Eistert, G. Bock, E. Kosch and F. Spalink, *Chem. Ber.*, **93**, 1451 (1960).
1733. B. Eistert, D. Greiber and I. Caspari, *Ann. Chem.*, **659**, 64 (1962).
1734. B. Eistert, D. Greiber and I. Caspari, *Ann. Chem.*, **659**, 79 (1962).
1735. B. Eistert, W. Schade and H. Selzer, *Chem. Ber.*, **97**, 1470 (1964).
1736. B. Eistert and G. Heck, *Ann. Chem.*, **681**, 138 (1965).
1737. B. Eistert, W. Schade and N. Mecke, *Ann. Chem.*, **717**, 80 (1968).
1738.* B. Eistert, M. Regitz, G. Heck and H. Schwall, in *Methoden der Organischen Chemie* (Houben–Weyl–Müller), Vol. X/4, 4th ed., Georg Thieme, Stuttgart, 1968, p. 714.
1739.* B. Eistert, M. Regitz, G. Heck and H. Schwall, *Methoden der Organischen Chemie* (Houben–Weyl–Müller), Vol. X/4, Georg Thieme Verlag, Stuttgart, 1968, p. 473.
1740. B. Eistert, H. Fink, J. Riedinger, H.-G. Hahn and H. Dürr, *Chem. Ber.*, **102**, 3111 (1969).
1741. B. Eistert and P. Donath, *Chem. Ber.*, **103**, 993 (1970).
1742. B. Eistert and H. Juraszyk, *Chem. Ber.*, **103**, 2707 (1970).
1743. B. Eistert, W. Kurze and G. W. Müller, *Ann. Chem.*, **732**, 1 (1970).
1744. B. Eistert and A. J. Thommen, *Chem. Ber.*, **104**, 3048 (1971).
1745. B. Eistert, K. Pfleger and P. Donath, *Chem. Ber.*, **105**, 3915 (1972).
1746. B. Eistert, J. Riedinger, G. Küffner and W. Lazik, *Chem. Ber.*, 106, 727 (1973).
1747. B. Eistert and P. Donath, *Chem. Ber.*, **106**, 1537 (1973).
1748. B. Eistert, K. Pfleger, T. J. Arackal and G. Holzer, *Chem. Ber.*, **108**, 693 (1975).
1749. B. Eistert, L. S. B. Goubran, C. Vamvakaris and T. J. Arackal, *Chem. Ber.*, **108**, 2941 (1975).
1750. B. Eistert, H. Juraszyk and T. J. Arackal, *Chem. Ber.*, **109**, 640 (1976).
1751. K. Eiter and O. Svierak, *Monatsh.*, **83**, 1474 (1952).
1752. M. A. F. Elkaschef, F. M. E. Abdel-Megeid and S. M. M. Elzein, *Acta Chim. Acad. Sci. Hung.*, **79**, 411 (1973).
1753. M. A. F. Elkaschef, F. M. E. Abdel-Megeid and S. M. A. Yassin, *J. prakt. Chem.*, **316**, 363 (1974).
1754. J. Engbersen and J. B. F. N. Engberts, *Syn. Commun.*, **1**, 121 (1971).
1755. J. B. F. N. Engberts and B. Zwanenburg, *Tetrahedron Letters*, 831 (1967).
1756. I. Ernst, *Chem. Listy*, **48**, 847 (1954).
1757. I. Ernst and J. Stanek, *Coll. Czech. Chem. Commun.*, **24**, 530 (1959).
1758. A. Eschenmoser, D. Felix and G. Ohloff, *Helv. Chim. Acta*, **50**, 708 (1967).
1758a. D. A. Evans and C. L. Sims, *Tetrahedron Letters*, 4691 (1973).
1759. E. Fahr and F. Scheckenbach, *Ann. Chem.*, **655**, 86 (1962).
1760. E. Fahr, K. Königsdorfer and F. Scheckenbach, *Ann. Chem.*, **690**, 138 (1965).
1761. E. Fahr, K. Döppert and F. Scheckenbach, *Ann. Chem.*, **696**, 136 (1966).
1762. E. Fahr, K. Döppert, K. Königsdorfer and F. Scheckenbach, *Tetrahedron*, **24**, 1011 (1968).
1763. E. Fahr and K. Königsdorfer, *Tetrahedron Letters*, 1873 (1966).
1764. E. Fahr, J. Markert and N. Pelz, *Ann. Chem.*, 2088 (1973).
1765. P. E. Fanta, R. M. W. Rickett and D. S. James, *J. Org. Chem.*, **26**, 938 (1961).
1766. H. Favre, B. Marinier and J. C. Richer, *Can. J. Chem.*, **34**, 1329 (1956).
1767. H. Favre and B. Marinier, *Can. J. Chem.*, **35**, 278 (1957).

1768. F. Feist, *Ann. Chem.*, **345**, 100 (1906).
1769.* H. Feuer (Ed.), *The Chemistry of the Nitro and Nitroso Groups*, Interscience, London, 1969.
1770. E. O. Fischer, W. Hafner and H. O. Stahl, *Z. anorg. Chem.*, **282**, 47 (1955).
1771. E. O. Fischer and A. Maasböl, *Angew. Chem.*, **76**, 645 (1964).
1772. E. O. Fischer and A. Maasböl, *Angew. Chem. Int. Ed.*, **3**, 580 (1964).
1773. N. Fischer and G. Opitz, *Org. Synth.*, **48**, 106 (1968).
1774.* N. H. Fischer, *Synthesis*, 393 (1970).
1775. L. Fitjer and J. M. Conia, *Angew. Chem.*, **85**, 349 (1973).
1776. I. Fleming and R. B. Woodward, *J. Chem. Soc. Perkin Trans. I*, 165 (1973).
1777.* I. Fleming, *Selected Organic Syntheses*, Wiley–Interscience, London, 1973.
1778. J. Font, J. Valls and F. Serratosa, *J. Chem. Soc. Chem. Commun.*, 721 (1970).
1779. J. Font, J. Valls and F. Serratosa, *Tetrahedron*, **30**, 455 (1974).
1780. M. O. Forster and F. P. Dunn, *J. Chem. Soc.*, **95**, 425 (1909).
1781. M. Franck-Neumann, *Angew. Chem. Int. Ed.*, **8**, 210 (1969).
1782. M. Franck-Neumann and G. Bucherer, *Tetrahedron Letters*, 15 (1969).
1783. M. Franck-Neumann and G. Leclerc, *Tetrahedron Letters*, 1063 (1969).
1784. M. Franck-Neumann and C. Buchecker, *Angew. Chem. Int. Ed.*, **9**, 526 (1970).
1785. M. Franck-Neumann and M. Sedrati, *Org. Magn. Resonance*, **5**, 217 (1973).
1786. M. Franck-Neumann and M. Sedrati, *Angew. Chem.*, **86**, 673 (1974).
1787. M. Franck-Neumann and M. Sedrati, *Angew. Chem. Int. Ed.*, **13**, 606 (1974).
1788. M. Franck-Neumann and D. Martina, *Tetrahedron Letters*, 1755 (1975).
1789. M. Franck-Neumann and D. Martina, *Tetrahedron Letters*, 1759 (1975).
1790. M. Franck-Neumann, D. Martina and C. Buchecker, *Tetrahedron Letters*, 1763 (1975).
1791. M. Franck-Neumann and D. Martina, *Tetrahedron Letters*, 1767 (1975).
1792. V. Franzen and H. Kuntze, *Ann. Chem.*, **627**, 15 (1959).
1793. V. Franzen, *Ann. Chem.*, **627**, 22 (1959).
1794. B. H. Freeman, G. S. Harris, B. W. Kennedy and D. Lloyd, *J. Chem. Soc., Chem. Commun.*, 912 (1972).
1795. B. H. Freeman, D. Lloyd and M. I. C. Singer, *Tetrahedron*, **30**, 211 (1974).
1796. B. H. Freeman and D. Lloyd, *Tetrahedron*, **30**, 2257 (1974).
1797. B. H. Freeman, J. M. F. Gagar and D. Lloyd, *Tetrahedron*, 4307 (1973).
1798. P. K. Freeman and D. G. Kuper, *Chem. and Ind.*, 424 (1965).
1799. D. M. Gale, W. J. Middleton and C. G. Krispan, *J. Amer. Chem. Soc.*, **87**, 657 (1965).
1800. D. M. Gale, W. J. Middleton and C. G. Krispan, *J. Amer. Chem. Soc.*, **88**, 3617 (1966).
1801. P. G. Gassman, F. J. Williams and J. Seter, *J. Amer. Chem. Soc.*, **90**, 6983 (1968).
1802. P. H. Gebert, R. W. King, R. A. Labor and W. M. Jones, *J. Amer. Chem. Soc.*, **95**, 2357 (1973).
1803. U. R. Ghatak, S. Chakrabarty and K. Rudra, *J. Chem. Soc., Perkin I*, 1957 (1974).
1804. V. A. Ginsburg, A. Ya. Yakubovich, A. S. Filatov, G. E. Zelinin, S. P. Makarov, V. A. Shpanskii, G. P. Kotel'nikov, L. F. Sergienko and L. L. Martynova, *Doklady Akad. Nauk S.S.S.R.*, **142**, 354 (1952).
1805. T. Gibson and W. F. Erman, *J. Org. Chem.*, **31**, 3028 (1966).
1806. T. Giraldi, G. Steppani and L. Baldini, *Biochem. Pharmacol.*, **21**, 3035 (1972).
1807. T. Giraldi, R. Della-Loggia and L. Baldini, *Pharmacol. Res. Commun.*, **4**, 237 (1972).
1808. T. Giraldi and L. Baldini, *Biochem. Pharmacol.*, **22**, 1793 (1973).
1809. T. Giraldi and L. Baldini, *Biochem. Pharmacol.*, **23**, 289 (1974).
1810. T. Giraldi, C. Monti-Bragadin and R. Della-Loggia, *Experientia*, **30**, 496 (1974).
1811. T. Giraldi and C. Nisi, *Chem. Biol. Interactions*, **11**, 59 (1975).
1812. T. Giraldi, L. Baldini and G. Saun, *Biochem. Pharmacol.*, in press (1976).
1813. T. Giraldi and C. Nisi, *Pharmacol. Res. Commun.*, in press (1976).
1814. J. Goerdeler and G. Gnad, *Tetrahedron Letters*, 795 (1964).
1815. R. Gompper and H. Herlinger, *Chem. Ber.*, **89**, 2824 (1956).
1816. I. Gosney and D. Lloyd, *Tetrahedron*, **29**, 1697 (1973).
1817. R. Grée and R. Carrié, *Tetrahedron Letters*, 4117 (1971).
1818. R. Grée and R. Carrié, *Tetrahedron Letters*, 2987 (1972).
1819. R. Grée, F. Tonnard and R. Carrié, *Tetrahedron Letters*, 453 (1973).
1820. R. Grée and R. Carrié, *Tetrahedron*, **32**, 683 (1976).

1821. J. A. Green and L. A. Singer, *Tetrahedron Letters*, 5094 (1969).
1822. R. M. Greene and D. M. Kochhar, *J. Embryol. exp Morph.*, **33**, 355 (1975).
1823. P. Grieco, *J. Amer. Chem. Soc.*, **91**, 5660 (1969).
1824. C. Grob and P. Schiess, *Angew. Chem.*, **70**, 502 (1958).
1825. A. de Groot, J. A. Boerma, J. de Valk and H. Wynberg, *J. Org. Chem.*, **33**, 4025 (1968).
1826. J. F. Grove, J. McMillan, T. P. C. Mulholland and M. A. T. Roger, *J. Chem. Soc.*, 3977 (1952).
1827. P. Grunanger and P. Vita-Finzi, *Atti Accad. Nazl. Lincei, Rend. Sci. fis mat. nat.*, **31**, 128 (1961).
1828. R. Grüning and J. Lorberth, *J. Organomet. Chem.*, **69**, 213 (1974).
1829. G. Guillerm, A. L'Honore, L. Veniard, G. Pourcelot and J. Benaum, *Bull. Soc. Chim. France*, 2739 (1973).
1830. G. Guillerm and M. Lequan, *C. R. Acad. Sci. Paris, Sér. C*, **269**, 853 (1969).
1831. C. D. Gutsche and K. L. Seligman, *J. Amer. Chem. Soc.*, **75**, 2579 (1953).
1832. C. D. Gutsche and F. A. Fleming, *J. Amer. Chem. Soc.*, **76**, 1771 (1954).
1833. C. D. Gutsche and M. Hillman, *J. Amer. Chem. Soc.*, **76**, 2236 (1954).
1834. F. F. Guzik and A. K. Colter, *Canad. J. Chem.*, **43**, 1441 (1965).
1835. G. Haberland and H. J. Siegert, *Chem. Ber.*, **71**, 2619 (1938).
1836. J. A. Haines, C. B. Reese and Lord Todd, *J. Chem. Soc.*, 1406 (1964).
1837. K. Haffner, *Angew. Chem.*, **70**, 419 (1958).
1838.* E. A. Halevi, R. Pauncz, I. Schek and H. Weinstein, in *Chemical and Biochemical Reactivity, The Jerusalem Symposia on Quantum Chemistry and Biochemistry*, Vol. VI, Jerusalem, 1974, p. 167.
1839. J. V. Halpern, *Rec. Trav. chim.*, **62**, 485 (1943).
1840. M. Hamaguchi and T. Ibata, *Tetrahedron Letters*, 4475 (1974).
1841. M. Hamaguchi and T. Ibata, *Chem. Letters*, 169 (1975).
1842. M. Hamaguchi and T. Ibata, *Chem. Letters*, 499 (1975).
1843. M. Hamaguchi and T. Ibata, *Chem. Letters*, 287 (1976).
1844. J. Hamelin, *C. R. Acad. Sci. Paris, Sér. C*, **261**, 4776 (1965).
1845. J. Hamelin and R. Carrié, *Bull. Soc. Chim. France*, 2162 (1968).
1846. J. Hamelin and R. Carrié, *Bull. Soc. Chim. France*, 2515 (1968).
1847. J. Hamelin and R. Carrié, *Bull. Soc. Chim. France*, 2521 (1968).
1848. J. Hamelin and R. Carrié, *Bull. Soc. Chim. France*, 3000 (1968).
1849. J. Hamelin and R. Carrié, *Bull. Soc. Chim. France*, 2054 (1972).
1850. W. Hammond and N. Turro, *J. Amer. Chem. Soc.*, **88**, 2880 (1966).
1851. S. H. Harper and K. C. Steep, *J. Sci. Food Agric.*, **6**, 116 (1955).
1852. S. C. Hartman and T. F. McGrath, *J. Biol. Chem.*, **248**, 8506 (1973).
1853. S. C. Hartman and E. M. Stochaj, *J. Biol. Chem.*, **248**, 8511 (1973).
1854. S. Hauptmann and K. Kirschberg, *J. prakt. Chem.*, **35**, 105 (1967).
1855. G. Hayes and G. Holt, *J. Chem. Soc. Perkin I*, 1206 (1973).
1856. J. Haywood-Farmer, R. E. Pincock and J. I. Wells, *Tetrahedron*, **22**, 2007 (1966).
1857. E. Heilbronner and H.-D. Martin, *Chem. Ber.*, **106**, 3376 (1973).
1858. L. Hellerman and R. L. Garner, *J. Amer. Chem. Soc.*, **57**, 139 (1935).
1859. M. E. Hendrick, W. J. Baron and M. Jones, Jr, *J. Amer. Chem. Soc.*, **93**, 1594 (1971).
1860. J. Hendrickson, C. Foote and N. Yoshimura, *Chem. Commun.*, 165 (1965).
1861. J. Hendrickson, T. Bogard and M. Fisch, *J. Amer. Chem. Soc.*, **92**, 5538 (1970).
1862. K. Henkel and F. Weigand, *Chem. Ber.*, **76**, 812 (1943).
1863. G. Hesse and E. Reichold, *Chem. Ber.*, **90**, 2101 (1957).
1864. G. Hesse, E. Reichold and S. Majmudar, *Chem. Ber.*, **90**, 2106 (1957).
1865. G. Hesse, *Angew. Chem.*, **70**, 134 (1958).
1866. G. Hesse and S. Majmudar, *Chem. Ber.*, **93**, 1129 (1960).
1867. D. P. Higley and R. W. Murray, *J. Amer. Chem. Soc.*, **96**, 3330 (1974).
1868. C. Hillhouse, D. S. Wulfman and B. Poling, unpublished work, 1975.
1869. G. Himbert and M. Regitz, *Ann. Chem.*, 1505 (1973).
1870. H. Hoberg, *Ann. Chem.*, **656**, 1 (1962).
1871. H. Hoberg, *Ann. Chem.*, **695**, 1 (1966).
1872. H. Hoberg, *Ann. Chem.*, **707**, 147 (1967).
1873. H. Hock and H. Kropf, *Chem. Ber.*, **88**, 1544 (1955).

1874. J. Hooz and S. Linke, *J. Amer. Chem. Soc.*, **90**, 5936 (1968).
1875. J. Hooz and S. Linke, *J. Amer. Chem. Soc.*, **90**, 6891 (1968).
1876. J. Hooz and D. M. Gunn, *J. Chem. Soc. Chem. Commun.*, 139 (1969).
1877. J. Hooz and D. M. Gunn, *J. Amer. Chem. Soc.*, **91**, 6195 (1969).
1878. L. Horner, E. Spietschka and A. Gross, *Ann. Chem.*, **573**, 17 (1951).
1879. L. Horner and E. Spietschka, *Chem. Ber.*, **88**, 934 (1955).
1880. L. Horner and H. Oediper, *Chem. Ber.*, **91**, 434 (1958).
1881. L. Horner, K. Muth and H. G. Schmelzer, *Chem. Ber.*, **92**, 2953 (1959).
1882. L. Horner, L. Hockenberger and W. Kirmse, *Chem. Ber.*, **94**, 290 (1961).
1883. H. O. House and C. J. Blankley, *J. Org. Chem.*, **33**, 47 (1968).
1884. W. Hughes, private communication, 1976.
1885. R. Huisgen, *Festschrift Zehnjahrenfeier Fonds der Chemischen Industrie, Düsseldorf*, p. 73 (1960).
1886. R. Huisgen, R. Fleischmann and A. Eckell, *Tetrahedron Letters*, 1 (1960).
1887. R. Huisgen and A. Eckell, *Tetrahedron Letters*, 5 (1960).
1888. R. Huisgen, H. Stangl, H. J. Sturm and H. Wagenhofer, *Angew. Chem.*, **73**, 170 (1961).
1889. R. Huisgen, H. S. Sturm and G. Binsch, *Chem. Ber.*, **97**, 2864 (1964).
1890. R. Huisgen and G. Juppe, *Chem. Ber.*, **94**, 2332 (1961).
1891. R. Huisgen, *Proc. Chem. Soc.*, 357 (1961).
1892. R. Huisgen, *Angew. Chem. Int. Ed.*, **2**, 565 (1963).
1893. R. Huisgen, *Angew. Chem.*, **75**, 604 (1963); *Int. Ed.*, **2**, 565 (1963).
1894. R. Huisgen, *Angew. Chem.*, **75**, 742 (1963); *Int. Ed.*, **2**, 633 (1963).
1895. R. Huisgen, R. Grashey and J. Sauer, in *The Chemistry of Alkenes* (Ed. S. Patai), Wiley–Interscience, London, 1964, p. 739.
1896. R. Huisgen, *J. Org. Chem.*, **33**, 2291 (1968).
1897. R. Huisgen, W. Scheer and H. Mader, *Angew. Chem. Int. Ed.*, **8**, 602 (1969).
1898. C. D. Hund and S. C. Liu, *Chem. Ber.*, **57**, 2656 (1935).
1899. R. Huttel, *Chem. Ber.*, **74**, 1680 (1941).
1900. R. Huttel and A. Gebhardt, *Ann. Chem.*, **558**, 34 (1948).
1901. R. Huttel, J. Riedl, H. Martin and K. Franke, *Chem. Ber.*, **93**, 1425 (1960).
1902. C. Huynh, S. Julia, R. Lorne and D. Michelot, *Bull. Soc. Chim. France*, 4057 (1972).
1903. T. Ibata, K. Veda and M. Takebayashi, *Bull. Chem. Soc. Japan*, **46**, 2897 (1973).
1904. T. Ibata, *Chem. Letters*, 233 (1974).
1905. T. Ibata, M. Hamaguchi and H. Kiyohara, *Chem. Letters*, 21 (1975).
1906. T. Ibata, *Chem. Letters*, 233 (1976).
1907. H. H. Inhoffen, R. Jonas, H. Krosche and U. Eder, *Ann. Chem.*, **694**, 19 (1966).
1908. S. Isshiki, *J. Pharm. Soc. Japan*, **65**, 10 (1945).
1909. H. Iwata, I. Yamamoto and E. Gohda, *Biochem. Pharmacol.*, **22**, 1845 (1973).
1910. R. A. Izydore and S. McLean, *J. Amer. Chem. Soc.*, **97**, 564 (1975).
1911. R. W. Jackson and R. H. Manske, *Can. J. Res.*, **13B**, 170 (1935).
1912. J. T. P. Jacobsen, K. Schaumburg and J. T. Nielsen, *J. Magn. Resonance*, **13**, 372 (1974).
1913. A. W. Johnson, A. Langemann and J. Murray, *J. Chem. Soc.*, 2136 (1953).
1914. W. S. Johnson, M. Neeman and S. P. Birkeland, *Tetrahedron Letters*, 1 (1960).
1915. W. S. Johnson, M. Neeman, S. P. Birkeland and N. H. Fedoruk, *J. Amer. Chem. Soc.*, **84**, 989 (1962).
1916. E. R. H. Jones, T. Y. Shem and M. C. Whitting, *J. Chem. Soc.*, 236 (1950).
1917. M. Jones, Jr and W. Ando, *J. Amer. Chem. Soc.*, **90**, 2200 (1968).
1918. M. Jones, Jr and W. Ando, *J. Amer. Chem. Soc.*, **90**, 2200 (1968).
1919. M. Jones, Jr and R. A. Moss, *Carbenes*, Vol. 1, Wiley–Interscience, New York, 1972.
1920. M. Jones, W. Ando, M. E. Hendrick, A. Kulczycki, P. M. Hawley, K. M. Hummel and D. S. Malament, *J. Amer. Chem. Soc.*, **94**, 7469 (1972).
1921. R. G. Jones, *J. Amer. Chem. Soc.*, **71**, 3994 (1949).
1922. V. K. Jones and A. J. Deutschman, Jr, *J. Org. Chem.*, **30**, 3978 (1965).
1923. W. M. Jones and J. M. Denham, *J. Amer. Chem. Soc.*, **86**, 944 (1964).
1924. W. M. Jones and C. L. Ennis, *J. Amer. Chem. Soc.*, **89**, 3069 (1967).

1925. W. M. Jones, M. E. Stowe, E. E. Wells and E. W. Lester, *J. Amer. Chem. Soc.*, **90**, 1849 (1968).
1926. W. M. Jones and C. L. Ennis, *J. Amer. Chem. Soc.*, **91**, 6391 (1969).
1927. W. Jugelt and D. Schmidt, *Tetrahedron*, **25**, 969 (1969).
1928. S. Julia, A. Constantino and G. Linstrumelle, *C. R. Acad. Sci. Paris, Sér. C*, **264**, 407 (1967).
1929. S. Julia, B. Cazes and C. Huynh, *C. R. Acad. Sci. Paris, Sér. C*, **274**, 2019 (1972).
1930. S. Julia, C. Huynh and D. Michelot, *Tetrahedron Letters*, 3587 (1972).
1931. M. I. Kabachnik, S. T. Ioffe and T. A. Mastryokava, *Zh. Obshch. Khim.*, **25**, 684 (1955).
1932. P. K. Kadaba and J. O. Edwards, *J. Org. Chem.*, **26**, 2331 (1961).
1933. P. K. Kadaba, *Tetrahedron*, **22**, 2453 (1966).
1934. P. K. Kadaba, *J. Hetero. Chem.*, **6**, 587 (1969).
1935. P. K. Kadaba and T. F. Colturi, *J. Hetero. Chem.*, **6**, 829 (1969).
1936. P. K. Kadaba, *Tetrahedron*, **25**, 3053 (1969).
1937. P. K. Kadaba, *Synthesis*, 71 (1973).
1938. P. K. Kadaba, *J. Hetero. Chem.*, **12**, 143 (1975).
1939. G. A. Karustes, *Ph.D. Thesis*, Yale University, 1967.
1940. A. R. Katritzky and S. Musierowicz, *J. Chem. Soc. C*, 78 (1966).
1941. A. S. Kende, *Chem. and Ind.*, 1053 (1956).
1942. A. S. Kende and E. D. Riecke, *J. Chem. Soc. Chem. Commun.*, 383 (1974).
1943. G. Kiehl, J. Streith and G. Taurand, *Tetrahedron Letters*, 2851 (1974).
1944. L. C. King and F. M. Miller, *J. Amer. Chem. Soc.*, **70**, 4154 (1948).
1945. L. C. King and F. Miller, *J. Amer. Chem. Soc.*, **71**, 367 (1949).
1946. W. Kirmse and L. Horner, *Ann. Chem.*, **614**, 1 (1958).
1947. W. Kirmse, *Chem. Ber.*, **93**, 2357 (1960).
1948. W. Kirmse and D. Grassman, *Chem. Ber.*, **99**, 1746 (1966).
1949. W. Kirmse, M. Kapps and R. B. Hager, *Chem. Ber.*, **99**, 2855 (1966).
1950. W. Kirmse and K. Horn, *Tetrahedron Letters*, 1827 (1967).
1951. K. Kitatani, T. Hiyama and H. Nozaki, *Tetrahedron Letters*, 1531 (1974).
1952. F. Klages and H. Meuresch, *Chem. Ber.*, **85**, 863 (1952).
1953. F. Klages and H. Meuresch, *Chem. Ber.*, **86**, 1322 (1953).
1954. F. Klages, H. Meuresch and W. Steppich, *Ann. Chem.*, **592**, 116 (1955).
1955. H. Kloosterziel, M. H. Deinema and H. J. Backer, *Rec. Trav. Chim.*, **71**, 1228 (1952).
1955a. H. Kloosterziel and H. J. Backer, *Rec. Trav. Chim.*, **71**, 1235 (1952).
1955b. H. Kloosterziel, J. S. Boerema and H. J. Backer, *Rec. Trav. chim.*, **72**, 612 (1953).
1956. E. P. Kohler, M. Tishler, H. Potter and H. T. Thompson, *J. Amer. Chem. Soc.*, **61**, 1057 (1939).
1957. M. I. Komendantov, V. Ya. Bespalov, O. A. Bezrukova and R. R. Bekmukhametov, *Zh. Org. Khim.*, **11**, 27 (1975).
1958. K. Kondo and I. Ojima, *J. Chem. Soc. Chem. Commun.*, 63 (1972).
1959. K. Kondo and I. Ojima, *Chem. Letters*, 771 (1972).
1960. N. Kornblum and R. A. Brown, *J. Amer. Chem. Soc.*, **86**, 2681 (1964).
1961. J. K. Korobitsyna and L. Rodina, *Zh. Org. Chem.*, **70**, 506 (1958).
1962. J. Kottwitz and H. Vorbruggen, *Synthesis*, 636 (1975).
1963. H. J. Kotzsch, *Chem. Ber.*, **99**, 1143 (1966).
1963a. A. P. Krapcho, M. P. S. Ivon, I. Goldberg and E. G. E. Jahngen, Jr, *J. Org. Chem.*, **39**, 860 (1974).
1964. H. Krzikalla and B. Eistert, *J. prakt. Chem.*, **2**, 143, 50 (1935).
1965. R. Kuhn and K. Henkel, *Ann. Chem.*, **549**, 279 (1941).
1966. R. Kuhn and H. H. Baer, *Chem. Ber.*, **86**, 724 (1953).
1967. R. A. Labor and W. M. Jones, *J. Amer. Chem. Soc.*, **95**, 2359 (1973).
1968. S. R. Lammert and S. Kukolja, *J. Amer. Chem. Soc.*, **97**, 5583 (1975).
1968a. L. Lardici, C. Battistini and R. Menicagli, *J. Chem. Soc. Perkin I*, 344 (1974).
1969. H. Ledon, *Dissertation*, Docteur Ingénieur, Paris, 1973.
1970. H. Ledon, G. Cannic, G. Linstrumelle and S. Julia, *Tetrahedron Letters*, 3971 (1970).
1971. H. Ledon, G. Linstrumelle and S. Julia, *Bull. Soc. Chim. France*, 2065 (1973).
1972. H. Ledon, G. Linstrumelle and S. Julia, *Bull. Soc. Chim. France*, 2071 (1973).

1973. H. Ledon, G. Linstrumelle and S. Julia, *Tetrahedron Letters*, 25 (1973).
1974. H. Ledon, G. Linstrumelle and S. Julia, *Tetrahedron*, **29**, 3609 (1973).
1975. D. M. Lemal, F. Menger and G. W. Clark, *J. Amer. Chem. Soc.*, **85**, 2529 (1963).
1976. R. M. Lemmon and W. Strohmeier, *J. Amer. Chem. Soc.*, **81**, 106 (1959).
1977. G. Leroy and M. Sana, *Tetrahedron*, **31**, 2091 (1975).
1978. G. Leroy and M. Sana, *Tetrahedron*, **32**, 709 (1976).
1979. G. Leroy and M. Sana, *Theor. Chim. Acta*, **33**, 329 (1974).
1980. A. L'Honoré, *Thèse*, Paris (1972).
1981. E. Lieber, N. Calvanico and C. N. R. Rao, *J. Org. Chem.*, **28**, 257 (1963).
1982. V. R. Likhterov, *Khim. Getero. Soed.*, 501 (1973).
1982a. V. R. Likhterov., *Khim. Getero. Soed.*, 545 (1973).
1982b. H. Lind and E. Fahr, *Tetrahedron Letters*, 4505 (1966).
1983. G. Linstrumelle, *Tetrahedron Letters*, 85 (1970).
1984. G. Linstrumelle, *Bull. Soc. Chim. France*, 919 (1970).
1985. G. Linstrumelle, *Bull. Soc. Chim. France*, 642 (1971).
1986. G. Linstrumelle, unpublished work.
1987. P. Lipp and R. Koster, *Chem. Ber.*, **64**, 2823 (1931).
1988. P. Lipp, J. Buchkremer and H. Seeles, *Ann. Chem.*, **499**, 1 (1932).
1989. O. Livi, P. L. Ferrarini, D. Bertini and I. Tonetti, *Fl. Farmaco Ed. Sci.*, **30**, 1017 (1975).
1990. D. Lloyd and F. I. Wasson, *Chem. and Ind.*, 1559 (1963).
1991. D. Lloyd and N. W. Preston, *Chem. and Ind.*, 1039 (1966).
1992. D. Lloyd and F. I. Wasson, *J. Chem. Soc. C*, 408 (1966).
1993. D. Lloyd and F. I. Wasson, *J. Chem. Soc. C*, 1086 (1966).
1994. D. Lloyd and M. I. C. Singer, *Chem. Commun.*, 390 (1967).
1995. D. Lloyd and M. I. C. Singer, *Chem. Commun.*, 1042 (1967).
1996. D. Lloyd and M. I. C. Singer, *Chem. and Ind.*, 118 (1967).
1997. D. Lloyd, M. I. C. Singer, M. Regitz and A. Liedhogener, *Chem. and Ind.*, 324 (1967).
1998. D. Lloyd and M. I. C. Singer, *Chem. and Ind.*, 510 (1967).
1999. D. Lloyd and M. I. C. Singer, *Chem. and Ind.*, 787 (1967).
2000. D. Lloyd and B. H. Freeman, *Chem. Commun.*, 924 (1970).
2001. D. Lloyd and M. I. C. Singer, *J. Chem. Soc. C*, 2939 (1971).
2002. E. T. McBee, J. A. Bosoms and C. J. Morton, *J. Org. Chem.*, **31**, 768 (1966).
2003. L. N. McCullagh and D. S. Wulfman, unpublished work.
2004. R. S. McDaniel, Jr, *Ph.D. Dissertation*, University of Missouri-Rolla, 1974.
2005. D. E. McGreer, P. Morris and G. Carmichael, *Can. J. Chem.*, **41**, 726 (1963).
2006. D. E. McGreer, N. W. K. Chiu and M. G. Vinje, *Can. J. Chem.*, **43**, 1398 (1965).
2007. D. E. McGreer, N. W. K. Chiu, M. G. Vinje and K. C. K. Wong, *Can. J. Chem.*, **43**, 1407 (1965).
2008. D. E. McGreer and W. S. Wu, *Can. J. Chem.*, **45**, 461 (1967).
2009. D. E. McGreer and N. W. K. Chiu, *Can. J. Chem.*, **46**, 2217 (1968).
2010. D. E. McGreer and N. W. K. Chiu, *Can. J. Chem.*, **46**, 2225 (1968).
2011. D. E. McGreer and Y. Y. Wigfield, *Can. J. Chem.*, **47**, 3965 (1969).
2012. D. E. McGreer and I. M. E. Masters, *Can. J. Chem.*, **47**, 3975 (1969).
2013. D. E. McGreer and J. W. McKinley, *Can. J. Chem.*, **49**, 105 (1971).
2014. M. M. McKown and R. I. Gregerman, *Life Sciences*, **16**, 71 (1975).
2014a. T. Machiguchi, Y. Yamamoto, M. Hoshino and Y. Kitihara, *Tetrahedron Letters*, 2627 (1973).
2015. S. P. Makarov, V. A. Shpanskii, V. A. Ginsburg, A. I. Shchekotikhin, A. S. Filatov, L. L. Martynova, I. V. Pavlovskaya, A. F. Golovaneva and A. Y. Yakubovich, *Doklady Akad. Nauk S.S.S.R.*, **142**, 596 (1962).
2016. G. Manecke and H. U. Schenck, *Tetrahedron Letters*, 2061 (1968).
2016a. G. Manecke and H. U. Schenck, *Tetrahedron Letters*, 617 (1969).
2017. A. P. Marchand and N. M. Brockway, *J. Amer. Chem. Soc.*, **92**, 5801 (1970).
2018. A. P. Marchand, in *Supplement A: The Chemistry of Double-bonded Functional Groups* (Ed. S. Patai), Chapter 7, Wiley–Interscience, London, 1977, p. 533.
2019. J. Markert and E. Fahr, *Tetrahedron Letters*, 4337 (1967).

2020. J. R. Marshall and J. Walker, *J. Chem. Soc.*, 467 (1952).
2021. J. Martelli and R. Carrié, *C. R. Acad. Sci. Paris, Sér. C*, **274**, 1222 (1972).
2022. J. Martelli, M. Bargain and R. Carrié, *C. R. Acad. Sci. Paris, Sér. C*, **276**, 523 (1973).
2022a. D. Martin and W. Mucke, *Z. Naturforsch.*, **3**, 347 (1963).
2022b. D. Martin and W. Mucke, *Ann. Chem.*, **682**, 90 (1965).
2023. D. Martin and A. Weise, *Chem. Ber.*, **99**, 317 (1966).
2024. T. Matsumoyo, H. Shirahama, A. Ichihara, H. Shin, S. Kagawa, F. Saken and K. Miyano, *Tetrahedron Letters*, 2049 (1971)
2025. S. Matsumura, T. Nagai and N. Tokura, *Bull. Chem. Soc. Japan*, **41**, 635 (1968).
2026. S. Matsumura, T. Nagai and N. Tokura, *Bull. Chem. Soc. Japan*, **41**, 2672 (1968).
2027. P. de Mayo (Ed.), *Molecular Rearrangements*, Wiley–Interscience, London, Part 1 (1963).
2028. P. de Mayo (Ed.), *Molecular Rearrangements*, Wiley–Interscience, London, Part 2 (1964).
2029. H. Meerwein, T. Bersen and W. Burneleit, *Chem. Ber.*, **61**, 1840 (1928).
2030. H. Meerwein, T. Bersen and W. Burneleit, *Chem. Ber.*, **62**, 999 (1929).
2031. H. Meerwein and G. Hinz, *Ann. Chem.*, **484**, 9 (1930).
2032. H. Meerwein and H. Sönke, *J. prakt. Chem.*, **2**, 137, 295 (1933).
2033. H. Meier, *Synthesis*, 235 (1972).
2034. H. Meier and K.-P. Zeller, *Angew. Chem. Int. Ed.*, **14**, 32 (1975); *Angew. Chem.*, **87**, 43 (1975).
2035. J. Meinwald, C. B. Jensen, A. Lewis and C. Swithenbank, *J. Org. Chem.*, **29**, 3469 (1964).
2036. I. Meinwald and G. H. Wahl, *Chem. and Ind.*, 424 (1965).
2037. J. Meinwald and J. K. Crandall, *J. Amer. Chem. Soc.*, **88**, 1292 (1966).
2038. K. Meischer and H. Kagi, *Helv. Chim. Acta*, **24**, 1471 (1941).
2039. A. Melzer and E. F. Jenny, *Tetrahedron Letters*, 4503 (1968).
2040. U. Mende, B. Raduchel, W. Skusalla and H. Vorbruggen, *Tetrahedron Letters*, 629 (1975).
2040a. P. Metzner, *Bull. Soc. Chim. France*, 2297 (1973).
2041. J. Meyer, *Monatsh.*, **26**, 1295 (1905).
2042. K. H. Meyer, *Chem. Ber.*, **52**, 1468 (1919).
2043. J. Meyer, *Helv. Chim. Acta*, **8**, 38 (1925).
2044. K. H. Meyer, *Chem. Ber.*, **52**, 1468 (1919).
2045. J. Michalsky, M. Holik and A. Podperova, *Monatsh.*, **90**, 814 (1959).
2046. D. Michelot, G. Linstrumelle and S. Julia, *J. Chem. Soc. Chem. Commun.*, 10 (1974).
2047. T. Mitsuhashi and W. M. Jones, *J. Chem. Soc. Chem. Commun.*, 103 (1974).
2048. T. Mitsuhashi, *Kagaku no Ryoiki*, **29**, 8 (1975).
2049. M. Mongrain, J. Lonfontaine, A. Belanger and P. Deslongchamps, *Can. J. Chem.*, **48**, 3273 (1970).
2050. C. Monti-Bragadin, M. Tamaro and E. Banfi, *Antimicrob. Agents Chemother.*, **6**, 655 (1974).
2051. R. Moore, A. Mishra and R. J. Crawford, *Can. J. Chem.*, **46**, 3305 (1968).
2052. K. Mori and M. Matsui, *Tetrahedron Letters*, 4435 (1969).
2053. K. Morita, I. Yamamoto and H. Iwata, *Biochem. Pharmacol.*, **22**, 1115 (1973).
2054. J. Moritani, T. Hosokawa and N. Obata, *J. Org. Chem.*, **34**, 670 (1969).
2055. W. R. Moser, *J. Amer. Chem. Soc.*, **91**, 1135 (1969).
2056. W. R. Moser, *J. Amer. Chem. Soc.*, **91**, 1141 (1969).
2057. E. Mosettig, *Chem. Ber.*, **61**, 1391 (1928).
2058. E. Mosettig and K. Czadek, *Monatsh.*, **57**, 291 (1931).
2059.* R. A. Moss, *Selective Organic Transformations*, Vol. 1 (Ed. B. S. Thyagarajan), Wiley–Interscience, 1970, p. 35.
2060. R. A. Moss, *J. Org. Chem.*, **31**, 3296 (1966).
2061. E. Mugnaini and P. Grunanger, *Atti Accad. naz. Lincei, Rend. Classe Sci. fis. mat. nat.*, **14**, 95 (1953).
2062. T. Mukai, H. Tsuruta, T. Nakzawa, K. Isabe and K. Kurabayashi, *Sci. Rept. Tohoku Univ. Ser. I*, **51**, 113 (1968).
2063. T. Mukai, T. Nakagawa and K. Isabe, *Tetrahedron Letters*, 565 (1968).

2064. T. Mukai, H. Tsuruta, K. Saitu, S. Mori and Y. Yamashita, *Sci. Rept. Tohoku Univ. Ser. I*, **57**, 131 (1974).
2065. T. P. C. Mulholland, R. I. W. Honeywood, H. D. Preston and D. T. Rosevear, *J. Chem. Soc.*, 4940 (1965).
2066. T. P. C. Mulholland, R. I. W. Honeywood, H. D. Preston and D. T. Rosevear, *J. Chem. Soc.* 4947 (1965).
2067. E. Müller, *Chem. Ber.*, **42**, 3270 (1909).
2068. E. Müller and A. Freytag, *J. prakt. Chem.*, [2], **146**, 56 (1936).
2069. E. Müller, H. Huber-Emden and W. Rundel, *Angew. Chem.*, **69**, 614 (1957).
2070. E. Müller and W. Rundel, *Angew. Chem.*, **70**, 105 (1958).
2071. E. Müller, H. Huber-Emden and W. Rundel, *Ann. Chem.*, **623**, 34 (1959).
2072. E. Müller, M. Bauer and W. Rundel, *Z. Naturforsch.*, **14b**, 209 (1959).
2073. E. Müller and H. Huber-Emden, *Ann. Chem.*, **649**, 81 (1961).
2074. E. Müller, H. Fricke and H. Kessler, *Tetrahedron Letters*, 1501 (1963).
2075. E. Müller, H. Fricke and H. Kessler, *Ann. Chem.*, **675**, 63 (1964).
2076. E. Müller, R. Heischkeil and M. Bauer, *Ann. Chem.*, **677**, 55 (1964).
2077. E. Müller and H. Kessler, *Ann. Chem.*, **692**, 58 (1966).
2078. E. Müller, H. Kessler and B. Zeeh, *Fortschr. Chem. Forsch.*, **7**, 128 (1966).
2078a. A. J. Mura, Jr, D. A. Bennett and T. Cohen, *Tetrahedron Letters*, 4433 (1975).
2078b. A. J. Mura, Jr, G. Majetich, P. A. Grieco and T. Cohen, *Tetrahedron Letters*, 4437 (1975).
2079. M. Muramatsu, N. Obata and T. Takizawa, *Tetrahedron Letters*, 2133 (1973).
2080. S. Murao, K. Oda and Y. Matsushita, *Agric. Biol. Chem.*, **37**, 1417 (1973).
2081. H. R. Musser, *Ph.D. Dissertation*, University of Missouri-Rolla, 1970.
2082. H. R. Musser and J. O. Stoffer, *J. Chem. Soc. Chem. Commun.*, 481 (1970).
2083. H. Musso and U. Beithan, *Chem. Ber.*, **97**, 2282 (1964).
2084. A. Mustafa, *J. Chem. Soc.*, 234 (1949).
2085. M. Neeman, M. C. Caserio, J. D. Roberts and W. S. Johnson, *Tetrahedron*, **6**, 36 (1959).
2086. C. D. Nenitzescu and E. Solomica, *Chem. Ber.*, **64**, 1924 (1931).
2087. N. P. Neureiter, *J. Amer. Chem. Soc.*, **88**, 558 (1966).
2088. M. S. Newman and P. F. Beal, *J. Amer. Chem. Soc.*, **72**, 5161 (1950).
2089. M. S. Newman and P. F. Beal, *J. Amer. Chem. Soc.*, **72**, 5163 (1950).
2090. M. S. Newman, G. Eglington and H. M. Grotta, *J. Amer. Chem. Soc.*, **75**, 349 (1953).
2091. A. T. Nielsen, unpublished work.
2092. G. A. Nikiforov, B. D. Sviridov, A. A. Volod'kin and V. V. Ershov, *Izvest. Akad. Nauk S.S.S.R. Ser. khim.*, 861 (1971).
2093. G. S. Nikol'skaya and A. T. Troshchenko, *Zh. Org. Khim.*, **3**, 498 (1967).
2094. M. Noel, Y. Vo-Quang and L. Vo-Quang, *C. R. Acad. Sci. Paris, Sér. C*, **270**, 80 (1970).
2095. J. Novaky, J. Batusky, V. Sneberk and F. Sarm, *Coll. Czech. Chem. Commun.*, **22**, 1836 (1957).
2096. R. Noyori, H. Takaya, Y. Nakanisi and H. Nozaki, *Can. J. Chem.*, **47**, 1242 (1969).
2097. H. Nozaki, H. Takaya and R. Noyori, *Tetrahedron Letters*, 2563 (1965).
2098. H. Nozaki, S. Moriuti, H. Takaya and R. Noyori, *Tetrahedron Letters*, 5239 (1966).
2099. H. Nozaki, H. Takaya, S. Moriuti and R. Noyori, *Tetrahedron*, **24**, 3655 (1968).
2100. K. Oda and S. Murao, *Agric. Biol. Chem.*, **38**, 2435 (1974).
2101. K. Oda, S. Murao, T. Oka and K. Morihara, *Agric. Biol. Chem.*, **39**, 477 (1975).
2102. I. Ojima and K. Kondo, *Bull. Chem. Soc. Japan*, **46**, 2571 (1973).
2103. G. Olah and S. Kuhn, *Chem. Ber.*, **89**, 864 (1956).
2104. E. Oliveri-Mandali, *Gazzetta*, **40**, I, 120 (1910).
2105. G. Opitz and K. Fischer, *Z. Naturforsch.*, **18**, 775 (1963).
2106. G. Opitz, *Angew. Chem.*, **79**, 161 (1967); *Angew Chem. Int. Ed.*, **6**, 107 (1967).
2107. G. Opitz and N. H. Tischer, cited in Reference 1774.
2108. G. Opitz and S. Mächtle, cited in Reference 1774.
2109. C. G. Overberger and J.-P. Anselme, *J. Org. Chem.*, **28**, 592 (1963).
2110. C. G. Overberger, N. Weinshenker and J.-P. Anselme, *J. Amer. Chem. Soc.*, **86**, 5364 (1964).

2111. C. G. Overberger, N. Weinshenker and J.-P. Anselme, *J. Amer. Chem. Soc.*, **87**, 4119 (1965).
2112. C. G. Overberger, R. Zangaro, R. Winter and J.-P. Anselme, *J. Org. Chem.*, **36**, 975 (1971).
2113. A. J. Owen, *Tetrahedron*, **17**, 237 (1961).
2114. W. E. Parham and J. L. Bleasdale, *J. Amer. Chem. Soc.*, **72**, 3843 (1950).
2115. W. E. Parham and J. L. Bleasdale, *J. Amer. Chem. Soc.*, **73**, 4664 (1951).
2116. W. E. Parham and W. R. Hasek, *J. Amer. Chem. Soc.*, **76**, 799 (1954).
2117. W. E. Parham and W. R. Hasek, *J. Amer. Chem. Soc.*, **76**, 935 (1954).
2118. W. E. Parham, C. Serres, Jr and P. R. O'Connor, *J. Amer. Chem. Soc.*, **80**, 588 (1958).
2119. W. E. Parham, H. G. Braxton and P. R. O'Connor, *J. Org. Chem.*, **26**, 1805 (1961).
2120. S. Parodi, A. Furlani, V. Scarcia, G. Brambilla, M. Carvana and M. De Barbieri, *Boll. Soc. ital. Biol. sper*, **48**, 871 (1972).
2121. D. J. Pasto and P. W. Wojtkowski, *Tetrahedron Letters*, 215 (1970).
2122. D. J. Pasto and P. W. Wojtkowski, *J. Org. Chem.*, **36**, 1790 (1971).
2123. H. Paul, I. Lange and A. Kausmann, *Chem. Ber.*, **98**, 1789 (1965).
2124. O. Pauli, *Inaug. Dissertation*, University of Marburg, Marburg, 1935.
2125. R. Paulissen, A. J. Hubert and Ph. Teyssie, *Tetrahedron Letters*, 1465 (1972).
2126. P. Pauls, *Inaug. Dissertation*, University of Marburg, Marburg, 1934.
2127. B. W. Peace, *Ph.D. Dissertation*, University of Missouri-Rolla, 1971.
2128. B. W. Peace and D. S. Wulfman, *J. Chem. Soc. D*, 1179 (1971).
2129. B. W. Peace, F. C. Carman and D. S. Wulfman, *Synthesis*, 658 (1971).
2130. B. W. Peace and D. S. Wulfman, *Tetrahedron Letters*, 3799 (1971).
2131. B. W. Peace and D. S. Wulfman, *Tetrahedron Letters*, 3903 (1972).
2132. B. W. Peace and D. S. Wulfman, *Synthesis*, 137 (1973).
2133. B. W. Peace, R. S. McDaniel, Jr and D. S. Wulfman, unpublished work.
2134. A. Peratoner and E. Azzarello, *Gazzetta*, **38**, I, 76 (1908).
2135. P. Pfeiffer and E. Enders, *Chem. Ber.*, **84**, 247 (1951).
2136. A. S. Pfau and P. A. Plattner, *Helv. Chim. Acta*, **19**, 858 (1936).
2137. D. D. Phillips and W. C. Champion, *J. Amer. Chem. Soc.*, **78**, 5452 (1956).
2138. D. D. Phillips and W. C. Champion, *J. Amer. Chem. Soc.*, **78**, 5452 (1956).
2139. J. P. Phillips and R. W. Keown, *J. Amer. Chem. Soc.*, **73**, 5483 (1951).
2140. E. Piers, R. Birttan and W. de Waal, *Can. J. Chem.*, **47**, 831 (1969).
2141. J. Plattner and H. Rapoport, *J. Amer. Chem. Soc.*, **93**, 1758 (1971).
2142. A. Plowman and M. A. Whitely, *J. Chem. Soc.*, **125**, 587 (1924).
2143. B. Poling and D. S. Wulfman, *Proc. Mo. Acad. Sci.*, 1976.
2144. H. Prinzbach, D. Stusche and R. Kitzing, *Angew. Chem.*, **82**, 393 (1970).
2145. E. Profft and W. Steinke, *J. prakt. Chem.* [4], **13**, 58 (1961).
2146. A. N. Pudovik, N. G. Khusainova, T. V. Tumoshina and O. E. Raevskaya, *Zhur. Obshch. Khim.*, **41**, 1476 (1971).
2147. J. Quintana, M. Torres and F. Serratosa, *Tetrahedron*, **29**, 2065 (1973).
2148. B. Raduchel, U. Mende, G. Cleve, G. A. Hoyer and H. Vorbruggen, *Tetrahedron Letters*, 633 (1975).
2149. E. T. Rakitzis, *Biochem. J.*, **141**, 601 (1974).
2150. J. Ramonczai and L. Vargha, *J. Amer. Chem. Soc.*, **72**, 2737 (1950).
2151. K. V. Rao and B. Ravindrath, *J. Hetero. Chem.*, **12**, 147 (1975).
2152. M. Reetz, U. Schollkopf and B. Banhidai, *Ann. Chem.*, 599 (1973).
2153. M. Regitz and F. Menz, *Chem. Ber.*, **101**, 2622 (1968).
2154. M. Regitz and J. Rüter, *Chem. Ber.*, **101**, 1263 (1968).
2155. M. Regitz, *Angew. Chem.*, *Int. Ed.*, **14**, 282 (1975).
2156. A. A. Reid, J. T. Sharp and H. R. Sood, *J. Chem. Soc.*, 2543 (1973).
2157. W. Reid and H. Mengler, *Ann. Chem.*, **678**, 113 (1964).
2158. H. Reimlinger, *Chem. Ber.*, **92**, 970 (1959).
2159. H. Reimlinger, *Angew. Chem.*, **72**, 33 (1960).
2160. H. Reimlinger, J. F. M. Oth and F. Billau, *Chem. Ber.*, **97**, 331 (1964).
2161. H. Reimlinger and C. H. Moussebois, *Chem. Ber.*, **98**, 1805 (1965).
2162. H. Reimlinger, *Ann. Chem.*, **713**, 113 (1968).

2163. H. Reimlinger, J. J. M. Vanderwalle and A. V. Overstraeten, *Ann. Chem.*, **720,** 124 (1968).
2164. A. G. Rendall and M. A. Whitely, *J. Chem. Soc.*, **121**, 2118 (1922).
2164a. G. A. Reynolds, *J. Org. Chem.*, **29**, 3733 (1964).
2165. W. Ried and H. Mengler, *Angew. Chem.*, **73**, 218 (1961).
2166. W. Ried and H. Mengler, *Ann. Chem.*, **651**, 54 (1962).
2167. W. Ried and J. Omran, *Ann. Chem.*, **666**, 144 (1963).
2168. W. Ried and B. M. Beck, *Ann. Chem.*, **673**, 124 (1964).
2169. W. Ried and B. M. Beck, *Ann. Chem.*, **673**, 128 (1964).
2170. W. Ried and H. Mengler, *Ann. Chem.*, **678**, 105 (1964).
2171. W. Ried and H. Mengler, *Ann. Chem.*, **678**, 113 (1964).
2172. W. Ried and R. Kraemer, *Ann. Chem.*, 1952 (1973).
2173. W. Ried, W. Kuhn and A. H. Schmidt, *Chem. Ber.*, **107**, 1147 (1974).
2174. V. Rosnati, G. Pagani and F. Sannicolo, *Tetrahedron Letters*, 1241 (1967).
2175. H. D. Roth, *J. Amer. Chem. Soc.*, **93**, 1527 (1971).
2176. H. D. Roth, *J. Amer. Chem. Soc.*, **93**, 4935 (1971).
2177. H. D. Roth, *J. Amer. Chem. Soc.*, **94**, 1761 (1972).
2178. H. D. Roth, *Mol. Photochem.*, **5**, 91 (1973).
2179. H. D. Roth and M. L. Manion, *J. Amer. Chem. Soc.*, **97**, 779 (1975).
2180. R. Rotter, *Monatsh.*, **47**, 353 (1926).
2181. R. Rotter and E. Schaudy, *Monatsh.*, **58**, 245 (1931).
2182. K.-E. Rozumek, H. Dürr and L. Schrader, *J. Chromatog.*, **48**, 53 (1970).
2183. H. Rupe and F. Hafliger, *Helv. Chim. Acta*, **23**, 139 (1940).
2184. H. Rupe and C. Frey, *Helv. Chim. Acta*, **27**, 627 (1944).
2185. C. Sabate-Alduy and J. Bastide, *Bull. Soc. chim. France*, 2764 (1972).
2186. T. Salgusa, Y. Ito, S. Kobayashi, K. Hirota, and T. Shimzu, *Tetrahedron Letters*, 6131 (1966).
2187. R. G. Salomon and J. K. Kochi, *J. Amer. Chem. Soc.*, **95**, 3300 (1973).
2188. R. G. Salomon, M. F. Salomon and T. R. Heyne, *J. Org. Chem.*, **40**, 756 (1975).
2189. T. Sato, *Tetrahedron Letters*, 835 (1968).
2190. F. Sauter and G. Büyük, *Monatsh.*, **105**, 550 (1974).
2191. S. E. Schaafsma, H. Steinberg and T. J. De Boer, *Rec. Trav. chim.*, **84**, 113 (1965).
2192. S. E. Schaafsma, H. Steinberg and T. J. De Boer, *Rec. Trav. chim.*, **86**, 651 (1967).
2193. G. O. Schenck and H. Ziegler, *Ann. Chem.*, **584**, 221 (1953).
2194. H. Schenkel-Rudin and M. Schenkel-Rudin, *Helv. Chim. Acta*, **27**, 1457 (1944).
2195. G. P. Schiemans, *Methoden der Organischen Chemie* (Houben–Weyl–Müller), Vol. IV/5b, Georg Thieme, Stuttgart, 1975, p. 1344.
2196. F. Schlotterbeck, *Chem. Ber.*, **40**, 479 (1907).
2197. A. Schmitz, U. Kraatz and F. Karte, *Chem. Ber.*, **108**, 1010 (1975).
2198. M. P. Schneider and R. J. Crawford, *Can. J. Chem.*, **48**, 628 (1970).
2199. A. Schönberg *et al.*, *J. Amer. Chem. Soc.*, **76**, 2273 (1954).
2200. A. Schönberg, A. E. K. Fateen and A. E. M. A. Sammour, *J. Amer. Chem. Soc.*, **79**, 6020 (1957).
2201. A. Schönberg and M. M. Sedky, *J. Amer. Chem. Soc.*, **81**, 2259 (1959).
2202. A. Schönberg, K. H. Brosowski and U. Singer, *Chem. Ber.*, **95**, 1910 (1962).
2203. A. Schönberg, E. Frese and K. H. Brosowski, *Chem. Ber.*, **95**, 3077 (1962).
2204. A. Schönberg and E. Frese, *Chem. Ber.*, **96**, 2420 (1963).
2205. A. Schönberg and K. Praefke, *Tetrahedron Letters*, 2043 (1964).
2206. A. Schönberg, B. König and E. Frese, *Chem. Ber.*, **98**, 3303 (1965).
2207. A. Schönberg and K. Praefke, *Chem. Ber.*, **99**, 196 (1966).
2208. A. Schönberg and K. Praefke, *Chem. Ber.*, **99**, 205 (1966).
2209. A. Schönberg and K. Praefke, *Chem. Ber.*, **99**, 2371 (1966).
2210. A. Schönberg, K. Praefke and J. Kohts, *Chem. Ber.*, **99**, 2433 (1966).
2211. A. Schönberg, E. Singer and W. Knöfel, *Chem. Ber.*, **99**, 3813 (1966).
2212. A. Schönberg, E. Singer, H. Schulze-Pannier and H. Schwarz, *Chem. Ber.*, **108**, 322 (1975).
2213. J. Schreiber, W. Leimgruber, M. Pesarv, P. Schudel, T. Threlfall and A. Eschenmoser, *Helv. Chim. Acta*, **44**, 540 (1961).

2214. G. Schröder, J. Oth and R. Meveny, *Angew. Chem. Int. Ed.*, **4**, 752 (1965).
2215. H. Schubert and J. Bleichert, *Z. Chem.*, **3**, 350 (1963).
2216. P. Schuster and O. E. Polansky, *MR. Chem. Bond.*, 396 (1965).
2217. E. E. Schweizer and C. S. Labaw, *J. Org. Chem.*, **38**, 3069 (1973).
2218. L. T. Scott, *J. Chem. Soc. Chem. Commun.*, 882 (1973).
2219. W. Scott and D. Evans, *J. Amer. Chem. Soc.*, **74**, 4780 (1972).
2220. A. Seetharmiah, *J. Chem. Soc.*, 894 (1948).
2221. F. Serratosa and J. Quintana, *Tetrahedron Letters*, 2245 (1967).
2222. F. Serratosa, F. Lopez and J. Font, *Anales real Soc. españ. Fis. Quim.*, **70**, 893 (1974).
2223. T. Severin and B. Bruck, *Chem. Ber.*, **98**, 3847 (1965).
2224. T. Severin, B. Bruck and P. Adhekary, *Chem. Ber.*, **99**, 3097 (1966).
2224a. D. Seyferth, H. D. Simmons, Jr and S. J. Todd, *J. Amer. Chem. Soc.*, **86**, 121 (1964); **91**, 5027 (1969).
2225. M. K. Shakhova, M. I. Budagyants, G. I. Samokhvalov and N. A. Preobrazhenskii, *Zh. Obshch. Khim.*, **32**, 2832 (1962).
2226. Y. F. Shealy and C. A. O'Dell, *J. Heterocyclic Chem.*, **10**, 839 (1973).
2227. H. Shechter and F. Conard, *J. Amer. Chem. Soc.*, **76**, 2716 (1954).
2228. J. C. Sheehan and P. T. Izzo, *J. Amer. Chem. Soc.*, **71**, 4059 (1949).
2229. J. C. Sheehan, *J. Amer. Chem. Soc.*, **71**, 4059 (1949).
2230. J. C. Sheehan and I. Lengyel, *J. Org. Chem.*, **28**, 3252 (1963).
2231. H. E. Sheffer and J A. Moore, *J. Org. Chem.*, **28**, 129 (1963).
2232. W. A. Sheppard and O. W. Webster, *J. Amer. Chem. Soc.*, **95**, 2695 (1973).
2233. W. A. Sheppard, Lecture, University of Missouri-Rolla, 1976.
2234. W. Shermer, *Doctoral Dissertation*, University of Illinois, 1970.
2235. J. S. Sherwell, J. R. Russell and D. Swern, *J. Org. Chem.*, **27**, 2853 (1962).
2236. T. Shirafuji, Y. Yamamoto and H. Nozaki, *Tetrahedron*, **27**, 5353 (1971).
2237. T. Shirafuji, K. Kitatani and H. Nozaki, *Bull. Chem. Soc. Japan*, **46**, 2249 (1973).
2238. O. Silberrad and C. S. Roy, *J. Chem. Soc.* (*B*), 646 (1971).
2239. H. E. Simmons and R. D. Smith, *J. Amer. Chem. Soc.*, **81**, 4256 (1959).
2240. H. E. Simmons, T. L. Cairns, S. A. Vladuchick and C. M. Hoiness, *Org. React.*, **20**, 1 (1973).
2241. P. S. Skell, S. J. Valenty and P. W. Hunter, *J. Amer. Chem. Soc.*, **95**, 5041 (1973).
2242. P. S. Skell and S. J. Valenty, *J. Amer. Chem. Soc.*, **95**, 5042 (1973).
2243. P. A. S. Smith, in Reference 1159, p. 370.
2244. L. I. Smith and P. O. Tawney, *J. Amer. Chem. Soc.*, **56**, 2167 (1934).
2245. L. I. Smith and C. L. Agre, *J. Amer. Chem. Soc.*, **60**, 648 (1938).
2246. M. B. Sohn and M. Jones, Jr, *J. Amer. Chem. Soc.*, **94**, 8280 (1972).
2246a. L. B. Sokolov, Y. I. Porfir'eva and A. A. Petrov, *Zh. Org. Khim.*, **1**, 610 (1965).
2246b. L. B. Sokolov, L. K. Vagina, V. N. Chistokletov and A. A. Petrov, *Zh. Org. Khim.*, **2**, 615 (1966).
2247. F. Sorm, *Coll. Czech. Chem. Commun.*, **12**, 245 (1947).
2248. J. H. Sperna-Weiland, *Rec. Trav. Chim.*, **83**, 81 (1964).
2249. H. Staudinger and O. Kupfer, *Chem. Ber.*, **44**, 2197 (1911).
2250. H. Staudinger, *Chem. Ber.*, **49**, 1884 (1916).
2251. H. Staudinger and J. Siegwart, *Chem. Ber.*, **49**, 1918 (1916).
2252. H. Staudinger, E. Anthes and F. Pfenninger, *Chem. Ber.*, **49**, 1928 (1916).
2253. H. Staudinger, E. Anthes and F. Pfenninger, *Chem. Ber.*, **49**, 1939 (1916).
2254. H. Staudinger and F. Pfenninger, *Chem. Ber.*, **49**, 1941 (1916).
2255. H. Staudinger and A. Gaule, *Chem. Ber.*, **49**, 1959 (1916).
2256. H. Staudinger and H. Hirzel, *Chem. Ber.*, **49**, 2526 (1917).
2257. H. Staudinger and J. Siegwart, *Helv. Chim. Acta*, **3**, 824 (1920).
2258. H. Staudinger and J. Siegwart, *Helv. Chim. Acta*, **3**, 848 (1920).
2259. H. Staudinger and T. Reber, *Helv. Chim. Acta*, **4**, 3 (1921).
2260. H. H. Stechl, *Chem. Ber.*, **97**, 2681 (1964).
2261. W. Steinkopf, *Ann. Chem.*, **434**, 21 (1923).
2262. E. Stephan, L. Vo-Quang and Y. Vo-Quang, *C. R. Acad. Sci. Paris, Sér. C*, **272**, 1731 (1971).
2263. E. Stephan, L. Vo-Quang and Y. Vo-Quang, *Bull. Soc. Chim. France*, 4781 (1972).

2264. E. Stephan, *Thèse de 3e Cycle*, Paris (1972).
2265. E. Stephan, L. Vo-Quang and Y. Vo-Quang, *Bull. Soc. Chim. France*, 2795 (1973).
2266. E. Stephan, L. Vo-Quang and Y. Vo-Quang, *Bull. Soc. Chim. France*, 1793 (1975).
2267. H. Stetter and H. Stark, *Chem. Ber.*, **92**, 732 (1959).
2268. J. M. Stewart, C. Carlisle, K. Kem and G. Lee, *J. Org. Chem.*, **35**, 2040, (1970).
2269. I. Strating and A. M. van Leusen, *Rec. Trav. chim.*, **81**, 966 (1962).
2270. G. Suchár and D. Kristian, *Chem. Zvesti*, **29** (2), 244 (1975).
2271. T. Sumner, L. E. Ball and J. Platner, *J. Org. Chem.*, **24**, 2017 (1959).
2272. H. de Suray, G. Leroy and J. Weiler, *Tetrahedron Letters*, 2209 (1974).
2273. I. Tabushi, K. Takagi, M. Ohano and R. Oda, *Tetrahedron*, **23**, 2621 (1967).
2273a. S. Tadashi and S. Katsuhiko, *Yuki Gosei Kagaku Kyoka Shi*, **26**, 432 (1968).
2274. K. Takagi and R. J. Crawford, *J. Amer. Chem. Soc.*, **93**, 5910 (1971).
2275. K. Takahashi, W.-J. Chang and J.-S. Ko, *J. Biochem.*, **76**, 897 (1974).
2276.* M. Takebayshi, T. Shingaki, N. Torimoto and M. Inagaki, *Kogyo Kagaku Zasshi*, **69**, 970 (1966).
2277. A. Tamburello and A. Millazzo, *Gazzetta*, **38, I**, 95 (1908).
2278. A. Tanaka, H. Uda and A. Yoshikoshi, *Chem. Commun.*, 308 (1969).
2279. T. Tanaka, T. Nagai and N. Tokura, *Chemistry Letters*, 1207 (1972).
2280. V. A. Tartakovskii, I. E. Chlenov, G. V. Lagodzinskaya and S. S. Novikov, *Izvest. Akad. Nauk S.S.S.R., Ser. Khim.*, 370 (1966).
2281. V. A. Tartakovskii, I. E. Chlenov, N. S. Morozova and S. S. Novikov, *Izvest. Akad. Nauk S.S.S.R., Ser. Khim.*, 370 (1966).
2282. Y. Tatsuno, A. Konishi, A. Nakamura and S. Otsuka, *J. Chem. Soc. Chem. Commun.*, 588 (1974).
2283. Y. Tatsuno, A. Konishi, A. Nakamura and S. Otsuka, *J. Chem. Soc., Chem. Commun.*, 588 (1974).
2284. B. Tchoubar, *Bull. Soc. Chim. France*, 164 (1949).
2285. A. P. Terbarg, H. Kloosterziel and N. Van Meurs, *Rec. Trav. chim.*, **82**, 717 (1963).
2286. P. B. Terent'ev, T. P. Mokvina, L. V. Moshentseva and A. N. Kost, *Khim. Getero. Soedin.*, **4**, 498 (1968).
2287. A. P. Terent'ev and A. A. Demidova, *Zh. Obshch. Khim.*, **7**, 2195 (1937).
2288. P. B. Terent'ev, T. P. Mokvina, L. V. Moshen'tseva and A. N. Kost, *Khim. Getero. Soedin.*, **4**, 498 (1968).
2289. F. Texier and R. Carrié, *Bull. Soc. Chim. France*, 3642 (1971).
2290. L. Thijs, A. Wagenaar, E. M. M. van Rens and B. Zwanenburg, *Tetrahedron Letters*, 3589 (1973).
2291. A. F. Thompson and M. Baer, *J. Amer. Chem. Soc.*, **62**, 2094 (1940).
2292. M. Tisler, M. Hrovat and N. Machiedo, *Croat. Chem. Acta*, **34**, 183 (1962).
2293. T. Toda, C. Tanigma, A. Yamae and T. Mukai, *Chem. Letters*, 447 (1972).
2294. N. Tokura, T. Nagai and S. Matsumura, *J. Org. Chem.*, **31**, 349 (1966).
2295. K. Torssell, *Arkiv Kemi*, **23**, 537 (1965).
2296. K. Torssell, *Arkiv Kemi*, **23**, 543 (1965).
2297. W. Triebs and M. Quang, *Annalen*, **598**, 38 (1956). J. M. J. Tronchet, B. Gentile and J. Tronchet, *Helv. Chim. Acta*, **58**, 1817 (1975).
2298. A. T. Troshchenko and A. A. Petrov, *Doklady Akad. Nauk S.S.S.R.*, **119**, 292 (1958).
2299. O. Tsuge and M. Koga, *Org. Prep. Proced. Intl.*, **7** (4), 173 (1975).
2300. D. Tsuru, K. Fujiwara, T. Yoshimoto, R. Watanabe, M. Tomumatsu and S. Hayashida, *Int. J. Peptide Protein Res.*, **5**, 293 (1973).
2301. D. Tsuru, K. Fujiwara, R. Watanabe, T. Yoshimoto, S. Hayashida, M. Tomimatsu and Y. Okoshi, *J. Biochem.*, **75**, 261 (1974).
2302. N. J. Turro and W. B. Hammond, *J. Amer. Chem. Soc.*, **88**, 3672 (1966).
2303. W. H. Urry and J. R. Eiszner, *J. Amer. Chem. Soc.*, **73**, 2977 (1951).
2304. W. H. Urry and J. R. Eiszner, *J. Amer. Chem. Soc.*, **74**, 5822 (1952).
2305. W. H. Urry, J. R. Eiszner and J. Wilt, *J. Amer. Chem. Soc.*, **79**, 918 (1957).
2306. W. H. Urry, H. W. Knuse and W. R. McBride, *J. Amer. Chem. Soc.*, **79**, 6568 (1957).
2307. W. H. Urry, P. Szecsi, C. Ikoku and D. W. Moore, *J. Amer. Chem. Soc.*, **88**, 2224 (1966).

2308. L. K. Vagina, V. N. Chistokletov, and A. A. Petrov, *Zh. Org. Khim.*, **2**, 417 (1966).
2309. S. J. Valenty and P. S. Skell, *J. Org. Chem.*, **38**, 3937 (1973).
2310. A. M. Van Leusen and J. Straling, *Quarterly Rep. Sulfur Chem.*, **5**, 67 (1970).
2311. E. E. Van Tamelen, T. Spencer, D. Allen and R. Orris, *Tetrahedron*, **14**, 8 (1961).
2312. L. V. Vargha and E. Kovacs, *Chem. Ber.*, **75**, 794 (1942).
2313. L. Veniard, *Thèse*, Paris, 1971; *Bull. Soc. Chim. France*, 2746 (1973).
2314. O. Vig, M. Bhatia, K. Gupta and K. Matta, *J. Indian Chem. Soc.*, **46**, 991 (1969).
2315. J. Vilarrasa and R. Granados, *J. Heterc. Chem.*, **11**, 867 (1974).
2316. A. G. Vitenberg, I. A. D'Yakonov and A. Zindel, *Zh. Obshch. Khim.*, **2**, 1532 (1966).
2316a. A. G. Vitenberg and I. A. D'Yakonov, *Zh. Org. Khim.*, **5**, 1036 (1969).
2317. E. Vogel, W. Wiedemann, H. Kiefer and W. F. Harrison, *Tetrahedron Letters*, 673 (1963).
2318. E. Vogel, A. Vogel, H. K. Kubbeler and W. Sturm, *Angew. Chem. Int. Ed.*, **9**, 514 (1970).
2319. E. Vogel and H. Reel, *J. Amer. Chem. Soc.*, **94**, 4388 (1972).
2320. K. von Auwers and R. Ottens, *Chem. Ber.*, **57**, 446 (1924).
2321. K. von Auwers and O. Ungemach, *Chem. Ber.*, **66**, 1205 (1933).
2322. K. von Auwers and O. Ungemach, *Chem. Ber.*, **66**, 1205 (1933).
2323. H. von Pechmann, *Chem. Ber.*, **28**, 855 (1895).
2324. H. von Pechmann, *Chem. Ber.*, **28**, 861 (1895).
2325. H. von Pechmann, *Chem. Ber.*, **28**, 1626 (1895).
2326. H. von Pechmann and A. Wold, *Chem. Ber.*, **29**, 2588 (1896).
2327. H. von Pechmann and A. Wold, *Chem. Ber.*, **31**, 557 (1898).
2328. L. Vo-Quang, *C. R. Acad. Sci. Paris, Sér. C*, **266**, 642 (1968).
2329. L. Vo-Quang and Y. Vo-Quang, *Bull. Soc. Chim. France*, 2575 (1974).
2330. E. E. Waali and W. M. Jones, *J. Amer. Chem. Soc.*, **95**, 8114 (1973).
2331. W. Walter, J. Voss, J. Curts and H. Pawelzik, *Ann. Chem.*, **660**, 60 (1962).
2332. V. Wannagat and R. Pfeffenschneider, *Naturwiss.*, **43**, 178 (1956).
2333. D. Wendisch, *Methoden der Organischen Chemie* (Houben–Weyl–Müller), Vol. IV/3, Georg Thieme, Stuttgart, 1971.
2334. H. Werner and J. H. Richards, *J. Amer. Chem. Soc.*, **90**, 4976 (1968).
2334a. E. A. Werner, *J. Chem. Soc.*, **115**, 1168 (1919).
2335. F. Weygand, W. Schwenke and H. J. Bestmann, *Angew. Chem.*, **70**, 506 (1958).
2336. F. Weygand and H. J. Bestmann, *Chem. Ber.*, **92**, 528 (1959).
2337. F. Weygand and H. J. Bestmann, *Angew. Chem.*, **72**, 539 (1960).
2338. D. H. White, P. B. Condit and R. G. Bergman, *J. Amer. Chem. Soc.*, **94**, 1348 (1972).
2339. J. E. White, *J. Chem. Educ.*, **44**, 128 (1967).
2340. H. W. Whitlock and H. A. Carlson, *Tetrahedron*, **20**, 2101 (1964).
2341. K. B. Wiberg, B. R. Lowry and T. H. Colby, *J. Amer. Chem. Soc.*, **83**, 3998 (1961).
2342. K. B. Wiberg and A. de Meijere, *Tetrahedron Letters*, 519 (1969).
2343. H. Wieland and S. Bloch, *Chem. Ber.*, **39**, 1488 (1906).
2344. A. L. Wilds and A. L. Meader, *J. Org. Chem.*, **13**, 763 (1948).
2345. C. J. Wilkerson and F. D. Greene, *J. Org. Chem.*, **40**, 3112 (1975).
2346. J. W. Wilt, T. P. Malloy, P. K. Mookerjee and D. R. Sullivan, *J. Org. Chem.*, **39**, 1327 (1974).
2347. J. W. Wilt and D. R. Sullivan, *J. Org. Chem.*, **40**, 1036 (1975).
2348. S. Winter and H. Pracejus, *Chem. Ber.*, **99**, 151 (1966).
2349. G. Wittig and H. Dürr, *Ann. Chem.*, **672**, 55 (1964).
2350. L. Wolff, *Ann. Chem.*, **325**, 169 (1902).
2351. L. Wolff, *Ann. Chem.*, **333**, 1 (1904).
2352. L. Wolff, *Ann. Chem.*, **394**, 23 (1912).
2352a. L. Wolff and R. Krüche, *Ann. Chem.*, **394**, 48 (1912).
2353. M. E. Wolff, D. Feldman, P. Catsoulacos, J. W. Funder, C. Hancock, Y. Amano and I. S. Edelman, *J. Biochem.*, **14**, 1750 (1975).
2354. R. B. Woodward and R. Hoffmann, *Angew. Chem. Int. Ed.*, **8**, 781 (1969).
2355. D. S. Wulfman, B. W. Peace and E. K. Steffen, *J. Chem. Soc. (D)*, 1360 (1971).
2356. D. S. Wulfman, F. C. Carman, B. G. McGibboney, E. K. Steffen and B. W. Peace, *Preprints, Div. Petro. Chem., Amer. Chem. Soc.*, **16** (B), L (1971).

2357. D. S. Wulfman, B. G. McGibboney and B. W. Peace, *Synthesis*, 49 (1972).
2358. D. S. Wulfman and B. W. Peace, *Tetrahedron Letters*, 3903 (1972).
2359. D. S. Wulfman, N. V. Thinh, R. S. McDaniel, Jr, B. W. Peace, M. Tom Jones and C. W. Heitsch, *J. Chem. Soc. Dalton Trans.*, 522 (1975).
2360. D. S. Wulfman, B. Poling and R. S. McDaniel, Jr, *Tetrahedron Letters*, 4519 (1975).
2361. D. S. Wulfman and R. S. McDaniel, Jr, *Tetrahedron Letters*, 4523 (1975).
2362. D. S. Wulfman and B. Poling, *Proc. Missouri Acad. Sci.*, in press.
2363. D. S. Wulfman, *Tetrahedron*, **32**, 1231 (1976).
2364. D. S. Wulfman, R. S. McDaniel, Jr and B. W. Peace, *Tetrahedron*, **32**, 1241 (1976).
2365. D. S. Wulfman, B. W. Peace and R. S. McDaniel, Jr, *Tetrahedron*, **32**, 1251 (1976).
2366. D. S. Wulfman, unpublished results.
2367. D. S. Wulfman; in footnotes to Reference 2187 this explanation (saturation of ligancy) was suggested as a rationale to overcome the mechanistic implications of the reported observation.
2368. D. S. Wulfman, B. G. McGibboney, E. K. Steffen, N. V. Thinh, R. S. McDaniel, Jr and B. W. Peace, *Tetrahedron*, **32**, 1257 (1976).
2369. L. A. Yanovskaya and V. A. Dombrovskii, *Russ. Chem. Rev.*, **44**, 154 (1975).
2370. P. Yates and E. W. Robb, *J. Amer. Chem. Soc.*, **79**, 5760 (1957).
2371. P. Yates and B. G. Christensen, *Chem. and Ind.*, 1441 (1958).
2372. P. Yates and R. J. Crawford, *J. Amer. Chem. Soc.*, **88**, 1562 (1966).
2373. L. G. Zaitseva, I. B. Avezov, O. A. Subbatin and I. G. Bolesov, *Zh. Org. Khim.*, **11**, 1415 (1975).
2374. E. Zbiral and E. Bauer, *Tetrahedron*, **28**, 4189 (1972).
2375. K.-P. Zeller, H. Meier and E. Müller, *Tetrahedron*, **28**, 5831 (1972).
2376. H. E. Zimmerman, H. G. C. Dürr, R. G. Lewis and S. Bram, *J. Amer. Chem. Soc.*, **84**, 4149 (1962).
2377. H. E. Zimmerman and D. H. Paskovich, *J. Amer. Chem. Soc.*, **86**, 2149 (1964).
2378. H. E. Zimmerman, H. G. C. Dürr, R. S. Givens and R. G. Lewis, *J. Amer. Chem. Soc.*, **89**, 1863 (1967).
2379. B. Zwanenburg, J. B. F. N. Engberts and I. Strating, *Tetrahedron Letters*, 543 (1964).
2380. B. Zwanenburg, A. Wagenaar, S. Thijs and J. Strating, *J. Chem. Soc. Perkin Trans. I*, 73, (1973).
2381. B. Zwanenburg and A. Wagenaar, *Tetrahedron Letters*, 5009 (1973).

Author Index

This author index is designed to enable the reader to locate an author's name and work with the aid of the reference numbers appearing in the text. The page numbers are printed in normal type in ascending numerical order, followed by the reference numbers in parentheses. The numbers in *italics* refer to the pages on which the references are actually listed.

Subject Index